J-3.152

Advanced
Calculus
for
Applications

Advanced
Calculus
for
Applications

SECOND EDITION

Francis B. Hildebrand

Professor of Mathematics
Massachusetts Institute of Technology

PRENTICE-HALL, INC.
Englewood Cliffs, New Jersey

Library of Congress Cataloging in Publication Data

HILDEBRAND, FRANCIS BEGNAUD.
   Advanced calculus for applications.

   Published in 1948 under title: Advanced calculus for engineers.
   Bibliography: p.
   Includes index.
   1. Calculus. I. Title.
QA303.H55 1976   515   75-34473
ISBN 0-13-011189-9

© 1976 by Prentice-Hall, Inc.
Englewood Cliffs, New Jersey

10 9 8 7 6 5 4 3

Printed in the United States of America

PRENTICE-HALL INTERNATIONAL, INC., *London*
PRENTICE-HALL OF AUSTRALIA, PTY. LTD., *Sydney*
PRENTICE-HALL OF CANADA, LTD., *Toronto*
PRENTICE-HALL OF INDIA PRIVATE LIMITED, *New Delhi*
PRENTICE-HALL OF JAPAN, INC., *Tokyo*
PRENTICE-HALL OF SOUTHEAST ASIA (PTE.) LTD., *Singapore*

Contents

5

Boundary-Value Problems and Characteristic-Function Representations   186

omit → (handwritten annotation pointing to 5.5)

6

Vector Analysis   269

10

Functions of a Complex Variable   **539**

11

Applications of Analytic Function Theory   **622**

Preface

The purpose of this text is to present an integrated treatment of a number of those topics in mathematics which can be made to depend only upon a sound course in elementary calculus, and which are of common importance in many fields of application.

An attempt is made to deal with the various topics in such a way that a student who may not proceed into the more profound areas of mathematics still may obtain an intelligent working knowledge of a substantial number of useful mathematical methods, together with an appropriate awareness of the foundations, interrelations, and limitations of these methods. At the same time, it is hoped that a student who is to progress, say, into a rigorous course in mathematical analysis will be provided, in addition, with increased incentive and motivation. For both of these purposes, the phrase "It can be shown" is used occasionally, not only to exhibit a generalization of an established conclusion or a useful related fact, but also to introduce a needed basic result for which a rigorous demonstration would require what is believed to be an inappropriately excessive amount of detailed analysis or of prerequisite preparation.

This revision incorporates a large number of relatively minor changes for the purpose of increased clarity or precision or to supply a previously omitted proof, a substantial amount of added textual material (particularly in the later chapters), and about 250 additional problems.

The first four chapters are concerned chiefly with ordinary differential equations, including analytical, operational, and numerical methods of solution, and with special functions generated as solutions of such equations. In particular, the material of the first chapter can be considered as either a systematic review or an initial introduction to the elementary concepts and techniques, associated with linear equations and with special solvable types of nonlinear equations, which are needed in subsequent chapters. The fifth chapter deals with boundary-value problems governed by ordinary differential equations, with the associated characteristic functions, and with series and integral representations of arbitrary functions in terms of these functions.

Chapter 6 develops the useful ideas and tools of vector analysis; Chapter 7 provides brief introductions to some special topics in higher-dimensional calculus which are rather frequently needed in applications. The treatment here occasionally consists essentially of indicating the plausibility and practical significance of a result and stating conditions under which its validity is rigorously established in listed references.

In Chapter 8, certain basic concepts associated with the simpler types of partial differential equations are introduced, after which, in Chapter 9, full use is made of most of the tools developed in earlier chapters for the purpose of formulating and solving a variety of typical problems governed by the partial differential equations of mathematical physics. A new section deals with the application of the so-called method of variation of parameters to such problems.

Chapter 10 treats the basic topics in the theory of analytic functions of a complex variable, including contour integration and residue calculus. Although certain developments in preceding chapters could be made more elegant and more complete if they were made to depend upon this treatment, introduced at an earlier stage, it is felt that, in some cases, the knowledge based on a brief initial study of analytic functions may not be sufficiently firm to support significantly dependent treatments of the other topics, but that such knowledge then may better serve to clarify the other topics when subsequently provided. However, since most of the treatments of Chapter 10, as well as most of those of Chapters 6 and 7, are independent of the content of preceding chapters, material from these chapters can indeed be introduced at an earlier stage in a given course, at the discretion of the instructor. It has been considered reasonable to assume knowledge of certain elementary properties of complex numbers in the earlier chapters, even though the solution of the equation $x^4 + 1 = 0$ then may occasion a personal review on the part of the reader.

A new Chapter 11 considers some applications of analytic function theory to other fields, including the derivation of methods for the inversion of Laplace transforms (an expansion of material previously presented in annotated problems), an indication of the properties and uses of conformal mapping (formerly included in Chapter 10), and a new brief treatment of Green's functions as related to partial differential equations.

Extensive sets of problems are included at the end of each chapter, grouped in correspondence with the respective sections with which they are associated. In addition to more-or-less routine exercises, there are numerous annotated problems which are intended to guide the reader in developing results or techniques which extend or complement treatments in the text, or in dealing with a particularly challenging application. Such problems may serve as focal points for extended discussions or for the introduction of additional (or alternative) material into a chapter, permitting the text to serve somewhat more flexibly in courses of varied types. New problems of this sort now permit the consideration of topics such as one-dimensional Green's functions and applications of elliptic integrals, Fourier transforms, and associated Legendre functions. Answers to all problems are either incorporated into the statement of the problem or listed at the end of the book.

The author is particularly indebted to Professor E. Reissner for valuable collaboration in the preliminary stages of the preparation of the original edition and for many ideas which contributed to whatever useful novelty some of the treatments may possess, and to Professor G. B. Thomas for additional advice and help, as well as to a rather long list of other colleagues and students who have offered criticisms and suggestions leading to many of the modifications incorporated into this revision.

F. B. HILDEBRAND

Advanced
Calculus
for
Applications

1

Ordinary Differential Equations

1.1. Introduction. A *differential equation* is an equation relating two or more variables in terms of derivatives or differentials. Thus the simplest differential equation is of the form

$$\frac{dy}{dx} = h(x), \tag{1}$$

where $h(x)$ is a given function of the independent variable x. The solution is obtained immediately by integration, in the form

$$y = \int h(x)\, dx + C, \tag{2}$$

where C is an arbitrary constant. Whether or not it happens that the integral can be expressed in terms of named or tabulated functions is incidental, in the sense that we accept as a *solution* of a differential equation any functional relation, *not involving derivatives or integrals of unknown functions*, the satisfaction of which implies the satisfaction of the differential equation. Similarly, in an equation of the form

$$F(x)G(y)\, dx + f(x)g(y)\, dy = 0, \tag{3}$$

we may *separate the variables* and obtain a solution by integration in the form

$$\int \frac{F(x)}{f(x)}\, dx + \int \frac{g(y)}{G(y)}\, dy = C, \tag{4}$$

if suitable account is taken of situations in which a divisor may vanish.

Usually we desire to obtain the *most general* solution of the differential equation; that is, we require *all* functional relations which imply the equation.

1

In the general case it may be difficult to determine when *all* such relations have indeed been obtained. Fortunately, however, this difficulty does not exist in the case of so-called *linear* differential equations, which are of most frequent occurrence in applications and which are to be of principal interest in what follows.

A differential equation of the form

$$a_0(x)\frac{d^n y}{dx^n} + a_1(x)\frac{d^{n-1}y}{dx^{n-1}} + \cdots + a_{n-1}(x)\frac{dy}{dx} + a_n(x)y = f(x) \tag{5}$$

is said to be a *linear* differential equation of *order n*. The distinguishing characteristic of such an equation is the absence of products or nonlinear functions of the *dependent variable* (unknown function) y and its derivatives, the highest derivative present being of order n. The coefficients $a_0(x), \ldots, a_n(x)$ may be arbitrarily specified functions of the independent variable x.

For a linear equation of the *first* order,

$$a_0(x)\frac{dy}{dx} + a_1(x)y = f(x),$$

it is shown in Section 1.4 that if both sides of the equation are multiplied by a certain determinable function of x (an "integrating factor"), the equation always can be put in an equivalent form

$$\frac{d}{dx}[p(x)y] = F(x),$$

where $p(x)$ and $F(x)$ are simply expressible in terms of a_0, a_1, and f, and hence then can be solved directly by integration.

Although no such simple general method exists for solving linear equations of higher order, there are two types of such equations which are of particular importance in applications and which can be completely solved by direct methods. These two cases are considered in Sections 1.5 and 1.6. In addition, this chapter presents certain techniques that are available for treatment of more general linear equations.

Many of the basically useful properties of linear differential equations do not hold for *nonlinear* equations, such as

$$\frac{dy}{dx} = x + y^2, \qquad \frac{d^2 y}{dx^2} + \sin y = 0, \qquad \frac{d^2 y}{dx^2} + x\left(\frac{dy}{dx}\right)^2 + y = e^x.$$

A few special types of solvable nonlinear equations are dealt with briefly in Section 1.12.

The equations to be considered in this chapter are known as *ordinary* differential equations, as distinguished from *partial* differential equations, which involve *partial derivatives* with respect to two or more independent variables. Equations of the latter type are treated in subsequent chapters.

Before proceeding to the study of linear ordinary differential equations, we next briefly introduce the notion of *linear dependence*, which is basic in this work.

1.2. Linear Dependence. By a *linear combination* of n functions $u_1(x)$, $u_2(x), \ldots, u_n(x)$ is meant an expression of the form

$$c_1 u_1(x) + c_2 u_2(x) + \cdots + c_n u_n(x) \equiv \sum_{k=1}^{n} c_k u_k(x), \tag{6}$$

where the c's are constants. When at least one c is not zero, the linear combination is termed *nontrivial*. The functions $u_1, u_2, \ldots, u_n$ are then said to be *linearly independent* over a given interval (say $a \le x \le b$) if over that interval no one of the functions can be expressed as a linear combination of the others, or, equivalently, if no *nontrivial* linear combination of the functions is identically zero over the interval considered. Otherwise, the functions are said to be *linearly dependent* over that interval.

As an example, the functions $\cos 2x$, $\cos^2 x$, and 1 are linearly dependent over *any* interval because of the identity

$$\cos 2x - 2 \cos^2 x + 1 \equiv 0.$$

It follows from the definition that *two* functions are linearly dependent over an interval if and only if one function is a constant multiple of the other over that interval. The necessity of the specification of the interval in the general case is illustrated by a consideration of the two functions x and $|x|$. In the interval $x > 0$ there follows $x - |x| \equiv 0$, whereas in the interval $x < 0$ we have $x + |x| \equiv 0$. Thus the two functions are linearly dependent over any interval not including the point $x = 0$; but they are linearly independent over any interval including $x = 0$, since no single linear combination of the two functions is identically zero over such an interval.

Although in practice the linear dependence or independence of a set of functions generally can be established by inspection, the following result is of some importance in theoretical discussions. We assume that each of a set of n functions $u_1, u_2, \ldots, u_n$ possesses n finite derivatives at all points of an interval I. Then, if a set of constants exists such that

$$c_1 u_1 + c_2 u_2 + \cdots + c_n u_n = 0$$

for all values of x in I, these same constants also satisfy the identities

$$c_1 \frac{du_1}{dx} + c_2 \frac{du_2}{dx} + \cdots + c_n \frac{du_n}{dx} = 0,$$

$$c_1 \frac{d^2 u_1}{dx^2} + c_2 \frac{d^2 u_2}{dx^2} + \cdots + c_n \frac{d^2 u_n}{dx^2} = 0,$$

$$\cdots \cdots \cdots \cdots \cdots \cdots \cdots \cdots \cdots \cdots \cdots \cdots$$

$$c_1 \frac{d^{n-1} u_1}{dx^{n-1}} + c_2 \frac{d^{n-1} u_2}{dx^{n-1}} + \cdots + c_n \frac{d^{n-1} u_n}{dx^{n-1}} = 0.$$

Thus the n constants must satisfy n homogeneous linear equations. However, such a set of equations can possess *nontrivial* solutions only if its coefficient determinant vanishes. Thus it follows that *if the functions* $u_1, u_2, \ldots, u_n$ *are linearly dependent over an interval I, then the determinant*

$$W(u_1, u_2, \ldots, u_n) = \begin{vmatrix} u_1 & u_2 & \cdots & u_n \\ \dfrac{du_1}{dx} & \dfrac{du_2}{dx} & \cdots & \dfrac{du_n}{dx} \\ \cdots\cdots\cdots\cdots\cdots\cdots \\ \dfrac{d^{n-1}u_1}{dx^{n-1}} & \dfrac{d^{n-1}u_2}{dx^{n-1}} & \cdots & \dfrac{d^{n-1}u_n}{dx^{n-1}} \end{vmatrix} \tag{7}$$

vanishes identically over I. This determinant appears frequently in theoretical work and is called the *Wronskian* (or Wronskian determinant) of the functions. Thus we see that *if the Wronskian of* $u_1, u_2, \ldots, u_n$ *is not identically zero over I, then the functions are linearly independent over I.*

To illustrate, since the value of the determinant

$$W(1, x, x^2, \ldots, x^n) = \begin{vmatrix} 1 & x & x^2 & x^3 & \cdots & x^n \\ 0 & 1! & 2x & 3x^2 & \cdots & nx^{n-1} \\ 0 & 0 & 2! & 6x & \cdots & n(n-1)x^{n-2} \\ 0 & 0 & 0 & 3! & \cdots & n(n-1)(n-2)x^{n-3} \\ \cdots\cdots\cdots\cdots\cdots\cdots\cdots\cdots\cdots \\ 0 & 0 & 0 & 0 & \cdots & n! \end{vmatrix}$$

is merely the product of the nonvanishing constants appearing in the principal diagonal and hence cannot vanish, it follows that the functions appearing in the first row are linearly independent (over *any* interval).

Unfortunately, the converse of the preceding theorem is *not* true since, in unusual cases, the Wronskian of a set of *linearly independent* functions also may vanish. That is, the vanishing of the Wronskian is *necessary* but not *sufficient* for linear dependence of a set of functions. (For an example establishing the insufficiency, see Problem 5.)

1.3. Complete Solutions of Linear Equations. The most general linear differential equation of the *n*th order can be written in the form

$$\frac{d^n y}{dx^n} + a_1(x)\frac{d^{n-1}y}{dx^{n-1}} + \cdots + a_{n-1}(x)\frac{dy}{dx} + a_n(x)y = h(x). \tag{8}$$

Here it is assumed that both sides of the equation have been divided by the coefficient of the highest derivative. We will speak of this form as the *standard form* of the equation. This equation is frequently written in the abbreviated form

$$Ly = h(x), \tag{9}$$

where *L* here represents the *linear differential operator*

$$L = \frac{d^n}{dx^n} + a_1(x)\frac{d^{n-1}}{dx^{n-1}} + \cdots + a_{n-1}(x)\frac{d}{dx} + a_n(x). \tag{10}$$

The problem of solving Equation (8) consists of determining the most general expression for y which, if substituted into the left-hand side of (8), or if operated on by (10), gives the prescribed right-hand side $h(x)$. When a relationship of the form $y = u(x)$ satisfies Equation (8), it is conventional to say that either the *relation* $y = u(x)$ or the *function* $u(x)$ is a *solution* of that equation.†

If all the coefficients $a_1(x), \ldots, a_n(x)$ were zero, the solution of Equation (8) would be accomplished directly by n successive integrations, each integration introducing an independent constant of integration. Thus it might be expected that the general solution of (8) also would contain n independent arbitrary constants. As a matter of fact, it is known that *in any interval I in which the coefficients are continuous, there exists a continuous solution to Equation (8) involving exactly n independent arbitrary constants; furthermore, there are no solutions of Equation (8) valid in I which cannot be obtained by specializing the constants in any such solution.*

It should be noticed that this is a property peculiar to *linear* differential equations. To illustrate, the nonlinear differential equation

$$\left(\frac{dy}{dx}\right)^2 - 2\frac{dy}{dx} + 4y = 4x - 1 \tag{11}$$

is of first order. A solution containing one arbitrary constant is of the form

$$y = x - (x - c)^2, \tag{12}$$

as can be verified by direct substitution. However, this is not the most general solution, since the function $y = x$ also satisfies the differential equation but cannot be obtained by specializing the arbitrary constant in the solution given. The additional solution $y = x$ is called a *singular solution*. Such solutions can occur only in the solution of nonlinear differential equations.

We consider first the result of replacing the function $h(x)$ by zero in Equation (8). The resulting differential equation, $Ly = 0$, is said to be *homogeneous*, since each term in the equation then involves the first power of y or of one of its derivatives. In this case, from the linearity of the equation, it is easily seen that any linear combination of individual solutions is also a solution. Thus, if n linearly independent solutions $u_1(x), u_2(x), \ldots, u_n(x)$ of the associated homogeneous equation

$$Ly_H = 0 \tag{13}$$

are known, the *general* solution of Equation (13) is of the form

$$y_H(x) = c_1 u_1(x) + c_2 u_2(x) + \cdots + c_n u_n(x) = \sum_{k=1}^{n} c_k u_k(x), \tag{14}$$

where the c's are the n required arbitrary constants. That is, all solutions of the homogeneous equation associated with (8) are obtained by suitably specializing the constants in Equation (14).

---

†Whereas a relationship of the implicit form $\varphi(x, y) = 0$ also would be acceptable as a solution, there is no need for this generalization when the equation is *linear*.

In this connection, it should be explained that we refer to a function as a solution of a differential equation in a given interval I if and only if that function satisfies the differential equation at *all points* of I. Thus, in the case of the homogeneous equation

$$\frac{d^2y}{dx^2} = 0$$

we say that the general solution is of the form $y = c_1 + c_2 x$. It may be argued that, since the function $y = |x|$ is a linear function over any interval not including the point $x = 0$, its second derivative is zero and hence it is a "solution" which cannot be obtained by specializing c_1 and c_2. However, it is clear that this function is not differentiable at $x = 0$. Consequently, since the left-hand side of the equation does not exist at $x = 0$, the equation is not satisfied at this point, and $y = |x|$ cannot be said to be a solution over any interval including $x = 0$. Over any interval not including $x = 0$, the function $y = |x|$ may be replaced by either $+x$ or $-x$ and hence is obtained from the general solution by setting $c_1 = 0$ and either $c_2 = 1$ or $c_2 = -1$.

Now suppose that one *particular solution* of Equation (8), say $y = y_P(x)$, can be obtained by inspection or otherwise, so that

$$Ly_P = h(x). \tag{15}$$

Then the *complete solution* of Equation (8) is of the form

$$y = y_H(x) + y_P(x) = \sum_{k=1}^{n} c_k u_k(x) + y_P(x), \tag{16}$$

since this expression contains n independent arbitrary constants and satisfies the differential equation

$$Ly = L(y_H + y_P) = Ly_H + Ly_P = h(x). \tag{17}$$

Thus it is seen that the process of solving an ordinary linear differential equation can be divided conveniently into two parts. First, n linearly independent solutions of the associated homogeneous equation may be obtained; then, if any one particular solution of the complete equation is found, the complete solution is given by Equation (16).

It is frequently convenient to say, "$y_H(x)$ is a *homogeneous solution* of $Ly = h$," in abbreviation of the statement, "$y = y_H(x)$ is a solution of the associated homogeneous equation $Ly = 0$." The term "complementary solution" also is used.

It will be shown in Section 1.9 that, if the general homogeneous solution of an nth-order linear equation is known, a particular solution can always be obtained by n integrations. In Sections 1.5 and 1.6 we consider important special cases in which the homogeneous solution is readily obtained.

1.4. The Linear Differential Equation of First Order. The linear equation of *first* order is readily solved in general terms, without determining separately homogeneous and particular solutions. For this purpose, we attempt to determine an *integrating factor* $p(x)$ such that the standard form

$$\frac{dy}{dx} + a_1(x)y = h(x) \qquad \text{LDE of} \qquad (18)$$
$$\text{Ist ORDER}$$

is equivalent to the equation

$$\frac{d}{dx}(py) = ph. \qquad (19)$$

Since Equation (19) can be written in the form

$$\frac{dy}{dx} + \left(\frac{1}{p}\frac{dp}{dx}\right)y = h(x),$$

it follows that Equations (18) and (19) are equivalent if p satisfies the equation

$$\frac{1}{p}\frac{dp}{dx} = a_1(x),$$

and hence an integrating factor is

$$p = e^{\int a_1(x)\,dx}. \qquad (20)$$

The solution of Equation (19) is obtained by integration:

$$py = \int ph\,dx + C,$$

so that the general solution of (18) is of the form

$$y = \frac{1}{p}\int ph\,dx + \frac{C}{p}, \qquad (21)$$

where p is the integrating factor defined by Equation (20), and C is an arbitrary constant.

In particular, that solution for which $y = y_0$ when $x = x_0$ is expressible in the form

$$y(x) = \int_{x_0}^{x} \frac{p(\xi)}{p(x)}h(\xi)\,d\xi + y_0\frac{p(x_0)}{p(x)}, \qquad (21')$$

where the dummy variable in the integrand is denoted by ξ in order to distinguish it from the free variable x, which is to be held fixed in the integration.

Example. To solve the differential equation

$$x\frac{dy}{dx} + (1-x)y = xe^x,$$

we first rewrite the equation in the standard form,

$$\frac{dy}{dx} + \left(\frac{1}{x} - 1\right)y = e^x.$$

An integrating factor is then

$$p = e^{\int [(1/x)-1]\,dx} = e^{\log x - x} = xe^{-x},$$

no constant being added in the integration, since only a particular integrating factor is needed. The solution is then given by Equation (21),

$$y = \frac{e^x}{x} \int x \, dx + C\frac{e^x}{x},$$

or

$$y = \frac{x}{2}e^x + C\frac{e^x}{x}. \qquad \blacksquare$$

It may be noticed that the general homogeneous solution of Equation (18) is $y_H = Cp^{-1}$, whereas a particular solution is $y_P = p^{-1}\int ph \, dx$.

1.5. Linear Differential Equations with Constant Coefficients. The simplest and perhaps the most important differential equation of higher order in practice is the linear equation

$$Ly = \frac{d^n y}{dx^n} + a_1 \frac{d^{n-1} y}{dx^{n-1}} + \cdots + a_{n-1}\frac{dy}{dx} + a_n y = h(x), \qquad (22)$$

in which the coefficients a_k are *constants*.

We first attempt to determine n linearly independent solutions of the corresponding homogeneous equation. The appearance of the equation suggests homogeneous solutions of the form e^{rx}, where r is a constant, since all derivatives of e^{rx} are constant multiples of the function itself,

$$\frac{d^m}{dx^m}e^{rx} = r^m e^{rx}.$$

We then have

$$Le^{rx} = (r^n + a_1 r^{n-1} + \cdots + a_{n-1}r + a_n)e^{rx}. \qquad (23)$$

This result shows that e^{rx} is a solution of the homogeneous equation associated with Equation (22) if r is one of the n roots $r_1, r_2, \ldots, r_n$ of the *characteristic equation*

$$r^n + a_1 r^{n-1} + \cdots + a_{n-1}r + a_n = 0. \qquad (24)$$

It should be noticed that this equation is obtained from the associated homogeneous differential equation by formally replacing $d^k y/dx^k$ by r^k, with the convention that $d^0 y/dx^0 \equiv y$. If the n roots of Equation (24) are distinct, exactly n independent solutions $e^{r_1 x}, \ldots, e^{r_n x}$ of the homogeneous equation are so obtained and the general homogeneous solution is

$$y_H = \sum_{k=1}^{n} c_k e^{r_k x}. \qquad (25)$$

However, if one or more of the roots is repeated, less than n independent solutions are obtained in this way. To find the missing solutions we may proceed as follows. Suppose that $r = r_1$ is a double root of Equation (24). Then Equation (23) is of the form

$$Le^{rx} = (r - r_1)^2(r - r_3) \cdots (r - r_n)e^{rx},$$

and it follows that not only the right-hand member itself but also its (partial) derivative with respect to r must vanish when $r = r_1$. The same then must be true for the left-hand member, and hence we conclude that in this case we have both

$$L[e^{rx}]_{r=r_1} = Le^{r_1x} = 0$$

and
$$\left[\frac{\partial}{\partial r}(Le^{rx})\right]_{r=r_1} = L\left[\frac{\partial}{\partial r}(e^{rx})\right]_{r=r_1} = Lxe^{r_1x} = 0,$$

when we make use of the fact that

$$\frac{\partial}{\partial r}\left(\frac{\partial^k e^{rx}}{\partial x^k}\right) = \frac{\partial^k}{\partial x^k}\left(\frac{\partial e^{rx}}{\partial r}\right).$$

Thus the part of the homogeneous solution corresponding to a double root r_1 can be written in the form

$$c_1 e^{r_1x} + c_2 x e^{r_1x} = e^{r_1x}(c_1 + c_2 x).$$

By a simple extension of this argument, it can be shown that the part of the homogeneous solution corresponding to an m-fold root r_1 is of the form

$$e^{r_1x}(c_1 + c_2 x + c_3 x^2 + \cdots + c_m x^{m-1}).$$

Hence, to each of the n roots of Equation (24), repeated roots being counted separately, there is a corresponding known homogeneous solution, and the general homogeneous solution is determined as a linear combination of these n independent solutions.

Example 1. For the equation

$$\frac{d^3y}{dx^3} - \frac{dy}{dx} = 0,$$

the characteristic equation is $r^3 - r = r(r + 1)(r - 1) = 0$, from which there follows $r = 0, \pm 1$. The general solution is then

$$y = c_1 + c_2 e^x + c_3 e^{-x}. \qquad \blacksquare$$

Example 2. For the differential equation

$$\frac{d^3y}{dx^3} - 5\frac{d^2y}{dx^2} + 8\frac{dy}{dx} - 4y = 0,$$

the characteristic equation is $(r - 1)(r - 2)^2 = 0$, from which there follows $r = 1, 2, 2$. The general solution is then

$$y = c_1 e^x + e^{2x}(c_2 + c_3 x). \qquad \blacksquare$$

If Equation (24) has *imaginary roots* and if the coefficients of Equation (24) are real, the roots must occur in conjugate pairs. Thus, if $r_1 = a + ib$ is one root, a second root must be $r_2 = a - ib$. The part of the solution corresponding to these two roots can be written in the form

$$Ae^{(a+ib)x} + Be^{(a-ib)x} = e^{ax}(Ae^{ibx} + Be^{-ibx}).$$

In order that this expression be real, the constants A and B must be imaginary.

By making use of *Euler's formula,*†

$$e^{i\theta} = \cos \theta + i \sin \theta, \tag{26}$$

we find that the solution becomes

$$e^{ax}[A (\cos bx + i \sin bx) + B (\cos bx - i \sin bx)]$$

and hence can be written in the more convenient form $r_1 = a + ib$

$$e^{ax}(c_1 \cos bx + c_2 \sin bx),   r_2 = a - bi$$

where c_1 and c_2 are new arbitrary constants replacing $A + B$ and $i(A - B)$, respectively. Accordingly, since $A = \frac{1}{2}(c_1 - ic_2)$ and $B = \frac{1}{2}(c_1 + ic_2)$, real values of c_1 and c_2 correspond to values of A and B which are conjugate complex. Similarly, if $a \pm ib$ are m-fold roots, the corresponding $2m$ terms in the homogeneous solution can be written in the real form

$$e^{ax}[(c_1 + c_2 x + \cdots + c_m x^{m-1}) \cos bx$$
$$+ (c_{m+1} + c_{m+2} x + \cdots + c_{2m} x^{m-1}) \sin bx].$$

Example 3. The equation

$$\frac{d^2 y}{dx^2} + 2 \frac{dy}{dx} + 5y = 0$$

                                                                        $a \quad b$

has the characteristic equation $r^2 + 2r + 5 = 0$, from which $r = -1 \pm 2i$; hence

$$y = e^{-x}(c_1 \cos 2x + c_2 \sin 2x). \qquad \blacksquare$$

Example 4. The equation

$$\frac{d^4 y}{dx^4} + 2 \frac{d^2 y}{dx^2} + y = 0$$

has the characteristic equation $(r^2 + 1)^2 = 0$, from which $r = \pm i, \pm i$; hence

$$y = (c_1 + c_2 x) \cos x + (c_3 + c_4 x) \sin x. \qquad \blacksquare$$

General methods for obtaining a particular solution of the complete non-homogeneous Equation (22) are given in Sections 1.7 and 1.9. A shorter method which can be applied in many practical cases is that of *undetermined coefficients*. This method may be used when the right-hand side of Equation (22) involves only terms of the form x^m, where m is an integer, terms of the form $\sin qx$, $\cos qx$, and e^{px}, and/or products of two or more such functions. The reason for the success of the method is the fact that each of these functions, or any product of a finite number of these functions, has only a finite number of linearly independent derivatives.

If we define the *family* of a function $f(x)$ as the set of linearly independent functions of which the function $f(x)$ and its derivatives with respect to x are

†Familiarity with this important relation, and with the elementary algebra of complex numbers, is assumed. Such topics are reviewed in the preliminary sections of Chapter 10.

linear combinations, the following families may be listed:

Term	Family
x^m	$x^m, x^{m-1}, x^{m-2}, \ldots, x^2, x, 1$
$\sin qx$	$\sin qx, \cos qx$
$\cos qx$	$\sin qx, \cos qx$
e^{px}	e^{px}

The family of a function consisting of a product of n terms of this type is readily seen to consist of all possible products of n factors, in which one factor in each product is taken from the family of each factor in the parent function. Thus it may be verified that the family of $x^2 \sin 3x$ is composed of two-factor products of terms in the families $\{x^2, x, 1\}$ and $\{\sin 3x, \cos 3x\}$, one term from each family appearing in each product:

$$\{x^2 \sin 3x, x \sin 3x, \sin 3x, x^2 \cos 3x, x \cos 3x, \cos 3x\}.$$

The method of undetermined coefficients now may be outlined as follows. It is assumed that the general homogeneous solution of the differential equation already has been obtained, and that any cosh or sinh functions occurring in it, or in the right-hand member $h(x)$, are replaced by equivalent linear combinations of exponential functions.

(1) Construct the family of each term (or product) of which $h(x)$ is a linear combination.

(2) If any family has a member which is a homogeneous solution of the differential equation, replace *that* family by a new family in which *each* member of the original family is multiplied by x, or by the lowest integral power of x for which no member of the new family is a homogeneous solution. Only members of the offending family are so modified. It should also be noticed, for example, that the presence of e^x or $\sin x$ in the homogeneous solution does not require modification of a family containing the *product* $e^x \sin x$ unless that product itself is also a homogeneous solution.

(3) Assume as a particular solution a linear combination of all members of the resultant families, with undetermined literal coefficients of combination, and determine these coefficients by requiring that the differential equation be identically satisfied by this assumed solution.

It will be found in all cases that the number of coefficients to be determined will equal the number of linearly independent functions whose coefficients must be matched, and that the resultant equations always have a solution. The detailed proof of this general statement is rather lengthy and is omitted. However, the relevant analysis of one typical case is presented, for the purpose of illustration, at the end of Section 1.7.

It should be emphasized that *this procedure does not generally apply unless the differential equation has constant coefficients and has a right-hand member possessing a finite family.*

Example 5. Consider the differential equation

$$\frac{d^3y}{dx^3} - \frac{dy}{dx} = 2x + 1 - 4\cos x + 2e^x.$$

The general homogeneous solution is

$$y_H = c_1 + c_2 e^x + c_3 e^{-x}.$$

The families of the terms x, 1, $\cos x$, and e^x on the right-hand side of the equation are, respectively,

$$\{x, 1\}, \qquad \{1\}, \qquad \{\cos x, \sin x\}, \qquad \{e^x\}.$$

The second family is contained in the first, and is discarded. Since the first family has the representative 1 in the homogeneous solution, it is replaced by the family $\{x^2, x\}$. Similarly, the last family is replaced by $\{xe^x\}$. A particular solution is then assumed in the form

$$y_P = Ax^2 + Bx + C\cos x + D\sin x + Exe^x.$$

When y is replaced by y_P, the differential equation becomes

$$-2Ax - B - 2D\cos x + 2C\sin x + 2Ee^x = 2x + 1 - 4\cos x + 2e^x$$

and, when the coefficients of x, 1, $\cos x$, $\sin x$, and e^x are equated, there follows

$$A = -1, \qquad B = -1, \qquad D = 2, \qquad C = 0, \qquad E = 1.$$

A particular solution thus is

$$y_P = -x^2 - x + 2\sin x + xe^x,$$

and the general solution is

$$y = c_1 + c_2 e^x + c_3 e^{-x} - x^2 - x + 2\sin x + xe^x. \qquad \blacksquare$$

1.6. The Equidimensional Linear Differential Equation. An equation of the form

$$Ly = x^n\frac{d^ny}{dx^n} + b_1 x^{n-1}\frac{d^{n-1}y}{dx^{n-1}} + \cdots + b_{n-1}x\frac{dy}{dx} + b_n y = f(x), \qquad (27)$$

where the b's are constants, has the property that each term on the left is unchanged when x is replaced by cx, where c is a nonzero constant. Thus the physical *dimension* of x is irrelevant in each term on the left and, if the b's are dimensionless, each term on the left has the dimensions of y. For this reason we shall refer to this equation as the *equidimensional linear equation.* The equation is also variously called "Euler's equation," "Cauchy's equation," and the "homogeneous linear equation," although each of these terms also has other connotations.

One method of solving this equation consists of introducing a new independent variable z by the substitution

$$x = e^z, \qquad z = \log x. \qquad (28)$$

There then follows

$$\frac{d}{dx} = \frac{dz}{dx}\frac{d}{dz} = \frac{1}{e^z}\frac{d}{dz},$$

and hence

$$x^m \frac{d^m}{dx^m} = e^{mz}\left(\frac{1}{e^z}\frac{d}{dz}\right)^m.$$

Thus, in particular, we obtain

$$x\frac{dy}{dx} = \frac{dy}{dz},$$

$$x^2\frac{d^2y}{dx^2} = \frac{d^2y}{dz^2} - \frac{dy}{dz} = \frac{d}{dz}\left(\frac{d}{dz} - 1\right)y,$$

$$x^3\frac{d^3y}{dx^3} = \frac{d^3y}{dz^3} - 3\frac{d^2y}{dz^2} + 2\frac{dy}{dz}$$

$$= \frac{d}{dz}\left(\frac{d}{dz} - 1\right)\left(\frac{d}{dz} - 2\right)y,$$

and, in general, it is found that

$$x^m\frac{d^my}{dx^m} = \frac{d}{dz}\left(\frac{d}{dz} - 1\right)\left(\frac{d}{dz} - 2\right)\cdots\left(\frac{d}{dz} - m + 1\right)y. \tag{29}$$

The transformed equation thus becomes linear with constant coefficients, and y then can be determined in terms of z by the methods of Section 1.5 if the new right-hand member is zero or if it has a finite family (with respect to z-differentiation). The final result is obtained by replacing z by log x.

Example 1. To solve the differential equation

$$x^2\frac{d^2y}{dx^2} - 2x\frac{dy}{dx} + 2y = x^2 + 2,$$

we make use of Equations (28) and (29) to obtain the transformed equation

$$\frac{d^2y}{dz^2} - 3\frac{dy}{dz} + 2y = e^{2z} + 2.$$

The solution is found, by the methods of Section 1.5, to be

$$y = c_1e^z + c_2e^{2z} + ze^{2z} + 1,$$

or, in terms of the original variable x,

$$y = c_1x + c_2x^2 + x^2 \log x + 1. \qquad\blacksquare$$

If the right-hand member is zero, a more convenient alternative procedure consists of directly assuming a homogeneous solution of the form

$$y_H = x^r,$$

corresponding to the assumption $y_H = e^{rz}$ in the transformed equation. By making use of the relationship

$$x^m \frac{d^m x^r}{dx^m} = r(r-1) \cdots (r-m+1)x^r,$$

there follows, with the notation of Equation (27),

$$Lx^r = \{[r(r-1) \cdots (r-n+1)]$$
$$+ b_1[r(r-1) \cdots (r-n+2)] + \cdots + b_{n-1}r + b_n\}x^r.$$

Hence x^r is a homogeneous solution if r satisfies the *characteristic equation*

$$[r(r-1) \cdots (r-n+1)] + b_1[r(r-1) \cdots (r-n+2)]$$
$$+ \cdots + b_{n-1}r + b_n = 0. \qquad (30)$$

This equation can be obtained from the left-hand side of Equation (27) by formally replacing $x^m(d^m y/dx^m)$ by the m-factor product

$$r(r-1) \cdots (r-m+1).$$

Let the n roots of Equation (30) be denoted by $r_1, r_2, \ldots, r_n$. If these roots are distinct, the general homogeneous solution is of the form

$$y_H = \sum_{k=1}^{n} c_k x^{r_k}. \qquad (31)$$

In analogy with the results of Section 1.5, we find that the second homogeneous solution corresponding to a double root r_1 is

$$\left[\frac{\partial}{\partial r}(x^r) \right]_{r=r_1} = x^{r_1} \log x$$

and the part of the homogeneous solution corresponding to an m-fold root r_1 is

$$x^{r_1}[c_1 + c_2 \log x + c_3 (\log x)^2 + \cdots + c_m (\log x)^{m-1}].$$

Further, to a conjugate pair of imaginary roots $r = a \pm ib$ there corresponds the solution

$$x^a[c_1 \cos (b \log x) + c_2 \sin (b \log x)].$$

The extension to the case of repeated imaginary roots is obvious.

Except in those cases in which the right-hand member is a linear combination of powers of x (and in certain other cases of little practical interest), particular solutions of nonhomogeneous equations of type (27) usually cannot be obtained by the method of undetermined coefficients. However, it is readily shown by using the substitution (28) that a particular solution corresponding to a right-hand member of the form x^s is given by $y_P = Ax^s$, where A is a constant to be determined by substitution, unless x^s is a homogeneous solution. If x^s is a homogeneous solution, the trial particular solution should be of the form $y_P = Ax^s(\log x)^k$, where k is the smallest positive integer for which this expression is not a homogeneous solution. In other cases, particular solutions can be obtained by the methods of Sections 1.7 and 1.9.

Example 2. For the equation

$$x^2 \frac{d^2y}{dx^2} - 2x \frac{dy}{dx} + 2y = x^2 + 2,$$

of Example 1, the characteristic equation (30) becomes

$$r(r-1) - 2r + 2 = r^2 - 3r + 2 = 0,$$

from which $r = 1, 2$. The homogeneous solution is thus

$$y_H = c_1 x + c_2 x^2.$$

Since x^2 is a homogeneous solution, we assume a solution $y_P = Ax^2 \log x$ corresponding to the right-hand term x^2 and a solution $y_P = B$ corresponding to the constant term, and hence write

$$y_P = Ax^2 \log x + B.$$

in this case
$k = 1$ which is the
smallest positive integer

Substitution into the given differential equation gives $A = B = 1$, and the complete solution is

$$y = c_1 x + c_2 x^2 + x^2 \log x + 1. \qquad \blacksquare$$

1.7. Properties of Linear Operators. We now consider more critically certain properties of *linear differential operators* of the *general* form

$$L = a_0(x) \frac{d^n}{dx^n} + a_1(x) \frac{d^{n-1}}{dx^{n-1}} + \cdots + a_{n-1}(x) \frac{d}{dx} + a_n(x). \qquad (32)$$

An expression of this sort has no intrinsic meaning by itself, but when it is followed by a function $u(x)$, the result Lu is defined to be a new function of x defined by the relationship

$$Lu \equiv \left(a_0 \frac{d^n}{dx^n} + a_1 \frac{d^{n-1}}{dx^{n-1}} + \cdots + a_{n-1} \frac{d}{dx} + a_n \right) u$$

$$\equiv a_0 \frac{d^n u}{dx^n} + a_1 \frac{d^{n-1} u}{dx^{n-1}} + \cdots + a_{n-1} \frac{du}{dx} + a_n u.$$

We speak of Lu as the result of *operating* on u by the *operator L*.

If L_1 and L_2 are two linear operators, we say that L_1 and L_2 are *equal* when $L_1 u = L_2 u$ for every function u for which the operations are defined. In addition, we write $L_2 L_1 u$ to indicate the operation $L_2(L_1 u)$, that is, the result of operating on $L_1 u$ by L_2, and similarly for three or more successive operations. The abbreviations $L^2 u \equiv LLu$, $L^3 u \equiv LLLu$, and so on, are frequently used. In particular, if the operator d/dx is written as D,

$$D = \frac{d}{dx}, \qquad (33)$$

there follows

$$D^2 = \frac{d}{dx}\left(\frac{d}{dx}\right) = \frac{d^2}{dx^2}, \qquad D^3 = \frac{d}{dx}\left(\frac{d^2}{dx^2}\right) = \frac{d^3}{dx^3}, \qquad \cdots,$$

and in general $D^m = d^m/dx^m$. Thus Equation (32) can be written in the equivalent form

$$L = a_0(x)D^n + a_1(x)D^{n-1} + \cdots + a_{n-1}(x)D + a_n(x)$$

$$= \sum_{k=0}^{n} a_{n-k}(x)D^k. \tag{34}$$

The operations L_2L_1u and L_1L_2u should be carefully distinguished from each other, since the two operations are not, in general, equivalent. To illustrate, let $L_1 = d/dx$ and $L_2 = x(d/dx)$. Then

$$L_2L_1u = x\frac{d}{dx}\left(\frac{du}{dx}\right) = x\frac{d^2u}{dx^2},$$

and

$$L_1L_2u = \frac{d}{dx}\left(x\frac{du}{dx}\right) = x\frac{d^2u}{dx^2} + \frac{du}{dx}.$$

If the order in which the operators L_1 and L_2 are applied is immaterial, that is, if $L_1L_2u = L_2L_1u$, the two operators are said to be *commutative*. Similarly, we say that a *set* of operators is commutative if each pair of operators in the set is commutative.

It is clear that any two operators of the form D^m and D^n are commutative; so also are two operators of the form a_mD^m and a_nD^n, where a_m and a_n are constant. From this fact it follows easily that *the set of linear operators with constant coefficients is commutative*.

The commutativity of two linear equidimensional operators, for which $a_k(x) = b_kx^k$, is seen to depend upon the commutativity of any two operators of the form $L_1 = b_mx^m(d^m/dx^m)$ and $L_2 = b_nx^n(d^n/dx^n)$. But with the substitution (28), Equation (29) shows that L_1 and L_2 become linear operators with constant coefficients, and hence are commutative. Thus it follows that *the set of equidimensional linear operators is commutative*.

It may be remarked, however, that commutativity is the exception rather than the rule. Thus, for example, the above illustration shows that linear operators with constant coefficients and equidimensional linear operators are not in general commutative with each other.

The *distributive* property of linear operators,

$$(c_1L_1 + c_2L_2 + \cdots + c_nL_n)u = c_1L_1u + c_2L_2u + \cdots + c_nL_nu, \tag{35}$$

as well as the distributive property of linear operations,

$$L(c_1u_1 + c_2u_2 + \cdots + c_nu_n) = c_1Lu_1 + c_2Lu_2 + \cdots + c_nLu_n, \tag{36}$$

of which use has already been made, is easily established.

In many cases it is possible to *factor* a linear operator into the product of n linear factors. If the factors are commutative, the result of factoring is unique, the *order* in which the factors are written being arbitrary. Otherwise, the component factors will differ in form according to the position that they occupy in the product.

To illustrate, the operator $D^2 - 3D + 2$ can be factored uniquely in the forms $(D - 2)(D - 1) = (D - 1)(D - 2)$. However, the operator $xD^2 + D$ factors in two ways, into the products $(D)(xD)$ and $(xD + 1)(D)$. By this statement we mean, of course, that

$$(D)(xD)u = (xD + 1)(D)u = (xD^2 + D)u$$

for any twice-differentiable u.

Use is frequently made of the factoring process in solving linear differential equations. Thus, in the case of the homogeneous linear equation with constant coefficients, the operator

$$L = D^n + a_1 D^{n-1} + \cdots + a_{n-1}D + a_n$$

can be factored uniquely into the linear factors

$$L = (D - r_1)(D - r_2) \cdots (D - r_n),$$

where $r_1, r_2, \ldots, r_n$ are roots of the formal equation $L = 0$. Thus the equation $Ly_H = 0$ becomes

$$(D - r_1)(D - r_2) \cdots (D - r_n)y_H = 0,$$

where each operator operates on the expression to its right. Hence the complete expression will be zero if the result of the first operation is zero. Since any one of the n operators can be written immediately before y, it follows that a solution of any one of the n equations

$$(D - r_k)y_H = 0 \qquad (k = 1, 2, \ldots, n)$$

is a solution of $Ly_H = 0$. But these equations are equivalent to

$$\frac{dy_H}{dx} - r_k y_H = 0 \qquad (k = 1, 2, \ldots, n)$$

and are readily solved to give the solutions

$$y_H = c_k e^{r_k x} \qquad (k = 1, 2, \ldots, n).$$

By superimposing these solutions, the general homogeneous solution is obtained in the case where the roots are distinct, in accordance with the results of Section 1.5. The part of the solution corresponding to m-fold roots can be obtained as the solution of $(D - r_1)^m y_H = 0$.

Analogous procedures can be applied in other cases. In particular, the general solution of any linear differential equation with constant coefficients and *arbitrary* right-hand side can be obtained by a method illustrated by the following example.

Example. To solve the differential equation

$$\frac{d^2y}{dx^2} - 3\frac{dy}{dx} + 2y = h(x),$$

we write the equation in the operational form

$$(D - 2)(D - 1)y = h(x).$$

With the definition

$$y_1 = (D - 1)y,$$

the differential equation becomes

$$(D - 2)y_1 = h(x) \quad \text{or} \quad \frac{dy_1}{dx} - 2y_1 = h(x).$$

This linear equation is of first order and is solved, by using the results of Section 1.4, in the form

$$y_1 = e^{2x} \int e^{-2x}h(x) \, dx + c_1 e^{2x}.$$

Next, replacing y_1 by $(D - 1)y$, we obtain a second first-order equation,

$$\frac{dy}{dx} - y = e^{2x} \int e^{-2x}h(x) \, dx + c_1 e^{2x},$$

with solution

$$y = e^x \int e^x \left[\int e^{-2x}h(x) \, dx \right] dx + c_1 e^{2x} + c_2 e^x. \qquad \blacksquare$$

Similarly, it can be shown that the linear equidimensional operator of Equation (27) can be factored into the commutative factors

$$(xD - r_1)(xD - r_2) \cdots (xD - r_n),$$

where $r_1, r_2, \ldots, r_n$ are the roots of the characteristic equation (30). In this connection it should be noticed that the operators

$$x^m \frac{d^m}{dx^m} \quad \text{and} \quad \left(x \frac{d}{dx} \right)^m$$

are *not* equivalent.

The notion of operators is useful in establishing the general validity of the method of *undetermined coefficients* described in Section 1.5. To illustrate the argument, we here consider an equation of the form $Ly = a \cos qx$, where L is a linear differential operator with constant coefficients. Since the operator $D^2 + q^2$ annihilates the right-hand member, it follows that all solutions of the given equation are included in the general solution of the equation $L^*y = 0$, where

$$L^* = (D^2 + q^2)L.$$

However, if $D^2 + q^2$ is not a factor of L, then the general solution of this equation is

$$y = y_H + A \cos qx + B \sin qx,$$

where $y = y_H$ is the general solution of $Ly = 0$. If $D^2 + q^2$ is an unrepeated factor of L, it is a *double* factor of L^*. Since $\cos qx$ and $\sin qx$ are already present in y_H, $A \cos qx + B \sin qx$ then must be replaced by $Ax \cos qx + Bx \sin qx$, and so forth, in accordance with the rules set down in Section 1.5.

1.8. Simultaneous Linear Differential Equations. Frequently, two or more unknown functions are related to a single independent variable by an equal number of linear differential equations. Thus, in the case of two unknown functions x and y and the independent variable t, we may have a pair of simultaneous equations of the form

$$L_1 x + L_2 y = h_1(t), \tag{37a}$$

$$L_3 x + L_4 y = h_2(t), \tag{37b}$$

where the L's are linear differential operators in t. The unknown functions x and y are to be determined as functions of t.

When the operators involved are commutative, all unknown functions except one can be eliminated successively from the given set of equations to give a new set of linear differential equations, each involving only one unknown function. We illustrate this procedure in the case of Equations (37a, b).

If Equation (37a) is operated on by L_4 and (37b) by $-L_2$, and if the resultant equations are added, there follows

$$(L_4 L_1 - L_2 L_3)x + (L_4 L_2 - L_2 L_4)y = L_4 h_1 - L_2 h_2,$$

or, if L_2 and L_4 are commutative,

$$(L_4 L_1 - L_2 L_3)x = L_4 h_1 - L_2 h_2. \tag{38a}$$

Similarly, to eliminate x we operate on Equation (37a) by $-L_3$ and on (37b) by L_1. If L_1 and L_3 are also commutative, we then obtain, by addition,

$$(L_1 L_4 - L_3 L_2)y = L_1 h_2 - L_3 h_1.$$

Finally, if L_1, L_4 and L_2, L_3 are commutative, the operators of x and y in the last equations are identical, and we have

$$(L_4 L_1 - L_2 L_3)y = L_1 h_2 - L_3 h_1. \tag{38b}$$

Equations (38a) and (38b) can be written formally in the determinantal form

$$\Delta x = \begin{vmatrix} h_1 & L_2 \\ h_2 & L_4 \end{vmatrix}, \tag{39a}$$

$$\Delta y = \begin{vmatrix} L_1 & h_1 \\ L_3 & h_2 \end{vmatrix}, \tag{39b}$$

where Δ is the operator

$$\Delta = \begin{vmatrix} L_1 & L_2 \\ L_3 & L_4 \end{vmatrix}, \tag{40}$$

if it is understood that in each term of the expansion of the right-hand sides of Equations (39a, b) the operator is to be written before the function operated upon. The formal analogy with *Cramer's rule* for solving linear equations by determinants should be noticed.

Since the same operator affects x and y in Equations (39a, b), it is seen that the homogeneous solutions of these equations are linear combinations of the same functions, the number n of independent constants in each linear combination being equal to the degree of the operator Δ. Thus the solutions of Equations (39a, b) will contain $2n$ independent constants.

At this stage of the solution a certain amount of care must be taken. It is clear that, since Equations (37a, b) imply (39a, b), all solutions of the original simultaneous equations are contained in the solutions of the final equations. However, since differentiation generally is involved in obtaining (39) from (37), the converse is not generally true, in the sense that the solutions of (39a, b) may satisfy (37a, b) only if certain relationships exist among the $2n$ constants. These relationships may be determined by substituting the solutions of (39) into (37a, b) and requiring that the resultant equations be identities. However, if the coefficients are *constants*, and if in one of Equations (37a, b) the two operators involved have no common factors, the relationships are completely determined by substitution into that single equation (see Problem 31). If x_P and y_P satisfy Equations (37), only the added x_H and y_H need be so checked. An alternative procedure consists of solving only one of Equations (39a, b) for one unknown function and of then substituting this result into whichever of (37a, b) is more convenient for the subsequent determination of the second unknown function. The expressions so obtained are then introduced into the remaining one of Equations (37a, b) to determine possible restrictions on the arbitrary constants.

The extension of this procedure to cases in which more than two unknown functions are present leads to results again completely analogous to the statement of Cramer's rule. Thus, assuming that all operators involved are commutative with each other, the solutions of the equations

$$L_1 x + L_2 y + L_3 z = h_1(t),$$
$$L_4 x + L_5 y + L_6 z = h_2(t),$$
$$L_7 x + L_8 y + L_9 z = h_3(t)$$

are also solutions of three linear differential equations each involving only one dependent variable, one of which can be written formally as

$$\Delta y = \begin{vmatrix} L_1 & h_1 & L_3 \\ L_4 & h_2 & L_6 \\ L_7 & h_3 & L_9 \end{vmatrix},$$

where

$$\Delta = \begin{vmatrix} L_1 & L_2 & L_3 \\ L_4 & L_5 & L_6 \\ L_7 & L_8 & L_9 \end{vmatrix},$$

if in each term of the expansion of the first determinant the function is written after the operators. If the operator Δ is of order n, the solutions of the three equations so obtained involve $3n$ arbitrary constants, and possible restriction on these constants must be obtained by substitution into the *original* equations. This procedure may be quite laborious if several unknown functions are present.

A useful check is provided by the known fact that, if all operators involved are commutative, *the total number of independent constants present in the solu-*

tion of a set of linear differential equations is equal to the order of the operator Δ. This order cannot exceed the sum of the orders of the several equations and in special cases may be less than this number.

A method of solving such sets of equations with reduced labor in certain problems, in cases when the operators have constant coefficients, is presented in Chapter 2.

To illustrate the preceding method, we solve the equations

$$\frac{d^2x}{dt^2} - x - 2y = t, \tag{41a}$$

$$\frac{d^2y}{dt^2} - 2y - 3x = 1. \tag{41b}$$

In operational form, these equations become

$$(D^2 - 1)x - 2y = t,$$
$$-3x + (D^2 - 2)y = 1.$$

Equation (40) gives

$$\Delta = \begin{vmatrix} D^2 - 1 & -2 \\ -3 & D^2 - 2 \end{vmatrix} = D^4 - 3D^2 - 4,$$

and Equations (39a, b) then become

$$(D^4 - 3D^2 - 4)x = \begin{vmatrix} t & -2 \\ 1 & D^2 - 2 \end{vmatrix} = (D^2 - 2)t + 2 = 2 - 2t \tag{42a}$$

and

$$(D^4 - 3D^2 - 4)y = \begin{vmatrix} D^2 - 1 & t \\ -3 & 1 \end{vmatrix} = (D^2 - 1)1 + 3t = 3t - 1. \tag{42b}$$

We notice that the characteristic equation for both x and y is obtained by formally replacing D by r in the expression for $\Delta = 0$,

$$r^4 - 3r^2 - 4 = 0,$$

from which $r = \pm 2, \pm i$. Hence we obtain

$$x_H = c_1 e^{2t} + c_2 e^{-2t} + c_3 \cos t + c_4 \sin t,$$
$$y_H = d_1 e^{2t} + d_2 e^{-2t} + d_3 \cos t + d_4 \sin t.$$

Particular solutions of Equations (42a, b) are readily found by inspection or by the method of undetermined coefficients,

$$x_P = \tfrac{1}{2}t - \tfrac{1}{2}, \qquad y_P = \tfrac{1}{4} - \tfrac{3}{4}t.$$

To determine the relationships which must exist among the c's and d's, we may first verify that x_P and y_P satisfy Equations (41a, b) as well as (42a, b). Hence we then introduce the expressions for x_H and y_H into the left-hand sides of Equations (41a, b) and require the results to be identically zero. Using first

Equation (41a), we find the conditions

$$d_1 = \tfrac{3}{2}c_1, \qquad d_2 = \tfrac{3}{2}c_2, \qquad d_3 = -c_3, \qquad d_4 = -c_4.$$

The same conditions are obtained by using Equation (41b) (see Problem 31). Thus four of the eight constants are truly arbitrary. Retaining the four c's, we write

$$x_H = c_1 e^{2t} + c_2 e^{-2t} + c_3 \cos t + c_4 \sin t,$$

$$y_H = \tfrac{3}{2}(c_1 e^{2t} + c_2 e^{-2t}) - (c_3 \cos t + c_4 \sin t),$$

and the final solutions are

$$x = c_1 e^{2t} + c_2 e^{-2t} + c_3 \cos t + c_4 \sin t + \tfrac{1}{2}t - \tfrac{1}{2},$$

$$y = \tfrac{3}{2}(c_1 e^{2t} + c_2 e^{-2t}) - (c_3 \cos t + c_4 \sin t) - \tfrac{3}{4}t + \tfrac{1}{4}.$$

It may be seen that Equation (42a) *could be obtained directly*, in this case, *by solving* (41a) *for y and substituting the result into* (41b). In more complicated cases the present procedure is generally preferable.

If the expression for x_H were introduced into the left-hand side of Equation (41a) and the right-hand side were replaced by zero, an expression for y_H would be found directly, in this case, in terms of the constants of x_H. Substitution of these results into the left-hand side of Equation (41b) would then show that no further restrictions on the constants were necessary. If Equation (41b) were used to determine y_H in terms of x_H, the two new constants introduced would be determined in terms of the c's by substitution into (41a).

The solutions of Equations (41a, b) can also be obtained by a slightly different but equivalent method which is of some practical interest. Since the coefficients in both linear equations are constants, it can be assumed initially that homogeneous solutions exist of the form

$$x_H = c_k e^{r_k t}, \qquad y_H = d_k e^{r_k t}.$$

By introducing these assumptions into Equations (41a, b) and replacing the right-hand sides by zeros, we obtain the conditions

$$(r_k^2 - 1)c_k - 2d_k = 0,$$

$$-3c_k + (r_k^2 - 2)d_k = 0.$$

In order that nontrivial solutions of these equations exist, it is necessary that the determinant of the coefficients of c_k and d_k vanish, giving the characteristic equation obtained previously. If r_k satisfies this equation, the coefficient d_k can be expressed in terms of c_k by either of the two equations. Thus, using the first equation, we have

$$d_k = \frac{r_k^2 - 1}{2} c_k.$$

For the roots $r = \pm 2$ there follows $d_k = \tfrac{3}{2}c_k$, whereas for $r = \pm i$ there follows $d_k = -c_k$. These results are easily shown to lead, by superposition, to the previously obtained homogeneous solutions.

In order to obtain particular solutions directly from Equations (41a, b), the method of undetermined coefficients can be applied if all terms on the right-hand sides of the equations are taken into account in constructing the families. Thus, from Equation (41a) we have the family $\{t, 1\}$ and from (41b) the family $\{1\}$, which is contained in the former family. Since there is no representative in either homogeneous solution, we assume particular solutions of the form

$$x_P = At + B, \qquad y_P = Ct + D.$$

Substitution into Equations (41a, b) gives

$$-(A + 2C)t - (B + 2D) = t,$$
$$-(3A + 2C)t - (3B + 2D) = 1.$$

In order that these be identities, we must have

$$-A - 2C = 1, \quad B + 2D = 0, \quad 3A + 2C = 0, \quad -3B - 2D = 1,$$

from which there follows

$$A = \tfrac{1}{2}, \qquad B = -\tfrac{1}{2}, \qquad C = -\tfrac{3}{4}, \qquad D = \tfrac{1}{4},$$

in accordance with the previously obtained results.

In order to illustrate a special situation, we consider also the set

$$2\frac{dx}{dt} - 3x + y = 4e^t,$$

$$x + 2\frac{dy}{dt} - 3y = 0. \tag{43}$$

If Equations (39) and (40) are applied to these equations, there follows

$$\frac{d^2x}{dt^2} - 3\frac{dx}{dt} + 2x = -e^t,$$

$$\frac{d^2y}{dt^2} - 3\frac{dy}{dt} + 2y = -e^t, \tag{44}$$

from which one obtains

$$x_H = c_1 e^{2t} + c_2 e^t,$$
$$y_H = d_1 e^{2t} + d_2 e^t \tag{45}$$

and

$$x_P = te^t,$$
$$y_P = te^t. \tag{46}$$

But here it happens that x_P and y_P do *not* satisfy the *original* Equations (43). Thus it is necessary to substitute the *sums* $x_H + x_P$ and $y_H + y_P$ into Equations (43) for the purpose of obtaining conditions on the constants in Equations (45). This process gives, finally,

$$d_2 = 2 + c_2, \qquad d_1 = -c_1. \tag{47}$$

1.9. Particular Solutions by Variation of Parameters. We next derive a method for determining the complete solution of *any* linear differential equation for which the general homogeneous solution is known.

Suppose that the general homogeneous solution of the equation

$$Ly = \frac{d^n y}{dx^n} + a_1(x)\frac{d^{n-1}y}{dx^{n-1}} + \cdots + a_{n-1}(x)\frac{dy}{dx} + a_n(x)y = h(x) \qquad (48)$$

has been obtained in the form

$$y_H = \sum_{k=1}^{n} c_k u_k(x), \qquad (49)$$

where the u's are n linearly independent homogeneous solutions and the c's are n arbitrary constants or "parameters." We will find that a particular solution of the complete equation can be obtained by replacing the constant parameters c_k in the solution of the associated homogeneous equation by certain *functions of x*. Thus we assume that

$$y_P = \sum_{k=1}^{n} C_k(x)u_k(x) \qquad (50)$$

is a solution of Equation (48) and attempt to choose the n functions C_k suitably. Since we have n functions to determine, and since the requirement that Equation (50) satisfy Equation (48) represents only one condition, we have $n-1$ additional conditions at our disposal.

Differentiating Equation (50) and using primes to denote differentiation with respect to x, we obtain

$$\frac{dy_P}{dx} = \sum_{k=1}^{n} C_k u_k' + \sum_{k=1}^{n} C_k' u_k.$$

In order to simplify this expression, we require as our first condition that the second summation vanish,

$$\sum_{k=1}^{n} C_k' u_k = 0. \qquad (51a)$$

There then follows

$$\frac{dy_P}{dx} = \sum_{k=1}^{n} C_k u_k'$$

and

$$\frac{d^2 y_P}{dx^2} = \sum_{k=1}^{n} C_k u_k'' + \sum_{k=1}^{n} C_k' u_k'.$$

As the second condition, we require again that the sum of the terms involving *derivatives* of the C's vanish,

$$\sum_{k=1}^{n} C_k' u_k' = 0. \qquad (51b)$$

Proceeding in this way through the $(n-1)$th derivative, we have as our $(n-1)$th condition the requirement

$$\sum_{k=1}^{n} C_k' u_k^{(n-2)} = 0 \qquad (51c)$$

and the $(n-1)$th derivative is

$$\frac{d^{n-1}y_P}{dx^{n-1}} = \sum_{k=1}^{n} C_k u_k^{(n-1)}.$$

The expression for the nth derivative is then

$$\frac{d^n y_P}{dx^n} = \sum_{k=1}^{n} C_k u_k^{(n)} + \sum_{k=1}^{n} C_k' u_k^{(n-1)}.$$

By introducing the expressions for y_P and its derivatives into the left-hand side of Equation (48), we find that the final condition, that Equation (50) satisfy (48), becomes

$$Ly_P = \sum_{k=1}^{n} C_k u_k^{(n)} + a_1(x) \sum_{k=1}^{n} C_k u_k^{(n-1)} + \cdots$$

$$+ a_{n-1}(x) \sum_{k=1}^{n} C_k u_k' + a_n(x) \sum_{k=1}^{n} C_k u_k + \sum_{k=1}^{n} C_k' u_k^{(n-1)} = h(x).$$

Combining the first summations, we obtain

$$\sum_{k=1}^{n} C_k \left[\frac{d^n u_k}{dx^n} + a_1(x) \frac{d^{n-1}u_k}{dx^{n-1}} + \cdots \right.$$

$$\left. + a_{n-1}(x) \frac{du_k}{dx} + a_n(x)u_k \right] + \sum_{k=1}^{n} C_k' u_k^{(n-1)} = h(x).$$

Now, since each function u_k satisfies Equation (48) with $h(x)$ replaced by zero, and since each bracket in the first summation is precisely the result of replacing y in the left-hand side of Equation (48) by a function u_k, the first summation vanishes identically, and the final condition becomes merely

$$\sum_{k=1}^{n} C_k' u_k^{(n-1)} = h(x). \tag{51d}$$

In summary, the n conditions imposed on the n unknown functions can be written in the expanded form

$$\begin{aligned}
C_1'(x)u_1(x) + C_2'(x)u_2(x) + \cdots + C_n'(x)u_n(x) &= 0, \\
C_1'(x)u_1'(x) + C_2'(x)u_2'(x) + \cdots + C_n'(x)u_n'(x) &= 0, \\
\cdots\cdots\cdots\cdots\cdots\cdots\cdots\cdots\cdots\cdots\cdots\cdots\cdots\cdots\cdots&\cdots \\
C_1'(x)u_1^{(n-2)}(x) + C_2'(x)u_2^{(n-2)}(x) + \cdots + C_n'(x)u_n^{(n-2)}(x) &= 0, \\
C_1'(x)u_1^{(n-1)}(x) + C_2'(x)u_2^{(n-1)}(x) + \cdots + C_n'(x)u_n^{(n-1)}(x) &= h(x).
\end{aligned} \tag{52}$$

If this set of equations is solved for $C_1', C_2', \ldots, C_n'$ by Cramer's rule, the common-denominator determinant is seen to be the Wronskian of $u_1, u_2, \ldots, u_n$.†

If the solutions $C_1', C_2', \ldots, C_n'$ are integrated and the results are introduced into Equation (50), the result is a particular solution of Equation (48) for any

---

†It is known that, if the coefficients $a_1, a_2, \ldots, a_n$ in the standard form (48) are continuous in an interval I, the indicated derivatives $u_k^{(r)}$ exist in I, and furthermore the Wronskian of the linearly independent functions cannot vanish in I. Hence a unique solution of (52) then exists.

choice of the n constants of integration. If the constants are left arbitrary, this procedure yields the complete solution of Equation (48).

It is important to notice that Equation (48) was written in "standard form." If the coefficient of $d^n y/dx^n$ in Equation (48) were $a_0(x)$, the last equation of (52) would be modified by replacing $h(x)$ by $h(x)/a_0(x)$.

In particular, for a second-order linear equation of the form

$$\frac{d^2 y}{dx^2} + a_1(x)\frac{dy}{dx} + a_2(x)y = h(x), \tag{53}$$

there follows

$$y = C_1(x)u_1(x) + C_2(x)u_2(x), \tag{54a}$$

where

$$C_1' = \frac{\begin{vmatrix} 0 & u_2 \\ h & u_2' \end{vmatrix}}{\begin{vmatrix} u_1 & u_2 \\ u_1' & u_2' \end{vmatrix}} = -\frac{h(x)u_2(x)}{W[u_1(x), u_2(x)]}$$

and, similarly,

$$C_2' = \frac{h(x)u_1(x)}{W[u_1(x), u_2(x)]}.$$

Thus we can write

$$\begin{aligned} C_1 &= -\int \frac{h(x)u_2(x)}{W[u_1(x), u_2(x)]}\, dx + c_1, \\ C_2 &= \int \frac{h(x)u_1(x)}{W[u_1(x), u_2(x)]}\, dx + c_2 \end{aligned} \tag{54b}$$

and the introduction of these results into Equation (54a) gives the required solution.

If $h(x)$ is not given explicitly, but the general solution is required for an *arbitrary* $h(x)$, we may combine the result of this substitution into a more compact form if, before substituting Equations (54b) into (54a), we denote the dummy variable of integration by a new symbol ξ, to distinguish it from the current variable x which now will appear in the integrand (as well as in the upper limit of integration). Substitution of Equations (54b) into (54a) then leads to the result

$$y = \int^x \frac{h(\xi)[u_1(\xi)u_2(x) - u_2(\xi)u_1(x)]}{W[u_1(\xi), u_2(\xi)]}\, d\xi + c_1 u_1(x) + c_2 u_2(x), \tag{55}$$

where a convenient lower limit is assumed in the integral. If $h(x)$ is given explicitly, the direct evaluation of Equations (54b) and subsequent substitution into (54a) is usually more convenient than the use of Equation (55).

Example 1. For the differential equation

$$\frac{d^2 y}{dx^2} + y = h(x),$$

two linearly independent homogeneous solutions are $u_1 = \cos x$, $u_2 = \sin x$. The Wronskian is

$$W(\cos x, \sin x) = \begin{vmatrix} \cos x & \sin x \\ -\sin x & \cos x \end{vmatrix} = \sin^2 x + \cos^2 x = 1.$$

Thus use of Equations (54a,b) gives the solution

$$y = -\cos x \left[\int h(x) \sin x\, dx + c_1 \right] + \sin x \left[\int h(x) \cos x\, dx + c_2 \right].$$

This form is usually most convenient for actual evaluation of the solution when $h(x)$ is given. The form (55), which is useful in more general considerations, here becomes

$$y = \int^x h(\xi)[\cos \xi \sin x - \sin \xi \cos x]\, d\xi + c_1 \cos x + c_2 \sin x$$

or $\qquad y = \int^x h(\xi) \sin (x - \xi)\, d\xi + c_1 \cos x + c_2 \sin x.$ ∎

Example 2. For the differential equation

$$\frac{d^3 y}{dx^3} - 3\frac{d^2 y}{dx^2} + 2\frac{dy}{dx} = h(x),$$

we may take $u_1 = 1$, $u_2 = e^x$, $u_3 = e^{2x}$. Equations (52) then become

$$C_1' + C_2' e^x + C_3' e^{2x} = 0,$$
$$C_2' e^x + 2C_3' e^{2x} = 0,$$
$$C_2' e^x + 4C_3' e^{2x} = h(x).$$

For the determinant of this system we find $W(1, e^x, e^{2x}) = 2e^{3x}$. Solving the three simultaneous equations, we obtain

$$C_1' = \tfrac{1}{2}h(x), \qquad C_2' = -e^{-x}h(x), \qquad C_3' = \tfrac{1}{2}e^{-2x}h(x).$$

The solution of the differential equation is then

$$y = 1\left[\tfrac{1}{2} \int h(x)\, dx + c_1 \right] + e^x\left[-\int e^{-x}h(x)\, dx + c_2 \right] + e^{2x}\left[\tfrac{1}{2} \int e^{-2x}h(x)\, dx + c_3 \right]$$

or, equivalently,

$$y = \tfrac{1}{2} \int^x h(\xi)[1 - 2e^{x-\xi} + e^{2(x-\xi)}]\, d\xi + c_1 + c_2 e^x + c_3 e^{2x}.$$ ∎

It will be shown in Section 1.10 that the Wronskian of two homogeneous solutions of Equation (53) is of the form

$$W(u_1, u_2) = Ae^{-\int a_1(x)\, dx}, \tag{56}$$

where A is a specific *constant* depending only on the choice of the arbitrary multiplicative constants involved in the homogeneous solutions u_1 and u_2. (See also Problem 8.) It follows (see Problem 45) that $W(u_1, u_2)$ can be determined if only the values of u_1 and u_2 and their first derivatives are known *at a single point*. This fact is useful in determining the integrand in Equation (55) if, for example, the solutions $u_1(x)$ and $u_2(x)$ are expressed in terms of power series (see Chapter 4).

It can be shown† that, more generally, the Wronskian of n homogeneous solutions of Equation (48) is also given by the right-hand member of Equation (56). The statement of this fact is known as *Abel's formula*. From the properties of the exponential function, it follows that, if $a_1(x)$ is continuous in an interval I, the Wronskian cannot vanish in I unless it vanishes *identically*.

1.10. Reduction of Order. One of the important properties of linear differential equations is the fact that, if one *homogeneous* solution of an equation of order n is known, a new linear differential equation of order $n - 1$, determining the remainder of the solution, can be obtained. This procedure is in a sense analogous to the reduction of the degree of an algebraic equation when one solution is known.

Suppose that one homogeneous solution $u_1(x)$ is known. We next write $y = v(x)u_1(x)$ and attempt to determine the function $v(x)$. Substituting vu_1 for y in the left-hand side of the differential equation, we obtain a new linear differential equation of order n to determine v. But since $y = cu_1(x)$ is a homogeneous solution of the original equation, $v = c$ must be a homogeneous solution of the new equation. Hence the new equation must lack the term of zero order in v; that is, the coefficient of v must be zero. Thus the new equation is of order $n - 1$ in the variable dv/dx.

We apply this procedure to the solution of the general second-order linear equation

$$\frac{d^2y}{dx^2} + a_1(x)\frac{dy}{dx} + a_2(x)y = h(x), \tag{57}$$

assuming that one homogeneous solution $u_1(x)$ is known. Writing

$$y = v(x)u_1(x), \tag{58}$$

and introducing Equation (58) into (57), there follows

$$v''u_1 + 2v'u_1' + a_1v'u_1 + v(u_1'' + a_1u_1' + a_2u_1) = h.$$

But since u_1 is a homogeneous solution of Equation (57), the expression in parentheses vanishes and the differential equation determining v becomes

$$v''u_1 + 2v'u_1' + a_1v'u_1 = h$$

or

$$(v')' + \left(2\frac{u_1'}{u_1} + a_1\right)v' = \frac{h}{u_1}. \tag{59}$$

This equation is of first order in v', with an integrating factor given by the results of Section 1.4 in the form

$$e^{2\log u_1 + \int a_1 dx} = pu_1^2,$$

where

$$p = e^{\int a_1 dx}. \tag{60}$$

---

†By differentiating the Wronskian determinant, one obtains the relation $dW/dx = -a_1 W$.

Hence there follows

$$v' = \frac{1}{pu_1^2} \int phu_1 \, dx + \frac{c_1}{pu_1^2}.$$

An integration then gives

$$v = \int \frac{\int phu_1 \, dx}{pu_1^2} \, dx + c_1 \int \frac{dx}{pu_1^2} + c_2 \tag{61}$$

and the introduction of Equation (61) into (58) yields the general solution

$$y = u_1(x) \int \frac{\int phu_1 \, dx}{pu_1^2} \, dx + c_1 u_1(x) \int \frac{dx}{pu_1^2} + c_2 u_1(x). \tag{62}$$

Thus, if u_1 is one homogeneous solution, the most general linearly independent solution of the associated homogeneous equation is of the form

$$u_2 = Au_1(x) \int \frac{dx}{pu_1^2} + Bu_1(x), \tag{63}$$

where A and B are arbitrary constants with $A \neq 0$, and a particular solution of the complete equation is

$$y = u_1(x) \int \frac{\int phu_1 \, dx}{pu_1^2} \, dx. \tag{64}$$

We remark that no constants of integration need be added in either of the integrations in Equation (64), since the additional terms thereby introduced can be absorbed into the homogeneous solution. However, if arbitrary constants *are* introduced in each of the integrations, Equation (64) represents the complete solution of Equation (57).

In view of Equation (63), the Wronskian of any two homogeneous solutions of (57) is given by

$$W(u_1, u_2) = \begin{vmatrix} u_1 & Au_1 \int \frac{dx}{pu_1^2} + Bu_1 \\ u_1' & \frac{A}{pu_1} + Au_1' \int \frac{dx}{pu_1^2} + Bu_1' \end{vmatrix}.$$

Expansion of this determinant gives Abel's result, in this case,

$$W(u_1, u_2) = \frac{A}{p(x)} = Ae^{-\int a_1(x) \, dx}. \tag{65}$$

Example. One homogeneous solution of the equation

$$x^2 \frac{d^2y}{dx^2} + (x - 1)\left(x \frac{dy}{dx} - y\right) = x^2 e^{-x}$$

is seen by inspection to be $y = x$. To find the complete solution, we first write the equation in the standard form of (57),

$$\frac{d^2y}{dx^2} + \left(1 - \frac{1}{x}\right)\frac{dy}{dx} + \frac{1-x}{x^2}y = e^{-x},$$

and find

$$a_1(x) = 1 - \frac{1}{x}, \qquad p(x) = \frac{1}{x}e^x.$$

With $u_1 = x$, a second linearly independent solution is obtained from Equation (63), taking $A = 1$ and $B = 0$ for convenience, in the form

$$u_2 = x \int \frac{e^{-x}}{x} \, dx.$$

The indefinite integral which appears here cannot be evaluated in terms of elementary functions. However, the function

$$Ei(-x) = \int_\infty^x \frac{e^{-\xi}}{\xi} \, d\xi = \int_{-\infty}^{-x} \frac{e^\xi}{\xi} \, d\xi \qquad (x > 0)$$

is a tabulated function, known as the "exponential-integral function." [See Reference 8 of Chapter 4, for example, where similar integrals involving $(\sin x)/x$ and $(\cos x)/x$, and known as "sine-integral" and "cosine-integral" functions, are also defined and tabulated.] Thus the second independent homogeneous solution can be taken to be

$$u_2 = xEi(-x),$$

and the general homogeneous solution is of the form

$$y_H = x[c_1 + c_2 Ei(-x)].$$

A particular solution is given by Equation (64),

$$y_P = x \int \frac{e^{-x} \int e^{-x}e^x \, dx}{x} \, dx = x \int e^{-x} \, dx = -xe^{-x},$$

if constants of integration are omitted. Thus the complete solution is

$$y = x[c_1 + c_2 Ei(-x) - e^{-x}] \qquad (x > 0). \qquad \blacksquare$$

1.11. Determination of Constants. The n arbitrary constants present in the general solution of a linear differential equation of order n are to be determined by n suitably prescribed supplementary conditions.

Frequently, these conditions consist of the requirement that the function and its first $n - 1$ derivatives take on prescribed values at a given point $x = a$,

$$y(a) = y_0, \quad y'(a) = y_0', \quad \ldots, \quad y^{(n-1)}(a) = y_0^{(n-1)}. \qquad (66)$$

When such conditions are prescribed, the problem is known as an *initial-value problem*. In this case it is true that *if the point $x = a$ is included in an interval where the coefficients $a_1(x), \ldots, a_n(x)$ and the right-hand side $h(x)$ of the differential equation, in standard form, are continuous, there exists a unique solution satisfying Equations* (66). Here, if the complete solution is written in the form

$$y = \sum_{k=1}^{n} c_k u_k(x) + y_P(x), \qquad (67)$$

the conditions of Equations (66) require that the constants c_k satisfy the n equations

$$\sum_{k=1}^{n} c_k u_k^{(m)}(a) = y_0^{(m)} - y_P^{(m)}(a) \qquad (m = 0, 1, 2, \ldots, n-1). \qquad (68)$$

We notice that the determinant of the coefficients of the constants c_k is the value of the Wronskian of the linearly independent functions u_k at the point $x = a$, which (see footnote on page 25) cannot vanish under the specified conditions. Thus a unique solution is assured.

Sets of conditions other than those of Equations (66) may, however, be prescribed. For example, values of the function and/or certain of the derivatives may be prescribed at *two* distinct points $x = a$ and $x = b$, the problem then being known as a *boundary-value problem*. In such cases a solution may or may not exist, and may or may not be unique.

Example. The general solution of the differential equation

$$\frac{d^2 y}{dx^2} + y = 0$$

is $y = c_1 \cos x + c_2 \sin x$. The initial-value problem with conditions $y(0) = y_0$, $y'(0) = y'_0$ has the unique solution $y = y_0 \cos x + y'_0 \sin x$. The conditions $y(0) = 1$, $y(\pi/2) = 1$ imply the unique solution $y = \cos x + \sin x$, while the only solution satisfying the conditions $y(0) = y(\pi/2) = 0$ is the trivial solution $y = 0$. However, the conditions $y(0) = y(\pi) = 0$ are both satisfied if we take $c_1 = 0$, and hence in this case there exist an infinite number of solutions of the form $y = A \sin x$, where A is arbitrary. ∎

1.12. Special Solvable Types of Nonlinear Equations. Although there exist no techniques of general applicability for the purpose of obtaining solutions of nonlinear differential equations in closed form, there are several special types of equations for which such solutions can be obtained, a few of which are treated very briefly in this section. Here, instead of seeking y as a function of x, we may be led to determine x as a function of y or to accept as a solution a functional relationship involving the two variables in a less simple way.

(1) *Separable Equations.* Separable first-order equations have already been mentioned in Section 1.1.

Example 1. The equation $(1 + x^2)\, dy + (1 + y^2)\, dx = 0$ is separable in the form

$$\frac{dx}{1 + x^2} + \frac{dy}{1 + y^2} = 0,$$

and integration gives the solution

$$\tan^{-1} x + \tan^{-1} y = c$$

or

$$x + y = c'(1 - xy) \qquad (c' = \tan c). \quad ∎$$

Example 2. The equation

$$\left(\frac{dy}{dx}\right)^2 - 4y + 4 = 0$$

yields *two* separable equations when solved algebraically for dy/dx,

$$\frac{dy}{dx} = \pm 2\sqrt{y-1},$$

from which

$$\pm \frac{dy}{\sqrt{y-1}} = 2\,dx,$$

provided that the division by $\sqrt{y-1}$ is legitimate, and hence there follows $\pm\sqrt{y-1}$ $= x - c$ or

$$y = 1 + (x - c)^2.$$

Since the relation $y - 1 = 0$ has been excluded in the derivation of this solution, the possibility that $y = 1$ may also be a solution must be explored separately. Direct substitution into the original differential equation shows that $y = 1$ *is* indeed a solution. Furthermore, it cannot be obtained by specializing the *constant* c in the relation $y = 1 + (x - c)^2$. Here the complete solution consists of the latter one-parameter solution *together with* the "singular solution," $y = 1$. It can be verified, in this case, that all the curves which represent particular solutions, in correspondence with particular choices of the constant c, are *tangent* to the straight line representing the singular solution, so that this line is the *envelope* of those curves.   ■

(2) *Exact First-order Equations.* A first-order equation, written in the form

$$P(x, y)\,dx + Q(x, y)\,dy = 0, \qquad (69)$$

where P and Q are assumed to have continuous first partial derivatives, is said to be *exact* when P and Q satisfy the condition

$$\frac{\partial P}{\partial y} = \frac{\partial Q}{\partial x}. \qquad (70)$$

In this case, and in this case only, there exists a function $u(x, y)$ such that

$$du = P\,dx + Q\,dy. \qquad (71)$$

Thus Equation (69) then is identical with the equation $du = 0$, whose general solution clearly is

$$u(x, y) = c, \qquad (72)$$

where c is an arbitrary constant.

Since Equation (71) implies

$$\frac{\partial u}{\partial x} = P, \qquad \frac{\partial u}{\partial y} = Q, \qquad (73)$$

the *necessity* of Equation (70) follows from the fact that

$$\frac{\partial}{\partial x}\left(\frac{\partial u}{\partial y}\right) = \frac{\partial}{\partial y}\left(\frac{\partial u}{\partial x}\right)$$

when the indicated derivatives are continuous. In order to obtain a function u satisfying the two relations of Equations (73), we may, for example, start with the first relation, integrating with respect to x *with y held constant* (since $\partial u/\partial x$ was formed in this way) to obtain†

$$u(x, y) = \int_{x_0}^{x} P(x, y) \, dx + f(y). \tag{74}$$

Here $f(y)$ is the added "constant of integration," to be determined by the second relation of Equations (73), which gives

$$\int_{x_0}^{x} \frac{\partial P}{\partial y} \, dx + f'(y) = Q$$

and hence

$$f'(y) = Q - \int_{x_0}^{x} \frac{\partial P}{\partial y} \, dx. \tag{75}$$

That the right-hand member is indeed only a function of y, so that $f(y)$ can be determined (with an irrelevant arbitrary additive constant) by direct integration, follows from the fact that its partial derivative with respect to x is zero, since

$$\frac{\partial}{\partial x}\left(Q - \int_{x_0}^{x} \frac{\partial P}{\partial y} \, dx\right) = \frac{\partial Q}{\partial x} - \frac{\partial P}{\partial y} = 0,$$

when Equation (70) is satisfied, so that the *sufficiency* of that condition is also established.

Example 3. The equation

$$\frac{dy}{dx} = \frac{1 + y^2 + 3x^2y}{1 - 2xy - x^3}$$

can be written in the form

$$(3x^2y + y^2 + 1) \, dx + (x^3 + 2xy - 1) \, dy = 0,$$

and the condition (70) of exactness is satisfied. From the relation

$$\frac{\partial u}{\partial x} = 3x^2y + y^2 + 1$$

there follows $u = x^3y + xy^2 + x + f(y)$. The relation

$$\frac{\partial u}{\partial y} = x^3 + 2xy - 1$$

then gives $x^3 + 2xy + f'(y) = x^3 + 2xy - 1$ or $f'(y) = -1$, from which $f(y) = -y$, apart from an irrelevant arbitrary additive constant. Hence the required solution $u = c$ is

$$x^3y + xy^2 + x - y = c. \qquad \blacksquare$$

†The special notation

$$\int_{x_0}^{x} P(x, y) \, \partial x$$

is sometimes used to indicate that "partial integration" with respect to x is to be effected, with y held constant in the process.

(3) *First-order Equations of Homogeneous Type.* A *function* $f(x, y)$ is said to be *homogeneous of degree n* if there exists a constant n such that, for every number λ, it is true that

$$f(\lambda x, \lambda y) = \lambda^n f(x, y).$$

Thus, for example, the functions $x^2 + xy$ and $\tan^{-1}(y/x)$ are both homogeneous, the first of degree two and the second of degree zero, whereas $x^2 + y$ is *not* homogeneous. The first-order *differential equation*

$$P(x, y)\, dx + Q(x, y)\, dy = 0 \qquad (76)$$

may be said to be *of homogeneous type* if P and Q are *both* homogeneous of degree n, for some constant n.†

Such an equation becomes *separable* upon the change of variables

$$y = vx, \qquad dy = v\, dx + x\, dv. \qquad (77)$$

For, since $P(x, vx) = x^n P(1, v)$ and $Q(x, vx) = x^n Q(1, v)$, this substitution reduces Equation (76) to the form

$$x^n P(1, v)\, dx + x^n Q(1, v)(v\, dx + x\, dv) = 0$$

or $$[P(1, v) + vQ(1, v)]\, dx + xQ(1, v)\, dv = 0, \qquad (78)$$

which is indeed separable.

The substitution

$$x = uy, \qquad dx = u\, dy + y\, du$$

also is appropriate and may or may not lead to a more convenient form after a separation of variables.

Example 4. The equation

$$(3y^2 - x^2)\, dx = 2xy\, dy$$

is of homogeneous type (with $n = 2$) and, with $y = vx$, it becomes

$$(v^2 - 1)\, dx - 2xv\, dv = 0$$

from which there follows

$$\frac{2v\, dv}{v^2 - 1} = \frac{dx}{x} \qquad (x \neq 0,\ v^2 \neq 1)$$

and hence $$\log |v^2 - 1| = \log |x| + \log |c| = \log |cx|$$

or $$v^2 - 1 = cx$$

or $$y^2 - x^2 = cx^3.$$

The temporarily excluded relations, $x = 0$ and $y = \pm x$, both are seen to satisfy the given equation and hence in fact are solutions, but they need not be appended to the solution obtained since they are included in it when $c = \infty$ and $c = 0$. ∎

---

†Such an equation is usually called a *homogeneous equation.* However, we will use this term only as it is defined in Section 1.8 and generalized in Section 5.1.

(4) *Miscellaneous First-order Equations.* Most of the other known techniques for solving special first-order equations in closed form consist of reductions to forms which are linear, separable, exact, or of homogeneous type. Some typical examples follow.

Example 5. *Bernoulli's equation.* The equation

$$\frac{dy}{dx} + P(x)y = Q(x)y^n \qquad (n \neq 1)$$

can be written in the form

$$y^{-n} \frac{dy}{dx} + P(x)y^{-n+1} = Q(x),$$

which clearly becomes *linear* under the substitution $v = y^{-n+1}$. ■

Example 6. The equation

$$\frac{dy}{dx} = \frac{ax + by + c}{Ax + By + C}$$

can be reduced to an equation of *homogeneous* type by writing $x = t + h$ and $y = s + k$ and determining suitable values of the constants h and k, provided that $aB \neq bA$. In that exceptional case, the substitution $s = ax + by + c$ renders the equation *separable.* ■

Example 7. Although the equation

$$y^2\, dx = x(x\, dy - y\, dx)$$

can be treated as an equation of homogeneous type or as a Bernoulli equation in y (Example 5), the combination $x\, dy - y\, dx$ tends to suggest division by y^2 or by x^2, for the purpose of forming $-d(x/y)$ or $d(y/x)$. Here a division by y^2, accompanied by a division by x, is clearly indicated, leading to the form

$$\frac{dx}{x} + \frac{y\, dx - x\, dy}{y^2} = 0$$

which is *exact*, since each term is an exact differential, and there follows

$$\log |x| + \frac{x}{y} = c$$

or

$$y = \frac{x}{c - \log |x|}.$$

In this case, multiplication of the equation by the *integrating factor* $x^{-1}y^{-2}$ makes the equation exact. In less contrived situations, the discovery of such a factor may be much less straightforward. (See Problem 63.) ■

Example 8. The nonlinear Equation (11),

$$\left(\frac{dy}{dx}\right)^2 - 2\frac{dy}{dx} + 4y = 4x - 1,$$

yields the two first-degree equations

$$\frac{dy}{dx} = 1 \pm 2\sqrt{x - y}.$$

The prominence of the expression $x - y$ may suggest the *substitution*

$$x - y = u, \qquad 1 - \frac{dy}{dx} = \frac{du}{dx},$$

which leads to the separable equations

$$\frac{du}{dx} = \pm\, 2\sqrt{\bar{u}}.$$

If it is noticed that the process of separation necessitates the special consideration of the relation $u = 0$, the one-parameter solution (12) and the singular solution $y = x$ then are easily obtained. Equally fortuitous substitutions may suggest themselves in other cases. ∎

(5) *Second-order Equations Lacking One Variable.* The general equation of second order is of the form

$$F\left(x, y, \frac{dy}{dx}, \frac{d^2y}{dx^2}\right) = 0, \tag{79}$$

containing d^2y/dx^2 explicitly. Any such equation can be written equivalently as a pair of simultaneous first-order equations, in various ways. In particular, we may write

$$\frac{dy}{dx} = p \tag{80}$$

and

$$\frac{d^2y}{dx^2} = \begin{cases} \dfrac{dp}{dx} \\[2mm] \dfrac{dp}{dy}\dfrac{dy}{dx} = p\dfrac{dp}{dy} \end{cases} \tag{81}$$

in accordance with which Equation (79) can be replaced by either the set

$$F\left(x, y, p, \frac{dp}{dx}\right) = 0,$$

$$\frac{dy}{dx} = p \tag{82}$$

or the set

$$F\left(x, y, p, p\frac{dp}{dy}\right) = 0,$$

$$\frac{dy}{dx} = p. \tag{83}$$

In the general case, both equations of either set involve the *three* variables x, y, and p, and hence no one of the equations can be solved independently of the associated equation. However, if y is not explicitly involved in F, then the first equation of (82) involves only x and p. If it can be solved, to provide a relation between x and p, and if the result can be used to eliminate p from the second relation in Equation (82), then y can be determined in terms of x by integration of the resulting equation.

On the other hand, if x is not explicitly involved in F, then the first equation of (83) involves only y and p. If it can be solved, to provide a relation between y and p, and if the result can be used to eliminate p from the second relation in Equation (83), then a separable first-order equation in y and x results.

In either case, it may be more feasible to obtain x and y in terms of p, with p retained as a parameter, than to eliminate p by using the solution of the first equation of the pair.

Example 9. The equation

$$\frac{d^2y}{dx^2} = x\left(\frac{dy}{dx}\right)^3$$

lacks the variable y. With $dy/dx = p$ and $d^2y/dx^2 = dp/dx$, there follows

$$\frac{dp}{dx} = xp^3$$

which separates to give

$$p = \frac{\pm 1}{\sqrt{c_1^2 - x^2}}.$$

Hence

$$dy = \pm\frac{dx}{\sqrt{c_1^2 - x^2}},$$

from which there follows

$$y = \pm\sin^{-1}\frac{x}{c_1} + c_2 \quad \text{or} \quad x = c_1' \sin(y - c_2),$$

where $c_1' = \pm c_1$. ∎

Example 10. The equation

$$y\frac{d^2y}{dx^2} = \left(\frac{dy}{dx}\right)^2$$

lacks the variable x. With $dy/dx = p$ and $d^2y/dx^2 = p(dp/dy)$, there follows

$$p\left(y\frac{dp}{dy} - p\right) = 0$$

and hence

$$y\frac{dp}{dy} = p \quad \text{or} \quad p = 0.$$

The first alternative gives

$$p = c_1 y,$$

which includes the second alternative $p = 0$ as a special case, and hence the general solution is

$$y = c_2 e^{c_1 x}.$$ ∎

Example 11. The equation

$$\frac{d^2y}{dx^2} - y = 0$$

lacks the variable x. Whereas it can be solved by the present method, it is *linear* in y and is much more easily solved by the methods of Section 1.5. ∎

Example 12. The equation

$$\left(\frac{dy}{dx}\right)^2 \frac{d^2y}{dx^2} = 1 + \left(\frac{dy}{dx}\right)^2$$

lacks both x and y. If the absence of y is exploited, there follows

$$p^2 \frac{dp}{dx} = 1 + p^2$$

and hence

$$x = p - \tan^{-1} p + c_1.$$

This relation can be used to eliminate x (rather than p) from the relation

$$dy = p \, dx = p \frac{dx}{dp} \, dp,$$

to give

$$dy = p\left(1 - \frac{1}{1 + p^2}\right) dp = \frac{p^3}{1 + p^2} \, dp.$$

Hence

$$y = \tfrac{1}{2}p^2 - \tfrac{1}{2} \log (p^2 + 1) + c_2.$$

Here the solution provides a parametric representation of x and y. The fact that the parameter p happens to be identifiable with dy/dx may afford a subsequent added convenience. ∎

REFERENCES

1. AGNEW, R. P., *Differential Equations*, 2nd ed., McGraw-Hill Book Company, Inc., New York, 1960.

2. BIRKHOFF, G., and G.-C. ROTA, *Ordinary Differential Equations*, 2nd ed., Ginn and Company, Boston, 1969.

3. CODDINGTON, E. A., and N. LEVINSON, *Theory of Ordinary Differential Equations*, McGraw-Hill Book Company, Inc., New York, 1955.

4. DANIEL, J. W., and R. E. MOORE, *Computation and Theory in Ordinary Differential Equations*, W. H. Freeman and Company, Publishers, San Francisco, 1970.

5. INCE, E. L., *Ordinary Differential Equations*, 4th ed., Dover Publications, Inc., New York, 1953.

6. PIAGGIO, H. T. H., *Differential Equations*, Bell Mathematical Series, Open Court Publishing Company, La Salle, Ill., 1952.

PROBLEMS

Section 1.1

1. Differentiate each of the following relations with respect to x and, by using the result to eliminate the constant c, obtain a first-order differential equation of which the given relation defines a one-parameter solution:

(a) $y = ce^x$,

(b) $y = ce^{x^2} + 2x$,

(c) $y = \dfrac{1}{c - x}$,

(d) $y = e^{cx}$,

(e) $x^2 + y^2 = c^2$,

(f) $y = cx + c^2 x^2$.

2. By differentiating each of the following relations twice, and eliminating the constants c_1 and c_2 from the three resultant equations, obtain a second-order differential equation of which the given relation defines a two-parameter solution:

(a) $y = c_1 e^x + c_2 e^{-x}$,

(b) $y = e^x(c_1 + c_2 x)$,

(c) $y = c_1 \cos x + c_2 \sin x$,

(d) $y = c_1 e^{c_2 x}$,

(e) $y = \dfrac{c_1 + x}{c_2 + x}$,

(f) $(x - c_1)^2 + (y - c_2)^2 = 1$.

3. Obtain the general solution of each of the following differential equations:

(a) $\dfrac{dy}{dx} = 2xy$,

(b) $y \, dy + \sqrt{1 - y^2} \, dx = 0$,

(c) $(1 - y^2) \, dx + (1 - x^2) \, dy = 0$,

(d) $\sqrt{1 - x^2} y \, dy = \sqrt{1 - y^2} x \, dx$.

Section 1.2

4. If $u_1(x)$ and $u_2(x)$ are linearly independent, prove that $A_1 u_1(x) + A_2 u_2(x)$ and $B_1 u_1(x) + B_2 u_2(x)$ are also linearly independent if $A_1 B_2 - A_2 B_1 \neq 0$.

5. By considering the functions $u_1 = x^3$ and $u_2 = x^2 |x|$ over an interval including the origin, show that identical vanishing of the Wronskian of u_1 and u_2 over an interval does *not* imply linear dependence of u_1 and u_2 over that interval. (Notice that $u_2 = x^3 \operatorname{sgn} x$ and $u_2' = 3x^2 \operatorname{sgn} x$ for all x, where $\operatorname{sgn} x = +1$ when $x > 0$, $\operatorname{sgn} x = -1$ when $x < 0$, and $\operatorname{sgn} 0 = 0$.)

6. (a) Prove that $e^{r_1 x}$, $e^{r_2 x}$, and $e^{r_3 x}$ are linearly independent over any interval if r_1, r_2, and r_3 are distinct constants.

  (b) Determine whether $\log x$, $\log x^2$, and $\log x^3$ are linearly independent when $x > 0$.

7. Determine whether each of the following sets of functions is linearly dependent over the specified interval:

(a) $1 + 2x$, $2 + 3x$, $3 + 4x$ $(-\infty < x < \infty)$,

(b) $0, 1, x$ $(0 < x < 1)$,

(c) $1, x, |x|$ $(-1 \le x \le 1)$,

(d) $x, \dfrac{(x - 1)^2}{x - 2}, \dfrac{1}{x - 2}$ $(2 < x < \infty)$.

8. Suppose that u_1 and u_2 both satisfy the linear differential equation

$$(py')' + qy = 0,$$

where p and q are functions of the independent variable x. Show that there follows

$$u_1(pu_2')' - u_2(pu_1')' = 0,$$

and that this equation is equivalent to the equation $(pW)' = 0$, where W is the Wronskian of u_1 and u_2. Hence deduce that $W = A/p$, where A is a constant.

Section 1.3

9. If $y = u_1(x)$ and $y = u_2(x)$ satisfy the homogeneous linear equation

$$\frac{d^n y}{dx^n} + a_1(x)\frac{d^{n-1}y}{dx^{n-1}} + \cdots + a_n(x)y = 0,$$

prove that $y = c_1 u_1(x) + c_2 u_2(x)$ is also a solution, for any constant values of c_1 and c_2.

10. Verify that $u_1 = 1$ and $u_2 = \log x$ each satisfy the *nonlinear* equation

$$y'' + y'^2 = 0,$$

but that $y = c_1 u_1 + c_2 u_2$ is *not* a solution unless either $c_2 = 0$ or $c_2 = 1$.

11. If $y_P^{(1)}$ and $y_P^{(2)}$ are two linearly independent particular solutions of the nonhomogeneous linear equation

$$\frac{d^2 y}{dx^2} + a_1(x)\frac{dy}{dx} + a_2(x)y = h(x),$$

show that the function $u_1 = y_P^{(1)} - y_P^{(2)}$ satisfies the associated homogeneous equation (in which h is replaced by 0).

12. Verify that $y = \log x$ reduces the nonlinear expression $y'' + y'^2$ to zero, and that $y = x$ reduces it to unity. To what does the *sum* $y = \log x + x$ reduce it?

Section 1.4

13. Solve the following differential equations:

(a) $x\dfrac{dy}{dx} - ky = x^2$ (k constant),

(b) $\dfrac{dy}{dx} - y\tan x = x,$

(c) $\dfrac{dy}{dx} + y\tan x = \sec x,$

(d) $\dfrac{dy}{dx} + (y - 2\sin x)\cos x = 0,$

(e) $(x^2 - 1)\dfrac{dy}{dx} + 2y = (x + 1)^2,$

(f) $x^2 \log x\, dy + (xy - 1)\, dx = 0,$

(g) $2xy\dfrac{dy}{dx} - y^2 = x^2,$

(h) $y\, dx = (x + y^2)\, dy.$

14. Obtain the solution of the problem

$$\frac{dy}{dx} - ky = h(x),$$

$$y(x_0) = y_0$$

in the form

$$y(x) = \int_{x_0}^{x} e^{k(x-\xi)}h(\xi)\,d\xi + y_0 e^{kx},$$

when k is a constant.

15. Show that the substitution $u = y^{1-n}$ reduces the nonlinear equation

$$\frac{dy}{dx} + a_1(x)y = f(x)y^n \qquad (n \neq 1)$$

to a linear equation. (This equation is often called *Bernoulli's equation*.)

Section 1.5

16. If two roots of the characteristic equation (24) are $r = \pm a$, show that the corresponding part of the homogeneous solution can be written in the form

$$y = c_1 \cosh ax + c_2 \sinh ax,$$

where $\sinh ax$ and $\cosh ax$ are the *hyperbolic functions* defined by the equations

$$\sinh ax = \frac{e^{ax} - e^{-ax}}{2}, \qquad \cosh ax = \frac{e^{ax} + e^{-ax}}{2}.$$

17. If two roots of Equation (24) are $r = a \pm b$, show that the corresponding part of the homogeneous solution can be written in the form

$$y = e^{ax}(c_1 \cosh bx + c_2 \sinh bx).$$

18. Solve the following differential equations:

(a) $\dfrac{d^2y}{dx^2} - \dfrac{dy}{dx} - 2y = 0,$ (b) $\dfrac{d^3y}{dx^3} - \dfrac{d^2y}{dx^2} - \dfrac{dy}{dx} + y = 0,$

(c) $\dfrac{d^2y}{dx^2} - 2\dfrac{dy}{dx} + 2y = 0,$ (d) $\dfrac{d^4y}{dx^4} - 4\dfrac{d^3y}{dx^3} + 7\dfrac{d^2y}{dx^2} - 6\dfrac{dy}{dx} + 2y = 0,$

(e) $\dfrac{d^3y}{dx^3} - y = 0,$ (f) $\dfrac{d^2y}{dx^2} - 2iy = 0 \qquad (i^2 = -1).$

19. The following differential equations arise in dealing with the problems noted. Find the general solution of each equation, assuming that k is a nonzero constant:

(a) $\dfrac{d^4y}{dx^4} - k^4y = 0$ (vibration of a beam,)

(b) $\dfrac{d^4y}{dx^4} + 4k^4y = 0$ (beam on an elastic foundation),

(c) $\dfrac{d^4y}{dx^4} - 2k^2\dfrac{d^2y}{dx^2} + k^4y = 0$ (bending of an elastic plate).

20. Use the method of undetermined coefficients to find the complete solution of each of the following differential equations:

(a) $\dfrac{d^2y}{dx^2} + k^2y = \sin x$ $(k^2 \neq 0, 1),$ (b) $\dfrac{d^2y}{dx^2} + y = \sin x,$

(c) $\dfrac{d^2y}{dx^2} - y = \sin x,$ (d) $\dfrac{d^2y}{dx^2} - y = e^x,$

(e) $\dfrac{d^2y}{dx^2} - y = xe^x,$ (f) $\dfrac{d^2y}{dx^2} - 2\dfrac{dy}{dx} + 2y = \sin x,$

(g) $\dfrac{d^2y}{dx^2} - 2\dfrac{dy}{dx} + 2y = e^x \sin x,$ (h) $\dfrac{d^2y}{dx^2} - 9\dfrac{dy}{dx} + 20y = 4x^2e^{3x},$

(i) $\dfrac{d^2y}{dx^2} - \dfrac{dy}{dx} - 2y = 3e^{-x} + 10 \sin x - 4x.$

Section 1.6

21. Solve the following differential equations:

(a) $x^2\dfrac{d^2y}{dx^2} + x\dfrac{dy}{dx} - k^2y = 0,$ $\bigcirc$ (b) $x^2\dfrac{d^2y}{dx^2} - x\dfrac{dy}{dx} + 2y = 0,$

$\bigcirc$ (c) $x^2\dfrac{d^2y}{dx^2} - 2y = 0,$ (d) $x^2\dfrac{d^2y}{dx^2} - x\dfrac{dy}{dx} + y = 0,$

(e) $x^3\dfrac{d^3y}{dx^3} + 2x^2\dfrac{d^2y}{dx^2} - x\dfrac{dy}{dx} + y = 0,$ (f) $x^2\dfrac{d^2y}{dx^2} + 2x\dfrac{dy}{dx} - n(n+1)y = 0,$

(g) $x^2\dfrac{d^2y}{dx^2} + x\dfrac{dy}{dx} - y = x^2,$ (h) $x^2\dfrac{d^2y}{dx^2} + x\dfrac{dy}{dx} - y = x,$

(i) $x^2\dfrac{d^2y}{dx^2} - 4x\dfrac{dy}{dx} + 6y = 6x + 12.$

22. Solve the differential equation

$$(a + x)^2\frac{d^2y}{dx^2} - 2y = h(x)$$

in each of the following cases:

(a) $h(x) = 0,$ (b) $h(x) = 3(a + x)^2 + 1,$ (c) $h(x) = x^2.$

Section 1.7

23. Show, by direct expansion, that

$$(x^2D^2)(xD) = (xD)(x^2D^2)$$

but that $(xD)^2 \neq x^2D^2.$

24. Use the method of the text Example to obtain the general solution of the equation

$$(D - r_1)(D - r_2)y = h(x) \qquad (r_1 \neq r_2),$$

where r_1 and r_2 are constants, in the form

$$y = e^{r_2x}\int e^{-r_2x}\left[\int e^{-r_1x}h(x)\,dx\right]e^{r_1x}\,dx + c_1e^{r_1x} + c_2e^{r_2x}.$$

25. (a) If the notation

$$y(x) = \frac{1}{D - r}h(x)$$

is used to indicate that $y(x)$ satisfies the equation $(D - r)y = h$, where $D \equiv d/dx$ and r is a constant, show that

$$\frac{1}{D - r}h(x) = e^{rx}\int e^{-rx}h(x)\,dx,$$

where an arbitrary additive constant of integration is implied in the integral.

(b) Verify that, with the notation of part (a), the expression

$$y = \frac{1}{(D - r_2)(D - r_1)}h(x) \equiv \frac{1}{D - r_2}\left[\frac{1}{D - r_1}h(x)\right]$$

satisfies the equation $(D - r_1)(D - r_2)y = h$. Hence obtain the general solution of that equation in the form

$$y = e^{r_2 x} \int e^{(r_1 - r_2)x} \left[\int e^{-r_1 x} h(x) \, dx \right] dx,$$

where an arbitrary additive constant is implied in each integration. (Compare Problem 24.)

26. Verify that, with the notation of Problem 25(a), the expression

$$y = \left[\frac{1}{r_1 - r_2} \left(\frac{1}{D - r_1} - \frac{1}{D - r_2} \right) \right] h(x)$$

$$= \frac{1}{r_1 - r_2} \left[\frac{1}{D - r_1} h(x) - \frac{1}{D - r_2} h(x) \right]$$

satisfies the equation $(D - r_1)(D - r_2)y = h$ where $r_1 \neq r_2$. Hence obtain the general solution of that equation in the form

$$y = \frac{1}{r_1 - r_2} \left[e^{r_1 x} \int e^{-r_1 x} h(x) \, dx - e^{r_2 x} \int e^{-r_2 x} h(x) \, dx \right]$$

when $r_1 \neq r_2$, where an arbitrary additive constant is implied in each integration.

27. Verify that the result of Problem 26 can be written in the form

$$y = \frac{1}{r_1 - r_2} \int^x [e^{r_1(x - \xi)} - e^{r_2(x - \xi)}] h(\xi) \, d\xi + c_1 e^{r_1 x} + c_2 e^{r_2 x}$$

when $r_1 \neq r_2$.

28. Solve the differential equation

$$\frac{d^2 y}{dx^2} - y = e^x$$

by use of each of the two formulas derived in Problems 25 and 26.

29. By proceeding as in Problems 25 and 26, obtain the general solution of the equation

$$(xD - r_1)(xD - r_2)y = f(x) \qquad (r_1 \neq r_2)$$

in each of the following forms:

(a) $y = x^{r_2} \int x^{r_1 - r_2} \left[\int x^{-r_1} f(x) \frac{dx}{x} \right] \frac{dx}{x} + c_1 x^{r_1} + c_2 x^{r_2}$,

(b) $y = \frac{1}{r_1 - r_2} \left[x^{r_1} \int x^{-r_1} f(x) \frac{dx}{x} - x^{r_2} \int x^{-r_2} f(x) \frac{dx}{x} \right] + c_1 x^{r_1} + c_2 x^{r_2}$,

(c) $y = \frac{1}{r_1 - r_2} \int^x \left[\left(\frac{x}{\xi} \right)^{r_1} - \left(\frac{x}{\xi} \right)^{r_2} \right] f(\xi) \frac{d\xi}{\xi} + c_1 x^{r_1} + c_2 x^{r_2}$.

30. Solve the differential equation

$$x^2 \frac{d^2 y}{dx^2} - 2y = x^3$$

by use of each of the two formulas derived in Problems 29(a, b).

Section 1.8

31. Suppose that the coefficients in Equations (37a,b) are constants. With the symbols $R_1 = L_1 x + L_2 y - h_1$ and $R_2 = L_3 x + L_4 y - h_2$, Equations (37a,b) become

$$R_1 = 0, \tag{a}$$

$$R_2 = 0, \tag{b}$$

and Equations (38a,b) or, equivalently, (39a,b) become

$$L_4 R_1 - L_2 R_2 = 0, \tag{c}$$

$$L_1 R_2 - L_3 R_1 = 0. \tag{d}$$

Show that if solutions of (c,d) also satisfy (b), so that $R_2 = 0$, there follows also $L_4 R_1 = 0$, $L_3 R_1 = 0$. Hence, noticing that these two equations can have a nontrivial common solution ($R_1 \neq 0$) only if L_4 and L_3 have a common factor, deduce that *if all the coefficients in Equations (37a, b) are constant and if, in either of Equations (37a, b), the two operators involved have no factors in common, the restrictions on the constants appearing in the solutions of Equations (39a, b) are completely determined by substitution into that equation. Otherwise, the solutions of Equations (39a, b) must be checked by substitution into both of Equations (37a, b).*

32. Illustrate the results of Problem 31 in the case of the simultaneous equations

$$\frac{d^2 y}{dt^2} + \frac{dx}{dt} = 0,$$

$$\frac{dy}{dt} + \frac{dx}{dt} + x = 0.$$

33. Find the solution of each of the following sets of equations:

(a) $\begin{cases} 2\dfrac{dx}{dt} = 3x - y, \\ 2\dfrac{dy}{dt} = 3y - x, \end{cases}$

O (b) $\begin{cases} \dfrac{dx}{dt} - \dfrac{dy}{dt} + x = 2e^t, \\ \dfrac{d^2 x}{dt^2} - \dfrac{dy}{dt} + 3x - y = 4e^t, \end{cases}$

(c) $\begin{cases} 2\dfrac{d^2 x}{dt^2} - \dfrac{dy}{dt} - 4x = 2t, \\ 2\dfrac{dx}{dt} - 4\dfrac{dy}{dt} - 3y = 0, \end{cases}$

(d) $\begin{cases} \dfrac{d^2 x}{dt^2} - \dfrac{1}{2}\dfrac{d^2 y}{dt^2} + k^2 x = 0, \\ \dfrac{d^2 y}{dt^2} - \dfrac{d^2 x}{dt^2} + k^2 y = 0, \end{cases}$

(e) $\begin{cases} \dfrac{d^2 x}{dt^2} + x + 2y = 7e^{2t} - 1, \\ \dfrac{d^2 x}{dt^2} + 3x + 2y = 9e^{2t} + 1, \end{cases}$

(f) $\begin{cases} \dfrac{d^2 x}{dt^2} + \dfrac{dx}{dt} + \dfrac{dy}{dt} + 3x + y = 0, \\ \dfrac{d^2 x}{dt^2} + \dfrac{dy}{dt} + x + y = 0, \end{cases}$

O (g) $\begin{cases} \dfrac{dx}{dt} = 2x, \\ \dfrac{dy}{dt} = 3x - 2y, \\ \dfrac{dz}{dt} = 2y + 3z. \end{cases}$

34. Find the solution of each of the following sets of equations:

(a) $\begin{cases} 2t\dfrac{dx}{dt} = 3x - y, \\ 2t\dfrac{dy}{dt} = 3y - x, \end{cases}$

(b) $\begin{cases} t\dfrac{dx}{dt} = 2x, \\ t\dfrac{dy}{dt} = 3x - 2y, \\ t\dfrac{dz}{dt} = 2y + 3z. \end{cases}$

Section 1.9

35. Solve by the method of variation of parameters:

(a) $\dfrac{dy}{dx} + a_1(x)y = h(x),$ (b) $\dfrac{d^2y}{dx^2} + y = \cot x,$

(c) $\dfrac{d^2y}{dx^2} + y = \sec x,$ (d) $\dfrac{d^2y}{dx^2} + y = \log x,$

(e) $\dfrac{d^2y}{dx^2} - 2\dfrac{dy}{dx} + 2y = e^x \tan x,$ (f) $x^2\dfrac{d^2y}{dx^2} - 4x\dfrac{dy}{dx} + 6y = x^2 \sin x,$

(g) $x^2\dfrac{d^2y}{dx^2} - 2x\dfrac{dy}{dx} + 2y = x \log x.$

36. If $y(x)$ satisfies Equation (55), show that

$$\frac{dy}{dx} = \int^x \frac{h(\xi)[u_1(\xi)u_2'(x) - u_2(\xi)u_1'(x)]}{W[u_1(\xi), u_2(\xi)]}\, d\xi + c_1u_1'(x) + c_2u_2'(x).$$

37. Obtain the general solution of the equation

$$\frac{d^2y}{dx^2} = h(x)$$

in the form

$$y = \int^x (x - \xi)h(\xi)\, d\xi + c_1x + c_2.$$

38. The differential equation

$$\frac{d^2y}{dx^2} + \frac{dy}{dx} - xy = 0$$

possesses solutions $u_1(x)$ and $u_2(x)$ which can be represented by series, valid near $x = 0$, the leading terms of which are as follows:

$$u_1(x) = 1 + \tfrac{1}{6}x^3 + \cdots, \qquad u_2(x) = x - \tfrac{1}{2}x^2 + \cdots.$$

Use Abel's formula to show that $W(u_1, u_2) = e^{-x}$, and hence deduce that the general solution of the equation

$$\frac{d^2y}{dx^2} + \frac{dy}{dx} - xy = h(x)$$

is of the form

$$y = \int^x h(\xi)[u_1(\xi)u_2(x) - u_2(\xi)u_1(x)]e^\xi\, d\xi + c_1u_1(x) + c_2u_2(x).$$

Section 1.10

39. Verify that $y = e^x$ satisfies the homogeneous equation associated with $(x - 1)y'' - xy' + y = 1$, and obtain the general solution.

40. Verify that $y = \tan x$ satisfies the equation $y'' \cos^2 x = 2y$, and obtain the general solution.

41. One homogeneous solution of the equation

$$(1 - x^2)\frac{d^2y}{dx^2} - 2x\frac{dy}{dx} + 2y = 6(1 - x^2)$$

is $y = x$. Find the complete solution.

42. One solution of *Legendre's equation*,

$$(1 - x^2)\frac{d^2y}{dx^2} - 2x\frac{dy}{dx} + n(n + 1)y = 0,$$

is called $P_n(x)$. Show that a second solution is of the form

$$P_n(x) \int \frac{dx}{(1 - x^2)[P_n(x)]^2}.$$

43. One solution of *Bessel's equation*,

$$x^2\frac{d^2y}{dx^2} + x\frac{dy}{dx} + (x^2 - p^2)y = 0,$$

is called $J_p(x)$. Show that a second solution is of the form

$$J_p(x) \int \frac{dx}{x[J_p(x)]^2}.$$

44. Given that the function $J_0(x)$ in Problem 43 has a Taylor series expansion with leading terms

$$J_0(x) = 1 - \frac{x^2}{4} + \frac{x^4}{64} - \cdots,$$

which converges for all values of x, show that any independent solution of the equation

$$x\frac{d^2y}{dx^2} + \frac{dy}{dx} + xy = 0$$

has an expansion of the form

$$y = J_0(x)\left[A(\log x + \frac{x^2}{4} + \cdots) + B\right],$$

where A and B are constants and $A \neq 0$, and hence becomes logarithmically infinite as $x \longrightarrow 0$.

45. Use Abel's formula (65) to show that if u_1 and u_2 are solutions of

$$\frac{d^2y}{dx^2} + a_1(x)\frac{dy}{dx} + a_2(x)y = 0$$

in an interval I, and if we write

$$p = e^{\int a_1\,dx},$$

then the Wronskian of u_1 and u_2 is given by the expression

$$W[u_1(x), u_2(x)] = \frac{p(a)}{p(x)} \cdot \begin{vmatrix} u_1(a) & u_2(a) \\ u_1'(a) & u_2'(a) \end{vmatrix},$$

where $x = a$ is a point in the interval I.

46. *Removal of first-derivative term.* In the differential equation

$$\frac{d^2y}{dx^2} + a_1(x)\frac{dy}{dx} + a_2(x)y = h(x),$$

make the substitution $y = u(x)v(x)$ and determine v so that the coefficient of du/dx in the resultant equation vanishes. Thus show that, if $p = e^{\int a_1\,dx}$, the substitution $y = u/\sqrt{p}$ reduces the differential equation to the form

$$\frac{d^2u}{dx^2} - \frac{1}{4}\left(a_1^2 + 2\frac{da_1}{dx} - 4a_2\right)u = \sqrt{p}\,h(x).$$

47. Use the result of Problem 46 to show that Bessel's equation (Problem 43) takes the form

$$x^2 \frac{d^2u}{dx^2} + \left(x^2 - p^2 + \frac{1}{4}\right)u = 0$$

with the substitution

$$y = \frac{u}{\sqrt{x}}.$$

Section 1.11

48. A *fundamental set* of solutions of a homogeneous linear differential equation $Ly = 0$, of order n, relative to a point $x = a$, may be defined as a set of n solutions $v_0(x), v_1(x), \ldots, v_{n-1}(x)$ such that

$$v_0(a) = 1, \qquad v_0'(a) = v_0''(a) = \cdots = v_0^{(n-1)}(a) = 0;$$
$$v_1(a) = 0, \qquad v_1'(a) = 1, \qquad v_1''(a) = \cdots = v_1^{(n-1)}(a) = 0;$$
$$\cdots\cdots\cdots\cdots\cdots\cdots\cdots\cdots\cdots\cdots\cdots\cdots\cdots\cdots\cdots$$
$$v_{n-1}(a) = v_{n-1}'(a) = \cdots = v_{n-1}^{(n-2)}(a) = 0, \qquad v_{n-1}^{(n-1)}(a) = 1.$$

(a) Deduce that the solution of the initial-value problem, in which $Ly = 0$ and the values $y(a) = y_0, y'(a) = y_0', \ldots, y^{(n-1)}(a) = y_0^{(n-1)}$ are prescribed, is then given by

$$y(x) = \sum_{k=0}^{n-1} y_0^{(k)} v_k(x).$$

(b) Show that the Wronskian of a fundamental set of solutions, relative to $x = a$, is *unity* at $x = a$.

49. Show that the equation

$$x^2 \frac{d^2y}{dx^2} + x \frac{dy}{dx} - y = 0$$

does not possess a fundamental set of solutions, relative to $x = 0$. Then determine a fundamental set relative to $x = a$, where $a \neq 0$.

50. Show that the functions $\cosh x$ and $\sinh x$ constitute a fundamental set of solutions relative to $x = 0$ for the equation

$$\frac{d^2y}{dx^2} - y = 0.$$

(See Problem 16.)

51. Obtain a fundamental set of solutions for the equation

$$\frac{d^3y}{dx^3} + \frac{dy}{dx} = 0,$$

relative to $x = 0$.

52. Obtain a fundamental set of solutions for the equation

$$\frac{d^4y}{dx^4} - y = 0,$$

relative to $x = 0$.

53. Let $v_0(x), \ldots, v_{n-1}(x)$ comprise a fundamental set of solutions of an nth-order homogeneous linear differential equation $Ly = 0$, relative to $x = a$, (Problem 48) and let $w(x)$ be a particular solution of the nonhomogeneous equation $Ly = h$ for which

$$w(a) = w'(a) = \cdots = w^{(n-1)}(a) = 0.$$

(a) Show that the solution of the equation

$$Ly = h$$

for which

$$y(a) = y_0, y'(a) = y_0', \ldots, y^{(n-1)}(a) = y_0^{(n-1)}$$

is

$$y(x) = w(x) + \sum_{k=0}^{n-1} y_0^{(k)} v_k(x).$$

(b) When $n = 2$, verify that the function

$$w(x) = \int_a^x \frac{h(\xi)[u_1(\xi)u_2(x) - u_2(\xi)u_1(\xi)]}{W[u_1(\xi), u_2(\xi)]} \, d\xi$$

has the required properties if $u_1(x)$ and $u_2(x)$ are any two linearly independent solutions of $Ly = 0$, provided that the integral exists.

54. (a) Determine those values of the constant k for which the differential equation $y'' + k^2 y = 0$ possesses a nontrivial solution which vanishes when $x = 0$ and when $x = a$, where a is a given positive constant.

(b) Determine those values of k for which the differential equation $y'' + k^2 y = 1$ possesses a solution which vanishes when $x = 0$ and when $x = a$.

55. Determine the solution of the problem

$$\frac{d^2 y}{dx^2} + y = h(x), \qquad y(0) = 0, \quad y(l) = 0,$$

when $\sin l \neq 0$, by the following steps:

(a) Use the result of text Example 1 to obtain

$$y(x) = \int_0^x h(\xi) \sin (x - \xi) \, d\xi + c_2 \sin x,$$

where c_2 is such that

$$0 = \int_0^l h(\xi) \sin (l - \xi) \, d\xi + c_2 \sin l.$$

(b) Show that, if $\sin l \neq 0$, there follows

$$y(x) = \frac{1}{\sin l} \left[\int_0^x h(\xi) \sin (x - \xi) \sin l \, d\xi - \int_0^l h(\xi) \sin (l - \xi) \sin x \, d\xi \right].$$

(c) By writing $\int_0^l = \int_0^x + \int_x^l$ in the second integral, deduce that the solution can be expressed in the form

$$y(x) = \int_0^l G(x, \xi) h(\xi) \, d\xi,$$

where

$$G(x, \xi) = \begin{cases} \dfrac{\sin \xi \sin (x - l)}{\sin l} & (\xi \leq x), \\[2mm] \dfrac{\sin (\xi - l) \sin x}{\sin l} & (x \leq \xi), \end{cases}$$

when $\sin l \neq 0$. [The function $G(x, \xi)$, or a constant multiple of it, is known as the *Green's function* for the expression $Ly = y'' + y$ and the conditions $y(0) = y(l) = 0$.]

56. If $\sin l = 0$, show that the problem

$$\frac{d^2y}{dx^2} + y = h(x), \qquad y(0) = 0, \quad y(l) = 0$$

has *no solution* unless $h(x)$ satisfies the condition

$$\int_0^l h(x) \sin (l - x) \, dx = 0,$$

in which case there are *infinitely many solutions*, each of the form

$$y(x) = \int_0^x h(\xi) \sin (x - \xi) \, d\xi + C \sin x,$$

where C is an arbitrary constant. [See Problem 55(a).]

57. *The Green's function.* Let L be the operator such that

$$Ly = a_0(x)\frac{d^2y}{dx^2} + a_1(x)\frac{dy}{dx} + a_2(x)y$$

and assume that the solution of the problem

$$Ly = h(x), \qquad y(a) = 0, \quad y(b) = 0$$

exists and can be expressed in the form

$$y(x) = \int_a^b G(x, \xi)h(\xi) \, d\xi,$$

in terms of the Green's function

$$G(x, \xi) = \begin{cases} \alpha(\xi)u(x) & (x \leq \xi), \\ \beta(\xi)v(x) & (x \geq \xi). \end{cases}$$

Determine G by the following steps:

(a) Show that the assumed solution is of the form

$$y(x) = v(x)\int_a^x \beta(\xi)h(\xi) \, d\xi + u(x)\int_x^b \alpha(\xi)h(\xi) \, d\xi$$

and deduce that

$$\frac{dy}{dx} = \int_a^b \frac{\partial G(x, \xi)}{\partial x} h(\xi) \, d\xi$$

if and only if the condition

$$\alpha(x)u(x) = \beta(x)v(x) \qquad \qquad \text{(A)}$$

is satisfied. [Recall that $(d/dx)\int_{x_0}^x \varphi(\xi) \, d\xi = \varphi(x)$ if $x_0 = $ constant.]

(b) Show that, when (A) is satisfied,

$$\frac{d^2y}{dx^2} = \int_a^b \frac{\partial^2 G(x, \xi)}{\partial x^2} h(\xi) \, d\xi + [\beta(x)v'(x) - \alpha(x)u'(x)]h(x),$$

and hence that

$$Ly = \int_a^b [LG(x, \xi)]h(\xi) \, d\xi + a_0(x)[\beta(x)v'(x) - \alpha(x)u'(x)]h(x),$$

so that $Ly = h(x)$ if

$$Lu(x) = 0, \qquad Lv(x) = 0 \qquad \qquad \text{(B)}$$

and

$$a_0(x)[\beta(x)v'(x) - \alpha(x)u'(x)] = 1. \qquad \qquad \text{(C)}$$

(c) Show that (A) implies

$$\frac{v(x)}{\alpha(x)} = \frac{u(x)}{\beta(x)} \equiv c(x),$$

so that

$$G(x, \xi) = \begin{cases} \dfrac{v(\xi)u(x)}{c(\xi)} & (x \leq \xi), \\ \dfrac{u(\xi)v(x)}{c(\xi)} & (x \geq \xi), \end{cases}$$

with c determined by (C) in the form

$$c(\xi) = a_0(\xi)[u(\xi)v'(\xi) - v(\xi)u'(\xi)].$$

(d) From the preceding results, deduce that $u(x)$ and $v(x)$ must be solutions of the equation $Ly = 0$, such that $u(a) = 0$ and $v(b) = 0$, and that G then is defined by the formula of part (c), provided that both a_0 and the *Wronskian* of u and v do not vanish in (a, b), and hence that u and v are not constant multiples of each other. Show also that if G is considered as a function of x, with ξ fixed, then G is *continuous* when $x = \xi$ and $\partial G/\partial x$ has a *jump* of $1/a_0(\xi)$ as x increases through ξ.

(e) Use Abel's formula to show that $c(\xi)$ is a *constant* if $a_1(x) = a_0'(x)$, so that $G(x, \xi)$ then is *symmetric* in x and ξ. [Notice that a_1 must be replaced by a_1/a_0 in (56).]

(f) Verify that the formula of part (c) correctly yields the result of Problem 55.

58. Generalize the results of Problem 57 as follows:

(a) Show that if the conditions $y(a) = 0$ and $y(b) = 0$ are replaced by more general homogeneous conditions, of the form

$$k_1 y(a) + k_2 y'(a) = 0, \qquad k_3 y(b) + k_4 y'(b) = 0,$$

then $u(x)$ must satisfy the condition imposed at $x = a$ and $v(x)$ that imposed at $x = b$. (Again the Wronskian of u and v must not vanish.)

(b) Show that, if the nonhomogeneous conditions

$$k_1 y(a) + k_2 y'(a) = g_1, \qquad k_3 y(b) + k_4 y'(b) = g_2$$

are prescribed instead, then the problem solution (if it exists) can be written in the form

$$y(x) = w(x) + \int_a^b G(x, \xi)h(\xi)\, d\xi$$

where G is the Green's function of part (a), corresponding to the "homogenized" end conditions, and where $w(x)$ is the solution of the equation $Ly = 0$ which satisfies the *prescribed* end conditions.

[See also Problem 80 of Chapter 11, for a different (but equivalent) approach to the Green's function.]

59. Use the results of Problems 57 and 58 to derive the solution exhibited for each of the following problems:

(a) $\dfrac{d^2y}{dx^2} = h(x), \qquad y(0) = A, \quad y(l) = B;$

$$y(x) = A\left(1 - \frac{x}{l}\right) + B\frac{x}{l} + \int_0^l G(x, \xi)h(\xi)\, d\xi,$$

where
$$G(x, \xi) = \begin{cases} \dfrac{(\xi - l)x}{l} & (x \leq \xi), \\ \dfrac{\xi(x - l)}{l} & (x \geq \xi). \end{cases}$$

(b) $\dfrac{d^2y}{dx^2} = h(x), \quad y(0) = A, \quad y'(l) = B;$

$$y(x) = A + Bx + \int_0^l G(x, \xi)h(\xi)\, d\xi,$$

where
$$G(x, \xi) = \begin{cases} -x & (x \leq \xi), \\ -\xi & (x \geq \xi). \end{cases}$$

(c) $\dfrac{d^2y}{dx^2} - y = h(x), \quad y(-\infty) = 0, \quad y(\infty) = 0;$

$$y(x) = \tfrac{1}{2} \int_{-\infty}^{\infty} e^{-|x - \xi|}h(\xi)\, d\xi.$$

(d) $\dfrac{d^2y}{dx^2} - \dfrac{1}{x}\dfrac{dy}{dx} = h(x), \quad y(0) = 0, \quad y(1) = 0;$

$$y(x) = \int_0^1 G(x, \xi)h(\xi)\, d\xi,$$

where
$$G(x, \xi) = \begin{cases} -\dfrac{(1 - \xi^2)x^2}{2\xi} & (x \leq \xi), \\ -\dfrac{\xi(1 - x^2)}{2} & (x \geq \xi). \end{cases}$$

Section 1.12

60. Solve the following:
  (a) $(2xy^3 + y)\, dx + (3x^2y^2 + x - 2y)\, dy = 0,$
  (b) $(3x^2y - y^3)\, dx - (3xy^2 - x^3)\, dy = 0,$
  (c) $e^x(y\, dx + dy) + e^y(dx + x\, dy) = 0.$

61. Solve the following:
  (a) $\dfrac{dy}{dx} = \dfrac{x - y}{x + y},$ (b) $(x^2 + 2xy)\, dy = y^2\, dx,$
  (c) $x\dfrac{dy}{dx} = y - \sqrt{x^2 + y^2},$ (d) $x\, dy - y\, dx = x \tan\left(\dfrac{y}{x}\right) dx.$

62. Solve the following:
  (a) $x\dfrac{dy}{dx} + y = y \log (xy),$
  (b) $x\, dy + (y - x^2)\, dx = 0,$
  (c) $y\, dx + (x - y^2)\, dy = 0,$
  (d) $(x + y - 3)\, dy = (x - y + 1)\, dx,$
  (e) $x\dfrac{dy}{dx} + y + x^2y^2 = 0,$
  (f) $(2xy^2 - x)\, dx + (x^2y + y)\, dy = 0.$

63. *Integrating factors.* Let q be an integrating factor for the differential equation

$$M\, dx + N\, dy = 0,$$

so that $qM\, dx + qN\, dy$ is an exact differential.

(a) Show that q must be such that

$$N\frac{\partial q}{\partial x} - M\frac{\partial q}{\partial y} = \left(\frac{\partial M}{\partial y} - \frac{\partial N}{\partial x}\right)q.$$

(b) If

$$\left(\frac{\partial M}{\partial y} - \frac{\partial N}{\partial x}\right)\bigg/ N \equiv \varphi$$

is independent of y, show that an integrating factor is

$$q(x) = e^{\int \varphi\, dx}.$$

(c) If

$$\left(\frac{\partial M}{\partial y} - \frac{\partial N}{\partial x}\right)\bigg/ M \equiv \psi$$

is independent of x, show that an integrating factor is

$$q(y) = e^{-\int \psi\, dy}.$$

64. Use the results of Problem 63 to solve the following differential equations:
(a) $(2x^2 - 4y^3)\, dx + 3xy^2\, dy = 0$,
(b) $mx^{m-1}y\, dx + [(n-p)y^n - px^m]\, dy = 0$.

65. Solve the following:

(a) $\dfrac{d^2y}{dx^2} + \left(\dfrac{dy}{dx}\right)^2 = 0$,

(b) $\dfrac{d^2y}{dx^2} + 2x\left(\dfrac{dy}{dx}\right)^2 = 0$,

(c) $y\dfrac{d^2y}{dx^2} + \left(\dfrac{dy}{dx}\right)^2 = 0$,

(d) $\dfrac{d^2y}{dx^2} = \left[1 + \left(\dfrac{dy}{dx}\right)^2\right]^{3/2}$,

(e) $\dfrac{d^2y}{dx^2} + y = 0$,

(f) $\left[6\left(\dfrac{dy}{dx}\right)^2 + 1\right]\dfrac{d^2y}{dx^2} = 1$.

66. Show that the substitution $y = u'/Qu$, where $u' = du/dx$, reduces the nonlinear first-order equation

$$\frac{dy}{dx} + P(x)y + Q(x)y^2 = R(x)$$

to the linear second-order equation

$$\frac{d^2u}{dx^2} + \left(P - \frac{Q'}{Q}\right)\frac{du}{dx} - RQu = 0,$$

and also that the substitution $y = y_1 + 1/v$, where $y_1(x)$ is any known solution of the given equation, leads to the linear first-order equation

$$\frac{dv}{dx} - (P + 2Qy_1)v = Q.$$

(The nonlinear form is known as *Riccati's equation*.)

67. Use the procedures suggested in Problem 66 to obtain the general solution of the equation

$$x^2\frac{dy}{dx} + xy + x^2y^2 = 1$$

in the form $xy = (x^2 - k)/(x^2 + k)$, where k is an arbitrary constant. (Take $y_1 = 1/x$.)

2

The Laplace Transform

2.1. An Introductory Example. If a function $f(t)$ is multiplied by e^{-st} and the result is integrated with respect to t from $t = 0$ to $t = \infty$, a new function of the new variable s is obtained when the integral exists.† This function (when it exists) is called the *Laplace transform* of $f(t)$. Before studying the properties of such transforms, we illustrate one of their most useful applications by considering a simple problem.

Suppose that we require the solution of the differential equation

$$\frac{dy}{dt} - y = e^{at}, \tag{1}$$

for positive values of t, which satisfies the initial condition

$$y(0) = -1. \tag{2}$$

In place of determining the general solution of Equation (1) by the methods of Chapter 1, and then determining the arbitrary constant by satisfying Equation (2), we proceed as follows.

We first take the Laplace transform of both sides of Equation (1), by multiplying both sides of the equation by e^{-st} and integrating the results with respect to t from zero to infinity, to obtain the equation

$$\int_0^\infty e^{-st} \frac{dy}{dt}\, dt - \int_0^\infty e^{-st} y\, dt = \int_0^\infty e^{-st} e^{at}\, dt. \tag{3}$$

---

†The variable s is considered to be a *real* variable in this chapter. However, all results which hold when s is *real* and $s > a$, for some real value of a, also hold when s is *complex* and (real part of s) $> a$.

It is assumed, of course, that the separate integrals exist for some range of values of s.

The integral on the right is readily evaluated,

$$\int_0^\infty e^{-st}e^{at}\,dt = -\frac{e^{-(s-a)t}}{s-a}\Big|_0^\infty = \frac{1}{s-a}, \tag{4}$$

the integral existing when $s > a$. The first integral on the left in Equation (3) can be integrated formally by parts to give

$$\int_0^\infty e^{-st}\frac{dy}{dt}\,dt = e^{-st}y(t)\Big|_0^\infty + s\int_0^\infty e^{-st}y\,dt$$

$$= -y(0) + s\int_0^\infty e^{-st}y\,dt$$

$$= 1 + s\int_0^\infty e^{-st}y\,dt, \tag{5}$$

assuming that $e^{-st}y(t)$ approaches zero, for sufficiently large values of s, as $t \to \infty$. Thus the transform of dy/dt is expressed in terms of the prescribed initial value of y and in terms of the transform of y itself.

If the results of Equations (4) and (5) are introduced into (3), there then follows

$$(s-1)\int_0^\infty e^{-st}y\,dt = \frac{1}{s-a} - 1$$

or

$$\int_0^\infty e^{-st}y\,dt = \frac{a+1-s}{(s-1)(s-a)}. \tag{6}$$

The original problem is now apparently reduced to the problem of determining a function $y(t)$ whose Laplace transform is given by the right-hand side of Equation (6). To determine such a function, we first expand this expression by the method of partial fractions, to obtain the equivalent form

$$\int_0^\infty e^{-st}y\,dt = \frac{1}{a-1}\frac{1}{s-a} - \frac{a}{a-1}\frac{1}{s-1}. \tag{6a}$$

Reference to Equation (4) then indicates that, since $1/(s-a)$ is the transform of e^{at}, the first term of (6a) is the transform of $e^{at}/(a-1)$ and the second term the transform of $-ae^t/(a-1)$. Thus (6a) will be satisfied if we write

$$y = \frac{1}{a-1}(e^{at} - ae^t), \tag{7}$$

when $a \neq 1$. A corresponding expression in the case $a = 1$ can be obtained by taking the limit as $a \to 1$, in the form $y = (t-1)e^t$.

With this expression for y, the validity of the transition from Equations (1) and (2) to (6a) is readily established when $s > a$. Still, it is by no means obvious that (7) is the *only* solution of (6a). That is, while (1) is known to have a unique solution satisfying (2), it is conceivable that (6a) could have several solutions, only one of which would then also satisfy (1) and (2).

However, direct substitution shows that Equation (7) actually does represent the solution of (1) and (2). Further, it can be shown also that (7) *is* the only continuous solution of (6a).

Although this procedure has the advantage that the *particular* solution required is obtained directly, without first obtaining the *general* solution, it is clearly desirable to simplify the procedure by eliminating the necessity of carrying out certain general integrations in each case, and to determine in what cases such a procedure is valid.

In the remainder of this chapter, certain properties of Laplace transforms are investigated and relevant formulas are tabulated in such a way that the solution of initial-value problems involving linear differential equations with constant coefficients, or sets of simultaneous equations of this type, can be conveniently obtained. Thus, for example, use of the tabulated formulas will permit immediate transition from Equations (1) and (2) to (6), and from (6a) to the solution (7). Use of the methods to be given will, in general, introduce a considerable saving in labor over the alternative procedures of Chapter 1.

Laplace transforms are also useful, for example, in connection with the solution of certain problems governed by partial differential equations (and integral equations), as will be shown in later chapters. Certain of the properties developed in this chapter are of principal use in these later applications.

2.2. Definition and Existence of Laplace Transforms. The Laplace transform of a function $f(t)$, defined for positive values of t, is frequently indicated by the notation $\mathcal{L}\{f(t)\}$ and is defined, as a function of the variable s, by the integral†

$$\mathcal{L}\{f(t)\} = \int_0^\infty e^{-st} f(t)\, dt \tag{8}$$

over that range of values of s for which the integral exists. The notation $\bar{f}(s)$, or merely $\bar{f}$, is often used in place of $\mathcal{L}\{f(t)\}$.

The integral (8) may fail to define a function of s, in particular, because of infinite discontinuities in $f(t)$ for certain positive values of t or because of failure of $f(t)$ to behave in a sufficiently regular way near $t = 0$ or for large values of t. However, the presence of a finite number of *finite* discontinuities or "jumps" will not, in itself, affect the existence of the integral.

A function $f(t)$ is said to be *piecewise continuous* in a finite range if it is possible to divide that range into a finite number of intervals such that $f(t)$ is continuous inside each interval and approaches finite values as either end of any interval is approached from the interior. Such functions may thus have finite jumps at points inside the range considered. At such a point, say $t = t_0$, different limits are approached by $f(t)$ as t approaches t_0 from the right (that is,

---

†Some authors replace definition (8) by the definition

$$\mathcal{L}\{f(t)\} = s \int_0^\infty e^{-st} f(t)\, dt.$$

from larger values of t) and from the left (from smaller values). These two limits are called *right-hand* and *left-hand* limits, respectively, and when necessary are conveniently indicated by the respective notations

$$\lim_{t \to t_0 +} f(t) = f(t_0 +)$$

and

$$\lim_{t \to t_0 -} f(t) = f(t_0 -).$$

In illustration, if $f(t)$ is defined to be unity when $0 < t < 1$ and zero elsewhere, then $f(t)$ is piecewise continuous over any range. There follows also, for example, $f(1+) = 0$ and $f(1-) = 1$. If $f(t)$ is $1/\sqrt{t}$ when $0 < t < 1$ and zero elsewhere, then $f(t)$ is piecewise continuous in any range not including $t = 0$ as an interior or end point.

In the developments of this chapter, *we consider only functions which are at least piecewise continuous in every positive range not including zero as an end point*. Then, if we write (8) as the sum of three integrals,

$$\mathcal{L}\{f(t)\} = \int_0^\infty e^{-st} f(t)\, dt = \left(\int_0^{t_1} + \int_{t_1}^T + \int_T^\infty \right) e^{-st} f(t)\, dt, \qquad (8a)$$

the second integral on the right exists for all positive finite values of t_1 and T.

If, in addition, $f(t)$ approaches a finite limit as $t \to 0+$ or if $|f(t)| \to \infty$ as $t \to 0+$ in such a way that *for some number* $n, n < 1$, *the product* $t^n f(t)$ *is bounded near* $t = 0$, then the first integral of Equation (8a) exists.

Finally, a sufficient additional condition to guarantee the existence of the third integral of (8a), at least for sufficiently large values of s, is the requirement that $f(t)$ belong to the rather extensive class of "functions of exponential order." A function $f(t)$ is said to be *of exponential order* if, for some number s_0, the product $e^{-s_0 t} |f(t)|$ is bounded for large values of t, say for $t > T$. If the bound is denoted by M, then there follows, when $t > T$,

$$e^{-s_0 t} |f(t)| < M \quad \text{or} \quad |f(t)| < M e^{s_0 t}. \qquad (9)$$

Thus, though $f(t)$ may become infinitely large as $t \to \infty$, we see that $|f(t)|$ must not "grow" more rapidly than a multiple of *some* exponential function of t. We say that $f(t)$ is "of the order" of $e^{s_0 t}$ and frequently write $f(t) = O(e^{s_0 t})$. In particular, if $\lim_{t \to \infty} e^{-s_0 t} |f(t)|$ exists (and is finite) for some $s_0 > 0$, then for sufficiently large values of t the product must be bounded, and hence $f(t)$ is of the order of $e^{s_0 t}$. The limit may, of course, be zero, in which case we also write $f(t) = o(e^{s_0 t})$.

We note that any bounded function is of exponential order with $s_0 = 0$. Other examples are e^{at} (with $s_0 = a$), $e^{at} \sin bt$ (with $s_0 = a$), and t^n (with s_0 *any* positive number no matter how small). The function e^{t^2} is *not* of exponential order, since $e^{-s_0 t} e^{t^2} = e^{t^2 - s_0 t}$ is unbounded as $t \to \infty$ for *all* values of s_0.

If $f(t)$ is piecewise continuous and of exponential order, then its integral $\int_0^t f(u)\, du$ is continuous and is *also* of exponential order. Although it *cannot* be said in general that derivatives of functions of exponential order have the same property, this is true in most practical cases.

As an example of a reasonably simple exceptional case, we notice that though $\sin(e^{t^2})$ is bounded, and hence of exponential order, its derivative $2te^{t^2}\cos(e^{t^2})$ is *not* of exponential order.

In case $f(t)$ is of exponential order, and hence satisfies Equation (9) when $t > T$, then there follows

$$|e^{-st}f(t)| < e^{-st} \cdot Me^{s_0 t} = Me^{-(s-s_0)t}.$$

Hence, since we have assumed $f(t)$ to be piecewise continuous and since $\int_T^\infty e^{-(s-s_0)t}\,dt$ exists if $s > s_0$, it follows that the third integral in (8a) then also exists when $s > s_0$.

Thus, in summary, the Laplace transform of $f(t)$ exists, when s is sufficiently large, if $f(t)$ satisfies the following conditions:

(1) $f(t)$ *is continuous or piecewise continuous in every finite interval* $t_1 \leq t \leq T$, *where* $t_1 > 0$.

(2) $t^n |f(t)|$ *is bounded near* $t = 0$ *for some number n, where* $n < 1$.

(3) $e^{-s_0 t}|f(t)|$ *is bounded for large values of t, for some number* s_0.

Although the transform may also exist in other cases, these conditions are sufficiently weak to include most functions occurring in practice.

For reference purposes, it is noted that whenever *any* integral of the form

$$\int_0^\infty e^{-st}f(t)\,dt$$

exists for $s = s_0$, it exists also for all s such that $s \geq s_0$. Also, it is then true that

$$\lim_{s \to c} \int_0^\infty e^{-st}f(t)\,dt = \int_0^\infty e^{-ct}f(t)\,dt \tag{10}$$

when $c \geq s_0$, that

$$\frac{d}{ds}\int_0^\infty e^{-st}f(t)\,dt = \int_0^\infty \left(\frac{d}{ds}e^{-st}\right)f(t)\,dt \tag{11}$$

when $s \geq s_0$, and that

$$\int_\alpha^\beta \left[\int_0^\infty e^{-st}f(t)\,dt\right]ds = \int_0^\infty \left[\int_\alpha^\beta e^{-st}f(t)\,ds\right]dt \tag{12}$$

when $s_0 \leq \alpha \leq \beta < \infty$. The remarkable fact that all these operations can be effected under the integral sign, *for any convergent Laplace transform*, is of frequent usefulness.

The direct calculation of Laplace transforms may be illustrated by the following simple cases:

$$\mathcal{L}\{1\} = \int_0^\infty e^{-st}\,dt = -\frac{e^{-st}}{s}\Big|_0^\infty = \frac{1}{s} \qquad (s > 0). \tag{13a}$$

$$\mathcal{L}\{e^{at}\} = \int_0^\infty e^{-(s-a)t}\, dt = -\left.\frac{e^{-(s-a)t}}{s-a}\right|_0^\infty = \frac{1}{s-a} \qquad (s > a). \tag{13b}$$

$$\mathcal{L}\{\sin at\} = \int_0^\infty e^{-st} \sin at\, dt = -\left.\frac{e^{-st}}{s^2+a^2}(s \sin at + a \cos at)\right|_0^\infty$$

$$= \frac{a}{s^2+a^2} \qquad (s > 0). \tag{13c}$$

2.3. Properties of Laplace Transforms. Among the most useful properties of Laplace transforms are the following:

$$\mathcal{L}\{af(t) + bg(t)\} = a\bar{f}(s) + b\bar{g}(s). \tag{14}$$

$$\mathcal{L}\left\{\frac{d^n f(t)}{dt^n}\right\} = s^n \bar{f}(s) - \left[s^{n-1} f(0+) + s^{n-2}\frac{df(0+)}{dt}\right.$$

$$\left. + s^{n-3}\frac{d^2 f(0+)}{dt^2} + \cdots + \frac{d^{n-1} f(0+)}{dt^{n-1}}\right]. \tag{15}$$

$$\mathcal{L}\left\{\int_0^t f(u)\, du\right\} = \frac{1}{s}\bar{f}(s). \tag{16}$$

$$\mathcal{L}\{e^{at} f(t)\} = \bar{f}(s-a). \tag{17}$$

$$\text{If } f(t) = \begin{cases} 0 & t < a \\ g(t-a) & t \geq a \end{cases} \text{ with } a \geq 0, \quad \text{then} \quad \bar{f}(s) = e^{-as}\,\bar{g}(s). \tag{18}$$

$$\mathcal{L}\{t^n f(t)\} = (-1)^n \frac{d^n \bar{f}(s)}{ds^n}. \tag{19}$$

$$\mathcal{L}\left\{\int_0^t f(t-u)g(u)\, du\right\} = \bar{f}(s)\bar{g}(s). \tag{20}$$

In these equations a and b are constants, and n is a positive integer.

In all cases except Equation (15), *we suppose that the functions* $f(t)$ *and* $g(t)$ *satisfy the conditions of page* 57. *In the case of* (15), *more stringent restrictions are imposed.* These conditions are stated in connection with the proof to be given in this section (see page 59).

Equation (*14*) expresses the *linear property* of Laplace transforms. Its proof follows directly from the definition, in view of the corresponding linear properties of the integral.

Example 1. By using Equations (14) and (13b), we obtain

$$\mathcal{L}\{\sinh at\} = \mathcal{L}\left\{\frac{1}{2}e^{at} - \frac{1}{2}e^{-at}\right\} = \frac{1}{2(s-a)} - \frac{1}{2(s+a)} = \frac{a}{s^2-a^2}. \qquad \blacksquare$$

Equation (*15*) states one of the most important properties of Laplace transforms. It expresses the transform of any derivative of a function in terms of the transform of the function itself and in terms of the values of the lower-order derivatives of the function at $t = 0$ (or, more precisely, the values approached by these derivatives as $t \to 0$ from positive values). We consider first the case when $n = 1$, for which, from the definition,

$$\mathcal{L}\left\{\frac{df(t)}{dt}\right\} = \int_0^\infty e^{-st} \frac{df(t)}{dt} dt.$$

An integration by parts gives

$$\int_0^\infty e^{-st} \frac{df(t)}{dt} dt = e^{-st} f(t) \Big|_0^\infty + s \int_0^\infty e^{-st} f(t) dt$$

if $f(t)$ is continuous and $df(t)/dt$ is piecewise continuous in every interval $(0, T)$.†
But since $f(t)$ is of exponential order, the integrated part vanishes as $t \to \infty$ (for $s > s_0$), and there follows

$$\mathcal{L}\left\{\frac{df(t)}{dt}\right\} = s\bar{f}(s) - f(0+). \tag{15a}$$

Similarly, in the case $n = 2$, integration by parts gives

$$\mathcal{L}\left\{\frac{d^2f(t)}{dt^2}\right\} = \int_0^\infty e^{-st} \frac{d^2f(t)}{dt^2} dt$$

$$= e^{-st} \frac{df(t)}{dt} \Big|_0^\infty + s \int_0^\infty e^{-st} \frac{df(t)}{dt} dt$$

$$= e^{-st} \frac{df(t)}{dt} \Big|_0^\infty + s\mathcal{L}\left\{\frac{df(t)}{dt}\right\},$$

if df/dt is continuous and d^2f/dt^2 piecewise continuous. If df/dt is also of exponential order, the integrated part again vanishes as $t \to \infty$ and, making use of Equation (15a), there then follows

$$\mathcal{L}\left\{\frac{d^2f(t)}{dt^2}\right\} = s^2 \bar{f}(s) - sf(0+) - \frac{df(0+)}{dt}. \tag{15b}$$

Equations (15a) and (15b) are special cases of (15). The general proof of (15) follows by induction from the general result

$$\mathcal{L}\left\{\frac{d^n f(t)}{dt^n}\right\} = s\mathcal{L}\left\{\frac{d^{n-1}f(t)}{dt^{n-1}}\right\} - \frac{d^{n-1}f(0+)}{dt^{n-1}},$$

if $d^{n-1}f(t)/dt^{n-1}$ is continuous and $d^n f(t)/dt^n$ is piecewise continuous, and if $f(t), df(t)/dt, \ldots, d^n f(t)/dt^n$ are *all* of exponential order. This result is obtained, as in the special cases $n = 1$ and 2, by an integration by parts.

We thus obtain the following result:
Equation (15) *is valid if* $f(t)$ *and its first* $n - 1$ *derivatives are continuous over every interval* $(0, T)$, *if* $d^n f(t)/dt^n$ *is* (at least) *piecewise continuous over every interval* $(0, T)$, *and if* $f(t)$ *and its first* n *derivatives are of exponential order.*

---

†Unless these conditions are satisfied, the formula for integration by parts may not be valid.

Example 2. If we take $f(t) = \sin at$, then, from (13c),

$$\bar{f}(s) = \frac{a}{s^2 + a^2}.$$

Equation (15a) then gives

$$\mathscr{L}\{a \cos at\} = s \frac{a}{s^2 + a^2} - \sin 0$$

or, using (14),

$$\mathscr{L}\{\cos at\} = \frac{s}{s^2 + a^2}. \qquad\blacksquare$$

In particular, for the class of functions considered, we see that *if a function and its first $n - 1$ derivatives vanish at $t = 0$ (or as $t \to 0+$), the transform of its nth derivative is obtained by multiplying the transform of the function by s^n.* Applications of this fact are closely related to the use of the operational notation $D^n f(t)$ to represent $d^n f(t)/dt^n$.

Equation (16) is established by similar methods. Again making use of integration by parts, and recalling that

$$\frac{d}{dt} \int_0^t f(u)\, du = f(t),$$

we obtain

$$\mathscr{L}\left\{ \int_0^t f(u)\, du \right\} = \int_0^\infty e^{-st} \left\{ \int_0^t f(u)\, du \right\} dt$$

$$= \left[\frac{e^{-st}}{-s} \int_0^t f(u)\, du \right]_0^\infty + \frac{1}{s} \int_0^\infty e^{-st} f(t)\, dt$$

$$= \frac{1}{s} \bar{f}(s),$$

the integrated part vanishing at the upper limit (for sufficiently large values of s), since $f(t)$, and hence $\int_0^t f(u)\, du$, is of exponential order. Thus, in general, *if a function is integrated over $(0, t)$, the transform of the integral is obtained by dividing the transform of the function by s.*

If the lower limit differs from zero, the formula

$$\mathscr{L}\left\{ \int_a^t f(u)\, du \right\} = \frac{1}{s} \bar{f}(s) - \frac{1}{s} \int_0^a f(u)\, du \qquad (21)$$

is easily established.

Equations (17) and (18) express the so-called *translation properties* of the Laplace transform. The proof of the former property follows immediately from the definition, since the transform of $e^{at} f(t)$ is given by

$$\int_0^\infty e^{-st} [e^{at} f(t)]\, dt = \int_0^\infty e^{-(s-a)t} f(t)\, dt,$$

and the last expression differs from $\bar{f}(s)$ only in that s is replaced by $s - a$.

Thus, *if a function is multiplied by e^{at}, the transform of the result is obtained by replacing s by $s - a$ in the transform of the original function.* It is seen that if $\bar{f}(s)$ is plotted as a function of s, the representation of the transform of $e^{at}f(t)$ is thus obtained by shifting or "translating" the transform of $f(t)$ through a units in the positive direction of s.

Example 3. If we take $f(t) = \sin bt$, Equation (13c) gives $\bar{f}(s) = b/(s^2 + b^2)$, and (17) then gives

$$\mathcal{L}\{e^{at}\sin bt\} = \frac{b}{(s - a)^2 + b^2}. \qquad \blacksquare$$

Suppose now that a function is defined to be $g(t)$ for $t \geq 0$ and to be zero for negative values of t. Its transform may be denoted by $\bar{g}(s)$. If the given function is translated through a units in the positive direction of t, and so becomes $g(t - a)$ when $t \geq a$ and zero otherwise, *Equation (18)* states that *the transform of the translated function is obtained by multiplying the transform of the original function by e^{-as}.* To establish this property, we notice that, since the translated function vanishes when $0 \leq t < a$, its transform is defined by the integral

$$\int_a^\infty e^{-st}g(t - a)\,dt.$$

If t is replaced by $t + a$, and the lower limit of the integral is changed accordingly, this integral becomes

$$\boxed{\int_0^\infty e^{-s(t+a)}g(t)\,dt = e^{-as}\int_0^\infty e^{-st}g(t)\,dt = e^{-as}\bar{g}(s),}$$

in accordance with Equation (18).

Example 4. If $f(t) = \sin(t - t_0)$ when $t \geq t_0$ and $f(t) = 0$ when $t < t_0$, there follows $g(t) = \sin t$, and, from (13c), $\bar{g}(s) = (s^2 + 1)^{-1}$. Thus Equation (18) gives

$$\mathcal{L}\{f(t)\} = \frac{e^{-st_0}}{s^2 + 1}.$$

Conversely, if this relationship is known, a reversed argument serves to determine $f(t)$.
$\blacksquare$

To establish the property of *Equation (19)*, we merely differentiate both sides of the equation

$$\bar{f}(s) = \int_0^\infty e^{-st}f(t)\,dt$$

n times with respect to s. Differentiation under the integral sign is valid for all values of s for which the transform exists, as was stated at the end of Section 2.2.

Example 5. To find the transform of t^n, Equation (19) states that we merely differentiate the transform of *unity* n times with respect to s and multiply the result by $(-1)^n$.

We thus obtain

$$\mathcal{L}\{t^n\} = (-1)^n \frac{d^n}{ds^n}\left(\frac{1}{s}\right) = \frac{n!}{s^{n+1}},$$

where n is a positive integer or zero, with the usual convention that $0! = 1$.   ■

2.4. The Inverse Transform. In applications of Laplace transforms we frequently encounter the inverse problem of determining a function which has a given transform. The notation $\mathcal{L}^{-1}\{F(s)\}$ is conventionally used for the *inverse Laplace transform* of $F(s)$; that is, if $F(s) = \mathcal{L}\{f(t)\}$, then we write also

$$f(t) = \mathcal{L}^{-1}\{F(s)\}.$$

The notation $f(t) \leftrightarrow F(s)$ is also frequently useful.

To determine the inverse transform of a given function $F(s)$ it is thus necessary to determine a function $f(t)$ which satisfies the equation

$$\int_0^\infty e^{-st} f(t)\, dt = F(s).$$

Since the unknown function $f(t)$ appears under an integral sign, an equation of this type is called an *integral equation*.

In more advanced works it is proved that, if this equation has a solution, then that solution is *unique*. Thus, *if one function having a given transform is known, it is the only possible one.* This result is known as *Lerch's theorem*.

More precisely, Lerch's theorem states that two functions having the same transform cannot differ throughout any interval of positive length. Thus, for example, Equation (13a) shows that the continuous solution of

$$\int_0^\infty e^{-st} f(t)\, dt = \frac{1}{s}$$

is $f(t) = 1$; that is, $\mathcal{L}^{-1}\{s^{-1}\} = 1$. However, it is clear that if we take $f(t)$ to be, say, zero at $t = 1$ and unity elsewhere, or otherwise redefine the function $f(t)$ at a finite number of points, the value of the integral is not changed. Hence the new function is also a solution. Such artificialities are, however, generally of no significance in applications.

Although the direct determination of inverse transforms involves methods outside the scope of this chapter,† extensive tables of corresponding functions and transforms are available in the literature, and their use (in conjunction with the use of the properties listed in Section 2.3) is sufficient for many purposes. A short table of this sort is presented on pages 67 and 68.

It should be pointed out that not all functions of s are transforms, but that the class of such functions is greatly restricted by requirements of continuity and satisfactory behavior as $s \to \infty$. A useful result in this connection is the following: *If $f(t)$ is piecewise continuous in every finite interval $0 \leq t \leq T$ and is of exponential order, then $\bar{f}(s) \to 0$ as $s \to \infty$; furthermore $s\bar{f}(s)$ is bounded*

†See, however, Sections 11.2 and 11.3.

as $s \longrightarrow \infty$. The proof follows from the fact that in such cases

$$|f(t)| < M e^{s_0 t} \quad \text{and} \quad |e^{-st} f(t)| < M e^{-(s-s_0)t}$$

for some fixed constants s_0 and M. Hence we have

$$|\bar{f}(s)| = \left| \int_0^\infty e^{-st} f(t) \, dt \right| \leq \int_0^\infty e^{-st} |f(t)| \, dt$$

$$\leq M \int_0^\infty e^{-(s-s_0)t} \, dt$$

$$\leq \frac{M}{s - s_0}.$$

The theorem stated then follows from the fact that $M/(s - s_0)$ approaches zero and $[s/(s - s_0)]M$ is bounded as $s \longrightarrow \infty$. Thus, such functions as 1, $s/(s + 1)$, $1/\sqrt{s}$, and $\sin s$ cannot be transforms of functions satisfying the conditions stated.

It should also be noted that *if $f(t)$ is continuous and $df(t)/dt$ is piecewise continuous in every finite interval $0 \leq t \leq T$, and if $f(t)$ and $df(t)/dt$ are of exponential order, then*

$$\lim_{s \to \infty} s\bar{f}(s) = f(0+). \tag{22}$$

This result follows from the fact that in this case the preceding theorem states that the left-hand side of Equation (15a) vanishes as $s \longrightarrow \infty$. It is useful in those cases when only the *initial value* of $f(t)$ is required and the *transform* of f is known.

2.5. The Convolution. It frequently happens that, although a given function $F(s)$ is not the transform of a known function, it can be expressed as the product of two functions, each of which *is* the transform of a known function. Thus, it may be possible to write

$$F(s) = \bar{f}(s)\bar{g}(s),$$

where $\bar{f}(s)$ and $\bar{g}(s)$ are known to be the transforms of the functions $f(t)$ and $g(t)$, respectively. We suppose that these functions satisfy the conditions of page 57. In this case, *Equation (20)* states that the product $\bar{f}(s)\bar{g}(s)$ is the transform of the function defined by the integral $\int_0^t f(t - u)g(u) \, du$. This integral is called the *convolution* of f and g, and may be denoted by the abbreviation $f*g$. It is indicated by the symmetry in $\bar{f}$ and $\bar{g}$ that f and g can be interchanged in the convolution, that is, $f*g = g*f$. Before outlining the proof of Equation (20), we illustrate its application.

Example. To determine $\mathcal{L}^{-1}\{a/s(s - a)\}$ we may refer to Equations (13a) and (13b) and write the given function of s in the form

$$\frac{a}{s} \frac{1}{s - a} = \bar{f}(s)\bar{g}(s),$$

where $f(t) = a$ and $g(t) = e^{at}$. Equation (20) states that the product is the transform of the function

$$f * g = \int_0^t a e^{au} \, du = e^{at} - 1.$$

If the functions f and g are interchanged, there follows alternatively

$$g * f = \int_0^t e^{a(t-u)} a \, du = e^{at} \int_0^t e^{-au} a \, du = e^{at}(1 - e^{-at}) = e^{at} - 1,$$

as before. The same result is obtained without making use of the convolution, in this case, by using (16) and (13b), or by expanding the product by the method of partial fractions in the form

$$\frac{1}{s - a} - \frac{1}{s}$$

and using (13a) and (13b).   ∎

   Equation (20) can be obtained formally as follows. From the definition, the right-hand side of (20) can be written in the form

$$\bar{f}(s)\bar{g}(s) = \left[\int_0^\infty e^{-sv} f(v) \, dv \right]\left[\int_0^\infty e^{-su} g(u) \, du \right]$$

$$= \int_0^\infty \int_0^\infty e^{-s(v+u)} f(v) g(u) \, dv \, du$$

$$= \int_0^\infty g(u) \left[\int_0^\infty e^{-s(v+u)} f(v) \, dv \right] du$$

if different "dummy variables" of integration (v and u) are used in defining the two transforms. If, in the inner integral of the last form, we replace v by a new variable t with the substitution

$$v = t - u, \qquad dv = dt,$$

there follows

$$\int_0^\infty e^{-s(v+u)} f(v) \, dv = \int_u^\infty e^{-st} f(t - u) \, dt,$$

and hence

$$\bar{f}(s)\bar{g}(s) = \int_0^\infty \left[\int_u^\infty e^{-st} f(t - u) g(u) \, dt \right] du.$$

Interchanging the order of integration in the double integral and changing the limits as indicated in Figure 2.1, we then obtain formally

$$\bar{f}(s)\bar{g}(s) = \int_0^\infty \left[\int_0^t e^{-st} f(t - u) g(u) \, du \right] dt$$

$$= \int_0^\infty e^{-st} \left[\int_0^t f(t - u) g(u) \, du \right] dt$$

$$= \mathcal{L}\left\{ \int_0^t f(t - u) g(u) \, du \right\}$$

in accordance with Equation (20). The interchange of order of integration can be shown to be legitimate, by using appropriate limiting processes, when f and g satisfy the assumed conditions.

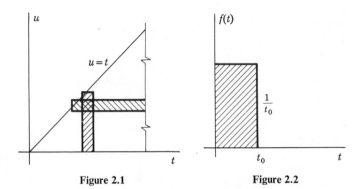

Figure 2.1   **Figure 2.2**

2.6. Singularity Functions. Consider the function $f(t)$ which has the value $1/t_0$ when $0 < t < t_0$ and is zero elsewhere (Figure 2.2). We then have

$$\int_0^\infty f(t)\, dt = \int_0^{t_0} f(t)\, dt = 1;$$

that is, the area under the graph representing $f(t)$ is unity. The transform of this function is found to be

$$\mathcal{L}\{f(t)\} = \frac{1}{t_0} \int_0^{t_0} e^{-st}\, dt = \frac{1 - e^{-st_0}}{st_0}.$$

Now as $t_0 \to 0$, the magnitude of $f(t)$ over $(0, t_0)$ increases without limit, while, at the same time, the length of the interval $(0, t_0)$ shrinks toward zero in such a way that the integral $\int_0^{t_0} f(t)\, dt$ retains the value of unity.

In the limit we have the occurrence of a "function" which is *infinite* at the point $t = 0$ and *zero* elsewhere, but which has the property that its integral across $t = 0$ is unity. Making use of L'Hospital's rule, we find

$$\lim_{t_0 \to 0} \frac{1 - e^{-st_0}}{st_0} = \lim_{t_0 \to 0} \frac{se^{-st_0}}{s} = 1.$$

That is, *the transform of $f(t)$ approaches unity as $t_0 \to 0$.*

If t represents time and $f(t)$ force, then $\int_0^{t_0} f(t)\, dt$ represents the *impulse* of the force $f(t)$ acting over the time interval $(0, t_0)$. Hence, as $t_0 \to 0$, we may speak loosely of a resulting "unit impulse" at $t = 0$, due to "an infinite force acting over a zero time interval." In view of this interpretation, the limiting form of $f(t)$ is frequently called the *unit impulse function*. If we denote it by $\delta(t)$, we are led to write

$$\mathcal{L}\{\delta(t)\} = 1. \tag{23}$$

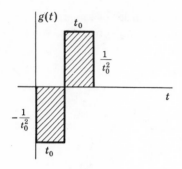

Figure 2.3

The "function" $\delta(t)$ is also often called the *Dirac delta function*. If, for example, t represented *distance* along, say, the centerline of a beam and $f(t)$ represented the intensity of a distributed *load*, then $\delta(t)$ could be considered as the formal representation of a *concentrated unit load* applied at the *point* $t = 0$, and analogous interpretations in other fields are frequently useful.†

In a similar way, if $g(t)$ has the value $-1/t_0^2$ when $0 < t < t_0$, the value $+1/t_0^2$ when $t_0 < t < 2t_0$, and the value zero elsewhere (Figure 2.3), there follows

$$\int_0^\infty (t - t_0)g(t)\,dt = \int_0^{2t_0} (t - t_0)g(t)\,dt = 1.$$

That is, the *moment* of the area under the graphical representation of $g(t)$, about the point t_0, is unity, whereas the (signed) area is zero. The transform of $g(t)$ is

$$\mathscr{L}\{g(t)\} = -\frac{1}{t_0^2}\left\{\int_0^{t_0} e^{-st}\,dt - \int_{t_0}^{2t_0} e^{-st}\,dt\right\}$$

$$= -\frac{1}{st_0^2}(1 - e^{-st_0})^2.$$

Repeated use of L'Hospital's rule shows that, as $t_0 \to 0$, $\mathscr{L}\{g(t)\} \to -s$. The limiting form of $g(t)$, as $t_0 \to 0$, is frequently referred to as the *doublet* (or *dipole*) *function*, because of certain interpretations relating to fluid flow and electric field theory. Denoting *the negative* of this limiting form by $\delta'(t)$, we write

$$\mathscr{L}\{\delta'(t)\} = s. \tag{24}$$

With the interpretation of t as *distance* and $g(t)$ as *load intensity*, $\delta'(t)$ could be considered as the formal representation of a *concentrated negative unit moment* applied at the *point* $t = 0$.

Such functions, often called *singularity functions*, are dealt with rigorously in the branch of mathematics known as the *theory of distributions*, and are of frequent use in physical applications. Although they do not conform to the restrictions of page 57, and although, in fact, *they are not true functions*, nevertheless *if formal use of them leads to a result which is capable of physical interpretation, then in practical cases the result may be accepted as correct.*

---

†In some situations it is necessary to replace the parent function $f(t)$ represented in Figure 2.2 by one which is *continuous* (and perhaps also has a certain number of continuous derivatives) for $0 \leq t \leq t_0$, before effecting the relevant limit operation as $t_0 \to 0$. Clearly, this can be done in many equivalent ways.

If the singularity occurs at $t = t_1$ rather than at $t = 0$, we denote the corresponding functions by $\delta(t - t_1)$ and $\delta'(t - t_1)$ and obtain (by limiting processes analogous to those given above) the formal results

$$\mathcal{L}\{\delta(t - t_1)\} = e^{-st_1}, \tag{25a}$$

$$\mathcal{L}\{\delta'(t - t_1)\} = se^{-st_1}. \tag{25b}$$

It may be noticed that these results are also obtained by formally applying the translation property (18) to Equations (23) and (24).

2.7. Use of Table of Transforms. A brief table of corresponding functions and transforms is presented in this section, to facilitate the determination of both direct and inverse transforms. The first ten pairs represent general relationships proved in Section 2.3. *In pairs (T3, 4, 5) the conditions of page 59 are assumed, whereas in the remaining pairs the less restrictive conditions of page 57 are implied.*

Table 1. Laplace Transforms

	Transform	Function
T1	$\bar{f}(s) = \mathcal{L}\{f(t)\} = \int_0^\infty e^{-st} f(t)\, dt$	$f(t)$
T2	$a\bar{f}(s) + b\bar{g}(s)$	$af(t) + bg(t)$
T3	$s\bar{f}(s) - f(0)$	$\dfrac{df(t)}{dt}$
T4	$s^2\bar{f}(s) - sf(0) - \dfrac{df(0)}{dt}$	$\dfrac{d^2 f(t)}{dt^2}$
T5	$s^n\bar{f}(s) - \displaystyle\sum_{k=1}^{n} s^{n-k} \dfrac{d^{k-1} f(0)}{dt^{k-1}}$	$\dfrac{d^n f(t)}{dt^n}$
T6	$\dfrac{1}{s^n}\bar{f}(s)$	$\overbrace{\int_0^t \cdots \int_0^t}^{n\ \text{times}} f(t)\, dt \overbrace{\cdots dt}^{n\ \text{times}}$
T7	$(-1)^n \dfrac{d^n \bar{f}(s)}{ds^n}$	$t^n f(t)$
T8	$\bar{f}(s - a)$	$e^{at} f(t)$
T9	$e^{-as}\bar{f}(s) \quad (a \geqq 0)$	$\begin{cases} f(t - a) & (t \geqq a) \\ 0 & (t < a) \end{cases}$
T10	$\bar{f}(s)\bar{g}(s)$	$\displaystyle\int_0^t f(t - u)g(u)\, du = \int_0^t f(u)g(t - u)\, du$
T11	$\dfrac{1}{s}$	1
T12	$\dfrac{1}{s + a}$	e^{-at}
T13	$\dfrac{1}{(s + a)(s + b)}$	$\dfrac{1}{a - b}(e^{-bt} - e^{-at})$
T14	$\dfrac{s}{(s + a)(s + b)}$	$\dfrac{1}{b - a}(be^{-bt} - ae^{-at})$
T15	$\dfrac{a}{s^2 + a^2}$	$\sin at$
T16	$\dfrac{s}{s^2 + a^2}$	$\cos at$

Table 1. (*Continued*)

	Transform	Function
T17	$\dfrac{a}{s^2 - a^2}$	$\sinh at$
T18	$\dfrac{s}{s^2 - a^2}$	$\cosh at$
T19	$\dfrac{2as}{(s^2 + a^2)^2}$	$t \sin at$
T20	$\dfrac{s^2 - a^2}{(s^2 + a^2)^2}$	$t \cos at$
T21	$\dfrac{2a^3}{(s^2 + a^2)^2}$	$\sin at - at \cos at$
T22	$\dfrac{2as}{(s^2 - a^2)^2}$	$t \sinh at$
T23	$\dfrac{s^2 + a^2}{(s^2 - a^2)^2}$	$t \cosh at$
T24	$\dfrac{2a^3}{(s^2 - a^2)^2}$	$at \cosh at - \sinh at$
T25	$\dfrac{a}{(s + b)^2 + a^2}$	$e^{-bt} \sin at$
T26	$\dfrac{s + b}{(s + b)^2 + a^2}$	$e^{-bt} \cos at$
T27	$\dfrac{4a^3}{s^4 + 4a^4}$	$\sin at \cosh at - \cos at \sinh at$
T28	$\dfrac{2a^2 s}{s^4 + 4a^4}$	$\sin at \sinh at$
T29	$\dfrac{2a^3}{s^4 - a^4}$	$\sinh at - \sin at$
T30	$\dfrac{2a^2 s}{s^4 - a^4}$	$\cosh at - \cos at$
T31	1	$\delta(t)$
T32	e^{-st_1}	$\delta(t - t_1)$
T33	s	$\delta'(t)$
T34	se^{-st_1}	$\delta'(t - t_1)$
T35	$\dfrac{1}{s^n}$	$\dfrac{t^{n-1}}{(n - 1)!} \qquad (n > 0)$
T35a	$\dfrac{n!}{s^{n+1}}$	$t^n \qquad (n > -1)$
T36	$\dfrac{1}{(s + a)^n}$	$\dfrac{t^{n-1}e^{-at}}{(n - 1)!} \qquad (n > 0)$
T37	$\dfrac{s}{(s + a)^n}$	$\dfrac{(n - 1) - at}{(n - 1)!} t^{n-2}e^{-at} \qquad (n > 1)$
T38	$\dfrac{a^{2n-1}}{(s^2 + a^2)^n}$	$\dfrac{1}{2^{n-1}(n - 1)!}\left[\sqrt{\dfrac{\pi}{2}}(at)^{n-1/2}J_{n-1/2}(at)\right] (n > 0)$
T39	$\dfrac{a^{2n-2}s}{(s^2 + a^2)^n}$	$\dfrac{at}{2^{n-1}(n - 1)!}\left[\sqrt{\dfrac{\pi}{2}}(at)^{n-3/2}J_{n-3/2}(at)\right] (n > \tfrac{1}{2})$
T40	$\dfrac{a^{2n-1}}{(s^2 - a^2)^n}$	$\dfrac{1}{2^{n-1}(n - 1)!}\left[\sqrt{\dfrac{\pi}{2}}(at)^{n-1/2}I_{n-1/2}(at)\right] (n > 0)$
T41	$\dfrac{a^{2n-2}s}{(s^2 - a^2)^n}$	$\dfrac{at}{2^{n-1}(n - 1)!}\left[\sqrt{\dfrac{\pi}{2}}(at)^{n-3/2}I_{n-3/2}(at)\right] (n > \tfrac{1}{2})$

Pairs (T11–30) either have been established in examples or can be easily obtained from established results by using certain of the general properties. Only frequently occurring basic forms are listed; other forms are readily deduced from those given. Pairs (T31–34) involve the singularity functions of the preceding section.

Pair (T35) was derived in Example 5 of Section 2.3 when n is a positive integer and (T36, 37) follow, in this case, by virtue of (T8) and (T3), respectively. The formulas are valid also (under the given restrictions on n) if n is *not* an integer, if $(n - 1)!$ is interpreted in a way to be defined in Section 2.9.

Pairs (T38–41) are included principally for future reference. In these pairs J_m and I_m are certain functions known as *Bessel functions of order m*. These functions are to be treated in Chapter 4. If, in these pairs, n is zero or a positive integer, then the order of the function involved is half an odd integer. In such cases the functions in brackets, in the right-hand column, can be expressed in terms of products of polynomials and either circular or hyperbolic functions. These expressions are given for $m = -\frac{1}{2}, \frac{1}{2}, \ldots, \frac{9}{2}$ in Table 2, page 70 (where x is written for at). The expressions for $m = \frac{11}{2}, \frac{13}{2}$, and so on, can be obtained in terms of these expressions by use of the recurrence formulas listed at the foot of the table.

Although the transforms of the simpler functions of frequent occurrence in practice can be obtained directly from the table, or by direct integration, the determination of *inverse transforms* may frequently involve a certain amount of manipulation. In this connection, it should be observed that *if n is a positive integer, all functions of t appearing in these tables* (except the singularity functions), *as well as all their derivatives, are continuous everywhere and are of exponential order.* Hence it follows that *if n is a positive integer, all properties* (T1–10) *can be applied to all succeeding pairs in the table, except for* (T31–34).

Pair (T3) is particularly useful in the determination of inverse transforms, *when $f(0) = 0$.* Reference to Equation (22) shows that this is so if $s\bar{f}(s)$ tends to zero as $s \to \infty$. Hence it follows that *if* $\lim_{s\to\infty} s\bar{f}(s) = 0$, *then*

$$\mathcal{L}^{-1}\{s\bar{f}(s)\} = \frac{df}{dt}. \tag{26}$$

Example 1. To determine $\mathcal{L}^{-1}\{s^2/(s^2 + 4)^2\}$, we first obtain from (T19) the result

$$\mathcal{L}^{-1}\left\{\frac{s}{(s^2 + 4)^2}\right\} = \frac{1}{4}t \sin 2t.$$

Hence, using Equation (26),

$$\mathcal{L}^{-1}\left\{\frac{s^2}{(s^2 + 4)^2}\right\} = \frac{d}{dt}\left(\frac{1}{4}t \sin 2t\right) = \frac{1}{4}\sin 2t + \frac{t}{2}\cos 2t. \qquad \blacksquare$$

When a given function $F(s)$, whose inverse transform is required, is the ratio of two polynomials in s, the method of partial fractions can be used to express $F(s)$ as the sum of a number of terms whose inverse transforms can be

Table 2. Bessel Functions of Order Half an Odd Integer

m	$\sqrt{\dfrac{\pi}{2}}\,x^m J_m(x)$	$\sqrt{\dfrac{\pi}{2}}\,x^m I_m(x)$
$-\tfrac{1}{2}$	$(\cos x)/x$	$(\cosh x)/x$
$\tfrac{1}{2}$	$\sin x$	$\sinh x$
$\tfrac{3}{2}$	$\sin x - x\cos x$	$x\cosh x - \sinh x$
$\tfrac{5}{2}$	$(3 - x^2)\sin x - 3x\cos x$	$(3 + x^2)\sinh x - 3x\cosh x$
$\tfrac{7}{2}$	$(15 - 6x^2)\sin x - (15 - x^2)x\cos x$	$(15 + 6x^2)\sinh x - (15 + x^2)x\sinh x$
$\tfrac{9}{2}$	$(105 - 45x^2 + x^4)\sin x - (105 - 10x^2)x\cos x$	$(105 + 45x^2 + x^4)\sinh x - (105 + 10x^2)x\cosh x$

Recurrence Formulas

$$x^{m+1}J_{m+1}(x) = 2m[x^m J_m(x)] - x^2[x^{m-1}J_{m-1}(x)]$$

$$x^{m+1}I_{m+1}(x) = -2m[x^m I_m(x)] + x^2[x^{m-1}I_{m-1}(x)]$$

determined from the table. In such cases, if the inverse transform does not involve the singularity functions of Section 2.6, the degree of the denominator must be greater than that of the numerator. In particular, if

$$F(s) = \frac{N(s)}{D(s)},$$

where $D(s)$ is a polynomial of degree n with n distinct real zeros $s = a_1, a_2, \ldots,$ a_n, and $N(s)$ is a polynomial of degree $n - 1$ or less, there follows (see Problem 17)

$$\frac{N(s)}{D(s)} = \frac{N(a_1)}{D'(a_1)} \frac{1}{s - a_1} + \frac{N(a_2)}{D'(a_2)} \frac{1}{s - a_2} + \cdots + \frac{N(a_n)}{D'(a_n)} \frac{1}{s - a_n}$$

$$= \sum_{m=1}^{n} \frac{N(a_m)}{D'(a_m)} \frac{1}{s - a_m},$$

and hence, from (T12),

$$\mathcal{L}^{-1}\left\{\frac{N(s)}{D(s)}\right\} = \sum_{m=1}^{n} \frac{N(a_m)}{D'(a_m)} e^{a_m t}. \tag{27}$$

If certain of the zeros of $D(s)$ are repeated or complex, recourse may be had to conventional methods of expansion in partial fractions.

Example 2. To determine $\mathcal{L}^{-1}\{(s^2 + 1)/(s^3 + 3s^2 + 2s)\}$, we write

$$N(s) = s^2 + 1, \qquad D(s) = s^3 + 3s^2 + 2s = s(s + 1)(s + 2).$$

With $a_1 = 0$, $a_2 = -1$, $a_3 = -2$, there follows

$$N(a_1) = 1, \qquad D'(a_1) = 2,$$
$$N(a_2) = 2, \qquad D'(a_2) = -1,$$
$$N(a_3) = 5, \qquad D'(a_3) = 2,$$

and Equation (27) gives

$$\mathcal{L}^{-1}\left\{\frac{s^2 + 1}{s^3 + 3s^2 + 2s}\right\} = \frac{1}{2} - 2e^{-t} + \frac{5}{2}e^{-2t}. \qquad \blacksquare$$

Example 3. To determine $\mathcal{L}^{-1}\{1/[(s + 1)(s^2 + 1)]\}$, we first assume an expansion of the form

$$\frac{1}{(s + 1)(s^2 + 1)} = \frac{A}{s + 1} + \frac{Bs + C}{s^2 + 1}.$$

After clearing fractions, we require the equation to be an identity and obtain $A = -B = C = \frac{1}{2}$. Hence

$$\frac{1}{(s + 1)(s^2 + 1)} = \frac{1}{2}\left[\frac{1}{s + 1} + \frac{1}{s^2 + 1} - \frac{s}{s^2 + 1}\right]$$

and the use of (T12, 15, 16) gives the inverse transform $\frac{1}{2}(e^{-t} + \sin t - \cos t)$. $\qquad \blacksquare$

The usefulness of (T25, 26) in determining inverse transforms should not be overlooked.

Example 4. To determine $\mathcal{L}^{-1}\{s/(s^2 + 4s + 5)\}$, we first write

$$\frac{s}{s^2 + 4s + 5} = \frac{s}{(s + 2)^2 + 1} = \frac{(s + 2) - 2}{(s + 2)^2 + 1}.$$

Pairs (T25, 26) then give the required inverse transform

$$e^{-2t}(\cos t - 2 \sin t).$$ ∎

2.8. Applications to Linear Differential Equations with Constant Coefficients.
It follows from the property (T5) and its special cases (T3, 4) that any ordinary
linear differential equation with constant coefficients, *with prescribed initial
conditions at $t = 0$,* can be transformed immediately to a *linear algebraic* equation determining the *transform* of the required solution, provided that the right-
hand member of the equation has a transform. The solution then is to be
obtained as the inverse of the transform so determined.

If the right-hand member is of exponential order, the same will be true of
the solution and of those derivatives whose transforms are involved, and all
the relations of Table 1 are appropriate.

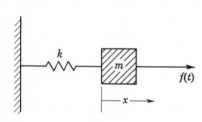

Figure 2.4

We take as a simple example the case
of *forced vibration* of a mass m attached
to a spring with spring constant k. (That
is, the force exerted on the free end of the
spring is assumed to be proportional to its
displacement x from the position of equilibrium, the constant of proportionality
being k.) If the applied force is $f(t)$ and if
no damping is present (Figure 2.4), the
differential equation of motion is

$$m\frac{d^2x}{dt^2} + kx = f(t). \tag{28}$$

Furthermore, if the mass is assumed to be at rest at equilibrium when $t = 0$,
the initial conditions

$$x(0) = \frac{dx(0)}{dt} = 0 \tag{29}$$

must be satisfied. With the transforms of $x(t)$ and $f(t)$ denoted by $\bar{x}(s)$ and
$\bar{f}(s)$, respectively, the transform of Equation (28) becomes merely

$$ms^2\bar{x} + k\bar{x} = \bar{f}.$$

Thus, if we write

$$\omega_0^2 = \frac{k}{m}, \tag{30}$$

the transform of the required solution is

$$\bar{x} = \frac{1}{m}\frac{\bar{f}}{s^2 + \omega_0^2}. \tag{31}$$

Since $1/(s^2 + \omega_0^2)$ is the transform of $(\sin \omega_0 t)/\omega_0$, this product can be considered as the product of the transforms of $(\sin \omega_0 t)/m\omega_0$ and $f(t)$, and hence use of the convolution (T10) gives the solution

$$x = \frac{1}{m\omega_0} \int_0^t f(u) \sin \omega_0(t - u) \, du, \tag{32}$$

in terms of an arbitrary force function $f(t)$.

However, in place of specializing this general form, it is often more convenient in specific cases to derive the required solution directly from Equation (31). We consider several cases of interest.

(1) Suppose that an instantaneous impulse of magnitude I is applied just after the time $t = 0$. Then $f(t) = I \cdot \delta(t)$ and $\bar{f}(s) = I \cdot 1 = I$. Hence we have

$$\bar{x} = \frac{I}{m} \frac{1}{s^2 + \omega_0^2}, \qquad x = \frac{I}{m\omega_0} \sin \omega_0 t \qquad (t > 0). \tag{33}$$

Thus the motion in this case is a sinusoidal vibration of amplitude $I/m\omega_0$ and angular frequency ω_0, following the application of the impulse; ω_0 is known as the *natural frequency* of the system. It should be noticed that here the initial condition $dx(0)/dt = 0$ apparently is not fulfilled. However, the velocity *cannot* vanish when the impulse is applied, since the momentum $mv = I$ must be imparted by the impulse, in accordance with Newton's laws of motion. Here we may suppose that x and $v = dx/dt$ are zero throughout an infinitesimal interval following the time $t = 0$, and that on the subsequent application of the impulse the velocity abruptly takes on the value I/m and a sinusoidal motion ensues. Interpretations of this general nature are frequently necessary in dealing with the idealized "singularity functions."

(2) If a sinusoidal force $f(t) = A \sin \omega t$ is applied, there follows

$$\bar{x} = \frac{A\omega}{m} \frac{1}{(s^2 + \omega_0^2)(s^2 + \omega^2)},$$

or, after expanding in terms of partial fractions,

$$\bar{x} = \frac{A\omega}{m(\omega^2 - \omega_0^2)} \left(\frac{1}{s^2 + \omega_0^2} - \frac{1}{s^2 + \omega^2} \right).$$

Hence

$$x = \frac{A\omega}{m(\omega^2 - \omega_0^2)} \left(\frac{\sin \omega_0 t}{\omega_0} - \frac{\sin \omega t}{\omega} \right).$$

or

$$x = \frac{A}{m\omega_0(\omega^2 - \omega_0^2)} (\omega \sin \omega_0 t - \omega_0 \sin \omega t). \tag{34}$$

Thus, if $\omega \neq \omega_0$, the motion is compounded of two *modes* of vibration, one (the *natural mode*) at the natural frequency ω_0, and the other (the *forced mode*) at the frequency of the imposed force. In case the system is excited at its natural frequency ($\omega = \omega_0$), the motion can be determined by considering the limiting form of Equation (34) as $\omega \to \omega_0$, or, more easily, by noticing that in this case

$$\bar{x} = \frac{A\omega_0}{m} \frac{1}{(s^2 + \omega_0^2)^2}.$$

Hence we obtain, from (T21),

$$x = \frac{A}{2m\omega_0^2}(\sin \omega_0 t - \omega_0 t \cos \omega_0 t). \tag{35}$$

Thus the last term of Equation (35) shows that, when the exciting frequency equals the natural frequency, the amplitude of the oscillations increases indefinitely with time. This is the case of *resonance*.

Similarly, if $f(t) = A \cos \omega_0 t$, there follows

$$x = \frac{A}{2m\omega_0} t \sin \omega_0 t.$$

(3) If a constant force $f(t) = A$ is applied when $t > 0$, there follows

$$\bar{x} = \frac{A}{m} \frac{1}{s(s^2 + \omega_0^2)}$$

$$= \frac{A}{m\omega_0^2}\left(\frac{1}{s} - \frac{s}{s^2 + \omega_0^2}\right),$$

and hence, from (T11, 16),

$$x = \frac{A}{m\omega_0^2}(1 - \cos \omega_0 t). \tag{36}$$

Thus, in this case, the mass oscillates with its natural frequency between the points $x = 0$ and $x = 2A/m\omega_0^2 = 2A/k$, when damping is absent.

(4) If constant force is applied only over the interval $0 < t < t_0$, and no force acts when $t > t_0$, there follows $\bar{f} = (A/s)(1 - e^{-st_0})$, and hence

$$\bar{x} = \frac{A}{m} \frac{1 - e^{-st_0}}{s(s^2 + \omega_0^2)} = \frac{A}{m}\left[\frac{1}{s(s^2 + \omega_0^2)} - \frac{e^{-st_0}}{s(s^2 + \omega_0^2)}\right].$$

The inverse transform of the first term is given by Equation (36), and, in view of (T9), the inverse of the second term is zero when $t < t_0$ and is obtained by replacing t by $t - t_0$ in (36) when $t > t_0$. Hence we have, *when* $0 < t < t_0$,

$$x = \frac{A}{m\omega_0^2}(1 - \cos \omega_0 t); \tag{37a}$$

and, *when* $t > t_0$,

$$x = \frac{A}{m\omega_0^2}\{(1 - \cos \omega_0 t) - [1 - \cos \omega_0(t - t_0)]\}$$

$$= \frac{A}{k}[\cos \omega_0(t - t_0) - \cos \omega_0 t]$$

$$= \frac{2A}{k}\left(\sin \frac{1}{2}\omega_0 t_0\right) \sin \omega_0\left(t - \frac{1}{2}t_0\right). \tag{37b}$$

Thus, while the force acts $(0 < t < t_0)$, the mass oscillates at its natural frequency, with amplitude A/k, about the point $x = A/k$; however, after the force is removed $(t > t_0)$, the mass oscillates about the point of equilibrium $(x = 0)$, at the same frequency, but with an amplitude $(2A/k) \sin \frac{1}{2}\omega_0 t_0$. If $t_0 = 2\pi/\omega_0 = T$, where T is the period of the natural mode of vibration, then

$x = 0$ when $t \geqq t_0$, so that the mass returns to its equilibrium position as the force is removed, and then remains at that position.

It is seen that in the preceding example, and in similar cases, the use of tables permits the determination of the transform of the solution by purely algebraical methods. This is true, however, only in cases when the coefficients of the linear differential equation are constants, and the usefulness of the present methods is mainly restricted, in such applications, to such cases. (See, however, Problems 46 to 49.)

The use of Laplace transforms is particularly advantageous in the solution of initial-value problems associated with sets of simultaneous linear equations. We illustrate the procedure by considering an example.

We require the solution of the simultaneous equations

$$\frac{dx}{dt} - y = e^t,$$
$$\frac{dy}{dt} + x = \sin t, \tag{38}$$

which satisfies the conditions

$$x(0) = 1, \qquad y(0) = 0. \tag{39}$$

The transforms of (38) satisfying (39) are

$$s\bar{x} - \bar{y} = \frac{1}{s - 1} + 1,$$
$$\bar{x} + s\bar{y} = \frac{1}{s^2 + 1},$$

from which we obtain, algebraically,

$$\bar{x} = \frac{s}{(s - 1)(s^2 + 1)} + \frac{s}{s^2 + 1} + \frac{1}{(s^2 + 1)^2},$$
$$\bar{y} = -\frac{1}{(s - 1)(s^2 + 1)} - \frac{1}{s^2 + 1} + \frac{s}{(s^2 + 1)^2}.$$

If the first terms on the right-hand sides of these equations are expanded in partial fractions, there follows

$$\bar{x} = \frac{1}{2}\left[\frac{1}{s - 1} + \frac{1}{s^2 + 1} + \frac{s}{s^2 + 1} + \frac{2}{(s^2 + 1)^2}\right],$$
$$\bar{y} = \frac{1}{2}\left[-\frac{1}{s - 1} - \frac{1}{s^2 + 1} + \frac{s}{s^2 + 1} + \frac{2s}{(s^2 + 1)^2}\right], \tag{40}$$

and reference to Table 1 gives the required solution,

$$x = \tfrac{1}{2}(e^t + 2 \sin t + \cos t - t \cos t),$$
$$y = \tfrac{1}{2}(-e^t - \sin t + \cos t + \sin t). \tag{41}$$

To illustrate the existence of exceptional cases which may arise in connection with simultaneous differential equations, we next attempt to find a solution of the equations

$$\frac{dx}{dt} + y = 0,$$

$$\frac{d^2x}{dt^2} + \frac{dy}{dt} + y = e^t,$$

(42)

satisfying the conditions

$$x(0) = 1, \qquad x'(0) = 0, \qquad y(0) = 0. \tag{43}$$

The transformed equations are

$$s\bar{x} + \bar{y} = 1,$$

$$s^2\bar{x} + (s + 1)\bar{y} = s + \frac{1}{s - 1},$$

from which there follow

$$\bar{x} = \frac{2}{s} - \frac{1}{s - 1}, \qquad \bar{y} = \frac{1}{s - 1}. \tag{44}$$

The inverse transforms of these expressions are then

$$x = 2 - e^t, \qquad y = e^t. \tag{45}$$

However, these solutions do not satisfy the last two of the prescribed initial conditions (43). It is readily shown, by methods of Chapter 1, that the most general solution of (42) is of the form

$$x = C - e^t, \qquad y = e^t$$

where C is an arbitrary constant. Hence only the initial value of x is arbitrary, and the problem as stated does not possess a solution.

This example shows that, *in the case of simultaneous equations*, although the method of Laplace transforms will yield the correct solution if it exists, it may also supply an erroneous solution (which fails to satisfy certain prescribed initial conditions) if no true solution exists. Thus, in doubtful cases, the satisfaction of initial conditions should be checked.

2.9. The Gamma Function. In calculating the transform of t^n, where $n > -1$ but n is not necessarily an integer, we encounter a function, known as the *Gamma function*, which also occurs frequently in many other applications. In this section we investigate certain properties of this function.

If, in the integral defining the transform of t^n,

$$\mathcal{L}\{t^n\} = \int_0^\infty e^{-st}t^n \, dt \qquad (n > -1),$$

we introduce a new variable of integration by setting $st = x$, there follows†

$$\mathcal{L}\{t^n\} = \frac{1}{s^{n+1}} \int_0^\infty e^{-x}x^n \, dx \qquad (n > -1). \tag{46}$$

---

†The restriction $n > -1$ is necessary to ensure the convergence of the integral. If $n \leq -1$, the function t^n does not have a Laplace transform, as here defined.

The integral appearing in Equation (46) depends only upon n. Although it cannot be expressed in terms of elementary functions of n, the same integral with n (inconveniently) replaced by $n - 1$ is a tabulated function which occurs frequently in practice and is known as the *Gamma function* of n, written $\Gamma(n)$:

$$\Gamma(n) = \int_0^\infty e^{-x} x^{n-1}\, dx \qquad (n > 0). \tag{47}$$

With this notation, Equation (46) can be written in the form

$$\mathscr{L}\{t^n\} = \frac{\Gamma(n+1)}{s^{n+1}} \qquad (n > -1). \tag{48}$$

By repeated use of integration by parts (or by comparison with the result of Example 5, Section 2.3), it follows that

$$\Gamma(n+1) = n! \tag{49}$$

if n is a positive integer, and that

$$\Gamma(1) = 0! = 1. \tag{50}$$

Thus it is seen that, if $n > -1$, $\Gamma(n+1)$ is a continuous function of n which takes on the value $n!$ when n is a positive integer or zero. For this reason, the Gamma function is often referred to as the *generalized factorial function*.

When n is not necessarily integral, an integration by parts leads to the result

$$\Gamma(n+1) = \int_0^\infty e^{-x} x^n\, dx = -x^n e^{-x}\Big|_0^\infty + n\int_0^\infty e^{-x} x^{n-1}\, dx$$

$$= n\int_0^\infty e^{-x} x^{n-1}\, dx \qquad (n > 0),$$

from which there follows

$$\Gamma(n+1) = n\Gamma(n) \qquad (n > 0). \tag{51}$$

This relation displays the characteristic property of the Gamma function. Inductive reasoning then leads to the formula

$$\Gamma(n+N) = (n+N-1)(n+N-2)\cdots(n+1)n\Gamma(n) \qquad (n > 0), \tag{52}$$

where N is any positive integer. Also, if n is replaced by $n - 1$, Equation (51) can be written in the alternative form

$$\Gamma(n-1) = \frac{\Gamma(n)}{n-1} \qquad (n > 1). \tag{53}$$

Four-place values of $\Gamma(x)$ over the interval $1 \le x \le 2$ are listed in Table 3. With such a table, Equation (52) can be used to evaluate the Gamma function for arguments greater than 2, since any such argument differs from a value in the tabulated range by some positive integer N. If we write $x = n + N$ and replace n by x_0, where $1 \le x_0 \le 2$, Equation (52) becomes

$$\Gamma(x) = (x-1)(x-2)\cdots(x_0+1)x_0\Gamma(x_0) \qquad (x > 2,\; 1 \le x_0 \le 2). \tag{54}$$

Also, Equation (53) serves to determine values of the Gamma function for arguments between zero and unity.

Table 3. Values of $\Gamma(x) = (x - 1)!$

x	.00	.01	.02	.03	.04	.05	.06	.07	.08	.09
1.0	1.0000	.9943	.9888	.9835	.9784	.9735	.9687	.9642	.9597	.9555
.1	.9514	.9474	.9436	.9399	.9364	.9330	.9298	.9267	.9237	.9209
.2	.9182	.9156	.9131	.9108	.9085	.9064	.9044	.9025	.9007	.8990
.3	.8975	.8960	.8946	.8934	.8922	.8912	.8902	.8893	.8885	.8879
.4	.8873	.8868	.8864	.8860	.8858	.8857	.8856	.8856	.8857	.8859
.5	.8862	.8866	.8870	.8876	.8882	.8889	.8896	.8905	.8914	.8924
.6	.8935	.8947	.8959	.8972	.8986	.9001	.9017	.9033	.9050	.9068
.7	.9086	.9106	.9126	.9147	.9168	.9191	.9214	.9238	.9262	.9288
.8	.9314	.9341	.9368	.9397	.9426	.9456	.9487	.9518	.9551	.9584
.9	.9618	.9652	.9688	.9724	.9761	.9799	.9837	.9877	.9917	.9958

For negative values of n, the function $\Gamma(n)$ is not defined by Equation (47), since the integral does not exist. However, it is conventional to extend the definition in such cases by requiring that the *recurrence formula* (51) hold also for negative values of n.† Since, for any negative value of n which is not an integer, there exists a positive integer N such that $n + N$ is in the tabulated range $(1 < n + N < 2)$, we may then replace $n + N$ by x_0 and n by x in (52), where $1 < x_0 < 2$, to obtain

$$\Gamma(x_0) = (x_0 - 1)(x_0 - 2)\cdots(x + 1)x\Gamma(x)$$

or, solving for $\Gamma(x)$,

$$\Gamma(x) = \frac{\Gamma(x_0)}{x(x + 1)(x + 2)\cdots(x_0 - 2)(x_0 - 1)} \qquad (x < 1, \; 1 < x_0 < 2). \qquad (55)$$

Equations (54) and (55) thus serve to determine values of the Gamma function for real arguments outside the tabulated range. It should be noticed, however, that since the denominator of Equation (55) vanishes when x is zero or a negative integer, the Gamma function is not defined for these values, and becomes infinite as these values are approached (see Figure 2.5).

It will be convenient in later work to use the notation $n!$ even in cases when n is not a positive integer or zero, with the convention that in such cases $n!$ is defined by $\Gamma(n + 1)$.

The value of $\Gamma(\frac{1}{2})$ is of particular interest. A well-known but quite indirect method of determining this value is now presented. From the definition (47), we have

$$\Gamma(\tfrac{1}{2}) = \int_0^\infty e^{-z}z^{-1/2} \, dz.$$

With the change in variables $z = x^2$, this integral becomes

$$\Gamma(\tfrac{1}{2}) = 2 \int_0^\infty e^{-x^2} \, dx. \qquad (56)$$

---

†A different method of definition yields $\Gamma(n)$ for all *complex* values of n except for zero and negative integers, at which points $\Gamma(n)$ does not remain finite.

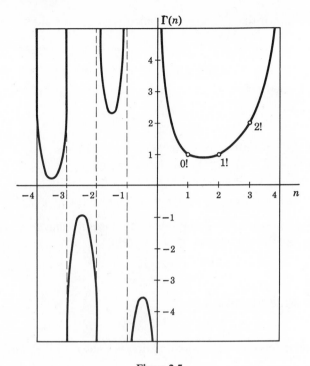

Figure 2.5

If the right-hand side of Equation (56) is multiplied by itself, and if the variable of integration is replaced by y in one factor, there follows

$$[\Gamma(\tfrac{1}{2})]^2 = 4\left(\int_0^\infty e^{-x^2}\, dx\right)\left(\int_0^\infty e^{-y^2}\, dy\right) = 4\int_0^\infty \int_0^\infty e^{-(x^2+y^2)}\, dx\, dy.$$

This double integral represents the volume under the surface $z = e^{-(x^2+y^2)}$ in the first quadrant ($x \geqq 0$, $y \geqq 0$). Changing to polar coordinates, we obtain

$$[\Gamma(\tfrac{1}{2})]^2 = 4\int_0^{\pi/2} \int_0^\infty e^{-r^2} r\, dr\, d\theta = 4\frac{\pi}{2}\frac{e^{-r^2}}{-2}\Big|_0^\infty = \pi.$$

Thus, finally,

$$\Gamma(\tfrac{1}{2}) = \sqrt{\pi}. \tag{57}$$

From Equation (56) we also have the useful result

$$\int_0^\infty e^{-x^2}\, dx = \tfrac{1}{2}\sqrt{\pi}. \tag{58}$$

We include without proof the formula

$$\Gamma(x)\Gamma(1-x) = \frac{\pi}{\sin \pi x}, \tag{59}$$

which can be shown to be valid for nonintegral values of x (see Problem 119 of Chapter 10) and which is of frequent use in applications of the Gamma function.

It can be shown that the Gamma function is defined alternatively by the limit

$$\Gamma(x) = \lim_{n \to \infty} \frac{n!}{x(x + 1) \cdots (x + n)} n^x.$$

Hence there follows also

$$\lim_{n \to \infty} \frac{n! n^x}{\Gamma(x + n + 1)} = 1. \tag{60}$$

Thus in limiting operations involving Gamma functions, when $n \to \infty$, the expression $\Gamma(x + n + 1)$ can be replaced by its approximation $n! n^x$. We indicate this fact by the notation

$$(n + x)! = \Gamma(x + n + 1) \sim n! n^x \qquad (n \to \infty), \tag{61}$$

where the symbol $\sim$ is to be interpreted as indicating that (61) implies (60).

A further limit of importance is of the form

$$\lim_{n \to \infty} \frac{\Gamma(n + 1)}{\sqrt{2\pi} \, n^{n + (1/2)} e^{-n}} = 1 \tag{62}$$

which we can write symbolically

$$n! = \Gamma(n + 1) \sim \sqrt{2\pi n} (n/e)^n \qquad (n \to \infty). \tag{63}$$

If n is a positive integer, this approximation is known as the *Stirling formula* for the factorial.

Proof of these relations is beyond the scope of this chapter.

REFERENCES

1. CAMPBELL, G. A., and R. M. FOSTER, *Fourier Integrals for Practical Applications*, Van Nostrand Reinhold Company, New York, 1947.†

2. CARSLAW, H. S., and J. C. JAEGER, *Operational Methods in Applied Mathematics*, 2nd ed., Oxford University Press, Inc., New York, 1948.

3. CHURCHILL, R. V., *Operational Mathematics*, 2nd ed., McGraw-Hill Book Company, Inc., New York, 1958.

4. ERDÉLYI, A., ed., *Tables of Integral Transforms*, 2 vols., McGraw-Hill Book Company, Inc., New York, 1954.

5. WIDDER, D. V., *The Laplace Transform*, Princeton University Press, Princeton, N.J., 1941.

†This reference contains 763 pairs, most of which can be interpreted as functions and their Laplace transforms. For this purpose, if the variable g is replaced by t, then the column headed "Coefficient $G(g)$" lists the function $f(t)$ in the present notation and the column headed "Coefficient $F(f)$" lists the corresponding Laplace transform $\mathcal{L}\{f(t)\} = \bar{f}(s)$.

PROBLEMS

Section 2.1

1. (a) Obtain the solution of the equation

$$\frac{dy}{dt} - \alpha y = e^{at}$$

for which $y(0) = y_0$, by the method of Section 2.1. Assume that $\alpha \neq a$.
(b) Verify the solution so obtained.

2. (a) Obtain the solution of the equation

$$\frac{dy}{dt} - ay = e^{at}$$

for which $y(0) = y_0$, by considering the limit of the solution of Problem 1 as $\alpha \longrightarrow a$.
(b) Verify the solution so obtained.

Section 2.2

3. Find the Laplace transform of each of the following functions, by direct integration:
(a) $e^{at} \cos kt$,
(b) $t^n e^{-at}$ (n a positive integer),
(c) $\begin{cases} \sin t & (0 < t < \pi), \\ 0 & (t > \pi), \end{cases}$
(d) $\begin{cases} 0 & (0 < t < a), \\ 1 & (a < t < b), \\ 0 & (t > b). \end{cases}$

Section 2.3

4. Find the Laplace transform of each of the following functions:
$\bigcirc$ (a) t^3,
$\bigcirc$ (b) $t^2 e^{-3t}$,
$\bigcirc$ (c) $\cos at \sinh at$,
$\bigcirc$ (d) $te^t \sin 2t$,
(e) $t^2 \sin at$,
(f) $e^{at} \cosh bt$.

5. Find the Laplace transform of each of the following functions:
(a) $\dfrac{d^3 f(t)}{dt^3}$,
(b) $te^t f(t)$,
(c) $\displaystyle\sum_{n=0}^{N} a_n t^n$,
(d) $\displaystyle\sum_{n=0}^{N} a_n \cos nt$.

6. The *Laguerre polynomial* of degree n is defined by the equation

$$L_n(t) = e^t \frac{d^n}{dt^n}(t^n e^{-t}).$$

Prove that

$$\mathcal{L}\{L_n(t)\} = \frac{n!}{s}\left(\frac{s-1}{s}\right)^n.$$

7. Prove that, if $\bar{f}(s) = \mathcal{L}\{f(t)\}$, and if $a > 0$, then also
(a) $\mathcal{L}\{f(at)\} = \dfrac{1}{a}\bar{f}\left(\dfrac{s}{a}\right)$,
(b) $\mathcal{L}^{-1}\{\bar{f}(as)\} = \dfrac{1}{a}f\left(\dfrac{t}{a}\right)$.

8. Prove that

$$\mathcal{L}\left\{\left(t\frac{d}{dt}\right)^n f(t)\right\} = (-1)^n \frac{d}{ds}\left(s\frac{d}{ds}\right)^{n-1} [s\bar{f}(s)]$$

when n is a positive integer. (Use induction.)

9. Let $f(t) = F(t)$ when $0 < t < a$, and let $f(t)$ be *periodic*, of period a, so that $f(t + a) = f(t)$. By writing

$$\mathcal{L}\{f(t)\} = \int_0^a e^{-st} f(t)\,dt + \int_a^{2a} e^{-st} f(t)\,dt + \int_{2a}^{3a} e^{-st} f(t)\,dt + \cdots$$

and transforming each integral in such a way that in each case the range of integration is $(0, a)$, show that

$$\mathcal{L}\{f(t)\} = \int_0^a e^{-st} F(t)\,dt\,[1 + e^{-as} + e^{-2as} + \cdots],$$

and hence, when $s > 0$,

$$\mathcal{L}\{f(t)\} = \frac{\int_0^a e^{-st} F(t)\,dt}{1 - e^{-as}}.$$

10. Apply the result of Problem 9 to the "square-wave function" for which $F(t) = 1$ when $0 < t < a/2$ and $F(t) = -1$ when $a/2 < t < a$. Show that the transform of this function is

$$\frac{(1 - e^{-as/2})^2}{s(1 - e^{-as})} = \frac{1}{s}\frac{1 - e^{-as/2}}{1 + e^{-as/2}} = \frac{1}{s}\tanh\frac{as}{4}.$$

11. Show that, if $f(t)$ is the "square-wave function" of Problem 10, then $\int_0^t f(u)\,du$ is a "triangular-wave function." Sketch it and give the expression for its transform.

12. If $f(t)$ is the "staircase function" such that $f(t) = b$ when $0 < x < a$, $f(t) = 2b$ when $a < x < 2a$, and so forth, show that

$$\mathcal{L}\{f(t)\} = \frac{b}{s(1 - e^{-as})}.$$

13. (a) Show that, if $\bar{f}(s) = \mathcal{L}\{f(t)\}$, and if an interchange of order of integrations is valid, then

$$\int_s^\infty \bar{f}(v)\,dv = \mathcal{L}\left\{\frac{f(t)}{t}\right\}.$$

[The result is valid, for s sufficiently large, whenever $f(t)/t$ has a transform.]
    (b) Use this result to deduce the transforms

$$\mathcal{L}\left\{\frac{\sin t}{t}\right\} = \cot^{-1} s, \qquad \mathcal{L}\left\{\frac{1 - e^{-t}}{t}\right\} = \log\left(\frac{s + 1}{s}\right).$$

14. (a) By formally setting $s = 0$ in the result of Problem 13(a), obtain the formula

$$\int_0^\infty \bar{f}(s)\,ds = \int_0^\infty \frac{f(t)}{t}\,dt.$$

(The result is valid *when the integral on the right exists.*)

(b) Use the result of part (a) to obtain the evaluations

$$\int_0^\infty \frac{\sin t}{t}\, dt = \frac{\pi}{2}, \qquad \int_0^\infty \frac{e^{-at} - e^{-bt}}{t}\, dt = \log \frac{b}{a} \qquad (a, b > 0).$$

(c) Show that, if $f(t) = e^t$, then neither side of the relation of part (a) exists, whereas if $f(t) = e^t \sin t$, then the left-hand member exists but the right-hand member does not.

15. By applying the property of Equation (16) to the result of Problem 13, obtain the formula

$$\mathcal{L}\left\{\int_0^t \frac{f(u)}{u}\, du\right\} = \frac{1}{s} \int_s^\infty \bar{f}(v)\, dv.$$

Also, assuming that the result of Problem 14 also applies, obtain the formula

$$\mathcal{L}\left\{\int_t^\infty \frac{f(u)}{u}\, du\right\} = \frac{1}{s} \int_0^s \bar{f}(v)\, dv.$$

[In each case, the given formula is valid, for s sufficiently large, whenever the left-hand member exists. The integral $\int_0^\infty f(t)/t\, dt$ need not exist.]

16. Assuming the results of Problem 15, show that

$$\mathcal{L}\{Si(t)\} = \frac{1}{s} \cot^{-1}s, \qquad \mathcal{L}\{Ci(t)\} = \frac{1}{s} \log \frac{1}{\sqrt{s^2 + 1}},$$

and

$$\mathcal{L}\{Ei(-t)\} = \frac{1}{s} \log \frac{1}{s + 1},$$

where

$$Si(t) = \int_0^t \frac{\sin u}{u}\, du, \qquad Ci(t) = -\int_t^\infty \frac{\cos u}{u}\, du, \qquad Ei(-t) = -\int_t^\infty \frac{e^{-u}}{u}\, du.$$

Section 2.4

17. Let $N(s)/D(s)$ denote the ratio of two polynomials, with no common factors, such that the degree n of $D(s)$ is greater than that of $N(s)$, and suppose that $D(s)$ has n distinct real zeros $s = a_1, a_2, \ldots, a_n$. Show that the coefficients in the *partial-fraction expansion*

$$\frac{N(s)}{D(s)} = \frac{A_1}{s - a_1} + \frac{A_2}{s - a_2} + \cdots + \frac{A_n}{s - a_n} = \sum_{m=1}^n \frac{A_m}{s - a_m}$$

are determined by the equations

$$A_m = \lim_{s \to a_m} (s - a_m) \frac{N(s)}{D(s)} = \frac{N(a_m)}{D'(a_m)}.$$

Hence show that in this case

$$\mathcal{L}^{-1}\left\{\frac{N(s)}{D(s)}\right\} = \sum_{m=1}^n \frac{N(a_m)}{D'(a_m)} e^{a_m t}.$$

18. If $\mathcal{L}\{f(t)\} = \bar{f}(s)$ and if

$$f(t) = A_0 + A_1 t + A_2 t^2 + \cdots + A_n t^n + \cdots$$

(in some interval about $t = 0$) and

$$\bar{f}(s) = \frac{B_0}{s} + \frac{B_1}{s^2} + \frac{B_2}{s^3} + \cdots + \frac{B_n}{s^{n+1}} + \cdots$$

(for sufficiently large values of s), show that

$$A_n = \frac{B_n}{n!}.$$

19. By using the ratio test, show that if $\lim_{n \to \infty} |B_{n+1}/B_n| = s_0$ the second series in Problem 18 converges when $s > s_0$. Deduce that in this case $\lim_{n \to \infty} |A_{n+1}/A_n| = 0$, so that the first series in Problem 18 then converges *for all values of t.*

20. Use the results of Problems 18 and 19 to find the inverse transform of each of the following functions as a power series in t:

(a) $\sin \dfrac{1}{s}$,
                         (b) $\dfrac{1}{\sqrt{s^2 + 1}} = \dfrac{1/s}{\sqrt{1 + (1/s)^2}}$.

21. (a) Show that if

$$\frac{d}{ds} F(s) = \mathcal{L}\{g(t)\}$$

then

$$\mathcal{L}^{-1}\{F(s)\} = -\frac{g(t)}{t}.$$

[Use Equation (19) with $n = 1$.]

    (b) Use the result of part (a) to deduce that

$$\mathcal{L}^{-1}\{\cot^{-1}s\} = \frac{\sin t}{t}, \qquad \mathcal{L}^{-1}\left\{\log \frac{s+1}{s}\right\} = \frac{1 - e^{-t}}{t}.$$

[Compare Problem 13(b).]

22. Use the result of Problem 21(a) to deduce the following formulas:

(a) $\mathcal{L}^{-1}\left\{\log \dfrac{s-b}{s-a}\right\} = \dfrac{e^{at} - e^{bt}}{t}$,
        (b) $\mathcal{L}^{-1}\left\{\tanh^{-1}\dfrac{s}{a}\right\} = \dfrac{\sinh at}{t}$.

23. Verify Equation (22) in each of the following cases:

(a) $f(t) = \cos at$,
                    (b) $f(t) = \sinh at$,

(c) $\bar{f}(s) = \dfrac{s}{s^2 - a^2}$,
                (d) $\bar{f}(s) = \dfrac{2s + 1}{s^2 + 2s + 2}$.

24. By starting with Equation (15a) and considering the limiting form as $s \to 0$, obtain the relation

$$\lim_{s \to 0} s\bar{f}(s) = f(0) + \lim_{s \to 0} \int_0^\infty e^{-st} f'(t) \, dt$$

and, by formally taking the limit on the right under the integral sign, obtain the result

$$\lim_{s \to 0} s\bar{f}(s) = f(\infty)$$

where $f(\infty) = \lim_{t \to \infty} f(t)$. [Compare Equation (22). The result is valid *when the integral* $\int_0^\infty f'(t) \, dt$ *exists. In particular,* $\lim_{t \to \infty} f(t)$ *must exist.*]

25. (a) Show that the result of Problem 24 is *not* valid, in particular, in the cases for which $\bar{f}(s)$ is given by

$$\frac{1}{s-1}, \qquad \frac{1}{s^2+1}, \qquad \frac{1}{s(s^2-3s+2)}.$$

(b) Show that the result of Problem 24 *is* valid, in particular, in the cases for which $\bar{f}(s)$ is given by

$$\frac{1}{s}, \qquad \frac{1}{s+1}, \qquad \frac{1}{s(s^2+3s+2)}.$$

[The result of Problem 24 occasionally is stated without qualifying restrictions. The above examples show that some care should be taken in applying it. In practical situations, it can be used if the existence of $\lim_{t\to\infty} f(t)$ is assured, from physical considerations or otherwise.]

Section 2.5

26. Determine the convolution of each of the following pairs of functions:
  ⓓ (a) 1, $\sin at$,   ⓓ (b) t, e^{at},
   (c) e^{at}, e^{bt},   (d) $\sin at$, $\sin bt$.

27. Verify Equation (20) in each of the cases considered in Problem 26.

28. If $\bar{f}(s) = \mathcal{L}\{f(t)\}$, express the inverse transform of each of the following functions as an integral:

(a) $\dfrac{\bar{f}(s)}{s+a}$,   (b) $\dfrac{\bar{f}(s)}{s^2+a^2}$,

(c) $\dfrac{\bar{f}(s)}{(s+a)^2}$,   (d) $\dfrac{\bar{f}(s)}{(s+a)(s+b)}$.

29. Suppose that $y(t)$ satisfies the *integral equation*

$$y(t) = F(t) + \int_0^t G(t-u)y(u)\,du,$$

where $F(t)$ and $G(t)$ are known functions, with Laplace transforms $\bar{F}(s)$ and $\bar{G}(s)$, respectively.

(a) Show that then

$$\bar{y}(s) = \frac{\bar{F}(s)}{1-\bar{G}(s)} = \bar{F}(s) + \frac{\bar{G}(s)}{1-\bar{G}(s)}\bar{F}(s).$$

(b) Deduce that the solution of the integral equation is

$$y(t) = F(t) + \int_0^t H(t-u)F(u)\,du,$$

where $H(t)$ is the function whose transform is given by

$$\bar{H}(s) = \frac{\bar{G}(s)}{1-\bar{G}(s)}.$$

(c) Illustrate this result in the special case when $G(t) = e^{-at}$.

30. (a) Show that $s^{-n}\bar{f}(s)$ is the transform of the function

$$F(t) = \frac{1}{(n-1)!} \int_0^t (t-u)^{n-1} f(u)\, du,$$

when n is a positive integer.

(b) Deduce that

$$\overbrace{\int_0^t \cdots \int_0^t}^{n \text{ times}} f(t)\, dt \cdots dt = \frac{1}{(n-1)!} \int_0^t (t-u)^{n-1} f(u)\, du.$$

Section 2.6

31. (a) Show that

$$\int_a^b \delta(u-t_1) f(u)\, du = f(t_1) \quad \text{if} \quad a < t_1 < b,$$

if the integral is defined as the limit (as $\epsilon \to 0$) of the integral in which $\delta(u-t_1)$ is replaced by a function equal to $1/(2\epsilon)$ when $t_1 - \epsilon < u < t_1 + \epsilon$ and equal to zero elsewhere.

(b) In a similar way, obtain the relation

$$\int_a^b \delta'(u-t_1) f(u)\, du = -f'(t_1) \quad \text{if} \quad a < t_1 < b.$$

(c) Show that the convolution of $\delta(t-t_1)$ and $f(t)$ is given by

$$\int_0^t \delta(u-t_1) f(t-u)\, du = \begin{cases} 0 & (t < t_1), \\ f(t-t_1) & (t > t_1), \end{cases}$$

when t and t_1 are positive.

32. (a) If the *Heaviside unit step function* $H(t)$ is defined such that $H(t) = 1$ when $t > 0$ and $H(t) = 0$ when $t < 0$, show that

$$\mathcal{L}\{H(t)\} = \frac{1}{s}, \qquad \mathcal{L}\{H(t-t_1)\} = \frac{e^{-st_1}}{s}$$

when $t_1 \geq 0$.

(b) Noticing that $\mathcal{L}\{\delta'(t)\} = s\, \mathcal{L}\{\delta(t)\} = s^2\, \mathcal{L}\{H(t)\}$, indicate a sense in which, correspondingly, we may be led to think of $\delta'(t)$ as a formal derivative of $\delta(t)$, and of $\delta(t)$ as a formal derivative of $H(t)$.

Section 2.7

33. Find the inverse Laplace transform of each of the following functions:

(a) $\dfrac{1}{s^2 - 3s + 2}$,   (b) $\dfrac{s}{s^2 - 2s + 5}$,   (c) $\dfrac{s+1}{s^4 + 1}$,

(d) $\dfrac{2s+1}{s(s+1)(s+2)}$,   (e) $\dfrac{1}{s^2(s^2+1)}$,   (f) $\dfrac{e^{-s}}{s+1}$.

34. Find the inverse Laplace transform of each of the following functions:

(a) $\dfrac{s^2}{(s^2+a^2)^2}$,   (b) $\dfrac{s^2}{(s^2-a^2)^2}$,   (c) $\dfrac{3s+1}{(s+1)(s+2)(s+3)}$,

(d) $\dfrac{1}{s^3 + a^3}$,   (e) $\dfrac{1 - e^{-s}}{s}$,   (f) $\dfrac{1}{(s^2 + a^2)(s^2 + b^2)}$.

35. Find the inverse Laplace transform of each of the following functions:

(a) $\dfrac{s + 1}{s^2 + 2s + 2}$,   (b) $\dfrac{s^2}{s^4 + 4a^4}$,   (c) $\dfrac{e^{-\pi s}}{s^2 + 1}$,

(d) $\dfrac{s^2}{(s + a)^3}$,   (e) $\dfrac{1}{(s^2 + 1)^3}$,   (f) $\dfrac{1}{(s^2 - 1)^3}$.

36. Determine the inverse transform of each of the following functions by making appropriate use of the expansion

$$\frac{1}{1 - \alpha} = 1 + \alpha + \alpha^2 + \cdots \qquad (|\alpha| < 1)$$

and of the property (18):

(a) $\dfrac{b}{s(1 - e^{-as})}$,   (b) $\dfrac{1}{s}\dfrac{1 - e^{-as/2}}{1 + e^{-as/2}}$,

(c) $\dfrac{\omega}{(1 - e^{-\pi s/\omega})(s^2 + \omega^2)}$,   (d) $\dfrac{\omega(1 + e^{-\pi s/\omega})}{(1 - e^{-\pi s/\omega})(s^2 + \omega^2)}$.

37. If $F(s) = \bar{g}(s)/(1 - e^{-as})$, where $g(t) = 0$ when $t \geq a$, show that $\mathcal{L}^{-1}\{F(s)\}$ is a *periodic function*, of period a, which agrees with $g(t)$ when $0 < t < a$.

38. Show that the transforms listed in parts (b), (c), and (d) of Problem 36 can be written in the form required by Problem 37, by taking $\bar{g}(s) = (1 - e^{-as/2})^2/s$ in part (b) and $\bar{g}(s) = (1 + e^{-\pi s/\omega})/(s^2 + \omega^2)$ in parts (c) and (d), with $a = 2\pi/\omega$ in part (c) and $a = \pi/\omega$ in part (d). Also verify the assertion of Problem 37 in those cases.

39. Show that the inverse of the transform

$$F(s) = \frac{1}{s} \frac{e^{-Ps/8} + e^{-3Ps/8}}{1 + e^{-Ps/2}}$$

is of period P and sketch its graph.

Section 2.8

40. Solve the following problems by the use of Laplace transforms:

(a) $\dfrac{dy}{dt} + ky = 0$, $y(0) = 1$.

(b) $\dfrac{dy}{dt} + ky = 1$, $y(0) = 0$.

(c) $\dfrac{dy}{dt} + ky = \delta(t - 1)$, $y(0) = 1$.

(d) $\dfrac{dy}{dt} + ky = f(t)$, $y(0) = y_0$.

41. Solve the following problems by the use of Laplace transforms:

(a) $\dfrac{d^2 y}{dt^2} + 2\dfrac{dy}{dt} + 2y = 0$, $y(0) = 1$, $\dfrac{dy(0)}{dt} = -1$.

(b) $\dfrac{d^2 y}{dt^2} + 2\dfrac{dy}{dt} + 2y = 2$, $y(0) = 0$, $\dfrac{dy(0)}{dt} = 1$.

(c) $\dfrac{d^2 y}{dt^2} + 2\dfrac{dy}{dt} + 2y = \delta(t - 1)$, $y(0) = 1$, $\dfrac{dy(0)}{dt} = -1$.

(d) $\dfrac{d^2y}{dt^2} + 2\dfrac{dy}{dt} + 2y = f(t)$, $y(0) = y_0$, $\dfrac{dy(0)}{dt} = y_0'$.

42. Solve the following problems by the use of Laplace transforms:

(a) $\dfrac{d^4y}{dt^4} + 4y = 0$, $y(0) = \dfrac{dy(0)}{dt} = \dfrac{d^2y(0)}{dt^2} = 0$, $\dfrac{d^3y(0)}{dt^3} = 1$.

(b) $\dfrac{d^4y}{dt^4} + 4y = 4$, $y(0) = \dfrac{dy(0)}{dt} = \dfrac{d^2y(0)}{dt^2} = \dfrac{d^3y(0)}{dt^3} = 0$.

43. Use Laplace transforms to solve the problem

$$\dfrac{d^2y}{dt^2} + y = \delta(t - a),   y(0) = 0,   y(b) = 0,$$

where $0 < a < b$, by first writing $dy(0)/dt = c$, and finally determining c so that the condition $y(b) = 0$ is satisfied by the inverse transform.

44. In Figure 2.6 a mass m is connected to an elastic spring with spring constant k, and to a dashpot which resists motion of the mass with a force numerically equal to c times the velocity of motion. The applied external force is indicated by $f(t)$ and the displacement of the mass from equilibrium position by x.

(a) Show that the differential equation of motion can be put into the form

$$\dfrac{d^2x}{dt^2} + 2\alpha\dfrac{dx}{dt} + (\alpha^2 + \beta^2)x = \dfrac{1}{m}f(t)$$

with the abbreviations

$$\dfrac{k}{m} = \omega_0^2,   \alpha = \dfrac{c}{2m},   \beta = \sqrt{\omega_0^2 - \alpha^2}.$$

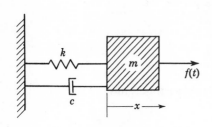

Figure 2.6

(b) Assuming that the mass starts from rest at its equilibrium position and is acted on by a uniform force f_0, find the resulting motion. Consider separately the cases when β is real and positive, $\beta = 0$, and $\beta = i\gamma$, where γ is real. Discuss the three cases and sketch typical curves representing the displacement as a function of time.

(c) Assuming that the mass starts from rest at the position $x = a$ and that no external force acts, investigate the resulting motion as in part (b).

45. (a) Use the method of Laplace transforms to obtain the solution of the simultaneous equations

$$\dfrac{dx}{dt} + \dfrac{dy}{dt} + x = -e^{-t},$$

$$\dfrac{dx}{dt} + 2\dfrac{dy}{dt} + 2x + 2y = 0,$$

which satisfies the initial conditions

$$x(0) = -1,   y(0) = +1.$$

(b) Solve the same problem by one of the methods of Chapter 1.

46. Establish the relation

$$\mathcal{L}\left\{t^k \frac{d^n y}{dt^n}\right\} = (-1)^k \frac{d^k}{ds^k}[s^n \bar{y} - s^{n-1} y(0) - \cdots - y^{(n-1)}(0)].$$

(Notice that by use of this relation a linear differential equation in y with *polynomial* coefficients can be transformed into a new linear differential equation in the transform $\bar{y}$, which may in certain cases be more tractable than the original equation. In particular, if the coefficients are *linear* functions of t, the transformed equation is of *first* order.)

47. Use the formula obtained in Problem 46 to show that, if $y(t)$ satisfies the differential equation

$$(at + b)\frac{d^2 y}{dt^2} + (ct + d)\frac{dy}{dt} + (et + f)y = h(t),$$

then its Laplace transform $\bar{y}(s)$ satisfies the equation

$$(as^2 + cs + e)\frac{d\bar{y}}{ds} - [bs^2 + (d - 2a)s + (f - c)]\bar{y}$$
$$= [(a - d) - bs]y(0) - by'(0) - \bar{h}(s).$$

48. Use the result of Problem 47 to transform the equation

$$t\frac{d^2 y}{dt^2} + 2\frac{dy}{dt} - \omega^2 ty = 0$$

to the equation

$$\frac{d\bar{y}}{ds} = \frac{y(0)}{\omega^2 - s^2}$$

and hence deduce that

$$y(t) = y(0)\frac{\sinh \omega t}{\omega t}.$$

[See Problem 22(b). Notice that $y'(0)$ cannot be assigned here and, indeed, that from the differential equation there follows $y'(0) = 0$ if y and y'' are finite at $t = 0$. Any solution other than the one obtained is not finite at $t = 0$, and does not possess a Laplace transform.]

49. Use the result of Problem 47 to transform the equation

$$t\frac{d^2 y}{dt^2} + \frac{dy}{dt} + ty = 0$$

to the equation

$$(s^2 + 1)\frac{d\bar{y}}{ds} + s\bar{y} = 0,$$

regardless of prescribed initial conditions. Hence deduce that

$$\bar{y}(s) = \frac{y(0)}{\sqrt{s^2 + 1}}$$

[see Equation (22)] and, by expanding $\bar{y}(s)$ in powers of $1/s$ [Problem 20(b)], obtain the solution

$$y(t) = y(0)\left[1 - \frac{(t^2/2)^2}{(1!)^2} + \frac{(t^2/2)^4}{(2!)^2} - \cdots\right].$$

[The series in brackets represents the Bessel function $J_0(t)$. Notice (T38). The comments on finiteness following Problem 48 also apply here.]

Section 2.9

50. Use Table 3 (when necessary) to evaluate the following to two decimal places:

(a) $\int_0^\infty e^{-x}\sqrt{x}\,dx$, (b) $\Gamma(2.7)$, (c) $\Gamma(-1.3)$,

(d) $(1.6)!$, (e) $(-1.3!)$, (f) $\Gamma(\tfrac{5}{2})$.

51. By making the indicated substitutions, transform the integral

$$\Gamma(n) = \int_0^\infty e^{-t}t^{n-1}\,dt$$

to the following equivalent forms:

(a) $\Gamma(n) = \int_0^1 \left(\log\dfrac{1}{x}\right)^{n-1} dx \qquad \left(t = \log\dfrac{1}{x}\right)$,

(b) $\Gamma(n) = 2\displaystyle\int_0^\infty e^{-r^2}r^{2n-1}\,dr \qquad (t = r^2)$.

52. (a) Obtain from Stirling's formula (63) the relation

$$\log_{10} n! \sim (n + \tfrac{1}{2})\log_{10} n - 0.4342945n + 0.39909 \qquad (n \longrightarrow \infty)$$

for use in numerical computation.

(b) Show that use of this formula gives the approximations

$$10! \approx 3.599 \times 10^6, \qquad 100! \approx 9.325 \times 10^{157}.$$

(The true values, to four figures, are 3.629×10^6 and 9.333×10^{157}. It can be shown, more generally, that Stirling's formula for $n!$ is accurate to within 1 percent when $n > 10$ and within 0.1 percent when $n > 100$.)

53. Use the Stirling formula to show that

$$\frac{a^n}{n!} \sim \frac{1}{\sqrt{2\pi n}}\left(\frac{ae}{n}\right)^n \qquad (n \longrightarrow \infty),$$

when a is any constant, and deduce that

$$\lim_{n\to\infty} \frac{a^n}{n!} = 0.$$

54. Investigate the following limits:

(a) $\displaystyle\lim_{n\to\infty} \frac{n!}{(\alpha + 1)(\alpha + 2)(\alpha + 3)\cdots(\alpha + n)}$,

(b) $\displaystyle\lim_{n\to\infty} \frac{(2n)!}{2^{2n}n!(n - \tfrac{1}{2})!}$.

55. (a) Starting with the relation

$$\frac{\Gamma(x + n + 1)}{\Gamma(x + 1)} = (x + 1)(x + 2)\cdots(x + n),$$

for any positive integer n, obtain the result

$$\frac{d}{dx}\log\Gamma(x + n + 1) = \frac{\Gamma'(x + n + 1)}{\Gamma(x + n + 1)}$$

$$= \frac{\Gamma'(x + 1)}{\Gamma(x + 1)} + \frac{1}{1 + x} + \frac{1}{2 + x} + \cdots + \frac{1}{n + x}.$$

(b) If we write $\Psi(z) = \Gamma'(z+1)/\Gamma(z+1)$, and set $x = 0$ in the result of part (a), show that there follows

$$\Psi(n) = \Psi(0) + 1 + \frac{1}{2} + \frac{1}{3} + \cdots + \frac{1}{n}.$$

[The function $\Psi(z)$ is often called the *Digamma function*. It takes on the value $\Psi(0) = -\gamma$, where $\gamma = 0.5772157 \ldots$ is a number known as *Euler's constant*.]

56. *The Beta function.* The Beta function of p and q is defined by the integral

$$B(p,q) = \int_0^1 t^{p-1}(1-t)^{q-1}\, dt \qquad (p, q > 0).$$

By writing $t = \sin^2 \theta$, obtain the equivalent form

$$B(p,q) = 2\int_0^{\pi/2} \sin^{2p-1}\theta \cos^{2q-1}\theta\, d\theta \qquad (p, q > 0).$$

57. Make use of the results of Problems 51(b) and 56 to verify the following development:

$$B(p,q)\Gamma(p+q) = 4\int_0^\infty e^{-r^2} r^{2p+2q-1}\, dr \int_0^{\pi/2} \sin^{2p-1}\theta \cos^{2q-1}\theta\, d\theta$$

$$= 4\int_0^\infty \int_0^\infty e^{-(x^2+y^2)} y^{2p-1} x^{2q-1}\, dx\, dy$$

$$= \Gamma(p)\Gamma(q) \qquad (p, q > 0).$$

Hence show that

$$B(p,q) = \frac{\Gamma(p)\Gamma(q)}{\Gamma(p+q)} \qquad (p, q > 0).$$

58. By writing $t = x/(x+a)$ in the definition of the Beta function, and using the result of Problem 57, obtain the result

$$\int_0^\infty \frac{x^{p-1}\, dx}{(x+a)^{p+q}} = a^{-q} B(p,q) = a^{-q}\frac{\Gamma(p)\Gamma(q)}{\Gamma(p+q)} \qquad (p, q, a > 0).$$

59. Use the results of preceding problems to verify the following evaluations:

(a) $\displaystyle \int_0^1 \frac{t^{p-1}}{\sqrt{1-t}}\, dt = B\!\left(p, \frac{1}{2}\right) = \sqrt{\pi}\,\frac{\Gamma(p)}{\Gamma(p+\frac{1}{2})} \qquad (p > 0),$

(b) $\displaystyle \int_0^{\pi/2} \sin^n \theta\, d\theta = \int_0^{\pi/2} \cos^n \theta\, d\theta = \frac{\sqrt{\pi}}{2}\,\frac{\Gamma\!\left(\dfrac{n+1}{2}\right)}{\Gamma\!\left(\dfrac{n+2}{2}\right)} \qquad (n > -1),$

(c) $\displaystyle \int_0^1 \frac{dt}{\sqrt{1-t^n}} = \frac{1}{n}\int_0^1 \frac{s^{(1/n)-1}\, ds}{\sqrt{1-s}} = \frac{\sqrt{\pi}}{n}\,\frac{\Gamma\!\left(\dfrac{1}{n}\right)}{\Gamma\!\left(\dfrac{1}{n}+\dfrac{1}{2}\right)} \qquad (n > 0),$

(d) $\displaystyle \int_0^\infty \frac{x^c\, dx}{(1+x)^2} = \Gamma(1-c)\Gamma(1+c) = \frac{\pi c}{\sin \pi c} \qquad (-1 < c < 1),$

(e) $\displaystyle \int_0^\infty \frac{dx}{1+x^n} = \frac{1}{n}\int_0^\infty \frac{s^{(1/n)-1}\, ds}{1+s} = \frac{\pi/n}{\sin (\pi/n)} \qquad (n > 1),$

(f) $\displaystyle \int_0^{\pi/2} \tan^n \theta\, d\theta = \frac{\pi/2}{\cos (n\pi/2)} \qquad (0 < n < 1).$

60. Verify the relationship

$$B(p, p) = 2 \int_0^{1/2} [t(1-t)]^{p-1} \, dt = \frac{\sqrt{\pi}}{2^{2p-1}} \frac{\Gamma(p)}{\Gamma(p + \frac{1}{2})},$$

by making use of the substitution $t = \frac{1}{2}(1 - \cos \theta)$.

61. By comparing the result of Problem 60 with the expression for $B(p, p)$ which follows from the result of Problem 57, deduce the *duplication formula* for the Gamma function:

$$\Gamma(2p) = \frac{2^{2p-1}}{\sqrt{\pi}} \Gamma(p)\Gamma(p + \frac{1}{2}).$$

Also show that this result can be written in the form

$$(2n)! = \frac{2^{2n}}{\sqrt{\pi}} n!(n - \frac{1}{2})!.$$

62. (a) By writing $x = u/t$ in the definition

$$B(m, n) = \int_0^1 x^{m-1}(1 - x)^{n-1} \, dx,$$

show that

$$B(m, n)t^{m+n-1} = \int_0^t u^{m-1}(t - u)^{n-1} \, du.$$

(b) By noticing that the right-hand member is the convolution of t^{m-1} and t^{n-1}, deduce that

$$B(m, n)\frac{\Gamma(m + n)}{s^{m+n}} = \frac{\Gamma(m)}{s^m}\frac{\Gamma(n)}{s^n},$$

and so obtain a simple derivation of the result of Problem 57.

3

Numerical Methods for Solving Ordinary Differential Equations

3.1. Introduction. In this chapter there are presented certain methods of numerically calculating particular solutions of ordinary differential equations which cannot be readily solved analytically. The methods given are, in general, *step-by-step* methods and are described initially for first-order equations; however, the extension of these methods to the solution of higher-order equations is also indicated.

Before considering these procedures, we outline a graphical method of solving first-order differential equations which is of some practical interest. If such a differential equation is solved for the derivative of the unknown function, the result consists of one or more relations of the form

$$\frac{dy}{dx} = F(x, y), \tag{1}$$

where it may be assumed that the function $F(x, y)$ is a single-valued function of x and y. Such an equation states that at any point (x, y) for which $F(x, y)$ is defined, the slope of any integral curve passing through that point is given by $F(x, y)$. If we plot the family of so-called *isocline* curves, defined by the equation

$$F(x, y) = C \tag{2}$$

for a series of values of the constant C, it then follows that all integral curves of (1) intersect a particular curve of the family (2) with the same slope angle φ, where $\tan \varphi$ is given by the value of C specifying the isocline. Thus, if on each isocline a series of short parallel segments having the required slope is drawn, an infinite number of integral curves can be sketched by starting in each case

93

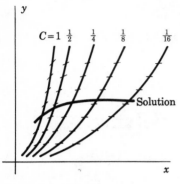

Figure 3.1

at a given point on one isocline and sketching a curve passing through that point with the indicated slope and crossing the successive isoclines with the slopes associated with them. This method can always be used to determine graphically the particular solution of (1) which passes through a prescribed point (x_0, y_0), when the function $F(x, y)$ is single-valued and continuous. The procedure is illustrated in Figure 3.1.

3.2. Use of Taylor Series. Suppose that the solution of (1) which passes through the point (x_0, y_0) is required. Knowing the value of y at $x = x_0$, we attempt, by a step-by-step method, to calculate successively approximate values of y at the points

$$x_1 = x_0 + h, \quad x_2 = x_0 + 2h, \quad \ldots, \quad x_k = x_0 + kh,$$

where h is a suitably chosen *spacing* along the x axis. For this purpose we now suppose that the value of y has been determined at $x = x_k$, and denote this value by y_k; that is, we write $y_k = y(x_k)$. We then make use of the Taylor series representation in the form

$$y(x + h) = y(x) + y'(x)\frac{h}{1!} + y''(x)\frac{h^2}{2!} + \cdots,$$

and, setting $x = x_k$, we obtain the formula

$$y_{k+1} = y_k + y'_k\frac{h}{1!} + y''_k\frac{h^2}{2!} + \cdots \tag{3}$$

for calculating the value of y_{k+1}. Primes are used to denote differentiation with respect to x. Since y' is given by (1) in terms of x and y, the coefficients in (3) can be determined from (1) by successive differentiation. Thus we obtain

$$y' = F(x, y) \qquad\qquad y'_k = F(x_k, y_k),$$

$$y'' = \frac{\partial F}{\partial x} + \frac{\partial F}{\partial y}\frac{dy}{dx}, \qquad y''_k = \frac{\partial F(x_k, y_k)}{\partial x} + \frac{\partial F(x_k, y_k)}{\partial y}y'_k,$$

and so forth, for the higher derivatives. Since y_0 is given, Equation (3) can be used to determine first y_1, with $k = 0$, then y_2 with $k = 1$, and so on, the number of terms being retained in (3) at each step depending upon the spacing and upon the accuracy desired. It is evident that a single series may, if preferred, be used for *several* calculations in some cases, by assigning successively increased values to h.

Example 1. To illustrate this procedure, we consider the solution of the differential equation

$$\frac{dy}{dx} = y - x.$$

with the initial condition that $y = 2$ at $x = 0$. We choose an interval $h = 0.1$, and hence calculate successively the approximate values of y at $x = 0.1, 0.2, 0.3$, and so on. By successive differentiation we obtain

$$y' = y - x, \qquad y'' = y' - 1, \qquad y''' = y'', \qquad \dots.$$

Hence, at $x = 0$, we have $y_0 = 2$ and

$$y_0' = 2, \qquad y_0'' = 1, \qquad y_0''' = 1, \qquad \dots,$$

and, with $k = 0$, Equation (3) gives

$$y_1 = y_0 + 2h + \frac{h^2}{2} + \frac{h^3}{6} + \cdots$$

$$\cong 2 + 0.2000 + 0.0050 + 0.0002 + \cdots = \mathbf{2.2052.}$$

Next, at $x = 0.1$, we have $y_1 \cong 2.2052$ and

$$y_1' \cong 2.1052, \qquad y_1'' \cong 1.1052, \qquad y_1''' \cong 1.1052, \qquad \dots,$$

and, with $k = 1$, Equation (3) gives

$$y_2 \cong y_1 + 2.1052h + 0.5526h^2 + 0.1842h^3 + \cdots$$

$$\cong 2.2052 + 0.2105 + 0.0055 + 0.0002 + \cdots = \mathbf{2.4214.}$$

The procedure may be repeated as often as is required. In this example the exact solution is readily found to be

$$y = e^x + x + 1,$$

and the above results are found to be accurate to the four decimal places retained. ∎

This procedure is readily generalized to the solution of initial-value problems involving differential equations of higher order, as may be seen from Example 2, below. For a second-order equation it is found to be necessary to calculate y_{k+1}' as well as y_{k+1} before proceeding to the calculation of y_{k+2}. For this purpose we may differentiate $y(x + h)$ with respect to x to obtain

$$y'(x + h) = y'(x) + y''(x)\frac{h}{1!} + y'''(x)\frac{h^2}{2!} + \cdots.$$

Setting $x = x_k$, there then follows

$$y_{k+1}' = y_k' + y_k''\frac{h}{1!} + y_k'''\frac{h^2}{2!} + \cdots. \tag{4}$$

This result can also be obtained by differentiating Equation (3) *with respect to h*, as may be seen directly from the fact that the derivatives of $y(x + h)$ with respect to x and h are identical.

Example 2. Consider the nonlinear differential equation

$$\frac{d^2y}{dx^2} - \frac{dy}{dx} + xy^2 = 0$$

with the initial conditions that $y = 1$ and $y' = -1$ when $x = 0$. We calculate the successive derivatives

$$y'' = y' - xy^2, \qquad y''' - y'' - 2xyy' - y^2,$$
$$y^{\text{iv}} = y''' - 2xy'^2 - 2xyy'' - 4yy', \qquad \dots$$

Hence, at $x = 0$, we have $y_0 = 1$, $y_0' = -1$, and

$$y_0'' = y_0' = -1, \qquad y_0''' = y_0'' - y_0^2 = -2, \qquad y_0^{\text{iv}} = 2, \qquad \dots$$

Then, with $k = 0$, Equation (3) gives

$$y_1 = 1 - h - \frac{h^2}{2} - \frac{h^3}{3} + \frac{h^4}{12} + \cdots,$$

and, taking $h = 0.1$,

$$y_1 \cong 1 - 0.1 - 0.005 - 0.0003 + \cdots = \mathbf{0.8947}.$$

Now, in order to calculate y_2 it will be necessary next to calculate y_1'', y_1''', and so on. However, the calculation of these values involves knowledge of the value y_1' in addition to the value of y_1, which is now known. The value of y_1' can be calculated by using the series

$$y_1' = -1 - h - h^2 + \frac{h^3}{3} + \cdots,$$

which is obtained by differentiating the series defining $y_1 = y(x_0 + h)$ with respect to h. Hence we obtain

$$y_1' \cong -1 - 0.1 - 0.01 + 0.0003 + \cdots = \mathbf{-1.1097}.$$

The values of y_1'', y_1''', and so on, can now be calculated from the forms given, and the calculation of y_2 and y_2' proceeds in the same way. ∎

A further generalization to the solution of two simultaneous equations of the form

$$\frac{dx}{dt} = F(x, y, t) \quad \text{and} \quad \frac{dy}{dt} = G(x, y, t),$$

with prescribed initial values of x and y, is readily devised.

3.3. The Adams Method.
Suppose again that the solution of the problem

$$\frac{dy}{dx} = F(x, y), \qquad y(x_0) = y_0 \tag{5}$$

has been determined up to the point $x_k = x_0 + kh$. If now we assume that over the interval $(x_k, x_k + h)$ the derivative dy/dx changes so slowly that it can be approximated by its value y_k' at the point x_k, then over that interval the approximate increase in y is given by hy_k', and we obtain

$$y_{k+1} \cong y_k + hy_k' \cong y_k + hF_k, \tag{6}$$

where F_k is written for $F(x_k, y_k)$. This formula obviously would give exact results if y were a linear function of x over the interval considered.

A more nearly exact formula is obtained, in general, if we assume that the derivative dy/dx is nearly *linear* over the interval $(x_k - h, x_k + h)$, and hence that the graph of y can be approximated by a parabola over this interval. We thus assume the approximation

$$F(x, y) \cong a + b(x - x_k) \qquad (x_k - h < x < x_k + h)$$

and determine the constants a and b in such a way that the approximation takes on the calculated values of $F(x, y)$ at the points $x_k - h$ and x_k. In this way we obtain

$$a = F_k, \qquad b = \frac{F_k - F_{k-1}}{h}.$$

Hence, integrating both sides of the differential equation (5) over the interval $(x_k, x_k + h)$, we find

$$y_{k+1} - y_k \cong \int_{x_k}^{x_k+h} \left[F_k + (F_k - F_{k-1})\frac{x - x_k}{h} \right] dx$$

$$\cong hF_k + \frac{h}{2}(F_k - F_{k-1})$$

or

$$y_{k+1} \cong y_k + hF_k + \frac{h}{2}(F_k - F_{k-1}). \tag{7}$$

The last term of this expression is seen to be a correction to the expression given by (6) and may be appreciable if the derivative dy/dx varies appreciably over the interval considered. It should be noticed that the first ordinate calculable by (7) is y_2, when $k = 1$, and that in this case the values of y_1 and F_1 are needed in addition to the known initial value F_0. The value of y_1 must be determined by another method, for example, by the use of Taylor series; the value of F_1 is then given as $F(x_1, y_1)$. With the notation

$$\Delta F_k = F_{k+1} - F_k,$$

Equation (7) can be written in the form

$$y_{k+1} \cong y_k + h(F_k + \tfrac{1}{2}\Delta F_{k-1}). \tag{8}$$

Still more accurate formulas can be obtained, in general, if the derivative of the unknown function y is approximated by a polynomial of higher degree n, taking on calculated values at $n + 1$ consecutive points. By an extension of the preceding method (see Section 3.7), a formula is obtained when $F(x, y)$ is approximated by a polynomial of *fourth degree*, in the form

$$y_{k+1} \cong y_k + h(F_k + \tfrac{1}{2}\Delta F_{k-1} + \tfrac{5}{12}\Delta^2 F_{k-2} + \tfrac{3}{8}\Delta^3 F_{k-3} + \tfrac{251}{720}\Delta^4 F_{k-4}) \tag{9}$$

with the notations

$$\Delta F_r = F_{r+1} - F_r,$$
$$\Delta^2 F_r = \Delta F_{r+1} - \Delta F_r,$$
$$\Delta^3 F_r = \Delta^2 F_{r+1} - \Delta^2 F_r, \tag{10}$$
$$\Delta^4 F_r = \Delta^3 F_{r+1} - \Delta^3 F_r.$$

These notations define the first, second, third, and fourth *forward differences* of the calculated values of F, which, as will be seen, are readily evaluated if the calculations are suitably tabulated.

Formula (9), which involves fourth differences, would give exact results if, over the interval $(x_k - 4h, x_k + h)$, the unknown function y were a polynomial of *fifth degree* in x. Formula (6) is obtained from (9) by neglecting all differeneces, and (7) or (8) is obtained by retaining only first differences. Formulas of intermediate accuracy can be obtained from (9) by retaining only terms through the second or third differences.

It is seen that, if, for example, it is decided that second differences are to be retained, the calculation of y_{k+1} makes use of the values of F_k, F_{k-1}, and F_{k-2}, and hence of the values y_k, y_{k-1}, and y_{k-2}. Thus the first ordinate calculable in this case would be y_3. Since only y_0 is prescribed at the start, the values of y_1 and y_2 must first be determined by another method, such as that of Taylor series, or the Runge–Kutta method of the following section when initial series developments are not feasible. Similarly, if third or fourth differences are retained, three or four additional ordinates, respectively, must be first calculated by another method before the present procedure can be applied.

Example 1 (continued). We apply this method to the continuation of the solution of the problem considered in Example 1 (Section 3.2), retaining *second differences*. The work may be tabulated as follows:

x	y	$F = y - x$	ΔF	$\Delta^2 F$
0	2	2	0.1052	0.0110
0.1	2.2052	2.1052	0.1162	0.0122
0.2	2.4214	2.2214	0.1284	
0.3	2.6498	2.3498		
0.4	2.8917			

Before the Adams formula is applied to calculate y_3, the two initial ordinates y_1 and y_2 are required in addition to the prescribed ordinate $y_0 = 2$. These ordinates are taken from the results of Example 1 and are entered into the second column of the table as shown. The corresponding values of F are then entered in the third column. Each entry in the fourth column is obtained by subtracting the corresponding entry in the F column from the succeeding entry in that column. *Algebraic signs must, of course, be retained.* The last column contains similar differences between successive entries in the ΔF column. In this way the entries above the division line in each column of the table are obtained. Now, to calculate the ordinate y_3, we make use of Formula (9), neglecting the last two terms, and notice that the quantities needed are exactly those which appear immediately above the division line in each column, that is, the last numbers entered in the columns at this stage of the computation. We thus obtain

$$y_3 \cong 2.4214 + 0.1[2.2214 + \tfrac{1}{2}(0.1162) + \tfrac{5}{12}(0.0110)] = \mathbf{2.6498}.$$

It is seen that the differences needed at a given step of the calculation recede along successive columns in the table. The value of y_3 is now entered in the second column, the corresponding value of F is calculated, and additional differences are determined. The next ordinate is then calculated as before,

$$y_4 \cong 2.6498 + 0.1[2.3498 + \tfrac{1}{2}(0.1284) + \tfrac{5}{12}(0.0122)] = \mathbf{2.8917},$$

and the process is continued in the same way. A rough check on the accuracy obtained at each step can be obtained by estimating the contribution of the neglected third difference. Thus retention of third differences in the calculation of y_4 would, in this case, increase the value obtained by $\tfrac{3}{8}(0.1)(0.0012) = 0.00005$, and hence the fourth places in the results are in doubt. The correct value is $e^{0.3} + 1.3 = 2.8918$.   ■

To apply the Adams method to a second-order equation of the form

$$\frac{d^2 y}{dx^2} = F\left(x, y, \frac{dy}{dx}\right), \tag{11}$$

we may first introduce the notation $p = dy/dx$ and hence replace Equation (11) by the simultaneous equations

$$\frac{dy}{dx} = p, \tag{12a}$$

$$\frac{dp}{dx} = F(x, y, p). \tag{12b}$$

The initial conditions at $x = x_0$,

$$y = y_0, \tag{13a}$$

$$\frac{dy}{dx} = p = p_0, \tag{13b}$$

are to be satisfied. Then, applying (9) separately to (12a) and (12b), we obtain the two formulas

$$y_{k+1} \cong y_k + h(p_k + \tfrac{1}{2}\Delta p_{k-1} + \cdots), \tag{14a}$$

$$p_{k+1} \cong p_k + h(F_k + \tfrac{1}{2}\Delta F_{k-1} + \cdots), \tag{14b}$$

and proceed step by step as before.

Example 2 (continued). We apply this method to the continuation of the solution of the problem considered in Example 2 (Section 3.2), retaining (for simplicity) only *first* differences. The work may be tabulated as follows:

x	y	p	Δp	$F = p - xy^2$	ΔF
0	1	-1	-0.110	-1	-0.190
0.1	0.895	-1.110		-1.190	
0.2	0.779	-1.238			

The values of y_1 and p_1, taken from Example 2, as well as the prescribed values of y_0 and p_0, are entered first. Next the corresponding values of F and the differences

Δp_0 and ΔF_0 are calculated and entered. Equations (14a, b) then give

$$y_2 \cong 0.895 + 0.1(-1.110 - 0.055) = \mathbf{0.779},$$

$$p_2 \cong -1.110 + 0.1(-1.190 - 0.095) = \mathbf{-1.238}.$$

At this stage a second difference $\Delta^2 p_0 = -0.018$ can be calculated. Since its contribution to the calculation of y_3 would be -0.00075, it may be presumed that the result for y_2 is also in doubt by about one unit in the third decimal place. ∎

The generalization of these methods to the solution of more general initial-value problems involving two simultaneous differential equations offers no difficulty.

3.4. The Modified Adams Method. A useful modification of the Adams method in the case of a first-order equation consists of using an appropriate truncation of the Adams formula

$$y_{k+1} \cong y_k + h(F_k + \tfrac{1}{2}\Delta F_{k-1} + \tfrac{5}{12}\Delta^2 F_{k-2} + \tfrac{3}{8}\Delta^3 F_{k-3} + \tfrac{251}{720}\Delta^4 F_{-4}) \tag{15}$$

only as a "predictor," to provide a first approximation to the value of y_{k+1}, and of using a corresponding truncation of the formula

$$y_{k+1} \cong y_k + h(F_{k+1} - \tfrac{1}{2}\Delta F_k - \tfrac{1}{12}\Delta^2 F_{k-1} - \tfrac{1}{24}\Delta^3 F_{k-2} - \tfrac{19}{720}\Delta^4 F_{k-3}) \tag{16}$$

as a "corrector." The derivation of the latter formula is indicated in Section 3.7.

The approximation yielded by (16) usually is better than that afforded by (15), when both are terminated with differences of like order, since the coefficients of the higher differences are smaller in (16). However, this advantage is partly offset by the fact that, since the right-hand member of (16) involves $F_{k+1} \equiv F(x_{k+1}, y_{k+1})$, this equation expresses y_{k+1} (approximately) in terms of itself. Thus (16) generally cannot be solved *analytically* for y_{k+1} except in special cases, such as those in which F is a linear function of y.

Fortunately, resort generally can be had to a simple iterative procedure for the solution of (16) when the spacing h is sufficiently small. For this purpose, (15) is first used to determine a "predicted" value of y_{k+1}, in correspondence with which the approximate value of F_{k+1} is calculated, together with the approximate new differences $\Delta F_k = F_{k+1} - F_k$, $\Delta^2 F_{k-1}$, and so on. Then (16) is used to determine a "corrected" value of y_{k+1}. If this value differs significantly from the predicted value, the entries F_{k+1}, ΔF_k, and so on, then may be correspondingly corrected and (16) used again, to provide a "recorrected" value of y_{k+1}. Generally, the need for this recorrection is avoided either by retaining a sufficiently large number of differences or by taking the spacing h to be sufficiently small.†

---

†When h is sufficiently small to avoid the need for *large* correction, it frequently happens, in fact, that the result of the *first* correction affords a better approximation to the true value than that which would be afforded by subsequent recorrections.

Example 1 (continued). In the preceding tabulation of Example 1, the entry 2.8917 here would be interpreted as the *predicted* value of y_4. When corresponding approximations to F_4, ΔF_3, and $\Delta^2 F_2$ are calculated, the tabulation appears as follows:

x	y	F	ΔF	$\Delta^2 F$
0	2	2	0.1052	0.0110
0.1	2.2052	2.1052	0.1162	0.0122
0.2	2.4214	2.2214	0.1284	0.0135
0.3	2.6498	2.3498	0.1419	
0.4	2.8917	2.4917		

From (16), truncated also to second differences, a *corrected* value is obtained,

$$y_4 \cong 2.6498 + 0.1[2.4917 - \tfrac{1}{2}(0.1419) - \tfrac{1}{12}(0.0135)] = \mathbf{2.8918},$$

in correspondence with which the tabulated values of F_4, ΔF_3, and $\Delta^2 F_2$ each are corrected by one unit in the last place retained. Clearly, no additional corrections will result from a repetition of the process, so that an advance to $x = 0.5$ appears to be in order. ∎

When both formulas are truncated with a difference of order r ($r = 2$ in Example 1), the error in the finally corrected value can be estimated by the formula

$$E_{k+1} \cong \frac{y_{k+1}^P - y_{k+1}^C}{2 + 4r}, \tag{17}$$

where y_{k+1}^P is the "predicted" value obtained from (15) and y_{k+1}^C is the "corrected" value obtained from (16). (See Reference 3.) This estimate applies only to the error introduced, by using (16) to approximate the differential equation, in progressing from x_k to x_{k+1}. It ignores the accumulated effects of errors of the same type introduced at earlier stages, as well as the effects of "roundoff" errors; also its validity essentially presumes that the spacing h is sufficiently small to obviate the need for "recorrection."

Thus, for example, when only second differences are retained, the error in the corrected value can be estimated as $(y_{k+1}^P - y_{k+1}^C)/10$, provided that this estimate is small, that the same is true of the corresponding error estimates at all preceding stages, that recorrection was not needed, and that roundoff errors have been controlled by the retention of an appropriate number of significant figures in all calculations.

It happens that the rate of convergence of the iterative process, in the determination of y_{k+1} from (16), depends upon the magnitude of the quantity

$$\rho_k = \beta h \frac{\partial F(x_k, y_k)}{\partial y}, \tag{18}$$

where β is the algebraic sum of the numerical coefficients of the differences retained in (16). (See Reference 3.) Thus, when *second* differences are retained,

$$\beta = 1 - \tfrac{1}{2} - \tfrac{1}{12} = \tfrac{5}{12}.$$

Unless $|\rho_k| \ll 1$, the process either will not converge or will converge so slowly that many recorrections will be needed, in general. In the case of the Example, the "convergence factor" ρ_3 is seen to be

$$\rho_3 = \tfrac{5}{12}h \approx 0.04.$$

Before *starting* a step-by-step calculation based on (16), it is desirable to verify that the spacing h is such that the magnitude of the initial convergence factor ρ_0 is small relative to unity.

The application of the modified Adams method to the second-order equation (11), or to a higher-order equation, is perfectly straightforward. In the case of Equation (11), $y'' = F(x, y, y')$, the approximate "convergence factor" is found to be

$$\rho_k = \beta h \left[\beta h \left(\frac{\partial F}{\partial y} \right)_k + \left(\frac{\partial F}{\partial y'} \right)_k \right]. \tag{19}$$

3.5. The Runge-Kutta Method. The method associated with the names of Runge and Kutta is a step-by-step process in which an approximation to y_{k+1} is obtained from y_k in such a way that the power series expansion of the approximation would coincide, up to terms of a certain order h^N in the spacing $h = x_{k+1} - x_k$, with the actual Taylor series development of $y(x_k + h)$ in powers of h. However, no preliminary differentiation is needed, and the method also has the advantage that no initial values are needed beyond the prescribed values. Such a method is particularly useful if certain coefficients in the differential equation are empirical functions for which analytical expressions are not known, and hence for which initial series developments are not feasible. In place of using values of N derivatives of y at one point, it uses values only of the *first* derivative at N suitably chosen points.

The derivation of the basic formulas may be illustrated by considering the special case when *second-order* accuracy in h is required. Again starting with the differential equation

$$\frac{dy}{dx} = F(x, y) \tag{20}$$

and the prescribed initial condition $y = y_0$ when $x = x_0$, the Taylor series expansion of $y_{k+1} = y(x_k + h)$ up to second powers of h is obtained, by introducing the two succeeding relations into (3), in the form

$$y_{k+1} = y_k + F_k h + \left(\frac{\partial F_k}{\partial x} + \frac{\partial F_k}{\partial y} F_k \right) \frac{h^2}{2} + \cdots \tag{21}$$

with the notations

$$F_k = F(x_k, y_k), \qquad \frac{\partial F_k}{\partial x} = \frac{\partial F(x_k, y_k)}{\partial x}, \qquad \frac{\partial F_k}{\partial y} = \frac{\partial F(x_k, y_k)}{\partial y}.$$

We assume an approximation of the form

$$y_{k+1} \cong y_k + \lambda_1 h F_k + \lambda_2 h F(x_k + \mu_1 h, \, y_k + \mu_2 h F_k) \tag{22}$$

and attempt to determine the constants λ_1, λ_2, μ_1, and μ_2 in such a way that the expansion of the right-hand side of (22) in powers of h agrees with the expression given by (21) *through terms of second order in h*. From the Taylor series expansion

$$f(x + H, y + K) = f(x, y) + H\frac{\partial f(x, y)}{\partial x} + K\frac{\partial f(x, y)}{\partial y} + \cdots,$$

for small values of H and K, where the following terms are of second order in H and K, we find

$$F(x_k + \mu_1 h, y_k + \mu_2 h F_k)$$
$$= F(x_k, y_k) + \mu_1 h\frac{\partial F(x_k, y_k)}{\partial x} + \mu_2 h\frac{\partial F(x_k, y_k)}{\partial y}F_k + \cdots$$
$$= F_k + h\left(\mu_1\frac{\partial F_k}{\partial x} + \mu_2\frac{\partial F_k}{\partial y}F_k\right) + \cdots,$$

where the omitted terms are of at least second order in h. Hence (22) becomes

$$y_{k+1} = y_k + (\lambda_1 + \lambda_2)F_k h + \lambda_2\left(\mu_1\frac{\partial F_k}{\partial x} + \mu_2\frac{\partial F_k}{\partial y}F_k\right)h^2 + \cdots. \qquad (23)$$

The terms retained in Equations (21) and (23) are brought into agreement if we take, in particular,

$$\lambda_1 = \lambda_2 = \tfrac{1}{2}, \qquad \mu_1 = \mu_2 = 1.$$

Thus Equation (22) becomes, in this case,

$$y_{k+1} \cong y_k + \frac{h}{2}[F(x_k, y_k) + F(x_k + h, y_k + hF_k)]$$

or, writing

$$K_1 = hF(x_k, y_k), \qquad K_2 = hF(x_k + h, y_k + K_1), \qquad (24)$$

this result can be put in the form

$$y_{k+1} \cong y_k + \tfrac{1}{2}(K_1 + K_2). \qquad (25)$$

It should be noticed that Equation (23) can be brought into agreement with (21) in infinitely many ways, by taking

$$\lambda_2 = 1 - \lambda_1, \qquad \mu_1 = \mu_2 = \frac{1}{2(1 - \lambda_1)}$$

where $\lambda_1 \neq 1$ but is otherwise arbitrary. Thus infinitely many other forms of (22) could be obtained, in addition to the rather symmetrical form chosen here.

By methods analogous to that just given, similar formulas giving higher-order accuracy in h may be obtained. We give without derivation two such procedures:

Third-order accuracy:

$$y_{k+1} \cong y_k + \tfrac{1}{6}(a_1 + 4a_2 + a_3), \qquad (26)$$
$$a_1 = hF(x_k, y_k),$$
$$a_2 = hF(x_k + \tfrac{1}{2}h, y_k + \tfrac{1}{2}a_1), \qquad (27)$$
$$a_3 = hF(x_k + h, y_k + 2a_2 - a_1).$$

Fourth-order accuracy:

$$y_{k+1} \cong y_k + \tfrac{1}{6}(b_1 + 2b_2 + 2b_3 + b_4), \tag{28}$$

$$
\begin{aligned}
b_1 &= hF(x_k, y_k), \\
b_2 &= hF(x_k + \tfrac{1}{2}h, y_k + \tfrac{1}{2}b_1), \\
b_3 &= hF(x_k + \tfrac{1}{2}h, y_k + \tfrac{1}{2}b_2), \\
b_4 &= hF(x_k + h, y_k + b_3).
\end{aligned}
\tag{29}
$$

The close relationship between Equations (26) and (27) and the formula of "Simpson's rule" (see Problem 27) may be noticed.

Let the *error* associated with using a procedure of Nth-order accuracy k times with spacing h be expressed in the form Kkh^{N+1}; that is, suppose that the correct result is given by *adding* Kkh^{N+1} to the calculated result. Then it can be shown that in ordinary cases the quantity K is not strongly dependent on k, h, and N. Thus, if the spacing were doubled, the error associated with the corresponding new calculation would be approximately

$$K \frac{k}{2}(2h)^{N+1} = 2^N Kkh^{N+1}.$$

Since the result of subtracting the ordinate determined by the second method from that determined by the first would then be $(2^N - 1) Kkh^{N+1}$, it follows that this difference is approximately $(2^N - 1)$ times the error in the first (more nearly exact) calculated value. In this way we are led to the following error estimate:

If a procedure of Nth-order accuracy gives an ordinate $y^{(1)}$ with spacing h and an ordinate $y^{(2)}$ with spacing $2h$, the error in $y^{(1)}$ is given approximately by $(y^{(1)} - y^{(2)})/(2^N - 1)$.

Thus the difference between the two results is divided by 3 if (24) and (25) are used, by 7 if (26) and (27) are used, and by 15 if (28) and (29) are used, to obtain an error estimate.

Example 1. We apply (26) and (27) to integrate the equation considered in Example 1 (Section 3.2). The work may be arranged as follows:

x	0	0.1	0.2
y	2	2.20517	2.42139
a_1	0.2	0.21052	
$x + \tfrac{1}{2}h$	0.05	0.15	
$y + \tfrac{1}{2}a_1$	2.1	2.31043	
a_2	0.205	0.21604	
$x + h$	0.1	0.2	
$y + 2a_2 - a_1$	2.21	2.42673	
a_3	0.211	0.22267	
$\tfrac{1}{6}(a_1 + 4a_2 + a_3)$	0.20517	0.21622	

If the spacing is doubled, the result $y_2 \cong 2.42133$ is obtained directly (with $h = 0.2$). Application of the error estimate gives the approximate error $(0.000006)/7 = 0.00001$, which indicates that the first value calculated for y_2 may be one unit too small in the last place retained. The exact value is $y_2 = 2.42140$ to five decimal places. ∎

To integrate a second-order differential equation of the form

$$\frac{d^2y}{dx^2} = F\left(x, y, \frac{dy}{dx}\right) \tag{30}$$

with the initial conditions $y = y_0$ and $y' = y'_0$ when $x = x_0$, with *third-order accuracy* in the spacing, we first calculate successively the values

$$k_1 = hy'_k,$$
$$k'_1 = hF(x_k, y_k, y'_k),$$
$$k_2 = h(y'_k + \tfrac{1}{2}k'_1),$$
$$k'_2 = hF(x_k + \tfrac{1}{2}h, y_k + \tfrac{1}{2}k_1, y'_k + \tfrac{1}{2}k'_1), \tag{31}$$
$$k_3 = h(y'_k + 2k'_2 - k'_1),$$
$$k'_3 = hF(x_k + h, y_k + 2k_2 - k_1, y'_k + 2k'_2 - k'_1).$$

The values y_{k+1} and y'_{k+1} are then given by

$$y_{k+1} \cong y_k + \tfrac{1}{6}(k_1 + 4k_2 + k_3) \tag{32}$$

and

$$y'_{k+1} \cong y'_k + \tfrac{1}{6}(k'_1 + 4k'_2 + k'_3). \tag{33}$$

A corresponding formula giving *fourth-order accuracy* can be written down by analogy from (28) and (29).

Example 2. Applying this method to calculate y_1 in the case of the problem of Example 2 (Section 3.2), we obtain successively

$$k_1 = -0.1, \qquad k_2 = -0.105, \qquad k_3 = -0.11190,$$
$$k'_1 = -0.1, \qquad k'_2 = -0.10951, \qquad k'_3 = -0.11982.$$

There then follows

$$y_1 \cong 1 - \tfrac{1}{6}(0.1 + 0.42 + 0.11190) = \mathbf{0.89468},$$
$$y'_1 \cong -1 - \tfrac{1}{6}(0.1 + 0.43804 + 0.11982) = -\mathbf{1.1096}. \qquad ∎$$

3.6. Picard's Method. In contrast with the step-by-step methods so far considered, in which successive ordinates are calculated point by point, the method of Picard is an *iterative method* that gives successive *functions* which, in favorable cases, tend as a whole (at least over a certain interval) toward the exact solution. Although the method is of limited practical usefulness, it illustrates a type of procedure which is of frequent use in other applications, and is in itself of theoretical importance.

Considering first an initial-value problem of first order,

$$\frac{dy}{dx} = F(x, y), \tag{34}$$

where $y = y_0$ when $x = x_0$, we formally integrate both sides of Equation (34) over the interval (x_0, x) to obtain an equivalent relation

$$y = y_0 + \int_{x_0}^{x} F(x, y)\, dx. \tag{35}$$

Now, to start the procedure, we take as an initial approximation to the function y (to be determined) a suitable *function of x*, say $y^{(1)}(x)$. If the general nature of the required solution of (34) is known, this initial function may be chosen on this basis. It is preferable that it satisfy the initial condition, although this is not necessary. In the absence of further information, the initial approximation function $y^{(1)}(x)$ may be taken as the constant y_0. With this assumed approximation for y, as a function of x, the function $F(x, y^{(1)})$ becomes a known function of x, and a second approximation to the function y, say $y^{(2)}(x)$, is given by

$$y^{(2)}(x) = y_0 + \int_{x_0}^{x} F[x, y^{(1)}(x)]\, dx. \tag{36}$$

A third approximation is obtained by replacing y by $y^{(2)}(x)$ in $F(x, y)$ and using Equation (35) to give

$$y^{(3)}(x) = y_0 + \int_{x_0}^{x} F[x, y^{(2)}(x)]\, dx. \tag{37}$$

In this way successive approximations are obtained *as functions of x*, according to the formula

$$y^{(n+1)}(x) = y_0 + \int_{x_0}^{x} F[x, y^{(n)}(x)]\, dx. \tag{38}$$

If $F(x, y)$ is sufficiently regular near the point (x_0, y_0), the successive approximations $y^{(1)}, y^{(2)}, \ldots, y^{(n)}$ will tend toward a limiting function $y(x)$ over some interval in x about $x = x_0$, and that function will satisfy the differential equation (34) as well as the prescribed initial condition.

Example. Applying this method to the solution of the equation of Example 1 (Section 3.2), we write

$$y^{(n+1)} = 2 + \int_0^x (y^{(n)} - x)\, dx.$$

Taking $y^{(1)} = 2$, we obtain successively

$$y^{(2)} = 2 + \int_0^x (2 - x)\, dx = 2 + 2x - \frac{x^2}{2},$$

$$y^{(3)} = 2 + \int_0^x \left(2 + x - \frac{x^2}{2}\right) dx = 2 + 2x + \frac{x^2}{2} - \frac{x^3}{6},$$

$$y^{(4)} = 2 + \int_0^x \left(2 + x + \frac{x^2}{2} - \frac{x^3}{6}\right) dx = 2 + 2x + \frac{x^2}{2} + \frac{x^3}{6} - \frac{x^4}{24},$$

and so forth. In this case it is readily shown by induction that

$$y^{(n+1)} = 2 + 2x + \frac{x^2}{2!} + \frac{x^3}{3!} + \cdots + \frac{x^n}{n!} - \frac{x^{n+1}}{(n+1)!},$$

and as n becomes large we are led to a consideration of the infinite series

$$y = 1 + x + \left(1 + \frac{x}{1!} + \frac{x^2}{2!} + \cdots + \frac{x^n}{n!} + \cdots\right),$$

which may be recognized as the expansion of the known exact solution

$$y = 1 + x + e^x,$$

the expansion converging for all values of x. ∎

In practice it is usually not feasible to obtain an explicit expression for the nth approximation and to proceed to the exact solution by a limiting process, as in the example given. Furthermore, the successive approximations need not be polynomials, as in this case. Still, if a particular approximation is used to calculate y, the accuracy at a given point can be estimated roughly by considering the deviation between this approximation and the preceding approximation in the neighborhood of the point. Unfortunately, it is true that in many cases the integrals defining successive approximations cannot be evaluated analytically, but in turn must be approximated by numerical methods.

Similar procedures can be devised for dealing with equations of higher order, the convergence of the successive approximations to the true solution depending upon the regularity of the coefficients.

3.7. Extrapolation with Differences. For purposes of simplicity, the details of the derivation of Equations (9) and (16), which specify the Adams method and its modification for the numerical integration of differential equations, were omitted in Sections 3.3 and 3.4. In this section we indicate this derivation and also point out the usefulness in other connections of certain intermediate and related results.

We again denote the value of a function $f(x)$ at one of $n + 1$ equally spaced points,

$$x_k, \quad x_{k-1} = x_k - h, \quad x_{k-2} = x_k - 2h, \quad \ldots, \quad x_{k-n} = x_k - nh, \quad (39)$$

by the abbreviation $f_r = f(x_r)$. In Section 3.3 the *forward differences* were defined by the equations

$$\Delta f_r = f_{r+1} - f_r, \quad \Delta^2 f_r = \Delta f_{r+1} - \Delta f_r = f_{r+2} - 2f_{r+1} + f_r, \quad (40)$$

and so forth. In some developments it is more convenient to use the so-called *backward difference* notation, according to which we write instead

$$\nabla f_r = f_r - f_{r-1}, \quad \nabla^2 f_r = \nabla f_r - \nabla f_{r-1} = f_r - 2f_{r-1} + f_{r-2}, \quad (41)$$

and so forth. These two notations clearly are related by the equations

$$\Delta f_r = \nabla f_{r+1}, \quad \Delta^2 f_r = \nabla^2 f_{r+2}, \quad \ldots, \quad \Delta^m f_r = \nabla^m f_{r+m}. \quad (42)$$

Thus, for example, we have

$$f_3 - f_2 = \Delta f_2 = \nabla f_3, \quad f_3 - 2f_2 + f_1 = \Delta^2 f_1 = \nabla^2 f_3.$$

In particular, Equations (9) and (16) take the forms

$$y_{k+1} \cong y_k + h(F_k + \tfrac{1}{2}\nabla F_k + \tfrac{5}{12}\nabla^2 F_k + \tfrac{3}{8}\nabla^3 F_k + \tfrac{251}{720}\nabla^4 F_k) \qquad (43)$$

and

$$y_{k+1} \cong y_k + h(F_{k+1} - \tfrac{1}{2}\nabla F_{k+1} - \tfrac{1}{12}\nabla^2 F_{k+1} \\ - \tfrac{1}{24}\nabla^3 F_{k+1} - \tfrac{19}{720}\nabla^4 F_{k+1}), \qquad (44)$$

with the notation (41) of backward differences, the subscripts on the right in these equations no longer varying from difference to difference.

In order to determine a *polynomial approximation* of degree n to a function $f(x)$, having the property that agreement is exact at the $n + 1$ points defined in (39), it is convenient for present purposes to write the approximation in the form

$$f(x) \cong a_0 + a_1(x - x_k) + a_2(x - x_k)(x - x_{k-1}) \\ + a_3(x - x_k)(x - x_{k-1})(x - x_{k-2}) + \cdots \\ + a_n(x - x_k)(x - x_{k-1}) \cdots (x - x_{k-n+1}). \qquad (45)$$

In this form the coefficients are readily determined by forming successive backward differences at $x = x_k$, as follows. First, setting $x = x_k$, and replacing the approximation by an equality, we have

$$a_0 = f_k. \qquad (46a)$$

Next, if we calculate the difference $\nabla f(x) = f(x) - f(x - h)$ and use (39), there follows

$$\nabla f(x) \cong h a_1 + 2h a_2(x - x_k) + 3h a_3(x - x_k)(x - x_{k-1}) + \cdots \\ + n h a_n(x - x_k)(x - x_{k-1}) \cdots (x - x_{k-n+2}).$$

Thus, setting $x = x_k$ and requiring equality, we obtain

$$a_1 = \frac{1}{h}\nabla f_k. \qquad (46b)$$

After calculating the second difference,

$$\nabla^2 f(x) \cong 2 \cdot 1 h^2 a_2 + 3 \cdot 2 h^2 a_3(x - x_k) + \cdots \\ + n(n - 1)h^2 a_n(x - x_k)(x - x_{k-1}) \cdots (x - x_{k-n+3}),$$

we determine a_2,

$$a_2 = \frac{1}{2! \, h^2}\nabla^2 f_k. \qquad (46c)$$

From the way in which these results were obtained, it can be seen (inductively) that the general result will be of the form

$$a_r = \frac{1}{r! \, h^r}\nabla^r f_k \qquad (k = 0, 1, 2, \ldots, n). \qquad (47)$$

The resultant approximation formula (45) can be put into a conveniently compact form if we write

$$x = x_k + sh, \qquad s = \frac{x - x_k}{h}, \qquad (48)$$

so that s is distance measured from the point x_k, in units of the spacing h. With this notation, the introduction of (47) into (45) then gives

$$f(x_k + sh) \cong f_k + s\nabla f_k + \frac{s(s+1)}{2!}\nabla^2 f_k + \frac{s(s+1)(s+2)}{3!}\nabla^3 f_k$$

$$+ \cdots + \frac{s(s+1)\cdots(s+n-1)}{n!}\nabla^n f_k, \qquad (49)$$

if use is made of (39). This result is known as *Newton's* (or *Gregory's*) *backward difference formula* for polynomial approximation.

Thus, if the values of a smooth function $f(x)$ are known at $n+1$ equally spaced points, we may suppose that the function is approximated by the nth-degree polynomial which agrees with $f(x)$ at these points and, accordingly, use (49) to determine approximate values of $f(x)$ at additional nearby points. This formula is particularly useful for extrapolation (prediction) beyond the point x_k, at the end of a range of tabulation, although it can also be used for interpolation, with negative values of s.

To illustrate the use of this formula in extrapolation, we consider the function $f(x) = \sqrt{x}$. Assuming known three-place values for $x = 2, 3, 4, 5$, and 6, we form the following difference table:

x	f	∇f	$\nabla^2 f$	$\nabla^3 f$
2.0	1.414			
		0.318		
3.0	1.732		−0.050	
		0.268		0.018
4.0	2.000		−0.032	
		0.236		0.009
5.0	2.236		−0.023	
		0.213		
6.0	2.449			

To extrapolate for $\sqrt{6.2}$, we set $s = 0.2$, since here $h = 1$, and retain third differences. Equation (49) then gives

$$f(6.2) \cong 2.449 + (0.2)(0.213) + \frac{(0.2)(1.2)}{2}(-0.023) + \frac{(0.2)(1.2)(2.2)}{6}(0.009)$$

$$= 2.449 + 0.0426 - 0.0028 + 0.0008$$

$$= \mathbf{2.490}.$$

Similarly, to interpolate for $\sqrt{5.8}$ we take $s = -0.2$ and, again retaining third differences, obtain the value 2.408. Both the values so obtained are correct to the same number of places as the given data. In general, however, extrapolation is less dependable than interpolation.

Equation (49) is also useful for *integrating* a tabular function over an interval near the end point x_k. Thus we have, for example,

$$\int_{x_k}^{x_k+h} f(x)\, dx = h \int_0^1 f(x_k + sh)\, ds$$

and (49) then gives

$$\int_{x_k}^{x_k+h} f(x)\, dx \cong h(f_k + \tfrac{1}{2}\nabla f_k + \tfrac{5}{12}\nabla^2 f_k + \tfrac{3}{8}\nabla^3 f_k + \tfrac{251}{720}\nabla^4 f_k + \cdots). \tag{50}$$

When $f(x)$ is a polynomial, the sum on the right terminates and yields an exact result. More generally, when $f(x)$ is *not* a polynomial, the result of retaining differences through the nth is identical with the result of replacing $f(x)$ by the polynomial of degree n which agrees with f at the $n + 1$ points $x_k, x_{k-1}, \ldots,$ x_{k-n} and integrating that polynomial over the interval $(x_k, x_k + h)$.

In illustration, if we use (50) to approximate $\int_6^7 \sqrt{x}\, dx$, with third differences, we obtain

$$\int_6^7 \sqrt{x}\, dx \cong 1[2.449 + \tfrac{1}{2}(0.213) + \tfrac{5}{12}(-0.023) + \tfrac{3}{8}(0.009)] = \mathbf{2.549}.$$

This is in agreement with the true three-place value.

Similarly, from the relation

$$\int_{x_k-h}^{x_k} f(x)\, dx = h \int_{-1}^0 f(x_k + sh)\, ds,$$

we obtain the formula

$$\int_{x_k-h}^{x_k} f(x)\, dx \cong h(f_k - \tfrac{1}{2}\nabla f_k - \tfrac{1}{12}\nabla^2 f_k - \tfrac{1}{24}\nabla^3 f_k - \tfrac{19}{720}\nabla^4 f_k + \cdots). \tag{51}$$

Equation (50), with f replaced by F, leads to the approximate relation (43) or, equivalently, (9) when $dy = F\, dx$, and hence

$$y_{k+1} - y_k = \int_{x_k}^{x_k+1} F\, dx.$$

Further, Equation (51) leads similarly to (44) or (16) when k is replaced by $k + 1$.

Finally, we may use (49) to obtain an approximate *derivative* formula. Since $f'(x) = h^{-1}(df/ds)$, we obtain the relation

$$f'(x_k + sh) \cong \frac{1}{h}\left(\nabla f_k + \frac{2s+1}{2}\nabla^2 f_k + \frac{3s^2+6s+2}{6}\nabla^3 f_k + \cdots\right). \tag{52}$$

Thus, to approximate $(d\sqrt{x}/dx)_{x=6}$ by using the tabulated data, we take $s = 0$, and hence, again using third differences, obtain

$$f'(6) \cong \tfrac{1}{1}[0.213 + \tfrac{1}{2}(-0.023) + \tfrac{1}{3}(0.009)]$$
$$= 0.213 - 0.0115 + 0.003 = \mathbf{0.2045}.$$

The true result is 0.204, to three places. Although satisfactory results were obtained here, it should be noted that, *in general, approximate differentiation is inaccurate.* Its accuracy may be very seriously impaired by small inaccuracies in the tabulated data. Still, there are occasions when there is no alternative but to use an approximate method of this sort.

The formulas and methods of this section illustrate the treatment of tabulated data near the end of the range over which a tabulation exists. These data are usually the most troublesome to deal with. This section may be considered also as illustrating the more general use of finite differences in approximate analysis.

Analogous formulas for interpolation and other operations near *interior* tabular points may be found in texts treating numerical methods. (See also Problems 23 to 26.) In such formulas the so-called *central difference* notation is particularly useful. In this notation one writes

$$f_r - f_{r-1} = \delta f_{r-1/2}, \qquad \delta f_{r+1/2} - \delta f_{r-1/2} = \delta^2 f_r, \qquad \dots, \tag{53}$$

so that the subscript of a difference is the *mean* of the subscripts of its parents.

Thus we have the general notational correspondence

$$f_{r+1} - f_r = \Delta f_r = \delta f_{r+1/2} = \nabla f_{r+1},$$

$$f_{r+2} - 2f_{r+1} + f_r = \Delta^2 f_r = \delta^2 f_{r+1} = \nabla^2 f_{r+2},$$

and so forth. The "forward" difference Δf is sometimes called a "descending" difference, whereas the "backward" difference ∇f sometimes called an "ascending" difference. Although this practice leads to a certain amount of confusion, its motivation may be realized by an inspection of the following table of difference notations.

f_0

$\Delta f_0 = \delta f_{1/2} = \nabla f_1$

f_1 $\qquad\qquad\qquad\qquad \Delta^2 f_0 = \delta^2 f_1 = \nabla^2 f_2$

$\Delta f_1 = \delta f_{3/2} = \nabla f_2$ $\qquad\qquad\qquad\qquad\qquad \Delta^3 f_0 = \delta^3 f_{3/2} = \nabla^3 f_3$

f_2 $\qquad\qquad\qquad\qquad \Delta^2 f_1 = \delta^2 f_2 = \nabla^2 f_3$

$\Delta f_2 = \delta f_{5/2} = \nabla f_3$ $\qquad\qquad\qquad\qquad\qquad \Delta^3 f_1 = \delta^3 f_{5/2} = \nabla^3 f_4$

f_3 $\qquad\qquad\qquad\qquad \Delta^2 f_2 = \delta^2 f_3 = \nabla^2 f_4$

$\Delta f_3 = \delta f_{7/2} = \nabla f_4$

f_4

It is apparent that the differences Δf_r, $\Delta^2 f_r$, $\Delta^3 f_r$, and so forth, "descend" to the right, whereas the differences ∇f_r, $\nabla^2 f_r$, $\nabla^3 f_r$, and so forth, "ascend" to the right. The central differences $\delta^2 f_r$, $\delta^4 f_r$, and so forth, remain in the same horizontal line as the entry f_r, whereas the differences $\delta f_{r+1/2}$, $\delta^2 f_r$, $\delta^3 f_{r+1/2}$, $\delta^4 f_r$, and so forth, form a "forward zigzag" set of differences in increasing order, remaining as close as possible to a horizontal line.

More detailed treatments of the errors associated with formulas for approximate interpolation, numerical integration and differentiation, and the numerical solution of differential equations may be found in many sources, including the references listed at the end of this chapter.

REFERENCES

1. COLLATZ, L., *The Numerical Treatment of Differential Equations*, 3rd ed., Springer-Verlag New York, Inc., New York, 1960.

2. HENRICI, P., *Discrete Variable Methods in Ordinary Differential Equations*, John Wiley & Sons, Inc., New York, 1962.

3. HILDEBRAND, F. B., *Introduction to Numerical Analysis*, 2nd ed., McGraw-Hill Book Company, Inc., New York, 1974.

4. MILNE, W. E., *Numerical Solution of Differential Equations*, 2nd ed., Peter Smith Publisher, Gloucester, Mass., 1970.

5. STEFFENSEN, J. F., *Interpolation*, 2nd ed., Chelsea Publishing Company, Inc., New York, 1950.

PROBLEMS

Section 3.1

1. Use the method of isoclines to sketch the integral curves of the equation

$$\frac{dy}{dx} - y = x$$

in the first quadrant.

2. Proceed as in Problem 1 with the equation

$$\frac{dy}{dx} - y^2 = x^2.$$

3. Use the method of isoclines to obtain a sketch, in the first quadrant, of the integral curve of the equation

$$\frac{dy}{dx} - x^2 y = x$$

which passes through the point $(0, 1)$.

Section 3.2

4. Use the method of Taylor series to determine to four places the values of the solution of the problem

$$\frac{dy}{dx} - x^2 y = x, \qquad y(0) = 1$$

at the points $x = 0.1, 0.2,$ and 0.3. (The true values at $x = 0.1, 0.2, 0.3, 0.4,$ and 0.5 round to 1.00533, 1.02270, 1.05428, 1.10260, and 1.17072.)

5. Use the method of Taylor series to determine to four places the values of the solution of the problem

$$\frac{dy}{dx} + xy^2 = 0, \qquad y(0) = 1$$

at the points $x = 0.1, 0.2$, and 0.3, (The true values at $x = 0.1, 0.2, 0.3, 0.4,$ and 0.5 round to $0.99502, 0.98039, 0.95694, 0.92593,$ and $0.88889.$)

6. Use the method of Taylor series to determine to four places the values of the solution of the problem

$$\frac{d^2y}{dx^2} - x^2\frac{dy}{dx} - 2xy = 1, \qquad y(0) = 1, \qquad y'(0) = 0$$

at the points $x = 0.1, 0.2$, and 0.3. (The true values are the same as those in Problem 4.)

Section 3.3

7. Assuming the values of y in Problem 4 for $x = 0, 0.1, 0.2,$ and 0.3, use the Adams method to calculate the values for $x = 0.4$ and 0.5, using third differences.

8. Assuming the values of y in Problem 5 for $x = 0, 0.1, 0.2,$ and 0.3, use the Adams method to calculate the values for $x = 0.4$ and 0.5, using third differences.

9. Assuming the results of Problem 6 for $x = 0, 0.1, 0.2,$ and 0.3, use the Adams method to calculate the values of y for $x = 0.4$ and 0.5, using third differences.

Section 3.4

10. Use the modified Adams method, retaining only second differences, to effect the determinations of (a) Problem 7, (b) Problem 8, and (c) Problem 9. At each stage, use Equation (17) to estimate the error introduced.

11. If no differences beyond the *second* are retained, show that the formulas (15) and (16), of the modified Adams method, can be written in the explicit forms

$$y_{k+1} \cong y_k + \frac{h}{12}(23F_k - 16F_{k-1} + 5F_{k-2})$$

and

$$y_{k+1} \cong y_k + \frac{h}{12}(5F_{k+1} + 8F_k - F_{k-1}),$$

respectively. (These formulas are easily used, since differencing is not involved. However, the use of differences is usually preferable in questionable cases, since then gross errors may be indicated by irregularities in the columns of differences, and also warnings may be served by trends observable in these columns.)

12. Making use of the fact that, for small values of the spacing h, there follows

$$\Delta^r f \approx \nabla^r f \approx \delta^r f \approx h^r \frac{d^r f}{dx^r},$$

show that the errors introduced into the formulas of Problem 11 by neglecting the *first* omitted difference are respectively approximated by

$$\frac{3h^4}{8}\left(\frac{d^4y}{dx^4}\right)_k, \qquad -\frac{h^4}{24}\left(\frac{d^4y}{dx^4}\right)_k.$$

[It is known that the true truncation error in each of these cases is given by the result of replacing $(d^4y/dx^4)_k$ by the value of d^4y/dx^4 at *some* point between x_{k-2} and x_{k+1} in the expression so obtained.]

Section 3.5

13. Use the Runge–Kutta method, with third-order accuracy, to determine the approximate values of y at $x = 0.1$ and 0.2 if y satisfies the conditions of (a) Problem 4, (b) Problem 5, and (c) Problem 6.

14. (a–c) Recalculate the value of $y(0.2)$ determined in the corresponding part of Problem 13 by taking only one step, with spacing $2h = 0.2$, and estimate the error in the two-step calculation by use of the procedure suggested in the text.

15. A function $y(x)$ satisfies the equation

$$\frac{d^2y}{dx^2} + y\varphi(x) = 0$$

and the initial conditions $y(0) = 1$ and $y'(0) = 0$. The following approximate values of the function $\varphi(x)$ are known:

x	0	0.05	0.10	0.15	0.20
$\varphi(x)$	1.000	1.032	1.115	1.249	1.434

Use the Runge–Kutta method, with third-order accuracy, to determine approximate values of y at $x = 0.1$ and 0.2.

16. Recalculate $y(0.2)$ in Problem 15 by taking one step, with spacing $2h$, and estimate the error in the two-step calculation.

Section 3.6

17. Apply Picard's method to the solution of the problem

$$\frac{dy}{dx} - x^2y = x, \qquad y(0) = 1,$$

taking $y^{(1)}(x) = 1$ and making two successive substitutions, and compare the approximations with the series expansion,

$$y = 1 + \tfrac{1}{2}x^2 + \tfrac{1}{3}x^3 + \tfrac{1}{10}x^5 + \tfrac{1}{18}x^6 + \cdots,$$

of the exact solution.

18. Apply Picard's method to the solution of the problem

$$\frac{dy}{dx} - y^2 = x, \qquad y(0) = 1,$$

taking $y^{(1)}(x) = 1$ and making two successive substitutions, and compare the approximations with the series expansion,

$$y = 1 + x + \tfrac{3}{2}x^2 + \tfrac{4}{3}x^3 + \tfrac{17}{12}x^4 + \cdots,$$

of the exact solution.

Section 3.7

19. Suppose that the following rounded values of a certain function $f(x)$ are known:

x	1.0	2.0	3.0	4.0	5.0
$f(x)$	1.2840	1.3499	1.4191	1.4918	1.5683

By making use of formulas derived in Section 3.7, obtain approximate values of the following quantities as accurately as possible:

(a) $f(4.8)$,   (b) $f(5.2)$,   (c) $f'(5.0)$,   (d) $\int_5^6 f(x)\,dx$.

20. Derive from Equation (49) the formula

$$\int_{x_k}^{x_k+th} f(x)\,dx \cong th\left[f_k + \frac{t}{2}\nabla f_k + \frac{t(3+2t)}{12}\nabla^2 f_k + \cdots \right].$$

21. Use the result of Problem 20 to obtain, from the data of Problem 19, approximate values of the following quantities:

(a) $\int_{5.0}^{5.2} f(x)\,dx$,   (b) $\int_{4.8}^{5.0} f(x)\,dx$.

22. Establish the following notational relations:

(a) $\Delta f_r = \delta f_{r+1/2}$,   (b) $\nabla\Delta f_r = \Delta\nabla f_r = \delta^2 f_r$,
(c) $\Delta\nabla\Delta f_r = \delta^3 f_{r+1/2}$.

23. Determine the coefficients specifying the polynomial approximation of degree $n = 2k$ of the form

$$f(x) \cong a_0 + a_1(x - x_0) + a_2(x - x_0)(x - x_1)$$
$$+ a_3(x - x_0)(x - x_1)(x - x_{-1})$$
$$+ a_4(x - x_0)(x - x_1)(x - x_{-1})(x - x_2) + \cdots$$
$$+ a_n(x - x_0)(x - x_1)(x - x_{-1}) \cdots (x - x_k)(x - x_{-k}),$$

so that the two members are equal at the $n + 1$ equally spaced points

$$x_{-k} = x_0 - kh, \quad \ldots, \quad x_0, \quad \ldots, \quad x_k = x_0 + kh.$$

by calculating the difference $\Delta f(x) = f(x + h) - f(x)$, then the differences $\nabla[\Delta f(x)]$, $\Delta[\nabla\Delta f(x)]$, and so on, and equating the two members at $x = x_0$ after each such differencing. Thus show that $a_0 = f_0$, that

$$\Delta f(x) \cong h[a_1 + 2a_2(x - x_0) + 3a_3(x - x_0)(x - x_{-1})$$
$$+ 4a_4(x - x_0)(x - x_{-1})(x - x_1) + \cdots],$$

and hence that $a_1 = \Delta f_0/h = \delta f_{1/2}/h$, and so forth. Make use of the results of Problem

22 to express the result in the form

$$f(x) \cong f_0 + (x - x_0)\frac{\delta f_{1/2}}{1! \, h} + (x - x_0)(x - x_1)\frac{\delta^2 f_0}{2! \, h^2}$$

$$+ (x - x_0)(x - x_1)(x - x_{-1})\frac{\delta^3 f_{1/2}}{3! \, h^3}$$

$$+ (x - x_0)(x - x_1)(x - x_{-1})(x - x_2)\frac{\delta^4 f_0}{4! \, h^4} + \cdots .$$

24. By writing $x = x_0 + sh$, express the result of Problem 23 in the form

$$f(x_0 + sh) \cong f_0 + s\delta f_{1/2} + \frac{s(s-1)}{2!}\delta^2 f_0 + \frac{s(s^2-1)}{3!}\delta^3 f_{1/2}$$

$$+ \frac{s(s^2-1)(s-2)}{4!}\delta^4 f_0 + \cdots .$$

(This is the interpolation formula of *Gauss*. The central differences involved remain as near as possible to the horizontal line through the tabular value f_0, comprising a "forward zigzag.")

25. By integrating the result of Problem 24 over appropriate intervals, obtain the integral formulas

(a) $\displaystyle\int_{x_0-h}^{x_0+h} f(x)\,dx \cong 2h(f_0 + \tfrac{1}{6}\delta^2 f_0 - \tfrac{1}{180}\delta^4 f_0 + \cdots)$,

(b) $\displaystyle\int_{x_0-2h}^{x_0+2h} f(x)\,dx \cong 4h(f_0 + \tfrac{2}{3}\delta^2 f_0 + \tfrac{7}{90}\delta^4 f_0 + \cdots)$,

26. Make appropriate use of the Gauss interpolation formula of Problem 24 to obtain, from the data given in Problem 19, approximate values of each of the following quantities:

(a) $f(2.8)$,   (b) $f(3.2)$,   (c) $f'(3.0)$,

(d) $f'(2.8)$,   (e) $\displaystyle\int_2^4 f(x)\,dx$,   (f) $\displaystyle\int_1^5 f(x)\,dx$.

27. By retaining only second differences in the result of Problem 25(a), deduce the formula

$$\int_{x_0-h}^{x_0+h} f(x)\,dx \cong \frac{h}{3}(f_{-1} + 4f_0 + f_1).$$

[This is the celebrated formula of *Simpson's rule*. Notice that, since the coefficient of the third difference in the more general formula is zero, the formula is exact when $f(x)$ is a *cubic* polynomial.]

28. Show (as in Problem 12) that the error associated with Simpson's rule is approximately

$$-\frac{h^5}{90}\left(\frac{d^4 f}{dx^4}\right)_{x_0} .$$

[The *true* error is known to be expressible in this form as well, but with $d^4 y/dx^4$ evaluated instead at some unknown point between $x_0 - h$ and $x_0 + h$.]

29. Show that, if Simpson's rule, of Problem 27, is applied successively over the adjacent double intervals $(a, a + 2h)$, $(a + 2h, a + 4h)$, ..., $(b - 2h, b)$ in the approximate evaluation of the integral $\int_a^b f(x)\,dx$, there follows

$$\int_a^b f(x)\,dx \cong \frac{h}{3}(f_0 + 4f_1 + 2f_2 + 4f_3 + \cdots + 2f_{n-2} + 4f_{n-1} + f_n),$$

where $h = (b - a)/n$ and n is an even integer. [This formula, often known as the *parabolic rule*, is probably the most widely used formula for numerical integration. It is *exact* when $f(x)$ is a polynomial of degree not greater than three. As is suggested by $n/2$ applications of the result of Problem 28, for a given interval (a, b) the error in the approximation is nearly proportional to h^4 and hence to $1/n^4$. From this fact it follows that, if two calculations are made, one (I_n) with n subdivisions and one (I_{2n}) with $2n$ subdivisions, the error in I_{2n} may be estimated by $(I_{2n} - I_n)/15$.]

30. Evaluate the following integrals approximately by use of the parabolic rule (Problem 29) first with $n = 2$ and then with $n = 4$, estimate the error in I_4 in each case, and compare the results with the given rounded true values:

(a) $\int_0^1 \frac{dx}{1 + x^2} = 0.7854,$ (b) $\int_0^\pi \frac{\sin x}{x}\,dx = 1.8518,$

(c) $\int_0^{\pi/2} \sqrt{1 - \tfrac{1}{2}\sin^2 x}\,dx = 1.3506.$

4

Series Solutions of Differential Equations: Special Functions

4.1. Properties of Power Series. A large class of ordinary differential equations possesses solutions expressible, over a certain interval, in terms of power series and related series. Before investigating methods of obtaining such solutions, we review without proof certain useful properties of power series.

An expression of the form

$$A_0 + A_1(x - x_0) + \cdots + A_n(x - x_0)^n + \cdots = \sum_{n=0}^{\infty} A_n(x - x_0)^n \qquad (1)$$

is called a *power series* and is defined as the limit

$$\lim_{N \to \infty} \sum_{n=0}^{N} A_n(x - x_0)^n$$

for those values of x for which the limit exists. For such values of x the series is said to *converge*. In this chapter we suppose that the variable x and the coefficients are *real*; *complex* power series are dealt with in Chapter 10.

To determine for what values of x the series (1) converges, we may make use of the *ratio test*, which states that, if the absolute value of the ratio of the $(n + 1)$th term to the nth term in *any* infinite series approaches a limit ρ as $n \to \infty$, then the series converges when $\rho < 1$ and diverges when $\rho > 1$. The test fails if $\rho = 1$. A more delicate test states that, if the absolute value of the same ratio is bounded by some number σ as $n \to \infty$, then the series converges when $\sigma < 1$. In the case of the power series (1) we obtain

$$\rho = \lim_{n \to \infty} \left| \frac{A_{n+1}}{A_n} \right| |x - x_0| = L |x - x_0|,$$

where

$$L = \lim_{n \to \infty} \left| \frac{A_{n+1}}{A_n} \right|, \qquad (2)$$

if the last limit exists. In this case it follows that (1) *converges* when

$$|x - x_0| < \frac{1}{L}$$

and *diverges* when

$$|x - x_0| > \frac{1}{L}.$$

Thus, when L exists and is finite, an *interval of convergence*

$$\left(x_0 - \frac{1}{L}, \ x_0 + \frac{1}{L}\right)$$

is determined symmetrically about the point x_0, such that inside the interval the series converges and outside the interval it diverges. The distance $R = 1/L$ is frequently called the *radius of convergence*.

The behavior of the series at the end points of the interval is not determined by the ratio test. Useful tests for investigating convergence of the two series of constants corresponding to the end points $x = x_0 \pm R$ are:

(1) If, at an end point, the successive terms of the series *alternate in sign* for sufficiently large values of n, the series converges if after a certain stage the successive terms always decrease in magnitude and if the nth term approaches zero, and the series diverges if the nth term does not tend to zero.

(2) If, at an end point, the successive terms of the series *are of constant sign*, and if the ratio of the $(n + 1)$th term to the nth term can be written in the form

$$1 - \frac{k}{n} + \frac{\theta_n}{n^2},$$

where k is independent of n and θ_n is bounded as $n \longrightarrow \infty$, then the series converges if $k > 1$ and diverges if $k \leq 1$. It should be noticed that this test is applicable even in the case when $k = 1$, so long as an expression of the indicated form can be obtained. We shall refer to this test as *Raabe's test*.†

Example. In illustration, we consider the series

$$\sum_{n=1}^{\infty} \frac{(1 \cdot 2 \cdot 3 \cdots n)(x - a)^n}{(\alpha + 1)(\alpha + 2)(\alpha + 3) \cdots (\alpha + n)},$$

where α is not a negative integer. In this case we obtain

$$L = \lim_{n \to \infty} \left| \frac{n + 1}{\alpha + n + 1} \right| = 1.$$

Hence the interval of convergence is given by $|x - a| < 1$ or

$$a - 1 < x < a + 1.$$

---

†This very useful test is also associated with the name of *Gauss*. The term θ_n/n^2 can in fact be replaced by θ_n/n^{1+p}, where it is required only that $p > 0$ and θ_n be bounded, and still more delicate modifications exist. (See, for example, Reference 10.)

When $x = a - 1$, the signs of successive terms alternate when $n > -\alpha$. Apart from algebraic sign, the nth term is then

$$\frac{n!}{(\alpha + 1)(\alpha + 2) \cdots (\alpha + n)} = \frac{n!\,\Gamma(\alpha + 1)}{\Gamma(\alpha + n + 1)}.$$

Reference to Equation (61) of Chapter 2 shows that this ratio is approximated by $\Gamma(\alpha + 1)n^{-\alpha}$ when n is large, and hence approaches zero as n increases only if $\alpha > 0$. Thus the series converges at $x = a - 1$ if $\alpha > 0$, and diverges at $x = a - 1$ otherwise. When $x = a + 1$, the terms are of constant sign when $n > -\alpha$. The ratio of consecutive terms is then

$$\frac{n + 1}{n + \alpha + 1} = 1 - \frac{\alpha}{n} + \frac{\alpha(\alpha + 1)}{n(n + \alpha + 1)}.$$

Hence, by Raabe's test with $\theta_n = [\alpha(\alpha + 1)n]/(n + \alpha + 1)$, the series converges at $x = a + 1$ if $\alpha > 1$ and diverges at $x = a + 1$ otherwise. ∎

It may be noticed that if L is zero the interval of convergence includes all values of x. However, if L is infinite, the series converges only *at the point* $x = x_0$. Whether or not the limit L exists, it is known that always, for the power series (1), either the series converges only when $x = x_0$, or the series converges everywhere, or there exists a positive number R such that the series converges when $|x - x_0| < R$ and diverges when $|x - x_0| > R$.

It may happen that the series (1) contains only terms for which the subscript n is an integral multiple of an integer $N > 1$, and hence is of the form

$$A_0 + A_N(x - x_0)^N + A_{2N}(x - x_0)^{2N} + \cdots = \sum_{k=0}^{\infty} A_{kN}(x - x_0)^{kN}, \qquad (3)$$

or is a product of a power of $x - x_0$ and such a series. Examples are afforded by the series

$$\cos x = 1 - \frac{x^2}{2!} + \frac{x^4}{4!} + \cdots = \sum_{k=0}^{\infty} (-1)^k \frac{x^{2k}}{(2k)!} \qquad (|x| < \infty)$$

and

$$\log \frac{1 + x^2}{1 - x^2} = 2\left(x^2 + \frac{x^6}{3} + \frac{x^{10}}{5} + \cdots\right)$$

$$= 2x^2 \sum_{k=0}^{\infty} \frac{x^{4k}}{2k + 1} \qquad (|x| < 1),$$

for which $N = 2$ and 4, respectively. In such cases, the ratio A_{n+1}/A_n is undefined for infinitely many values of n. The limiting absolute value of the ratio of successive terms is

$$\rho_N = \lim_{k \to \infty} \left|\frac{A_{(k+1)N}}{A_{kN}}\right| |x - x_0|^N = L_N |x - x_0|^N,$$

where

$$L_N = \lim_{k \to \infty} \left|\frac{A_{(k+1)N}}{A_{kN}}\right|, \qquad (4)$$

if that limit exists, and the series converges when $L_N |x - x_0|^N < 1$, or

$$|x - x_0| < \frac{1}{\sqrt[N]{L_N}}. \qquad (5)$$

A particularly useful property of power series is the fact that convergent power series can be treated, for many purposes, in the same way as polynomials. Inside its interval of convergence, a power series represents a continuous function of x with continuous derivatives of all orders. Inside this interval, a power series can be integrated or differentiated term by term, as in the case of a polynomial, and the resultant series will converge, *in the same interval*, to the integral or derivative of the function represented by the original series. Further, two power series in $x - x_0$ can be multiplied together term by term and the resultant series will converge to the product of the functions represented by the original series, *inside the common interval of convergence*. A similar statement applies to division of one series by another, provided that the denominator is not zero at x_0. Here the resultant series will converge to the ratio in *some* subinterval of the common interval of convergence.

By the statement on term-by-term multiplication of power series, we mean (taking $x_0 = 0$ for simplicity) that, if

$$f(x) = \sum_{i=0}^{\infty} a_i x^i \qquad (|x| < R_1)$$

and
$$g(x) = \sum_{j=0}^{\infty} b_j x^j \qquad (|x| < R_2),$$

where R_1 and R_2 are positive, then

$$f(x)g(x) = \left(\sum_{i=0}^{\infty} a_i x^i\right)\left(\sum_{j=0}^{\infty} b_j x^j\right) = \sum_{k=0}^{\infty} c_k x^k,$$

at least for $|x| < \min(R_1, R_2)$, where

$$c_k = a_0 b_k + a_1 b_{k-1} + \cdots + a_k b_0$$

$$= \sum_{i=0}^{k} a_i b_{k-i}. \tag{6}$$

With respect to *division* of power series, we mean that under the same assumptions, together with the requirement that $b_0 \neq 0$, we have

$$\frac{f(x)}{g(x)} = \frac{\sum_{i=0}^{\infty} a_i x^i}{\sum_{j=0}^{\infty} b_j x^j} = \sum_{k=0}^{\infty} d_k x^k,$$

in *some* interval $|x| < R$, where the d's are to be determined by the relations

$$a_i = \sum_{k=0}^{i} b_{i-k} d_k \qquad (i = 0, 1, 2, \dots). \tag{7}$$

Since these conditions take the form

$$a_0 = b_0 d_0,$$
$$a_1 = b_1 d_0 + b_0 d_1,$$
$$a_2 = b_2 d_0 + b_1 d_1 + b_0 d_2,$$
$$\cdots\cdots\cdots\cdots\cdots\cdots\cdots\cdots$$

it follows that, with $b_0 \neq 0$, the first equation yields d_0, the second yields d_1, and so forth.

Now suppose that the series $\sum_{n=0}^{\infty} A_n(x - x_0)^n$ converges in a nonzero interval about $x = x_0$ and hence represents a function, say $f(x)$, in that interval,

$$f(x) = \sum_{n=0}^{\infty} A_n(x - x_0)^n. \tag{8}$$

Then, differentiating both sides of Equation (8) k times and setting $x = x_0$ in the result, we obtain

$$f^{(k)}(x_0) = k! A_k \qquad (k = 0, 1, 2, \ldots),$$

and hence (8) becomes

$$f(x) = \sum_{n=0}^{\infty} \frac{f^{(n)}(x_0)}{n!}(x - x_0)^n. \tag{9}$$

This is the so-called *Taylor series* expansion of $f(x)$ near $x = x_0$.† It is clear that not all functions possess such expansions, since, in particular, in order that (9) be defined, *all derivatives* of $f(x)$ must exist at $x = x_0$. A function which possesses such an expansion is said to be *regular* at $x = x_0$. The above derivation shows that, if a function is regular at $x = x_0$, it has *only one* expansion in powers of $x - x_0$ and that expansion is given by (9).

If $f(x)$ and all its derivatives are continuous in an interval including $x = x_0$, then $f(x)$ can be expressed as a *finite* Taylor series plus a remainder, in the form

$$f(x) = \sum_{n=0}^{N-1} \frac{f^{(n)}(x_0)}{n!}(x - x_0)^n + R_N(x). \tag{10}$$

Here R_N, the remainder after N terms, is given by

$$R_N(x) = \frac{f^{(N)}(\xi)}{N!}(x - x_0)^N, \tag{11}$$

where ξ is some point in the interval (x_0, x). To show that the expansion (9) is valid, so that $f(x)$ is regular at x_0, we must show that $R_N(x) \to 0$ as $N \to \infty$ for values of x in an interval including $x = x_0$. A test which is much more easily applied and which is sufficient in the case of most functions occurring in practice consists of determining whether the formal series in Equation (9) converges in an interval about $x = x_0$.

It is apparent, in particular, that any *polynomial* in x is regular for *all* x. Further, any *rational function* (*ratio* of polynomials) is regular for all values of x which are not zeros of the denominator.

4.2. Illustrative Examples. To illustrate the use of power series in obtaining solutions of differential equations, we first consider the solution of three specific linear equations of second order.

(1) To solve the differential equation

$$Ly \equiv \frac{d^2 y}{dx^2} - y = 0, \tag{12}$$

---

†When $x_0 = 0$, the expansion is often called a *Maclaurin series*.

we assume a solution in the form

$$y = A_0 + A_1 x + A_2 x^2 + A_3 x^3 + A_4 x^4 + A_5 x^5 + \cdots$$

and assume that the series converges in an interval including $x = 0$. Differentiating twice term by term, we then obtain

$$\frac{d^2 y}{dx^2} = 2A_2 + 6A_3 x + 12A_4 x^2 + 20A_5 x^3 + \cdots.$$

With the assumed form for y there follows

$$Ly \equiv (2A_2 - A_0) + (6A_3 - A_1)x + (12A_4 - A_2)x^2$$
$$+ (20A_5 - A_3)x^3 + \cdots = 0.$$

In order that this equation be valid over an interval, it is necessary that the coefficients of all powers of x vanish independently, giving the equations

$$2A_2 = A_0, \qquad 6A_3 = A_1, \qquad 12A_4 = A_2, \qquad 20A_5 = A_3, \qquad \ldots,$$

from which there follows

$$A_2 = \tfrac{1}{2}A_0, \qquad A_3 = \tfrac{1}{6}A_1, \qquad A_4 = \tfrac{1}{12}A_2 = \tfrac{1}{24}A_0,$$
$$A_5 = \tfrac{1}{20}A_3 = \tfrac{1}{120}A_1, \qquad \cdots.$$

The solution then becomes

$$y = A_0(1 + \tfrac{1}{2}x^2 + \tfrac{1}{24}x^4 + \cdots) + A_1(x + \tfrac{1}{6}x^3 + \tfrac{1}{120}x^5 + \cdots).$$

It is seen that the coefficients A_0 and A_1 are undetermined, and hence arbitrary, but that succeeding coefficients are determined in terms of them. The general solution is thus of the form

$$y = A_0 u_1(x) + A_1 u_2(x),$$

where $u_1(x)$ and $u_2(x)$ are two linearly independent solutions, expressed as power series, of which the first three terms have been obtained. The terms found may be recognized as the first terms of the series representing the known solutions cosh x and sinh x, respectively.

A more compact and convenient procedure uses the summation notation in place of writing out a certain number of terms of the series. Thus we write the assumed solution in the form

$$y = \sum_{k=0}^{\infty} A_k x^k$$

and obtain, by differentiation,

$$\frac{d^2 y}{dx^2} = \sum_{k=0}^{\infty} k(k-1)A_k x^{k-2}.$$

There then follows

$$Ly \equiv \sum_{k=0}^{\infty} k(k-1)A_k x^{k-2} - \sum_{k=0}^{\infty} A_k x^k = 0.$$

In order to collect the coefficients of like powers of x, we next change the indices of summation in such a way that the exponents of x in the two summations are equal. For this purpose we may, for example, replace k by $k - 2$ in the second summation, so that it becomes

$$\sum_{k-2=0}^{\infty} A_{k-2}x^{k-2} = \sum_{k=2}^{\infty} A_{k-2}x^{k-2},$$

and hence

$$Ly \equiv \sum_{k=0}^{\infty} k(k - 1)A_k x^{k-2} - \sum_{k=2}^{\infty} A_{k-2}x^{k-2} = 0.$$

Since the first two terms ($k = 0, 1$) of the first summation are zero, we may replace the lower limit by $k = 2$ and then combine the summations to obtain

$$Ly \equiv \sum_{k=2}^{\infty} [k(k - 1)A_k - A_{k-2}]x^{k-2} = 0.$$

Equating to zero the coefficients of all powers of x involved in this sum, we obtain the condition

$$k(k - 1)A_k = A_{k-2} \qquad (k = 2, 3, \dots).$$

This condition is known as the *recurrence formula* for A_k. It expresses each coefficient A_k for which $k \geq 2$ as a multiple of the second preceding coefficient A_{k-2}, and reduces to the previously determined conditions when $k = 2, 3, 4$, and 5.

(2) As a second example we consider the equation

$$Ly \equiv x^2 \frac{d^2 y}{dx^2} + (x^2 + x)\frac{dy}{dx} - y = 0. \tag{13}$$

Assuming a solution of the form

$$y = \sum_{k=0}^{\infty} A_k x^k,$$

we obtain

$$\frac{dy}{dx} = \sum_{k=0}^{\infty} k A_k x^{k-1}, \qquad \frac{d^2 y}{dx^2} = \sum_{k=0}^{\infty} k(k - 1)A_k x^{k-2},$$

and hence

$$Ly \equiv \sum_{k=0}^{\infty} k(k - 1)A_k x^k + \sum_{k=0}^{\infty} k A_k x^{k+1} + \sum_{k=0}^{\infty} k A_k x^k - \sum_{k=0}^{\infty} A_k x^k.$$

The first, third, and fourth summations may be combined to give

$$\sum_{k=0}^{\infty} [k(k - 1) + k - 1]A_k x^k = \sum_{k=0}^{\infty} (k^2 - 1)A_k x^k,$$

and hence there follows

$$Ly \equiv \sum_{k=0}^{\infty} (k^2 - 1)A_k x^k + \sum_{k=0}^{\infty} k A_k x^{k+1}.$$

In order to combine these sums, we replace k by $k - 1$ in the second, to obtain

$$Ly \equiv \sum_{k=0}^{\infty} (k^2 - 1)A_k x^k + \sum_{k=1}^{\infty} (k - 1)A_{k-1} x^k.$$

Since the ranges of summation differ, the term corresponding to $k = 0$ must be extracted from the first sum, after which the remainder of the first sum can be combined with the second. In this way we find

$$Ly \equiv -A_0 + \sum_{k=1}^{\infty} [(k^2 - 1)A_k + (k - 1)A_{k-1}]x^k.$$

In order that Ly may vanish identically, the constant term, as well as the coefficients of the successive powers of x, must vanish independently, giving the condition

$$A_0 = 0$$

and the *recurrence formula*

$$(k - 1)[(k + 1)A_k + A_{k-1}] = 0 \qquad (k = 1, 2, 3, \dots).$$

The recurrence formula is *identically* satisfied when $k = 1$. When $k \geq 2$, it becomes

$$A_k = -\frac{A_{k-1}}{k + 1} \qquad (k = 2, 3, 4, \dots).$$

Hence we obtain

$$A_2 = -\frac{A_1}{3}, \quad A_3 = -\frac{A_2}{4} = \frac{A_1}{3 \cdot 4}, \quad A_4 = -\frac{A_3}{5} = -\frac{A_1}{3 \cdot 4 \cdot 5}, \quad \dots$$

Thus in this case $A_0 = 0$, A_1 is *arbitrary*, and all succeeding coefficients are determined in terms of A_1.† The solution becomes

$$y = A_1 \left(x - \frac{x^2}{3} + \frac{x^3}{3 \cdot 4} - \frac{x^4}{3 \cdot 4 \cdot 5} + \cdots \right).$$

If this solution is put in the form

$$y = \frac{2A_1}{x} \left(\frac{x^2}{2!} - \frac{x^3}{3!} + \frac{x^4}{4!} - \frac{x^5}{5!} + \cdots \right)$$

$$= \frac{2A_1}{x} \left[x - 1 + \left(1 - \frac{x}{1!} + \frac{x^2}{2!} - \frac{x^3}{3!} + \frac{x^4}{4!} - \cdots \right) \right],$$

the series in parentheses in the final form is recognized as the expansion of e^{-x}, and, writing $2A_1 = c$, the solution obtained may be put in the closed form

$$y = c \frac{e^{-x} - 1 + x}{x}.$$

In this case only one solution was obtained. This fact indicates that any linearly independent solution cannot be expanded in power series near $x = 0$; that is, it is *not regular* at $x = 0$. Although a second solution could be obtained by the method of Section 1.10, an alternative procedure given in a following section is somewhat more easily applied.

---

†Clearly, the temptation to "cancel the common factor" $k - 1$ before setting $k = 1$ in the recurrence formula must be resisted.

(3) As a final example we consider the equation

$$Ly \equiv x^3 \frac{d^2y}{dx^2} + y = 0. \tag{14}$$

Again assuming a solution of the form

$$y = \sum_{k=0}^{\infty} A_k x^k,$$

we obtain

$$Ly \equiv \sum_{k=0}^{\infty} k(k-1)A_k x^{k+1} + \sum_{k=0}^{\infty} A_k x^k.$$

If k is replaced by $k-1$ in the first sum, there follows

$$Ly \equiv \sum_{k=1}^{\infty} (k-1)(k-2)A_{k-1} x^k + \sum_{k=0}^{\infty} A_k x^k$$

$$\equiv A_0 + \sum_{k=1}^{\infty} [A_k + (k-1)(k-2)A_{k-1}]x^k.$$

The condition $Ly = 0$ then requires that

$$A_0 = 0$$

and that the succeeding coefficients satisfy the recurrence formula

$$A_k = -(k-1)(k-2)A_{k-1} \qquad (k = 1, 2, \dots).$$

For $k = 1$ and $k = 2$, the recurrence formula gives $A_1 = A_2 = 0$; and since, from this point, the remaining conditions express each A as a multiple of the preceding one, it follows that all the A's must be zero. Hence the only solution obtained is the trivial one $y = 0$. It thus follows that *the equation possesses no nontrivial solutions which are regular at $x = 0$.*

Next we proceed to a classification of types of linear differential equations of second order, and to a study of the basic differences among the three problems so far considered.

4.3. Singular Points of Linear Second-Order Differential Equations. If a homogeneous second-order linear differential equation is written in the *standard form*

$$\frac{d^2y}{dx^2} + a_1(x)\frac{dy}{dx} + a_2(x)y = 0, \tag{15}$$

the behavior of solutions of the equation near a point $x = x_0$ is found to depend upon the behavior of the coefficients $a_1(x)$ and $a_2(x)$ near $x = x_0$. The point $x = x_0$ is said to be an *ordinary point of the differential equation* if both $a_1(x)$ and $a_2(x)$ are regular *at $x = x_0$*, that is, if a_1 and a_2 can be expanded in power series in an interval including $x = x_0$. Otherwise, the point $x = x_0$ is said to be a *singular point of the differential equation*. In such a case, if the products $(x - x_0)a_1(x)$ and $(x - x_0)^2 a_2(x)$ are both regular at $x = x_0$, the point $x = x_0$ is said to be a *regular singular point;* otherwise, the point is called an *irregular singular point.*

In illustration, we notice that for the first differential equation of the preceding section, $a_1(x) = 0$ and $a_2(x) = -1$. Thus all points are ordinary points. In the second example the coefficients $a_1(x) = 1 + x^{-1}$ and $a_2(x) = -x^{-2}$ cannot be expanded in powers of x but can be expanded in power series near any other point $x = x_0$. Thus the point $x = 0$ is the only singular point. Since the products $xa_1(x) = x + 1$ and $x^2 a_2(x) = -1$ are regular at $x = 0$, it follows that the point is a regular singular point. In the third case the point $x = 0$ is readily seen to be an irregular singular point. Similarly, for the equation

$$x^3(1 - x)^2 \frac{d^2 y}{dx^2} - 2x^2(1 - x)\frac{dy}{dx} + 3y = 0,$$

it can be verified that $x = 0$ is an irregular singular point and $x = 1$ is a regular singular point. All other points are ordinary points.

When the coefficients $a_1(x)$ and $a_2(x)$ in the standard form (15) are *ratios of polynomials,* singular points can occur only when a denominator is zero, so that, unless the numerator also vanishes there, the corresponding coefficient is not *finite.* Most of the equations considered in this text will be of this type.

However, a singular point also may occur when $a_1(x)$ or $a_2(x)$ becomes infinite in some other way, or even in the absence of such behavior. For example, if $a_1(x) = (x - 1)^{5/3}$, it is seen that $a_1''(x)$ becomes infinite as $x \rightarrow 1$, so that $a_1(x)$ cannot be expanded in a series of powers of $x - 1$. Since the function $(x - 1)a_1(x) = (x - 1)^{7/3}$ also cannot be so expanded, it follows that the differential equation (15) has, in fact, an *irregular* singular point at $x = 1$ when $a_1(x) = (x - 1)^{5/3}$.

It will be shown that, if $x = x_0$ is an *ordinary* point of (15), then the equation possesses two linearly independent solutions which are regular at $x = x_0$, and hence are *both* expressible in the form $\sum_{k=0}^{\infty} A_k(x - x_0)^k$. If $x = x_0$ is a *regular singular point* of (15), it will be shown that the equation does not necessarily possess any nontrivial solution which is regular near $x = x_0$, but that at least one solution exists of the form

$$y = (x - x_0)^s \sum_{k=0}^{\infty} A_k(x - x_0)^k,$$

where s is a determinable number which may be real or imaginary. Such a solution is regular at $x = x_0$ only if s is zero or a positive integer. If $x = x_0$ is an *irregular singular point* of (15), the problem is more involved; a nontrivial solution of this type may or may not exist.†

In illustration, the equation

$$x^2 \frac{d^2 y}{dx^2} + (1 + 2x)\frac{dy}{dx} = 0$$

---

†It is known, however, that in this case the differential equation in fact cannot have *two* independent solutions of this type.

has an irregular singular point at $x = 0$. The general solution is

$$y = c_1 e^{1/x} + c_2;$$

thus the solution $y = $ constant is the only solution regular at $x = 0$. The equation

$$x^4 \frac{d^2y}{dx^2} + 2x^3 \frac{dy}{dx} + y = 0$$

also has an irregular singular point at $x = 0$ and has the general solution

$$y = c_1 \sin \frac{1}{x} + c_2 \cos \frac{1}{x}.$$

It follows from the nature of these functions that this equation has neither a nontrivial regular solution at $x = 0$ nor a solution expressible in the more general form

$$y = x^s \sum_{k=0}^{\infty} A_k x^k.$$

4.4. The Method of Frobenius. In this section we restrict attention to solutions valid in the neighborhood of the point $x = 0$. Solutions valid near a more general point $x = x_0$ may be obtained in an analogous way, although for this purpose it is frequently more convenient first to replace $x - x_0$ by a new variable t and then to determine solutions of the transformed differential equation near the point $t = 0$.

In place of reducing a second-order linear equation to the standard form (15), it is frequently more convenient to use a form which is, to some extent, cleared of fractions, particularly if $a_1(x)$ or $a_2(x)$ is the ratio of two polynomials. For this reason, to investigate the nature of solutions valid near $x = 0$, we suppose that the equation has been put in the form

$$Ly \equiv R(x) \frac{d^2y}{dx^2} + \frac{1}{x} P(x) \frac{dy}{dx} + \frac{1}{x^2} Q(x)y = 0, \tag{16}$$

where $R(x)$ does not vanish in some interval including $x = 0$. We also suppose that $P(x)$, $Q(x)$, and $R(x)$ are regular at $x = 0$. Then, with the notation of (15), the products

$$x a_1(x) = \frac{P(x)}{R(x)} \quad \text{and} \quad x^2 a_2(x) = \frac{Q(x)}{R(x)}$$

are regular at $x = 0$, and this point is either an ordinary point or, at worst, a regular singular point of the differential equation.

It is convenient to suppose also that the original equation has been divided through by a suitable constant so that $R(0) = 1$. Then we may write

$$P(x) = P_0 + P_1 x + P_2 x^2 + \cdots,$$
$$Q(x) = Q_0 + Q_1 x + Q_2 x^2 + \cdots, \tag{17}$$
$$R(x) = 1 + R_1 x + R_2 x^2 + \cdots,$$

the series converging in some interval including $x = 0$.

We attempt to find nontrivial solutions which are in the form of a power series in x multiplied by a power of x,

$$y = x^s \sum_{k=0}^{\infty} A_k x^k = A_0 x^s + A_1 x^{s+1} + A_2 x^{s+2} + \cdots, \qquad (18)$$

where s is to be determined. The number A_0 is now, by assumption, the coefficient of the *first term* in the series, and hence *must not vanish*. Substitution into the left-hand member of (16) then gives†

$$Ly \equiv (1 + R_1 x + R_2 x^2 + \cdots) \times [s(s-1)A_0 x^{s-2}$$
$$+ (s+1)sA_1 x^{s-1} + (s+2)(s+1)A_2 x^s + \cdots]$$
$$+ (P_0 + P_1 x + P_2 x^2 + \cdots) \times [sA_0 x^{s-2} + (s+1)A_1 x^{s-1}$$
$$+ (s+2)A_2 x^s + \cdots]$$
$$+ (Q_0 + Q_1 x + Q_2 x^2 + \cdots) \times [A_0 x^{s-2} + A_1 x^{s-1} + A_2 x^s + \cdots],$$

or, after multiplying term by term and collecting the coefficients of successive powers of x,

$$Ly \equiv [s(s-1) + P_0 s + Q_0]A_0 x^{s-2} + \{[(s+1)s + P_0(s+1) + Q_0]A_1$$
$$+ [R_1 s(s-1) + P_1 s + Q_1]A_0\}x^{s-1}$$
$$+ \{[(s+2)(s+1) + P_0(s+2) + Q_0]A_2$$
$$+ [R_1(s+1)s + P_1(s+1) + Q_1]A_1$$
$$+ [R_2 s(s-1) + P_2 s + Q_2]A_0\}x^s + \cdots. \qquad (19)$$

In order to abbreviate this relation, we next define the functions

$$f(s) = s(s-1) + P_0 s + Q_0 = s^2 + (P_0 - 1)s + Q_0 \qquad (20)$$

and
$$g_n(s) = R_n(s-n)(s-n-1) + P_n(s-n) + Q_n$$
$$= R_n(s-n)^2 + (P_n - R_n)(s-n) + Q_n. \qquad (21)$$

With this notation, Equation (19) then becomes

$$Ly \equiv f(s)A_0 x^{s-2} + [f(s+1)A_1 + g_1(s+1)A_0]x^{s-1}$$
$$+ [f(s+2)A_2 + g_1(s+2)A_1 + g_2(s+2)A_0]x^s + \cdots$$
$$+ \left[f(s+k)A_k + \sum_{n=1}^{k} g_n(s+k)A_{k-n} \right]x^{s+k-2} + \cdots. \qquad (22)$$

In order that (16) be satisfied in an interval including $x = 0$, this expression must vanish *identically*, in the sense that the coefficients of all powers of x in (22) must vanish independently. The vanishing of the coefficient of the *lowest* power x^{s-2} gives the requirement

$$f(s) = 0 \quad \text{or} \quad s^2 + (P_0 - 1)s + Q_0 = 0. \qquad (23)$$

This equation determines two values of s (which may however be equal) and is called the *indicial equation*. The two values of s, which specify the exponents

†A more compact development is obtained by using summation notation. (See Problem 10.)

of the leading terms of possible series solutions of the form (18), are called the *exponents of the differential equation* at $x = 0$.

For each such value of s, the vanishing of the coefficient of x^{s-1} in (22) gives the requirement

$$f(s + 1)A_1 = -g_1(s + 1)A_0$$

and hence determines A_1 in terms of A_0 if $f(s + 1) \neq 0$. Next, the vanishing of the coefficient of x^s in (22) determines A_2 in terms of A_1 and A_0,

$$f(s + 2)A_2 = -g_1(s + 2)A_1 - g_2(s + 2)A_0,$$

and hence in terms of A_0, if $f(s + 2) \neq 0$.

In general, the vanishing of the coefficient of x^{s+k-2} in Equation (22) gives the *recurrence formula*

$$f(s + k)A_k = -\sum_{n=1}^{k} g_n(s + k)A_{k-n} \qquad (k \geq 1), \tag{24}$$

which determines each A_k in terms of the preceding A's, and hence in terms of A_0, if for each positive integer k the quantity $f(s + k)$ is not zero.

Thus, if two distinct values of s are determined by (23), and if for each such value of s the quantity $f(s + k)$ is never zero for any positive integer k, the coefficients of two series of the form (18) are determined and these series are solutions of (16) in their interval of convergence. In such cases, if the solution obtained with $s = s_1$ is denoted by $A_0 u_1(x)$ and that with $s = s_2$ by $A_0 u_2(x)$, then the *general* solution can be expressed in the form $y = c_1 u_1(x) + c_2 u_2(x)$. We next investigate the *exceptional cases*.

Let the roots of Equation (23) be $s = s_1$ and $s = s_2$, where

$$s_1 = \frac{1 - P_0}{2} + \frac{1}{2}\sqrt{(1 - P_0)^2 - 4Q_0},$$

$$s_2 = \frac{1 - P_0}{2} - \frac{1}{2}\sqrt{(1 - P_0)^2 - 4Q_0}. \tag{25}$$

The first exceptional case is then the case when the exponents s_1 and s_2 are equal,

$$(1 - P_0)^2 - 4Q_0 = 0. \tag{26}$$

In this case only one solution of the form (18) can be obtained.

Now suppose that the two exponents are distinct. The second exceptional case may then arise if $f(s_1 + k)$ or $f(s_2 + k)$ vanishes for a positive integral value of k, say $k = K$, so that (24) cannot be solved for the coefficient A_K. With the notation of (25), we have

$$f(s) = (s - s_1)(s - s_2),$$

and hence

$$f(s + k) = (s + k - s_1)(s + k - s_2),$$

from which there follows

$$f(s_1 + k) = k[k + (s_1 - s_2)], \qquad f(s_2 + k) = k[k - (s_1 - s_2)]. \tag{27}$$

If s_1 is imaginary and the coefficients P_k, Q_k, and R_k are all real, then s_2 is the conjugate complex number. Hence in this case $s_1 - s_2$ is imaginary and the expressions in (27) cannot vanish for *any* real values of k except $k = 0$. Next suppose that s_1 and s_2 are *real* and distinct and that $s_1 > s_2$. Then, since $s_1 - s_2 > 0$, it follows that $f(s_1 + k)$ cannot vanish for $k \geq 1$ and that $f(s_2 + k)$ can vanish only when $k = s_1 - s_2$. Since k may take on only positive integral values, *this condition is possible only if $s_1 - s_2$ is a positive integer.* If $s_1 = s_2$, then $f(s_1 + k) = k^2$, and hence $f(s_1 + k)$ cannot vanish when $k \geq 1$.

We thus see that *if the two exponents s_1 and s_2 do not differ by zero or a positive integer, two distinct solutions of type* (18) *are obtained. If the exponents are equal, one such solution is obtained, whereas if the exponents differ by a positive integer, a solution of type* (18) *corresponding to the larger exponent is obtained.*

It is known (see, for example, Reference 6 of Chapter 1) that *the interval of convergence of each series so obtained is at least the largest interval, centered at $x = 0$, inside which the expansions of $xa_1(x)$ and $x^2a_2(x)$ in powers of x both converge,* with the natural understanding that the point $x = 0$ itself must be excluded when the exponent (s_1 or s_2) is negative or has a negative real part.

When x is *complex*, each infinite series converges to a solution in a *circle* in the complex plane, with center at the origin and radius at least the distance to the nearest singularity of $a_1(x)$ or $a_2(x)$, and with the center deleted when necessary. The solution then is said to have a *pole* at the origin when the associated exponent is a negative integer, and a *branch point* at the origin when the exponent is non-integral, as well as in the exceptional cases (Section 4.5) when the function $\log x$ is involved (see Chapter 10).

If $s_1 - s_2 = K$, where K is a positive integer, then when $k = K$ the recurrence formula (24) becomes

$$(s - s_2)(s - s_2 + K)A_K = -\sum_{n=1}^{K} g_n(s + K)A_{K-n}. \tag{28}$$

Thus, as we have seen, the left member vanishes when $s = s_2$, and the equation cannot be satisfied by any value of A_K unless it happens that the right member is also zero,† in which case the coefficient A_K is undetermined, and hence arbitrary. If this condition exists, a solution of type (18) is then obtained, *corresponding to the smaller exponent s_2,* which contains *two arbitrary constants A_0 and A_K,* and hence is the *complete solution.* Thus we conclude that, if the exponents differ by a positive integer, either *no solution* of type (18) is obtained for the smaller exponent or *two independent solutions* are obtained.

In the latter case the two solutions so obtained must then include the solution corresponding to the larger exponent as the coefficient of A_K. It is important to notice that this is the situation, for example, when $x = 0$ is an

---

†Here it is to be understood that the recurrence formula has been used to express $A_1, A_2, \ldots,$ A_{K-1} as multiples of A_0, so that the right-hand member of (28) is expressed as A_0 multiplied by a specific function of s.

ordinary point. For in this case one has $P_0 = Q_0 = Q_1 = 0$, and hence $s_1 = 1$, $s_2 = 0$, and $K = s_1 - s_2 = 1$. Thus Equation (28) here becomes

$$s(s + 1)A_1 = -g_1(s + 1)A_0$$
$$= -[R_1 s^2 + (P_1 - R_1)s]A_0$$
$$= -s[R_1 s + (P_1 - R_1)]A_0,$$

and when $s = s_2 = 0$ the recurrence formula is identically satisfied, leaving A_1 as well as A_0 arbitrary. Thus, *when $x = 0$ is an ordinary point, two linearly independent solutions which are regular at $x = 0$ are obtained.*

The preceding detailed derivation was intended for the purpose of investigating the *existence* of series solutions of the assumed type. Although the formulas obtained can be used directly for the determination of the coefficients, once the functions $f(s), g_1(s), g_2(s), \ldots$ are identified, it is usually preferable to obtain the indicial equation and the recurrence formulas in actual practice by direct substitution of the assumed series into the differential equation, written in any convenient form.

To illustrate the application of the preceding treatment, we consider again the second example, Equation (13), of Section 4.2. We thus seek solutions of the equation

$$Ly \equiv x^2 \frac{d^2y}{dx^2} + (x^2 + x)\frac{dy}{dx} - y = 0$$

of the form

$$y = x^s \sum_{k=0}^{\infty} A_k x^k = \sum_{k=0}^{\infty} A_k x^{k+s}.$$

By direct substitution in the differential equation, there follows

$$Ly \equiv (s^2 - 1)A_0 x^s + \sum_{k=1}^{\infty} \{[(s + k)^2 - 1]A_k + (s + k - 1)A_{k-1}\}x^{k+s}.$$

Hence the *indicial equation* is

$$s^2 - 1 = 0$$

and the *recurrence formula* is

$$[(s + k)^2 - 1]A_k + (s + k - 1)A_{k-1} = 0$$

or $$(s + k - 1)[(s + k + 1)A_k + A_{k-1}] = 0 \qquad (k \geq 1).$$

The exponents s_1 and s_2 are $+1$ and -1, respectively. Since they differ by an integer, a solution of the required type is *assured* only when s has the larger value $+1$.

With $s = +1$, the recurrence formula becomes

$$k[(k + 2)A_k + A_{k-1}] = 0,$$

or, since $k \neq 0$,

$$A_k = -\frac{A_{k-1}}{k + 2} \qquad (k \geq 1).$$

Thus one has

$$A_1 = -\frac{A_0}{3}, \qquad A_2 = \frac{A_0}{3 \cdot 4}, \qquad A_3 = -\frac{A_0}{3 \cdot 4 \cdot 5}, \qquad \cdots,$$

and hence the solution corresponding to $s = 1$ is

$$
\begin{aligned}
y &= x\left(A_0 - \frac{A_0}{3}x + \frac{A_0}{3 \cdot 4}x^2 - \frac{A_0}{3 \cdot 4 \cdot 5}x^3 + \cdots\right) \\
&= A_0\left(x - \frac{x^2}{3} + \frac{x^3}{3 \cdot 4} - \frac{x^4}{3 \cdot 4 \cdot 5} + \cdots\right) \\
&= 2A_0 \frac{e^{-x} - 1 + x}{x},
\end{aligned}
$$

in accordance with the result obtained previously.

With $s = -1$, the recurrence formula becomes

$$(k - 2)(kA_k + A_{k-1}) = 0 \qquad (k \geq 1).$$

It is important to notice that the factor $k - 2$ cannot be canceled except on the understanding that $k \neq 2$. That is, *when $k = 2$, the correct form of the recurrence formula is $0 = 0$*. If $k = 1$, there follows

$$A_1 \overset{\cdot}{=} -A_0.$$

If $k = 2$, the recurrence formula is identically satisfied, so that A_2 is *arbitrary*. If $k \geq 3$, the recurrence formula can be written in the form

$$A_k = -\frac{A_{k-1}}{k} \qquad (k \geq 3).$$

If we take $A_2 = 0$, then there follows $A_3 = A_4 = \cdots = 0$, and the solution corresponding to $s = -1$ becomes simply

$$y = x^{-1}(A_0 - A_0 x) = A_0 \frac{1 - x}{x}.$$

The general solution of the given equation is then a linear combination of the two solutions so obtained, and hence can be taken conveniently in the form

$$
\begin{aligned}
y &= c_1 x\left(1 - \frac{x}{3} + \frac{x^2}{3 \cdot 4} - \cdots\right) + c_2 x^{-1}(1 - x) \\
&= 2c_1 \frac{e^{-x} - 1 + x}{x} + c_2 \frac{1 - x}{x},
\end{aligned}
$$

or, alternatively,

$$y = C_1 \frac{e^{-x}}{x} + C_2 \frac{1 - x}{x},$$

where $C_1 = 2c_1$, and $C_2 = c_2 - 2c_1$. The only solution regular at $x = 0$ is the one formerly obtained,

$$y = c\frac{e^{-x} - 1 + x}{x}.$$

It can be verified that if A_2 is left arbitrary, the solution corresponding to the exponent -1 is the sum of A_0 times the two-term solution obtained and A_2 times the infinite series solution corresponding to the exponent $+1$.

For the purpose of computational efficiency, *when s_1 and s_2 differ by a positive integer K*, it usually is desirable to explore first the situation corresponding to the *smaller* exponent, since then if a nontrivial solution *is* obtained with $s = s_2$ it will be permissible to leave both A_K and A_0 arbitrary in that solution, and so to obtain the *general* solution without needing to proceed also to a determination with $s = s_1$. A frequently committed blunder, which vitiates this possibility, consists of dividing both members of the recurrence formula by $k - K$ and then setting $k = K$ in the result (and hence *dividing by zero*).

4.5. Treatment of Exceptional Cases.† We consider first the case of *equal exponents*, and attempt to determine a second solution which is independent of the one obtained by the method of Frobenius. Although this result could be accomplished by the methods of Section 1.10, the method to be given is usually more easily applied.

In place of first introducing the value of the repeated exponent s_1 into the recurrence formula (24) and then determining $A_1, A_2, \ldots, A_k, \ldots$ directly in terms of A_0, we suppose that the coefficients $A_1, A_2, \ldots, A_k, \ldots$ first are expressed, by use of the recurrence formula, in terms of A_0 *and s*. We indicate this fact by writing $A_1 = A_1(s)$, $A_2 = A_2(s)$, $\ldots$, where, in fact, each $A_k(s)$ will be of the form $A_0 c_k(s)$, with $c_k(s)$ a specific function of s. With these values of the A's, as functions of s, a function y depending upon s as well as x is determined and is denoted here by $y(x, s)$;

$$y(x, s) = x^s \sum_{k=0}^{\infty} A_k(s) x^k. \tag{29}$$

Reference to Equation (22) shows that satisfaction of the recurrence formula, for $k \geq 1$, brings about vanishing of all terms in (22) except the first, and so there follows

$$Ly(x, s) = A_0 f(s) x^{s-2}, \tag{30}$$

or, since in the case of a repeated exponent s_1 we have $f(s) = (s - s_1)^2$,

$$Ly(x, s) = A_0(s - s_1)^2 x^{s-2}. \tag{31}$$

The fact that the right-hand member of (31) vanishes when $s = s_1$ is in accordance with the known fact that (29) becomes a solution of (16), say $y_1(x)$, when $s = s_1$; that is,

$$y_1(x) = y(x, s_1).$$

However, since $s = s_1$ is a repeated zero of the right-hand member of (31), it

---

†Sections 4.5 and 4.6, together with the derivation of the series for $Y_0(x)$ in Section 4.8, can be omitted without logical difficulty. However, in this event a consideration of the last paragraph of Section 4.5 (and/or the working of Problems 16 and 17) is suggested.

follows also that the result of differentiating either member of (31) with respect to s (holding x constant),

$$\frac{\partial}{\partial s} Ly(x, s) = A_0[2(s - s_1) + (s - s_1)^2 \log x]x^{s-2},$$

is zero when $s = s_1$. But, since the operator $\partial/\partial s$ and the linear operator L are commutative, there follows also

$$\left[\frac{\partial}{\partial s} Ly(x, s)\right]_{s=s_1} = L\left[\frac{\partial y(x, s)}{\partial s}\right]_{s=s_1} = 0. \tag{32}$$

Hence a second solution of (16), when $s_1 = s_2$, is of the form

$$y_2(x) = \left[\frac{\partial y(x, s)}{\partial s}\right]_{s=s_1}. \tag{33}$$

The second exceptional case is that in which *the exponents differ by a positive integer K*,

$$s_1 - s_2 = K \geq 1,$$

but where the recurrence formula is not identically satisfied when $k = K$ and $s = s_2$, that is, when the right member of (28) does not vanish when $s = s_2$. In such a case, Equation (28) can be satisfied only if $A_0 = A_1 = \cdots = A_{K-1} = 0$, and hence Equation (16) does not possess a solution of type (18) beginning with a term of the form $A_0 x^{s_2}$.

In this case we suppose again that the recurrence formula is satisfied when $k \geq 1$ for all values of s, so that with each A_k expressed as the product of A_0 and a certain function of s, we again define a function $y(x, s)$ of form (29). In this case, however, it is clear from (28) and from the nature of the recurrence formula (24) that the expressions for the coefficients $A_K(s), A_{K+1}(s), \ldots$ now all will have a factor $s - s_2$ in a denominator, and hence will not approach finite limits as $s \to s_2$. If we consider the product $(s - s_2)y(x, s)$, we see that as $s \to s_2$ terms with coefficients A_k for which $k < K$ will vanish and the remaining terms will approach finite limits, thus giving rise to an infinite series of powers of x starting with a term involving $x^{s_2+K} = x^{s_1}$. Thus the limiting series must be proportional to the series for which $s = s_1$. In this case, however, since again satisfaction of the recurrence formula for $k \geq 1$ causes all terms of (22) except the first to vanish, we have

$$L\{(s - s_2)y(x, s)\} = A_0(s - s_2)^2(s - s_1)x^{s-2}. \tag{34}$$

But since the right member has a double zero $s = s_2$, the partial derivative of either member must vanish as $s \to s_2$, and, by an argument similar to that leading to (32), we conclude that

$$L\left\{\frac{\partial}{\partial s}[(s - s_2)y(x, s)]\right\}_{s=s_2} = 0$$

so that the function

$$y_2(x) = \left\{\frac{\partial}{\partial s}[(s - s_2)y(x, s)]\right\}_{s=s_2} \tag{35}$$

is a solution of (16), in addition to the solution $y_1(x) = [y(x, s)]_{s=s_1}$ corresponding to the larger exponent s_1.

From Equations (29) and (33) it follows that when $s_2 = s_1$ the second solution y_2 is expressible as

$$y_2(x) = \left[\sum_{k=0}^{\infty} A_k(s_1)x^{k+s_1} \right] \log x + \sum_{k=0}^{\infty} A'_k(s_1)x^{k+s_1}, \tag{36}$$

and the coefficient of $\log x$ is seen to be $y_1(x)$. Further, when s_2 and s_1 differ by a positive integer but there is no Frobenius series solution with $s = s_2$, Equations (29) and (35) show that the missing solution y_2 is expressible as

$$y_2(x) = \left[\sum_{k=0}^{\infty} (s - s_2)A_k(s)x^{k+s} \right]_{s=s_2} \log x$$

$$+ \sum_{k=0}^{\infty} \left\{ \frac{d}{ds}[(s - s_2)A_k(s)] \right\}_{s=s_2} x^{k+s_2}. \tag{37}$$

The coefficient of $\log x$ is $\lim_{s \to s_2} [(s - s_2)y(x, s)]$, which has been seen to be *proportional* to $y_1(x)$.

Hence it follows that, *in all cases when the differential equation,* having $x = 0$ *as a regular singular point with exponents* s_1 *and* s_2, *possesses only one solution*

$$y_1(x) = \sum_{k=0}^{\infty} A_k x^{k+s_1} \equiv A_0 u_1(x)$$

of the form (18), *any independent solution is of the form*

$$y_2(x) = Cu_1(x) \log x + \sum_{k=0}^{\infty} B_k x^{k+s_2}, \tag{38}$$

where C is a constant. Thus, *in place of using the results of Equations* (33) *or* (35) *in such cases, a second solution may be obtained by directly assuming a solution of this last form and by determining the necessary relationships between the coefficients* B_k *and an arbitrarily chosen constant* $C \neq 0$.

4.6. Example of an Exceptional Case. To illustrate the procedures developed in the preceding section, we consider the equation

$$Ly \equiv x\frac{d^2y}{dx^2} - y = 0. \tag{39}$$

With the notation

$$y(x, s) = \sum_{k=0}^{\infty} A_k(s)x^{k+s}, \tag{40}$$

there follows

$$Ly(x, s) \equiv s(s - 1)A_0 x^{s-1} + \sum_{k=1}^{\infty} [(k + s)(k + s - 1)A_k - A_{k-1}]x^{k+s-1}.$$

Hence we obtain the recurrence formula

$$(k + s)(k + s - 1)A_k = A_{k-1} \qquad (k \geq 1) \tag{41}$$

and the two indices

$$s_1 = 1, \qquad s_2 = 0. \tag{42}$$

Since the indices differ by unity, a solution of the form (40) is assured only for $s = 1$.

From 41 there follows

$$A_1 = \frac{A_0}{(s+1)s}, \qquad A_2 = \frac{A_0}{(s+2)(s+1)^2 s}, \qquad \cdots,$$

and in general, by inductive reasoning,

$$A_k(s) = \frac{A_0}{(s+k)[(s+k-1)\cdots(s+1)]^2 s} \qquad (k \geq 2). \tag{43}$$

The solution for which $s = s_1 = 1$ then becomes

$$y_1(x) = A_0 \sum_{k=0}^{\infty} \frac{x^{k+1}}{(k+1)!\,k!} \equiv A_0 u_1(x). \tag{44}$$

However, since $A_k(s) \to \infty$ when $s \to 0$, for all $k \geq 1$, there is indeed no solution of type (40) for which $s = 0$.

In order to obtain a second solution, we may refer to Equation (37). The coefficient of $\log x$ is seen to be

$$\sum_{k=0}^{\infty} [sA_k(s)]_{s=0} x^k = A_0 \sum_{k=1}^{\infty} \frac{x^k}{k!\,(k-1)!}$$

$$= A_0 \sum_{k=0}^{\infty} \frac{x^{k+1}}{(k+1)!\,k!} = A_0 u_1(x), \tag{45}$$

when account is taken of the vanishing of $(sA_0)_{s=0}$. The coefficient $d[sA_k(s)]/ds$, involved in the second series of Equation (37), is conveniently evaluated by logarithmic differentiation. For this purpose, we first deduce from (43) the relation

$$\log [sA_k(s)] = \log A_0 - \log (s+k) - 2 \sum_{m=1}^{k-1} \log (s+m),$$

then differentiate the equal members and resolve the result in the form

$$\frac{d}{ds}[sA_k(s)] = - \frac{\dfrac{1}{s+k} + 2 \displaystyle\sum_{m=1}^{k-1} \frac{1}{s+m}}{(s+k)[s+k-1)\cdots(s+1)]^2} A_0 \qquad (k \geq 2).$$

Thus there follows

$$\left\{ \frac{d}{ds}[sA_k(s)] \right\}_{s=0} = - \frac{\varphi(k) + \varphi(k-1)}{k!\,(k-1)!} A_0 \qquad (k \geq 2), \tag{46}$$

with the abbreviation

$$\varphi(k) = \begin{cases} \displaystyle\sum_{m=1}^{k} \frac{1}{m} = 1 + \frac{1}{2} + \frac{1}{3} + \cdots + \frac{1}{k} & (k = 1, 2, \ldots), \\ 0 & (k = 0). \end{cases} \tag{47}$$

It is found by direct calculation that Equation (46) also holds for $k = 1$, whereas the right-hand member must be replaced by A_0 when $k = 0$. Hence (37) gives

$$y_2(x) = A_0\left[u_1(x)\log x + 1 - \sum_{k=1}^{\infty}\frac{\varphi(k) + \varphi(k-1)}{k!(k-1)!}x^k\right] \qquad (48)$$

or, in expanded form,

$$y_2(x) = A_0[(x + \tfrac{1}{2}x^2 + \tfrac{1}{12}x^3 + \cdots)\log x + (1 - x - \tfrac{5}{4}x^2 - \tfrac{5}{18}x^3 - \cdots)].$$

The alternative procedure described at the end of the preceding section consists of substituting the relation of Equation (38) in the form

$$y_2(x) = Cu_1(x)\log x + \sum_{k=0}^{\infty}B_k x^k \qquad (49)$$

directly into (39), to obtain the condition

$$C\left[\left(x\frac{d^2u_1}{dx^2} - u_1\right)\log x + 2\frac{du_1}{dx} - \frac{1}{x}u_1\right]$$

$$+ \sum_{k=0}^{\infty}[(k+1)kB_{k+1} - B_k]x^k = 0. \qquad (50)$$

Since u_1 satisfies Equation (39), the coefficient of $\log x$ in (50) vanishes, and the introduction of (45) reduces (50) to the form

$$\sum_{k=0}^{\infty}[(k+1)kB_{k+1} - B_k]x^k + C\sum_{k=0}^{\infty}\left[\frac{2}{(k!)^2} - \frac{1}{(k+1)!k!}\right]x^k = 0.$$

The requirement that the coefficient of the general power x^k vanish becomes

$$(k+1)kB_{k+1} - B_k = -\frac{2k+1}{(k+1)!k!}C \qquad (k \geq 0).$$

By setting k successively equal to $0, 1, 2, \ldots$, we obtain

$$B_0 = C, \quad B_2 = \tfrac{1}{2}B_1 - \tfrac{3}{4}C, \quad B_3 = \tfrac{1}{12}B_1 - \tfrac{7}{36}C, \quad \ldots.$$

Here both B_1 and C are arbitrary and all the other coefficients are expressible in terms of these two constants, yielding the solution (49) in the form

$$y_2(x) = C[(x + \tfrac{1}{2}x^2 + \tfrac{1}{12}x^3 + \cdots)\log x + (1 - \tfrac{3}{4}x^2 - \tfrac{7}{36}x^3 + \cdots)]$$

$$+ B_1(x + \tfrac{1}{2}x^2 + \tfrac{1}{12}x^3 + \cdots). \qquad (51)$$

The coefficient of B_1 is seen to be $u_1(x)$. Thus, if C and B_1 are left as arbitrary constants, this expression for $y_2(x)$ in fact represents the *general* solution of (39). The particular expression for y_2 obtained in (48), by the first method, is obtained from (51) by choosing $C = -B_1 = A_0$.

4.7. A Particular Class of Equations.

Many important second-order equations, of frequent occurrence in practice, can be obtained by specializing the constants in the equation

$$(1 + R_M x^M)\frac{d^2y}{dx^2} + \frac{1}{x}(P_0 + P_M x^M)\frac{dy}{dx} + \frac{1}{x^2}(Q_0 + Q_M x^M)y = 0, \qquad (52)$$

where M is a positive integer. (In the case when $M = 0$, the equation is equidimensional.) Here the introduction of the assumption

$$y(x) = \sum_{k=0}^{\infty} A_k x^{k+s} \tag{53}$$

leads to the condition

$$\sum_{k=0}^{M-1} f(s+k)A_k x^{s+k-2} + \sum_{k=M}^{\infty} [f(s+k)A_k + g(s+k)A_{k-M}]x^{s+k-2} = 0,$$

where

$$f(s) = s^2 + (P_0 - 1)s + Q_0$$

and $$g(s) \equiv g_M(s) = R_M(s - M)^2 + (P_M - R_M)(s - M) + Q_M.$$

Thus, for each exponent satisfying the indicial equation $f(s) = 0$, the recurrence formula is

$$f(s+k)A_k = 0 \qquad (k = 1, 2, \ldots, M-1),$$
$$f(s+k)A_k = -g(s+k)A_{k-M} \qquad (k \geq M).$$

The first $M - 1$ conditions are satisfied by taking

$$A_1 = A_2 = \cdots = A_{M-1} = 0,$$

after which the recurrence formula for $k \geq M$ shows that *all coefficients A_k for which k is not an integral multiple of M can be taken to be zero.*

Accordingly, *it is convenient to write*

$$y(x) = \sum_{k=0}^{\infty} B_k x^{Mk+s} \tag{54}$$

when seeking a solution of a special case of Equation (52). Here k has been replaced by Mk in (53) and B_k has been written for A_{Mk}.

Further, it is seen that here *an exceptional case can occur only when the exponents s_1 and s_2 are equal or when $s_1 - s_2 = KM$, where K is a positive integer.*† In such a situation, when only one solution of type (54) is obtained, a second solution can be found, as usual, by use of Equations (33) or (35).

Since the expansion of $(1 + R_M x^M)^{-1}$ converges when

$$|x| < |R_M|^{-1/M}, \tag{55}$$

the solutions obtained for (52) also will converge in that interval. In particular, if $R_M = 0$, the series will converge for all finite values of x (the value $x = 0$ itself being excepted, as usual, when the real part of s is negative).

Among the many important specializations of Equation (52), we note *Bessel's equation,*

$$x^2 \frac{d^2y}{dx^2} + x\frac{dy}{dx} + (x^2 - p^2)y = 0, \tag{56}$$

---

†The present procedure clearly bypasses the possibility of determining a *two*-parameter solution with $s = s_2$ when $s_1 - s_2$ is a positive integer other than a multiple of M, but it provides *one* solution for *each* exponent in such special cases.

for which $M = 2$; *Legendre's equation,*

$$(1 - x^2)\frac{d^2y}{dx^2} - 2x\frac{dy}{dx} + p(p + 1)y = 0, \tag{57}$$

for which $M = 2$; and *Gauss's equation,*

$$x(1 - x)\frac{d^2y}{dx^2} + [\gamma - (\alpha + \beta + 1)x]\frac{dy}{dx} - \alpha\beta y = 0, \tag{58}$$

for which $M = 1$. The solutions of these equations, in the neighborhood of $x = 0$, are studied in the following sections.

Other notable special cases may be listed as follows.

(1) The equation

$$x\frac{d^2y}{dx^2} + (c - x)\frac{dy}{dx} - ay = 0, \tag{59}$$

for which $M = 1$, is satisfied by the *confluent hypergeometric function* of Kummer, $y = M(a, c, x)$ (see Problem 8). If $c = 1$ and $a = -n$, where n is a positive integer or zero, one solution is the nth *Laguerre polynomial,* $y = L_n(x)$. If $c = m + 1$ and $a = m - n$, where m and n are integral, one solution is the *associated Laguerre polynomial,*

$$y = L_n^m(x) = \frac{d^m}{dx^m}L_n(x),$$

if $m \leq n$.

(2) The equation

$$\frac{d^2y}{dx^2} - 2x\frac{dy}{dx} + 2ny = 0, \tag{60}$$

for which $M = 2$, is satisfied by the nth *Hermite polynomial,* $y = H_n(x)$, when n is a positive integer or zero.

(3) The equation

$$(1 - x^2)\frac{d^2y}{dx^2} - x\frac{dy}{dx} + n^2y = 0, \tag{61}$$

for which $M = 2$, is satisfied by the nth *Chebyshev polynomial,* $y = T_n(x)$, when n is a positive integer or zero.

(4) The equation

$$x(1 - x)\frac{d^2y}{dx^2} + [a - (1 + b)x]\frac{dy}{dx} + n(b + n)y = 0, \tag{62}$$

for which $M = 1$, is satisfied by the nth *Jacobi polynomial,* $y = J_n(a, b, x)$, when n is a positive integer or zero.

The functions mentioned are useful in many applications. It is a curious fact that they all satisfy equations which are special cases of (52),

4.8. Bessel Functions. Solutions of the differential equation

$$x^2 \frac{d^2y}{dx^2} + x \frac{dy}{dx} + (x^2 - p^2)y = 0, \tag{63}$$

or, equivalently,

$$x \frac{d}{dx}\left(x \frac{dy}{dx}\right) + (x^2 - p^2)y = 0, \tag{63a}$$

are known as *Bessel functions of order p*. These functions are of frequent use in the solution of many types of potential problems involving circular cylindrical boundaries, as well as in other applications, in such fields as elasticity, fluid flow, electrical field theory, and aerodynamic flutter analysis. We suppose that the constant p is *real*. Since only the quantity p^2 appears in Equation (63), we may also consider p to be *nonnegative* without loss of generality.

Since Equation (63) is of type (52), with $M = 2$, we may seek a solution of the form (54),

$$y(x) = \sum_{k=0}^{\infty} B_k x^{2k+s}. \tag{64}$$

Substitution into (63) yields the indices

$$s_1 = p, \qquad s_2 = -p \tag{65}$$

and the recurrence formula

$$(s + 2k + p)(s + 2k - p)B_k = -B_{k-1} \qquad (k \geqq 1). \tag{66}$$

The exceptional cases may arise only if $s_1 - s_2 = 2p$ is zero or an integral multiple of $M = 2$, that is, if p is zero or a positive integer. In such cases we can be certain only of one solution of type (64).

In correspondence with the exponent $s_1 = p$, repeated use of Equation (66) gives

$$B_k(p) = (-1)^k \frac{1}{(2 + 2p)(4 + 2p) \cdots (2k + 2p)} \frac{B_0}{2^k k!}$$

$$= (-1)^k \frac{1}{(1 + p)(2 + p) \cdots (k + p)} \frac{B_0}{2^{2k}k!} \qquad (k \geqq 1), \tag{67}$$

so that a series of type (64) is determined, for $s = p$, in the form

$$y_1(x) = B_0\left[x^p + \sum_{k=1}^{\infty} \frac{(-1)^k x^{2k+p}}{(1 + p)(2 + p) \cdots (k + p)2^{2k}k!}\right],$$

or, with use made of Equation (52) of Chapter 2,

$$y_1(x) = B_0\Gamma(1 + p) \sum_{k=0}^{\infty} \frac{(-1)^k x^{2k+p}}{2^{2k}k!\,\Gamma(k + p + 1)}.$$

This result is put into a more compact form if we use the abbreviation $\Gamma(k + p + 1) = (k + p)!$ and write

$$y_1(x) = 2^p p!\, B_0 \sum_{k=0}^{\infty} \frac{(-1)^k(x/2)^{2k+p}}{k!\,(k + p)!} \equiv B_0 u_1(x). \tag{68}$$

The series multiplying $2^p p! B_0$ in Equation (68) is known as the *Bessel function of the first kind, of order p*, and is denoted by $J_p(x)$,

$$J_p(x) = \sum_{k=0}^{\infty} \frac{(-1)^k (x/2)^{2k+p}}{k!(k+p)!}. \tag{69}$$

In particular, when $p = 0$ and 1, we obtain the series in the forms

$$J_0(x) = 1 - \frac{x^2}{2^2} + \frac{x^4}{2^4(2!)^2} - \frac{x^6}{2^6(3!)^2} + \cdots, \tag{70}$$

$$J_1(x) = \frac{x}{2} - \frac{x^3}{2^3 2!} + \frac{x^5}{2^5 2! 3!} - \frac{x^7}{2^7 3! 4!} + \cdots. \tag{71}$$

With $s = s_2 = -p$, Equation (66) yields the result of replacing p by $-p$ in (67),

$$B_k(-p) = (-1)^k \frac{1}{(1-p)(2-p)\cdots(k-p)} \frac{B_0}{2^{2k}k!} \qquad (k \geq 1). \tag{72}$$

Thus, if p is a positive integer, all coefficients B_k for which $k \geq p$ become infinite, and no Frobenius solution is obtained, in such a case, corresponding to the exponent $s = -p$. However, if p is not zero or a positive integer, a second solution is obtained by replacing p by $-p$ in the first solution, and hence may be taken in the form

$$J_{-p}(x) = \sum_{k=0}^{\infty} \frac{(-1)^k (x/2)^{2k-p}}{k!(k-p)!}. \tag{73}$$

Thus, if p is not zero or a positive integer, the complete solution of Bessel's equation (63) is a linear combination of the solutions (69) and (73), of the form

$$y = C_1 J_p(x) + C_2 J_{-p}(x). \tag{74}$$

If $p = 0$, the two solutions are identical. Moreover, if p is a positive integer, the second solution $J_{-p}(x)$ is not independent of $J_p(x)$. This statement is a consequence of the fact that if p is a positive integer n, the factor $1/(k-n)!$ in (73) is zero when $k < n$, and hence (73) is then equivalent to

$$J_{-n}(x) = \sum_{k=n}^{\infty} \frac{(-1)^k (x/2)^{2k-n}}{k!(k-n)!}$$

or, with the index k replaced by $k+n$,

$$J_{-n}(x) = \sum_{k=0}^{\infty} \frac{(-1)^{k+n}(x/2)^{2k+n}}{k!(k+n)!}.$$

Hence, *if n is an integer*, we obtain

$$J_{-n}(x) = (-1)^n J_n(x). \tag{75}$$

It should be noticed that, although the higher coefficients in the series $y_2(x)$ would become infinite as $p \to n$ ($n = 1, 2, 3, \ldots$) if the coefficient B_0 were held fixed, we have obtained (73) by setting $B_0(-p)! = 1$ or $B_0 = 1/\Gamma(1-p)$ in that series and, since this quantity tends to zero as $p \to n$, we have obtained a solution $J_{-p}(x)$ in which the coefficients which previously became infinite as $p \to n$ now approach finite limits, and the remaining coefficients tend to zero.

To find a second solution complementing $J_p(x)$ in the exceptional cases, recourse may be had to the methods of Section 4.5. We illustrate the procedure in the case of equal exponents, $p = 0$. In this case we obtain a function $y(x, s)$ by determining $B_k(s)$ from (66) and introducing the result into (64). The required second solution is then

$$y_2(x) = \left[\frac{\partial y(x, s)}{\partial s}\right]_{s=0},$$

where here

$$y(x, s) = \sum_{k=0}^{\infty} B_k(s)x^{2k+s},$$

so that

$$y_2(x) = \left[\sum_{k=0}^{\infty} B_k(0)x^{2k}\right] \log x + \sum_{k=1}^{\infty} B'_k(0)x^{2k}. \tag{76}$$

The recurrence formula (66) first gives

$$B_k(s) = (-1)^k \frac{B_0}{[(s+2)(s+4)\cdots(s+2k)]^2}.$$

To calculate $B'_k(s)$ it is again convenient first to take the logarithm of the two sides of the equation,

$$\log\left[(-1)^k \frac{B_k(s)}{B_0}\right] = -2 \log [(s+2)(s+4)\cdots(s+2k)]$$

$$= -2 \sum_{m=1}^{k} \log (s+2m).$$

Differentiation with respect to s then gives

$$\frac{B'_k(s)}{B_k(s)} = -2 \sum_{m=1}^{k} \frac{1}{s+2m}$$

and hence

$$\frac{B'_k(0)}{B_k(0)} = -2 \sum_{m=1}^{k} \frac{1}{2m} = -\sum_{m=1}^{k} \frac{1}{m}.$$

Thus, if we again introduce the abbreviation

$$\varphi(k) = \sum_{m=1}^{k} \frac{1}{m} = 1 + \frac{1}{2} + \cdots + \frac{1}{k} \qquad (k \geqq 1), \tag{77a}$$

with

$$\varphi(0) = 0, \tag{77b}$$

we obtain

$$B'_k(0) = -\varphi(k)B_k(0) = -\varphi(k)\left[(-1)^k \frac{B_0}{[2\cdot4\cdot6\cdots(2k)]^2}\right]$$

$$= -\frac{(-1)^k \varphi(k)}{2^{2k}(k!)^2} B_0,$$

and (76) then gives the required second solution in the case $p = 0$ in the form

$$y_2(x) = B_0\left[J_0(x) \log x + \sum_{k=1}^{\infty} (-1)^{k+1}\varphi(k)\frac{(x/2)^{2k}}{(k!)^2}\right].$$

The coefficient of B_0 is thus a second solution of Bessel's equation (63) when $p = 0$,

$$x\frac{d^2y}{dx^2} + \frac{dy}{dx} + xy = 0.$$

It was taken as the standard form of the second solution by *Neumann*, and is usually denoted by $Y^{(0)}(x)$. Thus any linear combination of $J_0(x)$ and $Y^{(0)}(x)$ is also a solution. The standard form chosen by *Weber* is defined in terms of J_0 and $Y^{(0)}$ by the equation

$$Y_0(x) = \frac{2}{\pi}[Y^{(0)}(x) + (\gamma - \log 2)J_0(x)], \tag{78}$$

where γ is *Euler's constant*, defined by the relation

$$\gamma = \lim_{k\to\infty} [\varphi(k) - \log k]$$

$$= \lim_{k\to\infty} \left(1 + \frac{1}{2} + \frac{1}{3} + \cdots + \frac{1}{k} - \log k\right) = 0.5772157\ldots. \tag{79}$$

We thus obtain a second solution in the form

$$Y_0(x) = \frac{2}{\pi}\left[\left(\log\frac{x}{2} + \gamma\right)J_0(x) + \sum_{k=0}^{\infty}(-1)^{k+1}\varphi(k)\frac{(x/2)^{2k}}{(k!)^2}\right]$$

$$= \frac{2}{\pi}\left[\left(\log\frac{x}{2} + \gamma\right)J_0(x) + \left\{\frac{x^2}{2^2} - \frac{x^4}{2^4(2!)^2}\left(1 + \frac{1}{2}\right)\right.\right.$$

$$\left.\left. + \frac{x^6}{2^6(3!)^2}\left(1 + \frac{1}{2} + \frac{1}{3}\right) - \cdots\right\}\right]. \tag{80}$$

The function $Y_0(x)$ is known as Weber's *Bessel function of the second kind, of order zero*. In German texts it is frequently denoted by $N_0(x)$. The complete solution thus can be written in the form

$$y = C_1J_0(x) + C_2Y_0(x). \tag{81}$$

Weber's definition of the function of the second kind [Equation (78)] is more convenient than that of Neumann because of the fact that the behavior of the function $Y_0(x)$ so defined, *for large values of x*, is more nearly comparable with the behavior of $J_0(x)$ [see Equation (88)].

A similar but more involved calculation leads to expressions for Weber's form of the *Bessel function of the second kind, of order n*,

$$Y_n(x) = \frac{2}{\pi}\left[\left(\log\frac{x}{2} + \gamma\right)J_n(x) - \frac{1}{2}\sum_{k=0}^{n-1}\frac{(n-k-1)!(x/2)^{2k-n}}{k!}\right.$$

$$\left. + \frac{1}{2}\sum_{k=0}^{\infty}(-1)^{k+1}[\varphi(k) + \varphi(k+n)]\frac{(x/2)^{2k+n}}{k!(n+k)!}\right] \tag{82}$$

when n is a positive integer. Thus, in particular,

$$Y_1(x) = \frac{2}{\pi}\left[\left(\log\frac{x}{2} + \gamma\right)J_1(x) - \frac{1}{x} - \frac{x}{4} + \left\{\frac{1 + (1 + \frac{1}{2})}{2}\right\}\frac{x^3}{2^3 2!}\right.$$

$$\left. - \left\{\frac{(1 + \frac{1}{2}) + (1 + \frac{1}{2} + \frac{1}{3})}{2}\right\}\frac{x^5}{2^5 2!3!} + \cdots\right]. \tag{83}$$

It follows that if $p = n$, where n is zero or a positive integer, the general solution of (63) can be taken in the form

$$y = C_1 J_n(x) + C_2 Y_n(x). \tag{84}$$

If p is not zero or a positive integer, the function $Y_p(x)$ is defined by the equation

$$Y_p(x) = \frac{(\cos p\pi) J_p(x) - J_{-p}(x)}{\sin p\pi}. \tag{85}$$

This definition can be shown to be consistent with Equation (82) as $p \to n$, and it defines $Y_p(x)$ as a linear combination of $J_p(x)$ and $J_{-p}(x)$ otherwise. It should be emphasized, however, that the second solution $Y_p(x)$ is not *needed* unless $p = n$.

The general solution of Bessel's equation is frequently abbreviated by use of the notation

$$y = Z_p(x), \tag{86}$$

with the convention that (86) stands for (74) unless p is zero or a positive integer, and for (84) in these cases.

The transformation $y = u(x)/\sqrt{x}$ transforms (63) to the equation

$$\frac{d^2u}{dx^2} + \left(1 - \frac{p^2 - \frac{1}{4}}{x^2}\right) u = 0 \tag{87}$$

(see Problem 47 of Chapter 1). For large values of x the term $(p^2 - \frac{1}{4})/x^2$ is negligible in comparison with unity. Thus it may be expected that for large values of x the behavior of solutions of (87) will be similar to that of corresponding solutions of the equation

$$\frac{d^2v}{dx^2} + v = 0.$$

Since such solutions can be written in the form $v = A \cos(x - \varphi)$, where A and φ are constants, we are led to the possibility that for large values of x any solution of (63) behaves like the function

$$\frac{v(x)}{\sqrt{x}} = \frac{A}{\sqrt{x}} \cos(x - \varphi),$$

for properly chosen values of A and φ. A rather involved analysis shows that for the function $J_p(x)$ one has

$$A_1 = \sqrt{\frac{2}{\pi}} \quad \text{and} \quad \varphi_1 = (2p + 1)\frac{\pi}{4},$$

whereas for $Y_p(x)$ there follows

$$A_2 = \sqrt{\frac{2}{\pi}} \quad \text{and} \quad \varphi_2 = \frac{\pi}{2} + \varphi_1.$$

Thus we may write

$$\left. \begin{aligned} J_p(x) &\sim \sqrt{\frac{2}{\pi x}} \cos(x - \alpha_p) \\[2ex] Y_p(x) &\sim \sqrt{\frac{2}{\pi x}} \sin(x - \alpha_p) \end{aligned} \right\} \quad (x \to \infty), \tag{88}$$

where $$\alpha_p = (2p + 1)\frac{\pi}{4}. \tag{89}$$

The notation of (88) denotes that the ratio of two expressions connected by the symbol $\sim$ approaches unity as $x \longrightarrow \infty$. We say that $J_p(x)$ *behaves asymptotically like* $\sqrt{2/\pi x} \cos (x - \alpha_p)$.

It follows from (88) that the complex function $J_p(x) + iY_p(x)$ has the asymptotic behavior

$$J_p(x) + iY_p(x) \sim \sqrt{\frac{2}{\pi x}} e^{i(x-\alpha_p)} \qquad (x \longrightarrow \infty), \tag{90}$$

whereas the conjugate complex function has the behavior

$$J_p(x) - iY_p(x) \sim \sqrt{\frac{2}{\pi x}} e^{-i(x-\alpha_p)} \qquad (x \longrightarrow \infty). \tag{91}$$

These complex functions are known as *Bessel functions of the third kind*, or, more generally, as the *Hankel functions of the first and second kinds*, respectively, and the abbreviations

$$\begin{aligned}
H_p^{(1)}(x) &= J_p(x) + iY_p(x), \\
H_p^{(2)}(x) &= J_p(x) - iY_p(x)
\end{aligned} \tag{92}$$

are conventional. The Hankel functions are particularly useful in studying certain types of wave propagation (see Section 9.13).

The differential equation

$$x^2 \frac{d^2y}{dx^2} + x\frac{dy}{dx} - (x^2 + p^2)y = 0, \tag{93}$$

which differs from Bessel's equation (63) only in the sign of x^2 in the coefficient of y, is transformed by the substitution $ix = t$ to the equation

$$t^2 \frac{d^2y}{dt^2} + t\frac{dy}{dt} + (t^2 - p^2)y = 0,$$

which is in the form of Bessel's equation (63). Hence, the general solution is of the form $y = Z_p(t)$, or, in terms of the original variable x, the general solution of (93) is

$$y = Z_p(ix). \tag{94}$$

That is, if p is not zero or a positive integer, the general solution is of the form

$$y = c_1 J_p(ix) + c_2 J_{-p}(ix),$$

whereas otherwise it may be taken in the form

$$y = c_1 J_n(ix) + c_2 Y_n(ix).$$

From Equation (69) we have

$$J_p(ix) = \sum_{k=0}^{\infty} \frac{(-1)^k i^{2k+p}(x/2)^{2k+p}}{k!(k+p)!} = i^p \sum_{k=0}^{\infty} \frac{(x/2)^{2k+p}}{k!(k+p)!}.$$

In place of using this function as a fundamental solution of (93), it is preferable to use the function $I_p(x) = i^{-p} J_p(ix)$,

$$I_p(x) = \sum_{k=0}^{\infty} \frac{(x/2)^{2k+p}}{k!(k+p)!},$$ (95)

since this function is *real* for real values of p. This function is known as the *modified Bessel function of the first kind, of order p*. The terms in the series representing $I_p(x)$ differ from those in the series for $J_p(x)$ only in that the terms are all positive in the I_p series, whereas they alternate in sign in the J_p series. Thus, if p is not zero or a positive integer, the general solution of (93) can be taken in the real form

$$y = Z_p(ix) = c_1 I_p(x) + c_2 I_{-p}(x).$$ (96)

As a second real fundamental solution of (93), in the case when $p = n$, where n is zero or a positive integer, it is rather conventional to define the function $K_n(x)$ by the equation†

$$K_n(x) = \frac{\pi}{2} i^{n+1} [J_n(ix) + i Y_n(ix)] = \frac{\pi}{2} i^{n+1} H_n^{(1)}(ix),$$ (97)

leading to the general solution

$$y = Z_n(ix) = c_1 I_n(x) + c_2 K_n(x),$$ (98)

when n is zero or a positive integer. The function $K_n(x)$ is known as the *modified Bessel function of the second kind, of order n*.

If p is not zero or a positive integer, the function $K_p(x)$ is defined by the equation

$$K_p(x) = \frac{\pi}{2} \frac{I_{-p}(x) - I_p(x)}{\sin p\pi},$$ (99)

which is consistent with (97) when $p \rightarrow n$.

For large values of x the modified functions have the asymptotic behavior

$$\left. \begin{array}{l} I_p(x) \sim \dfrac{e^x}{\sqrt{2\pi x}} \\[2ex] K_p(x) \sim \dfrac{e^{-x}}{\sqrt{\dfrac{2}{\pi} x}} \end{array} \right\} \quad (x \rightarrow \infty).$$ (100)

It is important to notice that the right members of (100) are independent of p.

4.9. Properties of Bessel Functions. It is readily verified directly that all *power series* involved in the definitions of all Bessel functions converge for all finite values of x. However, in consequence of the fact that these series are in

---

†Some references omit the factor π in (97) [and (99)], in order to equalize the two numerical factors in (100).

many cases multiplied by a negative power of x or by a logarithmic term, it is found that *only the functions $J_p(x)$ and $I_p(x)$ are finite at $x = 0$* (when $p \geq 0$).

For *small values of x*, retention of the leading terms in the respective series leads to the approximations

$$J_p(x) \sim \frac{1}{2^p p!} x^p, \quad J_{-p}(x) \sim \frac{2^p}{(-p)!} x^{-p} \quad (p \neq n), \tag{101}$$

$$Y_p(x) \sim -\frac{2^p(p-1)!}{\pi} x^{-p} \quad (p \neq 0), \quad Y_0(x) \sim \frac{2}{\pi} \log x, \tag{102}$$

$$I_p(x) \sim \frac{1}{2^p p!} x^p, \quad I_{-p}(x) \sim \frac{2^p}{(-p)!} x^{-p} \quad (p \neq n), \tag{103}$$

$$K_p(x) \sim 2^{p-1}(p-1)! \, x^{-p} \quad (p \neq 0), \quad K_0(x) \sim -\log x, \tag{104}$$

again with the usual implication that the ratio of two quantities connected by the symbol $\sim$ approaches unity as x tends to the relevant limit (here zero).

For *large values of x* $(x \longrightarrow \infty)$, we recapitulate the results listed in the preceding section:

$$J_p(x) \sim \sqrt{\frac{2}{\pi x}} \cos\left(x - \frac{\pi}{4} - \frac{p\pi}{2}\right), \quad Y_p(x) \sim \sqrt{\frac{2}{\pi x}} \sin\left(x - \frac{\pi}{4} - \frac{p\pi}{2}\right), \tag{105}$$

$$H_p^{(1)}(x) \sim \sqrt{\frac{2}{\pi x}} e^{i[x-(\pi/4)-(p\pi/2)]}, \quad H_p^{(2)}(x) \sim \sqrt{\frac{2}{\pi x}} e^{-i[x-(\pi/4)-(p\pi/2)]}, \tag{106}$$

$$I_p(x) \sim \frac{e^x}{\sqrt{2\pi x}}, \quad K_p(x) \sim \sqrt{\frac{\pi}{2x}} e^{-x}. \tag{107}$$

The following derivative formulas are of frequent use:

$$\frac{d}{dx}[x^p y_p(\alpha x)] = \begin{cases} \alpha x^p y_{p-1}(\alpha x) & (y = J, Y, I, H^{(1)}, H^{(2)}), \\ -\alpha x^p y_{p-1}(\alpha x) & (y = K); \end{cases} \tag{108}$$

$$\frac{d}{dx}[x^{-p} y_p(\alpha x)] = \begin{cases} -\alpha x^{-p} y_{p+1}(\alpha x) & (y = J, Y, K, H^{(1)}, H^{(2)}), \\ \alpha x^{-p} y_{p+1}(\alpha x) & (y = I). \end{cases} \tag{109}$$

These formulas are established for J_p and Y_p by considering their series definitions, and for the remaining functions by considering their definitions in terms of J_p and Y_p. Thus, to prove (108) for J_p, we note that from the definition (69) we have

$$\frac{d}{dx}[x^p J_p(\alpha x)] = \frac{d}{dx} \sum_{k=0}^{\infty} \frac{(-1)^k \alpha^{2k+p} x^{2k+2p}}{2^{2k+p} k!(k+p)!}$$

$$= \sum_{k=0}^{\infty} \frac{(-1)^k \alpha^{2k+p} x^{2k+2p-1}}{2^{2k+p-1} k!(k+p-1)!}$$

$$= \alpha x^p \sum_{k=0}^{\infty} \frac{(-1)^k (\alpha x/2)^{2k+p-1}}{k!(k+p-1)!}$$

$$= \alpha x^p J_{p-1}(\alpha x).$$

From (108) there follows

$$\frac{d}{dx}y_p(\alpha x) = \begin{cases} \alpha y_{p-1}(\alpha x) - \frac{p}{x}y_p(\alpha x) & (y = J, Y, I, H^{(1)}, H^{(2)}), \\[2mm] -\alpha y_{p-1}(\alpha x) - \frac{p}{x}y_p(\alpha x) & (y = K), \end{cases} \tag{110}$$

whereas (109) gives

$$\frac{d}{dx}y_p(\alpha x) = \begin{cases} -\alpha y_{p+1}(\alpha x) + \frac{p}{x}y_p(\alpha x) & (y = J, Y, K, H^{(1)}, H^{(2)}), \\[2mm] \alpha y_{p+1}(\alpha x) + \frac{p}{x}y_p(\alpha x) & (y = I). \end{cases} \tag{111}$$

By addition and subtraction of (110) and (111) we also obtain the relations

$$2\frac{d}{dx}y_p(\alpha x) = \alpha[y_{p-1}(\alpha x) - y_{p+1}(\alpha x)] \qquad (y = J, Y, H^{(1)}, H^{(2)}), \tag{112}$$

$$y_{p+1}(\alpha x) = \frac{2p}{\alpha x}y_p(\alpha x) - y_{p-1}(\alpha x) \qquad (y = J, Y, H^{(1)}, H^{(2)}), \tag{113}$$

$$2\frac{d}{dx}I_p(\alpha x) = \alpha[I_{p-1}(\alpha x) + I_{p+1}(\alpha x)], \tag{112a}$$

$$2\frac{d}{dx}K_p(\alpha x) = -\alpha[K_{p-1}(\alpha x) + K_{p+1}(\alpha x)], \tag{112b}$$

$$I_{p+1}(\alpha x) = -\frac{2p}{\alpha x}I_p(\alpha x) + I_{p-1}(\alpha x), \tag{113a}$$

$$K_{p+1}(\alpha x) = \frac{2p}{\alpha x}K_p(\alpha x) + K_{p-1}(\alpha x). \tag{113b}$$

Relations (108–112) are useful in evaluating certain integrals and derivatives involving Bessel functions. In particular, setting $p = 0$ in (109), we obtain the relations

$$\frac{d}{dx}y_0(\alpha x) = \begin{cases} -\alpha y_1(\alpha x) & (y = J, Y, K, H^{(1)}, H^{(2)}), \\[2mm] \alpha y_1(\alpha x) & (y = I). \end{cases} \tag{114}$$

The relations (113) are useful in expressing Bessel functions in terms of corresponding functions of lower order.

The Bessel functions of order p, where p is half an odd integer, can be expressed in closed form, in terms of elementary functions. To establish this fact, we consider first the case $p = \frac{1}{2}$ and denote the general solution of (63) by $y = Z_{1/2}(x)$. If we write

$$u(x) = \sqrt{x}Z_{1/2}(x),$$

Equation (87) shows that the function $u(x)$ satisfies the equation

$$\frac{d^2u}{dx^2} + u = 0$$

and hence is of the form $u = A \cos x + B \sin x$. Taking

$$u = \sqrt{x} J_{1/2}(x),$$

we have also, from (108),

$$\frac{du}{dx} = \sqrt{x} J_{-1/2}(x).$$

Hence, using (101), we find that

$$u(0) = 0 \quad \text{and} \quad u'(0) = \frac{2^{1/2}}{(-\frac{1}{2})!} = \frac{\sqrt{2}}{\Gamma(\frac{1}{2})} = \sqrt{\frac{2}{\pi}}.$$

Thus we must have $A = 0$, $B = \sqrt{2/\pi}$, and there follows

$$u = \sqrt{\frac{2}{\pi}} \sin x, \qquad \frac{du}{dx} = \sqrt{\frac{2}{\pi}} \cos x,$$

or, finally,

$$J_{1/2}(x) = \sqrt{\frac{2}{\pi x}} \sin x, \qquad J_{-1/2}(x) = \sqrt{\frac{2}{\pi x}} \cos x, \tag{115}$$

in accordance with (88), which here becomes an *equality*. Also, since $I_{1/2}(x) = i^{-1/2} J_{1/2}(ix)$ and $I_{-1/2}(x) = i^{1/2} J_{-1/2}(ix)$, we have

$$I_{1/2}(x) = \sqrt{\frac{2}{\pi x}} \sinh x, \qquad I_{-1/2}(x) = \sqrt{\frac{2}{\pi x}} \cosh x. \tag{116}$$

From (113), with $\alpha = 1$ and $p = n - \frac{1}{2}$, we obtain the recurrence formulas

$$J_{n+1/2}(x) = \frac{2n - 1}{x} J_{n-1/2}(x) - J_{n-3/2}(x) \tag{117}$$

and

$$I_{n+1/2}(x) = -\frac{2n - 1}{x} I_{n-1/2}(x) + I_{n-3/2}(x), \tag{118}$$

which permit the determination of $J_{n+1/2}(x)$ and $I_{n+1/2}(x)$ for all integral values of n, in terms of the functions in (115) and (116). Certain of those functions are listed in Table 2 of Chapter 2 (page 70).

As is indicated by the asymptotic approximations (105), the functions $J_p(x)$ and $Y_p(x)$ are oscillatory in nature, the amplitude of oscillation about a zero value tending to decrease with $\sqrt{2/\pi x}$ and the distance between successive zeros of the function decreasing toward π.† It can be shown that the zeros of $J_p(x)$ separate the zeros of $J_{p+1}(x)$; that is, between any two consecutive zeros of $J_{p+1}(x)$ there is one and only one zero of $J_p(x)$. (See Problem 36.) The same applies to the zeros of the functions of the second kind. The functions I_p and K_p are not oscillatory. It is found that the former function essentially increases exponentially with x, whereas the second essentially decreases exponentially. Sketches of these functions are presented in Figure 4.1.

†A brief table of values of zeros of certain Bessel functions is presented in Section 5.13.

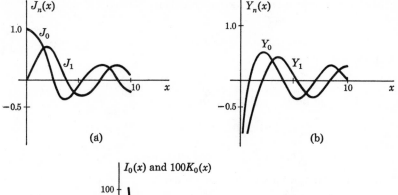

Figure 4.1

4.10. Differential Equations Satisfied by Bessel Functions.

The solution of the differential equation

$$X^2 \frac{d^2Y}{dX^2} + X\frac{dY}{dX} + (X^2 - p^2)Y = 0 \tag{119}$$

can be written in the form

$$Y = Z_p(X). \tag{120}$$

If we make the substitutions

$$Y = \frac{y}{g(x)}, \qquad X = f(x) \tag{121}$$

and notice that then

$$\frac{d}{dX} = \frac{1}{f'(x)}\frac{d}{dx}, \tag{122}$$

Equation (119) becomes

$$f^2\left\{\frac{1}{f'}\frac{d}{dx}\left[\frac{1}{f'}\frac{d}{dx}\left(\frac{y}{g}\right)\right]\right\} + f\left[\frac{1}{f'}\frac{d}{dx}\left(\frac{y}{g}\right)\right] + (f^2 - p^2)\frac{y}{g} = 0,$$

or

$$f\frac{d}{dx}\left[\frac{\dfrac{d}{dx}\left(\dfrac{y}{g}\right)}{f'}\right] + \frac{d}{dx}\left(\frac{y}{g}\right) + \frac{f'}{f}(f^2 - p^2)\frac{y}{g} = 0. \tag{123}$$

Reference to (120) and (121) shows that (123) is the differential equation satisfied by

$$y = g(x)Z_p[f(x)].\tag{124}$$

In particular, if we set

$$g(x) = x^A e^{-Bx^r}, \qquad f(x) = Cx^s,\tag{125}$$

Equation (123) can be reduced to the form

$$x^2 \frac{d^2y}{dx^2} + x[(1 - 2A) + 2rBx^r]\frac{dy}{dx}$$

$$+ [A^2 - p^2s^2 + s^2C^2x^{2s} - rB(2A - r)x^r + r^2B^2x^{2r}]y = 0.$$

This equation is somewhat simplified if we write

$$1 - 2A = a, \qquad rB = b, \qquad A^2 - p^2s^2 = c, \qquad s^2C^2 = d,$$

from which there follows

$$A = \frac{1 - a}{2}, \quad B = \frac{b}{r}, \quad C = \frac{\sqrt{d}}{s}, \quad p = \frac{1}{s}\sqrt{A^2 - c}.\tag{126}$$

With this notation, the differential equation takes the form

$$x^2 \frac{d^2y}{dx^2} + x(a + 2bx^r)\frac{dy}{dx} + [c + dx^{2s} - b(1 - a - r)x^r + b^2x^{2r}]y = 0,\tag{127}$$

and the solution of (127), obtained by introducing (125) and (126) into (124), is of the form

$$y = x^{\frac{1-a}{2}} e^{-\frac{bx^r}{r}} Z_p\left(\frac{\sqrt{d}}{s} x^s\right),\tag{128}$$

where

$$p = \frac{1}{s}\sqrt{\left(\frac{1 - a}{2}\right)^2 - c}.\tag{129}$$

If $\sqrt{d}/s$ is real, Z_p is to be interpreted by (74) or (84), whereas if $\sqrt{d}/s$ is imaginary, Z_p stands for (96) or (98), the choice in either case depending upon whether p *is not* or *is* zero or a positive integer, respectively.†

Thus, if it is possible to identify a particular differential equation with (127), by suitably choosing the constants in (127), the solution is given immediately by (128) in terms of Bessel functions of order p, where p is given by (129). We list here certain useful special forms with corresponding solutions readily obtainable in this way.

$$x^2y'' + xy' + (k^2x^2 - p^2)y = 0, \qquad y = Z_p(kx) \qquad (k \neq 0).\tag{130a}$$

$$x^2y'' + xy' - (k^2x^2 + p^2)y = 0, \qquad y = Z_p(ikx) \qquad (k \neq 0).\tag{130b}$$

†Because of the ambiguity of the signs of the radicals in (128) and (129), both the order p and the coefficient of x^s can be replaced by their negatives, and hence can be taken to be *nonnegative* when they are real, in (128) and in its specializations (130a–e).

$$x^2 y'' + xy' - (\beta^2 - \alpha x^{2s})y = 0, \qquad y = Z_{\beta/s}\left(\frac{\sqrt{\alpha}}{s}x^s\right) \qquad (\alpha s \neq 0). \qquad (130c)$$

$$y'' + kx^m y = 0, \qquad y = \sqrt{x}\, Z_{\frac{1}{m+2}}\left(\frac{2\sqrt{k}}{m+2}x^{\frac{m+2}{2}}\right) \qquad (mk \neq -2k). \qquad (130d)$$

$$\frac{d}{dx}\left(x^n \frac{dy}{dx}\right) + kx^m y = 0, \qquad y = x^{\frac{1-n}{2}} Z_p\left(\frac{\sqrt{k}}{s}x^s\right)$$

where
$$s = \frac{m-n}{2} + 1, \qquad p = \frac{1-n}{2s} \qquad (ks \neq 0). \qquad (130e)$$

The restricted cases in (130a–e) are all readily solvable as equidimensional equations (see Section 1.6).

4.11. Ber and Bei Functions. In certain problems it is convenient to obtain a desired solution as the real or imaginary part of a complex function, in terms of which a simplified formulation of the problem is possible. In such cases differential equations with complex coefficients may occur. In particular, equations are rather frequently obtained which are reducible to the form

$$x^2 \frac{d^2 y}{dx^2} + x \frac{dy}{dx} - (p^2 + ix^2)y = 0, \qquad (131)$$

where p is a real constant. Since (131) is equivalent to (130b) if $k^2 = i$ or $k = i^{1/2}$, the solution of (131) can be expressed in the form

$$y = Z_p(i^{3/2}x). \qquad (132)$$

The solutions of (131) which are finite at $x = 0$ are thus multiples of the function $J_p(i^{3/2}x)$. This function is a complex function whose real and imaginary parts are written $\text{ber}_p x$ and $\text{bei}_p x$, respectively,

$$\text{ber}_p x + i\, \text{bei}_p x = J_p(i^{3/2}x) = i^p I_p(i^{1/2}x). \qquad (133)$$

The *polar notation*

$$\text{ber}_p x + i\, \text{bei}_p x = M_p(x)e^{i\theta_p(x)} \qquad (134)$$

is also used.

We consider, in illustration, the case $p = 0$. In this case the zero subscripts are conventionally omitted in the notation for the real and imaginary parts,

$$J_0(i^{3/2}x) = \text{ber}\, x + i\, \text{bei}\, x. \qquad (135)$$

From the definition (69) we obtain, with $p = 0$,

$$J_0(i^{3/2}x) = \sum_{m=0}^{\infty} \frac{(-1)^m i^{3m}(x/2)^{2m}}{(m\,!)^2},$$

independently of the choice of the two possible interpretations of $i^{3/2}$. We now separate this series into two parts, in the first of which m takes on even values and in the second of which only odd values of m occur. If we replace m by $2k$

in the first series and m by $2k + 1$ in the second, we obtain

$$J_0(i^{3/2}x) = \sum_{k=0}^{\infty} \frac{i^{6k}(x/2)^{4k}}{[(2k)!]^2} - \sum_{k=0}^{\infty} \frac{i^{6k+3}(x/2)^{4k+2}}{[(2k+1)!]^2}.$$

Noticing that

$$i^{6k} = (-1)^{3k} = (-1)^k, \qquad i^{6k+3} = (-1)^k i^3 = (-1)^{k+1}i,$$

we obtain finally

$$J_0(i^{3/2}x) = \sum_{k=0}^{\infty} (-1)^k \frac{(x/2)^{4k}}{[(2k)!]^2} + i \sum_{k=0}^{\infty} (-1)^k \frac{(x/2)^{4k+2}}{[(2k+1)!]^2}. \tag{136}$$

A comparison of (135) and (136) gives the results

$$\text{ber } x = \sum_{k=0}^{\infty} (-1)^k \frac{(x/2)^{4k}}{[(2k)!]^2}$$

$$= 1 - \frac{x^4}{2^2 4^2} + \frac{x^8}{2^2 4^2 6^2 8^2} - \cdots \tag{137}$$

and

$$\text{bei } x = \sum_{k=0}^{\infty} (-1)^k \frac{(x/2)^{4k+2}}{[(2k+1)!]^2}$$

$$= \frac{x^2}{2^2} - \frac{x^6}{2^2 4^2 6^2} + \frac{x^{10}}{2^2 4^2 6^2 8^2 10^2} - \cdots. \tag{138}$$

Analogous series expansions defining the functions $\text{ber}_p x$ and $\text{bei}_p x$ for general values of p are obtained by similar methods.

Similar functions of the *second kind*, which are *not finite* at $x = 0$, are defined by the relation

$$\text{ker}_p x + i \, \text{kei}_p x = i^{-p} K_p(i^{1/2}x). \tag{139}$$

The general solution of (131) then can be written in the form

$$y = (c_1 \, \text{ber}_p x + c_2 \, \text{ker}_p x) + i(c_1 \, \text{bei}_p x + c_2 \, \text{kei}_p x). \tag{140}$$

To illustrate the occurrence of Bessel functions of order zero in practice, we consider particular solutions of the *partial differential equations*

$$\frac{\partial^2 U}{\partial x^2} + \frac{1}{x} \frac{\partial U}{\partial x} = \lambda \frac{\partial^2 U}{\partial t^2} \tag{141}$$

and

$$\frac{\partial^2 U}{\partial x^2} + \frac{1}{x} \frac{\partial U}{\partial x} = \mu \frac{\partial U}{\partial t}. \tag{142}$$

It is found that in many physical problems in various fields, a physical property U depending on a single distance variable x measured from a reference axis of symmetry, and on a time variable t, must satisfy one or the other of these equations, the quantities λ and μ involving known physical constants independent of time and position. It is frequently important to determine solutions of the form

$$f(x) \sin \omega t + g(x) \cos \omega t,$$

where ω is a constant. However, instead of proceeding directly to such a determination, it is more convenient to consider the required solution as *the real or imaginary part* of a *complex solution* of the form

$$U = y(x)e^{i\omega t}, \tag{143}$$

where $y(x)$ is a complex function of the form $y(x) = F(x) + iG(x)$.

Substitution of (143) into (141) and subsequent cancellation of the factor $e^{i\omega t}$ show that the function $y(x)$ must satisfy the differential equation

$$x\frac{d^2y}{dx^2} + \frac{dy}{dx} + \lambda\omega^2 xy = 0. \tag{144}$$

If $\lambda > 0$, the general solution of (144) is given by (130a) in the form $y = Z_0(\sqrt{\lambda}\omega x)$, whereas if $\lambda < 0$, the general solution is of the form $y = Z_0(i\sqrt{-\lambda}\omega x)$. In particular, if physical considerations require the solution to be finite when $x = 0$, the solutions must be of the form $CJ_0(\sqrt{\lambda}\omega x)$ when $\lambda > 0$ and $CI_0 (\sqrt{-\lambda}\omega x)$ when $\lambda < 0$, where in either case C is a constant (which may be complex).

In a similar way, substitution of (143) into (142) gives the equation

$$x\frac{d^2y}{dx^2} + \frac{dy}{dx} - i\mu\omega xy = 0 \tag{145}$$

to be satisfied by the complex amplitude function $y(x)$. If $\mu > 0$, comparison of (145) and (131) shows that the most general solution of (145) which is finite when $x = 0$ is of the form

$$y = CJ_0(i^{3/2}\sqrt{\mu\omega}\, x) = C(\text{ber }\sqrt{\mu\omega}\, x + i\,\text{bei }\sqrt{\mu\omega}\, x). \tag{146}$$

If $\mu < 0$, the corresponding solution is of the form

$$y = CJ_0(i^{1/2}\sqrt{-\mu\omega}\, x)$$

or, equivalently,

$$y = C(\text{ber }\sqrt{-\mu\omega}\, x - i\,\text{bei }\sqrt{-\mu\omega}\, x), \tag{147}$$

as may be seen by taking the complex conjugate of the equal members of (145).

Useful tables of Bessel functions are included in *Tables of Functions*, complied by Jahnke, Emde, and Lösch (Reference 8), the notation $N_p(x)$ being used in place of $Y_p(x)$. In addition, the real and imaginary parts of the functions $J_0(i^{1/2}x)$, $H_0^{(1)}(i^{1/2}x)$, and $i^{1/2}J_1(i^{1/2}x)$, $i^{1/2}H_1^{(1)}(i^{1/2}x)$ are tabulated therein. The first two functions are independent solutions of the equation

$$x^2\frac{d^2y}{dx^2} + x\frac{dy}{dx} - (p^2 - ix^2)y = 0 \tag{148}$$

when $p = 0$, and the last two functions are solutions when $p = 1$. Supplementary material, including tables of zeros of $J_p(x)$, is also included. Brief tables of the functions appearing in (140) are included in Dwight's *Tables* (Reference 3).

4.12. Legendre Functions.
Solutions of the differential equation

$$(1 - x^2)\frac{d^2y}{dx^2} - 2x\frac{dy}{dx} + p(p + 1)y = 0 \tag{149}$$

or, equivalently,

$$\frac{d}{dx}\left[(1 - x^2)\frac{dy}{dx}\right] + p(p + 1)y = 0 \tag{149a}$$

are known as *Legendre functions of order p*, where p is assumed here to be real and nonnegative.† They are of particular use in the solution of potential problems involving spherical boundaries, when rotational symmetry is present.

We may notice that $x = 0$ is an *ordinary* point for (149), so that one could assume $y = \sum A_k x^k$, with the knowledge that both A_0 and A_1 will be arbitrary. However, since (149) also is of type (52), with $M = 2$, it is somewhat more convenient to use the method of Section 4.7. Thus the introduction of

$$y(x) = \sum_{k=0}^{\infty} B_k x^{2k+s} \tag{150}$$

into Equation (149) readily yields the exponents

$$s_1 = 1, \qquad s_2 = 0 \tag{151}$$

(as was anticipated) and the recurrence formula

$$(s + 2k)(s + 2k - 1)B_k = -(p - s - 2k + 2)(p + s + 2k - 1)B_{k-1}. \tag{152}$$

With $s = 0$, this formula yields the result

$$B_k(0) = (-1)^k \frac{[p(p - 2)(p - 4) \cdots (p - 2k + 2)]}{[1 \cdot 3 \cdot 5 \cdots (2k - 1)]}$$

$$\times \frac{[(p + 1)(p + 3)(p + 5) \cdots (p + 2k - 1)]}{[2 \cdot 4 \cdot 6 \cdots (2k)]} B_0,$$

or

$$B_k(0) = \frac{(-1)^k B_0}{(2k)!}[p(p - 2) \cdots (p - 2k + 2)]$$

$$\times [(p + 1)(p + 3) \cdots (p + 2k - 1)]. \tag{153}$$

Similarly, when $s = 1$, Equation (152) gives

$$B_k(1) = \frac{(-1)^k B_0}{(2k + 1)!}[(p - 1)(p - 3) \cdots (p - 2k + 1)]$$

$$\times [(p + 2)(p + 4) \cdots (p + 2k)]. \tag{154}$$

The solutions corresponding to the exponents $s = 0$ and $s = 1$ then are of the respective forms

$$B_0 + \sum_{k=1}^{\infty} B_k(0)x^{2k}, \qquad B_0 x + \sum_{k=1}^{\infty} B_k(1)x^{2k+1}.$$

---

†Since $p(p + 1)$ is unchanged when p is replaced by $-(p + 1)$, the solutions for $p = -p_0$ are the same as those for $p = p_0 - 1$.

The coefficients of B_0 in these expressions are here denoted by $u_p(x)$ and $v_p(x)$, respectively, so that we write

$$u_p(x) = 1 - \frac{p(p+1)}{2!}x^2 + \frac{p(p-2)(p+1)(p+3)}{4!}x^4$$

$$- \frac{p(p-2)(p-4)(p+1)(p+3)(p+5)}{6!}x^6 + \cdots \quad (155)$$

and

$$v_p(x) = x - \frac{(p-1)(p+2)}{3!}x^3 + \frac{(p-1)(p-3)(p+2)(p+4)}{5!}x^5$$

$$- \frac{(p-1)(p-3)(p-5)(p+2)(p+4)(p+6)}{7!}x^7 + \cdots. \quad (156)$$

It may be seen that, if p is an *even positive integer n* (or zero) the series (155) terminates with the term involving x^n, and hence is a polynomial of degree n. Similarly, if p is an *odd positive integer n*, the series (156) terminates with the term involving x^n. Otherwise, the expressions are infinite series. The results of Section 4.4 show that the series converge when $-1 < x < 1$; they diverge otherwise (unless they terminate), as can be verified directly.

Thus the general solution of Equation (149) could be expressed in the form

$$y = c_1 u_p(x) + c_2 v_p(x)$$

when $-1 < x < 1$. However, a different terminology is conventional, for reasons which are now to be explained.

We consider first the cases when $p = n$, where n is a positive integer or zero. These are the cases commonly arising in practice. When $p = n$, one of the solutions (155) or (156) is a *polynomial* of degree n, whereas the other is an infinite series. That multiple of the polynomial of degree n which has the value unity when $x = 1$ is called the nth *Legendre polynomial* and is denoted by $P_n(x)$.

Thus we have, *when n is even*,

$$P_n(x) = \frac{u_n(x)}{u_n(1)} \quad (157a)$$

and, *when n is odd*,

$$P_n(x) = \frac{v_n(x)}{v_n(1)}. \quad (157b)$$

The first six Legendre polynomials are readily found to be

$$P_0(x) = 1, \quad P_1(x) = x, \quad P_2(x) = \tfrac{1}{2}(3x^2 - 1),$$

$$P_3(x) = \tfrac{1}{2}(5x^3 - 3x), \quad P_4(x) = \tfrac{1}{8}(35x^4 - 30x^2 + 3), \quad (158)$$

$$P_5(x) = \tfrac{1}{8}(63x^5 - 70x^3 + 15x).$$

For later reference, it is noted that the functions $u_n(x)$, with n even, and $v_n(x)$, with n odd, can be shown to have the following values at $x = 1$:

$$u_0(1) = 1, \qquad u_n(1) = (-1)^{n/2} \frac{2 \cdot 4 \cdot 6 \cdots n}{1 \cdot 3 \cdot 5 \cdots (n-1)} \quad (n = 2, 4, 6, \ldots),$$

$$\tag{159a}$$

$$v_1(1) = 1, \qquad v_n(1) = (-1)^{(n-1)/2} \frac{2 \cdot 4 \cdot 6 \cdots (n-1)}{1 \cdot 3 \cdot 5 \cdots n} \quad (n = 3, 5, 7, \ldots).$$

$$\tag{159b}$$

When p is an even integer n, the solution $v_n(x)$ is in the form of an infinite series, whereas if p is an odd integer n, the solution $u_n(x)$ is an infinite series. Suitable multiples of these solutions are called *Legendre functions of the second kind* and are denoted by $Q_n(x)$. It is conventional to take the multiplicative factors as $(-1)^n u_n(1)$ and $(-1)^n v_n(1)$, respectively, leading to the definition

$$Q_n(x) = \begin{cases} -v_n(1)u_n(x) & (n \text{ odd}), \\ u_n(1)v_n(x) & (n \text{ even}), \end{cases} \tag{160}$$

where the constants $u_n(1)$, n even, and $v_n(1)$, n odd, are defined by (159a, b). However, since the series appearing in (160) converge only when $|x| < 1$, the functions $Q_n(x)$ are defined by (160) only inside this interval.

Thus, when p is an integer n, a certain multiple of that solution (155) or (156) which is a polynomial is written as $P_n(x)$, and a certain multiple of the other (infinite series) solution is written as $Q_n(x)$, so that the general solution of (149) in this case is written in the form

$$y = c_1 P_n(x) + c_2 Q_n(x). \tag{161}$$

It can be shown (see Problem 64) that $P_n(x)$ and $Q_n(x)$ both satisfy the *recurrence formula*

$$ny_n(x) = (2n-1)xy_{n-1}(x) - (n-1)y_{n-2}(x). \tag{162}$$

Equation (162) permits the determination of expressions for Legendre functions in terms of corresponding functions of lower degree.

We next express $Q_0(x)$ and $Q_1(x)$ in closed form, by using the methods of Section 1.10. Reference to Problem 42 of Chapter 1 shows that the functions $Q_n(x)$ are expressible in the form

$$Q_n(x) = A_n P_n(x) \int \frac{dx}{(1-x^2)[P_n(x)]^2} + B_n P_n(x),$$

where A_n and B_n are suitably chosen constants. In particular, since $P_0(x) = 1$, there follows

$$Q_0(x) = A_0 \int \frac{dx}{1-x^2} + B_0,$$

or, if $|x| < 1$,

$$Q_0(x) = \frac{A_0}{2} \log \frac{1+x}{1-x} + B_0.$$

From (160) we obtain $Q_0(0) = u_0(1)v_0(0) = 0$ and $Q_0'(0) = u_0(1)v_0'(0) = 1$. Hence there follows $A_0 = 1$, $B_0 = 0$, and we have the result

$$Q_0(x) = \frac{1}{2} \log \frac{1 + x}{1 - x} = \tanh^{-1} x. \tag{163}$$

Similarly, when $n = 1$, since $P_1(x) = x$ there follows

$$Q_1(x) = A_1 x \int \frac{dx}{x^2(1 - x^2)} + B_1 x$$

or, if $|x| < 1$,

$$Q_1(x) = A_1 \left(\frac{x}{2} \log \frac{1 + x}{1 - x} - 1 \right) + B_1 x.$$

From (160) we have

$$Q_1(0) = -v_1(1)u_1(0) = -1 \quad \text{and} \quad Q_1'(0) = -v_1(1)u_1'(0) = 0.$$

Hence we must take $A_1 = 1$ and $B_1 = 0$, and so obtain

$$Q_1(x) = \frac{x}{2} \log \frac{1 + x}{1 - x} - 1 = xQ_0(x) - 1. \tag{164}$$

The series expansions of $Q_0(x)$ and $Q_1(x)$, in powers of x, are readily shown to be in agreement with the series indicated in (160).

Use of the recurrence formula (162) now permits the determination of $Q_n(x)$ for any positive integral value of n. In this way one obtains, in particular, the expressions

$$Q_2(x) = P_2(x)Q_0(x) - \frac{3}{2}x,$$

$$Q_3(x) = P_3(x)Q_0(x) - \frac{5}{2}x^2 + \frac{2}{3},$$

$$Q_4(x) = P_4(x)Q_0(x) - \frac{35}{8}x^3 + \frac{55}{24}x, \tag{165}$$

$$Q_5(x) = P_5(x)Q_0(x) - \frac{63}{8}x^4 + \frac{49}{8}x^2 - \frac{8}{15}.$$

For values of x such that $|x| > 1$, the integral $\int dx/(1 - x^2)$ takes the form

$$\frac{1}{2} \log \frac{x + 1}{x - 1} + C = \coth^{-1} x + C.$$

Thus, if $|x| > 1$, the function

$$Q_0(x) = \frac{1}{2} \log \frac{x + 1}{x - 1} = \coth^{-1} x \tag{166}$$

is a solution of (149) which complements the polynomial solution $P_0(x) = 1$. Corresponding solutions $Q_n(x)$ for integral values of n, *in the range* $|x| > 1$, are obtained by using the notation of (166) [in place of (163)] in Equations (164) and (165).

If p is *not* an integer, a certain combination of the series (155) and (156) can be determined so as to remain finite and take on the value unity at $x = 1$

(see Problem 65). This function is called $P_p(x)$, and a second independent combination is denoted by $Q_p(x)$, so that the general solution of (149) in the general case is written in the form

$$y = c_1 P_p(x) + c_2 Q_p(x). \tag{167}$$

The function $P_p(x)$ so defined, however, will not also remain finite at the point $x = -1$ unless p is integral, and the function $Q_p(x)$ cannot be finite at $x = 1$. Thus *the only Legendre functions which are finite at both $x = 1$ and $x = -1$ are the Legendre polynomials $P_p(x)$, for which p is integral.* (This fact will be of importance in Section 5.14.)

Rodrigues' formula, which expresses $P_n(x)$ in the alternative form

$$P_n(x) = \frac{1}{2^n n!} \frac{d^n (x^2 - 1)^n}{dx^n}, \tag{168}$$

is particularly useful in dealing with certain integrals involving Legendre polynomials. Proof that (168) is indeed consistent with (157) for all positive integral values of n is omitted here (see Problem 60), but it is readily verified that (168) reduces, in the special cases of (158), to the forms given.

From (168) it can be deduced that *all the zeros of $P_n(x)$ are real and unrepeated, and lie in the interval $-1 < x < 1$* (see Problem 61). Another important fact (Problem 62) is that, in the interval $-1 \leq x \leq 1$, the magnitude of each Legendre polynomial is maximum at the end points, so that

$$|P_n(x)| \leq 1 \quad when \quad |x| \leq 1,$$

Outside the interval $(-1, 1)$, each polynomial $P_n(x)$ increases or decreases steadily, without maxima or minima or turning points (see Figure 4.2).

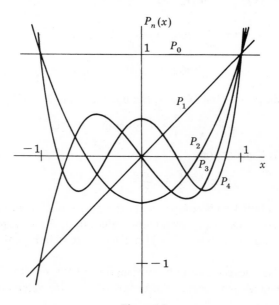

Figure 4.2

If $|x| < 1$, the substitution $x = \cos \varphi$ transforms Legendre's equation from the form (149a) into the form

$$\frac{1}{\sin \varphi} \frac{d}{d\varphi}\left(\sin \varphi \frac{dy}{d\varphi}\right) + n(n+1)y = 0 \tag{169}$$

or, equivalently,

$$\frac{d^2y}{d\varphi^2} + \frac{dy}{d\varphi} \cot \varphi + n(n+1)y = 0, \tag{169a}$$

when $p = n$, and hence (169) has the general solution

$$y = c_1 P_n(\cos \varphi) + c_2 Q_n(\cos \varphi). \tag{170}$$

Equations of such a form frequently arise when spherical coordinates are introduced in the solution of a potential problem with rotational symmetry. We note that $Q_n(\cos \varphi)$ is not finite when $\cos \varphi = \pm 1$, that is, when $\varphi = k\pi$, whereas $P_n(\cos \varphi)$ is merely a polynomial of degree n in $\cos \varphi$. In particular, we have the expressions

$$P_0(\cos \varphi) = 1, \qquad P_1(\cos \varphi) = \cos \varphi,$$
$$P_2(\cos \varphi) = \tfrac{1}{2}(3 \cos^2 \varphi - 1) = \tfrac{1}{4}(3 \cos 2\varphi + 1), \tag{171}$$
$$P_3(\cos \varphi) = \tfrac{1}{2}(5 \cos^3 \varphi - 3 \cos \varphi) = \tfrac{1}{8}(5 \cos 3\varphi + 3 \cos \varphi).$$

When $|x| < 1$, the functions

$$P_n^m(x) = (1 - x^2)^{m/2} \frac{d^m P_n(x)}{dx^m}, \qquad Q_n^m(x) = (1 - x^2)^{m/2} \frac{d^m Q_n(x)}{dx^m} \tag{172}$$

are called the *associated Legendre functions of degree n and order m, of the first and second kinds*, respectively. They can be shown to satisfy the differential equation

$$(1 - x^2)\frac{d^2y}{dx^2} - 2x\frac{dy}{dx} + \left[n(n+1) - \frac{m^2}{1 - x^2}\right]y = 0, \tag{173}$$

which differs from (149) only in the presence of the term involving m, and properly reduces to (149) when $m = 0$. When $|x| > 1$, the definitions (172) are modified by replacing $1 - x^2$ by $x^2 - 1$.

For any nonnegative integral value of m, the equation (173) *possesses a nontrivial solution which is finite at both $x = 1$ and $x = -1$ only when n is an integer, and that solution is a multiple of $P_n^m(x)$, with $n \geq m$.*

The substitution $x = \cos \varphi$ transforms (173) into the equation

$$\frac{d^2y}{d\varphi^2} + \frac{dy}{d\varphi} \cot \varphi + [n(n+1) - m^2 \csc^2 \varphi]y = 0, \tag{174}$$

which thus is satisfied by $P_n^m(\cos \varphi)$ and $Q_n^m(\cos \varphi)$. These functions are often useful in the solution of potential problems involving spherical boundaries, in the absence of rotational symmetry.

4.13. The Hypergeometric Function. Solutions of the differential equation

$$x(1 - x)\frac{d^2y}{dx^2} + [\gamma - (\alpha + \beta + 1)x]\frac{dy}{dx} - \alpha\beta y = 0 \qquad (175)$$

are generally called *hypergeometric functions*, since their series representations are, in a sense, generalizations of the elementary geometric series. Since (175) is of type (52), with $M = 1$, the series (54) reduces here to the usual form

$$y(x) = \sum_{k=0}^{\infty} B_k x^{k+s}. \qquad (176)$$

The exponents are found to be

$$s = 0, 1 - \gamma, \qquad (177)$$

so that only one solution of the assumed form can be expected when γ is integral. The recurrence formula is obtained in the form

$$(s + k)(s + k + \gamma - 1)B_k = (s + k + \alpha - 1)(s + k + \beta - 1)B_{k-1} \quad (k \geq 1), \qquad (178)$$

from which there follows

$$B_k(s) = \frac{[(s + \alpha)(s + \alpha + 1) \cdots (s + \alpha + k - 1)]}{[(s + 1)(s + 2) \cdots (s + k)]}$$

$$\times \frac{[(s + \beta)(s + \beta + 1) \cdots (s + \beta + k - 1)]}{[(s + \gamma)(s + \gamma + 1) \cdots (s + \gamma + k - 1)]}B_0 \qquad (k \geq 1). \qquad (179)$$

Corresponding to the exponent $s = 0$, we thus obtain the solution

$$y = B_0\left\{1 + \sum_{k=1}^{\infty} \frac{[\alpha(\alpha + 1) \cdots (\alpha + k - 1)][\beta(\beta + 1) \cdots (\beta + k - 1)]}{[1 \cdot 2 \cdots k][\gamma(\gamma + 1) \cdots (\gamma + k - 1)]}x^k\right\}. \qquad (180)$$

The coefficient of B_0 in (180) is written as $F(\alpha, \beta; \gamma; x)$, and the series is known as the *hypergeometric series* or *function*,

$$F(\alpha, \beta; \gamma; x) = 1 + \frac{\alpha \cdot \beta}{1 \cdot \gamma}x + \frac{\alpha(\alpha + 1)\beta(\beta + 1)}{1 \cdot 2 \cdot \gamma(\gamma + 1)}x^2 + \cdots. \qquad (181)$$

It is found that the series (181) converges in the interval $|x| < 1$ and also that, when $x = +1$, the series converges only if $\gamma - \alpha - \beta > 0$ and, when $x = -1$, the series converges only if $\gamma - \alpha - \beta + 1 > 0$.

It may be noticed that if $\alpha = 1$ and $\beta = \gamma$, the series becomes the elementary geometric series

$$F(1, \beta; \beta; x) = 1 + x + x^2 + \cdots + x^n + \cdots = \frac{1}{1 - x} \qquad (|x| < 1). \qquad (182)$$

It is seen also that, because of the symmetry in α and β, these parameters are interchangeable,

$$F(\alpha, \beta; \gamma; x) = F(\beta, \alpha; \gamma; x). \qquad (183)$$

The solution (181) does not exist (in general) when γ is zero or a negative integer.

Corresponding to the exponent $s = 1 - \gamma$, we obtain the solution

$$y = B_0 x^{1-\gamma}\left\{1 + \sum_{k=1}^{\infty} \frac{[(\alpha - \gamma + 1)(\alpha - \gamma + 2) \cdots (\alpha - \gamma + k)]}{[1 \cdot 2 \cdots k]}\right.$$

$$\left. \times \frac{[(\beta - \gamma + 1)(\beta - \gamma + 2) \cdots (\beta - \gamma + k)]}{[(2 - \gamma)(3 - \gamma) \cdots (k + 1 - \gamma)]} x^k\right\}. \qquad (184)$$

The series in braces in (184) is seen to differ from that in (180) only in that α, β, and γ in (180) are replaced by $(\alpha - \gamma + 1)$, $(\beta - \gamma + 1)$, and $(2 - \gamma)$, respectively, in (184). Hence (184) can be written in the form

$$y = B_0 x^{1-\gamma} F(\alpha - \gamma + 1, \beta - \gamma + 1; 2 - \gamma; x). \qquad (185)$$

The solution (185) does not exist (in general) when γ is a positive integer greater than unity. When $\gamma = 1$, the solution (185) becomes identical with (181).

Thus, *if γ is not zero or an integer*, the general solution of (175) can be expressed in the form

$$y = c_1 F(\alpha, \beta; \gamma; x) + c_2 x^{1-\gamma} F(\alpha - \gamma + 1, \beta - \gamma + 1; 2 - \gamma; x), \qquad (186)$$

when $|x| < 1$. The exceptional cases can be treated by the methods of Section 4.5.

Many elementary functions are expressible in terms of the hypergeometric function (181), including the following examples:

$$(1 - x)^{-\alpha} = F(\alpha, \beta; \beta; x),$$

$$(1 + x)^{-k} + (1 - x)^{-k} = 2F\left(\frac{k}{2}, \frac{k+1}{2}; \frac{1}{2}; x^2\right),$$

$$(1 + \sqrt{1 - x})^{-k} = 2^{-k}F\left(\frac{k}{2}, \frac{k+1}{2}; k + 1; x\right),$$

$$\log(1 - x) = xF(1, 1; 2; x),$$

$$\log \frac{1 + x}{1 - x} = 2xF\left(\frac{1}{2}, 1; \frac{3}{2}; x^2\right).$$

4.14. Series Solutions Valid for Large Values of x. In the preceding sections we have considered series solutions valid in an interval centered at the point $x = 0$, and have noticed that if solutions valid near a point $x = x_0$ were desired, such solutions could be conveniently obtained by first replacing $x - x_0$ by a new independent variable t and then seeking series solutions of the form

$$\sum A_n t^{n+s} = \sum A_n(x - x_0)^{n+s}$$

from the new equation. In such cases, the point $x = x_0$ naturally should be not worse than a regular singular point.

Thus, if series solutions of Bessel's equation of order zero in powers of $x - 1$ were required, we could set $t = x - 1$ and thus transform that equation to the form

$$(1 + t)\frac{d^2y}{dt^2} + \frac{dy}{dt} + (1 + t)y = 0.$$

Since the point $t = 0$ is an *ordinary point*, two solutions of the form $\sum A_n t^n$ can be obtained and rewritten finally in the desired form $\sum A_n (x - 1)^n$.

In order to investigate the behavior of solutions for *large* values of x, we are led to the possibility of replacing $1/x$ by a new independent variable t, and then of studying the behavior of solutions of the new equation for *small* values of t, since $t \to 0$ as $|x| \to \infty$. If the new equation has the point $t = 0$ as an *ordinary* point, we obtain two solutions of the form $\sum A_n t^n = \sum A_n x^{-n}$, whereas for a regular singular point at least one solution $\sum A_n t^{n+s} = \sum A_n x^{-n-s}$ is obtained.

With the substitution $x = 1/t$, the equation

$$\frac{d^2y}{dx^2} + a_1(x)\frac{dy}{dx} + a_2(x)y = 0 \tag{187}$$

becomes

$$\frac{d^2y}{dt^2} + \frac{1}{t^2}\left[2t - a_1\left(\frac{1}{t}\right)\right]\frac{dy}{dt} + \frac{1}{t^4}a_2\left(\frac{1}{t}\right)y = 0. \tag{188}$$

If the point $t = 0$ is an ordinary point (or a singular point) of (188), it is conventional to say that "the point $x = \infty$ is an ordinary point (or a singular point) of (187)." The use of such a phrase is motivated by the fact that, if $t = 0$ is an ordinary point (or *regular* singular point) of (188), then (187) possesses solutions of the form $\sum A_n x^{-n}$ (or $\sum A_n x^{-n-s}$).

Equation (188) shows that in order that $x = \infty$ be an *ordinary point* of (187) the functions

$$\frac{1}{t^2}\left[2t - a_1\left(\frac{1}{t}\right)\right] \quad \text{and} \quad \frac{1}{t^4}a_2\left(\frac{1}{t}\right) \tag{189}$$

must be regular at $t = 0$, whereas in order that $x = \infty$ be a *regular singular point* of (187) the functions

$$\frac{1}{t}a_1\left(\frac{1}{t}\right) \quad \text{and} \quad \frac{1}{t^2}a_2\left(\frac{1}{t}\right) \tag{190}$$

must be regular at $t = 0$.

To illustrate, we notice that, if a_1 and a_2 are *constants* in (187), the point $x = \infty$ is an *irregular singular point* unless $a_1 = a_2 = 0$, since the functions a_1/t and a_2/t^2 are not regular at $t = 0$.

Bessel's equation (63) also has an *irregular* singular point at $x = \infty$, since the function

$$\frac{1}{t^2}a_2\left(\frac{1}{t}\right) = \frac{1}{t^2} - p^2$$

is not regular at $t = 0$. Thus Bessel's equation has $x = 0$ as a regular singular point and $x = \infty$ as an irregular singular point. All other points are ordinary.

Legendre's equation (149), however, has a *regular* singular point at $x = \infty$, since

$$\frac{1}{t}a_1\left(\frac{1}{t}\right) = \frac{2}{1 - t^2} \quad \text{and} \quad \frac{1}{t^2}a_2\left(\frac{1}{t}\right) = \frac{p(p + 1)}{t^2 - 1}$$

are regular at $t = 0$, whereas the functions (189) are not regular at $t = 0$. Thus it is seen that Legendre's equation has regular singular points at $x = \pm 1$ and at $x = \infty$ and ordinary points elsewhere.

The expansion of $Q_n(x)$ in inverse powers of x when $|x| > 1$ can be expressed in terms of a hypergeometric series in the form

$$Q_n(x) = \frac{n!\sqrt{\pi}}{(n + \frac{1}{2})!\,(2x)^{n+1}}\, F\left(\frac{n+1}{2}, \frac{n+1}{2}; \frac{n+3}{2}; \frac{1}{x^2}\right).$$

For the *hypergeometric equation* (175) we find that $x = \infty$ is an *ordinary point* only if either $\alpha = 0$, $\beta = 1$ or $\alpha = 1$, $\beta = 0$, and is a *regular singular point* otherwise. Thus, except for the cases noted, the hypergeometric equation has regular singular points at $x = 0$, $x = 1$, and $x = \infty$ and ordinary points elsewhere. The general solution when $|x| > 1$ is of the form

$$y = c_1 x^{-\alpha} F\left(\alpha, \alpha - \gamma + 1; \alpha - \beta + 1; \frac{1}{x}\right)$$

$$+ c_2 x^{-\beta} F\left(\beta, \beta - \gamma + 1; \beta - \alpha + 1; \frac{1}{x}\right),$$

provided that $\beta - \alpha$ is nonintegral. (See Problem 76.)

If, in the hypergeometric equation (175), we replace x by a new independent variable x/β, we obtain the equation

$$x\left(1 - \frac{x}{\beta}\right)\frac{d^2 y}{dx^2} + \left[\gamma - x - (1 + \alpha)\frac{x}{\beta}\right]\frac{dy}{dx} - \alpha y = 0, \qquad (191)$$

which has regular singular points at $x = 0$, $x = \beta$, and $x = \infty$, and whose general solution, when $|x| < 1$, is given by (186) with x replaced by x/β. If now we let $\beta \longrightarrow \infty$, Equation (191) formally becomes

$$x\frac{d^2 y}{dx^2} + (\gamma - x)\frac{dy}{dx} - \alpha y = 0. \qquad (192)$$

In the transition from (191) to (192), we have moved the singular point at $x = \beta$ in (191) into coincidence or "confluence" with the second singular point at $x = \infty$. For this reason, (192) is known as the *confluent hypergeometric equation*. This equation is a special case of (52), as was pointed out at the end of Section 4.7, and is of some importance in applications. It is of interest to notice that $x = \infty$ is an *irregular* singular point of (192), formed by the confluence of two originally distinct regular singular points. (See also Problem 14.)

In cases where $x = \infty$ is an *irregular* singular point, it is frequently possible to obtain series of the type

$$y(x) \sim e^{rx} \sum_{k=0}^{\infty} \frac{A_k}{x^{k+s}}, \qquad (193)$$

such that *formal* substitution of the series into the differential equation reduces the equation to an identity. However, the series so obtained generally *do not converge for any finite values of x*. Still, they generally do have the property that if a finite number, N, of terms is retained, the sum, $S_N(x)$, of these terms approximates a solution $y(x)$ in such a way that not only the difference $y(x) - S_N(x)$ but also the product $x^N[y(x) - S_N(x)]$ approaches zero as $|x| \to \infty$. Such series, called *asymptotic expansions* of a solution $y(x)$, are of use not only in studying the *nature* of solutions for large values of $|x|$ but also in actually *calculating* values of such solutions to within a predictable accuracy. The reason is that *in such cases* the error associated with calculating $y(x)$ by using N terms is of the order of magnitude of the next following (neglected) term, *provided that that term is numerically smaller than the last term retained*. [This, of course, is not a general property of *all* divergent (or convergent) series.] Although this error eventually increases without limit as N increases, the first few successive terms frequently decrease rapidly in magnitude when x is large, so that it may be possible to stop with a term preceding a term of the order of magnitude of the tolerable error.†

The asymptotic approximations given in Equations (105–107) for the Bessel functions represent in each case the leading term of asymptotic expansions of this sort, which may be listed as follows:

$$J_p(x) \sim \sqrt{\frac{2}{\pi x}}\left[U_p(x) \cos\left(x - \frac{\pi}{4} - \frac{p\pi}{2}\right) - V_p(x) \sin\left(x - \frac{\pi}{4} - \frac{p\pi}{2}\right) \right],$$

$$(194)$$

$$Y_p(x) \sim \sqrt{\frac{2}{\pi x}}\left[U_p(x) \sin\left(x - \frac{\pi}{4} - \frac{p\pi}{2}\right) + V_p(x) \cos\left(x - \frac{\pi}{4} - \frac{p\pi}{2}\right) \right],$$

$$(195)$$

$$H_p^{(1)}(x) \sim \sqrt{\frac{2}{\pi x}}\,[U_p(x) + iV_p(x)]e^{i[x - (\pi/4) - (p\pi/2)]}, \qquad (196)$$

$$H_p^{(2)}(x) \sim \sqrt{\frac{2}{\pi x}}\,[U_p(x) - iV_p(x)]e^{-i[x - (\pi/4) - (p\pi/2)]}, \qquad (197)$$

$$I_p(x) \sim \frac{e^x}{\sqrt{2\pi x}}\,W_p(x), \qquad (198)$$

$$K_p(x) \sim \frac{e^{-x}}{\sqrt{\frac{2}{\pi}x}}\,W_p(-x), \qquad (199)$$

where U_p, V_p, and W_p denote the respective asymptotic series

$$U_p(x) = 1 - \frac{(4p^2 - 1^2)(4p^2 - 3^2)}{2!(8x)^2}$$

$$+ \frac{(4p^2 - 1^2)(4p^2 - 3^2)(4p^2 - 5^2)(4p^2 - 7^2)}{4!(8x)^4} - \cdots, \qquad (200)$$

$$V_p(x) = \frac{4p^2 - 1^2}{1!\,8x} - \frac{(4p^2 - 1^2)(4p^2 - 3^2)(4p^2 - 5^2)}{3!(8x)^3} + \cdots, \qquad (201)$$

†For further information on asymptotic expansions, see References 4 and 9.

$$W_p(x) = U_p(ix) - iV_p(ix)$$

$$= 1 - \frac{4p^2 - 1^2}{1!\,8x} + \frac{(4p^2 - 1^2)(4p^2 - 3^2)}{2!\,(8x)^2} - \cdots. \tag{202}$$

For both $U_p(x)$ and $V_p(x)$ it is known (Reference 17) that the error committed when the expansion is terminated with the kth term is not greater in magnitude than the absolute value of the $(k + 1)$th term, provided that $k > (2p - 1)/4$. The same statement applies to $W_p(x)$ if $k > (2p - 1)/2$. These expansions are frequently useful in the numerical evaluation of problem solutions involving Bessel functions of large argument, when the accuracy afforded by the leading term is insufficient or subject to question.

Asymptotic series, as defined above, are sometimes also called *semiconvergent series*.

REFERENCES

1. References at end of Chapter 1.
2. Davis, H. T., *Tables of Higher Mathematical Functions*, Principia Press, Bloomington, Ind., 1960.
3. Dwight, H. B., *Tables of Integrals and Other Mathematical Data*, 3rd ed., Dover Publications, Inc., New York, 1958.
4. Erdelyi, A., *Asymptotic Expansions*, Dover Publications, Inc., New York, 1956.
5. Fletcher, A., J. C. P. Miller, L. Rosenhead, and L. J. Comrie, *An Index of Mathematical Tables* 1, 2nd ed., Addison-Wesley Publishing Company, Inc., Reading, Mass., 1962.
6. Gray, A., G. R. Mathews, and T. M. MacRobert, *A Treatise on Bessel Functions*, 2nd ed., St. Martin's Press, Inc., New York, 1952.
7. Hobson, E. W., *Theory of Spherical and Ellipsoidal Harmonics*, Chelsea Publishing Company, Inc., New York, 1955.
8. Jahnke, E., F. Emde, and F. Lösch, *Tables of Higher Functions*, McGraw-Hill Book Company, Inc., New York, 1960.
9. Jeffreys, H., and B. S. Jeffreys, *Methods of Mathematical Physics*, Cambridge University Press, New York, 1956.
10. Knopp, K., *Infinite Sequences and Series*, Dover Publications, Inc., New York, 1956.
11. MacRobert, T. M., *Spherical Harmonics*, 2nd rev. ed., Dover Publications, Inc., New York, 1948.
12. Magnus, W., and F. Oberhettinger, *Special Functions of Mathematical Physics*, Chelsea Publishing Company, New York, 1949.
13. McLachlan, N. W., *Bessel Functions for Engineers*, 2nd ed., Oxford University Press, Inc., New York, 1955.
14. Rainville, E. D., *Special Functions*, Macmillan Publishing Co., Inc., New York, 1960.

15. Szego, O., *Orthogonal Polynomials*, 3rd ed., American Mathematical Society, Providence, R.I., 1967.

16. Watson, G. N., *A Treatise on the Theory of Bessel Functions*, 2nd ed., Macmillan Publishing Co., Inc., New York, 1945.

17. Whittaker, E. T., and G. N. Watson, *Modern Analysis*, 4th ed., Cambridge University Press, New York, 1958.

PROBLEMS

Section 4.1

1. Determine the interval of convergence for each of the following series, including consideration of the behavior of the series at each end point:

(a) $\displaystyle\sum_{n=0}^{\infty} \frac{2^n x^n}{n!}$,

(b) $\displaystyle\sum_{n=1}^{\infty} (-1)^n \frac{n(x+1)^n}{2^n}$,

(c) $\displaystyle\sum_{n=1}^{\infty} (-1)^n \frac{(x-1)^{2n}}{3^n n}$,

(d) $\displaystyle\sum_{n=1}^{\infty} \frac{k(k+1)\cdots(k+n-1)}{n!} x^n$,

(e) $\displaystyle\sum_{n=1}^{\infty} \frac{x^n}{n^n}$,

(f) $\displaystyle\sum_{n=1}^{\infty} n^n x^n$,

(g) $\displaystyle\sum_{n=1}^{\infty} \frac{(n-\alpha)!}{n!\,\alpha!}(x-a)^n$,

(h) $\displaystyle\sum_{n=0}^{\infty} \frac{(n!)^2}{(2n)!} x^n$,

(i) $\displaystyle\sum_{n=1}^{\infty} \frac{1}{nx^n}$,

(j) $\displaystyle\sum_{n=0}^{\infty} \left(\frac{x}{x+2}\right)^n$.

2. Starting with the expansions

$$f(x) = \sin x = x - \frac{x^3}{3!} + \frac{x^5}{5!} - \cdots, \qquad g(x) = \cos x = 1 - \frac{x^2}{2!} + \frac{x^4}{4!} - \cdots,$$

and assuming that

$$\tan x = x + \frac{x^3}{3} + \frac{2x^5}{15} + \cdots \qquad \left(|x| < \frac{\pi}{2}\right),$$

verify through three nonvanishing terms that term-by-term operations on the series properly yield the following results:

(a) $f'(x) = g(x)$,

(b) $\displaystyle\int_0^x g(u)\,du = f(x)$,

(c) $f(x)g(x) = \frac{1}{2}f(2x)$,

(d) $\displaystyle\frac{f(x)}{g(x)} = \tan x \qquad \left(|x| < \frac{\pi}{2}\right)$.

3. Show that

(a) $e^x = 1 + \dfrac{x}{1!} + \dfrac{x^2}{2!} + \cdots + \dfrac{x^n}{n!} + \dfrac{e^{\theta_1 x}}{(n+1)!} x^{n+1}$,

(b) $\cos x = 1 - \dfrac{x^2}{2!} + \dfrac{x^4}{4!} - \cdots + (-1)^m \dfrac{x^{2m}}{(2m)!} + (-1)^{m+1} \dfrac{\cos(\theta_2 x)}{(2m+2)!} x^{2m+2}$,

where $0 < \theta_{1,2} < 1$.

4. Prove that the remainder term in each of the representations in Problem 3 tends to zero as the number of terms increases, for *all* values of x. (See Problem 53 of Chapter 2.)

Section 4.2

5. Obtain the general solution of each of the following differential equations in terms of Maclaurin series:

(a) $\dfrac{d^2y}{dx^2} = xy$,

(b) $\dfrac{d^2y}{dx^2} + x\dfrac{dy}{dx} - y = 0$,

(c) $x\dfrac{d^2y}{dx^2} - \dfrac{dy}{dx} - 4x^3y = 0$.

6. For each of the following differential equations, obtain the most general solution which is representable by a Maclaurin series:

(a) $\dfrac{d^2y}{dx^2} + y = 0$,

(b) $\dfrac{d^2y}{dx^2} - (x - 3)y = 0$,

(c) $\left(1 - \dfrac{1}{2}x^2\right)\dfrac{d^2y}{dx^2} + x\dfrac{dy}{dx} - y = 0$,

(d) $x^2\dfrac{d^2y}{dx^2} - \dfrac{dy}{dx} + y = 0$,

(e) $(x^2 + x)\dfrac{d^2y}{dx^2} - (x^2 - 2)\dfrac{dy}{dx} - (x + 2)y = 0$,

(f) $x^2\dfrac{d^2y}{dx^2} - \dfrac{dy}{dx} = 0$.

Obtain three nonvanishing terms in each infinite series involved.

Section 4.3

7. (a–f) Locate and classify the singular points of the differential equations of Problem 6.

8. Locate and classify the singular points of the following differential equations:

(a) $(x - 1)\dfrac{d^2y}{dx^2} + \sqrt{x}\,y = 0$ $(x \geq 0)$,

(b) $\dfrac{d^2y}{dx^2} + \dfrac{dy}{dx}\log x + xy = 0$ $(x \geq 0)$,

(c) $x\dfrac{d^2y}{dx^2} + y\sin x = 0$,

(d) $\dfrac{d^2y}{dx^2} - |1 - x^2|y = 0$,

(e) $\dfrac{d^2y}{dx^2} + y\cos\sqrt{x} = 0$ $(x \geq 0)$.

9. Show that the change of variables $x = t^2$, with $t = \sqrt{x} \geq 0$, transforms the equation

$$\frac{d^2y}{dx^2} + \sqrt{x}\,y = 0 \quad (x \geq 0),$$

with an *irregular* singular point at $x = 0$, into the equation

$$t\frac{d^2y}{dt^2} - \frac{dy}{dt} + 4t^4y = 0 \quad (t \geq 0)$$

with a *regular* singular point at $t = 0$.

Section 4.4

10. Derive Equation (22) by using summation notation, as follows:

(a) Introduce the expansions

$$P(x) = \sum_{n=0}^{\infty} P_n x^n, \qquad Q(x) = \sum_{n=0}^{\infty} Q_n x^n, \qquad R(x) = \sum_{n=0}^{\infty} R_n x^n,$$

and

$$y(x) = \sum_{m=0}^{\infty} A_m x^{m+s},$$

with $R_0 = 1$, into the expression for Ly and obtain

$$Ly = \sum_{n=0}^{\infty} \sum_{m=0}^{\infty} [R_n(s+m)(s+m-1) + P_n(s+m) + Q_n] A_m x^{n+m+s-2}.$$

(b) Write $n + m = k$, using n and k as new summation indices and noticing that then n may vary from 0 through k as k varies from 0 to ∞. Thus transform the expression for Ly into

$$Ly = \sum_{k=0}^{\infty} \left[\sum_{n=0}^{k} g_n(s+k) A_{k-n} \right] x^{k+s-2},$$

where $g_n(s)$ is defined by (21) when $n = 1, 2, \ldots$ and $g_0(s) = f(s)$, and verify that this result is equivalent to (22).

11. Use the method of Frobenius to obtain the general solution of each of the following differential equations, valid near $x = 0$:

(a) $2x \dfrac{d^2 y}{dx^2} + (1 - 2x)\dfrac{dy}{dx} - y = 0,$

(b) $x^2 \dfrac{d^2 y}{dx^2} + x\dfrac{dy}{dx} + \left(x^2 - \dfrac{1}{4}\right)y = 0,$

(c) $x \dfrac{d^2 y}{dx^2} + 2\dfrac{dy}{dx} + xy = 0,$

(d) $x(1 - x)\dfrac{d^2 y}{dx^2} - 2\dfrac{dy}{dx} + 2y = 0.$

12. Use the method of Frobenius to obtain the general solution of each of the following differential equations, valid near $x = 0$:

(a) $x^2 \dfrac{d^2 y}{dx^2} - 2x\dfrac{dy}{dx} + (2 - x^2)y = 0,$

(b) $(x - 1)\dfrac{d^2 y}{dx^2} - x\dfrac{dy}{dx} + y = 0,$

(c) $x \dfrac{d^2 y}{dx^2} - \dfrac{dy}{dx} + 4x^3 y = 0,$

(d) $(1 - \cos x)\dfrac{d^2 y}{dx^2} - \dfrac{dy}{dx} \sin x + y = 0.$

Obtain three nonvanishing terms in each infinite series of parts (a–c). In part (d), obtain two such terms in each series.

13. (a–d) For each of the equations in Problem 12, give the largest interval inside which convergence of the Frobenius solutions is guaranteed (except possibly at $x = 0$) by the theorem stated on page 131.

14. Find the general solution of the differential equation

$$x\frac{d^2y}{dx^2} + (c - x)\frac{dy}{dx} - ay = 0,$$

valid near $x = 0$, assuming that c is nonintegral. The solution which is regular at $x = 0$ and which is unity at that point is called the *confluent hypergeometric function* and is usually denoted by $M(a, c; x)$. Show then that, if c is nonintegral, the general solution is of the form

$$y = c_1 M(a, c; x) + c_2 x^{1-c} M(1 + a - c, 2 - c; x).$$

15. Determine the two values of the constant α for which it is true that *all* solutions of the equation

$$x\frac{d^2y}{dx^2} + (x - 1)\frac{dy}{dx} - \alpha y = 0$$

are *regular* at $x = 0$, and obtain the general solution in each of these cases.

16. (a) Show that the equation

$$x\frac{d^2y}{dx^2} + \frac{dy}{dx} - y = 0$$

possesses equal exponents $s_1 = s_2 = 0$ at $x = 0$.

(b) Obtain the regular solution, and denote by $u_1(x)$ the result of setting the leading coefficient A_0 equal to unity.

(c) Assume a second solution of the form

$$y_2(x) = Cu_1(x) \log x + v(x),$$

where $C \neq 0$, and show that $v(x)$ must satisfy the equation

$$x\frac{d^2v}{dx^2} + \frac{dv}{dx} - v = -2C\frac{du_1}{dx}.$$

(d) Obtain one solution of this equation in the form

$$v(x) = \sum_{k=0}^{\infty} B_k x^{k+s_2} = \sum_{k=0}^{\infty} B_k x^k,$$

showing that C and B_0 are arbitrary, but taking $C = 1$ and $B_0 = 0$ for convenience. Hence obtain the general solution of the original equation in the form

$$y = c_1 u_1(x) + c_2[u_1(x) \log x + v(x)].$$

17. (a) Show that the equation

$$x\frac{d^2y}{dx^2} - x\frac{dy}{dx} - y = 0$$

possesses exponents $s_1 = 1$ and $s_2 = 0$ at $x = 0$.

(b) Obtain the regular solution, corresponding to $s_1 = 1$, and denote by $u_1(x)$ the result of setting the leading coefficient equal to unity.

(c) Assume a second solution in the form

$$y_2(x) = Cu_1(x) \log x + v(x),$$

where $C \neq 0$, and show that $v(x)$ must satisfy the equation

$$x\frac{d^2v}{dx^2} - x\frac{dv}{dx} - v = \frac{C}{x}\left[(1 + x)u_1 - 2x\frac{du_1}{dx}\right].$$

(d) Obtain one solution of this equation in the form

$$v(x) = \sum_{k=0}^{\infty} B_k x^{k+s_2} = \sum_{k=0}^{\infty} B_k x^k,$$

showing that C and B_1 are arbitrary, but taking $C = 1$ and $B_1 = 0$ for convenience. Hence obtain the general solution of the original equation in the form

$$y = c_1 u_1(x) + c_2[u_1(x)\log x + v(x)].$$

18. *Nonhomogeneous linear equations.*

(a) Show that a particular solution of the nonhomogeneous linear differential equation

$$x^2(1 + R_1 x + \cdots)\frac{d^2y}{dx^2} + x(P_0 + P_1 x + \cdots)\frac{dy}{dx} + (Q_0 + Q_1 x + \cdots)y$$

$$= x^r(d_0 + d_1 x + \cdots),$$

where r is any constant, and where all series converge in some interval about $x = 0$ or terminate, can be obtained in the form

$$y = x^r(A_0 + A_1 x + \cdots) = \sum_{k=0}^{\infty} A_k x^{k+r},$$

if neither of the exponents s_1 and s_2, for which

$$s^2 + (P_0 - 1)s + Q_0 = 0,$$

equals r or exceeds r by a positive integer.

(b) For an equation of the form

$$(1 + a_1 x + \cdots)\frac{d^2y}{dx^2} + (b_0 + b_1 x + \cdots)\frac{dy}{dx} + (c_0 + c_1 x + \cdots)y$$

$$= d_0 + d_1 x + \cdots,$$

which hence possesses an ordinary point at $x = 0$ and a regular right-hand member, deduce the existence of a particular solution of the form

$$y = v(x) \equiv x^2(A_0 + A_1 x + \cdots) = \sum_{k=0}^{\infty} A_k x^{k+2}.$$

[If the equation is written in the form $y'' + a_1 y' + a_2 y = x^{r-2}d$, the series solutions will converge at least inside the largest interval in which the series representing the resultant functions xa_1, $x^2 a_2$, and d would *all* converge. In the exceptional cases noted in part (a), logarithmic terms may be involved.]

19. Determine a particular solution of each of the following equations, in the form of a series valid near $x = 0$, by the method of Problem 18. In each case, obtain four nonvanishing terms.

(a) $\dfrac{d^2y}{dx^2} + y = e^x$

(b) $\dfrac{d^2y}{dx^2} + xy = 1,$

(c) $x\dfrac{d^2y}{dx^2} - y = x,$

(d) $x^2\dfrac{d^2y}{dx^2} + y = \dfrac{e^x}{\sqrt{x}}.$

Sections 4.5 and 4.6

20. Obtain the general solution of the equation

$$x\frac{d^2y}{dx^2} + \frac{dy}{dx} - y = 0$$

by use of the formula (36).

21. Obtain the general solution of the equation

$$x\frac{d^2y}{dx^2} - x\frac{dy}{dx} - y = 0$$

by use of the formula (37).

22. Obtain the general solution of the equation of Problem 14 when $a = c = 1$ by use of either of the two methods of Section 4.5.

23. Obtain the general solution of Problem 14 when $a = 1$ and $c = 0$ by use of either of the methods of Section 4.5.

Section 4.7

24. The differential equation

$$x\frac{d^2y}{dx^2} + (1 - x)\frac{dy}{dx} + ny = 0$$

is known as *Laguerre's equation*.

    (a) Verify that this equation is a special case of Equation (52), with $M = 1$, and show that the exponents at $x = 0$ are both zero.

    (b) Obtain the regular solution in the form

$$y_1(x) = B_0\left[1 + \sum_{k=1}^{\infty}(-1)^k\frac{n(n-1)(n-2)\cdots(n-k+1)}{(k!)^2}x^k\right].$$

    (c) Show that this solution is a polynomial of degree n when n is a nonnegative integer, and verify that the choice $B_0 = 1$ leads to the *Laguerre polynomial of degree n*, with the definition

$$L_n(x) = 1 - \binom{n}{1}\frac{x}{1!} + \binom{n}{2}\frac{x^2}{2!} - \cdots + \frac{(-x)^n}{n!},$$

where $\binom{n}{k}$ represents the binomial coefficient $n!/[(n-k)!k!]$.

25. The differential equation

$$\frac{d^2y}{dx^2} - 2x\frac{dy}{dx} + 2ny = 0$$

is known as *Hermite's equation*. Verify that this equation is a special case of Equation (52), with $M = 2$, and obtain the general solution in the form

$$y(x) = c_1u_n(x) + c_2v_n(x),$$

where

$$u_n(x) = 1 - \frac{n}{1}\frac{x^2}{1!} + \frac{n(n-2)}{1\cdot3}\frac{x^4}{2!} - \frac{n(n-2)(n-4)}{1\cdot3\cdot5}\frac{x^6}{3!} + \cdots$$

and

$$v_n(x) = x - \frac{n-1}{3}\frac{x^3}{1!} + \frac{(n-1)(n-3)}{3\cdot 5}\frac{x^5}{2!} - \frac{(n-1)(n-3)(n-5)}{3\cdot 5\cdot 7}\frac{x^7}{3!} + \cdots.$$

[Hence verify that the solution $u_n(x)$ is a polynomial of degree n when n is a positive even integer or zero, whereas $v_n(x)$ is a polynomial of degree n when n is a positive odd integer. That multiple of the nth-degree polynomial for which the coefficient of x^n is 2^n is called the nth *Hermite polynomial* and is often denoted by $H_n(x)$.]

26. The differential equation

$$(1 - x^2)\frac{d^2y}{dx^2} - cx\frac{dy}{dx} + n(n + c - 1)y = 0$$

is a specialization of *Jacobi's equation* (see Problem 68), referred to the interval $(-1, 1)$.

(a) Show that if n is zero or a positive integer it possesses a *polynomial* solution, of the form

$$u_n(x) = 1 - \frac{n(n+c-1)}{2!}x^2 + \frac{[n(n-2)][(n+c-1)(n+c+1)]}{4!}x^4 - \cdots$$

when n is even, and of the form

$$v_n(x) = x - \frac{(n-1)(n+2)}{3!}x^3 + \frac{[(n-1)(n-3)][(n+c)(n+c+2)]}{5!}x^5 - \cdots$$

when n is odd.

(b) When $c = 1$, that multiple of the nth-degree polynomial for which the coefficient of x^n is 2^{n-1} ($n \geq 1$) is called the nth *Chebyshev polynomial*, and is often denoted by $T_n(x)$. It is conventional to take $T_0(x) = 1$. Obtain the results

$$T_0(x) = 1, \quad T_1(x) = x, \quad T_2(x) = 2x^2 - 1, \quad T_3(x) = 4x^3 - 3x$$

and verify in these cases that

$$T_n(x) = \cos(n\cos^{-1}x).$$

(c) When $c = 3$, that multiple of the nth-degree polynomial for which the coefficient of x^n is 2^n is called the nth *Chebyshev polynomial of the second kind*, and is often denoted by $S_n(x)$. Obtain the results

$$S_0(x) = 1, \quad S_1(x) = 2x, \quad S_2(x) = 4x^2 - 1, \quad S_3(x) = 8x^3 - 4x$$

and verify in these cases that

$$S_n(x) = \frac{1}{n+1}T'_{n+1}(x) = \frac{\sin[(n+1)\cos^{-1}x]}{\sqrt{1-x^2}},$$

given that $T_4(x) = 8x^4 - 8x^2 + 1$.

(When $c = 2$, the *Legendre polynomials* are obtained. They are considered in some detail in Section 4.12.)

Section 4.8

27. Evaluate the following quantities, from the series definitions, to three-place accuracy:

(a) $J_1(0.3)$,

(b) $Y_0(0.2)$,

(c) $J_{0.75}(0.2)$,

(d) $I_2(1)$,

(e) $H_0^{(1)}(0.2)$,

(f) $J_1'(0.5)$.

28. Find the general solution of the simultaneous equations

$$y + t\frac{dx}{dt} = 0,$$

$$\frac{dy}{dt} - tx = 0.$$

29. By making an appropriate change of variables, obtain the general solution of the differential equation

$$(Ax + B)\frac{d^2y}{dx^2} + A\frac{dy}{dx} + A^2(Ax + B)y = 0.$$

Section 4.9

30. Use Equations (101–107) to evaluate the following limits:

(a) $\lim_{x \to 0} x^{1/3} J_{-1/3}(x)$,

(b) $\lim_{x \to 0} x Y_1(x)$,

(c) $\lim_{x \to 0} x^2 K_2(x)$,

(d) $\lim_{x \to 0} \frac{J_n(x)}{x^n}$,

(e) $\lim_{x \to \infty} x\{[J_p(x)]^2 + [Y_p(x)]^2\}$.

31. Show that the definitions (85) and (99) yield the special relations

$$Y_{1/2}(x) = -J_{-1/2}(x) = -\sqrt{\frac{2}{\pi x}}\cos x$$

and

$$K_{1/2}(x) = \frac{\pi}{2}[I_{-1/2}(x) - I_{1/2}(x)] = \sqrt{\frac{\pi}{2x}}e^{-x}$$

and verify that these results are consistent with (88) and (100).

32. Prove from the series definition that

$$\frac{d}{dx}[x^{-p}J_p(\alpha x)] = -\alpha x^{-p}J_{p+1}(\alpha x).$$

33. Use Equations (112) and (113) to prove that, if $G_p(x) = [J_p(x)]^2$, then

$$\frac{dG_p(x)}{dx} = \frac{x}{2p}[G_{p-1}(x) - G_{p+1}(x)].$$

34. Use Equations (108) and (109) to deduce (by induction) the formulas

$$\left(\frac{1}{x}\frac{d}{dx}\right)^m [x^p J_p(\alpha x)] = \alpha^m x^{p-m} J_{p-m}(\alpha x)$$

and

$$\left(\frac{1}{x}\frac{d}{dx}\right)^m [x^{-p} J_p(\alpha x)] = (-\alpha)^m x^{-p-m} J_{p+m}(\alpha x).$$

35. Use Equations (112) and (113) to show that if $J_p(\beta) = 0$ and $J'_p(\beta) = 0$, so that β is a zero of $J_p(x)$ of multiplicity at least two, then $J_{p+1}(\beta) = J_{p-1}(\beta)$ and hence, from (110) and (111), $J'_{p+1}(\beta) = J'_{p-1}(\beta)$. Then deduce that also $J''_p(\beta) = 0$, so that the multiplicity of β is at least three. Noticing that this argument can be repeated indefinitely, prove that *the zeros of $J_p(x)$ are all simple*; that is, if $J_p(\beta) = 0$, then $J'_p(\beta) \neq 0$.

36. Use Equation (111) and the result of Problem 35 to show that $J_{p+1}(x)$ has opposite signs at two consecutive zeros of $J_p(x)$, so that $J_{p+1}(x)$ has at least one zero between successive zeros of $J_p(x)$. Then use (110) with p replaced by $p + 1$, similarly, to show that $J_p(x)$ has at least one zero between successive zeros of $J_{p+1}(x)$. Hence prove that *there is exactly one zero of $J_p(x)$ between successive zeros of $J_{p+1}(x)$.*

37. Prove that $I_p(0) = 0$ when $p > 0$, but that $I_0(x)$ has no real zeros and $I_p(x)$ has no real zeros other than $x = 0$ when $p > 0$. [Inspect the power-series definition of $x^{-p}I_p(x)$.]

38. By using Equations (108) and (109), together with integration by parts, deduce the following reduction formulas:

(a) $\displaystyle\int x^m J_n(x)\, dx = x^m J_{n+1}(x) - (m - n - 1) \int x^{m-1} J_{n+1}(x)\, dx,$

(b) $\displaystyle\int x^m J_n(x)\, dx = -x^m J_{n-1}(x) + (m + n - 1) \int x^{m-1} J_{n-1}(x)\, dx.$

[Notice that the first reduction eventually yields a closed form, by iteration, when $m - n$ is an odd positive integer, whereas the second reduction does so when $m + n$ is an odd positive integer. When m and n are both even or both odd integers, appropriate use of one or both of the reductions will yield the sum of a closed form and a multiple of the term $\int J_0(x)\, dx$. Whereas this integral cannot be further simplified, the function $\int_0^x J_0(t)\, dt$ is a *tabulated* function.]

39. Use the results of Problem 38 to deduce the following formulas:

(a) $\displaystyle\int x^{p+1} J_p(\alpha x)\, dx = \frac{1}{\alpha} x^{p+1} J_{p+1}(\alpha x) + C,$

(b) $\displaystyle\int x^{1-p} J_p(\alpha x)\, dx = -\frac{1}{\alpha} x^{1-p} J_{p-1}(\alpha x) + C,$

(c) $\displaystyle\int x^3 J_0(x)\, dx = x^3 J_1(x) - 2x^2 J_2(x) + C,$

(d) $\displaystyle\int x^6 J_1(x)\, dx = x^6 J_2(x) - 4x^5 J_3(x) + 8x^4 J_4(x) + C,$

(e) $\displaystyle\int J_3(x)\, dx = -J_2(x) - \frac{2}{x} J_1(x) + C,$

(f) $\displaystyle\int x J_1(x)\, dx = -x J_0(x) + \int J_0(x)\, dx + C,$

(g) $\displaystyle\int x^{-1} J_1(x)\, dx = -J_1(x) + \int J_0(x)\, dx + C,$

(h) $\displaystyle\int J_2(x)\, dx = -2J_1(x) + \int J_0(x)\, dx + C.$

40. Establish the relation

$$J_0(x) = \frac{2}{\pi} \int_0^{\pi/2} \cos(x \sin \theta)\, d\theta$$

by verifying that the right-hand member satisfies Bessel's equation of order zero and investigating its value when $x = 0$. (The other Bessel functions have similar integral representations. See Problem 41, below, and Problem 65 of Chapter 5.)

41. (a) Deduce from Problem 40 that

$$J_1(x) = \frac{2}{\pi} \int_0^{\pi/2} \sin(x \sin \theta) \sin \theta \, d\theta.$$

(b) By integrating the right-hand member of this relation by parts, show that

$$\frac{1}{x} J_1(x) = \frac{2}{\pi} \int_0^{\pi/2} \cos(x \sin \theta) \cos^2 \theta \, d\theta.$$

(c) Deduce from Problems 40 and 41(b) that

$$J_2(x) = \frac{2}{\pi} \int_0^{\pi/2} \cos(x \sin \theta) \cos 2\theta \, d\theta.$$

42. Establish the relation

$$e^{\frac{x}{2}\left(r - \frac{1}{r}\right)} = \sum_{n=-\infty}^{\infty} r^n J_n(x)$$

by first considering the expansion

$$e^{\frac{x}{2}r} e^{-\frac{x}{2r}} = \left[\sum_{j=0}^{\infty} \frac{(xr/2)^j}{j!} \right] \left[\sum_{k=0}^{\infty} (-1)^k \frac{(x/2r)^k}{k!} \right] = \sum_{j=0}^{\infty} \sum_{k=0}^{\infty} (-1)^k \frac{(x/2)^{j+k}}{j!\,k!} r^{j-k},$$

replacing j by $n + k$, where n is a new index of summation, and then identifying the coefficient of r^n in the resultant series. [The function $e^{\frac{x}{2}\left(r - \frac{1}{r}\right)}$ is known as the *generating function* for $J_n(x)$.]

43. Let $U(x, r) = e^{\frac{x}{2}\left(r - \frac{1}{r}\right)}$.
   (a) Show that

$$\frac{\partial U}{\partial x} = \frac{1}{2}\left(r - \frac{1}{r}\right) U$$

and, by using the result of Problem 42, deduce the relation

$$\frac{d}{dx} J_n(x) = \frac{1}{2}[J_{n-1}(x) - J_{n+1}(x)].$$

   (b) Show that

$$\frac{\partial U}{\partial r} = \frac{x}{2}\left(1 + \frac{1}{r^2}\right) U$$

and, again using Problem 42, deduce that

$$(n + 1)J_{n+1}(x) = \frac{x}{2}[J_n(x) + J_{n+2}(x)].$$

[Note that these results are specializations of (112) and (113).]

44. (a) Use *Abel's formula*, Equation (65) of Chapter 1, to show that the *Wronskian* of any two solutions of Bessel's equation (63) or of the modified equation (93) is always of the form A/x, where A is an appropriate constant.
   (b) From this result deduce that

$$J_p(x) Y_p'(x) - Y_p(x) J_p'(x) = \frac{C}{x}$$

and, by considering the limiting form as $x \to 0$, show that $C = 2/\pi$ and hence

$$J_p(x)Y'_p(x) - Y_p(x)J'_p(x) = \frac{2}{\pi x}.$$

(c) In a similar way, show that

$$J_p(x)J'_{-p}(x) - J_{-p}(x)J'_p(x) = -\frac{2}{\pi x} \sin p\pi,$$

$$I_p(x)K'_p(x) - K_p(x)I'_p(x) = -\frac{1}{x},$$

$$I_p(x)I'_{-p}(x) - I_{-p}(x)I'_p(x) = -\frac{2}{\pi x} \sin p\pi.$$

[Use Equation (59) of Chapter 2.]

45. Use the result of Problem 44(b), and the methods of Section 1.9, to obtain the general solution of the equation

$$x^2 \frac{d^2y}{dx^2} + x\frac{dy}{dx} + (x^2 - p^2)y = f(x)$$

in the form

$$y = \frac{\pi}{2} \int^x [Y_p(x)J_p(\xi) - J_p(x)Y_p(\xi)]\frac{f(\xi)}{\xi} d\xi + c_1 J_p(x) + c_2 Y_p(x).$$

[Notice that here $h(x) = f(x)/x^2$, with the notation of Section 1.9.]

Section 4.10   use 06h 127

46. Obtain the general solution of each of the following equations in terms of Bessel functions or, if possible, in terms of elementary functions:

(a) $x\dfrac{d^2y}{dx^2} - 3\dfrac{dy}{dx} + xy = 0,$

(b) $x\dfrac{d^2y}{dx^2} - \dfrac{dy}{dx} + 4x^3y = 0,$

(c) $x^2\dfrac{d^2y}{dx^2} + x\dfrac{dy}{dx} - \left(x^2 + \dfrac{1}{4}\right)y = 0,$

(d) $x\dfrac{d^2y}{dx^2} + (2x + 1)\left(\dfrac{dy}{dx} + y\right) = 0,$

(e) $x\dfrac{d^2y}{dx^2} - \dfrac{dy}{dx} - xy = 0,$

(f) $x^4\dfrac{d^2y}{dx^2} + a^2y = 0,$

(g) $\dfrac{d^2y}{dx^2} - x^2y = 0,$

(h) $x\dfrac{d^2y}{dx^2} + (1 + 2x)\dfrac{dy}{dx} + y = 0,$

(i) $x\dfrac{d^2y}{dx^2} + (1 + 4x^2)\dfrac{dy}{dx} + x(5 + 4x^2)y = 0.$

47. The two following equations each have arisen in several physical investigations. Express the general solution of each equation in terms of Bessel functions and also show that it can be expressed in terms of elementary functions when m is an integer:

(a) $\dfrac{d^2y}{dx^2} - \alpha^2 y = \dfrac{m(m+1)}{x^2} y,$

(b) $\dfrac{d^2y}{dx^2} - \dfrac{2m}{x}\dfrac{dy}{dx} - \alpha^2 y = 0.$

48. Show that for the differential equation

$$x\frac{d^2y}{dx^2} + 3\frac{dy}{dx} + 4xy = 0$$

the condition $y(0) = 1$ determines a unique solution, and hence that $y'(0)$ cannot also be prescribed. Determine this solution.

49. Find the most general solution of the equation

$$x^2\frac{d^2y}{dx^2} + x\frac{dy}{dx} + (x^2 - 1)y = 0$$

for which

$$\lim_{x\to 0} 2\pi xy(x) = P,$$

where P is a given constant.

50. The differential equation for small deflections of a rotating string is of the form

$$\frac{d}{dx}\left(T\frac{dy}{dx}\right) + \rho\omega^2 y = 0.$$

Obtain the general solution of this equation under the following assumptions:
(a) $T = T_0 x^n,$ $\rho = \rho_0 x^n;$ $T_0 = l^2\rho_0\omega^2.$
(b) $T = T_0 x^n,$ $\rho = \rho_0,$ $n \neq 2;$ $T_0 = l^2\rho_0\omega^2.$
(c) $T = T_0 x^2,$ $\rho = \rho_0;$ $T_0 = 4\rho_0\omega^2.$

Section 4.11

51. From the series definitions (137) and (138), and the definition (134), calculate the value of each of the following quantities to four significant figures:
(a) ber 1,   (b) bei 1,   (c) $M_0(1),$ (d) $\theta_0(1).$

52. Establish the following relations, assuming that $i^{3/2} = e^{3\pi i/4}$:
(a) $\text{ber}'\, x + i\,\text{bei}'\, x = -i^{3/2}J_1(i^{3/2}x) = M_1(x)e^{i[\theta_1(x) - \pi/4]},$

(b) $\text{ber}_1\, x = \dfrac{\sqrt{2}}{2}(\text{ber}'\, x - \text{bei}'\, x),$

(c) $\text{bei}_1\, x = \dfrac{\sqrt{2}}{2}(\text{ber}'\, x + \text{bei}'\, x).$

53. (a) Use Equations (133) and (107) to obtain the results

$$\text{ber}_p\, x \sim \frac{e^{x/\sqrt{2}}}{\sqrt{2\pi x}}\cos\varphi_p, \qquad \text{bei}_p\, x \sim \frac{e^{x/\sqrt{2}}}{\sqrt{2\pi x}}\sin\varphi_p,$$

as $x \to \infty$, where

$$\varphi_p = \frac{x}{\sqrt{2}} - \frac{\pi}{8} + \frac{p\pi}{2}.$$

(b) Use Equations (139) and (107) to obtain the results

$$\ker_p x \sim \frac{e^{-x/\sqrt{2}}}{\sqrt{\frac{2}{\pi}x}}\cos\psi_p, \qquad \ker_p x \sim -\frac{e^{-x/\sqrt{2}}}{\sqrt{\frac{2}{\pi}x}}\sin\psi_p,$$

as $x \longrightarrow \infty$, where

$$\psi_p = \frac{x}{\sqrt{2}} + \frac{\pi}{8} + \frac{p\pi}{2}.$$

54. (a) Show that Equation (131) can be put into the form

$$\frac{d}{dx}(xy') = \left(\frac{p^2}{x} + ix\right)y.$$

(b) Noticing that this equation is satisfied by the complex function

$$y = \text{ber}_p x + i \, \text{bei}_p x,$$

deduce the results

$$\frac{d}{dx}(x \, \text{ber}'_p x) = -x \, \text{bei}_p x + p^2 \frac{\text{ber}_p x}{x},$$

$$\frac{d}{dx}(x \, \text{bei}'_p x) = x \, \text{ber}_p x + p^2 \frac{\text{bei}_p x}{x}.$$

(c) Show similarly that the functions $\ker_p x$ and $\ker_p x$ satisfy completely analogous equations.

Section 4.12

55. Determine particular solutions of the partial differential equation

$$\frac{1}{\sin\varphi}\frac{\partial}{\partial\varphi}\left(\sin\varphi\frac{\partial U}{\partial\varphi}\right) + \frac{\partial}{\partial r}\left(r^2\frac{\partial U}{\partial r}\right) = 0$$

of the form $U(r, \varphi) = r^n y(\varphi)$, where n is a constant and y is independent of r.

56. Establish the following properties of the Legendre polynomials:
  (a) $P_n(-x) = (-1)^n P_n(x), \qquad P'_n(-x) = (-1)^{n+1} P'_n(x),$
  (b) $P'_n(0) = \begin{cases} 0 & (n \text{ even}), \\ (-1)^{(n-1)/2}\dfrac{(1\cdot3\cdot5\,\cdots\,n)^2}{n!} & (n \text{ odd}), \end{cases}$
  (c) $P'_n(1) = (-1)^{n+1}P'_n(-1) = \frac{1}{2}n(n+1).$
[Use Equation (149) in part (c).]

57. By direct integration when $n = 0$, and otherwise by integrating the equal members of Equation (189) and using Problem 56(b) when necessary, establish the following results:

  (a) $\displaystyle\int_{-1}^{1} P_n(x)\,dx = \begin{cases} 2 & (n = 0), \\ 0 & (n \neq 0), \end{cases}$

  (b) $\displaystyle\int_{0}^{1} P_n(x)\,dx = \begin{cases} 1 & (n = 0), \\ 0 & (n = 2, 4, 6, \ldots), \\ (-1)^{(n-1)/2}\dfrac{1}{n(n+1)}\dfrac{(1\cdot3\cdot5\,\cdots\,n)^2}{n!} & (n = 1, 3, 5, \ldots). \end{cases}$

58. Show that the following differential equations have the indicated general solutions:

(a) $\dfrac{d^2y}{ds^2} + \dfrac{dy}{ds}\coth s - n(n+1)y = 0, \quad y = c_1 P_n(\cosh s) + c_2 Q_n(\cosh s);$

(b) $(1-x^2)\dfrac{d^2y}{dx^2} - 4x\dfrac{dy}{dx} + (n+2)(n-1)y = 0, \quad y = c_1 P_n'(x) + c_2 Q_n'(x);$

(c) $x(1-x)\dfrac{d^2y}{dx^2} + (1-2x)\dfrac{dy}{dx} + n(n+1)y = 0,$

$$y = c_1 P_n(1-2x) + c_2 Q_n(1-2x).$$

59. (a) Use Equations (157) and (159) to obtain an expression for $P_n(x)$ in descending powers of x in the form

$$P_n(x) = \frac{(2n-1)(2n-3)\cdots 3\cdot 1}{n!}\left[x^n - \frac{n(n-1)}{2(1)(2n-1)}x^{n-2}\right.$$

$$\left. + \frac{n(n-1)(n-2)(n-3)}{2^2(1\cdot 2)(2n-1)(2n-3)}x^{n-4} - \cdots\right].$$

(b) Verify that the coefficient of x^{n-2k} in the expression for $P_n(x)$ is

$$c_{nk} = (-1)^k \frac{(2n-2k)!}{2^n k!(n-k)!(n-2k)!},$$

where k may vary from 0 to $n/2$ or $(n-1)/2$, according as n is even or odd.

60. (a) Use the binomial expansion to obtain the result

$$(x^2-1)^n = \sum_{k=0}^{n}(-1)^k \frac{n!}{k!(n-k)!}x^{2n-2k}.$$

(b) Hence show that, with the notation of Problem 59(b),

$$\frac{d^n}{dx^n}(x^2-1)^n = 2^n n! \sum_{k=0}^{N} c_{nk}x^{n-2k} = 2^n n!\, P_n(x),$$

where $N = n/2$ or $(n-1)/2$, according as n is even or odd. [This establishes *Rodrigues' formula* (168).]

61. Use Rodrigues' formula (168) to prove that $P_n(x)$ has n distinct zeros inside the interval $(-1, 1)$. [Notice that $(x^2-1)^n$ is a polynomial of degree $2n$ with zeros of multiplicity n at the end points, so that $d(x^2-1)^n/dx$ has zeros of multiplicity $n-1$ at the end points and one interior zero, and so forth.]

62. Prove that $|P_n(x)| \leq 1$ when $|x| \leq 1$ by the following steps:
(a) With the definition

$$M(x) = [P_n(x)]^2 + \frac{1-x^2}{n(n+1)}[P_n'(x)]^2,$$

when $n = 1, 2, \ldots$, show that

$$\max_{|x|\leq 1}[P_n(x)]^2 \leq \max_{|x|\leq 1} M(x).$$

[Notice that $M(x) = [P_n(x)]^2$ when $P_n(x)$ takes on a relative maximum or minimum and when $x = \pm 1$.]
(b) Show that

$$M'(x) = \frac{2x}{n(n+1)}[P_n'(x)]^2.$$

(c) Deduce that in the interval $-1 \leqq x \leqq 1$ the function $M(x)$ takes on its maximum value at the end points, and hence complete the required proof.

63. Establish the relation

$$(1 - 2rx + r^2)^{-1/2} = \sum_{n=0}^{\infty} r^n P_n(x) \qquad (|x| \leqq 1, |r| < 1)$$

by first obtaining the binomial expansion

$$[1 - r(2x - r)]^{-1/2} = 1 + \frac{r}{2}(2x - r) + \frac{1 \cdot 3}{2^2 2!} r^2 (2x - r)^2 + \cdots$$

$$+ \frac{1 \cdot 3 \cdots (2n - 3)}{2^{n-1}(n - 1)!} r^{n-1}(2x - r)^{n-1}$$

$$+ \frac{1 \cdot 3 \cdots (2n - 1)}{2^n n!} r^n(2x - r)^n + \cdots,$$

when $|r(2x - r)| < 1$, and then picking out the coefficient of r^n in this expression and using the result of Problem 59(a). [Show that the coefficient of r^k in $(2x - r)^m$ is

$$(-1)^k \frac{m(m - 1) \cdots (m - k + 1)}{k!} (2x)^{m-k}$$

when $k \geq 1$ and is $(2x)^m$ when $k = 0$, and notice that no terms beyond those written can involve x^n. Since $|r(2x - r)| < 1$ when both $0 \leqq x \leqq 1$ and $0 \leqq r < 1$, and also when both $-1 \leqq x \leqq 0$ and $-1 < r \leqq 0$, in particular, the desired result is thus established in those cases. But since the series itself is a power series in r, the region inside which it converges must be symmetric in r, and hence must admit $|r| < 1$ when $|x| \leqq 1$. The function $(1 - 2rx + r^2)^{-1/2}$ is called the *generating function* for $P_n(x)$.]

64. Let $V(x, r) = (1 - 2rx + r^2)^{-1/2}$.
  (a) Show that

$$(1 - 2rx + r^2)\frac{\partial V}{\partial r} = (x - r)V$$

and, by using the result of Problem 63, deduce the relation

$$(n + 1)P_{n+1}(x) - 2nxP_n(x) + (n - 1)P_{n-1}(x) = xP_n(x) - P_{n-1}(x)$$

when $n \geq 1$, and hence establish the recurrence formula (162) for $P_n(x)$.
  (b) Show that

$$(1 - 2rx + r^2)\frac{\partial V}{\partial x} = rV$$

and, again using Problem 63, deduce the formula

$$\frac{d}{dx}P_n(x) - 2x\frac{d}{dx}P_{n-1}(x) + \frac{d}{dx}P_{n-2}(x) = P_{n-1}(x)$$

when $n \geq 2$.
  [By combining the results of parts (a) and (b) with the differential equation satisfied by $P_n(x)$, a variety of other differential recurrence formulas can be derived.]

65. (a) Show that the substitution $t = 1 - x$ transforms Legendre's equation to the form

$$t(2 - t)\frac{d^2 y}{dt^2} + 2(1 - t)\frac{dy}{dt} + p(p + 1)y = 0,$$

that this equation has one solution regular at $t = 0$, with exponent zero, and that all other solutions become logarithmically infinite at $t = 0$.

(b) Determine the regular solution, and hence show that the solution $P_p(x)$ of Legendre's equation which is finite at $x = 1$ and which is unity at that point is given by

$$P_p(x) = 1 + \sum_{k=1}^{\infty} [(p + 1)(p + 2) \cdots (p + k)]$$

$$\times [(-p)(1 - p) \cdots (k - 1 - p)]\frac{(1 - x)^k}{2^k (k\,!)^2}$$

near $x = 1$.

(c) Verify that this result is in accordance with Equations (158) when $p = 0, 1$, and 2 by writing out the terms in the series.

Section 4.13

66. Show that

$$\frac{d}{dx} F(\alpha, \beta\,;\gamma\,;x) = \frac{\alpha\beta}{\gamma} F(\alpha + 1, \beta + 1\,;\gamma + 1\,;x).$$

67. (a) Show that the change in variables $x = 1 - t$ transforms the hypergeometric equation (175) to the form

$$t(1 - t)\frac{d^2 y}{dt^2} + [(\alpha + \beta - \gamma + 1) - (\alpha + \beta + 1)t]\frac{dy}{dt} - \alpha\beta y = 0.$$

(b) By comparing the coefficients in this equation with those in Equation (175), deduce that the general solution of (175) can be written in the form

$$y = c_1 F(\alpha, \beta\,;\alpha + \beta - \gamma + 1\,;1 - x)$$

$$+ c_2 x^{\gamma - \alpha - \beta} F(\gamma - \beta, \gamma - \alpha\,;1 - \alpha - \beta + \gamma\,;1 - x)$$

near the second singular point at $x = 1$, if $\alpha + \beta - \gamma$ is nonintegral.

68. (a) Show that *Jacobi's equation*,

$$x(1 - x)\frac{d^2 y}{dx^2} + [a - (1 + b)x]\frac{dy}{dx} + n(b + n)y = 0,$$

is a hypergeometric equation.

(b) Deduce that the general solution, near $x = 0$, can be expressed in the form

$$y = c_1 F(n + b, -n\,;a\,;x) + c_2 x^{1-a} F(n + b - a + 1, 1 - n - a\,;2 - a\,;x),$$

if a is nonintegral.

69. Show that a solution of Jacobi's equation (Problem 68) which is regular at $x = 0$ can be written in the explicit form

$$J_n(a, b, x) = 1 - \frac{(n + b)n}{a}\frac{x}{1!} + \frac{(n + b)(n + b + 1)n(n - 1)}{a(a + 1)}\frac{x^2}{2!} + \cdots,$$

if a is not zero or a negative integer, where

$$J_n(a, b, x) = F(n + b, -n\,;a\,;x).$$

(Notice that this solution is a polynomial of degree n when n is a positive integer or zero.)

70. (a) Show that the substitution $x = 1 - 2t$ transforms the equation

$$(1 - x^2)\frac{d^2y}{dx^2} - cx\frac{dy}{dx} + n(n + c - 1)y = 0$$

(see Problem 26) to the special Jacobi equation

$$t(1 - t)\frac{d^2y}{dt^2} + \frac{c}{2}(1 - 2t)\frac{dy}{dt} + n(n + c - 1)y = 0.$$

(b) Deduce that, when n is a nonnegative integer, the polynomial solution $\varphi_n(x)$ of the original equation can be expressed in the form

$$\varphi_n(x) = \varphi_n(1)F\left(n + c - 1; -n; \frac{c}{2}; \frac{1 - x}{2}\right).$$

(c) Use this result and the results of Problems 26 and 69 to obtain the following relations:

$$T_n(x) = F\left(n, -n; \frac{1}{2}; \frac{1 - x}{2}\right) = J_n\left(\frac{1}{2}, 0, \frac{1 - x}{2}\right),$$

$$P_n(x) = F\left(n + 1, -n; 1; \frac{1 - x}{2}\right) = J_n\left(1, 1, \frac{1 - x}{2}\right),$$

$$S_n(x) = (n + 1)F\left(n + 2, -n; \frac{3}{2}; \frac{1 - x}{2}\right) = (n + 1)J_n\left(\frac{3}{2}, 2, \frac{1 - x}{2}\right).$$

Section 4.14

71. Show that the following differential equations each have an ordinary point at $x = \infty$ and in each case obtain two independent solutions expressed in the form $y = \sum A_k x^{-k}$:

(a) $x^2\frac{d^2y}{dx^2} + (1 + 2x)\frac{dy}{dx} = 0,$ (b) $x^4\frac{d^2y}{dx^2} + 2x^3\frac{dy}{dx} + y = 0.$

72. (a) Show that the differential equation

$$x^4\frac{d^2y}{dx^2} + y = 0$$

has a regular singular point at $x = \infty$ and obtain two independent solutions directly in the form $y = \sum A_k x^{-k-s}$.

(b) Obtain the same solutions by first making the change in variables $x = 1/t$.

73. Verify that the substitution $x = 1/t$ transforms the hypergeometric equation (175) to the form

$$t(1 - t)\frac{d^2y}{dt^2} + [(1 - \alpha - \beta) - (2 - \gamma)t]\frac{dy}{dt} + \frac{\alpha\beta}{t}y = 0,$$

and that this equation possesses a regular singular point at $t = 0$ unless $\alpha\beta = 0$ and $\alpha + \beta = 1$, in which case $t = 0$ is an ordinary point.

74. Show that the exponents of the differential equation obtained in Problem 73 are α and β at $t = 0$ and deduce that the transformed equation possesses solutions of the forms $t^\alpha u(t)$ and $t^\beta v(t)$, where $u(t)$ and $v(t)$ are regular at $t = 0$, unless α and β are equal or differ by an integer.

75. Verify that the substitution $y = t^\alpha u(t)$ transforms the equation of Problem 73 to the form

$$t(1 - t)\frac{d^2u}{dt^2} + [(1 + \alpha - \beta) - (2\alpha - \gamma + 2)t]\frac{du}{dt} - \alpha(1 + \alpha - \gamma)u = 0.$$

76. Show that the equation obtained in Problem 75 is of the form of Equation (175) if β is replaced by $1 + \alpha - \gamma$ and γ is replaced by $1 + \alpha - \beta$ in (175), and hence deduce that the functions

$$x^{-\alpha}F\left(\alpha, 1 + \alpha - \gamma; 1 + \alpha - \beta; \frac{1}{x}\right), \qquad x^{-\beta}F\left(\beta, 1 + \beta - \gamma; 1 - \alpha + \beta; \frac{1}{x}\right)$$

are independent solutions of the hypergeometric equation (175) for large values of x unless either α and β differ by an integer, in which case one of these expressions is undefined, or $\alpha = \beta$, in which case the two expressions are identical.

77. Use Equations (194) and (195) to verify the three-place values

(a) $J_0(10) = -0.246$,   (b) $Y_0(10) = 0.056$.

78. Show that the relations (194–199) are equalities for $p = \frac{1}{2}$.

79. Assume as a formal solution of the modified Bessel equation of order zero,

$$\frac{d^2y}{dx^2} + \frac{1}{x}\frac{dy}{dx} - y = 0,$$

an expression of the form

$$y \sim e^{rx}\sum_{k=0}^{\infty}A_k x^{-(k+s)}$$

and obtain the formal requirement

$$(r^2 - 1)A_0 x^{-s} + [(r^2 - 1)A_1 + r(1 - 2s)A_0]x^{-s-1}$$

$$+ \sum_{k=0}^{\infty}[(r^2 - 1)A_{k+2} - r(1 + 2k + 2s)A_{k+1} + (k + s)^2 A_k]x^{-s-k-2} = 0.$$

Hence deduce that, if $A_0 \neq 0$, there must follow either

$$r = 1, \qquad s = \frac{1}{2}, \qquad A_{k+1} = \frac{(2k + 1)^2}{8(k + 1)}A_k \qquad (k \geq 0)$$

or $$r = -1, \quad s = \frac{1}{2}, \qquad A_{k+1} = -\frac{(2k + 1)^2}{8(k + 1)}A_k \qquad (k \geq 0).$$

Thus obtain the formal solutions

$$y_1(x) \sim c_1\frac{e^x}{\sqrt{x}}\left[1 + \frac{1^2}{1!(8x)} + \frac{1^2 \cdot 3^2}{2!(8x)^2} - \cdots\right]$$

$$y_2(x) \sim c_2\frac{e^{-x}}{\sqrt{x}}\left[1 - \frac{1^2}{1!(8x)} + \frac{1^2 \cdot 3^2}{2!(8x)^2} - \cdots\right] \Bigg\} \qquad (x \to \infty).$$

[With $c_1 = 1/\sqrt{2\pi}$, $c_2 = \sqrt{\pi/2}$, these results are respectively the asymptotic (but *divergent*) expansions of $I_0(x)$ and $K_0(x)$, in agreement with (198), (199), and (202). In accordance with the statement at the end of Section 4.14, retention of only the first term would introduce an error of less than approximately 1 percent for values of x larger than about 12.]

5

Boundary-Value Problems and Characteristic-Function Representations

5.1. Introduction. In many problems the solution of an ordinary differential equation must satisfy certain conditions which are specified for *two or more* values of the independent variable. Such problems are called *boundary-value problems*, as distinct from *initial-value problems*, wherein all conditions are specified at one point.

A linear condition or equation is said to be *homogeneous* if, when it is satisfied by a particular function F, it is also satisfied by cF, where c is an arbitrary constant. For example, the requirement that a function or one of its derivatives (or some linear combination of the function and/or certain of its derivatives) *vanish* at a point is a homogeneous condition. In the present chapter we are mainly concerned with homogeneous linear ordinary differential equations and associated homogeneous boundary conditions.

In illustration, we may require a solution of a homogeneous linear equation of second order, of the form

$$\frac{d^2y}{dx^2} + a_1(x)\frac{dy}{dx} + a_2(x)y = 0 \tag{1}$$

which vanishes at the two points $x = a$ and $x = b$,

$$y(a) = 0, \qquad y(b) = 0. \tag{2}$$

Since the general solution of the differential equation is of the form

$$y = c_1 u_1(x) + c_2 u_2(x), \tag{3}$$

where u_1 and u_2 are linearly independent solutions and c_1 and c_2 are constants,

186

the boundary conditions constitute the requirements

$$c_1 u_1(a) + c_2 u_2(a) = 0,$$
$$c_1 u_1(b) + c_2 u_2(b) = 0. \tag{4}$$

One solution of these equations is $c_1 = c_2 = 0$, leading to the *trivial solution* $y \equiv 0$. If the determinant of the coefficients of c_1 and c_2 does not vanish, then (by Cramer's rule) this is *the only solution*. Hence, in order that nontrivial solutions exist, it is necessary that the determinant of coefficients vanish,

$$\begin{vmatrix} u_1(a) & u_2(a) \\ u_1(b) & u_2(b) \end{vmatrix} = 0. \tag{5}$$

If this condition exists, the two equations in (4) are in general equivalent and one constant can be expressed as a multiple of the second by use of either equation, the second constant then being arbitrary. Thus, *if* (5) *is satisfied*, the second equation of (4) may be discarded and the first equation then gives $c_2 u_2(a) = -c_1 u_1(a)$. If we write $c_1 = C u_2(a)$, there follows $c_2 = -C u_1(a)$, and the solution (3) becomes

$$y = C[u_2(a)u_1(x) - u_1(a)u_2(x)], \tag{6}$$

with C arbitrary.

It should be noticed that (6) is a nontrivial solution only if $u_1(a)$ and $u_2(a)$ are not both zero. If $u_1(a) = u_2(a) = 0$, the first equation of (4) is the trivial identity, and the *second* equation must be used to relate c_1 and c_2. This process leads to a nontrivial solution of the form

$$y = C[u_2(b)u_1(x) - u_1(b)u_2(x)],$$

assuming that $u_1(b)$ and $u_2(b)$ are not also both zero. If $u_1(x)$ and $u_2(x)$ should both vanish at $x = a$ and at $x = b$, then (3) would satisfy (2) for arbitrary values of both c_1 and c_2.

In many cases one or both of the coefficients $a_1(x)$ and $a_2(x)$ in (1), and hence the solutions $u_1(x)$ and $u_2(x)$, depend upon a constant parameter λ which may take on various constant values in a particular discussion. In such cases the determinant (5) may vanish for certain specific values of λ, say $\lambda = \lambda_1$, $\lambda_2, \ldots$. For each such value of λ a solution of type (6), involving an arbitrary multiplicative factor, is then obtained. Problems of this sort are known as *characteristic-value problems*; the values of λ for which nontrivial solutions exist are called the *characteristic values* of λ, and the corresponding solutions (with convenient choices of the arbitrary multiplicative constants) are called the *characteristic functions* of the problem. The terms "eigenvalues" and "eigenfunctions" are also frequently used.

In the remainder of this chapter we first consider certain examples of such problems, and then establish and investigate certain rather general properties of characteristic functions which are extremely useful in a wide class of related problems.

5.2. The Rotating String. We consider the problem of determining the form assumed by a tightly stretched flexible string of length l and linear density ρ, rotating with uniform angular velocity ω about its equilibrium position along the x axis. It is assumed that the initial tension in the string is so large that additional nonuniform stress introduced by the curvature of the string is relatively negligible. Denoting the displacement from the axis of rotation by $y(x)$ and the uniform tensile force by T, and considering only small displacements and slopes, the condition of force equilibrium is of the form

$$T\frac{d^2y}{dx^2} + \rho\omega^2 y = 0. \tag{7}$$

If we consider a small element of the deformed string, projecting into the interval $(x, x + \Delta x)$, the y component of the tensile force is given by $-T\,dy/ds$ at the end (x, y) and by $(T\,dy/ds)_{x+\Delta x}$ at the end $(x + \Delta x, y + \Delta y)$, where s is arc length measured along the string. The differential resultant force on the element is thus given by $(d/dx)(T\,dy/ds)\,\Delta x$. If the distributed external force, per unit distance along the string, in the y direction is denoted by Y, then the requirement of differential force equilibrium is

$$\frac{d}{dx}\left(T\frac{dy}{ds}\right)\Delta x + Y\Delta s = 0.$$

By dividing by Δx, letting $\Delta x \longrightarrow 0$, and replacing Y in the present case by the inertia force $\rho\omega^2 y$, we thus obtain the equation

$$\frac{d}{ds}\left(T\frac{dy}{ds}\right) + \rho\omega^2 y = 0.$$

For small slopes we have

$$\frac{d(\cdots)}{ds} = \frac{\dfrac{d(\cdots)}{dx}}{\sqrt{1 + \left(\dfrac{dy}{dx}\right)^2}} \approx \frac{d(\cdots)}{dx} \qquad \left(\left|\frac{dy}{dx}\right| \ll 1\right).$$

With this approximation and the assumption of nearly uniform tensile force, we obtain Equation (7).

Restricting attention to the case of a string of *uniform* density, we define the constant parameter

$$\lambda = \frac{\rho\omega^2}{T} \tag{8}$$

and write (7) in the form

$$\frac{d^2y}{dx^2} + \lambda y = 0. \tag{9}$$

If one end of the string is attached to the axis of rotation at the point $x = 0$, the end condition

$$y(0) = 0 \tag{10}$$

must be satisfied. If the other end is also attached to the axis, the second end condition is

$$y(l) = 0. \tag{11a}$$

If, however, the second end is attached to a yielding support equivalent to an elastic spring which exerts a restoring force proportional to its stretch, toward the axis of rotation, the second end condition is of the form $T\,dy/dx = -ky$, where k is the "spring constant." With the abbreviation

$$\alpha = \frac{T}{kl},$$

where α is a *dimensionless* constant inversely proportional to the elastic modulus of the spring support, this end condition can be written in the more convenient form

$$\alpha l y'(l) = -y(l). \tag{11b}$$

In the limiting case where $\alpha = 0$, Equation (11b) reduces to the condition of fixity (11a); in the limiting case $\alpha = \infty$, the end at $x = l$ is not restrained from motion normal to the axis of rotation, and we are led to the "free-end" condition

$$y'(l) = 0. \tag{11c}$$

In order that the hypothesis of nearly uniform tension be realized, it must be supposed that the "free end" is, however, restricted from appreciable movement in the direction of the axis of rotation.

The most general solution of Equation (9) satisfying (10) is of the form

$$y = C \sin \sqrt{\lambda}\, x. \tag{12}$$

If the end $x = l$ is *fixed*, the condition of (11a) becomes

$$C \sin \sqrt{\lambda}\, l = 0. \tag{13}$$

Hence nontrivial solutions of this problem exist only if λ has a value such that

$$\sqrt{\lambda}\, l = n\pi \qquad (n = 1, 2, \dots). \tag{14}$$

If these values of λ are ordered as $\lambda_1, \lambda_2, \dots, \lambda_n, \dots$, with respect to their magnitude, we can then write

$$\lambda_n = \frac{n^2 \pi^2}{l^2}, \tag{15}$$

where n is an integer. For each such value of λ, the solution (12) can be written in the form

$$y = C\varphi_n(x),$$

where

$$\varphi_n(x) = \sin \frac{n\pi x}{l}. \tag{16}$$

It is clear that only *positive* integral values of n need be considered.

Thus the boundary-value problem consisting of the differential equation (9) and the homogeneous boundary conditions (10) and (11a) has *no solution* other than the trivial solution $y \equiv 0$, unless λ has one of the *characteristic values* given by (15). Corresponding to each characteristic value of λ, there exists a characteristic function $\varphi_n(x)$, given by (16), such that any constant multiple of this function is a solution of the problem.

Each characteristic value λ_n corresponds to a specific value of the angular velocity ω (for given T and ρ), according to the notation of (8),

$$\omega_n = \frac{n\pi}{l}\sqrt{\frac{T}{\rho}} \qquad (n = 1, 2, \dots).$$

(17)

These velocities are known as the *critical speeds* of the rotating string.

The present analysis indicates that for speeds smaller than ω_1,

$$0 < \omega < \frac{\pi}{l}\sqrt{\frac{T}{\rho}},$$

the only stable position of the string is its undeformed position along the axis of rotation. However, if the speed is continuously increased until the first critical speed is attained ($\omega = \omega_1$), a possible new equilibrium form or *deflection mode*

$$y = C \sin \pi \frac{x}{l}$$

may exist. Corresponding to the second critical speed ($\omega = \omega_2 = 2\omega_1$) another mode

$$y = C \sin 2\pi \frac{x}{l}$$

may exist, and so on.

It may be seen that only the *shapes* of the string modes are determined, the amplitudes being apparently arbitrary. This indeterminacy is a result of the approximations made in deriving the linearized formulation of the problem and can be removed by a more nearly exact analysis of the true nonlinear problem.

Thus, if use is made of the additional relation $T\,dx/ds = T_0 = $ constant, which follows from the condition of force equilibrium in the x direction (in the absence of *external* force components in that direction), the tension T can be eliminated from the relation

$$\frac{d}{ds}\left(T\frac{dy}{ds}\right) + \rho\omega^2 y = 0$$

to yield the nonlinear equation

$$T_0\frac{d^2 y}{dx^2} + \rho\omega^2 y \sqrt{1 + \left(\frac{dy}{dx}\right)^2} = 0.$$

When ρ is constant, the appropriate solution of this equation can be obtained by use of so-called *elliptic integrals*,† after which T is given by the relation

$$T = T_0\sqrt{1 + \left(\frac{dy}{dx}\right)^2}.$$

---

†The integrals

$$F(k, \varphi) = \int_0^{\varphi} \frac{d\theta}{\sqrt{1 - k^2 \sin^2 \theta}} \quad \text{and} \quad E(k, \varphi) = \int_0^{\varphi} \sqrt{1 - k^2 \sin^2 \theta}\, d\theta,$$

where $0 < k < 1$, are known as *elliptic integrals of the first and second kind*, respectively, and are tabulated. For reductions expressing related integrals in terms of these integrals, see Reference 4 of Chapter 7.

If the end $x = l$ is attached to the axis of rotation by an elastic spring, the requirement that (12) satisfy (11b) becomes

$$-\alpha\sqrt{\lambda}l \cos \sqrt{\lambda}l = \sin \sqrt{\lambda}l.$$

With the introduction of a dimensionless parameter μ, of the form

$$\mu = \sqrt{\lambda}l, \qquad (18)$$

this condition can be written

$$\tan \mu = -\alpha\mu. \qquad (19)$$

That there are an infinite number of values of μ which satisfy (19), for a given value of α, is readily seen if the two curves $y = \tan \mu$ and $y = -\alpha\mu$ are plotted together and the desired roots are recognized as the values of μ corresponding to the intersections.† If we order the positive roots with respect to magnitude as $\mu_1, \mu_2, \ldots, \mu_n, \ldots$, the corresponding *characteristic values* of λ are given by

$$\lambda_n = \frac{\mu_n^2}{l^2} \qquad (20)$$

and the corresponding *deflection modes* are of the form

$$\varphi_n(x) = \sin \mu_n \frac{x}{l}. \qquad (21)$$

Finally, the corresponding *critical speeds* are given by

$$\omega_n = \frac{\mu_n}{l}\sqrt{\frac{T}{\rho}}. \qquad (22)$$

In the limiting case $\alpha = \infty$, when the end $x = l$ is not restrained from moving normal to the axis of rotation, Equation (19) gives

$$\cos \mu = 0, \qquad \mu_n = \frac{2n-1}{2}\pi. \qquad (23)$$

The characteristic values of λ are thus

$$\lambda_n = \left(\frac{2n-1}{2}\right)^2 \frac{\pi^2}{l^2} \qquad (n = 1, 2, \ldots), \qquad (24)$$

with corresponding characteristic modes

$$\varphi_n(x) = \sin \frac{2n-1}{2}\pi \frac{x}{l}, \qquad (25)$$

and the critical speeds are given by

$$\omega_n = \frac{2n-1}{2}\frac{\pi}{l}\sqrt{\frac{T}{\rho}}. \qquad (26)$$

---

†Once approximate values of roots of transcendental equations such as (19) have been obtained graphically, improved values often can be obtained by a "feedback" method, in which an approximation is inserted in *one side* of the equality and the resultant equation is then solved for the next approximation. *Newton's method* (see Section 7.10) is also useful.

5.3. The Rotating Shaft. We next consider the determination of the possible deflection modes of an originally straight shaft of length l, rotating with uniform angular velocity ω about its equilibrium position along the x axis. According to the elementary theory of bending of beams, the deflection y from the straight form is determined (approximately) by the differential equation

$$\frac{d^2}{dx^2}\left(EI\frac{d^2y}{dx^2}\right) - \rho\omega^2 y = 0, \tag{27}$$

where E is Young's modulus, I is the moment of inertia of a cross section about an axis perpendicular to the xy plane, and ρ is the linear density of the shaft material.

If we consider the portion of the shaft to the left of an arbitrary cross section (x positive to the right, y positive upward), the influence of the right-hand portion of the shaft on the portion considered can be resolved into a vertical *shearing force S* and a *bending moment M*. We choose the convention that S is then positive *downward* and M is positive when *counterclockwise* (see Figure 5.1). Thus, if the portion considered is to be in equilibrium, S must equal the algebraic sum of the actual *upward* forces (distributed and concentrated) acting to the left of the section considered, and M must equal the resultant *clockwise* moment, about the section chosen, of all forces acting on the portion of the

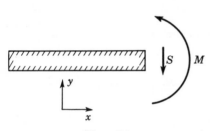

Figure 5.1

shaft to the left of the section. With these conventions, there follows also

$$M = EI\frac{d^2y}{dx^2} \tag{28}$$

and

$$S = \frac{d}{dx}\left(EI\frac{d^2y}{dx^2}\right). \tag{29}$$

If we consider a small element of the shaft, in the interval $(x, x + \Delta x)$, and assume small deflections and slopes, the differential unbalanced vertical force acting on the element is given by $-(dS/dx)\,\Delta x$ and the differential unbalanced counterclockwise moment is $(dM/dx)\,\Delta x - S\,\Delta x$. With the distributed external load intensity (positive in the positive y direction) denoted by $Y(x)$, the conditions of differential force and moment equilibrium become

$$-\frac{dS}{dx}\Delta x + Y\Delta x = 0, \qquad \frac{dM}{dx}\Delta x - S\Delta x = 0,$$

and there follows

$$\frac{dM}{dx} = S, \qquad \frac{dS}{dx} = Y.$$

The elementary theory of bending leads to the approximate relationship

$$\frac{EI}{R} = M,$$

where $1/R$ is the curvature of the shaft and is approximated by d^2y/dx^2 if the slope of the shaft is small. We thus obtain the relations (28) and (29), in addition to the equation

$$\frac{d^2}{dx^2}\left(EI\frac{d^2y}{dx^2}\right) = Y(x),$$

which reduces to (27) if Y is replaced by the inertia loading $\rho\omega^2 y$.

In addition to the differential equation (27), four boundary conditions must be prescribed, two conditions being imposed at each end of the shaft. Types of homogeneous end conditions of particular interest are the following:

(1) *Hinged End.* In this case the lateral displacement y and the bending moment M must vanish, at the end considered, giving the two conditions

$$y = 0, \qquad EI\frac{d^2y}{dx^2} = 0. \tag{30a}$$

(2) *Fixed End.* Here the displacement y and the slope dy/dx must vanish,

$$y = 0, \qquad \frac{dy}{dx} = 0. \tag{30b}$$

(3) *Free End.* If no moment or shearing force is applied at an end, there follows

$$EI\frac{d^2y}{dx^2} = 0, \qquad \frac{d}{dx}\left(EI\frac{d^2y}{dx^2}\right) = 0. \tag{30c}$$

(4) *Sliding Clamped End.* If one end of the shaft is constrained to retain zero slope but is completely free to move in the y direction, the slope dy/dx and the shear S must vanish at that end,

$$\frac{dy}{dx} = 0, \qquad \frac{d}{dx}\left(EI\frac{d^2y}{dx^2}\right) = 0. \tag{30d}$$

(5) *Elastically Supported End.* If the motion of one end in the y direction is partially restrained as by an elastic spring, with modulus k_1, the magnitude of the shearing force must be k_1 times the displacement,

$$\frac{d}{dx}\left(EI\frac{d^2y}{dx^2}\right) = \pm k_1 y. \tag{30e}$$

If the change in slope of the end is also partially restrained as by a spring system such that the restraining moment is proportional to the slope, the second condition at that end is of the form

$$EI\frac{d^2y}{dx^2} = \pm k_2\frac{dy}{dx}. \tag{30f}$$

By taking the four possible combinations of the limiting forms of these conditions when k_1 and k_2 are zero or infinitely large, the four preceding cases are attained. Specifically, $k_1 = \infty$, $k_2 = 0$ gives (30a); $k_1 = k_2 = \infty$ gives (30b); $k_1 = k_2 = 0$ gives (30c); and $k_1 = 0$, $k_2 = \infty$ gives (30d).

We restrict attention to the case of a *uniform* shaft, for which EI and ρ are constant, and introduce the constant parameter

$$\lambda = \frac{\rho\omega^2}{EI}. \tag{31}$$

The differential equation (27) then becomes

$$\frac{d^4y}{dx^4} - \lambda y = 0, \tag{32}$$

with general solution of the form

$$y = c_1 \sinh \sqrt[4]{\lambda}x + c_2 \cosh \sqrt[4]{\lambda}x + c_3 \sin \sqrt[4]{\lambda}x + c_4 \cos \sqrt[4]{\lambda}x. \tag{33}$$

In the special case of a rotating shaft *hinged at both ends*, we take the origin at one end of the shaft and impose the boundary conditions

$$y(0) = y''(0) = 0, \qquad y(l) = y''(l) = 0. \tag{34}$$

From the conditions at $x = 0$ we obtain

$$c_2 + c_4 = 0, \qquad c_2 - c_4 = 0,$$

from which there follows

$$c_2 = c_4 = 0. \tag{35}$$

To simplify the notation in the remaining work, it is convenient to introduce the dimensionless parameter μ, where

$$\mu = \sqrt[4]{\lambda}\, l, \qquad \mu^2 = \omega l^2 \sqrt{\frac{\rho}{EI}}. \tag{36}$$

Making use of (35) and (36), we then write (33) in the form

$$y = c_1 \sinh \mu\frac{x}{l} + c_3 \sin \mu\frac{x}{l}. \tag{37}$$

The last two conditions of (34) then become

$$c_1 \sinh \mu + c_3 \sin \mu = 0, \tag{38a}$$

$$c_1 \sinh \mu - c_3 \sin \mu = 0. \tag{38b}$$

These equations are compatible only if the determinant of coefficients vanishes,

$$\begin{vmatrix} \sinh \mu & \sin \mu \\ \sinh \mu & -\sin \mu \end{vmatrix} = -2 \sinh \mu \sin \mu = 0.$$

Since the value $\mu = 0$ leads to a trivial solution $y = 0$, the only permissible values of μ which are of interest are the positive roots of the equation $\sin \mu = 0$,

$$\mu_n = n\pi \qquad (n = 1, 2, \ldots). \tag{39}$$

For such values of μ, Equations (38a, b) require that

$$c_1 = 0. \tag{40}$$

Hence the characteristic functions corresponding to permissible values μ_n are of the form $y = c_3 \varphi_n(x)$, where

$$\varphi_n(x) = \sin \frac{n\pi x}{l}. \tag{41}$$

These *deflection modes* are of the same form as the modes of a rotating flexible string with fixed ends. The characteristic values of λ in the present case are given by (36),

$$\lambda_n = \frac{n^4 \pi^4}{l^4}, \tag{42}$$

and the critical speeds are given by

$$\omega_n = \frac{n^2 \pi^2}{l^2} \sqrt{\frac{EI}{\rho}} \qquad (n = 1, 2, \ldots). \tag{43}$$

The smallest critical speed corresponds to $n = 1$. If we denote this *fundamental critical speed by* Ω,

$$\Omega = \frac{\pi^2}{l^2} \sqrt{\frac{EI}{\rho}}, \tag{44}$$

then the nth critical speed is given by

$$\omega_n = n^2 \Omega. \tag{45}$$

Thus, whereas Equation (17) shows that for a rotating string with fixed ends the successive critical speeds increase in proportion to the integers, Equation (45) shows that for a shaft hinged at both ends the speeds increase in proportion to the *squares* of the integers.

5.4. Buckling of Long Columns under Axial Loads. As a final example, we next consider a long shaft or column subjected to an axial compressive force P applied at an end and investigate possible modes of lateral deflection from the initial equilibrium position along the x axis. The differential equation determining the deflection y, according to the elementary theory, is then of the form

$$\frac{d^2}{dx^2}\left(EI \frac{d^2 y}{dx^2}\right) + P \frac{d^2 y}{dx^2} = 0. \tag{46}$$

The bending moment M_P at an arbitrary section x, due to a compressive force P applied in the negative x direction at the end $x = L$, is given by $P[y(L) - y(x)]$. Hence (see Section 5.3) the effect of the axial load can be taken into account if a fictitious distributed lateral loading of intensity

$$Y_P(x) = \frac{d^2 M_P}{dx^2} = -P \frac{d^2 y}{dx^2}$$

is introduced. If the *actual* distributed lateral load intensity is denoted by $Y(x)$, the differential equation governing lateral deflection of the column is then

$$\frac{d^2}{dx^2}\left(EI \frac{d^2 y}{dx^2}\right) = Y(x) - P \frac{d^2 y}{dx^2}.$$

This equation reduces to (46) in the absence of distributed lateral loading.

The boundary conditions to be imposed depend upon the nature of the end restraints, as was outlined in Section 5.3.

Restricting attention to a *uniform* column, we write

$$\lambda = \frac{P}{EI}, \tag{47}$$

and Equation (46) then becomes

$$\frac{d^4 y}{dx^4} + \lambda \frac{d^2 y}{dx^2} = 0, \tag{48}$$

with general solution of the form

$$y = c_1 \sin \sqrt{\lambda} x + c_2 \cos \sqrt{\lambda} x + c_3 x + c_4. \tag{49}$$

In the special case when the two ends $x = 0$ and $x = l$ are *both hinged*, the conditions

$$y(0) = y''(0) = 0$$

at the end $x = 0$ give $c_2 + c_4 = 0$ and $c_2 = 0$, from which there follows

$$c_2 = c_4 = 0, \tag{50}$$

and Equation (49) takes the form

$$y = c_1 \sin \sqrt{\lambda} x + c_3 x. \tag{51}$$

The conditions

$$y(l) = y''(l) = 0$$

at the end $x = l$ then give

$$c_1 \sin \sqrt{\lambda} l + c_3 l = 0, \qquad c_1 \sin \sqrt{\lambda} l = 0 \tag{52}$$

so that nontrivial solutions of the problem exist only if $\sin \sqrt{\lambda} l = 0$ and $c_3 = 0$. Thus the characteristic values of λ are of the form

$$\lambda_n = \frac{n^2 \pi^2}{l^2} \qquad (n = 1, 2, \ldots), \tag{53}$$

and the corresponding characteristic functions are the deflection modes

$$\varphi_n(x) = \sin \frac{n \pi x}{l}. \tag{54}$$

The values of the axial load P corresponding to these modes are known as the *critical buckling loads*, and are given by (47) and (53) in the form

$$P_n = n^2 \pi^2 \frac{EI}{l^2} \qquad (n = 1, 2, \ldots). \tag{55}$$

For axial loads smaller in magnitude than the smallest critical value

$$P_1 = \pi^2 \frac{EI}{l^2}, \tag{56}$$

the column is stable only in its unbent position. That is, if the column is artificially imparted a small amount of bending, it will tend to return to its initial straight form. When $P = P_1$, however, this theory predicts that under a small disturbance it will assume a form

$$y = C \sin\frac{\pi x}{l}, \tag{57}$$

known as the *fundamental buckling mode*. Although the linear theory does not predict the amplitude C of the sinusoidal deformation, a more nearly exact analysis shows that the amplitude will increase with a small increase in the axial load P above the critical value P_1, until bending failure of the column occurs. The critical value (56) is known as the *Euler load* of a column hinged at both ends.

In practice, failure is not caused entirely by bending except for very long columns, and the structure may fail before the Euler load is attained. However, the Euler load does give an *upper limit* of stability in the undeformed position.

The higher modes cannot be attained in practice unless additional constraints are introduced.

If *the end $x = 0$ is hinged and the end $x = l$ is fixed*, we find, by imposing the conditions

$$y(l) = y'(l) = 0$$

on Equation (51) and introducing the dimensionless parameter

$$\mu = \sqrt{\lambda}\, l = \sqrt{\frac{P}{EI}}\, l, \tag{58}$$

that the critical loads are then of the form

$$P_n = \mu_n^2 \frac{EI}{l^2}, \tag{59}$$

where μ_n is a positive solution of the transcendental equation

$$\tan \mu = \mu. \tag{60}$$

The corresponding deflection modes are of the form

$$\varphi_n(x) = \frac{\sin(\mu_n x/l)}{\sin \mu_n} - \frac{x}{l}. \tag{61}$$

Since the smallest positive root of (60) is approximately

$$\mu_1 = 1.43\pi,$$

the Euler load in this case is approximately

$$P_1 = 2.05\pi^2 \frac{EI}{l^2}. \tag{62}$$

5.5. The Method of Stodola and Vianello. An iterative procedure, which is sometimes known as the *method of Stodola and Vianello*, often is useful for the *approximate* determination of the characteristic numbers and functions of a boundary-value problem. Before outlining the application of this procedure to problems of a rather general type, we first illustrate its use by considering a previously solved problem.

In the case of a rotating string fixed at its ends (Section 5.2), the boundary-value problem consisting of the differential equation

$$\frac{d^2 y}{dx^2} + \lambda y = 0 \tag{63}$$

and the end conditions

$$y(0) = 0, \qquad y(l) = 0 \tag{64}$$

was found to have the characteristic values

$$\lambda_n = \frac{n^2 \pi^2}{l^2} \qquad (n = 1, 2, \ldots), \tag{65}$$

with corresponding characteristic functions

$$\varphi_n = \sin \frac{n\pi x}{l}. \tag{66}$$

The method to be presented is a method of successive approximations leading to approximate expressions for the *smallest* characteristic value and the corresponding characteristic function, although the method can be modified in such a way that the higher modes are obtained.†

We first transpose the term involving λ to the right in Equation (63), to obtain

$$\frac{d^2 y}{dx^2} = -\lambda y. \tag{67}$$

Next, we replace the unknown function y on the right-hand side of (67) by a conveniently chosen first approximation $y_1(x)$. It is preferable that this function satisfy the prescribed end conditions, although this is not necessary. The resulting equation,

$$\frac{d^2 y}{dx^2} = -\lambda y_1(x), \tag{68}$$

then can be solved directly and the two constants of integration can be determined so that the solution satisfies the prescribed end conditions. It is clear that the solution so obtained will contain the parameter λ as a factor, and hence can be written in the form

$$y = \lambda f_1(x). \tag{69}$$

Now, if the originally assumed function $y_1(x)$ were actually a characteristic function of the problem, the function $y(x)$ given by (69) would necessarily become identical with the assumed function $y_1(x)$ if λ were assigned the corresponding characteristic value. That is, the corresponding characteristic value of λ would be given by the constant ratio $y_1(x)/f_1(x)$.

†See Problem 47 and Reference 1.

Thus, if we take

$$y_1(x) = \sin\frac{\pi x}{l},$$

the solution of (68) for which $y(0) = y(l) = 0$ is readily found to be

$$y(x) = \lambda\frac{l^2}{\pi^2}\sin\frac{\pi x}{l}.$$

In order that $y_1(x) = y(x)$, we must then have $\lambda = \pi^2/l^2$, in accordance with (65) and (66).

However, since $y_1(x)$ is generally not a characteristic function, the functions $y_1(x)$ and $f_1(x)$ will in general not be in a constant ratio. Still, as will be shown in Section 5.9, if a convenient multiple of $f_1(x)$ is taken as a new approximation $y_2(x)$ and the process is repeated indefinitely, the ratio $y_n(x)/f_n(x)$ tends toward a constant value over the interval $(0, l)$ as the number n of cycles is increased, and this constant value is exactly the *smallest* characteristic value λ_1. Also, the sequence of successive functions $y_1(x), y_2(x), \ldots, y_n(x), \ldots$ converges to the corresponding characteristic function.

Successive estimates of the characteristic value λ_1 may be obtained after each cycle by requiring that the functions $y_n(x)$ and $y(x) = \lambda f_n(x)$ agree as well as possible (in some sense) over the interval $(0, l)$. In particular, a simple procedure consists of determining λ so that the integral, over $(0, l)$, of the difference between the functions be zero. Thus the nth approximation to the smallest permissible value of λ is given by

$$\lambda_1^{(n)} = \frac{\int_0^l y_n(x)\,dx}{\int_0^l f_n(x)\,dx}. \tag{70}$$

To illustrate this procedure in the present problem, we choose as our first approximation the function

$$y_1(x) = x(l - x),$$

which satisfies the end conditions (64). If y is replaced by $y_1(x)$ on the right-hand side of (67), there results

$$\frac{d^2y}{dx^2} = \lambda(x^2 - lx),$$

and the solution of this equation satisfying (64) is found by direct integration in the form

$$y = \frac{\lambda}{12}(x^4 - 2lx^3 + l^3x) \equiv \lambda f_1(x).$$

If the integrals of $y_1(x)$ and $y(x) = \lambda f_1(x)$ over $(0, l)$ are equated, there follows

$$\frac{l^3}{6} = \lambda\frac{l^5}{60},$$

from which the first approximation to the smallest characteristic value of λ is

$$\lambda_1^{(1)} = \frac{10}{l^2}.$$

Comparison with the exact value $\pi^2/l^2 = 9.870/l^2$ shows that the error is about 1.3 percent.

If now we take

$$y_2(x) = x^4 - 2lx^3 + l^3x$$

as the initial approximation of the next cycle, we obtain

$$y = \lambda f_2(x) = -\frac{\lambda}{30}(x^6 - 3lx^5 + 5l^3x^3 - 3l^5x),$$

and Equation (70) gives the second approximation to λ_1,

$$\lambda_1^{(2)} = \frac{9.882}{l^2},$$

which differs from the exact value by about 0.12 percent.

The procedure as illustrated in this problem applies without essential modification to somewhat more involved boundary-value problems. Thus, for example, we may have differential equations of the form

$$\frac{d}{dx}\left[p(x)\frac{dy}{dx}\right] + \lambda r(x)y = 0 \tag{71a}$$

or

$$\frac{d^2}{dx^2}\left[p(x)\frac{d^2y}{dx^2}\right] + \lambda r(x)y = 0, \tag{71b}$$

with appropriate homogeneous end conditions. If the term involving λ is transposed to the right and if the factor y in its coefficient is then replaced by a suitable approximation $y_1(x)$, the resultant differential equation can be solved by direct integration for a new function of the form $y = \lambda f_1(x)$ and the remaining procedure outlined above is directly applicable.

In place of using Equation (70) to determine successive approximations to λ_1, one may make use of either of the following formulas (see Section 5.9):

$$\lambda_1^{(n)} = \frac{\int_a^b r(x)[y_n(x)]^2\,dx}{\int_a^b r(x)f_n(x)y_n(x)\,dx} \tag{72a}$$

or

$$\lambda_1^{(n)} = \frac{\int_a^b r(x)f_n(x)y_n(x)\,dx}{\int_a^b r(x)[f_n(x)]^2\,dx}. \tag{72b}$$

Either of these formulas will in general give a better approximation to the true value of λ_1, the last form being in general the most accurate.† In illustration,

†It can be shown that, in the general case of (71a) or (71b), the result of using (72a) in the nth cycle is comparable with the result of using (70) in the $(2n)$th cycle, while the result of using (72b) in the nth cycle is comparable with that of using (70) in the $(2n + 1)$th cycle. In the case of (71a), if y_1 is such that $(py_1')' = $ constant, these pairs of approximations are respectively equal. (See Problem 43.) The same is true of (71b) if $(py_1')' = $ constant.

we list the results of using the three formulas to estimate $l^2 \lambda_1$ in the preceding example.

n	(70)	(72a)	(72b)
1	10.00000	9.88235	9.87096
2	9.88235	9.86975	
∞	9.86960		

The tabulated values of $l^2 \lambda_1^{(n)}$ show that in this case only one approximation, using (72b), is needed to obtain an approximate value of λ_1 correct to within 2 parts in 10,000.

As a further application we investigate the Euler load of a column of length l, so constructed that its bending stiffness factor EI is of the form

$$EI(x) = C \frac{x}{l}, \tag{73}$$

where C is the maximum stiffness factor (at $x = l$). We suppose that both ends of the column are *hinged*. Then the bending moment $M(x)$ at a section x is due entirely to the axial load P and is given by $M = -Py(x)$. Hence, from Equation (28), the relevant differential equation is of the form

$$C \frac{x}{l} \frac{d^2 y}{dx^2} + Py = 0. \tag{74}$$

(See also Problem 15.) If we introduce the dimensionless parameter

$$\mu^2 = \frac{Pl^2}{C}, \tag{75}$$

Equation (74) becomes

$$x \frac{d^2 y}{dx^2} + \frac{\mu^2}{l} y = 0. \tag{76}$$

The end conditions are of the form

$$y(0) = 0, \qquad y(l) = 0. \tag{77}$$

According to Equation (130d) of Chapter 4, the general solution of (76) is of the form

$$y = \sqrt{x} Z_1 \left(2\mu \sqrt{\frac{x}{l}} \right)$$

or

$$y = c_1 \sqrt{x} J_1 \left(2\mu \sqrt{\frac{x}{l}} \right) + c_2 \sqrt{x} Y_1 \left(2\mu \sqrt{\frac{x}{l}} \right).$$

Since, for small values of x, we have

$$\sqrt{x} J_1 \left(2\mu \sqrt{\frac{x}{l}} \right) \sim \frac{\mu}{\sqrt{l}} x, \qquad \sqrt{x} Y_1 \left(2\mu \sqrt{\frac{x}{l}} \right) \sim -\frac{\sqrt{l}}{\pi \mu},$$

the condition $y(0) = 0$ requires that

$$c_2 = 0,$$

and the remaining condition $y(l) = 0$ requires that μ be a solution of the equation

$$J_1(2\mu) = 0. \tag{78}$$

A rounded value of the smallest positive root of this equation (see Table 1 of Section 5.13) is

$$\mu_1 = 1.916. \tag{79}$$

Correspondingly, the Euler load is given by

$$P_1 = \mu_1^2 \frac{C}{l^2} = 3.670 \frac{C}{l^2}, \tag{80}$$

and the fundamental buckling mode is given by

$$\varphi(x) = \sqrt{x} J_1 \left(3.832 \sqrt{\frac{x}{l}} \right). \tag{81}$$

To find an *approximate* value of μ, we take as our first approximation the function

$$y_1(x) = x(l - x). \tag{82}$$

From (76) we then have

$$\frac{d^2 y}{dx^2} = \frac{\mu^2}{l}(x - l),$$

from which there follows

$$y = \frac{\mu^2}{6l}(x^3 - 3lx^2 + 2l^2 x) \equiv \mu^2 f_1(x). \tag{83}$$

The condition

$$\int_0^l y_1(x)\,dx = \mu^2 \int_0^l f_1(x)\,dx \tag{84}$$

then gives, as a first approximation,

$$\mu_1^{(1)} = 2, \tag{85}$$

leading to an approximation

$$P_1^{(1)} = 4\frac{C}{l^2} \tag{86}$$

for the Euler load. A second approximation gives

$$\mu_1^{(2)} = 1.936, \qquad P_1^{(2)} = 3.750\frac{C}{l^2}. \tag{87}$$

If the condition

$$\int_0^l r(x)[y_n(x)]^2\,dx = \mu^2 \int_0^l r(x)f_n(x)y_n(x)\,dx$$

with

$$r(x) = \frac{1}{x},$$

corresponding to (72a), is used in place of (84) to determine approximations to μ_1, the improved values $\mu_1^{(1)} = 1.936$ and $\mu_1^{(2)} = 1.917$ are obtained. Use of (72b) leads to the successive approximations 1.922 and 1.916.

An investigation of the convergence of the iterative procedure to the lowest characteristic mode, as well as an indication of modifications leading to the determination of higher modes, is presented in Section 5.9.

5.6. Orthogonality of Characteristic Functions. Two functions $\varphi_m(x)$ and $\varphi_n(x)$ are said to be *orthogonal* on an interval (a, b) if the integral of the product $\varphi_m \varphi_n$ over that interval vanishes:

$$\int_a^b \varphi_m(x)\varphi_n(x)\,dx = 0. \tag{88}$$

More generally, the functions $\varphi_m(x)$ and $\varphi_n(x)$ are said to be *orthogonal with respect to a weighting function* $r(x)$, on an interval (a, b), if

$$\int_a^b r(x)\varphi_m(x)\varphi_n(x)\,dx = 0. \tag{89}$$

Finally, a *set* of functions is said to be orthogonal on (a, b) if all pairs of distinct functions in the set are orthogonal on (a, b).

An extremely useful property of boundary-value problems of a rather general type consists of the fact that the sets of characteristic functions corresponding to such problems are orthogonal with respect to a weighting function. In order to establish this fact, we consider first the boundary-value problem consisting of the linear homogeneous second-order differential equation

$$\frac{d}{dx}\left[p(x)\frac{dy}{dx}\right] + [q(x) + \lambda r(x)]y = 0 \tag{90}$$

and suitably prescribed homogeneous boundary conditions, to be specified presently, at the ends of an interval (a, b). The functions p, q, and r are assumed to be *real*. In terms of the operator

$$L = \frac{d}{dx}\left(p\frac{d}{dx}\right) + q = p\frac{d^2}{dx^2} + \frac{dp}{dx}\frac{d}{dx} + q, \tag{91}$$

Equation (90) can be written in the abbreviated form

$$Ly + \lambda r(x)y = 0. \tag{92}$$

We notice that *any equation of the form*

$$a_0(x)\frac{d^2y}{dx^2} + a_1(x)\frac{dy}{dx} + [a_2(x) + \lambda a_3(x)]y = 0 \tag{93}$$

can be written in the form of (90) by setting

$$p = e^{\int a_1/a_0\,dx}, \qquad q = \frac{a_2}{a_0}p, \qquad r = \frac{a_3}{a_0}p. \tag{94}$$

Suppose now that λ_1 and λ_2 are any two different characteristic values of the problem considered and that the corresponding characteristic functions are

$y = \varphi_1(x)$ and $y = \varphi_2(x)$, respectively. There then follows

$$\frac{d}{dx}\left(p\frac{d\varphi_1}{dx}\right) + (q + \lambda_1 r)\varphi_1 = 0,$$

$$\frac{d}{dx}\left(p\frac{d\varphi_2}{dx}\right) + (q + \lambda_2 r)\varphi_2 = 0. \tag{95}$$

If the first of these equations is multiplied by $\varphi_2(x)$ and the second by $\varphi_1(x)$, and the resultant equations are subtracted from each other, there follows

$$\varphi_2\frac{d}{dx}\left(p\frac{d\varphi_1}{dx}\right) - \varphi_1\frac{d}{dx}\left(p\frac{d\varphi_2}{dx}\right) + (\lambda_1 - \lambda_2)r\varphi_1\varphi_2 = 0,$$

and hence

$$(\lambda_2 - \lambda_1)\int_a^b r\varphi_1\varphi_2 \, dx = \int_a^b \left[\varphi_2\frac{d}{dx}\left(p\frac{d\varphi_1}{dx}\right) - \varphi_1\frac{d}{dx}\left(p\frac{d\varphi_2}{dx}\right)\right] dx. \tag{96}$$

Integrating the right member by parts, we obtain

$$(\lambda_2 - \lambda_1)\int_a^b r\varphi_1\varphi_2 \, dx = \left[\varphi_2\left(p\frac{d\varphi_1}{dx}\right) - \varphi_1\left(p\frac{d\varphi_2}{dx}\right)\right]_a^b$$

$$- \int_a^b \left[\frac{d\varphi_2}{dx}\left(p\frac{d\varphi_1}{dx}\right) - \frac{d\varphi_1}{dx}\left(p\frac{d\varphi_2}{dx}\right)\right] dx.$$

Since the last integrand vanishes identically, there follows finally

$$(\lambda_2 - \lambda_1)\int_a^b r\varphi_1\varphi_2 \, dx = \left[p(x)\left\{\varphi_2(x)\frac{d\varphi_1(x)}{dx} - \varphi_1(x)\frac{d\varphi_2(x)}{dx}\right\}\right]_a^b. \tag{97}$$

In view of the fact that both $\varphi_1(x)$ and $\varphi_2(x)$ satisfy the conditions prescribed in connection with (90) at the points $x = a$ and $x = b$, the right-hand member of (97) clearly vanishes if at *each* end point a prescribed condition is of one of the forms

$$y = 0 \tag{98a}$$

or

$$\frac{dy}{dx} = 0 \tag{98b}$$

or

$$y + \gamma\frac{dy}{dx} = 0 \tag{98c}$$

when $x = a$ or $x = b$. Here the vanishing of the right-hand member of (97) at an end point for which $\varphi_1(x)$ and $\varphi_2(x)$ satisfy (98c), for any value of γ, follows from the identity

$$\varphi_2\varphi_1' - \varphi_1\varphi_2' \equiv (\varphi_2 + \gamma\varphi_2')\varphi_1' - (\varphi_1 + \gamma\varphi_1')\varphi_2'.$$

Further, if it happens that

$$p(x) = 0 \quad \text{when } x = a \text{ or } x = b, \tag{99}$$

then vanishing of the right-hand member of (97) at $x = a$ or $x = b$ is assured if only y is required to be finite at that point and if either dy/dx is finite or $p\,dy/dx$ tends to zero at that point.

Finally, if

$$p(b) = p(a), \tag{100}$$

the right-hand member of (97) vanishes if the conditions

$$y(b) = y(a), \qquad y'(b) = y'(a) \tag{101}$$

are satisfied.† It should be noticed that (101) will be satisfied, in particular, if the solutions $\varphi_1(x)$ and $\varphi_2(x)$ are required to be *periodic*, of period $b - a$.

In the cases listed, there follows

$$\int_a^b r(x)\varphi_1(x)\varphi_2(x)\,dx = 0 \qquad \text{if } \lambda_2 \neq \lambda_1; \tag{102}$$

that is, if the characteristic functions correspond to *different* characteristic numbers, then they are orthogonal with respect to the function $r(x)$.

A boundary-value problem consisting of a differential equation of type (90), together with two homogeneous conditions of the types noted, is called a *Sturm–Liouville problem*. In all the cases considered here, except only when the periodicity conditions (100) and (101) are involved, it can be shown that to each characteristic number λ_n there corresponds *only one* characteristic function $\varphi_n(x)$. (See Problem 27.) From the homogeneity of the problem it is clear, however, that each such function involves an arbitrary multiplicative factor.

Example 1. We have seen that the problem

$$\frac{d^2y}{dx^2} + \lambda y = 0, \qquad y(0) = y(l) = 0$$

has the characteristic numbers $\lambda_n = n^2\pi^2/l^2$, with corresponding characteristic functions proportional to the functions $\varphi_n = \sin(n\pi x/l)$. Since in this case $r(x) = 1$, there follows from (102)

$$\int_0^l \varphi_m(x)\varphi_n(x)\,dx = \int_0^l \sin\frac{m\pi x}{l}\sin\frac{n\pi x}{l}\,dx = 0 \qquad (m \neq n),$$

when m and n are positive integers. This fact is readily verified independently by direct integration. ∎

In most applications, the functions $p(x)$ and $r(x)$ in Equation (90) are *positive* throughout the interval (a, b), including the end points, while the function $q(x)$ is *nonpositive* in that interval. Furthermore, when one or both of the end conditions is of type (98c), with $\gamma \neq 0$, normally γ is *negative* when the condition is imposed at the lower limit $x = a$ but is *positive* when the condition is imposed at the upper limit $x = b$. We will speak of a Sturm–Liouville problem satisfying these restrictions as a *proper* problem.

---

†Boundary conditions such as those of (101), which each involve data at *two* boundaries, are sometimes called *mixed boundary conditions*. [The same phrase is also used to describe a condition such as (98c), which relates *the function and its derivative* at *one* boundary.] More general mixed conditions of two-point type, which also lead to orthogonality when (100) holds, are considered in Problem 26.

For a proper Sturm–Liouville problem, it is found (see Problems 28 and 29) that *all characteristic numbers are real and nonnegative*, and that *the corresponding characteristic functions are real* (or can be made real by rejecting a possible complex constant multiplicative factor). In addition, it can be shown in this case that there are *infinitely many* characteristic numbers, that they are *discretely* distributed (and hence do not fill out any interval), and that their array is *unbounded*.

Furthermore, the integral of the weighted *square* of a characteristic function $\varphi_n(x)$,

$$C_n = \int_a^b r(x)[\varphi_n(x)]^2 \, dx, \tag{103}$$

then has a positive numerical value, known as the *norm* (or sometimes as the *square* of the norm) of φ_n, with respect to the weighting function r. If the arbitrary multiplicative factor involved in the definition of $\varphi_n(x)$ is so chosen that this integral has the value *unity*, the function $\varphi_n(x)$ is said to be *normalized* with respect to $r(x)$. A set of normalized orthogonal functions is said to be *orthonormal*.

Example 2. In the above Example 1, we find by direct integration that

$$C_n = \int_0^l \phi_n^2 \, dx = \int_0^l \sin^2 \frac{n\pi x}{l} \, dx = \frac{l}{2}.$$

Hence, in order to normalize the functions $\sin (n\pi x/l)$ over $(0, l)$, we would divide them by the common *normalizing factor* $\sqrt{l/2}$. The set of functions

$$\varphi_n^*(x) = \sqrt{\frac{2}{l}} \sin \frac{n\pi x}{l} \qquad (n = 1, 2, \ldots)$$

is thus an *orthonormal set* on $(0, l)$.   ∎

Orthogonality properties of sets of characteristic functions generated by certain boundary-value problems of higher order are readily established. Thus, in the case of a problem involving the *fourth-order* differential equation

$$\frac{d^2}{dx^2}\left[s(x)\frac{d^2y}{dx^2} \right] + \frac{d}{dx}\left[p(x)\frac{dy}{dx} \right] + [q(x) + \lambda r(x)]y = 0 \tag{104}$$

and appropriate homogeneous boundary conditions, the methods used above lead to the equation

$$(\lambda_2 - \lambda_1)\int_a^b r\varphi_1\varphi_2 \, dx = [\{\varphi_2(s\varphi_1'')' - \varphi_1(s\varphi_2'')'\}$$
$$- s(\varphi_2'\varphi_1'' - \varphi_1'\varphi_2'') + p(\varphi_2\varphi_1' - \varphi_1\varphi_2')]_a^b, \tag{105}$$

where $\varphi_1(x)$ and $\varphi_2(x)$ are characteristic functions corresponding to distinct characteristic numbers λ_1 and λ_2, and primes indicate differentiation with respect to x. It follows, in particular, that the functions φ_1 and φ_2 are orthogonal with respect to $r(x)$ on (a, b) if at *each* end of the interval there is prescribed one of the following pairs of homogeneous conditions:

$$y = 0, \qquad \frac{dy}{dx} = 0 \tag{106a}$$

or
$$y = 0, \qquad s(x)\frac{d^2y}{dx^2} = 0 \tag{106b}$$

or
$$\frac{dy}{dx} = 0, \qquad \frac{d}{dx}\left[s(x)\frac{d^2y}{dx^2}\right] = 0 \tag{106c}$$

when $x = a$ or $x = b$.

5.7. Expansion of Arbitrary Functions in Series of Orthogonal Functions. Suppose that we have a set of functions $\{\varphi_n(x)\}$, orthogonal on a given interval (a, b) with respect to a certain known weighting function $r(x)$, and desire to expand a given function $f(x)$ in terms of a series of these functions, of the form

$$f(x) = A_0\varphi_0(x) + A_1\varphi_1(x) + A_2\varphi_2(x) + \cdots$$
$$= \sum_{n=0}^{\infty} A_n\varphi_n(x). \tag{107}$$

If we *assume* that such an expansion exists, and multiply both sides of (107) by $r(x)\varphi_k(x)$, where $\varphi_k(x)$ is the kth function in the set, we have

$$r(x)f(x)\varphi_k(x) = \sum_{n=0}^{\infty} A_n r(x)\varphi_n(x)\varphi_k(x).$$

Next, if we integrate both sides of this last equation over the interval (a, b) and *assume* that the integral of the infinite sum is equivalent to the sum of the integrals,† there follows formally

$$\int_a^b r(x)f(x)\varphi_k(x)\,dx = \sum_{n=0}^{\infty} A_n \int_a^b r(x)\varphi_n(x)\varphi_k(x)\,dx. \tag{108}$$

But by virtue of the orthogonality of the set $\{\varphi_n(x)\}$, *all terms in the sum on the right are zero except that one for which $n = k$*, and hence (108) reduces to the equation

$$A_n \int_a^b r(x)[\varphi_n(x)]^2\,dx = \int_a^b r(x)f(x)\varphi_n(x)\,dx. \tag{109}$$

Thus Equation (109) determines each constant involved in the required series (107) as the ratio of integrals of known functions.

With these values of the constants, a formal series $\sum A_n\varphi_n(x)$ is determined. It should be emphasized, however, that we have not established the fact that this series actually does *represent* the function $f(x)$ in the interval (a, b). In fact, we have not even shown that the series *converges* in (a, b) and hence represents *any* function in that interval. We may, however, speak of the series so obtained as the *formal representation* of the function $f(x)$.

---

†This assumption is justified, in particular, if the series (107) is *uniformly convergent* in (a, b).

Example 1. In illustration, we have seen that the set of functions $\{\sin (n\pi x/l)\}$ is orthogonal on $(0, l)$, with the weighting function $r(x) = 1$. In the formal representation of the function $f(x) = x$ in a series of these functions, the constants A_n are given, in accordance with (109), by the equation

$$A_n \int_0^l \sin^2 \frac{n\pi x}{l} dx = \int_0^l x \sin \frac{n\pi x}{l} dx$$

or

$$A_n = -\frac{2l}{n\pi} \cos n\pi = \frac{2l}{\pi} \frac{(-1)^{n+1}}{n}.$$

Hence we obtain the formal expansion

$$x = \frac{2l}{\pi} \sum_{n=1}^{\infty} \frac{(-1)^{n+1}}{n} \sin \frac{n\pi x}{l} = \frac{2l}{\pi} \left(\frac{1}{1} \sin \frac{\pi x}{l} - \frac{1}{2} \sin \frac{2\pi x}{l} + \cdots \right). \qquad ■$$

The general problem of determining whether the formal expansion obtained actually converges to the given function in the interval involved is a relatively difficult one and is beyond the scope of this work. In connection with this problem, it may be noticed that if the function $f(x)$ to be expanded were not identically zero in (a, b), but were orthogonal to *all* the functions φ_n of the set, with respect to the weighting function $r(x)$, all coefficients in the formal expansion would vanish and no expansion would be obtained. However, it can be shown that, except for artificial functions of no physical interest, there are no nontrivial functions which are orthogonal to *all* members of any set generated as characteristic functions of a proper Sturm–Liouville problem. In this sense we say that the sets so generated are *complete*.

If the Sturm–Liouville problem is proper, and if also the relevant functions $p(x)$, $q(x)$, and $r(x)$ are *regular* in the interval (a, b), then it is known that *the formal representation of a piecewise differentiable† function $f(x)$ in a series of the generated characteristic functions converges to $f(x)$ inside (a, b) at all points where $f(x)$ is continuous, and converges to the mean value $\frac{1}{2}[f(x+) + f(x-)]$ at points where finite jumps occur.* In the sequel, to simplify the notation, we will tacitly assume that, when a function $f(x)$ has a finite jump at an interior point $x = c$ in (a, b), the mean of the right- and left-hand limits at that point is *defined* to be $f(c)$, so that the series also converges to $f(x)$ at such points.

Certain cases in which one or both of the functions p and r vanish at an end point of the interval are also of importance, as has been noted; although they require individual treatment, it has been shown that in the special cases to be treated here the convergence of the formal representation is again described by the above statement.

When the conditions imposed on the generating Sturm–Liouville problem require that the characteristic functions vanish at one or both end points of the

---

†A function $f(x)$ is said to be *piecewise differentiable* in (a, b) if that interval can be divided into a finite number of subintervals in such a way that, in each subinterval, $f(x)$ has a derivative at each interior point, a right-hand derivative at the initial point, and a left-hand derivative at the terminal point. Such a function thus may have a finite number of finite "jumps" in (a, b).

interval (a, b), or relate their values at the two ends [as in (101)], convergence to $f(x)$ at the end point or points in question will ensue only if $f(x)$ also satisfies the responsible conditions.

Example 2. In the case of Example 1, the theorem stated asserts that the expansion obtained converges to the function $f(x) = x$ at all points *inside* the interval $(0, l)$. Since all terms in the series vanish at *both* end points $x = 0$ and $x = l$, whereas $f(x)$ vanishes only at $x = 0$, it is seen that the series also represents $f(x) = x$ at $x = 0$, but *does not* represent it at the second end point $x = l$. From the convergence at the point $x = l/2$ we obtain, in particular,

$$\frac{l}{2} = \frac{2l}{\pi}\left(1 - \frac{1}{3} + \frac{1}{5} - \frac{1}{7} + \cdots\right)$$

and hence establish the useful result

$$1 - \frac{1}{3} + \frac{1}{5} - \frac{1}{7} + \cdots = \frac{\pi}{4}. \quad\blacksquare$$

It is important to notice the very great generality of such expansions. In the case of the power series expansions of Taylor and Maclaurin, a function cannot be representable over an interval unless *that function and all its derivatives* are continuous throughout that interval, and even these stringent conditions are not sufficient to ensure representation. However, in the present case of expansions in series of characteristic functions, a representable function may itself possess a finite number of finite discontinuities, and may even be defined by different analytical expressions over different parts of the interval of representation.

On the other hand, whereas it is true that a power series representation of a function $f(x)$ can be *differentiated* term by term with the result assuredly converging to $f'(x)$ at all points inside the interval of convergence of the parent series, the same statement does not apply to the present expansions. If the generating Sturm–Liouville problem is proper, it is known, however, that *if $f(x)$ is continuous and has a derivative $f'(x)$ which is piecewise differentiable in (a, b), and if $f(x)$ satisfies the boundary conditions relevant to the generating characteristic-value problem, then the result of differentiating the series* (107) *term by term will converge to $f'(x)$ inside (a, b) at all points where $f'(x)$ is continuous.* When the conditions specified in this statement are not all satisfied, term-by-term differentiation may or may not be legitimate. (See also Section 5.12.)

Example 3. If the series representation

$$x = \frac{2l}{\pi} \sum_{n=1}^{\infty} \frac{(-1)^{n+1}}{n} \sin\frac{n\pi x}{l} \quad (0 \leq x < l),$$

obtained in Example 1, is formally differentiated term by term, the queried result

$$1 \overset{?}{=} 2 \sum_{n=1}^{\infty} (-1)^{n+1} \cos\frac{n\pi x}{l}$$

is *not valid anywhere*, since the nth term of the series does not tend to zero as $n \rightarrow \infty$ and hence the series diverges everywhere. Here the function $f(x) = x$ violates only the requirement that it vanish at $x = l$.   ∎

An operation which is rather frequently needed (as in the two following sections), and which involves *two* differentiations, is the term-by-term application of the operator L defined by (91) to the associated series (107). In order to investigate its justification, we notice that the result of *formally* evaluating Lf, by allowing L to operate on the series (107) term by term, is the expression

$$Lf(x) \sim \sum_{n=0}^{\infty} A_n L\varphi_n(x) = -r(x) \sum_{n=0}^{\infty} \lambda_n A_n \varphi_n(x), \tag{110}$$

since $L\varphi_n(x) + \lambda_n r(x)\varphi_n(x) = 0$. On the other hand, if $[Lf(x)]/r(x)$ is piecewise differentiable in (a, b), then its true representation as a series of the φ's is of the form

$$\frac{Lf(x)}{r(x)} = \sum_{n=0}^{\infty} B_n \varphi_n(x), \tag{111}$$

where

$$B_n \int_a^b r\varphi_n^2 \, dx = \int_a^b r \frac{Lf}{r} \varphi_n \, dx = \int_a^b \varphi_n Lf \, dx$$

or

$$B_n \int_a^b r\varphi_n^2 \, dx = \int_a^b \varphi_n (pf')' \, dx + \int_a^b \varphi_n qf \, dx. \tag{112}$$

If the first term on the right is integrated twice by parts, the result can be written in the form

$$B_n \int_a^b r\varphi_n^2 \, dx = [p(\varphi_n f' - f\varphi_n')]_a^b + \int_a^b fL\varphi_n \, dx$$

or, equivalently,

$$B_n \int_a^b r\varphi_n^2 \, dx = [p(\varphi_n f' - f\varphi_n')]_a^b - \lambda_n \int_a^b rf\varphi_n \, dx. \tag{113}$$

Hence it follows that $B_n = -\lambda_n A_n$, where A_n is defined by (109), and accordingly that (110) properly agrees with (111), if and only if $f(x)$ is such that

$$[p(\varphi_n f' - f\varphi_n')]_a^b = 0. \tag{114}$$

Thus we conclude that *if $[Lf(x)]/r(x)$ is piecewise differentiable in (a, b), and if $f(x)$ satisfies the end conditions relevant to the generating proper characteristic-value problem, then the operator L can be applied term by term to the series* (107):

$$L \sum_{n=0}^{\infty} A_n \varphi_n(x) = \sum_{n=0}^{\infty} A_n L\varphi_n(x) = -r(x) \sum_{n=0}^{\infty} \lambda_n A_n \varphi_n(x). \tag{115}$$

The series (115) then represents $Lf(x)$ in the usual sense inside the interval (a, b) and may or may not converge to $Lf(x)$ at the end points.

Expansions of the type considered here will be of particular usefulness in the solution of boundary-value problems involving *partial differential equations*, as will be seen in Chapter 9. We next point out certain other useful applications of the theory.

5.8. Boundary-Value Problems Involving Nonhomogeneous Differential Equations. Suppose that we require the solution of the *nonhomogeneous* differential equation

$$\left[\frac{d}{dx}\left(p\frac{dy}{dx}\right) + qy\right] + \Lambda ry = h(x) \tag{116}$$

which satisfies prescribed *homogeneous* boundary conditions of the types listed in Section 5.7, with Λ a prescribed constant. Using the operational notation of (91), we may write (116) in the form

$$Ly + \Lambda ry = h(x). \tag{117}$$

The most important special case is that in which $\Lambda = 0$, so that we require the solution of the equation $Ly = h(x)$, in which case the function $r(x)$ can be chosen at our convenience unless other considerations dictate the choice.

With the given problem we may associate the boundary-value problem consisting of the *homogeneous* equation

$$Ly + \lambda ry = 0 \tag{118}$$

and the prescribed boundary conditions. This problem determines a set of orthogonal characteristic functions $\{\varphi_n(x)\}$, such that the nth function $\varphi_n(x)$ corresponds to a characteristic number λ_n,

$$L\varphi_n(x) + \lambda_n r(x)\varphi_n(x) = 0. \tag{119}$$

Assuming that the required solution of Equation (117) exists, we express it as a series of the form

$$y = \sum a_n\varphi_n(x) \tag{120}$$

and attempt to determine the coefficients a_n. For this purpose we introduce (120) into the differential equation (117). Then, since (119) gives

$$L\varphi_n = -\lambda_n r\varphi_n,$$

Equation (117) becomes formally

$$r(x) \sum (\Lambda - \lambda_n)a_n\varphi_n(x) = h(x). \tag{121}$$

Hence, if we use (109) to calculate the coefficients A_n in the expansion

$$f(x) = \frac{h(x)}{r(x)} = \sum A_n\varphi_n(x), \tag{122}$$

assuming that $h(x)/r(x)$ is piecewise differentiable, a comparison of Equations (121) and (122) determines the required coefficients a_n as solutions of the equations

$$(\Lambda - \lambda_n)a_n = A_n, \tag{123}$$

so that, provided that Λ is not one of the characteristic numbers, the solution (120) of the nonhomogeneous problem becomes simply

$$y = \sum \frac{A_n}{\Lambda - \lambda_n}\varphi_n(x) = \frac{A_0}{\Lambda - \lambda_0}\varphi_0(x) + \frac{A_1}{\Lambda - \lambda_1}\varphi_1(x) + \cdots. \tag{124}$$

The term-by-term differentiation of (120) which leads to (121) is justified, when the Sturm–Liouville problem is proper, by reference to the result obtained at the end of the preceding section, since the function y in (120) satisfies the relevant boundary conditions as well as the equation

$$\frac{Ly}{r} = -\Lambda y + \frac{h}{r},$$

where h/r has been assumed to be piecewise differentiable. The particular exceptional cases which will be encountered in later sections require special treatment, but justification again is possible. In fact, because of the growth of the factor $\Lambda - \lambda_n$ in the denominator of (124), as $n \to \infty$, the solution series (124) often will converge even when the series (122) does not.

It is immediately verified that the above derivation, beginning with Equation (117), is unchanged if the operator L in (117) is interpreted as the *fourth-order operator*

$$L = \frac{d^2}{dx^2}\left[s(x)\frac{d^2}{dx^2}\right] + \frac{d}{dx}\left[p(x)\frac{d}{dx}\right] + q(x) \tag{125}$$

in Equation (104), and if boundary conditions of the type listed in (106) are then assumed.

From these results one may draw certain important conclusions. We have seen that *if $h(x)$ is identically zero*, the problem consisting of (116) and the prescribed homogeneous boundary conditions has nontrivial solutions only if Λ has a characteristic value $\Lambda = \lambda_k$. Equation (124) shows that if $h(x)$ is *not* identically zero, the corresponding problem has a solution in general only if Λ *does not* take on one of the characteristic values of the homogeneous problem. More specifically, the solution (124) becomes nonexistent as $\Lambda \to \lambda_k$ unless it happens that $A_k = 0$, that is, unless

$$\int_a^b r(x)f(x)\varphi_k(x)\,dx = \int_a^b h(x)\varphi_k(x)\,dx = 0,$$

so that $h(x)$ is simply orthogonal to $\varphi_k(x)$. In that special case, Equation (123) shows that the coefficient of $\varphi_k(x)$ in (120) is *arbitrary*.

In connection with the physical problems of rotating strings and shafts (Sections 5.2 and 5.3), the presence of a prescribed function of the axial distance x on the right-hand side of the relevant differential equation would correspond to the presence of a distribution of transverse load. The above conclusions indicate that, for noncritical rotation speeds, definite deflection shapes are determined, but that (according to the linearized theory) as a critical speed is approached, the amplitude of the deflection will in general increase without limit.

5.9. Convergence of the Method of Stodola and Vianello. We may also make use of these developments in investigating the convergence of the iterative methods of Stodola and Vianello (Section 5.5), as applied to the Sturm–Liouville problem (with $q = 0$) consisting of equation (71a),

$$\frac{d}{dx}\left[p(x)\frac{dy}{dx}\right] + \lambda r(x)y = 0, \tag{126}$$

and appropriate homogeneous boundary conditions. We suppose that the problem is a *proper* one. In addition, we assume that $\lambda = 0$ is not a characteristic number. The first cycle in the procedure consists of assuming an initial approximation $y_1(x)$ and determining the solution of the differential equation

$$\frac{d}{dx}\left[p(x)\frac{dy}{dx}\right] = -\lambda r(x)y_1(x), \qquad (127)$$

which satisfies the boundary conditions. The initial approximation may be imagined as expressed in terms of a series of the exact characteristic functions of the problem, say

$$y_1(x) = \sum_{n=1}^{\infty} A_n\varphi_n(x), \qquad (128)$$

where

$$\frac{d}{dx}\left[p(x)\frac{d\varphi_n}{dx}\right] = -\lambda_n r(x)\varphi_n(x) \qquad (n = 1, 2, 3, \ldots) \qquad (129)$$

and where $0 < \lambda_1 < \lambda_2 < \cdots$.

The solution of (127) can also be written in the form

$$y(x) = \sum_{n=1}^{\infty} a_n\varphi_n(x). \qquad (130)$$

To determine the coefficients in (130), we introduce (130) into (127) and obtain†

$$\sum_{n=1}^{\infty} a_n\frac{d}{dx}\left[p(x)\frac{d\varphi_n}{dx}\right] = -\lambda r(x)\sum_{n=1}^{\infty} A_n\varphi_n(x),$$

or, using (129),

$$-\sum_{n=1}^{\infty} a_n\lambda_n r(x)\varphi_n(x) = -\lambda r(x)\sum_{n=1}^{\infty} A_n\varphi_n(x).$$

Hence the coefficients in (130) are determined in the form

$$a_n = \frac{\lambda}{\lambda_n}A_n \qquad (131)$$

and (130) becomes

$$y(x) = \lambda \sum_{n=1}^{\infty} \frac{A_n}{\lambda_n}\varphi_n(x)$$

or

$$y(x) = \lambda\left[\frac{A_1}{\lambda_1}\varphi_1(x) + \frac{A_2}{\lambda_2}\varphi_2(x) + \frac{A_3}{\lambda_3}\varphi_3(x) + \cdots\right]$$

$$= \frac{\lambda}{\lambda_1}\left[A_1\varphi_1(x) + \frac{\lambda_1}{\lambda_2}A_2\varphi_2(x) + \frac{\lambda_1}{\lambda_3}A_3\varphi_3(x) + \cdots\right]. \qquad (132)$$

If the last expression in brackets is taken as the initial approximation $y_2(x)$ in the second cycle, the resulting solution is found, by a repetition of the above process, in the form

$$y(x) = \frac{\lambda}{\lambda_1}\left[A_1\varphi_1(x) + \left(\frac{\lambda_1}{\lambda_2}\right)^2 A_2\varphi_2(x) + \left(\frac{\lambda_1}{\lambda_3}\right)^2 A_3\varphi_3(x) + \cdots\right],$$

†Term-by-term application of the operator $\frac{d}{dx}\left(p\frac{d}{dx}\right)$ is justified by the result obtained at the end of Section 5.7.

and, similarly, in the Nth cycle we have the initial approximation

$$y_N(x) = A_1\varphi_1(x) + \left(\frac{\lambda_1}{\lambda_2}\right)^{N-1} A_2\varphi_2(x) + \left(\frac{\lambda_1}{\lambda_3}\right)^{N-1} A_3\varphi_3(x) + \cdots \quad (133)$$

and the corresponding solution

$$y(x) \equiv \lambda f_N(x) = \frac{\lambda}{\lambda_1}\left[A_1\varphi_1(x) + \left(\frac{\lambda_1}{\lambda_2}\right)^{N} A_2\varphi_2(x) + \left(\frac{\lambda_1}{\lambda_3}\right)^{N} A_3\varphi_3(x) + \cdots\right]. \quad (134)$$

Since λ_1 is the *smallest* characteristic number, the ratios $\lambda_1/\lambda_2, \lambda_1/\lambda_3, \ldots,$ are smaller than unity and increasing powers of these ratios tend to zero as $N \longrightarrow \infty$, showing that successive approximations always tend to a multiple of the *characteristic function corresponding to the smallest characteristic number* λ_1 unless $A_1 = 0$, that is, *unless the initial approximation* $y_1(x)$ *is orthogonal to* $\varphi_1(x)$ *with respect to* $r(x)$. It is also clear that the ratio $y_N(x)/f_N(x)$, where $f_N(x) = y(x)/\lambda$, then tends to the limit λ_1, as was stated in Section 5.5.

It is seen further that, if the condition

$$\int_a^b r(x)\varphi_1(x)y(x)\,dx = \int_a^b r(x)\varphi_1(x)y_N(x)\,dx \quad (135)$$

is imposed in the Nth cycle, there follows from the orthogonality of the system

$$\frac{\lambda}{\lambda_1}A_1 \int_a^b r(x)[\varphi_1(x)]^2\,dx = A_1 \int_a^b r(x)[\varphi_1(x)]^2\,dx$$

or
$$\lambda = \lambda_1.$$

Hence, if the characteristic function $\varphi_1(x)$ were known exactly in any cycle of the calculation, Equation (135) would identify λ with the exact value of λ_1. Although $\varphi_1(x)$ is not known exactly, an approximate determination of λ_1 may be accomplished by replacing $\varphi_1(x)$ by its approximation $y_N(x)$ or $f_N(x)$ in (135). This consideration is the motivation of Equations (72a, b).

Modifications of the iterative procedure which permit numerical determination of the higher characteristic modes depend essentially on approximate methods of "subtracting off" the terms involving the lower modes in successive cycles.

5.10. Fourier Sine Series and Cosine Series. Since the boundary-value problem

$$\frac{d^2y}{dx^2} + \lambda y = 0, \qquad y(0) = 0, \quad y(l) = 0 \quad (136)$$

has the characteristic functions

$$\varphi_n(x) = \sin\frac{n\pi x}{l} \qquad (n = 1, 2, 3, \ldots) \quad (137)$$

corresponding to the characteristic numbers $\lambda_n = n^2\pi^2/l^2$, the results of the

preceding section 5.7 state that, if a function $f(x)$ can be represented by a series
of the form

$$f(x) = A_1 \sin \frac{\pi x}{l} + A_2 \sin \frac{2\pi x}{l} + \cdots$$

$$= \sum_{n=1}^{\infty} A_n \sin \frac{n\pi x}{l} \qquad (0 < x < l), \qquad (138)$$

then the coefficients A_n are given, in accordance with (109) with $r(x) = 1$, by

$$A_n \int_0^l \sin^2 \frac{n\pi x}{l} \, dx = \int_0^l f(x) \sin \frac{n\pi x}{l} \, dx. \qquad (139)$$

This formula is a consequence of the orthogonality of the characteristic functions $\varphi_n = \sin (n\pi x/l)$ with respect to the weighting function $r(x) = 1$,

$$\int_0^l \sin \frac{m\pi x}{l} \sin \frac{n\pi x}{l} \, dx = 0 \qquad (m \neq n). \qquad (140)$$

Since, by direct integration,

$$\int_0^l \sin^2 \frac{n\pi x}{l} \, dx = \frac{l}{2} \qquad (n = 1, 2, 3, \ldots), \qquad (141)$$

there follows

$$A_n = \frac{2}{l} \int_0^l f(x) \sin \frac{n\pi x}{l} \, dx. \qquad (142)$$

It may be noticed that *the coefficient* A_n *in* (142) *is given by twice the average value of the product* $f(x) \sin (n\pi x/l)$ *in the interval* (0, *l*).

The average value (with respect to x) of a function $F(x)$ in (a, b) is defined by the integral

$$\bar{F} = \frac{1}{b - a} \int_a^b F(x) \, dx.$$

If $f(x)$ *is piecewise differentiable in the interval* (0, *l*), *the series* (138) *converges to* $f(x)$ *at points of continuity, and to the mean of the two values approached from the right and left at a finite discontinuity.*

The series (138) is known as the *Fourier sine series* representation of $f(x)$ in the interval (0, *l*). It is seen that all terms or "*harmonics*" in (138) are *periodic* and have the common period 2*l*; that is, *the common period is twice the length of the interval of representation* (0, *l*). Also, if x is replaced by $-x$, the algebraic sign of each harmonic is merely reversed. Functions having this last property are known as *odd* functions of x.

In general, a function $F(x)$ is said to be an *odd* function if $F(-x) = -F(x)$ and an *even* function if $F(-x) = F(x)$. In a graphical representation, $y = F(x)$, an even function is *symmetrical* with respect to the y axis, whereas an odd function is *antisymmetrical*; that is, the graph of an odd function is symmetrical with respect to the origin of the coordinate system. Thus the functions $\sin mx$ and x^k, where k is an odd integer, are odd functions, and the functions $\cos mx$ and x^k, where k is an even integer, are even functions. Such functions as e^x and $\log x$ are *neither even nor odd*. It is clear that the product of two even functions or of two odd functions is an even function, whereas the product of an even function and an odd function is an odd function. Further, the derivative of an even function is an odd function and conversely; the integral of an odd function is an even function, whereas the integral of an even function is the sum of a constant and an odd function.

It follows that in the interval $(-l, 0)$ the series (138) represents the function $-f(-x)$. Since all terms have the common period $2l$, the behavior of the series in $(-l, l)$ is repeated periodically for all values of x.

If $f(x)$ *is an odd function of x*, the series (138) accordingly represents $f(x)$ not only in the interval $(0, l)$ but also in the larger interval $(-l, l)$. *If, in addition, $f(x)$ is periodic, of period $2l$, the series represents $f(x)$ everywhere* (with the usual convention regarding points of discontinuity).

Thus the sine series representation of the odd function $f(x) = x$, given as an illustration in Section 5.9, represents x in the larger interval $(-l, l)$ except at the end points $x = \pm l$, and repeats this behavior *periodically* for all values of x (see Figure 5.2).

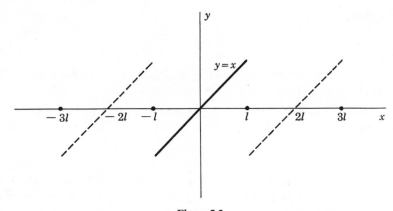

Figure 5.2

Example 1. To obtain the Fourier sine series representing $f(x) = e^x$ in the interval $(0, \pi)$, we set $l = \pi$ in Equations (138) and (142) and obtain

$$A_n = \frac{2}{\pi} \int_0^\pi e^x \sin nx \, dx = \frac{2}{\pi} \frac{n}{n^2 + 1} (1 - e^\pi \cos n\pi).$$

Thus the desired expansion is of the form

$$e^x = \frac{2}{\pi} \left[\frac{e^\pi + 1}{2} \sin x - \frac{2(e^\pi - 1)}{5} \sin 2x + \frac{3(e^\pi + 1)}{10} \sin 3x + \cdots \right] \quad (0 < x < \pi).$$
(143)

The series (143) represents e^x when $0 < x < \pi$, and vanishes at the end points $x = 0$

and $x = \pi$. In the interval $-\pi < x < 0$ the series represents the function $-e^{-x}$, and, in general, the series represents an *odd* periodic function which coincides with e^x when $0 < x < \pi$, with $-e^{-x}$ when $-\pi < x < 0$, which is zero when $x = -\pi, 0, \pi$, and which is of period 2π. ∎

Similar series developments involving cosine terms, rather than sine terms, may be obtained by considering the boundary-value problem

$$\frac{d^2y}{dx^2} + \lambda y = 0, \qquad y'(0) = 0, \quad y'(l) = 0, \tag{144}$$

which has the characteristic functions

$$\varphi_n(x) = \cos\frac{n\pi x}{l} \qquad (n = 0, 1, 2, \ldots) \tag{145}$$

corresponding to the characteristic numbers $\lambda_n = n^2\pi^2/l^2$. It is important to notice that $\varphi_0(x) = 1$ is a member of the set (145), corresponding to $\lambda_0 = 0$. The functions (145) possess the orthogonality property

$$\int_0^l \cos\frac{m\pi x}{l}\cos\frac{n\pi x}{l}\,dx = 0 \qquad (m \neq n), \tag{146}$$

as can be independently verified by direct integration. The constants in a series representation of the form

$$f(x) = A_0 + A_1\cos\frac{x\pi}{l} + A_2\cos\frac{2\pi x}{l} + \cdots$$

$$= A_0 + \sum_{n=1}^{\infty} A_n\cos\frac{n\pi x}{l} \qquad (0 \leq x \leq l) \tag{147}$$

are given by (109), with $r(x) = 1$, in the forms

$$A_0\int_0^l dx = \int_0^l f(x)\,dx, \qquad A_n\int_0^l \cos^2\frac{n\pi x}{l}\,dx = \int_0^l f(x)\cos\frac{n\pi x}{l}\,dx.$$

Since
$$\int_0^l \cos^2\frac{n\pi x}{l}\,dx = \begin{cases} l/2 & (n = 1, 2, 3, \ldots), \\ l & (n = 0), \end{cases} \tag{148}$$

there follows

$$A_0 = \frac{1}{l}\int_0^l f(x)\,dx, \qquad A_n = \frac{2}{l}\int_0^l f(x)\cos\frac{n\pi x}{l}\,dx \quad (n = 1, 2, 3, \ldots). \tag{149}$$

Hence *the coefficient A_0 in (147) is the average value of $f(x)$ in $(0, l)$, whereas the coefficient A_n, when $n \neq 0$, is twice the average value of $f(x)\cos(n\pi x/l)$ in $(0, l)$*.

With these values of the coefficients, the series (147) is known as the *Fourier cosine series* representation of $f(x)$ in $(0, l)$. It is seen that all harmonics in (147) are *even* functions of x and are again *periodic*, of period $2l$. Thus, if $f(x)$ is piecewise differentiable in $(0, l)$, the series (147) will converge to a function $f^*(x)$ which is periodic, with period $P = 2l$, and which is such that $f^*(x) = f(x)$

for $0 \leq x \leq l$ and $f^*(x) = f(-x)$ for $-l \leq x \leq 0$, again with the adopted convention regarding discontinuities. Since $f^*(x) = f(x)$ when $x = 0$ and when $x = l$, the series will in fact converge to $f(x)$ at those end points [as is asserted in (147)]. In addition, *if $f(x)$ is an even function*, that is, if $f(-x) = f(x)$, *the series will converge to $f(x)$ in the double interval $-l \leq x \leq l$*. Finally, if $f(x)$ also is periodic, with period $2l$, the series will converge to $f(x)$ *everywhere*.

Thus the cosine series representation of the *odd* function $f(x) = x$ in $(0, l)$ would represent x *only* for $0 \leq x \leq l$ and would represent $f(-x) = -x$ for $-l \leq x \leq 0$ (see Figure 5.3).

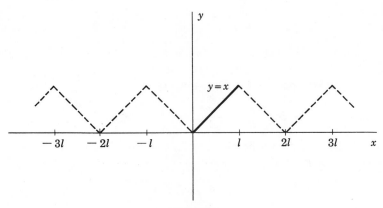

Figure 5.3

Example 2. To obtain the Fourier cosine series representing $f(x) = e^x$ in the interval $(0, \pi)$, we set $l = \pi$ in (147) and (149) and obtain

$$A_0 = \frac{1}{\pi} \int_0^\pi e^x \, dx = \frac{e^\pi - 1}{\pi},$$

$$A_n = \frac{2}{\pi} \int_0^\pi e^x \cos nx \, dx = \frac{2}{\pi} \frac{1}{n^2 + 1} (e^\pi \cos n\pi - 1) \qquad (n \neq 0).$$

Thus the desired expansion is of the form

$$e^x = \frac{2}{\pi} \left(\frac{e^\pi - 1}{2} - \frac{e^\pi + 1}{2} \cos x + \frac{e^\pi - 1}{5} \cos 2x - \frac{e^\pi + 1}{10} \cos 3x + \cdots \right)$$

$$(0 \leq x \leq \pi). \qquad (150)$$

The series (150) represents an even function of period 2π which is identified with e^x when $0 \leq x \leq \pi$ and with e^{-x} when $-\pi \leq x \leq 0$. ∎

We notice that the two expansions (143) and (150) *both* represent e^x in the interval $(0, \pi)$ but represent entirely distinct functions outside this interval. It may be remarked that half the sum of the two series will represent a periodic function which is *zero* when $-\pi < x < 0$ and which coincides with e^x when $0 < x < \pi$. At $x = 0$ this composite series, *consisting of both sine and cosine terms*, converges to the value $\frac{1}{2}$, whereas at the points $x = \pm\pi$ the series converges to the value $\frac{1}{2}e^{\pi}$.

5.11. Complete Fourier Series. It has now been seen that any reasonably well-behaved function $f(x)$ can be represented in the interval $(0, l)$ by a series consisting either of sines or of cosines with common period $P = 2l$, and that for an *odd* function $f(x)$ the *sine* series representation is valid in $(-l, l)$, whereas for an *even* function $f(x)$ the *cosine* series representation is valid in $(-l, l)$. We next show that, if *both sines and cosines* of common period $2l$ are used, a representation valid in the interval $(-l, l)$ can be obtained for *any* function $f(x)$ which is piecewise differentiable in $(-l, l)$.

It is clear that any function $f(x)$ can be expressed as the sum of an even function $f_e(x)$ and an odd function $f_o(x)$,

$$f(x) = f_e(x) + f_o(x), \tag{151}$$

since if we write

$$f(x) = \tfrac{1}{2}[f(x) + f(-x)] + \tfrac{1}{2}[f(x) - f(-x)], \tag{152}$$

the first bracket is an even function of x, whereas the second bracket is an odd function. The preceding theory states that the expansions

$$f_e(x) = A_0 + \sum_{n=1}^{\infty} A_n \cos\frac{n\pi x}{l},$$
$$f_o(x) = \sum_{n=1}^{\infty} B_n \sin\frac{n\pi x}{l} \tag{153}$$

are valid in the interval $(-l, l)$, if the coefficients are calculated by the formulas

$$A_0 = \frac{1}{l}\int_0^l f_e(x)\,dx,$$

$$A_n = \frac{2}{l}\int_0^l f_e(x)\cos\frac{n\pi x}{l}dx,$$

$$B_n = \frac{2}{l}\int_0^l f_o(x)\sin\frac{n\pi x}{l}dx.$$

Since all integrands are *even* functions of x, we clearly may replace $\int_0^l$ by $\frac{1}{2}\int_{-l}^l$ in each case. Finally, if we use (151) to replace $f_e(x)$ by $f(x) - f_o(x)$ in the first two integrals and replace $f_o(x)$ by $f(x) - f_e(x)$ in the third integral, and notice that the integral of an odd function over a symmetrical interval $(-l, l)$ vanishes, there follows

$$A_0 = \frac{1}{2l}\int_{-l}^l f(x)\,dx - \frac{1}{2l}\int_{-l}^l f_o(x)\,dx = \frac{1}{2l}\int_{-l}^l f(x)\,dx,$$

$$A_n = \frac{1}{l}\int_{-l}^l f(x)\cos\frac{n\pi x}{l}dx - \frac{1}{l}\int_{-l}^l f_o(x)\cos\frac{n\pi x}{l}dx$$

$$= \frac{1}{l}\int_{-l}^l f(x)\cos\frac{n\pi x}{l}dx,$$

$$B_n = \frac{1}{l} \int_{-l}^{l} f(x) \sin \frac{n\pi x}{l} dx - \frac{1}{l} \int_{-l}^{l} f_e(x) \sin \frac{n\pi x}{l} dx$$

$$= \frac{1}{l} \int_{-l}^{l} f(x) \sin \frac{n\pi x}{l} dx.$$

Hence, by introducing (153) into (151), there follows

$$f(x) = A_0 + \sum_{n=1}^{\infty} \left(A_n \cos \frac{n\pi x}{l} + B_n \sin \frac{n\pi x}{l} \right) \qquad (-l < x < l), \qquad (154)$$

where

$$A_0 = \frac{1}{2l} \int_{-l}^{l} f(x)\, dx, \qquad A_n = \frac{1}{l} \int_{-l}^{l} f(x) \cos \frac{n\pi x}{l} dx,$$

$$(155)$$

$$B_n = \frac{1}{l} \int_{-l}^{l} f(x) \sin \frac{n\pi x}{l} dx.$$

The series (154) is the complete *Fourier series* representation of $f(x)$ in the interval $(-l, l)$. It is seen that *again A_0 is the average value of $f(x)$*, the average now being calculated *over the interval of representation $(-l, l)$, whereas A_n and B_n are twice the average values of $f(x) \cos (n\pi x/l)$ and $f(x) \sin (n\pi x/l)$, respectively, over the same interval.* For an even function there follows $B_n = 0$ and the cosine series previously obtained results; for an odd function $A_0 = A_n = 0$ and the previously obtained sine series results. If $f(x)$ is neither even nor odd, the series (154) will contain both sines and cosines of period $2l$ and will represent $f(x)$, in the sense described, in the interval $(-l, l)$ of length $2l$. Because of the periodicity of all terms in the series, this representation is repeated *periodically* for all values of x.

Example 1. To obtain a Fourier series of period 2π, representing $f(x) = e^x$ in the interval $(-\pi, \pi)$, we set $l = \pi$ in (154) and (155) and obtain

$$A_0 = \frac{1}{2\pi} \int_{-\pi}^{\pi} e^x\, dx = \frac{1}{2\pi}(e^\pi - e^{-\pi}) = \frac{1}{\pi} \sinh \pi,$$

$$A_n = \frac{1}{\pi} \int_{-\pi}^{\pi} e^x \cos nx\, dx = \frac{2}{\pi} \frac{\cos n\pi}{n^2 + 1} \sinh \pi \qquad (n \neq 0),$$

$$B_n = \frac{1}{\pi} \int_{-\pi}^{\pi} e^x \sin nx\, dx = -\frac{2}{\pi} \frac{n \cos n\pi}{n^2 + 1} \sinh \pi.$$

The desired expansion is thus of the form

$$e^x = \frac{\sinh \pi}{\pi} \left(1 - \cos x + \frac{2}{5} \cos 2x - \frac{1}{5} \cos 3x + \cdots \right.$$

$$\left. + \sin x - \frac{4}{5} \sin 2x + \frac{3}{5} \sin 3x + \cdots \right) \qquad (-\pi < x < \pi)$$

or

$$e^x = \frac{\sinh \pi}{\pi} \left[1 - 2 \sum_{n=1}^{\infty} \frac{(-1)^{n+1}}{n^2 + 1} (\cos nx - n \sin nx) \right] \qquad (-\pi < x < \pi). \qquad (156)$$

This series converges to e^x at all points inside the interval $(-\pi, \pi)$. Since

$$e^x = \cosh x + \sinh x,$$

it follows that the even terms represent the function $\cosh x$ in $(-\pi, \pi)$, whereas the odd terms represent $\sinh x$:

$$\cosh x = \frac{\sinh \pi}{\pi}\left[1 - 2 \sum_{n=1}^{\infty} \frac{(-1)^{n+1}}{n^2 + 1} \cos nx\right] \qquad (-\pi \leqq x \leqq \pi),$$

$$\sinh x = 2\frac{\sinh \pi}{\pi} \sum_{n=1}^{\infty} \frac{(-1)^{n+1}n}{n^2 + 1} \sin nx \qquad (-\pi < x < \pi). \qquad \blacksquare$$

If we consider the *periodic extension* of $f(x)$, of period $2l$, that is, the function $f^*(x)$ which is of period $2l$ and which agrees with $f(x)$ when $-l < x < l$, we see that at both the points $x = \pm l$ the right- and left-hand limits of $f^*(x)$ are $f(-l)$ and $f(l)$, respectively. Since the series (154) converges everywhere to $f^*(x)$ when $f(x)$ is piecewise differentiable in $(-l, l)$, with the usual convention regarding points of discontinuity, we deduce that *the series* (154) *converges at both end points* $x = \pm l$ *to the value* $\frac{1}{2}[f(l) + f(-l)]$, *that is, to the mean of the two values of* $f(x)$ *at the end points.* We recall, in summary, that also *the sine series* (138) *vanishes identically at the end points of the interval* $(0, l)$, *and hence converges to* $f(x)$ *at those points only if* $f(x)$ *vanishes there*, and that *the cosine series* (147) *converges to* $f(x)$ *at both end points of the interval* $(0, l)$.

We next show that the constants in (154) can be determined in such a way that the given function $f(x)$ is represented in *any* specified interval of length $2l$, that is, of length equal to the common period of the trigonometric functions involved. Suppose that a function $f(x)$ is specified in an interval $(a, a + 2l)$ and that it is required that the coefficients in (154) be determined so that (154) represents $f(x)$ in the usual way in that interval. We next again consider the periodic extension $f^*(x)$ defined for all values of x in such a way that $f^*(x)$ coincides with $f(x)$ in the given interval, and such that $f^*(x)$ is a periodic function, of period $2l$. Then if $f(x)$ is replaced by $f^*(x)$ in the definitions of the constants (155), the corresponding series (154) clearly will represent the periodic function $f^*(x)$ for all values of x (in the usual sense), and hence, in particular, will represent $f(x)$ in the interval $(a, a + 2l)$. But since the integrands in (155) are then periodic functions, of period $2l$, the range of integration may equally well be taken as *any* interval of length $2l$. In particular, we may choose as that range the interval over which representation is desired, where $f^*(x) = f(x)$.

It follows that *if a representation of the form*

$$f(x) = A_0 + \sum_{n=1}^{\infty} \left(A_n \cos \frac{2n\pi x}{P} + B_n \sin \frac{2n\pi x}{P}\right) \qquad (157)$$

is required over the interval $(a, a + P)$, *the coefficients are to be determined by the equations*

$$A_0 = \frac{1}{P}\int_a^{a+P} f(x)\,dx, \qquad A_n = \frac{2}{P}\int_a^{a+P} f(x) \cos \frac{2n\pi x}{P}\,dx,$$

$$B_n = \frac{2}{P}\int_a^{a+P} f(x) \sin \frac{2n\pi x}{P}\,dx. \qquad (158)$$

In particular, representation in the interval $(0, P)$ is obtained by setting $a = 0$.

These general results are also obtainable directly by considering the boundary-value problem

$$\frac{d^2y}{dx^2} + \lambda y = 0, \qquad y(a) = y(a + P), \quad y'(a) = y'(a + P) \tag{159}$$

for which the characteristic values $\lambda_n = 4n^2\pi^2/P^2$ correspond to the two-parameter characteristic functions

$$\varphi_n = c_n \sin \frac{2n\pi x}{P} + d_n \cos \frac{2n\pi x}{P}. \tag{160}$$

The orthogonality of these functions for arbitrarily assigned values of both c_n and d_n, on the interval $(a, a + P)$, corresponds to the relations

$$\int_a^{a+P} \sin \frac{2m\pi x}{P} \sin \frac{2n\pi x}{P} dx = 0 \quad (m \neq n),$$

$$\int_a^{a+P} \cos \frac{2m\pi x}{P} \cos \frac{2n\pi x}{P} dx = 0 \quad (m \neq n), \tag{161}$$

$$\int_a^{a+P} \sin \frac{2m\pi x}{P} \cos \frac{2n\pi x}{P} dx = 0,$$

where the validity of the third relation when $m = n$ is not guaranteed by the results of Section 5.6 (since the two functions involved then correspond to the *same* characteristic number), but is directly verified.

Example 2. We suppose that it is required to find a Fourier series of period 6 which in the interval $(1, 7)$ represents a function $f(x)$ taking on the constant value $+1$ when $1 < x < 4$ and the constant value -1 when $4 < x < 7$. With $a = 1$, $P = 6$, the series (157) becomes

$$f(x) = A_0 + \sum_{n=1}^{\infty} \left(A_n \cos \frac{n\pi x}{3} + B_n \sin \frac{n\pi x}{3} \right),$$

where, in accordance with (158),

$$A_0 = \frac{1}{6} \int_1^7 f(x)\, dx = \frac{1}{6} \left[\int_1^4 1\, dx + \int_4^7 (-1)\, dx \right] = 0,$$

$$A_n = \frac{1}{3} \int_1^7 f(x) \cos \frac{n\pi x}{3}\, dx = \frac{1}{3} \left[\int_1^4 \cos \frac{n\pi x}{3}\, dx - \int_4^7 \cos \frac{n\pi x}{3}\, dx \right]$$

$$= \frac{1}{n\pi} \left[2 \sin \frac{4n\pi}{3} - \sin \frac{n\pi}{3} - \sin \frac{7n\pi}{3} \right] \quad (n \neq 0),$$

$$B_n = \frac{1}{3} \int_1^7 f(x) \sin \frac{n\pi x}{3}\, dx = \frac{1}{3} \left[\int_1^4 \sin \frac{n\pi x}{3}\, dx - \int_4^7 \sin \frac{n\pi x}{3}\, dx \right]$$

$$= -\frac{1}{n\pi} \left[2 \cos \frac{4n\pi}{3} - \cos \frac{n\pi}{3} - \cos \frac{7n\pi}{3} \right].$$

By virtue of the relations

$$\sin \frac{4n\pi}{3} = \cos n\pi \sin \frac{n\pi}{3}, \qquad \sin \frac{7n\pi}{3} = \sin \frac{n\pi}{3},$$

$$\cos \frac{4n\pi}{3} = \cos n\pi \cos \frac{n\pi}{3}, \qquad \cos \frac{7n\pi}{3} = \cos \frac{n\pi}{3},$$

these results take the form

$$A_0 = 0, \qquad A_n = -2\frac{1 - \cos n\pi}{n\pi} \sin \frac{n\pi}{3},$$

$$B_n = 2\frac{1 - \cos n\pi}{n\pi} \cos \frac{n\pi}{3}.$$

Since the factor $1 - \cos n\pi$ is zero when n is even and 2 when n is odd, it follows that the required series contains only harmonics corresponding to the *odd* integers. The series can be written in the form†

$$f(x) = \frac{4}{\pi} \sum_{n \text{ odd}} \frac{1}{n}\left(\cos \frac{n\pi}{3} \sin \frac{n\pi x}{3} - \sin \frac{n\pi}{3} \cos \frac{n\pi x}{3}\right).$$

We notice that this series is equivalent to the form

$$f(x) = \frac{4}{\pi} \sum_{n \text{ odd}} \frac{1}{n} \sin \frac{n\pi(x - 1)}{3}$$

which could have been obtained more directly, in this case, by first transforming the variable by moving the origin to the point $x = 1$. ∎

Any function $f(x)$ which is of period P and which also satisfies the relation

$$f\left(x + \frac{P}{2}\right) = -f(x) \tag{162}$$

is called an *odd-harmonic function*, since it is readily shown that the Fourier series representation of such a function possesses only harmonics corresponding to odd integers. The function in the preceding example is of this type.

5.12. Term-by-Term Differentiation of Fourier Series. Because of the fact that the formal (term-by-term) derivative of a Fourier series is another Fourier series, and also that the formal derivative of a sine series is a cosine series, and conversely, it is possible to obtain conditions permitting term-by-term differentiation of those series by use of a relatively simple argument. Since the need for such operations arises fairly often, we now deal briefly with their justification.

Suppose that $f(x)$ possesses a Fourier series of period P which converges to $f(x)$ inside the interval $(a, a + P)$. Then that series is specified by Equation (157), where the coefficients are determined by (158). The result of *formally* differentiating that series term by term is then

$$\sum_{n=1}^{\infty} \left(\frac{2n\pi}{P} B_n \cos \frac{2n\pi x}{P} - \frac{2n\pi}{P} A_n \sin \frac{2n\pi x}{P}\right). \tag{163}$$

†The phrase "n odd" is used in summations to indicate that n is to take on the values $n = 1$, 3, 5, ...

On the other hand, if $f'(x)$ exists and is piecewise differentiable, its Fourier series of period P in $(a, a + P)$ is

$$f'(x) = a_0 + \sum_{n=1}^{\infty} \left(a_n \cos \frac{2n\pi x}{P} + b_n \sin \frac{2n\pi x}{P} \right), \tag{164}$$

where

$$a_0 = \frac{1}{P} \int_a^{a+P} f'(x)\, dx, \qquad a_n = \frac{2}{P} \int_a^{a+P} f'(x) \cos \frac{2n\pi x}{P}\, dx,$$

$$b_n = \frac{2}{P} \int_a^{a+P} f'(x) \sin \frac{2n\pi x}{P}\, dx. \tag{165}$$

If $f(x)$ is continuous and $f'(x)$ is (at least) piecewise continuous, we may transform the formulas (165) to the forms

$$a_0 = \frac{1}{P}[f(a + P) - f(a)],$$

$$a_n = \frac{2}{P} \left\{ \left[f(x) \cos \frac{2n\pi x}{P} \right]_a^{a+P} + \frac{2n\pi}{P} \int_a^{a+P} f(x) \sin \frac{2n\pi x}{P}\, dx \right\}$$

$$= \frac{2}{P}[f(a + P) - f(a)] \cos \frac{2n\pi a}{P} + \frac{2n\pi}{P} B_n,$$

$$b_n = \frac{2}{P} \left\{ \left[f(x) \sin \frac{2n\pi x}{P} \right]_a^{a+P} - \frac{2n\pi}{P} \int_a^{a+P} f(x) \cos \frac{2n\pi x}{P}\, dx \right\}$$

$$= \frac{2}{P}[f(a + P) - f(a)] \sin \frac{2n\pi a}{P} - \frac{2n\pi}{P} A_n,$$

by use of integration by parts. Consequently, the formal series (163) then is identified with (164) if and only if the additional condition

$$f(a + P) = f(a) \tag{166}$$

is satisfied.

Furthermore, if $f'(x)$ is piecewise differentiable in $(a, a + P)$, its expansion (164) *will* in fact converge to $f'(x)$ at each interior point of that interval at which $f'(x)$ is continuous. When these facts are summarized, we are led to conclude that *if $f(x)$ is continuous and $f'(x)$ is piecewise differentiable in $(a, a + P)$, and if $f(a + P) = f(a)$, then the result of term-by-term differentiation of the Fourier series of period P representing $f(x)$ inside $(a, a + P)$ converges to $f'(x)$ at each interior point at which $f'(x)$ is continuous.*

This result is in accordance with the more general result quoted in Section 5.7, except that here $f(x)$ need satisfy only the *first* of the two conditions involved in the generating Sturm–Liouville problem (159). It may be seen that, if the condition $f'(a + P) = f'(a)$ is also satisfied, then the differentiated series also will converge to $f'(x)$ at the two end points $x = a$ and $x = a + P$.

Analogous considerations of the sine series (138) and the cosine series (147) lead easily to the conclusion that, *if $f(x)$ is continuous and $f'(x)$ is piecewise* differentiable in $(0, l)$, then the result of differentiating the sine series (138) term by term converges to $f'(x)$ at each interior point of $(0, l)$ at which $f'(x)$ is continuous, as well as at the end points $x = 0$ and $x = l$, provided that $f(x)$ also satisfies the end conditions $f(0) = 0$ and $f(l) = 0$. For the cosine series (147), the same statement applies, with two exceptions: convergence of the differentiated series to $f'(x)$ at the end points generally does not follow and also no supplementary end conditions need be satisfied by $f(x)$. In the case of the cosine series, it is seen that each term of the differentiated series vanishes at both end points and, accordingly, that satisfaction of the end conditions $f'(0) = 0$ and/or $f'(l) = 0$ would in fact bring about convergence of the differentiated series to $f'(x)$ at one or both of the end points as well.

Example 1. From the readily established representation

$$x = \frac{l}{2} - \frac{4l}{\pi^2} \sum_{n \text{ odd}} \frac{1}{n^2} \cos \frac{n\pi x}{l} \qquad (0 \leq x \leq l), \tag{167}$$

we deduce the representation

$$1 = \frac{4}{\pi} \sum_{n \text{ odd}} \frac{1}{n} \sin \frac{n\pi x}{l} \qquad (0 < x < l), \tag{168}$$

by differentiation. The validity of this result is easily confirmed independently by obtaining (168) directly, by use of (138). ∎

Example 2. The function

$$f(x) = \begin{cases} x & (0 \leq x \leq l/2), \\ l - x & (l/2 \leq x \leq l) \end{cases}$$

has the sine-series expansion

$$f(x) = \frac{4l}{\pi^2} \sum_{n=1}^{\infty} \frac{\sin (n\pi/2)}{n^2} \sin \frac{n\pi x}{l}$$

$$= \frac{4l}{\pi^2} \left(\frac{1}{1^2} \sin \frac{\pi x}{l} - \frac{1}{3^2} \sin \frac{3\pi x}{l} + \cdots \right) \qquad (0 \leq x \leq l)$$

[see Problem 49(b)]. Since $f(x)$ satisfies the stated conditions [in spite of the discontinuity in $f'(x)$], we may differentiate term by term, when $0 \leq x < l/2$ and when $l/2 < x \leq l$, to produce the verifiable representation

$$g(x) = \frac{4}{\pi} \sum_{n=1}^{\infty} \frac{\sin (n\pi/2)}{n} \cos \frac{n\pi x}{l}$$

$$= \frac{4}{\pi} \left(\frac{1}{1} \cos \frac{\pi x}{l} - \frac{1}{3} \cos \frac{3\pi x}{l} + \cdots \right),$$

where

$$g(x) = f'(x) = \begin{cases} 1 & (0 \leq x < l/2), \\ -1 & (l/2 < x \leq l), \end{cases}$$

and $g(l/2) = 0$. ∎

Since, for Fourier series, the associated operator L is such that $Ly = y''$, the result obtained at the end of Section 5.7 applies to this operation, with respect to the interior of the relevant interval. (See also Problems 67 and 68.)

5.13. Fourier–Bessel Series. Expansions in terms of Bessel functions arise most frequently in connection with a boundary-value problem involving the differential equation

$$x^2 \frac{d^2y}{dx^2} + x \frac{dy}{dx} + (\mu^2 x^2 - p^2)y = 0 \tag{169}$$

or, equivalently,

$$\frac{d}{dx}\left(x \frac{dy}{dx}\right) + \left(-\frac{p^2}{x} + \mu^2 x\right)y = 0. \tag{170}$$

Reference to Equation (130a), Chapter 4, shows that the general solution of this equation is of the form $y = Z_p(\mu x)$ or

$$y = \begin{cases} c_1 J_p(\mu x) + c_2 J_{-p}(\mu x) & (p \text{ not an integer}), \\ c_1 J_p(\mu x) + c_2 Y_p(\mu x) & (p \text{ an integer}). \end{cases} \tag{171}$$

Equation (170) is identified with (90) if we write

$$p(x) = x, \qquad q(x) = -\frac{p^2}{x}, \qquad r(x) = x, \qquad \lambda = \mu^2. \tag{172}$$

Restricting attention to the interval $(0, l)$, we see that since $p(0) = 0$, the results of Section 5.6 show that the characteristic functions of the problem are orthogonal on $(0, l)$, with respect to the weighting function $r(x) = x$, if at $x = 0$ it is required merely that y be finite and $x\, dy/dx$ be zero and if at $x = l$ a homogeneous condition is prescribed.

The conditions at $x = 0$ require that we take

$$c_2 = 0 \tag{173}$$

in (171). If at $x = l$ the condition

$$y(l) = 0 \tag{174a}$$

is prescribed, the characteristic values of μ are determined as the roots of the equation $J_p(\mu_n l) = 0$ and hence, if we write $\mu_n l = \alpha_n$, so that α_n is dimensionless, there follows

$$\mu_n = \frac{\alpha_n}{l} \qquad \text{with } J_p(\alpha_n) = 0, \tag{174b}$$

so that α_n is the nth zero of $J_p(x)$. Similarly, the end condition

$$y'(l) = 0 \tag{175a}$$

requires that the characteristic numbers μ_n satisfy

$$\mu_n = \frac{\alpha_n}{l} \qquad \text{with } J_p'(\alpha_n) = 0, \tag{175b}$$

whereas the more general condition

$$ky(l) + ly'(l) = 0 \qquad (k \geqq 0), \tag{176a}$$

where k is a prescribed dimensionless constant, determines μ_n in the form

$$\mu_n = \frac{\alpha_n}{l} \quad \text{with } kJ_p(\alpha_n) + \alpha_n J'_p(\alpha_n) = 0. \tag{176b}$$

In all these cases the characteristic functions are of the form

$$\phi_n(x) = J_p(\mu_n x) \equiv J_p\left(\alpha_n \frac{x}{l}\right), \tag{177}$$

where μ_n and α_n are defined by (174b), (175b), or (176b), and the results of Section 5.6 show that these functions are orthogonal on $(0, l)$ with the weighting function $r(x) = x$,

$$\int_0^l x J_p(\mu_m x) J_p(\mu_n x)\, dx = 0 \qquad (m \neq n). \tag{178}$$

Although Equation (170) is one of the exceptional cases described in Section 5.7 in that $p(x)$ and $r(x)$ vanish at the end point $x = 0$, and also $q(x)$ is not regular at $x = 0$ unless the order p of the Bessel functions is zero, still it can be shown that the formal series (107), with coefficients determined by (109), will converge to $f(x)$ inside the interval $(0, l)$ under the same conditions as those stated with reference to the Fourier series.

We must, of course, be careful to include *all* the characteristic functions of the problem involved. From the definition of the Bessel functions $J_p(x)$, it is clear that the relation

$$J_p(-x) = (-1)^p J_p(x)$$

is satisfied. Thus solutions of (174b), (175b), and (176b) exist in pairs, symmetrically located with respect to the point $x = 0$. However, since replacing μ_n by $-\mu_n$ in (177) either does not change $\varphi_n(x)$ or merely multiplies it by a numerical factor, the negative values of μ_n need not be considered. If $\mu_0 = 0$ were a characteristic number, the corresponding function $\varphi_0(x)$, as given by (177), would be identically zero (and hence not a characteristic function) *except in the case when the order p of the relevant Bessel functions is zero.* When $p = 0$, only (175b) permits the solution $\mu_0 = 0$. Thus it is necessary to consider only the set of functions $\{\varphi_n(x)\}$ corresponding to *positive* characteristic numbers μ_n ($n = 1$, 2, 3, . . .) in all cases *except in the case of (175b) when $p = 0$, in which case the function $\varphi_0 = 1$ corresponding to $\mu_0 = 0$ must be added to the set.*

Temporarily excluding this exceptional case, the coefficients in the series

$$f(x) = \sum_{n=1}^{\infty} A_n J_p(\mu_n x) \equiv \sum_{n=1}^{\infty} A_n J_p\left(\alpha_n \frac{x}{l}\right) \qquad (0 < x < l), \tag{179}$$

where the numbers μ_n and α_n are the positive quantities defined by one of Equations (174b), (175b), or (176b), are given, by virtue of (109), by the equation

$$A_n \int_0^l x[J_p(\mu_n x)]^2\, dx = \int_0^l x f(x) J_p(\mu_n x)\, dx.$$

The coefficient of A_n is independent of $f(x)$ and can be determined once and for all. Denoting it by C_n, we then have

$$A_n = \frac{1}{C_n} \int_0^l x f(x) J_p(\mu_n x)\, dx, \qquad (180)$$

where

$$C_n = \int_0^l x [J_p(\mu_n x)]^2\, dx. \qquad (181)$$

To determine the factors C_n, we proceed by an indirect method as follows. With the notation of (177) we have

$$\frac{d}{dx}\left(x \frac{d\varphi_n}{dx}\right) + \left(\mu_n^2 x - \frac{p^2}{x}\right)\varphi_n = 0, \qquad (182)$$

since the function $\varphi_n(x)$ satisfies (170) when $\mu = \mu_n$. If both sides of (182) are multiplied by $2x\, d\varphi_n/dx$, the resulting equation can be written in the form

$$(\mu_n^2 x^2 - p^2)\frac{d}{dx}(\varphi_n^2) = -\frac{d}{dx}\left(x \frac{d\varphi_n}{dx}\right)^2.$$

When both sides of this equation are integrated over $(0, l)$, and the resulting left-hand side is simplified by an integration by parts, there follows

$$[\mu_n^2 x^2 - p^2)\varphi_n^2]_0^l - 2\mu_n^2 \int_0^l x \varphi_n^2\, dx = -\left[x^2\left(\frac{d\varphi_n}{dx}\right)^2\right]_0^l.$$

If we notice that $\varphi_n(0) = J_p(0) = 0$ when $p > 0$, that $J_0(0) = 1$, and that $x J_p'(x) = 0$ at $x = 0$, it is readily seen that the integrated terms vanish at the lower limit. Thus we obtain the important relation

$$C_n = \int_0^l x[J_p(\mu_n x)]^2\, dx$$

$$= \frac{1}{2\mu_n^2}\left\{(\mu_n^2 l^2 - p^2)[\varphi_n(l)]^2 + l^2\left[\frac{d\varphi_n(l)}{dx}\right]^2\right\} \qquad (183)$$

where

$$\varphi_n(x) = J_p(\mu_n x). \qquad (184)$$

We also have

$$\varphi_n(l) = J_p(\mu_n l)$$

and, making use of Equation (111), Section 4.9,

$$\frac{d\varphi_n}{dx} = \frac{d}{dx} J_p(\mu_n x) = -\mu_n J_{p+1}(\mu_n x) + \frac{p}{x} J_p(\mu_n x),$$

so that there follows

$$\frac{d\varphi_n(l)}{dx} = -\mu_n J_{p+1}(\mu_n l) + \frac{p}{l}\varphi_n(l).$$

Thus, if μ_n satisfies (174b), there follows $\varphi_n(l) = 0$ and

$$C_n = \frac{l^2}{2}[J_{p+1}(\mu_n l)]^2 \equiv \frac{l^2}{2}[J_{p+1}(\alpha_n)]^2. \qquad (185a)$$

If μ_n satisfies (175b), there follows $d\varphi_n(l)/dx = 0$ and

$$C_n = \frac{\mu_n^2 l^2 - p^2}{2\mu_n^2}[J_p(\mu_n l)]^2 \equiv \frac{l^2}{2}\frac{\alpha_n^2 - p^2}{\alpha_n^2}[J_p(\alpha_n)]^2. \tag{185b}$$

Finally, if μ_n satisfies (176b), there follows $l\,d\varphi_n(l)/dx = -k\varphi_n(l)$ and

$$C_n = \frac{\mu_n^2 l^2 - p^2 + k^2}{2\mu_n^2}[J_p(\mu_n l)]^2 \equiv \frac{l^2}{2}\frac{\alpha_n^2 - p^2 + k^2}{\alpha_n^2}[J_p(\alpha_n)]^2. \tag{185c}$$

In the exceptional case noted above, when an expansion in terms of the functions $J_0(\mu_n x)$ is required, where $\mu_n = \alpha_n/l$ is a solution of the equation

$$J_0'(\mu_n l) = -\mu_n J_1(\mu_n l) = 0 \quad or \quad \alpha_n J_1(\alpha_n) = 0, \tag{186}$$

the presence of the additional characteristic number $\mu_0 = 0$ introduces a constant term A_0, leading to the series

$$f(x) = A_0 + \sum_{n=1}^{\infty} A_n J_0(\mu_n x) \equiv A_0 + \sum_{n=1}^{\infty} A_n J_0\left(\alpha_n \frac{x}{l}\right), \tag{187}$$

where

$$A_0 = \frac{2}{l^2}\int_0^l x f(x)\,dx \tag{188}$$

and the remaining coefficients are given by Equations (180) and (185b) with $p = 0$.

Although the explicit evaluation of the integral appearing in (180) often is not feasible, even when $f(x)$ is of a simple form, in practice the integral can be evaluated approximately for as many values of n as are required, by numerical or graphical methods.

Expansions of the types considered are of frequent use in connection with the solution of certain boundary-value problems involving partial differential equations, the particular type of expansion to be used being determined by the nature of the problem.

Example. As an application, we determine the expansion of the function $f(x) = 1$ in the interval $(0, L)$, in a series of the form

$$1 = A_1 J_0(\mu_1 x) + A_2 J_0(\mu_2 x) + \cdots$$

$$= \sum_{n=1}^{\infty} A_n J_0(\mu_n x) \quad (0 \leq x < l), \tag{189}$$

where $J_0(\mu_n l) = 0$, so that

$$\mu_n = \frac{\alpha_n}{l} \quad \text{with } J_0(\alpha_n) = 0. \tag{190}$$

In this case Equations (180) and (185a) give

$$A_n = \frac{1}{C_n}\int_0^l x J_0(\mu_n x)\,dx, \tag{191}$$

where

$$C_n = \frac{L^2}{2}[J_1(\mu_n L)]^2. \tag{192}$$

Also, making use of Equation (108), Section 4.9, we have

$$\mu_n x J_0(\mu_n x) = \frac{d}{dx}[x J_1(\mu_n x)]$$

and hence

$$\int_0^l x J_0(\mu_n x)\, dx = \frac{1}{\mu_n}[x J_1(\mu_n x)]_0^l = \frac{l}{\mu_n} J_1(\mu_n l). \tag{193}$$

With the results of (192) and (193), Equation (191) becomes

$$A_n = \frac{2}{(\mu_n l) J_1(\mu_n l)}, \tag{194}$$

and the desired series is of the form

$$1 = \frac{2}{\mu_1 l}\frac{J_0(\mu_1 x)}{J_1(\mu_1 l)} + \frac{2}{\mu_2 l}\frac{J_0(\mu_2 x)}{J_1(\mu_2 l)} + \cdots = \sum_{n=1}^{\infty} \frac{2}{\mu_n l}\frac{J_0(\mu_n x)}{J_1(\mu_n l)}, \tag{195}$$

where $J_0(\mu_n l) = 0$, or, equivalently,

$$1 = \frac{2}{\alpha_1}\frac{J_0\left(\alpha_1 \frac{x}{l}\right)}{J_1(\alpha_1)} + \frac{2}{\alpha_2}\frac{J_0\left(\alpha_2 \frac{x}{l}\right)}{J_1(\alpha_2)} + \cdots, \tag{196}$$

where α_n is the nth zero of $J_0(x)$.

Since the Bessel functions involved are *even* functions of x, the series must represent 1 not only in $(0, l)$ but also in the symmetrical interval $(-l, l)$. At the end points $x = \pm l$ the series clearly does not converge to the given function, since all terms in the series vanish. ∎

Three-place values of the first five zeros of some Bessel functions are included in Table 1. Much more elaborate tabulations appear in the literature.

Table 1. Zeros of Bessel Functions: $J_p(\alpha_n) = 0$

n	$p = 0$	$p = 1$	$p = 2$	$p = 3$	$p = 4$	$p = 5$
1	2.405	3.832	5.135	6.379	7.586	8.780
2	5.520	7.016	8.417	9.760	11.064	12.339
3	8.654	10.173	11.620	13.017	14.373	15.700
4	11.792	13.323	14.796	16.224	17.616	18.982
5	14.931	16.470	17.960	19.410	20.827	22.220

5.14. Legendre Series. Legendre's differential equation,

$$(1 - x^2)\frac{d^2 y}{dx^2} - 2x\frac{dy}{dx} + p(p + 1)y = 0, \tag{197}$$

or, equivalently,

$$\frac{d}{dx}\left[(1 - x^2)\frac{dy}{dx}\right] + p(p + 1)y = 0, \tag{198}$$

can be identified with (90) if we write

$$p(x) = 1 - x^2, \qquad q(x) = 0, \qquad r(x) = 1, \qquad \lambda = p(p + 1). \tag{199}$$

Since the function $p(x)$ vanishes at the points $x = \pm 1$, the results of Section 5.6 state that any two distinct solutions of (197) which are *finite* and have a finite derivative at $x = \pm 1$ are orthogonal, with respect to the weighting function $r(x) = 1$, on the interval $(-1, 1)$. Since (197) possesses solutions which are finite at $x = \pm 1$ only if p is a positive integer or zero (see Section 4.12), this condition of finiteness determines permissible values of p in the form

$$p = n \qquad (n = 0, 1, 2, 3, \ldots). \tag{200}$$

The corresponding solutions which are finite when $x = \pm 1$ are proportional to the Legendre polynomials

$$\varphi_n = P_n(x), \tag{201}$$

and the above reasoning leads to the conclusion

$$\int_{-1}^{1} P_m(x) P_n(x)\, dx = 0 \qquad (m \neq n). \tag{202}$$

Just as in the preceding cases, a function $f(x)$ which is piecewise differentiable in the interval $(-1, 1)$ can be represented by a series of the functions φ_n,

$$f(x) = A_0 P_0(x) + A_1 P_1(x) + A_2 P_2(x) + \cdots$$

$$= \sum_{n=0}^{\infty} A_n P_n(x) \qquad (-1 < x < 1). \tag{203}$$

The coefficients A_n are given by the equation

$$A_n \int_{-1}^{1} [P_n(x)]^2\, dx = \int_{-1}^{1} f(x) P_n(x)\, dx. \tag{204}$$

In order to evaluate the integrals appearing in (204), it is frequently useful to express $P_n(x)$ by *Rodrigues' formula* [Equation (168), Section 4.12],

$$P_n(x) = \frac{1}{2^n n!} \frac{d^n}{dx^n} (x^2 - 1)^n. \tag{205}$$

Hence we may write

$$\int_{-1}^{1} f(x) P_n(x)\, dx = \frac{1}{2^n n!} \int_{-1}^{1} f(x) \frac{d^n}{dx^n} (x^2 - 1)^n\, dx. \tag{206}$$

If the right-hand member of (206) is integrated by parts N times, where $N \leq n$, *assuming that the first N derivatives of $f(x)$ are continuous in $(-1, 1)$*,[†] and if it is noticed that the first $n - 1$ derivatives of $(x^2 - 1)^n$ vanish when $x = \pm 1$, Equation (206) becomes

$$\int_{-1}^{1} f(x) P_n(x)\, dx = \frac{(-1)^N}{2^n n!} \int_{-1}^{1} \left[\frac{d^N}{dx^N} f(x)\right]\left[\frac{d^{n-N}}{dx^{n-N}} (x^2 - 1)^n\right] dx \qquad (N \leq n). \tag{207}$$

---

†Otherwise, integration by parts may not be valid.

In particular, when $N = n$ there follows

$$\int_{-1}^{1} f(x) P_n(x)\, dx = \frac{(-1)^n}{2^n n!} \int_{-1}^{1} (x^2 - 1)^n \frac{d^n f(x)}{dx^n}\, dx \qquad (208)$$

if $f(x)$ and its first n derivatives are continuous in the interval $(-1, 1)$.

Replacing $f(x)$ by $P_n(x)$ in (208) and noticing that, from (205),

$$\frac{d^n P_n(x)}{dx^n} = \frac{1}{2^n n!} \frac{d^{2n}}{dx^{2n}}(x^2 - 1)^n$$

$$= \frac{1}{2^n n!} \frac{d^{2n}}{dx^{2n}}(x^{2n} - nx^{2n-2} + \cdots) = \frac{(2n)!}{2^n n!}, \qquad (209)$$

we obtain, from (208),

$$\int_{-1}^{1} [P_n(x)]^2\, dx = \frac{(2n)!}{2^{2n}(n!)^2} \int_{-1}^{1} (1 - x^2)^n\, dx. \qquad (210)$$

The integral on the right can be evaluated by successive reductions (involving integration by parts) in the form

$$\int_{-1}^{1} (1 - x^2)^n\, dx = 2\frac{(2n)(2n - 2)\cdots 4\cdot 2}{(2n + 1)(2n - 1)\cdots 5\cdot 3} = \frac{2^{2n+1}(n!)^2}{(2n + 1)!}. \qquad (211)$$

Thus Equation (210) takes the simple form

$$\int_{-1}^{1} [P_n(x)]^2\, dx = \frac{2}{2n + 1} \qquad (212)$$

and Equations (204) and (208) give for the coefficients in the expansion

$$f(x) = A_0 P_0(x) + A_1 P_1(x) + \cdots$$

$$= \sum_{n=0}^{\infty} A_n P_n(x) \qquad (-1 < x < 1) \qquad (213a)$$

the alternative forms

$$A_n = \begin{cases} \dfrac{2n + 1}{2} \displaystyle\int_{-1}^{1} f(x) P_n(x)\, dx, \\[2ex] \dfrac{2n + 1}{2^{n+1} n!} \displaystyle\int_{-1}^{1} (1 - x^2)^n \dfrac{d^n f(x)}{dx^n}\, dx, \end{cases} \qquad (213b)$$

where, again, the second form in (213b) is to be used only if $f(x)$ and its first n derivatives are continuous in $(-1, 1)$.

Since $P_n(x)$ is an *even* function of x when n is *even*, and an *odd* function when n is *odd*, it follows that *if $f(x)$ is an even function of x*, the coefficients A_n will vanish when n is odd; whereas *if $f(x)$ is an odd function of x*, the coefficients A_n will vanish when n is even.

Thus, for an *even* function $f(x)$, there follows

$$A_n = \begin{cases} 0 \qquad (n\ \text{odd}), \\[2ex] (2n + 1) \displaystyle\int_{0}^{1} f(x) P_n(x)\, dx \qquad (n\ \text{even}), \end{cases} \qquad (214)$$

whereas for an *odd* function $f(x)$ we have

$$A_n = \begin{cases} (2n + 1) \int_0^1 f(x)P_n(x)\,dx & (n \text{ odd}), \\ 0 & (n \text{ even}). \end{cases} \qquad (215)$$

If we are interested in representing a given function $f(x)$ merely over the interval $(0, 1)$, it is then clear that a series containing only *even* polynomials can be obtained by using (214), or, if preferred, a series containing only *odd* polynomials can be obtained by using (215). Either of these series will then represent $f(x)$ inside the interval $(0, 1)$, and the first series will represent $f(-x)$ and the second $-f(-x)$ in the interval $(-1, 0)$.

Finally, we remark that expansions valid in the more general interval $(-l, l)$ are readily obtained by replacing x by x/l in the preceding developments. This procedure leads to the expansion

$$f(x) = A_0 P_0\left(\frac{x}{l}\right) + A_1 P_1\left(\frac{x}{l}\right) + \cdots$$

$$= \sum_{n=0}^{\infty} A_n P_n\left(\frac{x}{l}\right) \qquad (-l < x < l), \qquad (216)$$

where

$$A_n = \begin{cases} \dfrac{2n + 1}{2l} \displaystyle\int_{-l}^{l} f(x)P_n\left(\frac{x}{l}\right)dx, \\[2ex] \dfrac{2n + 1}{2^{n+1}n!\,l^{n+1}} \displaystyle\int_{-l}^{l} (l^2 - x^2)^n \frac{d^n f(x)}{dx^n}dx, \end{cases} \qquad (217)$$

which generalizes (213).

It is clear that if $f(x)$ is a *polynomial* of degree k, all derivatives of $f(x)$ of order n vanish identically when $n > k$. Hence it follows from (213) or (217) that *any polynomial of degree k can be expressed as a linear combination of the first $k + 1$ Legendre polynomials.*

Example 1. As an application, we determine A_0, A_1, and A_2 so that

$$x^2 = A_0 P_0(x) + A_1 P_1(x) + A_2 P_2(x),$$

the higher coefficients vanishing by virtue of the above remark. Since x^2 is an even function of x, the remaining odd coefficient A_1 must also vanish:

$$A_1 = 0.$$

To calculate A_0 and A_2, we use the second form of (213b) and obtain

$$A_0 = \frac{1}{2}\int_{-1}^{1} 1 \cdot x^2\,dx = \frac{1}{3}, \qquad A_2 = \frac{5}{8 \cdot 2}\int_{-1}^{1} (1 - x^2)^2 \cdot 2\,dx = \frac{2}{3},$$

and hence

$$x^2 = \tfrac{1}{3}[P_0(x) + 2P_2(x)]. \qquad (218)$$

The truth of this identity is verified by reference to the expressions for $P_0(x)$ and $P_2(x)$ in Section 4.12 [Equation (158)]. ∎

Example 2. We determine the Legendre expansion in $(-1, 1)$ of the unit step function $f(x)$ which vanishes when $-1 < x < 0$ and which takes on the value unity when $0 < x < 1$,

$$f(x) = \begin{cases} 0 & (-1 < x < 0), \\ 1 & (0 < x < 1). \end{cases} \tag{219}$$

Since $f(x)$ is discontinuous at $x = 0$, the second expression in (213b) cannot be used. The basic relation becomes, in this case,

$$A_n = \frac{2n + 1}{2} \int_0^1 P_n(x)\, dx.$$

Hence, referring to Equation (158), Section 4.12, we find that

$$A_0 = \tfrac{1}{2} \int_0^1 1\, dx = \tfrac{1}{2},$$

$$A_1 = \tfrac{3}{2} \int_0^1 x\, dx = \tfrac{3}{4},$$

$$A_2 = \tfrac{5}{2} \cdot \tfrac{1}{2} \int_0^1 (3x^2 - 1)\, dx = 0,$$

$$A_3 = \tfrac{7}{2} \cdot \tfrac{1}{2} \int_0^1 (5x^3 - 3x)\, dx = -\tfrac{7}{16},$$

and so forth. Since $f(x)$ can be considered as the sum of the constant $\tfrac{1}{2}$ and an odd function of x, it is seen that A_0 is the only nonvanishing *even* coefficient. The first terms of the desired series are of the form

$$f(x) = \tfrac{1}{2}P_0(x) + \tfrac{3}{4}P_1(x) - \tfrac{7}{16}P_3(x) + \tfrac{11}{32}P_5(x) + \cdots \qquad (-1 < x < 1). \tag{220}$$

The determination of the general coefficient requires knowledge of the formula

$$\int_0^1 P_n(x)\, dx = \begin{cases} 1 & (n = 0), \\ 0 & (n = 2, 4, 6, \ldots), \\ (-1)^{(n-1)/2} \dfrac{1}{n(n+1)} \dfrac{(1 \cdot 3 \cdot 5 \cdots n)^2}{n!} & (n = 1, 3, 5, \ldots) \end{cases} \tag{221}$$

(see Problem 57 of Chapter 4). By use of this result, the expansion (220) can be written in the form

$$f(x) = \frac{1}{2} + \sum_{n \text{ odd}} (-1)^{(n-1)/2} \frac{2n + 1}{2n + 2} \frac{(1 \cdot 3 \cdot 5 \cdots n)^2}{n \cdot n!} P_n(x) \qquad (-1 < x < 1). \tag{222}$$

5.15. The Fourier Integral.

We have seen in Section 5.10 that the functions

$$\phi_n(x) = \sin \frac{n\pi}{l} x \qquad (n = 1, 2, 3, \ldots)$$

form an orthogonal set in the interval $0 < x < l$ and that a piecewise differentiable function $f(x)$ can be represented as a series of these functions in that interval. The question arises as to the possibility of a similar representation in the *semi-infinite* interval $0 < x < \infty$. However, since the functions $\varphi_n(x)$ vanish when $l \rightarrow \infty$, it is clear that such a representation cannot be obtained by merely replacing $1/l$ by zero in the series (138).

If we consider the relation

$$\frac{2}{l} \int_0^{l} \sin u_1 x \sin u_2 x \, dx = \begin{cases} \dfrac{\sin (u_2 - u_1)l}{(u_2 - u_1)l} - \dfrac{\sin (u_2 + u_1)l}{(u_2 + u_1)l} & (u_2 \neq u_1), \\ 1 - \dfrac{\sin 2u_1 l}{2u_1 l} & (u_2 = u_1), \end{cases}$$

(223)

when u_1 and u_2 are positive, we notice that the right-hand member *vanishes* if u_1 and u_2 are *distinct integral multiples of* π/l and is equal to *unity* if u_1 and u_2 both take on the *same* integral multiple of π/l. This is, of course, in accordance with the results of Section 5.10. In addition, however, we notice that as $l \to \infty$ the right-hand member *approaches* zero for *any* positive values of u_1 and u_2, *so long as* $u_2 \neq u_1$, but that if $u_2 = u_1$, the right-hand member approaches unity. That is, if we write

$$\phi_u(x) = \sin ux,$$

(224)

we obtain

$$\lim_{l \to \infty} \frac{2}{l} \int_0^{l} \phi_{u_1}(x) \phi_{u_2}(x) \, dx = \begin{cases} 0 & (u_2 \neq u_1), \\ 1 & (u_2 = u_1), \end{cases}$$

(225)

when u_1 and u_2 are positive. Thus two functions of form (224), corresponding to *any* two positive values of u, have a sort of orthogonality property in the semi-infinite interval $(0, \infty)$, and we are led to suspect that a representation of a function $f(x)$ in that interval generally must involve *all possible* functions of type (224), where u is *any* positive number and is not restricted, as in the finite case, to a set of *discrete* values.

In the finite case $f(x)$ is expressed as a linear combination of functions of the latter type, that is, as an *infinite series*,

$$f(x) = \sum_{n=1}^{\infty} A_n \sin \frac{n\pi x}{l} \qquad (0 < x < l),$$

where the constants A_n specify the contributions of the successive harmonics in the series. In the limiting case $l \to \infty$, if we write the contribution of $\sin ux$ to the representation of $f(x)$ in the form $A(u) \sin ux$, we may expect that the complete representation of $f(x)$ will be obtained by superimposing these contributions by means of an *infinite integral*,

$$f(x) = \int_0^{\infty} A(u) \sin ux \, du \qquad (0 < x < \infty).$$

(226)

Assuming that such a representation exists, for a sufficiently well-behaved function $f(x)$, it remains to determine $A(u)$ for *all* positive values of u, that is, as a function of u. In the following developments we indicate a formal method of determining $A(u)$, and afterward state conditions under which the result obtained is correct.

In determining the function $A(u)$, we are guided by the analogous procedure of Section 5.10. Thus we first multiply both sides of (226) by $\sin u_0 x$, where u_0

is any positive value of u,

$$f(x) \sin u_0 x = \sin u_0 x \int_0^\infty A(u) \sin ux \, du.$$

We next integrate both sides of the result with respect to x over the interval $(0, l)$, where l is destined to become infinite, and so obtain

$$\int_0^l f(x) \sin u_0 x \, dx = \int_0^l \sin u_0 x \left[\int_0^\infty A(u) \sin ux \, du \right] dx$$

or, assuming that the order of integration may be interchanged,

$$\int_0^l f(x) \sin u_0 x \, dx = \int_0^\infty A(u) \left(\int_0^l \sin ux \sin u_0 x \, dx \right) du. \qquad (227)$$

We denote the right-hand member of (227) by R_l; then, making use of (223), we obtain

$$R_l = \frac{1}{2} \int_0^\infty \frac{A(u)}{u - u_0} \sin l(u - u_0) \, du - \frac{1}{2} \int_0^\infty \frac{A(u)}{u + u_0} \sin l(u + u_0) \, du. \qquad (228)$$

With the substitution $l(u - u_0) = t$ in the first integral of (228) and the substitution $l(u + u_0) = t$ in the second integral, this expression takes the form

$$R_l = \frac{1}{2} \int_{-lu_0}^\infty A\left(u_0 + \frac{t}{l}\right) \frac{\sin t}{t} \, dt - \frac{1}{2} \int_{lu_0}^\infty A\left(-u_0 + \frac{t}{l}\right) \frac{\sin t}{t} \, dt. \qquad (229)$$

Now, if $u_0 > 0$, we see that as $l \rightarrow \infty$ the second integral *formally* approaches zero,[†] whereas the first integral takes the value

$$\frac{1}{2} \int_{-\infty}^\infty A(u_0) \frac{\sin t}{t} \, dt = \frac{A(u_0)}{2} \int_{-\infty}^\infty \frac{\sin t}{t} \, dt = \frac{\pi}{2} A(u_0).[‡]$$

Thus, as $l \rightarrow \infty$, Equation (227) gives formally

$$\int_0^\infty f(x) \sin u_0 x \, dx = \lim_{l \to \infty} R_l = \frac{\pi}{2} A(u_0) \qquad (u_0 > 0). \qquad (230)$$

Replacing u_0 by u in (230), we have determined the function $A(u)$ required in (226) in the form

$$A(u) = \frac{2}{\pi} \int_0^\infty f(x) \sin ux \, dx. \qquad (231)$$

To distinguish the dummy variable x in (231) from the current variable x in (226), we replace x by t in (231) before introducing (231) into (226), and so obtain

$$f(x) = \frac{2}{\pi} \int_0^\infty \sin ux \int_0^\infty f(t) \sin ut \, dt \, du \qquad (0 < x < \infty). \qquad (232)$$

[†]Since $A(u)$ is not yet known, the vanishing of this term as $l \rightarrow \infty$ cannot be *guaranteed* at this stage. However, the plausibility of this occurrence follows from the fact that $(\sin t)/t$ executes damped oscillations over the range of integration.
[‡]See Problem 14(b) of Chapter 2.

This expression is known as the *Fourier sine integral* representation of $f(x)$ and can be shown, by more rigorous analysis, to be valid when $x > 0$ if $f(x)$ is piecewise differentiable in every finite positive interval *and if the integral* $\int_0^\infty |f(x)|\, dx$ *exists.* As in the Fourier series theory, the integral represents $f(x)$ at points of continuity and, in order that the same be true at a point $x = c$ at which $f(x)$ has a finite jump, we again assume the definition $f(c) \equiv \frac{1}{2}[f(c+) + f(c-)]$.

In a similar way, the *Fourier cosine integral* representation

$$f(x) = \frac{2}{\pi} \int_0^\infty \cos ux \int_0^\infty f(t) \cos ut\, dt\, du \qquad (0 < x < \infty) \qquad (233)$$

can be obtained, subject to the same restrictions.

It is clear that (232) represents $-f(-x)$ when $x < 0$, whereas (233) represents $f(-x)$ when $x < 0$. Thus, if $f(x)$ is an *odd* function, the representation (232) includes all values of x, whereas if $f(x)$ is an *even* function, the same is true of (233). As in the finite case, we can make use of these facts to obtain a representation valid for all values of x by using both sine and cosine harmonics. Thus, if we write $f(x) = f_e(x) + f_o(x)$, where f_e is even and f_o is odd, we verify that

$$\frac{1}{\pi} \int_0^\infty \cos ux \int_{-\infty}^\infty f(t) \cos ut\, dt\, du = \frac{1}{\pi} \int_0^\infty \cos ux \int_{-\infty}^\infty f_e(t) \cos ut\, dt\, du$$

$$= \frac{2}{\pi} \int_0^\infty \cos ux \int_0^\infty f_e(t) \cos ut\, dt\, du$$

$$= f_e(x) \qquad (-\infty < x < \infty), \qquad (234a)$$

and, similarly,

$$\frac{1}{\pi} \int_0^\infty \sin ux \int_{-\infty}^\infty f(t) \sin ut\, dt\, du = f_o(x) \qquad (-\infty < x < \infty), \qquad (234b)$$

under the assumption that both $\int_0^\infty |f_e(x)|\, dx$ and $\int_0^\infty |f_o(x)|\, dx$ exist or, equivalently, that $\int_{-\infty}^\infty |f(x)|\, dx$ exists. Hence, adding (234a) and (234b), we obtain the representation

$$f(x) = \int_0^\infty [A(u) \cos ux + B(u) \sin ux]\, du \qquad (-\infty < x < \infty), \qquad (235)$$

where $A(u)$ and $B(u)$ are defined by the expressions

$$A(u) = \frac{1}{\pi} \int_{-\infty}^\infty f(t) \cos ut\, dt, \qquad B(u) = \frac{1}{\pi} \int_{-\infty}^\infty f(t) \sin ut\, dt. \qquad (236)$$

By introducing (236) into (235), we obtain also

$$f(x) = \frac{1}{\pi} \int_0^\infty \left[\int_{-\infty}^\infty f(t) \cos u(t - x)\, dt \right] du \qquad (-\infty < x < \infty), \qquad (237a)$$

or, equivalently,

$$f(x) = \frac{1}{2\pi} \int_{-\infty}^\infty \int_{-\infty}^\infty f(t) \cos u(t - x)\, dt\, du \qquad (-\infty < x < \infty). \qquad (237b)$$

This expression is known as the complete *Fourier integral* representation of $f(x)$ and represents $f(x)$ for all values of x in the usual sense if $f(x)$ is piecewise differentiable in every finite interval, *and if the integral $\int_{-\infty}^{\infty} |f(x)|\, dx$ exists.*
   By writing

$$\cos u(t - x) = \tfrac{1}{2}e^{iu(t-x)} + \tfrac{1}{2}e^{-iu(t-x)},$$

we can show that (237a) also can be written in the alternative *complex form*

$$f(x) = \frac{1}{2\pi} \int_{-\infty}^{\infty} \int_{-\infty}^{\infty} f(t)e^{-iu(t-x)}\, dt\, du \qquad (-\infty < x < \infty), \qquad (238a)$$

or, equivalently,

$$f(x) = \frac{1}{2\pi} \int_{-\infty}^{\infty} e^{iux} \int_{-\infty}^{\infty} e^{-iut} f(t)\, dt\, du \qquad (-\infty < x < \infty). \qquad (238b)$$

Finally, if we write $u = 2\pi\omega$, Equation (238b) becomes

$$f(x) = \int_{-\infty}^{\infty} e^{2\pi i\omega x} \int_{-\infty}^{\infty} e^{-2\pi i\omega t} f(t)\, dt\, d\omega \qquad (-\infty < x < \infty). \qquad (239)$$

Equations (235), (237), (238), and (239) are all equivalent forms of the Fourier integral.
   If we denote the inner integral in (238b) by $\bar{f}(u) \equiv \mathfrak{F}\{f(x)\}$, there follows

$$\bar{f}(u) \equiv \mathfrak{F}\{f(x)\} = \int_{-\infty}^{\infty} e^{-iux} f(x)\, dx \qquad (-\infty < u < \infty), \qquad (240a)$$

$$f(x) = \frac{1}{2\pi} \int_{-\infty}^{\infty} e^{iux}\bar{f}(u)\, du \qquad (-\infty < x < \infty). \qquad (240b)$$

The function $\bar{f}(u)$ defined by (240a) is called the *Fourier transform* of f. If the Fourier transform of f is known, Equation (240b) permits the determination of the function f in terms of an integral involving the transform $\bar{f}$.
   For a function $f(x)$ which *vanishes* when $x < 0$, the Fourier transform becomes formally identical with the Laplace transform (Chapter 2) if we replace iu by s and x by t.
   In a similar way, we obtain from (232) and (233) the inversion formulas

$$f_s(u) \equiv \mathfrak{S}\{f(x)\} = \int_0^{\infty} f(x) \sin ux\, dx \qquad (0 < u < \infty), \qquad (241a)$$

$$f(x) = \frac{2}{\pi} \int_0^{\infty} f_s(u) \sin ux\, du \qquad (0 < x < \infty) \qquad (241b)$$

and

$$f_c(u) \equiv \mathfrak{C}\{f(x)\} = \int_0^{\infty} f(x) \cos ux\, dx \qquad (0 < u < \infty), \qquad (242a)$$

$$f(x) = \frac{2}{\pi} \int_0^{\infty} f_c(u) \cos ux\, du \qquad (0 < x < \infty). \qquad (242b)$$

The functions defined by (241a) and (242a) are called the *Fourier sine and cosine transforms* of f, respectively.†

Some of the useful properties of these transforms are exhibited in Problems 89–92.

Example. To illustrate the definitions, we consider the function

$$f(x) = \begin{cases} 0 & (x < 0), \\ 1 & (0 < x < a), \\ 0 & (x > a). \end{cases} \tag{243}$$

From (232) we obtain the Fourier sine integral representation,

$$f(x) = \frac{2}{\pi} \int_0^\infty \frac{1 - \cos au}{u} \sin ux \, du \qquad (0 < x < \infty); \tag{244a}$$

Equation (233) gives the Fourier cosine integral representation,

$$f(x) = \frac{2}{\pi} \int_0^\infty \frac{\sin au \cos ux}{u} du \qquad (0 < x < \infty); \tag{244b}$$

and Equations (235), (238b), and (239) give three equivalent forms of the complete Fourier integral representation,

$$f(x) = \frac{1}{\pi} \int_0^\infty \left(\frac{\sin au}{u} \cos xu + \frac{1 - \cos au}{u} \sin xu \right) du \qquad (-\infty < x < \infty), \tag{245a}$$

$$f(x) = \frac{1}{2\pi i} \int_{-\infty}^\infty \frac{1 - e^{-iau}}{u} e^{iux} \, du \qquad (-\infty < x < \infty), \tag{245b}$$

and

$$f(x) = \frac{1}{2\pi i} \int_{-\infty}^\infty \frac{1 - e^{-2\pi ia\omega}}{\omega} e^{2\pi ix\omega} \, d\omega \qquad (-\infty < x < \infty). \tag{245c}$$

Also, from (240a), (241a), and (242a) we obtain the transforms of $f(x)$,

$$\bar{f}(u) = \frac{1 - e^{-iau}}{iu}, \tag{246a}$$

$$f_S(u) = \frac{1 - \cos au}{u} \qquad (u > 0), \tag{246b}$$

$$f_C(u) = \frac{\sin au}{u} \qquad (u > 0). \tag{246c}$$

∎

When $x < 0$, the integrals appearing in (244a) and (244b) represent $-f(-x)$ and $f(-x)$, respectively. Thus, for example, taking into account the discon-

---

†An important property of the transforms $f_S(u)$ and $f_C(u)$ is that *both tend to zero* as $u \longrightarrow \infty$ if $f(x)$ is such that the integral $\int_0^\infty |f(x)| \, dx$ exists. In addition, $\bar{f}(u)$ *tends to zero* as $u \longrightarrow \pm\infty$ if $\int_{-\infty}^\infty |f(x)| \, dx$ exists. Clearly, corresponding statements apply to integrals of the form $\int_0^\infty F(u) \cos ux \, du$, $\int_0^\infty F(u) \sin ux \, du$, and $\int_{-\infty}^\infty F(u) e^{iux} \, du$ as $x \longrightarrow \infty$ or $x \longrightarrow \pm\infty$.

tinuities in $f(x)$, we have from (244b) the useful result

$$\int_0^\infty \frac{\sin au \cos ux}{u} du = \begin{cases} \dfrac{\pi}{2} & (|x| < a), \\[2mm] \dfrac{\pi}{4} & (|x| = a), \\[2mm] 0 & (|x| > a). \end{cases} \qquad (247)$$

It should be noted that minor variations of the definitions used here also appear in the literature. Specifically, e^{-iux} and e^{iux} sometimes are interchanged in (240a, b) and the factor $1/(2\pi)$ in (240b) may appear instead in (240a), or the factor $1/\sqrt{2\pi}$ may appear in *both* definitions, for the purpose of increased symmetry. Similarly, in (241a, b) and (242a, b), the factor $2/\pi$ may be transferred to the definition of the transform or the factor $\sqrt{2/\pi}$ may appear in both the transform and its inverse, to bring about *complete* symmetry.

In this text we denote by $\bar{f}$ both the *Laplace* transform (Chapter 2) and the *Fourier* transform, continuing to use s to denote the transform variable in the former function and using a different letter (usually u, as here) for that purpose in the latter one. The remaining notational inconsistency is not troublesome since the two transforms are not used together.

Except when $f(x)$ vanishes outside a finite interval, as in (243), the actual evaluation of an integral specifying a transform (or an inverse transform) usually must be effected by methods of so-called *residue calculus* (see Sections 10.13–10.16), or by reference to tables.

Analogous integral representations involving Bessel functions can be obtained. In particular, it can be shown† that the equation

$$f(x) = \int_0^\infty \int_0^\infty ut\, f(t) J_p(ut) J_p(ux)\, dt\, du \qquad (0 < x < \infty) \qquad (248)$$

is valid when $p > -1$, with the usual convention regarding discontinuities, if $f(x)$ is piecewise differentiable in every finite positive interval and if the integral $\int_0^\infty x\,|f(x)|\, dx$ exists. This expression is known as the *Fourier–Bessel integral* representation of $f(x)$, of order p.

The *Hankel (or Fourier–Bessel) transform, of order p*, of a function $f(x)$ may be defined by the relation

$$f_{H_p}(u) = \mathcal{B}_p\{f(x)\} = \int_0^\infty x f(x) J_p(ux)\, dx \qquad (0 < u < \infty), \qquad (249a)$$

after which the function $f(x)$ is expressible in terms of its transform by use of the equation

$$f(x) = \int_0^\infty u f_{H_p}(u) J_p(ux)\, du \qquad (0 < x < \infty), \qquad (249b)$$

by virtue of (248).

†See Reference 5 of Chapter 4.

REFERENCES

1. BESKIN, L., and R. M. ROSENBERG, "Higher Modes of Vibration by a Method of Sweeping," *J. Aero. Sci.*, **13**: 597–604, 1946.
2. CARSLAW, H. S., *Introduction to the Theory of Fourier's Series and Integrals,* 3rd ed., Dover Publications, Inc., New York, 1952.
3. CHURCHILL, R. V., *Fourier Series and Boundary Value Problems,* McGraw-Hill Book Company, Inc., New York, 1941.
4. JACKSON, D., *Fourier Series and Orthogonal Polynomials,* Sixth Carus Mathematical Monograph, Open Court Publishing Company, La Salle, Ill., 1941.
5. SNEDDON, I. N., *Fourier Transforms,* McGraw-Hill Book Company, Inc., New York, 1951.
6. TITCHMARSH, E. C., *Eigenfunction Expansions Associated with Second-Order Differential Equations,* Oxford University Press, Inc., New York, 1946.
7. WHITTAKER, E. T., and G. N. WATSON, *Modern Analysis,* 4th ed., Cambridge University Press, New York, 1958.

PROBLEMS

Section 5.1

1. Show that the boundary-value problem

$$\frac{d^2y}{dx^2} - k^2y = 0, \qquad y(0) = y(l) = 0$$

cannot have a nontrivial solution for real values of k.

2. Determine those values of k for which the partial differential equation

$$\frac{\partial^2 T}{\partial x^2} + \frac{\partial^2 T}{\partial y^2} = 0$$

possesses nontrivial solutions of the form $T(x, y) = f(x) \sinh ky$ which vanish when $x = 0$ and when $x = l$.

3. Show that the equation

$$\frac{d^2y}{dx^2} + \lambda x^2 y = 0$$

can possess a nontrivial solution which vanishes at $x = 0$ and at $x = 1$ only if λ is such that

$$J_{1/4}(\tfrac{1}{2}\lambda^{1/2}) = 0$$

and that, corresponding to such a characteristic number λ_k, any multiple of the function

$$\varphi_k(x) = x^{1/2} J_{1/4}(\tfrac{1}{2}\lambda_k^{1/2}x^2)$$

is a solution with the required properties.

4. Show that the equation

$$x\frac{d^2y}{dx^2} + \frac{dy}{dx} + \lambda xy = 0$$

can possess a nontrivial solution which is *finite* at $x = 0$ and which vanishes when $x = a$ only if λ is such that $J_0(\lambda^{1/2}a) = 0$, and obtain the form of the solution in this case.

Section 5.2

5. Suppose that the uniform string considered in Section 5.2 is unrestrained from transverse motion at the end $x = 0$, and is attached at the end $x = l$ to a yielding support of modulus k, the ends of the string being constrained against appreciable movement parallel to the axis of rotation. Show that the nth critical speed ω_n is given by

$$\omega_n = \frac{\mu_n}{l}\sqrt{\frac{T}{\rho}}$$

where $\cot \mu_n = \alpha \mu_n$ and $\alpha = T/kl$.

6. Suppose that a uniform rotating string has both ends attached to yielding supports, the modulus at $x = 0$ being k_1 and that at $x = l$ being k_2, so that the end conditions

$$\alpha_1 ly'(0) = y(0), \qquad \alpha_2 ly'(l) = -y(l)$$

are to be satisfied, where $\alpha_1 = T/(k_1 l)$ and $\alpha_2 = T/(k_2 l)$. Show that the nth deflection mode is a multiple of the function

$$\varphi_n(x) = \sin \mu_n \frac{x}{l} + \alpha_1 \mu_n \cos \mu_n \frac{x}{l},$$

where μ_n is the nth positive solution of the equation

$$(\alpha_1 \alpha_2 \mu_n^2 - 1) \tan \mu_n = (\alpha_1 + \alpha_2)\mu_n$$

and that the nth corresponding critical speed is

$$\omega_n = \frac{\mu_n}{2}\sqrt{\frac{T}{\rho}}.$$

(Notice that combinations of "fixed" or "free" ends correspond to combinations of α_1 and/or α_2 zero or infinite.)

7. The linear density of a flexible string of length l varies according to the law $\rho = \rho_0(1 + x/l)^2$, where ρ_0 is a constant and x is measured from one end. The ends of the string are attached to an axis rotating with angular velocity ω.

(a) Show that the governing differential equation can be written in the form

$$\frac{d^2y}{dr^2} + 4\mu^2 r^2 y = 0,$$

where $r = 1 + x/l$ and $\mu^2 = (\rho_0\omega^2 l^2)/(4T)$, and that the end conditions then require that y vanish when $r = 1$ and when $r = 2$.

(b) Show that the nth critical speed ω_n is given by

$$\omega_n = 2\sqrt{\frac{T}{\rho_0}}\frac{\mu_n}{l}$$

where μ_n is the nth solution of the equation

$$J_{1/4}(\mu)\,Y_{1/4}(4\mu) - J_{1/4}(4\mu)\,Y_{1/4}(\mu) = 0.$$

8. (a) If a flexible string is caused to vibrate transversely in a plane with respect to a straight form along the x axis, the displacement w of a point is a function of x and of time t. Noticing that the inertia force per unit length is then given by $-\rho(\partial^2 w/\partial t^2)$, and denoting the actual transverse external force by $f(x, t)$, show that w satisfies the partial differential equation

$$\frac{\partial}{\partial x}\left(T\frac{\partial w}{\partial x}\right) - \rho\frac{\partial^2 w}{\partial t^2} + f = 0$$

when small slopes and displacements are considered.

(b) In the case of free periodic vibrations of circular frequency ω, with $f = 0$, we must have

$$w(x, t) = y(x)\sin(\omega t + \varphi),$$

where φ is constant and the amplitude y depends only on x. Show that $y(x)$ then satisfies the same ordinary differential equation as the displacement of a string rotating about the x axis with angular velocity ω (Section 5.2), and hence that the results of Section 5.2 can be interpreted in terms of natural vibration modes and natural frequencies in the analysis of periodic vibrations.

9. Deal with the problem

$$T_0\frac{d^2 y}{dx^2} + \rho\omega^2 y\sqrt{1 + \left(\frac{dy}{dx}\right)^2} = 0,$$

$$y(0) = 0, \qquad y(l) = 0$$

(see page 190) as follows:

(a) By writing $dy/dx = p$ and $d^2 y/dx^2 = p\,dp/dy$ (see Section 1.12), separating, and integrating the resultant equal members, deduce the relation

$$\sqrt{1 + p^2} = 1 + \alpha\left(1 - \frac{y^2}{y_m^2}\right),$$

where y_m is the value of y when $p = 0$ and

$$\alpha = \frac{\rho\omega^2 y_m^2}{2T_0}.$$

(b) By solving for p and defining a parameter k such that

$$k^2 = \frac{\alpha}{2 + \alpha}, \qquad \alpha = \frac{2k^2}{1 - k^2},$$

obtain the equation

$$\frac{dy}{dx} = \pm\frac{2k}{1 - k^2}\sqrt{\left(1 - \frac{y^2}{y_m^2}\right)\left(1 - k^2\frac{y^2}{y_m^2}\right)}.$$

(c) Restricting attention to the case when y has only one extreme value, and noticing from the symmetry that then $y = y_m$ when $x = l/2$, show that

$$\int_{y_m}^{y}\frac{dy}{\sqrt{\left(1 - \frac{y^2}{y_m^2}\right)\left(1 - k^2\frac{y^2}{y_m^2}\right)}} = \pm\frac{2k}{1 - k^2}\left(x - \frac{l}{2}\right).$$

(d) Make the substitution

$$y = y_m \sin \varphi, \qquad \varphi = \sin^{-1} \frac{y}{y_m}$$

and obtain the relation

$$y_m \int_{\pi/2}^{\varphi} \frac{d\varphi}{\sqrt{1 - k^2 \sin^2 \varphi}} = \pm \frac{2k}{1 - k^2} \left(x - \frac{l}{2} \right).$$

(e) By expressing y_m in terms of k and using the notation

$$F(k, \varphi) = \int_0^{\varphi} \frac{d\varphi}{\sqrt{1 - k^2 \sin^2 \varphi}},$$

rewrite this result in the form

$$x = \frac{l}{2} \pm \sqrt{\frac{T_0}{\rho \omega^2}} \sqrt{1 - k^2} \left[F\left(k, \frac{\pi}{2}\right) - F(k, \varphi) \right],$$

where

$$\varphi = \sin^{-1} \frac{y}{y_m}, \qquad y_m = 2 \sqrt{\frac{T_0}{\rho \omega^2}} \frac{k}{\sqrt{1 - k^2}}.$$

(f) Show that the end condition $y(0) = 0$ requires that k satisfy the equation

$$\sqrt{1 - k^2} K(k) = \sqrt{\frac{\rho \omega^2 l^2}{4 T_0}},$$

where

$$K(k) = F\left(k, \frac{\pi}{2}\right) = \int_0^{\pi/2} \frac{d\varphi}{\sqrt{1 - k^2 \sin^2 \varphi}}.$$

[Thus the required solution is defined by the equation obtained in part (e), with the constant k determined by the equation in part (f). The parameter k is called the *modulus* of the elliptic integral $F(k, \varphi)$; $K(k)$ is called the *complete* elliptic integral of the first kind.]

(g) Verify that, if we assume that $y_m/l \ll 1$, it follows that $k \ll 1$, and also verify that, if k is replaced by *zero* in the relations obtained in parts (e) and (f), the resultant approximate relations are

$$x = \frac{l}{2} \pm \sqrt{\frac{T_0}{\rho \omega^2}} \left(\frac{\pi}{2} - \sin^{-1} \frac{y}{y_m} \right),$$

where

$$\sqrt{\frac{\rho \omega^2 l^2}{4 T_0}} = \frac{\pi}{2}$$

or, equivalently,

$$y = y_m \sin \frac{\pi x}{l}$$

and

$$\omega = \frac{\pi}{l} \sqrt{\frac{T_0}{\rho}},$$

in accordance with (16) and (17) when $n = 1$.

Section 5.3

10. Determine expressions for the critical speeds and deflection modes of a rotating uniform shaft of length l fixed at both ends, taking the origin *at one end* of the shaft.

11. Solve Problem 10, taking the origin *at the center* of the shaft. (Express the general solution in terms of circular and hyperbolic functions and, by suitably combining the four equations determining the constants of integration, obtain an equivalent set consisting of two pairs of equations each involving coefficients of either only even or only odd functions. Notice that symmetrical and antisymmetrical modes are thus readily distinguished.)

12. Proceeding as in Problem 8, show that the displacement $w(x, t)$ of a shaft or beam executing small transverse vibrations in a plane satisfies the equation

$$\frac{\partial^2}{\partial x^2}\left(EI\frac{\partial^2 w}{\partial x^2}\right) + \rho\frac{\partial^2 w}{\partial t^2} - f = 0.$$

Show also that in the case of free periodic vibration the amplitude function satisfies Equation (27).

13. A uniform beam is hinged at the end $x = 0$, whereas the end $x = l$ is free. Show that the natural frequencies of free transverse vibrations of the beam are given by $\omega_n = \mu_n^2 l^{-2}\sqrt{EI/\rho}$, where μ_n is a root of the equation $\tanh \mu = \tan \mu$, and the deflection in a natural mode is given by

$$w_n(x, t) = C_n\left[\frac{\sinh (\mu_n x/l)}{\cosh \mu_n} + \frac{\sin (\mu_n x/l)}{\cos \mu_n}\right] \sin \omega_n t.$$

(See Problem 12.) Notice that the problem also admits the solution $w = Cx$ when $\omega = 0$, corresponding to a rigid-body rotation about the end $x = 0$.

14. Suppose that a mass $k\rho l$, equal to k times the mass of the entire beam, is attached to the free end of the beam of Problem 13.

   (a) Verify that the condition of vanishing shear at $x = l$ then is replaced by the requirement $l^3 y'''(l) + k\mu^4 y(l) = 0$, where $\mu^4 = \rho\omega^2 l^4/EI$, and where $y(x)$ is the amplitude function.

   (b) Show that μ then must satisfy the equation

$$\tanh \mu = \tan \mu + 2k\mu \tan \mu \tanh \mu$$

or

$$\cot \mu = \coth \mu + 2k\mu.$$

(Notice that this condition reduces to that obtained in Problem 13 when $k = 0$.)

Section 5.4

15. By integrating Equation (46) twice and determining the two constants of integration, show that the differential equation for buckling of a long column *hinged at both ends* can be taken in the form

$$EI\frac{d^2 y}{dx^2} + Py = 0,$$

with the two end conditions $y(0) = y(l) = 0$, if no transverse loads are acting.

16. An axial load P is applied to a column of circular cross section with linear taper, so that $I(x) = I_0(x/b)^4$, where x is measured from the point at which the column would taper to a point if it were extended, and I_0 is the value of I at the end $x = b$.

   (a) If the column is hinged at the ends $x = a$ and $x = b$, show that the governing differential equation can be put in the form

$$x^4\frac{d^2 y}{dx^2} + \mu^2 y = 0,$$

where $\mu^2 = Pb^4/EI_0$, and where the conditions $y(a) = y(b) = 0$ are to be satisfied.

(b) Show that the general solution of this equation can be expressed first in terms of Bessel functions and finally in the elementary form

$$y = x\left(c_1 \cos \frac{\mu}{x} + c_2 \sin \frac{\mu}{x}\right).$$

(c) Show that the critical loads are of the form

$$P_n = n^2\pi^2 \left(\frac{a}{b}\right)^2 \frac{EI_0}{l^2},$$

where $l = b - a$ is the length of the column, and that the buckling modes are given by

$$\varphi_n(x) = x \sin\left[n\pi \frac{b}{l}\left(1 - \frac{a}{x}\right)\right].$$

17. An originally vertical column of length l is assumed to bend under its own weight. If its mass per unit length is denoted by $\rho(x)$, where x is measured from the top of the column downward, show that the shearing force S at distance x from the top is given by the component of the load $w(x) = g \int_0^x \rho \, dx$ in the direction normal to the deflected axis of the column, and hence is of the form $S = -w \tan \theta$, where θ is the angle of deflection from the vertical. Noticing that the bending moment is given by $M = EI \, d\theta/dx$, and that $S = dM/dx$, obtain the governing differential equation in the form

$$\frac{d}{dx}\left(EI \frac{d\theta}{dx}\right) + w(x)\theta = 0,$$

when small deflections and slopes are assumed, where, if the end $x = 0$ is unrestrained whereas the end $x = l$ is built in, the end conditions $EI\theta'(0) = 0$ and $\theta(l) = 0$ are to be satisfied.

18. In the case of a *uniform* column, for which $EI = $ constant and $w' = g\delta A = $ constant, where δ is the volume mass density and A the cross-sectional area, show that the differential equation of Problem 17 takes the form

$$\frac{d^2\theta}{dx^2} + \frac{g\delta A}{EI} x\theta = 0,$$

and that the solution of this equation for which $\theta'(0) = 0$ is of the form

$$\theta(x) = Cx^{1/2} J_{-1/3}\left(\frac{2}{3}\sqrt{\frac{g\delta A}{EI}} x^{3/2}\right),$$

where C is an arbitrary constant. By imposing the condition $\theta(l) = 0$, and making use of the fact that the smallest zero of $J_{-1/3}(z)$ is $z \doteq 1.866$, deduce that the smallest value of l for which buckling may occur (the "critical length" of the column) is given by

$$l_{cr} \doteq 1.986\left(\frac{EI}{g\delta A}\right)^{1/3}.$$

Section 5.5

19. Determine an approximation to the Euler load P_1 of a column of length l, hinged at both ends, with bending stiffness

$$EI(x) = C\frac{l}{l+x},$$

where C is a constant, by the method of Stodola and Vianello. Take Equation (82) as a first approximation to the deflection curve and compare the results of using Equations (70) and (72a) or (72b) in calculating $\lambda_1^{(1)} = l^2 P_1^{(1)}/C$.

20. Find approximately the smallest characteristic value of λ for the problem

$$\frac{d^2y}{dx^2} + \lambda y = 0, \qquad y'(0) = y(1) = 0$$

by using the method of Vianello and Stodola as follows:

(a) Determine a first approximation, by first replacing the term λy by a convenient constant in the given differential equation, in the form $y_1 = 1 - x^2$.

(b) Show that there follows $f_1 = \frac{1}{12}(5 - 6x^2 + x^4)$.

(c) Show that the use of Equation (70) gives $\lambda_1 \approx 2.5$, whereas the use of (72a) gives $\lambda_1 \approx \frac{42}{17} \doteq 2.471$.

(d) Show that the true value of λ_1 is $\pi^2/4 \doteq 2.467$.

21. Find approximately the smallest value of μ for which the problem

$$x\frac{d^2y}{dx^2} - \frac{dy}{dx} + \mu^2 xy = 0, \qquad y(0) = y(1) = 0$$

possesses a nontrivial solution, by using the method of Stodola and Vianello as follows:

(a) Show that the differential equation can be written in the form

$$\frac{d}{dx}\left(\frac{1}{x}\frac{dy}{dx}\right) + \frac{\mu^2}{x}y = 0.$$

(b) Determine the first approximation $y_1(x)$ in such a way that

$$\frac{d}{dx}\left(\frac{1}{x}\frac{dy_1}{dx}\right) = \text{constant},$$

and also $y_1(0) = y_1(1) = 0$, and show that there follows $y_1 = x^2(1 - x)$, with a convenient choice of the constant.

(c) Show that then $f_1(x) = \frac{1}{120}(7x^2 - 15x^4 + 8x^5)$.

(d) Verify that the use of Equation (70), with $\lambda = \mu^2$, gives

$$\mu_1 \approx \sqrt{15} \doteq 3.873,$$

whereas the use of (72a), with $p = 1/x$, gives

$$\mu_1 \approx \sqrt{\frac{280}{19}} \doteq 3.839.$$

(e) Show that the true value of μ_1 is the smallest zero of $J_1(x)$ and hence (see Table 1 of Section 5.13) is given by $\mu_1 \doteq 3.832$.

Section 5.6

22. Reduce each of the following differential equations to the standard form

$$\frac{d}{dx}\left(p\frac{dy}{dx}\right) + (q + \lambda r)y = 0:$$

(a) $x\frac{d^2y}{dx^2} + 2\frac{dy}{dx} + (x + \lambda)y = 0,$

(b) $\frac{d^2y}{dx^2} + \frac{dy}{dx}\cot x + \lambda y = 0,$

$$\begin{cases} p = e^{\int \frac{a_1}{a_0} dx} \\ q = \frac{a_2}{a_0} p \\ r = \frac{a_3}{a_0} p \end{cases}$$

(c) $\dfrac{d^2 y}{dx^2} + a \dfrac{dy}{dx} + (b + \lambda)y = 0$ (a, b constants),

(d) $x \dfrac{d^2 y}{dx^2} + (c - x)\dfrac{dy}{dx} - ay + \lambda y = 0$ (a, c constants).

23. By considering the characteristic functions of the problem

$$\frac{d^2 y}{dx^2} + \mu^2 y = 0, \qquad y(0) = 0, \quad \alpha l y'(l) + y(l) = 0,$$

with $\alpha \geqq 0$, and using the results of Section 5.6, show that

$$\int_0^l \sin \mu_1 x \sin \mu_2 x \, dx = 0$$

when μ_1 and μ_2 are distinct positive solutions of the equation $\tan \mu l + \alpha \mu l = 0$.

24. By considering the characteristic functions of the problem

$$x \frac{d^2 y}{dx^2} + \frac{dy}{dx} + \mu^2 x y = 0, \qquad y(0) \text{ finite}, \quad y(l) = 0,$$

and using the results of Section 5.6, show that

$$\int_0^l x J_0(\mu_1 x) J_0(\mu_2 x) \, dx = 0$$

when μ_1 and μ_2 are distinct positive roots of the equation $J_0(\mu l) = 0$.

25. By considering the characteristic functions of the problem

$$(1 - x^2)\frac{d^2 y}{dx^2} - 2x \frac{dy}{dx} + \lambda y = 0, \qquad y(\pm 1) \text{ finite}$$

and using the results of Section 5.6, show that

$$\int_{-1}^1 P_r(x) P_s(x) \, dx = 0$$

when r and s are distinct nonnegative integers. [Notice that the equation is Legendre's equation, and that λ must be of the form $n(n + 1)$, *where n is an integer*, in order that solutions finite at both $x = 1$ and $x = -1$ exist, according to Section 4.12.]

26. If $p(b) = p(a)$, and if the conditions

$$y(b) = \alpha_{11} y(a) + \alpha_{12} y'(a),$$
$$y'(b) = \alpha_{21} y(a) + \alpha_{22} y'(a)$$

are associated with the differential equation (90), show that the orthogonality property (102) holds for two characteristic functions, corresponding to distinct characteristic numbers, provided that

$$\begin{vmatrix} \alpha_{11} & \alpha_{12} \\ \alpha_{21} & \alpha_{22} \end{vmatrix} = 1.$$

27. Prove that *if an end condition imposed in a proper Sturm–Liouville problem requires that*

$$p(\varphi_1 \varphi_2' - \varphi_2 \varphi_1') = 0 \qquad \text{at an end point,}$$

when φ_1 and φ_2 are characteristic functions, then no characteristic number can correspond to two linearly independent characteristic functions. [*Suggestion:* Assume, on the contrary, that φ_1 and φ_2 are linearly independent and that both correspond to the same characteristic number (and hence both satisfy the same differential equation). Show that then $p(x)[\varphi_1(x)\varphi_2'(x) - \varphi_2(x)\varphi_1'(x)]$ is a *constant* (Abel's theorem), so that its vanishing at a point implies its *identical* vanishing in (a, b). Then prove that $\varphi_2(x)$ is a constant multiple of $\varphi_1(x)$ in (a, b) and note the resultant contradiction.]

28. Prove, by the following steps, that *the characteristic numbers of a real Sturm–Liouville problem with $r(x) > 0$ are all real*:

(a) Show that if λ_1 is *assumed* to be nonreal, with an associated characteristic function $\varphi_1(x)$, then $\bar\lambda_1$ also is a characteristic number with an associated characteristic function $\overline{\varphi_1(x)}$, which is the complex conjugate of $\varphi_1(x)$.

(b) Deduce from (97) that if $\varphi_1(x) = u_1(x) + iv_1(x)$, where u_1 and v_1 are real, there follows

$$(\lambda_1 - \bar\lambda_1) \int_a^b r\varphi_1\bar\varphi_1 \, dx \equiv (\lambda_1 - \bar\lambda_1) \int_a^b r(u_1^2 + v_1^2) \, dx = 0,$$

and hence, since $r(x) > 0$, conclude that λ_1 is in fact real. [Thus $\varphi_1(x)$ also can be taken to be real, by suppressing a possible imaginary *constant* multiplicative factor.]

29. Prove, by the following steps, that *the characteristic numbers of a proper Sturm–Liouville problem are real and nonnegative*:

(a) Let λ_1 and $\varphi_1(x)$ be corresponding real characteristic quantities (see Problem 28) and show that

$$\lambda_1 \int_a^b r\varphi_1^2 \, dx = -\left[p\varphi_1 \frac{d\varphi_1}{dx} \right]_a^b + \int_a^b p\left(\frac{d\varphi_1}{dx}\right)^2 dx - \int_a^b q\varphi_1^2 \, dx.$$

(b) Show that the first term on the right vanishes unless a condition of type (98c) is imposed at $x = a$ and/or $x = b$, that a condition $y + \gamma_1 y' = 0$ at $x = a$ would introduce the term $-\gamma_1 p(a)[\varphi_1'(a)]^2$, and that a condition $y + \gamma_2 y' = 0$ at $x = b$ would introduce the term $+\gamma_2 p(b)[\varphi'(b)]^2$. Thus, by imposing the requirements that the problem be proper, deduce that $\lambda_1 \geqq 0$. Show also that $\lambda_1 = 0$ *is a characteristic number of a proper problem if and only if $q(x) = 0$ and the boundary conditions are satisfied by a constant.*

30. Derive Equation (105) from (104).

Section 5.7

31. Determine the coefficients in the representation

$$f(x) = \sum_{n=1}^\infty A_n \sin nx \qquad (0 < x < \pi)$$

in the following cases:

(a) $f(x) = 1$,   (b) $f(x) = x$,   (c) $f(x) = \begin{cases} 1 & (x < \pi/2), \\ \frac{1}{2} & (x = \pi/2), \\ 0 & (x > \pi/2). \end{cases}$

32. (a–c) Show that the series obtained in Problem 31 cannot be differentiated term by term. (Investigate the convergence of the result in each case.)

33. (a) If the expansion $f(x) = \sum_{n=1}^\infty A_n \sin nx$ is valid in $(0, \pi)$, show formally that

$$\int_0^\pi [f(x)]^2 \, dx = \frac{\pi}{2} \sum_{n=1}^\infty A_n^2.$$

(b) From this result, and from the results of Problems 31(a, b), deduce the following relations:

$$\frac{1}{1^2} + \frac{1}{3^2} + \frac{1}{5^2} + \cdots = \frac{\pi^2}{8}, \qquad \frac{1}{1^2} + \frac{1}{2^2} + \frac{1}{3^2} + \cdots = \frac{\pi^2}{6}.$$

34. Expand the function $f(x) = 1$ in a series of the characteristic functions of the boundary-value problem

$$\frac{d^2 y}{dx^2} + \lambda y = 0,$$

$$y(0) = 0, \qquad ly'(l) + ky(l) = 0 \quad (k \geqq 0),$$

over the interval $(0, l)$.

35. Consider the Sturm–Liouville problem

$$x^2 \frac{d^2 y}{dx^2} + x \frac{dy}{dx} + \lambda y = 0, \qquad y(1) = y(b) = 0,$$

where $b > 1$.

(a) Show that the characteristic numbers and functions are

$$\lambda_n = \frac{n^2 \pi^2}{(\log b)^2}, \qquad \varphi_n(x) = \sin\left(n\pi \frac{\log x}{\log b}\right) \qquad (n = 1, 2, \ldots).$$

(b) If $f(x)$ is piecewise differentiable in the interval $(1, b)$, and if

$$f(x) = \sum_{n=1}^{\infty} A_n \sin\left(n\pi \frac{\log x}{\log b}\right) \qquad (1 < x < b),$$

show that

$$A_n = \frac{\displaystyle\int_1^b f(x) \sin\left(n\pi \frac{\log x}{\log b}\right) \frac{dx}{x}}{\displaystyle\int_1^b \sin^2\left(n\pi \frac{\log x}{\log b}\right) \frac{dx}{x}}$$

and, by introducing an appropriate change of variables, reduce this result to the form

$$A_n = 2 \int_0^1 f(b^t) \sin n\pi t \, dt.$$

(c) Evaluate A_n when $f(x) = 1$ and when $f(x) = x$.

Section 5.8

36. Obtain the solution of the problem

$$\frac{d^2 y}{dx^2} + \Lambda y = h(x), \qquad y(0) = y(l) = 0$$

in the form

$$y(x) = \sum_{n=1}^{\infty} \frac{A_n}{\Lambda - n^2 \pi^2 / l^2} \sin \frac{n\pi x}{l} \qquad (0 \leqq x \leqq l)$$

when $\Lambda \neq (n\pi/l)^2$ $(n = 1, 2, \ldots)$, where A_n is the nth coefficient in the expansion

$$h(x) = \sum_{n=1}^{\infty} A_n \sin \frac{n\pi x}{l} \qquad (0 < x < l),$$

assuming that $h(x)$ is piecewise differentiable in $(0, l)$.

37. Obtain the condition which must be satisfied by a piecewise differentiable function $h(x)$ in order that the problem

$$\frac{d^2y}{dx^2} + \frac{p^2\pi^2}{l^2}y = h(x), \qquad y(0) = y(l) = 0$$

possesses a solution, when p is a positive integer, and express the corresponding most general solution as a series $\sum a_n \sin(n\pi x/l)$.

38. Let $h(x) = \sin(p\pi x/l)$ in Problem 36, where p is a positive integer.
    (a) Show that the solution of that problem then is

$$y = \frac{\sin(p\pi x/l)}{\Lambda - (p^2\pi^2/l^2)}$$

if $\Lambda \neq (n\pi/l)^2$ $(n = 1, 2, \ldots)$.
    (b) If $\Lambda = (r\pi/l)^2$, where r is a positive integer but $r \neq p$, show that the solution of part (a) becomes

$$y = \frac{l^2}{\pi^2} \frac{\sin(p\pi x/l)}{r^2 - p^2}$$

and also show that to this solution can be added any constant multiple of $\sin(r\pi x/l)$.
    (c) Account for the nonuniqueness of the solution in part (b).

39. By introducing the value of A_n (as an integral) into the result of Problem 36 and interchanging the summation and integration (this can be shown to be legitimate), show that the solution of the problem

$$\frac{d^2y}{dx^2} + \Lambda y = h(x), \qquad y(0) = y(l) = 0$$

can be expressed in the form

$$y(x) = \int_0^l G(x, \xi) h(\xi) \, d\xi$$

where $G(x, \xi)$ possesses the expansion

$$G(x, \xi) = \frac{2}{l} \sum_{n=1}^{\infty} \frac{\sin(n\pi x/l) \sin(n\pi \xi/l)}{\Lambda - (n^2\pi^2/l^2)} \qquad (0 \leq x \leq l, \ \ 0 \leq \xi \leq l),$$

provided that $\Lambda \neq n^2\pi^2/l^2$ $(n = 1, 2, \ldots)$. [Here $G(x, \xi)$ is the *Green's function* for the expression $d^2y/dx^2 + \Lambda y$ and the conditions $y(0) = y(l) = 0$. It is given in closed form when $\Lambda = 1$ and when $\Lambda = 0$ in Problems 55 and 59 of Chapter 1.]

40. Obtain the solution of the problem

$$\frac{d}{dx}\left(x\frac{dy}{dx}\right) + \Lambda xy = h(x), \qquad y(0) \text{ finite}, \quad y(l) = 0$$

in the form

$$y(x) = \sum_{n=1}^{\infty} \frac{A_n}{\Lambda - \mu_n^2} J_0(\mu_n x) \qquad (0 \leq x \leq l),$$

when $\Lambda \neq \mu_1^2, \mu_2^2, \ldots$, where μ_n is the nth root of the equation $J_0(\mu l) = 0$, and where A_n is the nth coefficient in the expansion

$$\frac{h(x)}{x} = \sum_{n=1}^{\infty} A_n J_0(\mu_n x) \qquad (0 < x < l),$$

assuming that $h(x)/x$ is piecewise differentiable in $(0, l)$. Show also that the coefficient A_n is to be determined by the equation

$$A_n \int_0^l x[J_0(\mu_n x)]^2 \, dx = \int_0^l h(x) J_0(\mu_n x) \, dx.$$

Section 5.9

41. With the abbreviation

$$\gamma_k = \int_a^b r\varphi_k^2 \, dx,$$

show formally that, with the notation of Sections 5.7 and 5.9,

$$\int_a^b r y_N^2 \, dx = \sum_{k=1}^\infty \left(\frac{\lambda_1}{\lambda_k}\right)^{2N-2} \gamma_k A_k^2,$$

$$\int_a^b r f_N y_N \, dx = \frac{1}{\lambda_1} \sum_{k=1}^\infty \left(\frac{\lambda_1}{\lambda_k}\right)^{2N-1} \gamma_k A_k^2,$$

$$\int_a^b r f_N^2 \, dx = \frac{1}{\lambda_1^2} \sum_{k=1}^\infty \left(\frac{\lambda_1}{\lambda_k}\right)^{2N} \gamma_k A_k^2,$$

and hence deduce that, if we write

$$\mu_m = \sum_{k=1}^\infty \left(\frac{\lambda_1}{\lambda_k}\right)^m \gamma_k A_k^2,$$

then, in the Nth cycle, the approximation to λ_1 yielded by Equation (72a) is $(\mu_{2N-2}/\mu_{2N-1})\lambda_1$, whereas that yielded by (72b) is $(\mu_{2N-1}/\mu_{2N})\lambda_1$. (These results are valid when the generating Sturm–Liouville is proper, as is assumed in Section 5.9.)

42. With the abbreviation

$$\alpha_k = \int_a^b \varphi_k \, dx$$

and the notation of Problem 41 and Sections 5.7 and 5.9, show formally that, in the nth cycle,

$$\int_a^b y_n \, dx = \sum_{k=1}^\infty \left(\frac{\lambda_1}{\lambda_k}\right)^{n-1} \alpha_k A_k,$$

$$\int_a^b f_n \, dx = \frac{1}{\lambda_1} \sum_{k=1}^\infty \left(\frac{\lambda_1}{\lambda_k}\right)^n \alpha_k A_k,$$

and hence, with the additional abbreviation

$$v_m = \sum_{k=1}^\infty \left(\frac{\lambda_1}{\lambda_k}\right)^m \alpha_k A_k,$$

deduce that the approximation to λ_1 yielded by (70) in the nth cycle is $(v_{n-1}/v_n)\lambda_1$.

43. With the notation of Problems 41 and 42, suppose that the initial approximation $y_1(x)$ is such that

$$\frac{d}{dx}\left(p \, \frac{dy_1}{dx}\right) = 1.$$

(a) Show that Equations (128) and (129) then imply also that

$$\frac{d}{dx}\left(p\frac{dy_1}{dx}\right) = -\sum_{n=1}^{\infty} \lambda_n A_n r(x)\varphi_n(x),$$

and hence also

$$r(x)\sum_{n=1}^{\infty} \lambda_n A_n \varphi_n(x) = -1,$$

when $a < x < b$. Thus deduce that in this case there follows

$$\alpha_k = -\lambda_k \gamma_k A_k,$$

and hence also

$$v_m = -\lambda_1 \mu_{m-1}.$$

(b) Deduce that, with the assumed determination of $y_1(x)$, the use of Equation (72a) gives the same approximation to λ_1 in the Nth cycle as does the use of (70) in the $(2N)$th cycle, and also the use of (72b) in the Nth cycle gives the same approximation as does the use of (70) in the $(2N+1)$th cycle.

44. Use results of Problems 41 and 42 to show that, when $A_1 \neq 0$ and $A_2 \neq 0$ in a Stodola–Vianello sequence, the error in the approximation afforded to a multiple of $\varphi_1(x)$ by a multiple of $f_N(x)$ in the Nth cycle is small of order $(\lambda_1/\lambda_2)^N$, whereas the error in the approximation afforded to λ_1 by use of either Equation (72a) or (72b) in that cycle is small of order $(\lambda_1/\lambda_2)^{2N}$. (If $A_2 = 0$ but $A_3 \neq 0$, the ratio λ_1/λ_2 is to be replaced by λ_1/λ_3, and so forth.)

45. If the initial approximation $y_1(x)$ is orthogonal to the first characteristic function $\varphi_1(x)$ with respect to $r(x)$, show that the method of Stodola and Vianello leads to the *second* characteristic quantities λ_2 and $\varphi_2(x)$ unless $y_1(x)$ is also orthogonal to $\varphi_2(x)$.

46. Show that zero is a characteristic number of the problem

$$\frac{d^2y}{dx^2} + \lambda y = 0, \qquad y'(0) = y'(1) = 0,$$

corresponding to the characteristic function $\varphi(x) = 1$, in addition to the characteristic numbers $\lambda_n = n^2\pi^2$ $(n = 1, 2, \ldots)$, corresponding to the respective characteristic functions $\varphi_n(x) = \cos n\pi x$. Hence deduce that the method of Stodola and Vianello would lead to $\lambda = 0$ and $y(x) = 1$ in this case unless the initial approximation $y_1(x)$ were orthogonal to the function $y = 1$ over $(0, 1)$, that is, unless

$$\int_0^1 y_1(x)\,dx = 0.$$

47. Apply the method of Stodola and Vianello to the approximate determination of the smallest nonzero characteristic number of the problem considered in Problem 46, assuming $y_1(x)$ in the form

$$y_1(x) = c_0 + c_1 x + c_2 x^2 + c_3 x^3,$$

and determining the constants such that

$$y_1'(0) = 0, \qquad y_1'(1) = 0,$$

and

$$\int_0^1 y_1(x)\,dx = 0.$$

Show that the resultant initial approximation $y_1 = 1 - 6x^2 + 4x^3$, corresponding to a convenient choice of the arbitrary multiplicative factor, leads to the result

$$f_1 = \tfrac{1}{10}(1 - 5x^2 + 5x^4 - 2x^5)$$

if the same three conditions are imposed on f_1. Notice that Equation (70) cannot be applied here, but that the simple process of equating $y_1(x)$ to $\lambda f_1(x)$ at either $x = 0$ or $x = 1$ gives $\lambda_1 \approx 10$, whereas the use of (72a) gives $\lambda_1 \approx \tfrac{306}{31} \doteq 9.871$. (The exact result is $\lambda_1 = \pi^2 \doteq 9.870$.)

Section 5.10

48. Establish, by direct integration, the truth of (a) Equation (140), (b) Equation (146), and (c) Equations (161).

49. Expand each of the following functions in a Fourier sine series of period $2l$, over the interval $(0, l)$, and in each case sketch the function represented by the series in the interval $(-3l, 3l)$:

(a) $f(x) = \begin{cases} 0 & (x < 0), \\ x(l - x) & (x > 0), \end{cases}$

(b) $f(x) = \begin{cases} 0 & (x < 0), \\ x & (0 < x < l/2), \\ l - x & (x > l/2), \end{cases}$

(c) $f(x) = \begin{cases} 1 & (x < l/2), \\ 0 & (x > l/2), \end{cases}$

(d) $f(x) = \sin \dfrac{\pi x}{2l}$,

(e) $f(x) = \begin{cases} \sin \dfrac{\pi x}{l} & (0 < x < l/2), \\ 0 & (\text{otherwise}), \end{cases}$

(f) $f(x) = \begin{cases} \dfrac{1}{\epsilon} & (0 < x < \epsilon < l), \\ 0 & (\text{otherwise}). \end{cases}$

50. (a–f) Expand each of the functions listed in Problem 49 in a Fourier cosine series of period $2l$, over the interval $(0, l)$, and in each case sketch the function represented by the series in the interval $(-3l, 3l)$.

51. (a) Obtain the expansion

$$\cos \alpha x = \frac{\sin \pi \alpha}{\pi \alpha} + \sum_{n=1}^{\infty} (-1)^n \frac{2\alpha \sin \pi \alpha}{\pi(\alpha^2 - n^2)} \cos nx \qquad (-\pi \leq x \leq \pi),$$

when α is nonintegral.

(b) Deduce from this result that

$$\cot \pi \alpha = \frac{1}{\pi} \left(\frac{1}{\alpha} - \sum_{n=1}^{\infty} \frac{2\alpha}{n^2 - \alpha^2} \right),$$

when α is nonintegral.

52. Plot the first three partial sums $s_n(x)$, for which

$$s_1(x) = \sin x, \qquad s_2(x) = \sin x - \tfrac{1}{2} \sin 2x,$$

and

$$s_3(x) = \sin x - \tfrac{1}{2} \sin 2x + \tfrac{1}{3} \sin 3x,$$

of the right-hand member of the relation

$$\frac{x}{2} = \sum_{k=1}^{\infty} \frac{(-1)^{k+1}}{k} \sin kx \qquad (0 \leq x < \pi),$$

over the interval $(0, \pi)$, and compare these plots with a plot of the limit function.

53. Plot the first three distinct partial sums of the series representation

$$x(\pi - x) = \frac{8}{\pi}\left(\sin x + \frac{1}{3^3}\sin 3x + \frac{1}{5^3}\sin 5x + \cdots\right) \qquad (0 \leq x \leq \pi),$$

over the interval $(0, \pi)$, and compare these plots with a plot of the limit function.

54. Use (124) to solve the problem

$$\frac{d^2y}{dx^2} + \Lambda y = \sin \omega x, \qquad y(0) = y(l) = 0$$

when $\Lambda \neq p^2\pi^2/l^2$ and $\omega^2 \neq p^2\pi^2/l^2$ ($p = 0, 1, 2, \ldots$).

55. Investigate the excluded cases in Problem 54, assuming that $\omega \geq 0$.

56. Show that, if we determine the coefficients in an N-term approximation of the form

$$f(x) \approx \sum_{n=1}^{N} a_n \sin \frac{n\pi x}{l} \qquad (0 \leq x \leq l)$$

in such a way that

$$\int_0^l \left[f(x) - \sum_{n=1}^{N} a_n \sin \frac{n\pi x}{l} \right]^2 dx = \text{minimum}$$

(that is, by the method of *least squares*), the coefficients are determined in the form

$$a_k = \frac{2}{l} \int_0^l f(x) \sin \frac{k\pi x}{l} dx \qquad (k = 1, 2, \ldots, N),$$

that is, as the first N coefficients in the Fourier sine series representation. (Differentiate the quantity to be minimized with respect to a general coefficient a_k and equate the result to zero.)

Section 5.11

57. (a–f) Expand each of the functions listed in Problem 49 in a Fourier series of period $2l$, over the interval $(-l, l)$, and in each case sketch the function represented by the series in the interval $(-3l, 3l)$.

58. Expand each of the following functions in a Fourier series of period equal to the length of the indicated interval of representation:

(a) $f(x) = a + bx$ $\qquad$ $(0 < x < P)$,

(b) $f(x) = \begin{cases} 0 & (x < 0) \\ 1 & (x > 0) \end{cases}$ $\qquad$ $(-1 < x < 1)$,

(c) $f(x) = \sin x$ $\qquad$ $(0 \leq x \leq \pi)$,

(d) $f(x) = x$ $\qquad$ $(1 < x < 2)$.

59. Prove that the Fourier series of period P which represents an odd-harmonic function of period P [satisfying (162)] involves only the odd harmonics.

60. Obtain the Fourier series of period 2π which represents the solution of the problem

$$\frac{d^2y}{dx^2} + \Lambda y = h(x), \qquad y(-\pi) = y(\pi), \quad y'(-\pi) = y'(\pi)$$

when

$$h(x) = \begin{cases} 0 & (-\pi < x < 0), \\ 1 & (0 < x < \pi/2), \\ 0 & (\pi/2 < x < \pi), \end{cases}$$

assuming that $\Lambda \neq p^2$ $(p = 0, 1, 2, \ldots)$.

61. (a) If the representation

$$f(x) = A_0 + \sum_{n=1}^{\infty} A_n \cos \frac{n\pi x}{l} \qquad (0 \leq x \leq l)$$

is valid, show formally that

$$\frac{2}{l} \int_0^l [f(x)]^2 \, dx = 2A_0^2 + \sum_{n=1}^{\infty} A_n^2.$$

(b) If the representation

$$f(x) = \sum_{n=1}^{\infty} B_n \sin \frac{n\pi x}{l} \qquad (0 < x < l)$$

is valid, show formally that

$$\frac{2}{l} \int_0^l [f(x)]^2 \, dx = \sum_{n=1}^{\infty} B_n^2.$$

(c) If the representation

$$f(x) = A_0 + \sum_{n=1}^{\infty} \left(A_n \cos \frac{n\pi x}{l} + B_n \sin \frac{n\pi x}{l} \right) \qquad (-l < x < l)$$

is valid, show formally that

$$\frac{1}{l} \int_{-l}^l [f(x)]^2 \, dx = 2A_0^2 + \sum_{n=1}^{\infty} (A_n^2 + B_n^2).$$

[It is a remarkable fact that the *results* of parts (a), (b), and (c) are valid whenever $f(x)$ is bounded and integrable over the relevant interval, whether or not the associated Fourier representations referred to are valid. These results are often called the *Parseval equalities*.]

62. (a) Show that the Fourier series expansion described by Equations (154) and (155) can be written in the form

$$f(x) = \frac{1}{2l} \int_{-l}^l f(t) \, dt + \sum_{n=1}^{\infty} \frac{1}{l} \int_{-l}^l f(t) \left(\cos \frac{n\pi x}{l} \cos \frac{n\pi t}{l} + \sin \frac{n\pi x}{l} \sin \frac{n\pi t}{l} \right) dt$$

$$= \frac{1}{2l} \int_{-l}^l f(t) \, dt + \sum_{n=1}^{\infty} \frac{1}{l} \int_{-l}^l f(t) \cos \frac{n\pi(x - t)}{l} \, dt$$

when $-l < x < l$.

(b) By making use of Euler's formula

$$\cos \frac{n\pi(x - t)}{l} = \frac{1}{2} [e^{\frac{in\pi(x-t)}{l}} + e^{-\frac{in\pi(x-t)}{l}}],$$

show that the last form can be transformed to

$$f(x) = \sum_{n=-\infty}^{\infty} \frac{1}{2l} \int_{-l}^l f(t) e^{\frac{in\pi(x-t)}{l}} \, dt$$

or, equivalently,

$$f(x) = \sum_{n=-\infty}^{\infty} a_n e^{\frac{in\pi x}{l}} \qquad (-l < x < l)$$

where

$$a_n = \frac{1}{2l} \int_{-l}^{l} f(x) e^{-\frac{in\pi x}{l}} \, dx.$$

(This is known as the *complex form of the Fourier series*.)

63. Determine the coefficients in the complex form of the Fourier series of period $P = 2l$ (Problem 62) representing each of the following functions in $(-l, l)$:

(a) $f(x) = 1$,

(b) $f(x) = \cos x$,

(c) $f(x) = e^{ax}$,

(d) $f(x) = \begin{cases} -1 & (-l < x < 0), \\ 1 & (0 < x < l). \end{cases}$

64. (a) Assuming the relation

$$e^{\frac{x}{2}\left(r - \frac{1}{r}\right)} = \sum_{n=-\infty}^{\infty} r^n J_n(x)$$

(see Problem 42 of Chapter 4), replace r by $e^{i\theta}$ and deduce the Fourier expansion

$$e^{ix \sin \theta} = \sum_{n=-\infty}^{\infty} J_n(x) e^{in\theta}.$$

(b) By equating separately the real and imaginary parts of the two sides of this equation, and recalling that $J_{-n}(x) = (-1)^n J_n(x)$, show that

$$\cos (x \sin \theta) = J_0(x) + 2[J_2(x) \cos 2\theta + J_4(x) \cos 4\theta + \cdots]$$

and $$\sin (x \sin \theta) = 2[J_1(x) \sin \theta + J_3(x) \sin 3\theta + \cdots].$$

65. (a) From the result of Problem 64(a) deduce the integral formula

$$J_n(x) = \frac{1}{2\pi} \int_{-\pi}^{\pi} e^{i(x \sin \theta - n\theta)} \, d\theta$$

$$= \frac{1}{\pi} \int_{0}^{\pi} \cos (x \sin \theta - n\theta) \, d\theta,$$

when n is an integer.

(b) From this result or from the result of Problem 64(b), also obtain the relations

$$J_n(x) = \frac{1}{\pi} \int_{0}^{\pi} \cos (x \sin \theta) \cos n\theta \, d\theta$$

$$= \frac{2}{\pi} \int_{0}^{\pi/2} \cos (x \sin \theta) \cos n\theta \, d\theta \qquad (n \text{ even});$$

$$J_n(x) = \frac{1}{\pi} \int_{0}^{\pi} \sin (x \sin \theta) \sin n\theta \, d\theta$$

$$= \frac{2}{\pi} \int_{0}^{\pi/2} \sin (x \sin \theta) \sin n\theta \, d\theta \qquad (n \text{ odd}).$$

[Notice that the integrands in part (b) are symmetrical with respect to $\theta = \pi/2$, with the stated restrictions on n.]

(c) In particular, show that

$$J_0(x) = \frac{2}{\pi} \int_0^{\pi/2} \cos(x \sin \theta) \, d\theta$$

$$= \frac{2}{\pi} \int_0^1 \frac{\cos xt}{\sqrt{1 - t^2}} \, dt.$$

Section 5.12

66. Derive the statements made in Section 5.12 relative to term-by-term differentiation of (a) the Fourier cosine series and (b) the Fourier sine series.

67. If $f(x)$ and $f'(x)$ are continuous and $f''(x)$ is piecewise differentiable in $(a, a + P)$, show that the series obtained by two term-by-term differentiations of the complete Fourier series (157) converges to $f''(x)$ at each interior point of $(a, a + P)$ at which $f''(x)$ is continuous, provided that

$$f(a + P) = f(a), \qquad f'(a + P) = f'(a),$$

and also converges to $f''(x)$ at the end points if also $f''(a + P) = f''(a)$.

68. If $f(x)$ and $f'(x)$ are continuous and $f''(x)$ is piecewise differentiable in $(0, l)$, derive the following statements:

(a) The series obtained by twice differentiating the Fourier sine series (138) term by term converges to $f''(x)$ at each interior point of $(0, l)$ at which $f''(x)$ is continuous provided that

$$f(0) = f(l) = 0,$$

and also converges to $f''(x)$ at an end point if $f''(x) = 0$ at that point.

(b) The series obtained by twice differentiating the Fourier cosine series (147) term by term converges to $f''(x)$ at each interior point of $(0, l)$ at which $f''(x)$ is continuous, and at both end points, provided that

$$f'(0) = f'(l) = 0.$$

69. Show that the results of Problem 68 do not permit two term-by-term differentiations of the expansions of text Examples 1 and 2 and verify that in fact the resultant series would diverge at all interior points in both cases.

70. *Term-by-term integration of Fourier series.* Show that the result of term-by-term integration of the formal series

$$f(x) = A_0 + \sum_{n=1}^{\infty} A_n \cos \frac{n\pi x}{l}$$

over the interval $(0, x)$,

$$\int_0^x f(t) \, dt = A_0 x + \sum_{n=1}^{\infty} \frac{l}{n\pi} A_n \sin \frac{n\pi x}{l},$$

is valid when $0 \leq x \leq l$ provided that the A's are calculated from (148) and that $f(x)$ is piecewise continuous in $(0, l)$. [Let $F(x) = \int_0^x f(t) \, dt - (x/l) \int_0^l f(t) \, dt$ and expand $F(x)$ in a Fourier sine series over $(0, l)$, noticing that $F(0) = F(l) = 0$ and that $F'(x) = f(x) - (1/l) \int_0^l f(t) \, dt$, where $f(x)$ is continuous. Here the *original* cosine series need not converge to $f(x)$. Similar conclusions follow for the sine series and for the complete series.]

Section 5.13

71. Expand the function $f(x) = x^p$ in a series of the characteristic functions of the boundary-value problem

$$x\frac{d}{dx}\left(x\frac{dy}{dx}\right) + (\mu^2 x^2 - p^2)y = 0, \qquad y(0) \text{ finite}, \quad y(l) = 0$$

over the interval $0 < x < l$, where p is a given nonnegative constant. [Make use of Equation (108), Section 4.9.]

72. Expand $f(x) = x^p$ in a series of the characteristic functions of the boundary-value problem

$$x\frac{d}{dx}\left(x\frac{dy}{dx}\right) + (\mu^2 x^2 - p^2)y = 0, \qquad y(0) \text{ finite}, \quad y'(l) = 0$$

over the interval $0 < x < l$, where p is a given nonnegative constant. (Consider the case $p = 0$ separately.)

73. Obtain the solution of the problem

$$\frac{d}{dx}\left(x\frac{dy}{dx}\right) + \Lambda xy = x, \qquad y(0) \text{ finite}, \quad y(l) = 0$$

in the form

$$y = \sum_{n=1}^{\infty} \frac{2}{\Lambda - \mu_n^2} \frac{J_0(\mu_n x)}{\mu_n l J_1(\mu_n l)} \qquad (0 \le x \le l),$$

where $J_0(\mu_n l) = 0$ $(n = 1, 2, \ldots)$, if $J_0(\sqrt{\Lambda}\, l) \ne 0$.

74. If $\Lambda = 0$ in Problem 73, show that the solution is of the form $y = \frac{1}{4}(x^2 - l^2)$. Hence deduce the representation

$$1 - \left(\frac{x}{l}\right)^2 = \sum_{n=1}^{\infty} \frac{8}{\alpha_n^3} \frac{J_0\left(\alpha_n \frac{x}{l}\right)}{J_1(\alpha_n)} \qquad (0 \le x \le l),$$

where α_n is the nth zero of $J_0(x)$.

Section 5.14

75. Show that the coefficients in the expansion

$$F(\varphi) = \sum_{n=0}^{\infty} A_n P_n(\cos \varphi) \qquad (0 < \varphi < \pi)$$

are of the form

$$A_n = \frac{2n+1}{2} \int_0^\pi F(\varphi) P_n(\cos \varphi) \sin \varphi \, d\varphi \qquad (n = 0, 1, \ldots).$$

76. Find the first three coefficients in the expansion of the function

$$f(x) = \begin{cases} 0 & (-1 < x < 0), \\ x & (0 < x < 1) \end{cases}$$

in a series of Legendre polynomials over the interval $(-1, 1)$.

77. Find the first three coefficients in the expansion of the function

$$F(\varphi) = \begin{cases} \cos \varphi & (0 < \varphi < \pi/2), \\ 0 & (\pi/2 < \varphi < \pi) \end{cases}$$

in a series of the form

$$F(\varphi) = \sum_{n=0}^{\infty} A_n P_n(\cos \varphi) \qquad (0 < \varphi < \pi).$$

78. Show that if the coefficients $A_0, A_1, \ldots, A_n$ are determined in such a way that

$$I \equiv \int_{-1}^{1} \{f(x) - [A_0 P_0(x) + A_1 P_1(x) + \cdots + A_n P_n(x)]\}^2 \, dx = \text{minimum},$$

so that the integral of the squared error in the approximation

$$f(x) \approx \sum_{k=0}^{n} A_k P_k(x) \qquad (-1 \le x \le 1)$$

is least, the coefficients are obtained (by requiring that $\partial I/\partial A_k = 0$ for $k = 0, 1, \ldots,$ n and making use of the relevant orthogonality) in the form

$$A_k \int_{-1}^{1} [P_k(x)]^2 \, dx = \int_{-1}^{1} f(x) P_k(x) \, dx \qquad (k = 0, 1, \ldots, n).$$

Hence, recalling that any polynomial of degree n can be expressed as a linear combination of $P_0(x), P_1(x), \ldots, P_n(x)$, deduce that *the polynomial of degree n which best approximates a function over* $(-1, 1)$ *in the least-squares sense consists of the sum of the terms of degree not greater than n in the Legendre expansion of that function over that interval.*

79. Determine the first three nonvanishing terms in the Legendre expansion, over the interval $(-1, 1)$, of the function

$$h(x) = \begin{cases} \dfrac{1}{2\epsilon} & (|x| < \epsilon), \\ 0 & (\epsilon < |x| < 1), \end{cases}$$

in the form

$$h(x) = A_0 P_0(x) + A_2 P_2(x) + A_4 P_4(x) + \cdots \qquad (-1 < x < 1)$$

where $\quad A_0 = \frac{1}{2}, \quad A_2 = -\frac{5}{4}(1 - \epsilon^2), \quad A_4 = \frac{9}{16}(3 - 10\epsilon^2 + 7\epsilon^4), \quad \ldots.$

80. Use the results of Section 5.8 to obtain the solution of the problem

$$\frac{d}{dx}\left[(1 - x^2)\frac{dy}{dx}\right] + ky = h(x), \qquad y(\pm 1) \text{ finite},$$

where $h(x)$ is defined in Problem 79, in the form

$$y = \sum_{n=0}^{\infty} \frac{A_n}{k - n(n + 1)} P_n(x),$$

if $k \ne 0, 6, 20, \ldots.$

81. As $\epsilon \longrightarrow 0$ in Problems 79 and 80, the function $h(x)$ tends toward the "unit impulse function." It can be shown that the series in Problem 79 does not converge when $\epsilon = 0$. However, the series solution in Problem 80 *does* then converge.

(a) Show that if $h(x)$ tends toward the unit impulse function $\delta(x)$, the solution of Problem 80 becomes

$$y = \frac{\frac{1}{2}}{k}P_0(x) - \frac{\frac{5}{4}}{k-6}P_2(x) + \frac{\frac{27}{16}}{k-20}P_4(x) + \cdots.$$

(b) Show that the same result is obtained if the *formal* expansion

$$\delta(x) \sim \tfrac{1}{2}P_0(x) - \tfrac{5}{4}P_2(x) + \tfrac{27}{16}P_4(x) + \cdots$$

is dealt with as though it were truly a convergent representation.

[In the procedure of part (a), the limit is taken *after* the coefficients of the expansion have been substituted into the general solution, whereas in that of part (b) the limit is taken formally *before* substitution. The former is mathematically sound; the second, strictly speaking, is not. However, both give the same final result. Since in many similar problems the second procedure may be much more simply applied than the first, it is rather frequently used in practice and can be justified from a practical point of view by an analysis similar to the one just considered.]

82. (a) With the notation of Section 4.12, show that the solution of Legendre's equation of order p which vanishes when $x = 0$ must be a multiple of $v_p(x)$. Making use of the fact that $v_p(x)$ is not finite at $x = 1$ unless p is an odd integer, deduce that the problem

$$\frac{d}{dx}\left[(1 - x^2)\frac{dy}{dx}\right] + \lambda y = 0, \qquad y(0) = 0, \quad y(1) \text{ finite}$$

possesses the characteristic numbers $\lambda_n = 2n(2n - 1)$ and the corresponding characteristic functions $\varphi_n(x) = P_{2n-1}(x)$, where $n = 1, 2, \ldots$.

(b) In a similar way, deduce that the problem

$$\frac{d}{dx}\left[(1 - x^2)\frac{dy}{dx}\right] + \lambda y = 0, \qquad y'(0) = 0, \quad y(1) \text{ finite}$$

possesses the characteristic numbers $\lambda_n = 2n(2n + 1)$ and the corresponding characteristic functions $\varphi_n(x) = P_{2n}(x)$, where $n = 0, 1, 2, \ldots$.

83. (a) Obtain the solution of the problem

$$\frac{d}{dx}\left[(1 - x^2)\frac{dy}{dx}\right] + \Lambda y = h(x), \qquad y(0) = 0, \quad y(1) \text{ finite}$$

in the form

$$y = \sum_{n=1}^{\infty} \frac{C_n}{\Lambda - 2n(2n - 1)}P_{2n-1}(x), \quad C_n = (4n - 1)\int_0^1 h(x)P_{2n-1}(x)\, dx,$$

if $\Lambda \neq 2, 12, 30, \ldots$.

(b) If the condition at $x = 0$ in part (a) is replaced by the condition $y'(0) = 0$, obtain the solution in the form

$$y = \sum_{n=0}^{\infty} \frac{C_n}{\Lambda - 2n(2n + 1)}P_{2n}(x), \quad C_n = (4n + 1)\int_0^1 h(x)P_{2n}(x)\, dx,$$

if $\Lambda \neq 0, 6, 20, \ldots$.

Section 5.15

84. With the symbols

$$u_n = \frac{n\pi}{l} = n \, \Delta u, \qquad \Delta u = \frac{\pi}{l}$$

where u_n is a function of the integer n, increasing by jumps of magnitude Δu, show that the complex form of the Fourier series representation of $f(x)$, as obtained in Problem 62, can be expressed in the form

$$f(x) = \frac{1}{2\pi} \sum_{n=-\infty}^{\infty} \Delta u \int_{-l}^{l} f(t) e^{-iu_n(t-x)} \, dt.$$

[Notice the formal similarity between this form (as $l \longrightarrow \infty$) and the Fourier integral representation expressed in the form (238a).]

85. Use (244a) to evaluate the integral

$$\int_0^\infty \frac{1 - \cos au}{u} \sin ux \, du$$

for all real values of x when $a > 0$.

86. (a) If α is a positive real constant, determine the Fourier sine and cosine integral representations of $e^{-\alpha x}$ in the forms

$$e^{-\alpha x} = \int_0^\infty \left(\frac{2}{\pi} \frac{u}{\alpha^2 + u^2} \right) \sin ux \, du = \int_0^\infty \left(\frac{2}{\pi} \frac{u}{\alpha^2 + u^2} \right) \cos ux \, du \qquad (\alpha > 0, \ x > 0).$$

(b) Use these results to determine functions $A(u)$ and $B(u)$ such that

$$\frac{x}{\alpha^2 + x^2} = \int_0^\infty A(u) \sin ux \, du, \qquad \frac{\alpha}{\alpha^2 + x^2} = \int_0^\infty B(u) \cos ux \, du \qquad (\alpha > 0, \ x > 0).$$

(c) Deduce from preceding results that

$$\mathcal{S}\left\{ \frac{x}{\alpha^2 + x^2} \right\} = \frac{\pi}{2} e^{-\alpha u}, \qquad \mathcal{S}\{e^{-\alpha x}\} = \frac{u}{\alpha^2 + u^2}$$

and

$$\mathcal{C}\left\{ \frac{\alpha}{\alpha^2 + x^2} \right\} = \frac{\pi}{2} e^{-\alpha u}, \qquad \mathcal{C}\{e^{-\alpha x}\} = \frac{\alpha}{\alpha^2 + u^2}$$

when $\alpha > 0$.

87. Show that α can be replaced by $a + ib$, where a and b are real and $a > 0$, in Problem 86(a). In particular, show that

$$e^{-ax}(\cos bx - i \sin bx) = \frac{2}{\pi} \int_0^\infty \frac{u \sin ux}{(a^2 - b^2 + u^2) + 2iab} \, du$$

when $a > 0$ and $x > 0$ and, by equating real and imaginary parts of the equal members, deduce the sine integral representations of $e^{-ax} \cos bx$ and $e^{-ax} \sin bx$ when $a > 0$.

88. (a) If α is a positive real constant, determine the complex form of the Fourier integral representation of $e^{-\alpha|x|}$ in the form

$$e^{-\alpha|x|} = \int_{-\infty}^\infty \left(\frac{1}{\pi} \frac{\alpha}{\alpha^2 + u^2} \right) e^{iux} \, du \qquad (\alpha > 0).$$

(b) Deduce from this result the function $C(u)$ such that

$$\frac{\alpha}{\alpha^2 + x^2} = \int_{-\infty}^{\infty} C(u)e^{iux}\,du \qquad (\alpha > 0).$$

(c) Deduce that

$$\mathcal{F}\left\{\frac{\alpha}{\alpha^2 + x^2}\right\} = \pi e^{-\alpha|u|}, \qquad \mathcal{F}\{e^{-\alpha|x|}\} = \frac{2\alpha}{\alpha^2 + u^2}$$

when $\alpha > 0$.

89. Assuming that $f'(x)$ and $f''(x)$ are continuous (or, more generally, that integration by parts is permissible when needed), and that the relevant transforms exist, obtain the following relations:

(a) $\mathcal{F}\{f'(x)\} = iu\bar{f}(u), \qquad \mathcal{F}\{f''(x)\} = -u^2\bar{f}(u)$

if $f(\pm\infty) = f'(\pm\infty) = 0$.

(b) $\mathcal{S}\{f'(x)\} = -uf_C(u), \qquad \mathcal{S}\{f''(x)\} = uf(0+) - u^2 f_S(u)$

if $f(\infty) = f'(\infty) = 0$.

(c) $\mathcal{C}\{f'(x)\} = -f(0+) + uf_S(u), \qquad \mathcal{C}\{f''(x)\} = -f'(0+) - u^2 f_C(u)$

if $f(\infty) = f'(\infty) = 0$.

90. *Transforms of the delta function.*

(a) By replacing $f(x)$ by $f(x)/a$ in the text Example and considering the limit of the Fourier transform of the result as $a \longrightarrow 0$, indicate the sense in which it is said that "the Fourier transform of the delta function $\delta(x)$ is 1."

(b) Generalize the result of part (a) to show that

$$\mathcal{F}\{\delta(x - a)\} = e^{-iau}$$

and, similarly, that

$$\mathcal{S}\{\delta(x - a)\} = \sin au \ (a > 0), \qquad \mathcal{C}\{\delta(x - a)\} = \cos au \ (a > 0).$$

91. *The convolution.* Let the convolution of $f(x)$ and $g(x)$ in $(-\infty, \infty)$ be defined as the function

$$F(x) = \int_{-\infty}^{\infty} f(x - \xi)g(\xi)\,d\xi,$$

assuming the existence of the integral.

(a) Assuming also that $F(x)$, $f(x)$, and $g(x)$ have the Fourier transforms $\bar{F}(u)$, $\bar{f}(u)$, and $\bar{g}(u)$, respectively, show that

$$\bar{F}(u) = \bar{f}(u)\bar{g}(u),$$

and hence that *if $\bar{f}$ and $\bar{g}$ are the Fourier transforms of f and g, then $\bar{f}\bar{g}$ is the Fourier transform of the convolution of f and g.* (Notice that f and g can be interchanged in the definition of the convolution F.)

(b) If $f(x)$ and $g(x)$ both vanish when $x < 0$, and $F(x)$ is their convolution, show that also $F(x)$ vanishes when $x < 0$ and that

$$F(x) = \int_{0}^{x} f(x - \xi)g(\xi)\,d\xi \qquad (x > 0).$$

(Compare the definition in Section 2.9.)

92. *Convolution properties of sine and cosine transforms.*

(a) Show that the result of Problem 91(a) can be written in the form

$$\int_{-\infty}^{\infty} f(x - \xi)g(\xi)\, d\xi = \frac{1}{2\pi} \int_{-\infty}^{\infty} e^{iux} \bar{f}(u)\bar{g}(u)\, du.$$

(b) When $f(x)$ and $g(x)$ are both *even* functions of x, show that $\bar{f}$ and $\bar{g}$ are even functions of u, and also that $\bar{f}(u) = 2f_C(u)$ and $\bar{g}(u) = 2g_C(u)$, where f_C and g_C are the cosine transforms of f and g. Thus deduce that, when $x > 0$,

$$\int_{-\infty}^{0} f(x - \xi)g(-\xi)\, d\xi + \int_{0}^{x} f(x - \xi)g(\xi)\, d\xi + \int_{x}^{\infty} f(\xi - x)g(\xi)\, d\xi$$

$$= \frac{4}{\pi} \int_{0}^{\infty} \cos ux\, f_C(u)g_C(u)\, du$$

and hence

$$\frac{1}{2} \int_{0}^{\infty} [f(|x - \xi|) + f(x + \xi)]g(\xi)\, d\xi = \frac{2}{\pi} \int_{0}^{\infty} \cos ux\, f_C(u)g_C(u)\, du,$$

so that, *if f_C and g_C are the cosine transforms of f and g, then $f_C g_C$ is the cosine transform of the function*

$$\frac{1}{2} \int_{0}^{\infty} [f(|x - \xi|) + f(x + \xi)]g(\xi)\, d\xi.$$

(c) By supposing that $f(x)$ is *even* and $g(x)$ is *odd*, and proceeding as in part (b), show that *if f_C is the cosine transform of f and g_S is the sine transform of g, then $f_C g_S$ is the sine transform of the function*

$$\frac{1}{2} \int_{0}^{\infty} [f(|x - \xi|) - f(x + \xi)]g(\xi)\, d\xi.$$

[Notice that the sine and cosine transforms require the definition of f and g only for *positive arguments*, and that the extended definitions for negative arguments in parts (b) and (c) are merely for the purpose of appropriately using the result of part (a).]

93. Solve the problem

$$\frac{d^2 y}{dx^2} - k^2 y = h(x), \qquad y(\pm\infty) = 0$$

by introducing the Fourier transform, using Problem 89 to show that

$$\bar{y}(u) = -\frac{\bar{h}(u)}{u^2 + k^2},$$

and using Problems 88 and 91 to deduce that

$$y(x) = -\frac{1}{2k} \int_{-\infty}^{\infty} e^{-k|x-\xi|} h(\xi)\, d\xi$$

when $k > 0$.

94. Solve the problem

$$\frac{d^2 y}{dx^2} - k^2 y = h(x), \qquad y'(0) = y'_0, \quad y'(\infty) = 0$$

by introducing the Fourier cosine transform, using Problem 89 to show that

$$y_C(u) = -\frac{y_0' + h_C(u)}{u^2 + k^2},$$

and using Problems 86 and 92 to deduce that

$$y(x) = -\frac{y_0'}{k} e^{-kx} - \frac{1}{2k} \int_0^\infty [e^{-k|x-\xi|} + e^{-k(x+\xi)}] h(\xi)\, d\xi$$

when $k > 0$.

95. Solve the problem

$$\frac{d^2 y}{dx^2} - k^2 y = h(x), \quad y(0) = y_0, \quad y(\infty) = 0$$

by introducing the Fourier sine transform, using Problem 89 to show that

$$y_S(u) = \frac{u y_0 - h_S(u)}{u^2 + k^2},$$

and using Problems 86 and 92 to deduce that

$$y(x) = y_0 e^{-kx} - \frac{1}{2k} \int_0^\infty [e^{-k|x-\xi|} - e^{-k(x+\xi)}] h(\xi)\, d\xi$$

when $k > 0$.

96. Solve the problem

$$\frac{d}{dx}\left(x \frac{dy}{dx}\right) - k^2 x y = x h(x), \quad y(0) \text{ finite}, \quad y(\infty) = 0$$

in the form

$$y(x) = -\int_0^\infty \frac{u h_{H_0}(u)}{u^2 + k^2} J_0(ux)\, du,$$

where

$$h_{H_0}(u) = \int_0^\infty x h(x) J_0(ux)\, dx.$$

97. It is required to determine $C(u)$ in such a way that the representation

$$f(x) = \frac{2}{\pi} \int_0^\infty C(u)(\alpha \sin ux + \beta u \cos ux)\, du$$

is valid when $0 < x < \infty$, where α and β are prescribed constants and $\alpha\beta \geq 0$.

   (a) Assuming that such a representation exists, and that differentiation under the integral sign is permissible, show that, if we write

$$F(x) = \frac{2}{\pi} \int_0^\infty C(u) \sin ux\, du,$$

there follows

$$\beta F'(x) + \alpha F(x) = f(x), \quad F(0) = 0$$

and

$$C(u) = \int_0^\infty F(x) \sin ux\, dx.$$

(b) Deduce that

$$F(x) = \frac{1}{\beta} \int_0^x f(t) e^{-(\alpha/\beta)(x-t)}\, dt$$

and, by formally interchanging the order of integration in the result of introducing this expression into the expression for $C(u)$, and changing the dummy variable of integration in the result, obtain the evaluation

$$C(u) = \int_0^\infty f(x) \frac{\alpha \sin ux + \beta u \cos ux}{\alpha^2 + \beta^2 u^2}\, dx.$$

(c) Verify that the representation obtained reduces to the Fourier cosine and sine representations when $\alpha = 0$ and when $\beta = 0$, respectively. [The result can be shown to be valid under the same restrictions on $f(x)$ as those which apply in those special cases when $\alpha\beta > 0$. An additional term is needed when $\alpha\beta < 0$.]

(d) Show that the representation obtained expresses $f(x)$ as a superposition of the characteristic functions of the problem

$$\frac{d^2y}{dx^2} + \lambda y = 0, \qquad \beta y'(0) = \alpha y(0), \quad y(\infty) \text{ finite}$$

when $\alpha\beta > 0$, but that when $\alpha\beta < 0$ there is a missing characteristic function $e^{-|\alpha/\beta|x}$, corresponding to the characteristic number $\lambda = -\alpha^2/\beta^2$.

(e) Verify that, with the definition

$$C(u) = \frac{f_M(u)}{\alpha + \beta u},$$

there follows

$$f_M(u) = \frac{\alpha + \beta u}{\alpha^2 + \beta^2 u^2} \int_0^\infty f(x)(\alpha \sin ux + \beta u \cos ux)\, dx$$

and

$$f(x) = \frac{2}{\pi} \int_0^\infty f_M(u) \frac{\alpha \sin ux + \beta u \cos ux}{\alpha + \beta u}\, du.$$

Show also that $f_M(u)$ reduces to $f_C(u)$ when $\alpha = 0$ and to $f_S(u)$ when $\beta = 0$. [Since $f_M(u)$ corresponds to the "mixed" end condition $\beta y'(0) - \alpha y(0) = 0$, it might be called a Fourier *mixed transform*.]

98. Solve the problem

$$\frac{d^2y}{dx^2} - k^2 y = h(x), \qquad \alpha y(0) - \beta y'(0) = \gamma, \quad y(\infty) = 0,$$

when $\alpha\beta > 0$, by use of the Fourier "mixed transform" (Problem 97), as follows:

(a) Show that the mixed transform of $f''(x)$ is

$$\frac{(\alpha + \beta u)u}{\alpha^2 + \beta^2 u^2}[\alpha f(0+) - \beta f'(0+)] - u^2 f_M(u).$$

(b) Use this result to show that

$$y_M(u) = \frac{(\alpha + \beta u)u}{\alpha^2 + \beta^2 u^2} - \frac{h_M(u)}{u^2 + k^2}.$$

(c) Deduce that the solution can be written in the form

$$y(x) = \frac{2\gamma}{\pi} \int_0^\infty \frac{u(\alpha \sin ux + \beta u \cos ux)}{(\alpha^2 + \beta^2 u^2)(u^2 + k^2)} \, du - \frac{2}{\pi} \int_0^\infty \frac{h_M(u)(\alpha \sin ux + \beta u \cos ux)}{(\alpha + \beta u)(u^2 + k^2)} du,$$

where

$$h_M(u) = \frac{\alpha + \beta u}{\alpha^2 + \beta^2 u^2} \int_0^\infty h(x)(\alpha \sin ux + \beta u \cos ux) \, dx.$$

[See footnote on page 239 with respect to the condition $y(\infty) = 0$. The result of part (c) can be reduced to the form

$$y(x) = \frac{\gamma}{\alpha + \beta k} e^{-kx} - \frac{1}{2k} \int_0^\infty \left[e^{-k|x-\xi|} - \frac{\alpha - \beta k}{\alpha + \beta k} e^{-k(x+\xi)} \right] h(\xi) \, d\xi$$

when $k > 0$. This form can be deduced directly (and more easily) from the result of part (b), by making use of results corresponding to those of Problems 86 and 92 relevant to the mixed transform.]

99. Show formally that the function

$$T(x, y) = \int_0^\infty e^{-yu} A(u) \sin ux \, du \qquad (y > 0)$$

satisfies the partial differential equation

$$\frac{\partial^2 T}{\partial x^2} + \frac{\partial^2 T}{\partial y^2} = 0$$

and vanishes when $x = 0$ and when $y \to \infty$. Determine $A(u)$ so that $T(x, y)$ formally reduces to a prescribed function $f(x)$ when $y = 0$ and x is positive.

100. Show formally that the function

$$T(x, y) = \frac{1}{\pi} \int_0^\infty \int_{-\infty}^\infty e^{-yu} f(t) \cos u(t - x) \, dt \, du \qquad (y > 0)$$

satisfies the partial differential equation

$$\frac{\partial^2 T}{\partial x^2} + \frac{\partial^2 T}{\partial y^2} = 0,$$

vanishes when $y \to \infty$, and reduces to $f(x)$ when $y = 0$.

101. Show formally that the function

$$\varphi(r, z) = \int_0^\infty \int_0^\infty e^{-zu} u t f(t) J_0(ut) J_0(ur) \, dt \, du$$

satisfies the partial differential equation

$$\frac{\partial^2 \varphi}{\partial r^2} + \frac{1}{r} \frac{\partial \varphi}{\partial r} + \frac{\partial^2 \varphi}{\partial z^2} = 0,$$

vanishes when $z \to \infty$, and reduces to $f(r)$ when $z = 0$.

102. (a) If $f(x) = x^p$ when $0 < x < a$ and $f(x) = 0$ when $a < x < \infty$, show that the Hankel transform of $f(x)$, of order p, is

$$\mathcal{B}_p\{x^p\} = \frac{1}{u} a^{p+1} J_{p+1}(au).$$

(b) Deduce that

$$\int_0^\infty J_{p+1}(au) J_p(ux)\, du = \begin{cases} \dfrac{1}{a}\left(\dfrac{x}{a}\right)^p & (0 < x < a), \\[2mm] \dfrac{1}{2a}\left(\dfrac{x}{a}\right)^p & (x = a), \\[2mm] 0 & (a < x < \infty). \end{cases}$$

103. Verify that the result of Problem 102(b) becomes equivalent to Equation (247) when $p = -1/2$.

6

Vector Analysis

6.1. Elementary Properties of Vectors. A *vector* quantity is distinguished from a *scalar* quantity by the fact that a scalar quantity possesses only *magnitude*, whereas a vector quantity possesses both *magnitude* and (except for the zero vector) *direction*. It is conventional to represent a vector geometrically as an arrow, pointing in the direction associated with the vector, and having a length proportional to the associated magnitude. Thus, in particular, the position of a point B relative to a point A can be completely described by a vector $\overrightarrow{AB}$ from A to B, in the sense that the vector $\overrightarrow{AB}$ specifies both the distance and direction from A to B and, in fact, indicates the displacement from A to B. Since motion from A to a third point C can be accomplished along the vector $\overrightarrow{AC}$ or, alternatively, along the vector $\overrightarrow{AB}$ to B and thence along $\overrightarrow{BC}$ to C (Figure 6.1), it is natural to extend the concept of *addition* to vector quantities by writing

$$\overrightarrow{AC} = \overrightarrow{AB} + \overrightarrow{BC}.$$

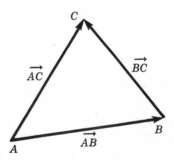

Figure 6.1

We say that two vectors are *equal* if they have the same direction and the same magnitude. In accordance with this definition a vector is unchanged if it is moved parallel to itself in any way; that is, the actual *position* of the vector in space can be assigned at pleasure. It should be noted, however, that in some applications it may be *necessary* to specify the position of a vector (as, for example, when a *moment* associated with a *force* vector is to be determined).

With this convention,† any two vectors **a** and **b** can be represented as arrows so placed that the terminal point of **a** coincides with the initial point of **b**. The sum of **a** and **b**,

$$c = a + b,$$

is then defined as the vector extending from the initial point of **a** to the terminal point of **b**. From the definition it follows readily that vector addition is *commutative*,

$$a + b = b + a \tag{1}$$

and also that the *associative* law

$$(a + b) + c = a + (b + c) \tag{2}$$

is satisfied. The common value of the two expressions in (2) is written as the sum **a** + **b** + **c** of the three vectors, and the definition is readily extended to the sum of any number of vectors.

If the initial and terminal points of the arrow representing a vector are interchanged, the resulting vector is called the *negative* of the given vector. Thus, for example $\overrightarrow{BA} = -\overrightarrow{AB}$. To *subtract* a vector **b** from a vector **a**, we add the negative of **b** to **a**,

$$a - b = a + (-b). \tag{3}$$

If a vector **a** is multiplied by a scalar m, the result is a vector $m\mathbf{a}$ whose magnitude is the arithmetic product of m and the magnitude of **a** and whose direction is that of **a** if m is positive or that of $-\mathbf{a}$ if m is negative.

A vector of unit length is called a *unit vector*. It is clear that any vector can be written as the product of its length and a unit vector.

A vector of zero length (and *arbitrary* direction) is called a *zero vector* and is denoted by **0**. The difference between two equal vectors is a zero vector.

We now consider a right-handed rectangular coordinate system, in which the coordinates $x, y,$ and z are measured along three mutually perpendicular axes such that rotation of the x axis into the y axis about the z axis is accomplished by the right-hand rule, and define *unit vectors* **i, j,** and **k** having the directions of the positive $x, y,$ and z axes, respectively (Figure 6.2). Then it is readily verified that any vector **v** whose projections on these axes are $v_x, v_y,$ and v_z, respectively, can be written as the vector sum

$$v = v_x i + v_y j + v_z k. \tag{4}$$

The numbers $v_x, v_y,$ and v_z are called the scalar *components* of **v** in the $x, y,$ and z directions (Figure 6.3). If, when the initial point of **v** coincides with the origin of the system, the angles measured to **v** from the positive $x, y,$ and z axes

---

†In printed work a vector quantity is often denoted (as here) by a boldface letter. Thus **a** represents a vector, and a represents a scalar quantity. In written work, the use of an arrow ($\vec{a}$) or of an underline ($\underline{a}$) is convenient; the notation $\hat{u}$ (or $\hat{\mathbf{u}}$) is often used to denote a vector of *unit length*.

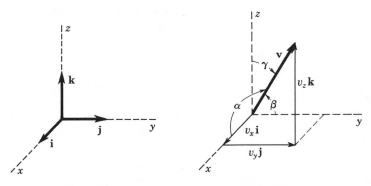

Figure 6.2   Figure 6.3

are denoted by α, β, and γ, respectively, there follows

$$v_x = v \cos \alpha, \qquad v_y = v \cos \beta, \qquad v_z = v \cos \gamma, \tag{5}$$

where v is written for the length or magnitude of the vector $\mathbf{v}$. The notation $|\mathbf{v}|$ is also frequently used in place of v. The numbers $\cos \alpha$, $\cos \beta$, and $\cos \gamma$ are called the *direction cosines* of $\mathbf{v}$ and are frequently denoted by l, m, and n, respectively. Any three numbers (A, B, C) proportional to (l, m, n) are known as "direction ratios." From the geometrical relationship

$$v = \sqrt{v_x^2 + v_y^2 + v_z^2} \tag{6}$$

relating v to the scalar components of $\mathbf{v}$, it follows that the direction cosines satisfy the equation

$$l^2 + m^2 + n^2 \equiv \cos^2 \alpha + \cos^2 \beta + \cos^2 \gamma = 1. \tag{7}$$

We see also that if we write Equation (4) in the form

$$\mathbf{v} = v(\mathbf{i} \cos \alpha + \mathbf{j} \cos \beta + \mathbf{k} \cos \gamma), \tag{8}$$

the coefficient of v is a unit vector, and hence the vector

$$\mathbf{v}_1 = \mathbf{i} \cos \alpha + \mathbf{j} \cos \beta + \mathbf{k} \cos \gamma \tag{9}$$

is a unit vector in the direction of $\mathbf{v}$. Combining Equations (5) and (6), we have

$$\cos \alpha = \frac{v_x}{\sqrt{v_x^2 + v_y^2 + v_z^2}}, \qquad \cos \beta = \frac{v_y}{\sqrt{v_x^2 + v_y^2 + v_z^2}},$$

$$\cos \gamma = \frac{v_z}{\sqrt{v_x^2 + v_y^2 + v_z^2}}. \tag{10}$$

6.2. The Scalar Product of Two Vectors. Two types of products of two vectors $\mathbf{a}$ and $\mathbf{b}$ are conventionally defined. The first type of product is called the *scalar, dot,* or *inner product* and is written as $\mathbf{a} \cdot \mathbf{b}$ or $(\mathbf{a}\,\mathbf{b})$. This product is defined to be a *scalar* equal to the product of the lengths of the two vectors and the cosine of the angle θ between the *positive* directions of the two vectors

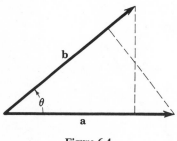

Figure 6.4

(see Figure 6.4), and hence is given by the equation

$$\mathbf{a} \cdot \mathbf{b} = ab \cos \theta. \qquad (11)$$

Since $b \cos \theta$ is the projection of $\mathbf{b}$ on $\mathbf{a}$, whereas $a \cos \theta$ is the projection of $\mathbf{a}$ on $\mathbf{b}$, it follows that $\mathbf{a} \cdot \mathbf{b}$ is numerically equal to the length of $\mathbf{b}$ times the projection of $\mathbf{a}$ on $\mathbf{b}$, and also is equal to the length of $\mathbf{a}$ times the projection of $\mathbf{b}$ on $\mathbf{a}$. If θ is obtuse, the projections considered are to be taken as negative.

In particular, *if two vectors are perpendicular, their dot product is zero.* From the geometrical definition it is readily seen that the dot product is commutative,

$$\mathbf{a} \cdot \mathbf{b} = \mathbf{b} \cdot \mathbf{a}, \qquad (12)$$

and distributive,

$$\mathbf{a} \cdot (\mathbf{b} + \mathbf{c}) = \mathbf{a} \cdot \mathbf{b} + \mathbf{a} \cdot \mathbf{c}. \qquad (13)$$

For the unit vectors $\mathbf{i}, \mathbf{j},$ and $\mathbf{k}$ there follows immediately

$$\mathbf{i} \cdot \mathbf{i} = \mathbf{j} \cdot \mathbf{j} = \mathbf{k} \cdot \mathbf{k} = 1,$$
$$\mathbf{i} \cdot \mathbf{j} = \mathbf{j} \cdot \mathbf{i} = \mathbf{j} \cdot \mathbf{k} = \mathbf{k} \cdot \mathbf{j} = \mathbf{i} \cdot \mathbf{k} = \mathbf{k} \cdot \mathbf{i} = 0, \qquad (14)$$

and hence, if

$$\mathbf{a} = a_x \mathbf{i} + a_y \mathbf{j} + a_z \mathbf{k}, \qquad \mathbf{b} = b_x \mathbf{i} + b_y \mathbf{j} + b_z \mathbf{k}, \qquad (15)$$

we have, using Equation (13),

$$\mathbf{a} \cdot \mathbf{b} = a_x b_x + a_y b_y + a_z b_z. \qquad (16)$$

The cosine of the angle θ between $\mathbf{a}$ and $\mathbf{b}$ is given by Equations (11) and (16) in the form

$$\cos \theta = \frac{a_x b_x + a_y b_y + a_z b_z}{\sqrt{a_x^2 + a_y^2 + a_z^2} \sqrt{b_x^2 + b_y^2 + b_z^2}}, \qquad (17)$$

or, if we denote the direction cosines of $\mathbf{a}$ and $\mathbf{b}$ by (l_1, m_1, n_1) and (l_2, m_2, n_2),

$$\cos \theta = l_1 l_2 + m_1 m_2 + n_1 n_2. \qquad (18)$$

The abbreviation $\mathbf{a}^2$ is sometimes used to indicate $\mathbf{a} \cdot \mathbf{a}$. Then we also have

$$\mathbf{a}^2 = \mathbf{a} \cdot \mathbf{a} = |\mathbf{a}|^2 = a^2. \qquad (19)$$

The dot product is particularly useful in expressing a given vector $\mathbf{v}$ as a linear combination of three mutually perpendicular unit vectors $\mathbf{u}_1, \mathbf{u}_2,$ and $\mathbf{u}_3$. For if we write

$$\mathbf{v} = c_1 \mathbf{u}_1 + c_2 \mathbf{u}_2 + c_3 \mathbf{u}_3, \qquad (20)$$

we may successively take the dot product of $\mathbf{u}_1, \mathbf{u}_2,$ and $\mathbf{u}_3$ into both sides of (20) and so obtain

$$c_1 = \mathbf{v} \cdot \mathbf{u}_1, \qquad c_2 = \mathbf{v} \cdot \mathbf{u}_2, \qquad c_3 = \mathbf{v} \cdot \mathbf{u}_3.$$

Thus Equation (20) becomes

$$v = (v \cdot u_1)u_1 + (v \cdot u_2)u_2 + (v \cdot u_3)u_3. \tag{21}$$

It is seen that this representation could fail only if a component of v were perpendicular to u_1, u_2, and u_3. This condition, however, is clearly impossible in space of three dimensions, since u_1, u_2, and u_3 are assumed to be mutually perpendicular.

6.3. The Vector Product of Two Vectors. The second conventional type of product of two vectors a and b is called the *vector, cross,* or *outer product* and is written as $a \times b$ or $[a\,b]$. It is defined to be a *vector* having the properties that (1) the length of $a \times b$ is the product of the lengths of a and b and the numerical value of the *sine* of the angle θ between the vectors and (2) the vector $a \times b$ is perpendicular to the plane of a and b and is so oriented that a is rotated into b about $a \times b$ by the right-hand rule, through not more than $180°$.

According to the definition, there follows (see Figure 6.5)

$$|a \times b| = ab\,|\sin \theta|. \tag{22}$$

Thus *the length of $a \times b$ is equal to twice the area of the triangle of which a and b form coterminous sides.* It is seen that the cross product is *not commutative,* since, from the definition,

$$b \times a = -(a \times b). \tag{23}$$

However, the cross product is distributive, so that

$$a \times (b + c) = a \times b + a \times c. \tag{24}$$

This relation can be established by geometrical considerations. It is seen that *the cross product of two parallel vectors is zero.*

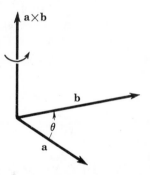

Figure 6.5

For the unit vectors i, j, and k there follows, from the definition,

$$i \times i = j \times j = k \times k = 0,$$
$$i \times j = -j \times i = k, \quad j \times k = -k \times j = i, \quad k \times i = -i \times k = j. \tag{25}$$

These relations are easily remembered in terms of the cyclic arrangement

$$\ldots \quad i \quad j \quad k \quad i \quad j \quad k \quad \ldots$$

if we notice that the cross product of a vector into its neighbor is the following vector when reading to the right and is the negative of the following vector when reading to the left.

To calculate the cross product of two vectors a and b in terms of their components as given in (15), we make use of Equations (24) and (25) and obtain

$$a \times b = (a_y b_z - a_z b_y)i + (a_z b_x - a_x b_z)j + (a_x b_y - a_y b_x)k. \tag{26}$$

This result can be written conveniently as the determinant

$$\mathbf{a} \times \mathbf{b} = \begin{vmatrix} \mathbf{i} & \mathbf{j} & \mathbf{k} \\ a_x & a_y & a_z \\ b_x & b_y & b_z \end{vmatrix}. \tag{27}$$

The usefulness of the cross product may be illustrated by two physical applications. First, we consider a point P in a rigid body rotating with angular velocity of magnitude ω about a fixed axis. Let O be a point on the axis of rotation and represent the angular velocity by a vector $\boldsymbol{\omega}$ of length ω, extending along the axis of rotation in the sense determined by the right-hand rule (Figure 6.6). Then, if the position vector from O to P is written as $\mathbf{r}$, the velocity vector associated with the point P is given by the equation

$$\mathbf{v} = \boldsymbol{\omega} \times \mathbf{r}, \tag{28}$$

since the magnitude of this vector is $\omega r |\sin \theta|$, where $r |\sin \theta|$ is the distance from the axis of rotation to the point P, and the direction of the vector is as indicated in the figure.

As a second application, we consider the moment vector $\mathbf{M}$ at a point O, associated with a force $\mathbf{F}$ acting at a point P (Figure 6.7). If the vector from O

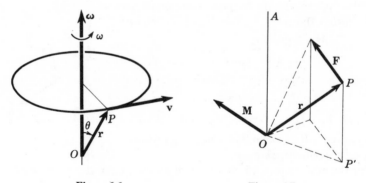

Figure 6.6   Figure 6.7

to P is denoted by $\mathbf{r}$, the moment of $\mathbf{F}$ about the *point* O is defined as the vector

$$\mathbf{M} = \mathbf{r} \times \mathbf{F}. \tag{29}$$

We see that the vector $\mathbf{M}$ is perpendicular to the plane of $\overline{OP}$ and the vector $\mathbf{F}$ and has as its magnitude

$$|\mathbf{M}| = M = Fr |\sin \theta|,$$

where θ is the angle between $\mathbf{r}$ and $\mathbf{F}$, and hence $r |\sin \theta|$ is the perpendicular distance from O to the line of action of $\mathbf{F}$. We may thus speak of M as the scalar moment of the force about the axis of $\mathbf{M}$. Geometrically, M is seen to be twice the area of the triangle determined by $\mathbf{r}$ and $\mathbf{F}$. Further, the scalar moment M_{OA} of the force about *any* axis OA through O is seen to be twice the area of the projection of this triangle on the plane through O perpendicular to OA. But since the angle between the two planes considered is the same as the angle between the corresponding axes, the projected area must be equal to the scalar projection of **M** on OA. That is, *the scalar moment of* **F** *about any axis through* O *is numerically equal to the length of the projection of* **M** *on that axis.*

6.4. Multiple Products. Three type of products involving three vectors are of importance, namely, those of the respective forms $(\mathbf{a} \cdot \mathbf{b})\mathbf{c}$, $(\mathbf{a} \times \mathbf{b}) \cdot \mathbf{c}$, and $(\mathbf{a} \times \mathbf{b}) \times \mathbf{c}$. The first type,

$$(\mathbf{a} \cdot \mathbf{b})\mathbf{c},$$

is merely the product of the scalar $\mathbf{a} \cdot \mathbf{b}$ and the vector **c**.

The second type,

$$(\mathbf{a} \times \mathbf{b}) \cdot \mathbf{c},$$

which is called the *triple scalar product*, is seen to be the dot product of the vector $\mathbf{a} \times \mathbf{b}$ and the vector **c**, and hence is a scalar quantity. The value of this product is given by the product of the length of $\mathbf{a} \times \mathbf{b}$ and the projection of **c** on $\mathbf{a} \times \mathbf{b}$ (Figure 6.8). But since $\mathbf{a} \times \mathbf{b}$ is a vector perpendicular to the plane of **a** and **b**, having a length numerically equal to the area of the parallelogram of which **a** and **b** form coterminous sides, and since the projection of **c** on $\mathbf{a} \times \mathbf{b}$ is the altitude of the parallelepiped with **a**, **b**, and **c** as coterminous edges, it follows that $(\mathbf{a} \times \mathbf{b}) \cdot \mathbf{c}$ *is numerically equal to the volume of this parallelepiped.* Alternatively, we see that $(\mathbf{a} \times \mathbf{b}) \cdot \mathbf{c}$ is numerically equal to six times the volume of the tetrahedron determined by **a**, **b**, and **c** as edges. The sign of the product depends upon the relative orientation of the three vectors,

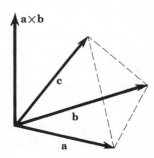

Figure 6.8

and is positive if and only if **a**, **b**, and **c** form a right-handed system, in the sense that **c** and $\mathbf{a} \times \mathbf{b}$ lie on the same side of the plane determined by **a** and **b**. From these facts it follows easily that

$$(\mathbf{a} \times \mathbf{b}) \cdot \mathbf{c} = (\mathbf{b} \times \mathbf{c}) \cdot \mathbf{a} = (\mathbf{c} \times \mathbf{a}) \cdot \mathbf{b} \tag{30}$$

and that the other products

$$(\mathbf{a} \times \mathbf{c}) \cdot \mathbf{b} = (\mathbf{b} \times \mathbf{a}) \cdot \mathbf{c} = (\mathbf{c} \times \mathbf{b}) \cdot \mathbf{a} \tag{31}$$

have the opposite algebraic sign. We see that *the triple scalar product is not changed by a cyclic permutation of the three elements.* Since also, from Equations (30) and (12),

$$(\mathbf{a} \times \mathbf{b}) \cdot \mathbf{c} = (\mathbf{b} \times \mathbf{c}) \cdot \mathbf{a} = \mathbf{a} \cdot (\mathbf{b} \times \mathbf{c}),$$

it follows that *the dot and cross can be interchanged in a triple scalar product.* For this reason, the notation $(\mathbf{a}\,\mathbf{b}\,\mathbf{c})$ is frequently used to indicate the common value of the products listed in Equation (30).

If we write

$$\mathbf{a} = a_x\mathbf{i} + a_y\mathbf{j} + a_z\mathbf{k}, \quad \mathbf{b} = b_x\mathbf{i} + b_y\mathbf{j} + b_z\mathbf{k}, \quad \mathbf{c} = c_x\mathbf{i} + c_y\mathbf{j} + c_z\mathbf{k},$$

we find that

$$(\mathbf{a}\,\mathbf{b}\,\mathbf{c}) = \mathbf{a} \cdot (\mathbf{b} \times \mathbf{c}) = \mathbf{a} \cdot \begin{vmatrix} \mathbf{i} & \mathbf{j} & \mathbf{k} \\ b_x & b_y & b_z \\ c_x & c_y & c_z \end{vmatrix}$$

$$= a_x(b_yc_z - b_zc_y) + a_y(b_zc_x - b_xc_z) + a_z(b_xc_y - b_yc_x),$$

or, in determinant form,

$$(\mathbf{a}\,\mathbf{b}\,\mathbf{c}) = \begin{vmatrix} a_x & a_y & a_z \\ b_x & b_y & b_z \\ c_x & c_y & c_z \end{vmatrix}. \tag{32}$$

From Equation (32) we see that *if two vectors in the product* $(\mathbf{a}\,\mathbf{b}\,\mathbf{c})$ *are parallel, then the product vanishes,* since in this case the corresponding elements in two rows of the determinant (32) are proportional.

The third type of triple product,

$$(\mathbf{a} \times \mathbf{b}) \times \mathbf{c},$$

is clearly a vector. Since it is perpendicular to $\mathbf{a} \times \mathbf{b}$, which is itself perpendicular to the plane of $\mathbf{a}$ and $\mathbf{b}$, and is also perpendicular to $\mathbf{c}$, it follows that $(\mathbf{a} \times \mathbf{b}) \times \mathbf{c}$ *is a vector which is in the plane of* $\mathbf{a}$ *and* $\mathbf{b}$ *and perpendicular to* $\mathbf{c}$. Thus this product must be expressible as a linear combination of $\mathbf{a}$ and $\mathbf{b}$,

$$(\mathbf{a} \times \mathbf{b}) \times \mathbf{c} = m\mathbf{a} + n\mathbf{b}.$$

In order to determine the scalars m and n, we first form the dot product of $\mathbf{c}$ into the equal members of this relation and deduce that

$$m\mathbf{a} \cdot \mathbf{c} + n\mathbf{b} \cdot \mathbf{c} = 0.$$

If we then write $n = \lambda\mathbf{a} \cdot \mathbf{c}$, there follows also $m = -\lambda\mathbf{b} \cdot \mathbf{c}$, and hence we now know that

$$(\mathbf{a} \times \mathbf{b}) \times \mathbf{c} = \lambda[(\mathbf{a} \cdot \mathbf{c})\mathbf{b} - (\mathbf{b} \cdot \mathbf{c})\mathbf{a}], \tag{33}$$

for some scalar λ. Substitution of the relations $\mathbf{a} = a_1\mathbf{i} + a_2\mathbf{j} + a_3\mathbf{k}$, and so forth, finally yields the determination

$$\lambda = 1,$$

and hence there follows

$$(\mathbf{a} \times \mathbf{b}) \times \mathbf{c} = (\mathbf{a} \cdot \mathbf{c})\mathbf{b} - (\mathbf{b} \cdot \mathbf{c})\mathbf{a}. \tag{34a}$$

(A somewhat less tedious determination of λ is indicated in Problem 22.) In a similar way, the vector $\mathbf{a} \times (\mathbf{b} \times \mathbf{c})$ is seen to be expressible as a linear combination of $\mathbf{b}$ and $\mathbf{c}$, and the identity

$$\mathbf{a} \times (\mathbf{b} \times \mathbf{c}) = (\mathbf{a} \cdot \mathbf{c})\mathbf{b} - (\mathbf{b} \cdot \mathbf{a})\mathbf{c} \tag{34b}$$

can be deduced from (34a).

The two formulas (34a, b) are easily remembered if it is noticed that in each expansion the *middle factor* on the left is multiplied by the dot product of the other two factors, whereas in the term with the negative sign the *other factor in parentheses* on the left is multiplied by the dot product of the remaining factors.

It should be noticed that such combinations as **ab**, **(a × b)c**, and **a × b × c** are here left undefined.

Vector products involving more than three vectors are readily evaluated in terms of the products considered above. For example, the product

$$(\mathbf{a} \times \mathbf{b}) \cdot (\mathbf{c} \times \mathbf{d})$$

can be considered, say, as the triple scalar product of **a**, **b**, and **(c × d)** and hence, if we write temporarily $\mathbf{u} = \mathbf{c} \times \mathbf{d}$, there follows

$$\mathbf{a} \times \mathbf{b} \cdot \mathbf{u} = \mathbf{a} \cdot \mathbf{b} \times \mathbf{u}$$
$$= \mathbf{a} \cdot [\mathbf{b} \times (\mathbf{c} \times \mathbf{d})]$$
$$= \mathbf{a} \cdot [(\mathbf{b} \cdot \mathbf{d})\mathbf{c} - (\mathbf{b} \cdot \mathbf{c})\mathbf{d}],$$

and hence

$$(\mathbf{a} \times \mathbf{b}) \cdot (\mathbf{c} \times \mathbf{d}) = (\mathbf{a} \cdot \mathbf{c})(\mathbf{b} \cdot \mathbf{d}) - (\mathbf{a} \cdot \mathbf{d})(\mathbf{b} \cdot \mathbf{c}). \tag{35}$$

In particular, if we take

$$\mathbf{c} = \mathbf{a}, \qquad \mathbf{d} = \mathbf{b},$$

Equation (35) becomes

$$(\mathbf{a} \times \mathbf{b}) \cdot (\mathbf{a} \times \mathbf{b}) = (\mathbf{a} \cdot \mathbf{a})(\mathbf{b} \cdot \mathbf{b}) - (\mathbf{a} \cdot \mathbf{b})^2. \tag{36}$$

This relationship is known as the *identity of Lagrange*. The truth of Equation (36) follows also from the fact that it can be written in the form

$$|\mathbf{a} \times \mathbf{b}|^2 = a^2b^2 - a^2b^2 \cos^2 \theta = (ab \sin \theta)^2,$$

which result is a consequence of the definition of **a × b**.

6.5. Differentiation of Vectors. If the definition of a vector quantity **v** involves a parameter t, the derivative of the vector $\mathbf{v}(t)$ with respect to t is defined as the limit

$$\frac{d\mathbf{v}(t)}{dt} = \lim_{\Delta t \to 0} \frac{\mathbf{v}(t + \Delta t) - \mathbf{v}(t)}{\Delta t}, \tag{37}$$

when that limit exists.

From this definition it follows that the derivative of the product of a scalar $s(t)$ and a vector $\mathbf{v}(t)$ is given by the familiar product law

$$\frac{d}{dt} s\mathbf{v} = s\frac{d\mathbf{v}}{dt} + \frac{ds}{dt}\mathbf{v}. \tag{38}$$

Hence, if a vector is expressed in terms of its components along the fixed coordinate axes,

$$\mathbf{v} = f(t)\mathbf{i} + g(t)\mathbf{j} + h(t)\mathbf{k},$$

there follows

$$\frac{d\mathbf{v}}{dt} = \frac{df}{dt}\mathbf{i} + \frac{dg}{dt}\mathbf{j} + \frac{dh}{dt}\mathbf{k}, \tag{39}$$

since $\mathbf{i}, \mathbf{j}$, and $\mathbf{k}$ are constant vectors.

It follows also, from the definition, that the derivative of a product involving two or more vectors is defined as in the corresponding scalar case if the order of the factors is retained. Thus, for example, we obtain the formulas

$$\frac{d}{dt}(\mathbf{a} \cdot \mathbf{b}) = \mathbf{a} \cdot \frac{d\mathbf{b}}{dt} + \frac{d\mathbf{a}}{dt} \cdot \mathbf{b},$$

$$\frac{d}{dt}(\mathbf{a} \times \mathbf{b}) = \mathbf{a} \times \frac{d\mathbf{b}}{dt} + \frac{d\mathbf{a}}{dt} \times \mathbf{b},$$

$$\frac{d}{dt}(\mathbf{a} \cdot \mathbf{b} \times \mathbf{c}) = \frac{d\mathbf{a}}{dt} \cdot \mathbf{b} \times \mathbf{c} + \mathbf{a} \cdot \frac{d\mathbf{b}}{dt} \times \mathbf{c} + \mathbf{a} \cdot \mathbf{b} \times \frac{d\mathbf{c}}{dt}.$$

In the first case the order of factors in the separate terms is irrelevant. This is, however, not true in the second and third cases.

The derivative of a vector of constant length, but changing direction, is perpendicular to the vector. This may be seen by noticing that if $\mathbf{a}$ has constant length there follows

$$\frac{d}{dt}(\mathbf{a} \cdot \mathbf{a}) = \frac{d}{dt}a^2 = 0$$

and also

$$\frac{d}{dt}(\mathbf{a} \cdot \mathbf{a}) = 2\mathbf{a} \cdot \frac{d\mathbf{a}}{dt}.$$

These results are compatible only if either $d\mathbf{a}/dt$ is zero or $d\mathbf{a}/dt$ is perpendicular to $\mathbf{a}$.

6.6. Geometry of a Space Curve. The equations

$$x = x(t), \qquad y = y(t), \qquad z = z(t) \tag{40}$$

define a curve in space as the parameter t varies over a specified range. We denote by $\mathbf{r}$ the *position vector* from the origin O to the point $P(x, y, z)$, corresponding to a specified value of t,

$$\mathbf{r} = x\mathbf{i} + y\mathbf{j} + z\mathbf{k}. \tag{41}$$

If t is increased by Δt, and x, y, z increase accordingly, the vector from O to the new point Q is given by

$$\mathbf{r} + \Delta\mathbf{r} = (x + \Delta x)\mathbf{i} + (y + \Delta y)\mathbf{j} + (z + \Delta z)\mathbf{k}$$

and hence there follows

$$\Delta\mathbf{r} = (\Delta x)\mathbf{i} + (\Delta y)\mathbf{j} + (\Delta z)\mathbf{k}. \tag{42}$$

This vector is clearly the vector $\overrightarrow{PQ}$, and hence is of length equal to the chord $\overline{PQ}$. (See Figure 6.9.) If both sides of Equation (42) are divided by the increment

Δt, there follows

$$\frac{\Delta \mathbf{r}}{\Delta t} = \frac{\Delta x}{\Delta t}\mathbf{i} + \frac{\Delta y}{\Delta t}\mathbf{j} + \frac{\Delta z}{\Delta t}\mathbf{k}.$$

If s represents arc length along the curve, we can artificially rewrite this equation in the form

$$\frac{\Delta \mathbf{r}}{\Delta t} = \left(\frac{\Delta x}{\Delta s}\mathbf{i} + \frac{\Delta y}{\Delta s}\mathbf{j} + \frac{\Delta z}{\Delta s}\mathbf{k}\right)\frac{\Delta s}{\Delta t}. \qquad (43)$$

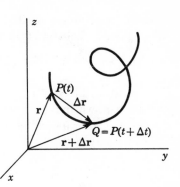

Figure 6.9

The vector $\Delta \mathbf{r}/\Delta s$ in parentheses is in the direction $\overrightarrow{PQ}$ if s increases with t, and is in the opposite direction otherwise; it has a length equal to the ratio of chord length to arc length. Thus, as $\Delta t \rightarrow 0$, we have in the limit

$$\frac{d\mathbf{r}}{dt} = \left(\frac{dx}{ds}\mathbf{i} + \frac{dy}{ds}\mathbf{j} + \frac{dz}{ds}\mathbf{k}\right)\frac{ds}{dt}, \qquad (44)$$

assuming that the derivatives exist, where the expression in parentheses is clearly a unit vector tangent to the space curve at the point P. Since this expression is equivalent to $d\mathbf{r}/ds$, we have the result that *the derivative of a position vector to a space curve, with respect to arc length along the curve, is a unit vector tangent to the curve, pointing in the direction of increasing arc length.* We denote this unit tangent vector by $\mathbf{u}$,

$$\mathbf{u} = \frac{d\mathbf{r}}{ds} = \frac{dx}{ds}\mathbf{i} + \frac{dy}{ds}\mathbf{j} + \frac{dz}{ds}\mathbf{k}. \qquad (45)$$

Since $\mathbf{u}$ is a unit vector, there follows

$$\left(\frac{dx}{ds}\right)^2 + \left(\frac{dy}{ds}\right)^2 + \left(\frac{dz}{ds}\right)^2 = 1.$$

This equation can be written in the differential form

$$ds = \sqrt{dx^2 + dy^2 + dz^2} = \sqrt{\left(\frac{dx}{dt}\right)^2 + \left(\frac{dy}{dt}\right)^2 + \left(\frac{dz}{dt}\right)^2}\ dt. \qquad (46)$$

giving the element of arc length for a space curve.

If t represents *time*, the vector $d\mathbf{r}/dt$ is the *velocity vector* associated with a point moving with speed ds/dt along the curve. Thus Equation (44) then becomes

$$\mathbf{v} = \mathbf{u}\frac{ds}{dt} = v\mathbf{u}, \qquad (47)$$

where $v = ds/dt$.

If Equation (47) is differentiated with respect to t, the *acceleration vector* $\mathbf{a}$ is obtained in the form

$$\mathbf{a} = \frac{dv}{dt}\mathbf{u} + v\frac{d\mathbf{u}}{dt} = \frac{dv}{dt}\mathbf{u} + v^2\frac{d\mathbf{u}}{ds}. \qquad (48)$$

Since **u** is a unit vector, the derivative $d\mathbf{u}/ds$ is perpendicular to the tangent vector. Making use of Equation (45), we obtain

$$\frac{d\mathbf{u}}{ds} = \frac{d^2\mathbf{r}}{ds^2} = \frac{d^2x}{ds^2}\mathbf{i} + \frac{d^2y}{ds^2}\mathbf{j} + \frac{d^2z}{ds^2}\mathbf{k}.$$

The length of this vector is called the *curvature* of the curve and is a measure of the rate at which the tangent vector changes its direction with distance along the curve. The reciprocal of this value is the *radius of curvature*, denoted by ρ. Hence we may write

$$\frac{d\mathbf{u}}{ds} = \frac{1}{\rho}\mathbf{n}. \tag{49}$$

where **n** is a unit vector, perpendicular to the tangent vector **u** at P, and known as the *principal normal vector*.† There follows also

$$\frac{1}{\rho} = \sqrt{\left(\frac{d^2x}{ds^2}\right)^2 + \left(\frac{d^2y}{ds^2}\right)^2 + \left(\frac{d^2z}{ds^2}\right)^2}. \tag{50}$$

By introducing Equation (49) into (48), we may express the acceleration vector in the form

$$\mathbf{a} = \frac{v^2}{\rho}\mathbf{n} + \frac{dv}{dt}\mathbf{u}. \tag{51}$$

Thus, if a particle moves along a space curve with speed v, its acceleration can be resolved into a component dv/dt along the curve and a component v^2/ρ along the principal normal to the curve.

In addition to the two perpendicular vectors **u** and **n** at a point P of a curve, a third vector, known as the *binormal vector*, is defined by the equation

$$\mathbf{b} = \mathbf{u} \times \mathbf{n}, \tag{52}$$

and hence is a unit vector perpendicular to both **u** and **n**. Thus, with each point P moving along a space curve, we may associate a moving and rotating triad of mutually orthogonal unit vectors. For a *plane curve*, the tangent vector **u** and the principal normal vector **n** lie in the plane of the curve, whereas the binormal vector is a constant unit vector perpendicular to that plane.

If we differentiate (52) and make use of (49), there follows

$$\frac{d\mathbf{b}}{ds} = \frac{d\mathbf{u}}{ds} \times \mathbf{n} + \mathbf{u} \times \frac{d\mathbf{n}}{ds} = \mathbf{u} \times \frac{d\mathbf{n}}{ds}.$$

Thus $d\mathbf{b}/ds$ is perpendicular to **u** and, since **b** is a unit vector, $d\mathbf{b}/ds$ is also perpendicular to **b**. Hence $d\mathbf{b}/ds$ is a scalar multiple of **n** and can be written in the form

$$\frac{d\mathbf{b}}{ds} = -\frac{1}{\tau}\mathbf{n}. \tag{53}$$

---

†Equation (49) states that **n** and $d\mathbf{u}/ds$ point either in the same direction or in opposite directions, depending upon the *sign* of ρ. We here define ρ to be always positive, so that **n** and $d\mathbf{u}/ds$ always point in the same direction. However, other conventions also appear in the literature. For a straight line, $1/\rho = 0$ and **n** can be taken to be *any* unit vector normal to **u**.

The scalar $1/\tau$ is called the *torsion* of the curve, the negative sign having been introduced so that the torsion is positive when the vector triad rotates in a right-handed sense about the tangent as it progresses along the curve. The length $|\tau|$ is called the *radius of torsion*. A curve whose torsion is not identically zero, and which hence does not lie in a plane, is often called a *twisted curve*.

Equations (49) and (53) give the derivatives of $\mathbf{u}$ and $\mathbf{b}$ with respect to arc length. To calculate $d\mathbf{n}/ds$, we write $\mathbf{n} = \mathbf{b} \times \mathbf{u}$ and differentiate, making use of (49) and (53), to obtain

$$\frac{d\mathbf{n}}{ds} = \frac{d\mathbf{b}}{ds} \times \mathbf{u} + \mathbf{b} \times \frac{d\mathbf{u}}{ds} = -\frac{1}{\tau}(\mathbf{n} \times \mathbf{u}) + \frac{1}{\rho}(\mathbf{b} \times \mathbf{n}),$$

or
$$\frac{d\mathbf{n}}{ds} = \frac{1}{\tau}\mathbf{b} - \frac{1}{\rho}\mathbf{u}. \tag{54}$$

Equations (49), (53), and (54) are known as *Frenet's formulas*.

If we form the dot product of $\mathbf{b}$ with the two sides of (54) and use (52) and (49), we obtain

$$\frac{1}{\tau} = \mathbf{b} \cdot \frac{d\mathbf{n}}{ds} = \mathbf{u} \times \mathbf{n} \cdot \frac{d\mathbf{n}}{ds} = \rho^2 \left(\mathbf{u} \times \frac{d\mathbf{u}}{ds} \cdot \frac{d^2\mathbf{u}}{ds^2}\right). \tag{55a}$$

This result can be written in the determinant form

$$\frac{1}{\tau} = \rho^2 \begin{vmatrix} \dfrac{dx}{ds} & \dfrac{dy}{ds} & \dfrac{dz}{ds} \\[2mm] \dfrac{d^2x}{ds^2} & \dfrac{d^2y}{ds^2} & \dfrac{d^2z}{ds^2} \\[2mm] \dfrac{d^3x}{ds^3} & \dfrac{d^3y}{ds^3} & \dfrac{d^3z}{ds^3} \end{vmatrix}, \tag{55b}$$

where ρ is defined by Equation (50).

Formulas expressing ρ and τ in terms of a general parameter t, rather than arc length s, are deduced in Problems 36 to 39.

The plane determined at a point P_0 on a curve by $\mathbf{u}$ and $\mathbf{n}$ is called the *osculating plane*, that determined by $\mathbf{n}$ and $\mathbf{b}$ the *normal plane*, and that determined by $\mathbf{u}$ and $\mathbf{b}$ the *rectifying plane*. Hence, if $\mathbf{r}_0$ is the position vector to P_0 and $\mathbf{r} = x\mathbf{i} + y\mathbf{j} + z\mathbf{k}$, the equations of these planes are

$$\mathbf{b}_0 \cdot (\mathbf{r} - \mathbf{r}_0) = 0, \qquad \mathbf{u}_0 \cdot (\mathbf{r} - \mathbf{r}_0) = 0, \qquad \mathbf{n}_0 \cdot (\mathbf{r} - \mathbf{r}_0) = 0,$$

respectively, where $\mathbf{b}_0$, $\mathbf{u}_0$, and $\mathbf{n}_0$ are evaluated at P_0.

6.7. The Gradient Vector. If φ is a scalar function of the coordinates x, y, and z, the values of the partial derivatives $\partial\varphi/\partial x$, $\partial\varphi/\partial y$, and $\partial\varphi/\partial z$ at a point represent the rates of change of φ with respect to distance in the $x, y,$ and z directions, respectively, at that point. If we consider the vector

$$\mathbf{V} = \mathbf{i}\frac{\partial\varphi}{\partial x} + \mathbf{j}\frac{\partial\varphi}{\partial y} + \mathbf{k}\frac{\partial\varphi}{\partial z}, \tag{56}$$

we see that the scalar components of this vector in the $x, y,$ and z directions at any point P are then exactly the respective rates at which φ is changing with respect to distance in those directions at P. To determine the component of $\mathbf{V}$ in *any* direction at P, we consider the position vector $\mathbf{r}$ from the origin to P and indicate a differential displacement from P in any chosen direction by $d\mathbf{r}$,

$$d\mathbf{r} = \mathbf{i}\, dx + \mathbf{j}\, dy + \mathbf{k}\, dz. \tag{57}$$

The scalar component of $\mathbf{V}$ in this direction is then obtained as the dot product of $\mathbf{V}$ and the unit vector $\mathbf{u} = d\mathbf{r}/ds$, where $ds = |d\mathbf{r}|$, and hence has the value

$$\mathbf{V} \cdot \mathbf{u} = \frac{\partial \varphi}{\partial x}\frac{dx}{ds} + \frac{\partial \varphi}{\partial y}\frac{dy}{ds} + \frac{\partial \varphi}{\partial z}\frac{dz}{ds} = \frac{d\varphi}{ds}. \tag{58}$$

Thus we see that the component of $\mathbf{V}$ in *any* direction is the rate of change of φ with respect to distance in that direction, so that, in fact, $\mathbf{V}$ is a vector function which is associated with the scalar function φ in a way which is *independent of the coordinate system* employed for their specification. It is called the *gradient* of φ, and accordingly we may write

$$\operatorname{grad} \varphi = \mathbf{i}\frac{\partial \varphi}{\partial x} + \mathbf{j}\frac{\partial \varphi}{\partial y} + \mathbf{k}\frac{\partial \varphi}{\partial z} \tag{59}$$

when we employ a rectangular coordinate system. In the general case, if $\mathbf{r}$ represents the position vector to a point P and if $d\mathbf{r}$ represents differential displacement of length ds from P, Equation (58) provides the basic definition

$$\frac{d\mathbf{r}}{ds} \cdot \operatorname{grad} \varphi = \frac{d\varphi}{ds} \tag{60a}$$

or, in terms of differentials,

$$d\mathbf{r} \cdot \operatorname{grad} \varphi = d\varphi. \tag{60b}$$

Since *the component of the vector grad φ in any direction is the derivative of φ in that direction*, it follows that grad φ *must point in the direction in which the derivative of φ is numerically greatest, and must have a length numerically equal to that maximum derivative.* In particular, grad φ must have no component in the directions in which the derivative of φ is zero, and hence *must be perpendicular to the surfaces* $\varphi(x, y, z) = $ constant.

Example. If the temperature at any point $P(x, y, z)$ is given by

$$T = C(x^2 + y^2 + z^2),$$

the gradient vector has the form

$$\operatorname{grad} T = 2C(x\mathbf{i} + y\mathbf{j} + z\mathbf{k}).$$

This vector is clearly a radial vector pointed outward from the origin, if $C > 0$, and hence is perpendicular to the spherical equithermal surfaces

$$x^2 + y^2 + z^2 = \text{constant}.$$

The magnitude of grad T is given by

$$|\operatorname{grad} T| = 2Cr,$$

where $r = \sqrt{x^2 + y^2 + z^2}$, and, since $T = Cr^2$, we have

$$|\operatorname{grad} T| = \frac{dT}{dr}.$$

This result is, of course, in accordance with the fact that the prescribed temperature changes most rapidly in the r direction. The rate of change of the temperature at P with respect to distance in a direction specified by the direction cosines (l, m, n) is given by

$$(l\mathbf{i} + m\mathbf{j} + n\mathbf{k}) \cdot \operatorname{grad} T = 2C(lx + my + nz). \qquad \blacksquare$$

6.8. The Vector Operator ∇. It is conventional to write

$$\operatorname{grad} \varphi = \nabla\varphi, \qquad (61)$$

where the symbol ∇, called *del*, represents a vector operator which, accordingly, is of the form

$$\nabla = \mathbf{i}\frac{\partial}{\partial x} + \mathbf{j}\frac{\partial}{\partial y} + \mathbf{k}\frac{\partial}{\partial z} \qquad (62)$$

in rectangular coordinates.† That is, we adopt the convention

$$\nabla\varphi = \left(\mathbf{i}\frac{\partial}{\partial x} + \mathbf{j}\frac{\partial}{\partial y} + \mathbf{k}\frac{\partial}{\partial z}\right)\varphi = \mathbf{i}\frac{\partial\varphi}{\partial x} + \mathbf{j}\frac{\partial\varphi}{\partial y} + \mathbf{k}\frac{\partial\varphi}{\partial z}, \qquad (63)$$

where φ is any differentiable scalar function of x, y, and z. In this notation, Equations (60) become

$$\frac{d\mathbf{r}}{ds} \cdot \nabla\varphi = \frac{d\varphi}{ds}, \qquad d\mathbf{r} \cdot \nabla\varphi = d\varphi, \qquad (64)$$

independently of the coordinate system.

If $\mathbf{F}$ is a vector function of x, y, and z, we may define the dot and cross products of the operator ∇ into $\mathbf{F}$. These products are called the *divergence* and *curl* of $\mathbf{F}$, respectively, and are of the form

$$\nabla \cdot \mathbf{F} = \mathbf{i} \cdot \frac{\partial \mathbf{F}}{\partial x} + \mathbf{j} \cdot \frac{\partial \mathbf{F}}{\partial y} + \mathbf{k} \cdot \frac{\partial \mathbf{F}}{\partial z} = \operatorname{div} \mathbf{F}, \qquad (65)$$

$$\nabla \times \mathbf{F} = \mathbf{i} \times \frac{\partial \mathbf{F}}{\partial x} + \mathbf{j} \times \frac{\partial \mathbf{F}}{\partial y} + \mathbf{k} \times \frac{\partial \mathbf{F}}{\partial z} = \operatorname{curl} \mathbf{F}. \qquad (66)$$

If $\mathbf{F} = F_x\mathbf{i} + F_y\mathbf{j} + F_z\mathbf{k}$, we then have

$$\nabla \cdot \mathbf{F} = \operatorname{div} \mathbf{F} = \frac{\partial F_x}{\partial x} + \frac{\partial F_y}{\partial y} + \frac{\partial F_z}{\partial z} \qquad (67)$$

and

$$\nabla \times \mathbf{F} = \operatorname{curl} \mathbf{F} = \begin{vmatrix} \mathbf{i} & \mathbf{j} & \mathbf{k} \\ \dfrac{\partial}{\partial x} & \dfrac{\partial}{\partial y} & \dfrac{\partial}{\partial z} \\ F_x & F_y & F_z \end{vmatrix} \qquad (68a)$$

†Expressions for the operator ∇, and for the related quantities introduced in this section, are obtained in terms of other coordinate systems in Section 6.17.

or

$$\mathbf{V} \times \mathbf{F} = \mathbf{i}\left(\frac{\partial F_z}{\partial y} - \frac{\partial F_y}{\partial z}\right) + \mathbf{j}\left(\frac{\partial F_x}{\partial z} - \frac{\partial F_z}{\partial x}\right) + \mathbf{k}\left(\frac{\partial F_y}{\partial x} - \frac{\partial F_x}{\partial y}\right). \tag{68b}$$

Just as the product **ab** has not been assigned a meaning, the combination **VF** is here left undefined.

The total differential of a vector function **F** is given in rectangular coordinates by

$$d\mathbf{F} = \frac{\partial \mathbf{F}}{\partial x} dx + \frac{\partial \mathbf{F}}{\partial y} dy + \frac{\partial \mathbf{F}}{\partial z} dz,$$

or, in operational form,

$$d\mathbf{F} = \left(dx\frac{\partial}{\partial x} + dy\frac{\partial}{\partial y} + dz\frac{\partial}{\partial z}\right)\mathbf{F}.$$

Since the operator is the dot product of the differential vector $d\mathbf{r} = \mathbf{i}\, dx + \mathbf{j}\, dy + \mathbf{k}\, dz$ and the operator **V**, we may write this relation in the invariant form

$$d\mathbf{F} = (d\mathbf{r} \cdot \mathbf{V})\mathbf{F}. \tag{69}$$

The derivative of **F** in the direction of $d\mathbf{r}$ is then

$$\frac{d\mathbf{F}}{ds} = \left(\frac{d\mathbf{r}}{ds} \cdot \mathbf{V}\right)\mathbf{F}. \tag{70}$$

More generally, *if* **u** *is any unit vector, then the derivative of* **F** *in the direction of* **u** *is given by*

$$\frac{d\mathbf{F}}{ds} = (\mathbf{u} \cdot \mathbf{V})\mathbf{F}. \tag{71}$$

Since **VF** is undefined, the parentheses in Equation (71) are generally omitted and we follow the convention

$$\mathbf{v} \cdot \mathbf{V}\mathbf{F} \equiv (\mathbf{v} \cdot \mathbf{V})\mathbf{F}, \tag{72}$$

where, if **v** is any vector with rectangular components $v_x, v_y,$ and v_z, the product $\mathbf{v} \cdot \mathbf{V}$ is the scalar operator

$$\mathbf{v} \cdot \mathbf{V} = v_x\frac{\partial}{\partial x} + v_y\frac{\partial}{\partial y} + v_z\frac{\partial}{\partial z}. \tag{73}$$

6.9. Differentiation Formulas. The following identities are of frequent use:

$$\mathbf{V} \cdot \varphi\mathbf{u} = \varphi\mathbf{V} \cdot \mathbf{u} + \mathbf{u} \cdot \mathbf{V}\varphi, \tag{74a}$$

$$\mathbf{V} \times \varphi\mathbf{u} = \varphi\mathbf{V} \times \mathbf{u} + \mathbf{V}\varphi \times \mathbf{u}, \tag{74b}$$

$$\mathbf{V} \cdot \mathbf{u} \times \mathbf{v} = \mathbf{v} \cdot \mathbf{V} \times \mathbf{u} - \mathbf{u} \cdot \mathbf{V} \times \mathbf{v}, \tag{74c}$$

$$\mathbf{V} \times (\mathbf{u} \times \mathbf{v}) = \mathbf{v} \cdot \mathbf{V}\mathbf{u} - \mathbf{u} \cdot \mathbf{V}\mathbf{v} + \mathbf{u}(\mathbf{V} \cdot \mathbf{v}) - \mathbf{v}(\mathbf{V} \cdot \mathbf{u}), \tag{74d}$$

$$\mathbf{V}(\mathbf{u} \cdot \mathbf{v}) = \mathbf{u} \cdot \mathbf{V}\mathbf{v} + \mathbf{v} \cdot \mathbf{V}\mathbf{u} + \mathbf{u} \times (\mathbf{V} \times \mathbf{v}) + \mathbf{v} \times (\mathbf{V} \times \mathbf{u}), \tag{74e}$$

$$\mathbf{V} \times (\mathbf{V}\varphi) = \text{curl grad } \varphi = 0, \tag{74f}$$

$$\mathbf{V} \cdot (\mathbf{V} \times \mathbf{u}) = \text{div curl } \mathbf{u} = 0, \tag{74g}$$

$$\mathbf{V} \times (\mathbf{V} \times \mathbf{u}) = \text{curl curl } \mathbf{u} = \mathbf{V}(\mathbf{V} \cdot \mathbf{u}) - \mathbf{V} \cdot \mathbf{V}\mathbf{u}$$

$$= \text{grad div } \mathbf{u} - \mathbf{V}^2\mathbf{u}, \tag{74h}$$

$$\mathbf{V} \cdot (\mathbf{V}\varphi_1 \times \mathbf{V}\varphi_2) = 0. \tag{74i}$$

In these formulas $\mathbf{u}$ and $\mathbf{v}$ are arbitrary vectors and φ, φ_1, and φ_2 arbitrary scalars for which the indicated derivatives exist. *In formulas* (74f) *and* (74g), *we assume in addition that φ and $\mathbf{u}$ have continuous partial derivatives of the second order.*

These identities can all be verified by direct expansion in terms (say) of components along $\mathbf{i}, \mathbf{j}$, and $\mathbf{k}$. In some cases, however, the proof can be considerably shortened by making use of the fact that *the operator $\mathbf{V}$ may be substituted for a vector in any vector identity provided that it operates on the same factors in all terms.*

It should be noticed that the operator $\mathbf{V}$ is a *distributive* operator, in the sense that the equations

$$\mathbf{V}(\varphi_1 + \varphi_2) = (\mathbf{V}\varphi_1) + (\mathbf{V}\varphi_2), \qquad \mathbf{V} \cdot (\mathbf{u} + \mathbf{v}) = (\mathbf{V} \cdot \mathbf{u}) + (\mathbf{V} \cdot \mathbf{v}),$$

$$\mathbf{V} \times (\mathbf{u} + \mathbf{v}) = (\mathbf{V} \times \mathbf{u}) + (\mathbf{V} \times \mathbf{v})$$

are true.

Also, direct expansion shows that the identity

$$(\mathbf{u} \cdot \mathbf{V})\varphi = \mathbf{u} \cdot (\mathbf{V}\varphi) \tag{75}$$

is valid. For this reason the parentheses are conventionally omitted in this case and the product is written in the form $\mathbf{u} \cdot \mathbf{V}\varphi$. Further, as has been noted, no parentheses are needed in a product such as $\mathbf{u} \cdot \mathbf{V}\mathbf{v}$, since $\mathbf{V}\mathbf{v}$ is undefined and hence this notation can only mean $(\mathbf{u} \cdot \mathbf{V})\mathbf{v}$.

In general, we adopt the convention that in a multiple product $\mathbf{V}$ operates on all terms to its right which are not specifically excluded from its influence by the use of parentheses, brackets, or braces. If this is not convenient, we will here indicate a term which is to be treated as a constant in the differentiation, and hence is *not* to be operated on by $\mathbf{V}$, by underlining that term. Thus we use the notations $\mathbf{V} \cdot (\varphi\underline{\mathbf{u}})$ and $\mathbf{V} \times (\varphi\underline{\mathbf{u}})$ to indicate that $\mathbf{u}$ is to be treated as a constant vector in the differentiation implied in $\mathbf{V}$, so that $\mathbf{V}$ operates only on φ. It is readily verified that we then have

$$\mathbf{V} \cdot (\varphi\underline{\mathbf{u}}) = \mathbf{u} \cdot \mathbf{V}\varphi, \qquad \mathbf{V} \times (\varphi\underline{\mathbf{u}}) = -\mathbf{u} \times \mathbf{V}\varphi. \tag{76}$$

From the nature of $\mathbf{V}$ we see that

$$\mathbf{V} \cdot (\varphi\mathbf{u}) = \mathbf{V} \cdot (\underline{\varphi}\mathbf{u}) + \mathbf{V} \cdot (\varphi\underline{\mathbf{u}}),$$

$$\mathbf{V} \times (\varphi\mathbf{u}) = \mathbf{V} \times (\underline{\varphi}\mathbf{u}) + \mathbf{V} \times (\varphi\underline{\mathbf{u}}).$$

That is, the derivative of the product is the sum of the two derivatives obtained by holding one of the factors constant and allowing the other to be operated on by $\mathbf{V}$. These considerations lead directly to formulas (74a) and (74b).

Formula (74c) is obtained in a similar way if we first notice that

$$\nabla \cdot \mathbf{u} \times \mathbf{c} = -\nabla \cdot \mathbf{c} \times \mathbf{u} = \mathbf{c} \cdot \nabla \times \mathbf{u}$$

when $\mathbf{c}$ is a constant vector.

We see that if the product

$$\nabla \times (\mathbf{u} \times \mathbf{v})$$

is *formally* expanded by substituting ∇, $\mathbf{u}$, and $\mathbf{v}$ for $\mathbf{a}$, $\mathbf{b}$, and $\mathbf{c}$ in Equation (34), the result

$$(\nabla \cdot \mathbf{v})\mathbf{u} - (\mathbf{u} \cdot \nabla)\mathbf{v}$$

is *not* necessarily equivalent to the given product, since there ∇ operates on both $\mathbf{u}$ and $\mathbf{v}$, whereas in the two terms of the formal expansion, ∇ operates only on $\mathbf{v}$. However, if $\mathbf{u}$ is treated as a constant vector in the given product, the expansion is valid. That is, the expansion represents $\nabla \times (\underline{\mathbf{u}} \times \mathbf{v})$. By adding to this expansion the corresponding expression for $\nabla \times (\mathbf{u} \times \underline{\mathbf{v}})$, we obtain formula (74d).

Formula (74e) may be established by an indirect method in which we first obtain the results

$$\mathbf{u} \times (\nabla \times \mathbf{v}) = \nabla(\underline{\mathbf{u}} \cdot \mathbf{v}) - \mathbf{u} \cdot \nabla\mathbf{v},$$

$$\mathbf{v} \times (\nabla \times \mathbf{u}) = \nabla(\mathbf{u} \cdot \underline{\mathbf{v}}) - \mathbf{v} \cdot \nabla\mathbf{u},$$

and then form the sum of these equations.

Equations (74f) and (74g) are easily established by direct calculation. Thus, to establish (74f), we write

$$\nabla \times \nabla\varphi = \begin{vmatrix} \mathbf{i} & \mathbf{j} & \mathbf{k} \\ \dfrac{\partial}{\partial x} & \dfrac{\partial}{\partial y} & \dfrac{\partial}{\partial z} \\ \dfrac{\partial\varphi}{\partial x} & \dfrac{\partial\varphi}{\partial y} & \dfrac{\partial\varphi}{\partial z} \end{vmatrix} = \mathbf{i}\left(\dfrac{\partial^2\varphi}{\partial y\,\partial z} - \dfrac{\partial^2\varphi}{\partial z\,\partial y}\right) + \mathbf{j}\left(\dfrac{\partial^2\varphi}{\partial z\,\partial x} - \dfrac{\partial^2\varphi}{\partial x\,\partial z}\right)$$
$$+ \mathbf{k}\left(\dfrac{\partial^2\varphi}{\partial x\,\partial y} - \dfrac{\partial^2\varphi}{\partial y\,\partial x}\right).$$

If the order of differentiation is immaterial, all components on the right vanish. This situation exists, in particular, *if the second crossed partial derivatives of φ are continuous*. Similarly, Equation (74g) is readily shown to be valid if the corresponding derivatives of the components of $\mathbf{u}$ are continuous.

Equation (74h) is obtained by replacing $\mathbf{a}$, $\mathbf{b}$, $\mathbf{c}$, by ∇, ∇, $\mathbf{u}$ in Equation (34). The scalar operator

$$\nabla^2 = \nabla \cdot \nabla = \frac{\partial^2}{\partial x^2} + \frac{\partial^2}{\partial y^2} + \frac{\partial^2}{\partial z^2} \tag{77}$$

is of frequent occurrence, and is known as the *Laplacian* operator. Equation (74i) is established by making use of (74c) and (74f).

If $\mathbf{r}$ is the position vector, so that

$$\mathbf{r} = x\mathbf{i} + y\mathbf{j} + z\mathbf{k}$$

in rectangular coordinates, we find by direct calculation that

$$\nabla \cdot \mathbf{r} = 3, \qquad (78a)$$

$$\nabla \times \mathbf{r} = \mathbf{0}, \qquad (78b)$$

and also that

$$\mathbf{u} \cdot \nabla \mathbf{r} = \mathbf{u}, \qquad (79)$$

where $\mathbf{u}$ is any vector.

6.10. Line Integrals. Let $\mathbf{F}$ represent a vector function of position in a space of three dimensions and let C represent a space curve, each point P of which is specified by its position vector $\mathbf{r}$ from an origin O to the point P. With each point P of the curve we associate a *differential distance vector* $d\mathbf{r}$. Then, since

$$d\mathbf{r} = \frac{d\mathbf{r}}{ds} ds = \mathbf{u} \, ds, \qquad (80)$$

where $\mathbf{u}$ is a unit tangent vector to C at P, the differential vector $d\mathbf{r}$ has as its length the differential arc length ds along C at P and as its direction the direction of the curve at P. The scalar differential $\mathbf{F} \cdot d\mathbf{r} = (\mathbf{F} \cdot \mathbf{u}) \, ds$ is then numerically equal to the product of the component of $\mathbf{F}$ in the direction of C at P and the differential length ds. Finally, the integral

$$\int_C \mathbf{F} \cdot d\mathbf{r} = \int_C (\mathbf{F} \cdot \mathbf{u}) \, ds, \qquad (81)$$

taken along the curve between two specified points P_0 and P_1, is known as the *line integral* of $\mathbf{F}$ along C.

In particular, if $\mathbf{F}$ represents *force* acting on a particle, the line integral (81) clearly represents the *work* done by the force in moving the particle along C from P_0 to P_1.

We assume here, and in the sequel, that any curve C along which a line integral is to be evaluated is made up of a finite number of arcs, along each of which the tangent vector $\mathbf{u}$ not only exists, but also varies continuously with s, so that the integrand $\mathbf{F} \cdot \mathbf{u}$ is continuous when $\mathbf{F}$ is continuous. Such a curve is said to be *piecewise smooth*. In particular, when $\mathbf{u}$ is continuous over *all* of C, we say that C is a *smooth curve*. For some purposes, we will require also that a curve not intersect itself, that is, that no point be encountered twice as the curve is traced out. Such a curve is called a *simple curve*.

If we write

$$\mathbf{F} = P(x, y, z)\mathbf{i} + Q(x, y, z)\mathbf{j} + R(x, y, z)\mathbf{k},$$
$$\mathbf{r} = x\mathbf{i} + y\mathbf{j} + z\mathbf{k}, \qquad (82)$$

the line integral (81) takes the form

$$\int_C \mathbf{F} \cdot d\mathbf{r} = \int_C (P \, dx + Q \, dy + R \, dz). \qquad (83)$$

On each arc of the curve C, the variables x, y, and z, as well as the corresponding differentials, are to be expressed in terms of an appropriate single variable, and the limits of the integral along that arc must be determined accordingly.

It should be pointed out that the line integral $\int_C \mathbf{F} \cdot d\mathbf{r}$, taken along the portion of a curve between two points A and B, may be defined more fundamentally as the limit of a sum of the form

$$\sum_{k=0}^{n} \mathbf{F}_k \cdot (\overrightarrow{\Delta r})_k = \mathbf{F}_0 \cdot (\overrightarrow{\Delta r})_0 + \mathbf{F}_1 \cdot (\overrightarrow{\Delta r})_1 + \cdots + \mathbf{F}_n \cdot (\overrightarrow{\Delta r})_n,$$

where $\mathbf{F}_0, \mathbf{F}_1, \ldots, \mathbf{F}_n$ are the values of $\mathbf{F}$ at A and at n points $P_1, P_2, \ldots, P_n$, arbitrarily chosen along the arc of C between A and B, and where $(\overrightarrow{\Delta r})_0 = \overrightarrow{AP_1}$, $(\overrightarrow{\Delta r})_1 = \overrightarrow{P_1 P_2}, \ldots, (\overrightarrow{\Delta r})_n = \overrightarrow{P_n B}$. The limit is taken as $n \to \infty$ in such a way that all the chords tend to zero. This definition is analogous to the definition of the ordinary integral. When $\mathbf{F}$ is continuous and C is piecewise smooth, the integral so defined can be evaluated by the method described above.

Example. As an illustration, we take

$$\mathbf{F} = yz\mathbf{i} + xy\mathbf{j} + xz\mathbf{k}$$

and first calculate the line integral of $\mathbf{F}$ from $(0, 0, 0)$ to $(1, 1, 1)$ along the path C consisting of the curve $x = y^2$, $z = 0$ in the xy plane from $(0, 0, 0)$ to $(1, 1, 0)$ and the line $x = 1$, $y = 1$ perpendicular to the xy plane from $(1, 1, 0)$ to $(1, 1, 1)$. In the first part of the path, C_1, we have

$$x = y^2, \quad z = 0; \qquad dx = 2y\, dy, \quad dz = 0,$$

and hence

$$\mathbf{F} \cdot d\mathbf{r} = yz\, dx + xy\, dy + xz\, dz = y^3\, dy.$$

Thus there follows

$$\int_{C_1} \mathbf{F} \cdot d\mathbf{r} = \int_0^1 y^3\, dy = \tfrac{1}{4}.$$

In the second part of the path, C_2, we have

$$x = 1, \quad y = 1; \qquad dx = dy = 0,$$

and hence

$$\mathbf{F} \cdot d\mathbf{r} = z\, dz.$$

Thus there follows

$$\int_{C_2} \mathbf{F} \cdot d\mathbf{r} = \int_0^1 z\, dz = \tfrac{1}{2},$$

and, finally,

$$\int_C \mathbf{F} \cdot d\mathbf{r} = \tfrac{1}{4} + \tfrac{1}{2} = \tfrac{3}{4}.$$

If we integrate instead along the path C' consisting of the straight line $x = y = z$, directly from $(0, 0, 0)$ to $(1, 1, 1)$, we have, on C',

$$x = z, \quad y = z; \qquad dx = dz, \quad dy = dz$$

and hence

$$\int_{C'} \mathbf{F} \cdot d\mathbf{r} = \int_0^1 3z^2\, dz = 1. \qquad \blacksquare$$

In this example the value of the line integral between the given points depends upon the *path* chosen. However, in certain cases this is not so. For if the expression $P\,dx + Q\,dy + R\,dz$ is the differential of a function $\varphi(x, y, z)$,

$$P\,dx + Q\,dy + R\,dz = d\varphi, \tag{84}$$

then the integral (83) becomes merely

$$\int_C \mathbf{F} \cdot d\mathbf{r} = \int_C d\varphi \tag{85}$$

and its value accordingly is *the change in φ along the curve C.* Thus the line integral depends only on the location of the end points P_0 and P_1 *if the function φ is single-valued*, so that the value approached by φ at a point in space does not depend upon the manner of approach.

Since the differential of φ is also given by

$$d\varphi = \frac{\partial \varphi}{\partial x}\,dy + \frac{\partial \varphi}{\partial y}\,dx + \frac{\partial \varphi}{\partial z}\,dz, \tag{86}$$

there follows, by comparing (84) and (86) and noticing that $x, y,$ and z are independent variables,

$$\frac{\partial \varphi}{\partial x} = P, \qquad \frac{\partial \varphi}{\partial y} = Q, \qquad \frac{\partial \varphi}{\partial z} = R. \tag{87}$$

By appropriately differentiating the first two equations, we find also that

$$\frac{\partial^2 \varphi}{\partial y\,\partial x} = \frac{\partial P}{\partial y}, \qquad \frac{\partial^2 \varphi}{\partial x\,\partial y} = \frac{\partial Q}{\partial x},$$

and hence, if the derivatives involved are continuous, we conclude that P and Q must satisfy the condition

$$\frac{\partial P}{\partial y} = \frac{\partial Q}{\partial x}. \tag{88a}$$

In a similar way we find that the two conditions

$$\frac{\partial P}{\partial z} = \frac{\partial R}{\partial x}, \tag{88b}$$

$$\frac{\partial Q}{\partial z} = \frac{\partial R}{\partial y}. \tag{88c}$$

must also be satisfied.

Hence, if a function φ exists so that Equation (84) is true, that is, if $\mathbf{F} \cdot d\mathbf{r} = P\,dx + Q\,dy + R\,dz$ is an exact differential, the functions $P, Q,$ and R must satisfy (88a, b, c). *Conversely*, when $P, Q,$ and R and their first partial derivatives are continuous in a region $\Re$, it can be shown (see Problem 63) that, if these conditions are satisfied, a function φ exists for which (84) is true in that region.

We notice next that since

$$\nabla \times \mathbf{F} = \mathbf{i}\left(\frac{\partial R}{\partial y} - \frac{\partial Q}{\partial z}\right) + \mathbf{j}\left(\frac{\partial P}{\partial z} - \frac{\partial R}{\partial x}\right) + \mathbf{k}\left(\frac{\partial Q}{\partial x} - \frac{\partial P}{\partial y}\right), \qquad (89)$$

the satisfaction of (88a, b, c) is equivalent to the vanishing of $\nabla \times \mathbf{F}$. Thus we conclude that, if $\mathbf{F}$ is continuously differentiable in a region $\mathcal{R}$ and if $\nabla \times \mathbf{F} = \mathbf{0}$ everywhere in $\mathcal{R}$, then a scalar function φ exists such that $d\varphi = \mathbf{F} \cdot d\mathbf{r}$. Further, if C is any curve lying in $\mathcal{R}$ and joining the points P_0 and P_1, then

$$\int_C \mathbf{F} \cdot d\mathbf{r} = \int_C d\varphi = \Delta_C\varphi,$$

where $\Delta_C\varphi$ is the change in φ corresponding to a transition along C from P_0 to P_1. However, if φ is not single-valued in $\mathcal{R}$, this change still may depend upon C.

To exclude the possibility that φ be multiple-valued, we henceforth require that the region $\mathcal{R}$ have the property that an arbitrary closed curve lying in $\mathcal{R}$ can be shrunk continuously to a point in $\mathcal{R}$ without passing outside of $\mathcal{R}$. Such a region is called a *simply connected region* or, more briefly, a *simple region*. Thus the plane annular region included between two concentric circles is *not* a simple region, since a curve which surrounds the inner circle and lies in the annulus cannot be shrunk to a point without passing outside the annulus, and the same conclusion can be drawn for the region between two *coaxial cylinders* in space. The region between two *concentric spheres* is simple; the interior of a *torus* is not.

To show rigorously the relationship between simplicity of $\mathcal{R}$ and single-valuedness of φ in $\mathcal{R}$ is beyond the scope of this work. (See also Section 6.16.) It may be remarked, however, that the proof depends upon the fact that by restricting $\mathcal{R}$ to be a simple region we ensure that any closed curve C in $\mathcal{R}$ is the complete boundary of some open surface S in $\mathcal{R}$, to which *Stokes's theorem* (Section 6.16) can be applied.

For any curve C joining the points P_0 and P_1 and lying in a simple region $\mathcal{R}$ throughout which $\mathbf{F}$ is continuously differentiable and also $\nabla \times \mathbf{F} = \mathbf{0}$, we may write

$$\int_C \mathbf{F} \cdot d\mathbf{r} = \int_{P_0}^{P_1} \mathbf{F} \cdot d\mathbf{r} = \int_{P_0}^{P_1} d\varphi = \varphi(P_1) - \varphi(P_0), \qquad (90)$$

where φ is single-valued in $\mathcal{R}$. That is, in such a case *the line integral from P_0 to P_1 is independent of the path*, so long as that path remains in $\mathcal{R}$. In particular, for a *closed* path lying in $\mathcal{R}$ the initial and terminal points coincide and the line integral vanishes. We may denote the line integral once around a closed curve by the symbol $\oint$ and write

$$\oint_C \mathbf{F} \cdot d\mathbf{r} = 0 \qquad (91)$$

in this situation.

If the conditions stated are not satisfied, the integral around C may or may not vanish. The value of that integral is called the *circulation* of $\mathbf{F}$ around C.

When a closed curve C lies on a surface S, one side of which is regarded as its positive side, it is conventional to define the *positive direction* on C (that is, the direction of $\mathbf{u} = d\mathbf{r}/ds$) as the direction along which an observer, traveling on the positive side of S, would move while keeping the area enclosed by C to his *left*. In this connection, for a coordinate plane or surface (on which a coordinate is constant) it is to be understood in what follows that the positive side is the one from which the relevant coordinate increases, unless the contrary is stated.

6.11. The Potential Function. Since the existence of a single-valued function φ such that

$$d\varphi = \mathbf{F} \cdot d\mathbf{r} \qquad (92)$$

at all points of a simple region is guaranteed by the condition $\nabla \times \mathbf{F} = \mathbf{0}$, and since Equation (64) gives also

$$d\varphi = \nabla\varphi \cdot d\mathbf{r},$$

there follows by subtraction

$$(\mathbf{F} - \nabla\varphi) \cdot d\mathbf{r} = 0.$$

Hence the vector $\mathbf{F} - \nabla\varphi$ must be perpendicular to $d\mathbf{r}$; but since the direction of $d\mathbf{r}$ is arbitrary, we conclude that $\mathbf{F} - \nabla\varphi$ must vanish,

$$\mathbf{F} = \nabla\varphi. \qquad (93)$$

Thus, if $\nabla \times \mathbf{F} = \mathbf{0}$ *in a simple region, then* $\mathbf{F}$ *is the gradient of a single-valued scalar function* φ *in that region.* Conversely, Equation (74f) states that if $\mathbf{F} = \nabla\varphi$ and if φ has continuous second partial derivatives, then $\nabla \times \mathbf{F} = \mathbf{0}$.

The function φ defined by Equation (93) is known as the *potential* of $\mathbf{F}$. If $\mathbf{F}$ represents *force*, the *negative* of φ is called the *potential energy* associated with $\mathbf{F}$. When such a function φ exists, and is single-valued in a region, the force $\mathbf{F}$ is said to be *conservative*, since in this case the total work done in moving a particle around a closed contour in that region is zero.

To illustrate the determination of the potential, we consider the force field

$$\mathbf{F} = y^2\mathbf{i} + 2(xy + z)\mathbf{j} + 2y\mathbf{k}. \qquad (94)$$

It is readily verified that $\nabla \times \mathbf{F} = \mathbf{0}$ everywhere. Hence the force is conservative and the work done by $\mathbf{F}$ in moving a particle between two points is independent of the path. Further, a scalar potential function φ exists such that

$$\mathbf{F} \cdot d\mathbf{r} = d\varphi \qquad (95)$$

and also

$$\mathbf{F} = \nabla\varphi. \qquad (96)$$

To determine φ, we write Equation (95) in the form

$$d\varphi = y^2\, dx + 2(xy + z)\, dy + 2y\, dz.$$

Comparison of this equation with (86) shows that φ must satisfy the three conditions

$$\frac{\partial \varphi}{\partial x} = y^2, \tag{97a}$$

$$\frac{\partial \varphi}{\partial y} = 2(xy + z), \tag{97b}$$

$$\frac{\partial \varphi}{\partial z} = 2y. \tag{97c}$$

If we integrate (97a), holding y and z constant, we obtain

$$\varphi = xy^2 + f(y, z), \tag{98}$$

where $f(y, z)$ is an arbitrary function of y and z. In order that (98) satisfy (97b), we must then have

$$2xy + \frac{\partial f(y, z)}{\partial y} = 2xy + 2z \quad \text{or} \quad \frac{\partial f(y, z)}{\partial y} = 2z.$$

Integration gives

$$f(y, z) = 2yz + g(z), \tag{99}$$

where $g(z)$ is an arbitrary function of z. Introducing (98) and (99) into (97c), we then obtain the final condition

$$2y + \frac{dg(z)}{dz} = 2y,$$

from which there follows

$$g(z) = c, \tag{100}$$

where c is an arbitrary constant. Thus we have finally

$$\varphi = xy^2 + 2yz + c, \tag{101}$$

where the constant c can be chosen arbitrarily, so that the potential at a convenient reference point is zero. (See also Problem 64.)

The surfaces $\varphi = $ constant are called *equipotential surfaces*. We see that the work done in moving a particle from a point on the surface $\varphi = c_1$ to a point on the surface $\varphi = c_2$ is merely $c_2 - c_1$.

To illustrate the existence of unusual cases, we notice that if

$$\mathbf{F} = -\frac{y}{x^2 + y^2}\mathbf{i} + \frac{x}{x^2 + y^2}\mathbf{j},$$

there then follows

$$\mathbf{F} \cdot d\mathbf{r} = \frac{x\,dy - y\,dx}{x^2 + y^2} = d\tan^{-1}\frac{y}{x} = d\theta,$$

where θ is angular displacement about the z axis. Thus it seems that we have

$$\mathbf{F} = \nabla\theta,$$

and hence that, from (74f), there follows

$$\nabla \times \mathbf{F} = 0.$$

Hence it might appear that the circulation of **F** around any closed curve C would be zero. However, we notice that **F** is discontinuous at points on the z axis, in the sense that the magnitude of **F**,

$$|\mathbf{F}| = \frac{1}{\sqrt{x^2 + y^2}},$$

becomes infinite as $(x, y) \to (0, 0)$. Thus $\nabla \times \mathbf{F}$ is *not defined* along the z axis. At all other points, however, $\nabla \times \mathbf{F}$ *is* defined and is zero. Any closed curve *which does not surround the z axis* can be included in a simple region where **F** is continuously differentiable and $\nabla \times \mathbf{F} = \mathbf{0}$, and hence the circulation around any such path must vanish. To verify this statement, we consider an arbitrary closed path C beginning and ending at a point P. Since

$$\int_C \mathbf{F} \cdot d\mathbf{r} = \int_C d\theta = \Delta_C \theta,$$

we see that if C does not enclose the z axis, the angle θ may alternately increase and decrease along C until a maximum value is reached, after which θ eventually returns to its original value at P and the total net increase in θ is indeed zero. That is, over such a closed path the circulation is zero.

However, if the path C surrounds the z axis once, and the path is described in the positive (counterclockwise) direction, the angle θ experiences a net increase of 2π when the complete path is described, and the circulation is 2π. Similarly, the circulation along a closed curve encircling the z axis k times in the positive direction is $2k\pi$.

It is important to notice here that θ is, in fact, *not* a single-valued function, since at any point P not on the z axis θ has an *infinite* number of admissible values, differing from each other by integral multiples of 2π. (On the z axis, θ may have *any* value.) In a *simple* region $\mathfrak{R}$ not including any point on the z axis (and hence also not including any closed curve which surrounds that axis), we may *define* a single-valued *interpretation* of θ such that the interpretation so defined is continuous (and differentiable) along any smooth curve C in $\mathfrak{R}$, so that $d\theta$ exists everywhere on C, and so that $\int_C d\theta$ then is dependent only upon the *end points* of C. If there were any closed curve in $\mathfrak{R}$ which surrounded the z axis, such a definition no longer would be possible. It should be noted that the shaded two-dimensional region indicated in Figure 6.10 is an example of a *permissible* region, for present purposes.

If we delete the region in the neighborhood of the z axis by cutting out an infinitely extended right circular cylinder of arbitrarily small radius, with its axis along the z axis, the remainder of space is a region $\mathfrak{R}'$ inside which $\nabla \times \mathbf{F} = \mathbf{0}$ at all points. However, the region $\mathfrak{R}'$ clearly is not a *simple* region. Those closed curves in $\mathfrak{R}'$ which *can* be shrunk to a point in $\mathfrak{R}'$ without passing outside $\mathfrak{R}'$ are said to be *reducible*. Thus any closed curve in $\mathfrak{R}'$ which does not enclose the z axis is reducible, and we see that in the present example the circulation around any *reducible* closed curve in $\mathfrak{R}'$ is zero.

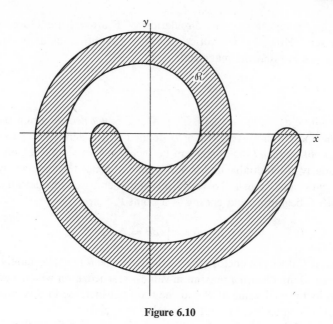

Figure 6.10

It can be shown, more generally, that if $\mathbf{F}$ is piecewise differentiable and $\mathbf{\nabla} \times \mathbf{F} = \mathbf{0}$ in *any* region $\mathfrak{R}$, whether simply connected or not, the circulation of $\mathbf{F}$ around any *reducible* closed curve C in that region is zero.†

6.12. Surface Integrals. A surface S in three dimensions is usually specified by an equation of the form

$$g(x, y, z) = 0, \tag{102}$$

where restrictions may be imposed on x, y, and z. If we think of S as one of the family of surfaces $g(x, y, z) = c$, as c takes on all possible values, we see that at any point where the vector $\mathbf{\nabla}g$ exists, it is normal to the particular surface of the family which passes through that point, and hence, in particular, $\mathbf{\nabla}g$ is normal to S at points of S. At such points $\mathbf{\nabla}g$ can then be expressed as the product of a scalar and a *unit vector* $\mathbf{n}$ normal to S. We can thus write

$$\mathbf{n} = \pm \frac{\mathbf{\nabla}g}{|\mathbf{\nabla}g|}, \tag{103}$$

where the choice of the ambiguous sign depends upon the desired orientation of $\mathbf{n}$.

For a *closed surface*, it is conventional to define the orientation of $\mathbf{n}$ at any point as *outward*. In any case, we suppose that S actually *possesses two sides*,

---

†This result can be established by applying Stokes's theorem (Section 6.16) to a surface S which is traced out as the reducible curve C is shrunk to a point in $\mathfrak{R}$, while remaining in $\mathfrak{R}$.

which can be distinguished from each other, and that *one* side has been selected as the positive side, from which **n** is to point.

We will suppose also that S is a so-called *simple surface*, having the property that it does not intersect itself, so that there cannot be more than one vector **n** at a single point.

Further, we suppose here, and in the remainder of the text, that any surface S over which an integration is to be effected can be subdivided, by smooth curves, into a finite number of parts such that on each part the normal **n** exists and also varies continuously with position. Such a surface is said to be *piecewise smooth*. When **n** is continuous over *all* of S, we say that S is a *smooth surface*. In the cases when S is described by an equation of form (102), the preceding requirement is merely that $\partial g/\partial x$, $\partial g/\partial y$, and $\partial g/\partial z$ exist and be continuous over S, or over each of a finite number of parts into which S is divided.

Thus, for the unit sphere $x^2 + y^2 + z^2 = 1$, we can write

$$g(x, y, z) = x^2 + y^2 + z^2 - 1,$$

and hence

$$\nabla g = 2x\mathbf{i} + 2y\mathbf{j} + 2z\mathbf{k}.$$

It follows that at any point (x_0, y_0, z_0) on the sphere we have

$$\mathbf{n} = \pm \frac{x_0\mathbf{i} + y_0\mathbf{j} + z_0\mathbf{k}}{\sqrt{x_0^2 + y_0^2 + z_0^2}} = \pm(x_0\mathbf{i} + y_0\mathbf{j} + z_0\mathbf{k}).$$

The *positive* sign must be chosen if **n** is to point outward from the surface of the sphere.

At a point $P(x, y, z)$ on S we define a *differential surface area vector* $d\boldsymbol{\sigma}$ whose length is numerically equal to the element of surface area $d\sigma$ associated with P, and whose direction coincides with the normal vector **n** at P. If the element $d\boldsymbol{\sigma}$ is constructed by projecting a rectangular element $dA = dx\,dy$ in the xy plane vertically onto the surface S at P, it is seen that the projection of $d\boldsymbol{\sigma}$ on the xy plane is numerically equal to the projection of **n** $d\sigma$ on the vector which is normal to the xy plane. (See Figure 6.11.) Hence we have

$$|\mathbf{n} \cdot \mathbf{k}|\, d\sigma = dx\,dy.$$

If the direction cosines of **n** are denoted by $(\cos \alpha, \cos \beta, \cos \gamma)$, this equation becomes

$$d\sigma = |\sec \gamma|\, dx\,dy \qquad (104)$$

and from Equation (103) we obtain

$$\cos \gamma = \mathbf{n} \cdot \mathbf{k} = \pm \frac{\partial g/\partial z}{\sqrt{(\partial g/\partial x)^2 + (\partial g/\partial y)^2 + (\partial g/\partial z)^2}}. \qquad (105)$$

In particular, if the surface is given in the special form

$$z = f(x, y), \qquad (106)$$

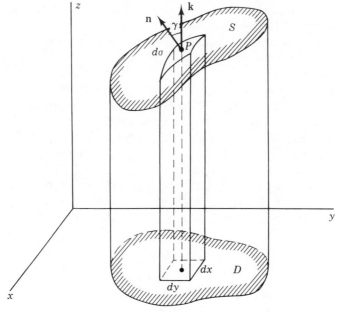

Figure 6.11

there follows $g(x, y, z) = z - f(x, y)$, and Equation (105) becomes

$$\cos \gamma = \pm \frac{1}{\sqrt{1 + (\partial f/\partial x)^2 + (\partial f/\partial y)^2}}. \tag{107}$$

With the definition $d\boldsymbol{\sigma} = \mathbf{n}\, d\sigma$, the integral

$$\iint_S \mathbf{F} \cdot d\boldsymbol{\sigma} = \iint_S \mathbf{F} \cdot \mathbf{n}\, d\sigma, \tag{108}$$

carried out over a surface S, is called the *surface integral* of $\mathbf{F}$ over S.† We notice that the element of integration is the product of the component of $\mathbf{F}$ normal to the surface at a point P and the scalar element of surface area associated with P.

Example. We calculate the surface integral of the vector function

$$\mathbf{F} = x\mathbf{i} + y\mathbf{j}$$

over the portion of the surface of the unit sphere

$$S: \quad x^2 + y^2 + z^2 = 1$$

above the xy plane, $z \geqq 0$. The unit normal at a point of S is given by

$$\mathbf{n} = x\mathbf{i} + y\mathbf{j} + z\mathbf{k}.$$

Also, we obtain

$$\cos \gamma = \mathbf{n} \cdot \mathbf{k} = z,$$

†Similarly defined integrals of the more general form $\iint_S h\, d\sigma$ (where the scalar function h is not necessarily expressed as $\mathbf{F} \cdot \mathbf{n}$) or of the vector form $\iint_S \mathbf{V}\, d\sigma$ also may arise in practice.

and hence Equation (104) gives

$$d\sigma = \frac{dx\,dy}{z}.$$

Thus the required surface integral is of the form

$$\iint_S \mathbf{F} \cdot d\boldsymbol{\sigma} = \iint_D (x^2 + y^2) \frac{dx\,dy}{z},$$

where D denotes the projection of S on the xy plane, and hence D is the interior of the unit circle $x^2 + y^2 = 1$ in the xy plane. In this integral, the variable z must be expressed in terms of x and y by using the equation of S. Hence we obtain the form

$$\iint_S \mathbf{F} \cdot d\boldsymbol{\sigma} = 4 \int_0^1 \int_0^{\sqrt{1-x^2}} \frac{x^2 + y^2}{\sqrt{1 - (x^2 + y^2)}}\,dy\,dx,$$

or, after changing for convenience to polar coordinates and evaluating the result as a repeated integral,

$$\iint_S \mathbf{F} \cdot d\boldsymbol{\sigma} = 4 \int_0^{\pi/2} \int_0^1 \frac{r^2}{\sqrt{1 - r^2}} r\,dr\,d\theta = \frac{4}{3}\pi. \qquad \blacksquare$$

It should be noticed that, if lines parallel to the z axis intersect the surface S in more than one point, the surface must be considered in two or more parts, the formulas resulting from the use of (104) being applied separately to the individual parts. In such cases it may be more convenient to project the element $d\sigma$ onto the xz or yz plane, or to proceed by a more direct method.

6.13. Interpretation of Divergence. The Divergence Theorem. Let $\mathbf{V}$ represent the velocity of flow of a fluid in three dimensions, and denote the mass density of the fluid at a point (x, y, z) by $\rho(x, y, z)$. Then the vector $\mathbf{Q} = \rho\mathbf{V}$ points in the direction of flow and has a magnitude Q numerically equal to the rate of flow of fluid mass through unit area perpendicular to the direction of flow. The differential rate of flow through a directed element of surface area $d\boldsymbol{\sigma} = \mathbf{n}\,d\sigma$ is then given by $\mathbf{Q} \cdot d\boldsymbol{\sigma} = \mathbf{Q} \cdot \mathbf{n}\,d\sigma$, this quantity being positive if the projection of $\mathbf{Q}$ on the vector $\mathbf{n}$ is positive. In particular, if $d\boldsymbol{\sigma}$ is an element of a closed surface, then $\mathbf{Q} \cdot d\boldsymbol{\sigma}$ is positive if the flow is *outward* from the surface. We write

$$\mathbf{Q} = Q_x\mathbf{i} + Q_y\mathbf{j} + Q_z\mathbf{k}. \qquad (109)$$

Now consider a small closed differential element of volume consisting of a rectangular parallelepiped with one vertex at $P(x, y, z)$ and with edges dx, dy, dz parallel to the coordinates axes. Then the left-hand face is represented by the differential surface vector $-\mathbf{j}\,dx\,dz$ and the differential rate of flow through this face is given by

$$\mathbf{Q} \cdot (-\mathbf{j}\,dx\,dz) = -Q_y\,dx\,dz,$$

the negative sign indicating that, if Q_y is positive, the flow through this face is *into* the volume element. Similarly, the differential rate of flow through the right-hand face is given by

$$\mathbf{Q}\Big|_{y+dy} \cdot (\mathbf{j} \, dx \, dz) = Q_y\Big|_{y+dy} dx \, dz = \left(Q_y + \frac{\partial Q_y}{\partial y} \, dy\right) dx \, dz$$

within infinitesimals of higher order. If the remaining four faces are treated similarly, we find that the *net* differential rate of flow dF outward from the volume element $d\tau = dx \, dy \, dz$ is given by

$$dF = \left[\left(Q_x + \frac{\partial Q_x}{\partial x} \, dx\right) - Q_x\right] dy \, dz + \left[\left(Q_y + \frac{\partial Q_y}{\partial y} \, dy\right) - Q_y\right] dx \, dz$$

$$+ \left[\left(Q_z + \frac{\partial Q_z}{\partial z} \, dz\right) - Q_z\right] dx \, dy$$

$$= \left(\frac{\partial Q_x}{\partial x} + \frac{\partial Q_y}{\partial y} + \frac{\partial Q_z}{\partial z}\right) dx \, dy \, dz,$$

or $$dF = (\mathbf{\nabla} \cdot \mathbf{Q}) \, d\tau. \tag{110}$$

Thus we may say that the divergence of $\mathbf{Q}$ at a point P represents the rate of fluid flow, per unit volume, outward from a differential volume associated with P, or that *the divergence of* $\mathbf{Q}$ *is the rate of decrease of mass per unit volume* in the neighborhood of the point. Thus, if no mass is added to or withdrawn from the element $d\tau$, we have the relation

$$\mathbf{\nabla} \cdot \mathbf{Q} = -\frac{\partial \rho}{\partial t}, \tag{111}$$

which can also be written in the form

$$\mathbf{\nabla} \cdot (\rho \mathbf{V}) + \frac{\partial \rho}{\partial t} = 0$$

or, using Equation (74a),

$$\rho \mathbf{\nabla} \cdot \mathbf{V} = -\left(\mathbf{V} \cdot \mathbf{\nabla}\rho + \frac{\partial \rho}{\partial t}\right). \tag{112}$$

For an *incompressible* fluid, $\rho = $ constant and there follows

$$\mathbf{\nabla} \cdot \mathbf{Q} = \rho \mathbf{\nabla} \cdot \mathbf{V} = 0. \tag{113}$$

Thus *the velocity of an incompressible fluid has zero divergence.*

It has been assumed here that no mass is introduced into the system or taken from the system, that is, that there are no points in the element $d\tau$ where fluid is added to or withdrawn from the system. A vector $\mathbf{V}$ with nonzero divergence can be considered as a velocity vector of an incompressible fluid in a region only if such points are assumed to be present. We speak of points where fluid is added to or taken from the system as *sources* and *sinks*, respectively. Thus the presence of a *source* distribution is associated with a *positive* divergence of $\mathbf{V}$.

Consider now a closed region $\Re$ in three dimensions and suppose that in each elementary volume $d\tau$ of this region there are present sources through which an amount $dF = \rho\mathbf{V} \cdot \mathbf{V}\, d\tau$ of incompressible fluid is introduced into the region per unit time. A sink is considered to be a negative source. Then the net mass of fluid introduced into $\mathfrak{R}$ in unit time through sources and sinks is given by the integral

$$\rho \iiint_{\mathfrak{R}} \mathbf{V} \cdot \mathbf{V}\, d\tau, \tag{114}$$

where the integration is carried out over the volume of the region. If the total mass in $\mathfrak{R}$ is conserved, this fluid clearly must escape from the region through the surface S which bounds it. If we consider a vector surface element $d\boldsymbol{\sigma} = \mathbf{n}\, d\sigma$, where $\mathbf{n}$ is the *outward* unit normal, the rate of mass flow outward through the element $d\sigma$ is given by $\rho\mathbf{V} \cdot \mathbf{n}\, d\sigma = \rho\mathbf{V} \cdot d\boldsymbol{\sigma}$. Thus the total rate of flow outward through S is given by the surface integral of $\mathbf{V}$ over S,

$$\rho \oiint_{S} \mathbf{V} \cdot d\boldsymbol{\sigma} = \rho \oiint_{S} \mathbf{V} \cdot \mathbf{n}\, d\sigma, \tag{115}$$

where the symbol $\oiint$ indicates that the integration is carried out over a closed surface.

The statement that (114) and (115) must be equivalent,

$$\iiint_{\mathfrak{R}} \mathbf{V} \cdot \mathbf{V}\, d\tau = \oiint_{S} \mathbf{V} \cdot \mathbf{n}\, d\sigma, \tag{116}$$

that is, the consequence of conservation of mass in $\mathfrak{R}$, is known as the *divergence theorem* (or as *Gauss's theorem*). This theorem can be established without reference to the physical considerations used here,† and *is true if* $\mathbf{V}$ *and its partial derivatives are continuous in* $\mathfrak{R}$ *and on* S *and if* S *is simple and piecewise smooth.* The region $\mathfrak{R}$ need not be simply connected, so long as its *complete* boundary is taken into account in the surface integral.

If we write

$$\mathbf{V} = P\mathbf{i} + Q\mathbf{j} + R\mathbf{k}, \qquad \mathbf{n} = \mathbf{i}\cos\alpha + \mathbf{j}\cos\beta + \mathbf{k}\cos\gamma,$$

Equation (116) takes the form

$$\iiint_{\mathfrak{R}} \left(\frac{\partial P}{\partial x} + \frac{\partial Q}{\partial y} + \frac{\partial R}{\partial z} \right) dx\, dy\, dz$$
$$= \oiint_{S} (P\cos\alpha + Q\cos\beta + R\cos\gamma)\, d\sigma. \tag{117}$$

This theorem is useful in many applications. In particular, it serves the purpose of avoiding the direct evaluation of a surface integral of a vector over a closed surface S, by replacing that integral by a volume integral over the interior $\mathfrak{R}$, if the divergence of the vector is of a convenient form in $\mathfrak{R}$. When the original integration is over an *open* surface S_1, one may first introduce an additional surface S_2 which, together with S_1, forms a *closed* surface S bounding a convenient region $\mathfrak{R}$. The surface integral over S_1 then can be evaluated as a

---

†For an indication of the analytical proof, see Problem 76.

volume integral over $\mathfrak{R}$ reduced by the surface integral over S_2, if the latter computation is preferable to the former.

Example. In illustration, we consider the example treated in Section 6.12. There we calculated directly the surface integral

$$\iint_S \mathbf{F} \cdot d\boldsymbol{\sigma},$$

where
$$\mathbf{F} = x\mathbf{i} + y\mathbf{j}$$

and S is the upper half of the unit sphere $x^2 + y^2 + z^2 = 1$. If we close the surface of integration by adding the portion of the xy plane which spans the hemisphere, we notice that the surface integral of $\mathbf{F}$ over the added surface is *zero*, since

$$\mathbf{F} \cdot \mathbf{n} = \mathbf{F} \cdot (-\mathbf{k}) = 0$$

over this area. Thus the divergence theorem states that we may calculate the required surface integral of $\mathbf{F}$ by evaluating

$$\iiint_\mathfrak{R} \mathbf{V} \cdot \mathbf{F} \, d\tau$$

throughout the interior of the hemisphere. Since $\mathbf{V} \cdot \mathbf{F} = 2$, the result is merely twice the volume of the unit hemisphere, or $4\pi/3$, as was obtained by direct integration. ∎

The surface integral of a vector $\mathbf{F}$ over a closed surface S is sometimes called the *flux* of $\mathbf{F}$ through S. Thus the divergence theorem states that the flux of $\mathbf{F}$ through S is a measure of the divergence of $\mathbf{F}$ inside S. In particular, *if $\mathbf{F}$ is continuously differentiable and $\mathbf{V} \cdot \mathbf{F} = 0$ in the region bounded by S, the flux of $\mathbf{F}$ through S is zero,*

$$\oiint_S \mathbf{F} \cdot d\boldsymbol{\sigma} = \oiint_S \mathbf{F} \cdot \mathbf{n} \, d\sigma = 0.$$

Thus, remembering the result of Section 6.10, we see that, for a continuously differentiable function $\mathbf{F}$, *the circulation $\oint \mathbf{F} \cdot d\mathbf{r}$ of $\mathbf{F}$ around a closed curve C is zero if C is in a simple region throughout which $\mathbf{V} \times \mathbf{F} = 0$, and also the flux $\oiint \mathbf{F} \cdot d\boldsymbol{\sigma}$ of $\mathbf{F}$ through a closed surface S is zero if $\mathbf{V} \cdot \mathbf{F} = 0$ in the region bounded by S.*

We may expect that the circulation of $\mathbf{F}$ around C will be in some sense a measure of the curl of $\mathbf{F}$ at points on a surface of which C is a boundary, in analogy with the relation (116) which connects flux through a surface to divergence inside the surface. A relationship of this sort, known as *Stokes's theorem*, is considered in Section 6.16.

The two-dimensional form of the divergence theorem is

$$\iint_D \mathbf{V} \cdot \mathbf{V} \, dA = \oint_C \mathbf{V} \cdot \mathbf{n} \, ds, \tag{118}$$

where D is a region in the xy plane, C is its complete boundary, and $\mathbf{n}$ is the unit *outward* normal along C (see Problem 78). In particular, for a simply connected region D we may write

$$\mathbf{n} = \frac{dy}{ds}\mathbf{i} - \frac{dx}{ds}\mathbf{j} \tag{119}$$

along C and, with

$$\mathbf{V} = P\mathbf{i} + Q\mathbf{j}, \tag{120}$$

(118) then takes the form

$$\int\int_D \left(\frac{\partial P}{\partial x} + \frac{\partial Q}{\partial y}\right) dx\, dy = \oint_C (P\, dy - Q\, dx). \tag{121}$$

6.14. Green's Theorem. If, in the divergence theorem (116), we write

$$\mathbf{V} = \varphi_1 \nabla \varphi_2,$$

where φ_1 and φ_2 are scalar functions of position, we obtain the result

$$\int\int\int_{\mathfrak{R}} \mathbf{V} \cdot \varphi_1 \nabla \varphi_2 \, d\tau = \oiint_S \mathbf{n} \cdot \varphi_1 \nabla \varphi_2 \, d\sigma.$$

By making use of (74a), this result takes the form

$$\int\int\int_{\mathfrak{R}} [\varphi_1 \nabla^2 \varphi_2 + (\nabla \varphi_1) \cdot (\nabla \varphi_2)] \, d\tau = \oiint_S \mathbf{n} \cdot \varphi_1 \nabla \varphi_2 \, d\sigma. \tag{122}$$

This equation is known as the *first form of Green's theorem*. A more symmetrical form is obtained if φ_1 and φ_2 are interchanged in Equation (122) and the resultant equation is subtracted from (122) to give

$$\int\int\int_{\mathfrak{R}} (\varphi_1 \nabla^2 \varphi_2 - \varphi_2 \nabla^2 \varphi_1) \, d\tau = \oiint_S \mathbf{n} \cdot (\varphi_1 \nabla \varphi_2 - \varphi_2 \nabla \varphi_1) \, d\sigma. \tag{123}$$

This equation is known as the *second form of Green's theorem*, or frequently merely as *Green's theorem*, and is frequently useful in applications. In both (122) and (123), the region $\mathfrak{R}$ is the region bounded by the closed surface S. It is assumed that φ_1 and φ_2 are twice continuously differentiable in $\mathfrak{R}$.

Two special cases of these theorems are of particular interest. If we take $\varphi_1 = \varphi_2 = \varphi$ in (122), there follows

$$\int\int\int_{\mathfrak{R}} [\varphi \nabla^2 \varphi + (\nabla \varphi)^2] \, d\tau = \oiint_S \varphi \mathbf{n} \cdot \nabla \varphi \, d\sigma, \tag{124}$$

where $(\nabla \varphi)^2 = (\nabla \varphi) \cdot (\nabla \varphi)$. We notice that the product $\mathbf{n} \cdot \nabla \varphi$ is, according to (64), the derivative of φ in the direction of $\mathbf{n}$, that is, in the direction of the outward normal to S at a point on S. Hence we may write

$$\frac{\partial \varphi}{\partial n} = \mathbf{n} \cdot \nabla \varphi \tag{125}$$

and speak of this quantity as the *normal derivative* of φ at a point on S. With this notation, Equation (124) becomes

$$\int\int\int_{\mathfrak{R}} [\varphi \nabla^2 \varphi + (\nabla \varphi)^2] \, d\tau = \oiint_S \varphi \frac{\partial \varphi}{\partial n} \, d\sigma, \tag{126}$$

and Equations (122) and (123) can be rewritten in a similar way.

A second important special case of Green's theorem is obtained by taking $\varphi_1 = \varphi$ and $\varphi_2 = 1$ in (123). Since then $\nabla\varphi_2 = 0$ and $\nabla^2\varphi_2 = 0$, there follows, with the notation of (125),

$$\iiint_R \nabla^2\varphi \, d\tau = \oiint_S \frac{\partial\varphi}{\partial n} \, d\sigma. \tag{127}$$

The relations (126) and (127) will be useful in certain applications which follow (Section 9.2). Their two-dimensional specializations are considered in Problem 80.

6.15. Interpretation of Curl. Laplace's Equation. Suppose that the motion of a fluid is simply a *rotation* about a given axis fixed in space. Then we may represent the angular velocity by a constant vector $\boldsymbol{\omega}$, as in Section 6.3, where the length ω is the scalar angular velocity and the direction of $\boldsymbol{\omega}$ is along the direction of the axis of rotation in accordance with the right-hand rule. If we take the origin of a rectangular coordinate system on this axis and denote the position vector to a point $P(x, y, z)$ by $\mathbf{r}$, then the results of Section 6.3 show that the velocity of P is given by the vector

$$\mathbf{V} = \boldsymbol{\omega} \times \mathbf{r}. \tag{128}$$

We have then, using Equation (74c),

$$\nabla \cdot \mathbf{V} = \nabla \cdot \boldsymbol{\omega} \times \mathbf{r} = \mathbf{r} \cdot (\nabla \times \boldsymbol{\omega}) - \boldsymbol{\omega} \cdot (\nabla \times \mathbf{r}),$$

and hence, since $\boldsymbol{\omega}$ is constant and $\nabla \times \mathbf{r} = \mathbf{0}$, from (78b), there follows

$$\nabla \cdot \mathbf{V} = 0. \tag{129}$$

We also have, using Equation (74d) and again noticing that $\boldsymbol{\omega}$ is constant,

$$\nabla \times \mathbf{V} = \nabla \times (\boldsymbol{\omega} \times \mathbf{r}) = \boldsymbol{\omega}(\nabla \cdot \mathbf{r}) - (\boldsymbol{\omega} \cdot \nabla)\mathbf{r},$$

or, by virtue of (78a) and (79),

$$\nabla \times \mathbf{V} = 2\boldsymbol{\omega}. \tag{130}$$

Thus, *if a fluid experiences a pure rotation, the divergence of its velocity vector is zero, and the curl of the velocity vector is a constant vector equal to twice the angular velocity vector.*

More generally if $\mathbf{V}$ is the velocity vector of a fluid flow, and if $\nabla \times \mathbf{V} = 2\boldsymbol{\omega}_0 \neq \mathbf{0}$ at a point P_0, then a component of $\mathbf{V}$ corresponds to a local tendency of the fluid to rotate about the vector $\boldsymbol{\omega}_0$, with angular velocity $|\boldsymbol{\omega}_0|$.

If $\nabla \times \mathbf{V} = \mathbf{0}$ in a region, we say that the flow is *irrotational* in that region. The results of Section 6.10 state that the circulation around a closed curve in a simple region where the flow is irrotational is zero. If the fluid is *incompressible* and there is no distribution of sources or sinks in the region, we have also $\nabla \cdot \mathbf{V} = 0$. Since the condition $\nabla \times \mathbf{V} = \mathbf{0}$ implies the existence of a potential φ such that

$$\mathbf{V} = \nabla\varphi, \tag{131}$$

we see that if also $\mathbf{V} \cdot \mathbf{V} = 0$ there follows $\mathbf{V} \cdot \mathbf{V}\varphi = \mathbf{V}^2\varphi = 0$. That is, *in the flow of an incompressible irrotational fluid without distributed sources or sinks, the velocity vector is the gradient of a potential φ which satisfies the equation*

$$\mathbf{V}^2\varphi = 0 \quad \text{or} \quad \frac{\partial^2 \varphi}{\partial x^2} + \frac{\partial^2 \varphi}{\partial y^2} + \frac{\partial^2 \varphi}{\partial z^2} = 0. \tag{132}$$

This equation is known as *Laplace's equation.*

We see, more generally, that, in *any* continuously differentiable vector field $\mathbf{F}$ with zero divergence and curl in a simple region, the vector $\mathbf{F}$ is the gradient of a solution of Laplace's equation. Solutions of this equation are called *harmonic functions*, and their determination is studied in Chapter 9.

6.16. Stokes's Theorem. We next indicate that if S is a two-sided surface in three dimensions having a closed curve C as its boundary, then the circulation of a respectable vector $\mathbf{V}$ around C is equal to the flux of the curl of $\mathbf{V}$ over S. It is assumed throughout that S and C are simple and piecewise smooth.

The relation to be considered is thus of the form

$$\iint_S \mathbf{n} \cdot (\mathbf{V} \times \mathbf{V}) \, d\sigma = \oint_C \mathbf{V} \cdot d\mathbf{r}, \tag{133}$$

where $\mathbf{n}$ is the unit vector normal to S on that side of S which is arbitrarily taken as the *positive side*. The positive direction along C is then defined as the direction along which an observer, traveling on the positive side of S, would proceed in keeping the enclosed area to his left. Equation (133) is known as *Stokes's theorem*, and is true if S is contained in a simple region where $\mathbf{V}$ is continuously differentiable. In the following considerations, however, we assume also the existence of continuous partial derivatives of the *second* order.

Suppose that Equation (133) has been established for *one* surface S_1 having C as its boundary, and consider any other surface S_2 having the same boundary. If we denote the closed region included between S_1 and S_2 by $\Re$, the divergence theorem states that

$$\iint_{S_1} \mathbf{n}_1 \cdot (\mathbf{V} \times \mathbf{V}) \, d\sigma + \iint_{S_2} \mathbf{n}_2 \cdot (\mathbf{V} \times \mathbf{V}) \, d\sigma = \iiint_\Re \mathbf{V} \cdot \mathbf{V} \times \mathbf{V} \, d\tau = 0, \tag{134}$$

if $\mathbf{V}$ is twice continuously differentiable in $\Re$, according to Equation (74g). In this equation the vectors $\mathbf{n}_1$ and $\mathbf{n}_2$ are headed *outward* from the enclosed region $\Re$. Hence if we choose a positive direction around the common boundary C so that $\mathbf{n}_1$ is on the positive side of S_1, and hence $\mathbf{n}_1 = \mathbf{n}$ on S_1, it follows that $\mathbf{n}_2$ is then on the negative side of S_2, and hence $\mathbf{n}_2 = -\mathbf{n}$ on S_2. Thus (134) becomes

$$\iint_{S_1} \mathbf{n} \cdot (\mathbf{V} \times \mathbf{V}) \, d\sigma = \iint_{S_2} \mathbf{n} \cdot (\mathbf{V} \times \mathbf{V}) \, d\sigma. \tag{135}$$

Accordingly, if Equation (133) is true for one surface having C as its boundary, it is also true for any other such surface, if both surfaces lie in a simple region where $\mathbf{V}$ is twice continuously differentiable.

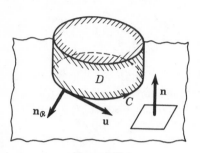

We prove (133) in the special case when C is a plane curve by taking S as the plane region D bounded by C in its plane. The vector $\mathbf{n}$ normal to D is then a constant vector. The proof is obtained most readily by an indirect method in which we apply the divergence theorem to a vector $\mathbf{V} \times \mathbf{n}$, which is thus parallel to the plane of D, in a closed three-dimensional region $\mathfrak{R}$ which is the interior of a right cylinder of constant height h having D as its lower base (Figure 6.12). We then have

Figure 6.12

$$\iiint_{\mathfrak{R}} \mathbf{V} \cdot (\mathbf{V} \times \mathbf{n}) \, d\tau = \iint_{S_{\mathfrak{R}}} \mathbf{n}_{\mathfrak{R}} \cdot (\mathbf{V} \times \mathbf{n}) \, d\sigma_{\mathfrak{R}}, \qquad (136)$$

where the surface integral is extended over the complete boundary $S_{\mathfrak{R}}$ of $\mathfrak{R}$, and $\mathbf{n}_{\mathfrak{R}}$ is the outward unit normal vector on $S_{\mathfrak{R}}$. On the upper and lower faces of $S_{\mathfrak{R}}$ we have $\mathbf{n}_{\mathfrak{R}} = \pm\mathbf{n}$, and hence the integrand in the surface integral vanishes. On the lateral boundary of $S_{\mathfrak{R}}$ we have $\mathbf{n}_{\mathfrak{R}} = \mathbf{u} \times \mathbf{n}$, where $\mathbf{u}$ is the unit tangent vector to the curve C which bounds D, and hence here the integrand becomes

$$(\mathbf{u} \times \mathbf{n}) \cdot (\mathbf{V} \times \mathbf{n}) = (\mathbf{V} \cdot \mathbf{u})(\mathbf{n} \cdot \mathbf{n}) - (\mathbf{u} \cdot \mathbf{n})(\mathbf{n} \cdot \mathbf{V}) = \mathbf{V} \cdot \mathbf{u},$$

by virtue of (35) and the fact that $\mathbf{u} \cdot \mathbf{n} = 0$. Also, since $\mathbf{n}$ is constant, we have $\mathbf{V} \cdot (\mathbf{V} \times \mathbf{n}) = \mathbf{n} \cdot \mathbf{V} \times \mathbf{V}$, from (74c), and hence Equation (136) becomes

$$\int_0^h \left(\iint_D \mathbf{n} \cdot \mathbf{V} \times \mathbf{V} \, d\sigma \right) dh = \int_0^h \left(\oint_C \mathbf{V} \cdot \mathbf{u} \, ds \right) dh.$$

Since this relationship is true for arbitrary values of h, we conclude that

$$\iint_D \mathbf{n} \cdot \mathbf{V} \times \mathbf{V} \, d\sigma = \oint_C \mathbf{V} \cdot d\mathbf{r}, \qquad (137)$$

as was to be shown.

To extend this result to the case of an arbitrary closed curve C in space, we first approximate C by a space polygon C_n whose n sides $\overline{P_1 P_2}, \overline{P_2 P_3}, \ldots, \overline{P_n P_1}$ are formed by joining n successive points $P_1, P_2, \ldots, P_n$ on C by straight lines. A surface S_n having C_n as its boundary can be defined as the polyhedron whose triangular faces are determined by the n sides of C_n and the $n - 3$ straight lines joining P_1 to the points $P_3, \ldots, P_{n-1}$. If we apply Equation (137) to each of the triangular faces of S_n and add the results, we notice that the line integrals along the lines $\overline{P_1 P_3}, \overline{P_1 P_4}, \ldots, \overline{P_1 P_{n-1}}$ are taken twice in opposite directions and hence cancel, leaving only the line integral around the polygon C_n. Thus (137) is also true for S_n and its boundary C_n. As n becomes infinite in such a way that all the chords $P_k P_{k+1}$ approach zero, the polygon C_n approaches C, and S_n approaches a surface S for which the relation (133) is true.

From (135) we conclude finally that Stokes's theorem is valid for an arbitrary piecewise smooth surface S with C as its boundary, if S is in a simple region where $\mathbf{V}$ is twice continuously differentiable. If we write

$$\mathbf{V} = P\mathbf{i} + Q\mathbf{j} + R\mathbf{k}, \qquad \mathbf{n} = \mathbf{i}\cos\alpha + \mathbf{j}\cos\beta + \mathbf{k}\cos\gamma,$$

this theorem takes the form

$$\oint_C (P\,dx + Q\,dy + R\,dz) = \int\!\!\int_S \left[\left(\frac{\partial R}{\partial y} - \frac{\partial Q}{\partial z}\right)\cos\alpha \right.$$

$$\left. + \left(\frac{\partial P}{\partial z} - \frac{\partial R}{\partial x}\right)\cos\beta + \left(\frac{\partial Q}{\partial x} - \frac{\partial P}{\partial y}\right)\cos\gamma \right] d\sigma. \qquad (138)$$

In addition to serving as an important theoretical tool, Stokes's theorem can be used to avoid the calculation of a surface integral by substituting an equivalent line integral. However, it is perhaps more frequently employed in the opposite direction, in cases when the line integral is not easily evaluated but the quantity $\mathbf{n} \cdot \mathbf{V} \times \mathbf{V}$ is of relatively simple form on some open surface which spans C.

Example 1. In order to evaluate the integral

$$I = \oint_C [(e^{-x^2} - yz)\,dx + (e^{-y^2} + xz + 2x)\,dy + e^{-z^2}\,dz]$$

when C is the circle

$$x = \cos\theta, \qquad y = \sin\theta, \qquad z = 2,$$

oriented in the direction of increasing θ, we may note that

$$\mathbf{V} \times \mathbf{V} = -x\mathbf{i} - y\mathbf{j} + (2 + 2z)\mathbf{k}.$$

Hence, if we take S to be the circular disk† bounded by C in the plane $z = 2$, so that $\mathbf{n} = \mathbf{k}$, there follows $\mathbf{n} \cdot \mathbf{V} \times \mathbf{V} = 6$ on S and accordingly

$$I = 6\int\!\!\int_S d\sigma = 6\pi. \qquad \blacksquare$$

For a *plane region* D in the xy plane, there follows $dz = 0$, $\cos\alpha = \cos\beta = 0$, $\cos\gamma = 1$, $d\sigma = dx\,dy$, and hence Stokes's theorem reduces to the two-dimensional form

$$\oint_C (P\,dx + Q\,dy) = \int\!\!\int_D \left(\frac{\partial Q}{\partial x} - \frac{\partial P}{\partial y}\right) dx\,dy. \qquad (139)$$

Equation (139) is also frequently referred to as *Green's theorem*, although we here reserve this name for the theorem of Section 6.14. It may be noticed that in fact the two-dimensional form (139) of Stokes's theorem is equivalent to the two-dimensional form (121) of the *divergence theorem*, since (139) becomes (121) with a notational change in which P is replaced by $-Q$ and Q by P in (139).

---

†A *circular disk* is a plane region bounded by a circle.

As a special case of (139) we take $P = -y$ and $Q = x$. The result,

$$\tfrac{1}{2} \oint_C (x \, dy - y \, dx) = A, \tag{140}$$

is useful in finding the area enclosed by a simple plane curve C whose equation is given in parametric form.

Example 2. For the ellipse $(x^2/a^2) + (y^2/b^2) = 1$, we can write

$$x = a \cos \theta, \qquad y = b \sin \theta, \qquad (0 \leq \theta \leq 2\pi),$$

and Equation (140) gives for its area

$$A = \tfrac{1}{2} \int_0^{2\pi} ab \, (\cos^2 \theta + \sin^2 \theta) \, d\theta = \pi ab. \qquad \blacksquare$$

A useful interpretation of the curl vector, supplementing that suggested in Section 6.15, can be obtained from Stokes's theorem (133) by taking C to be a circle with center at a point P and S to be the circular disk bounded by C, and by then letting the radius of C tend to zero. Since at each stage of the limiting process the left-hand member of (133) is equal to the product of the area of the disk and the value of $\mathbf{n} \cdot \nabla \times \mathbf{V}$ at some interior point, we may deduce in the limit that *the component of* $\nabla \times \mathbf{V}$ *in the direction of a vector* $\mathbf{n}$, *at a point* P, *is the circulation per unit area of* $\mathbf{V}$ *around* P, *in a plane normal to* $\mathbf{n}$.

The rigorous proof of Stokes's theorem (133) in full generality involves subtleties which are beyond the level of the present treatments. However, it is of some importance here to notice that from that theorem one may deduce, in particular, that *if* $\mathbf{V}$ *is continuously differentiable in a simply connected region* $\mathfrak{R}$, *and if* $\nabla \times \mathbf{V} = \mathbf{0}$ *everywhere in* $\mathfrak{R}$, *then*

$$\oint_C \mathbf{V} \cdot d\mathbf{r} = 0$$

for any piecewise smooth closed curve C *in* $\mathfrak{R}$. This is a result made plausible in Section 6.10 [Equation (91)].

From that result, in turn, one can reason in a direction opposite to that followed in Section 6.10 to deduce that, *under the conditions just stated, the line integral* $\int_C \mathbf{V} \cdot d\mathbf{r}$ *between two points in* $\mathfrak{R}$ *is independent of the path* C *joining those points, provided only that* C *is piecewise smooth and lies in* $\mathfrak{R}$, *and it then follows also that* $\mathbf{V} \cdot d\mathbf{r} = d\varphi$, *where* φ *is single-valued in* $\mathfrak{R}$.

The specialization of these results to the case when $\mathfrak{R}$ is a two-dimensional region, in the xy plane, and in which Stokes's theorem consequently reduces to (139), will be of particular significance in Section 10.5.

6.17. Orthogonal Curvilinear Coordinates. Suppose that the rectangular coordinates x, y, z are expressed in terms of new coordinates u_1, u_2, u_3 by the equations

$$x = x(u_1, u_2, u_3), \qquad y = y(u_1, u_2, u_3), \qquad z = z(u_1, u_2, u_3), \tag{141}$$

and that, conversely, these relations can be inverted to express u_1, u_2, u_3 in terms of x, y, z, when x, y, z and/or u_1, u_2, u_3 are suitably restricted. Then, at least in some region, any point with coordinates (x, y, z) has corresponding coordinates (u_1, u_2, u_3). We assume that the correspondence is unique. If a particle moves from a point P in such a way that u_2 and u_3 are held constant and only u_1 varies, a curve in space is generated. We speak of this curve as the u_1 curve. Similarly, two other coordinate curves, the u_2 and u_3 curves, are determined at each point (Figure 6.13). Further, if only one coordinate is held constant, we determine successively three surfaces passing through a point of space, these surfaces intersecting in the coordinate curves. It is often convenient to choose the new coordinates in such a way that the coordinate curves are mutually perpendicular at each point in space. Such coordinates are called *orthogonal curvilinear coordinates*.

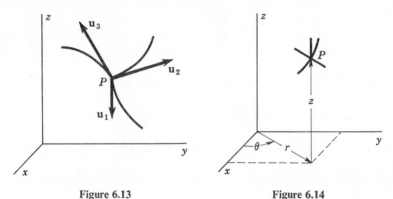

Figure 6.13   Figure 6.14

As an example (see Figure 6.14), in circular cylindrical coordinates (r, θ, z) we have

$$x = r \cos \theta, \qquad y = r \sin \theta, \qquad z = z,$$

where $r \geqq 0$ and $0 \leqq \theta < 2\pi$. If we take

$$u_1 = r, \qquad u_2 = \theta, \qquad u_3 = z,$$

then at any point P the u_2 curve is a circle of radius r and the u_1 and u_3 curves are straight lines.

Let $\mathbf{r}$ represent the position vector of a point P in space,

$$\mathbf{r} = x\mathbf{i} + y\mathbf{j} + z\mathbf{k}. \tag{142}$$

Then a tangent vector to the u_1 curve at P is given by

$$\mathbf{U}_1 = \frac{\partial \mathbf{r}}{\partial u_1} = \frac{\partial \mathbf{r}}{\partial s_1} \frac{ds_1}{du_1}, \tag{143}$$

where s_1 is arc length along the u_1 curve. Since $\partial \mathbf{r}/\partial s_1$ is a *unit vector*, we can write

$$\mathbf{U}_1 = h_1 \mathbf{u}_1, \tag{144}$$

where $\mathbf{u}_1$ is the unit vector tangent to the u_1 curve in the direction of increasing arc length and $h_1 = ds_1/du_1$ is the length of $\mathbf{U}_1$. Considering the other coordinate curves similarly, we thus write

$$\mathbf{U}_1 = h_1\mathbf{u}_1, \qquad \mathbf{U}_2 = h_2\mathbf{u}_2, \qquad \mathbf{U}_3 = h_3\mathbf{u}_3, \tag{145}$$

where $\mathbf{u}_k\ (k = 1, 2, 3)$ is the unit vector tangent to the u_k curve, and

$$h_1 = \frac{ds_1}{du_1} = \left|\frac{\partial\mathbf{r}}{\partial u_1}\right|, \qquad h_2 = \frac{ds_2}{du_2} = \left|\frac{\partial\mathbf{r}}{\partial u_2}\right|, \qquad h_3 = \frac{ds_3}{du_3} = \left|\frac{\partial\mathbf{r}}{\partial u_3}\right|. \tag{146}$$

These equations can be written in the differential form

$$ds_1 = h_1\,du_1, \qquad ds_2 = h_2\,du_2, \qquad ds_3 = h_3\,du_3. \tag{147}$$

We thus see that h_1, h_2, h_3 are of the nature of *scale factors*, giving the ratios of differential distances to the differentials of the coordinate parameters. It should be noticed that the calculation of $\mathbf{U}_1$, from Equation (143), determines both the scale factor $h_1 = |\mathbf{U}_1|$ and the unit vector $\mathbf{u}_1 = \mathbf{U}_1/h_1$.

We adopt the convention that $\mathbf{u}_1, \mathbf{u}_2$, and $\mathbf{u}_3$ form a right-handed system in the order written, in consequence of an appropriate numbering of the coordinates.

If s is arc length along a curve in *any* direction, the vector

$$\frac{d\mathbf{r}}{ds} = \frac{\partial\mathbf{r}}{\partial u_1}\frac{du_1}{ds} + \frac{\partial\mathbf{r}}{\partial u_2}\frac{du_2}{ds} + \frac{\partial\mathbf{r}}{\partial u_3}\frac{du_3}{ds}$$
$$= \mathbf{U}_1\frac{du_1}{ds} + \mathbf{U}_2\frac{du_2}{ds} + \mathbf{U}_3\frac{du_3}{ds} \tag{148}$$

has unit length. Thus, *if the coordinate curves are orthogonal*, so that

$$\mathbf{U}_1 \cdot \mathbf{U}_2 = \mathbf{U}_2 \cdot \mathbf{U}_3 = \mathbf{U}_3 \cdot \mathbf{U}_1 = 0, \tag{149}$$

there follows

$$\mathbf{U}_1 \cdot \mathbf{U}_1\,du_1^2 + \mathbf{U}_2 \cdot \mathbf{U}_2\,du_2^2 + \mathbf{U}_3 \cdot \mathbf{U}_3\,du_3^2 = ds^2$$

or, using Equation (145),

$$ds^2 = h_1^2\,du_1^2 + h_2^2\,du_2^2 + h_3^2\,du_3^2. \tag{150}$$

To find the element of volume, we notice that the vectors $\mathbf{U}_1\,du_1, \mathbf{U}_2\,du_2$, and $\mathbf{U}_3\,du_3$ are mutually perpendicular vectors having as their lengths the arc-length differentials ds_1, ds_2, and ds_3. Thus the element of volume $d\tau$ is given by the volume of the rectangular parallelepiped determined by these vectors,

$$d\tau = \mathbf{U}_1\,du_1 \times \mathbf{U}_2\,du_2 \cdot \mathbf{U}_3\,du_3 = (\mathbf{u}_1 \times \mathbf{u}_2 \cdot \mathbf{u}_3)h_1h_2h_3\,du_1\,du_2\,du_3,$$

and hence

$$d\tau = h_1h_2h_3\,du_1\,du_2\,du_3. \tag{151}$$

The fact that $\mathbf{U}_1 \times \mathbf{U}_2 \cdot \mathbf{U}_3$ is *positive* and that $\mathbf{u}_1 \times \mathbf{u}_2 \cdot \mathbf{u}_3 = +1$ is a consequence of the assumed right-handedness.

On the surface $u_1 = $ constant, the vector element of surface area $d\boldsymbol{\sigma}_1$ is, in a similar way, given by the vector product $\mathbf{U}_2\,du_2 \times \mathbf{U}_3\,du_3$, and hence

$$d\boldsymbol{\sigma}_1 = \mathbf{u}_1 h_2 h_3\,du_2\,du_3. \tag{152}$$

Analogous expressions are obtained on the other coordinate surfaces. In particular, the *scalar* surface elements *on the coordinate surfaces* are of the form

$$d\sigma_1 = h_2 h_3 \, du_2 \, du_3, \qquad d\sigma_2 = h_3 h_1 \, du_3 \, du_1, \qquad d\sigma_3 = h_1 h_2 \, du_1 du_2. \tag{153}$$

We next proceed to the determination of expressions for the gradient, divergence, and curl, and for the Laplacian operator $\mathbf{V}^2$ in the present coordinates. To determine the gradient of a scalar function f, we make use of the relation (64),

$$df = \mathbf{V}f \cdot d\mathbf{r} \tag{154}$$

which specifies the basic geometrical property of the gradient. In terms of the coordinates u_1, u_2, u_3 we have

$$df = \frac{\partial f}{\partial u_1} \, du_1 + \frac{\partial f}{\partial u_2} \, du_2 + \frac{\partial f}{\partial u_3} \, du_3$$

and

$$d\mathbf{r} = \mathbf{U}_1 \, du_1 + \mathbf{U}_2 \, du_2 + \mathbf{U}_3 \, du_3$$
$$= h_1 \mathbf{u}_1 \, du_1 + h_2 \mathbf{u}_2 \, du_2 + h_3 \mathbf{u}_3 \, du_3.$$

If we now write

$$\mathbf{V}f = \lambda_1 \mathbf{u}_1 + \lambda_2 \mathbf{u}_2 + \lambda_3 \mathbf{u}_3, \tag{155}$$

where the λ's are to be determined, the substitution of these expressions into (154) gives

$$\frac{\partial f}{\partial u_1} \, du_1 + \frac{\partial f}{\partial u_2} \, du_2 + \frac{\partial f}{\partial u_3} \, du_3 = h_1 \lambda_1 \, du_1 + h_2 \lambda_2 \, du_2 + h_3 \lambda_3 \, du_3,$$

and hence, since the variables u_1, u_2, u_3 are independent, we have

$$\lambda_k = \frac{1}{h_k} \frac{\partial f}{\partial u_k} \qquad (k = 1, 2, 3). \tag{156}$$

Thus Equation (155) becomes

$$\mathbf{V}f = \frac{\mathbf{u}_1}{h_1} \frac{\partial f}{\partial u_1} + \frac{\mathbf{u}_2}{h_2} \frac{\partial f}{\partial u_2} + \frac{\mathbf{u}_3}{h_3} \frac{\partial f}{\partial u_3} \tag{157}$$

and the operator $\mathbf{V}$ accordingly has the form

$$\mathbf{V} = \frac{\mathbf{u}_1}{h_1} \frac{\partial}{\partial u_1} + \frac{\mathbf{u}_2}{h_2} \frac{\partial}{\partial u_2} + \frac{\mathbf{u}_3}{h_3} \frac{\partial}{\partial u_3}. \tag{158}$$

Equations (157) and (158) can be written in the equivalent forms

$$\mathbf{V}f = \mathbf{u}_1 \frac{\partial f}{\partial s_1} + \mathbf{u}_2 \frac{\partial f}{\partial s_2} + \mathbf{u}_3 \frac{\partial f}{\partial s_3}, \tag{159}$$

$$\mathbf{V} = \mathbf{u}_1 \frac{\partial}{\partial s_1} + \mathbf{u}_2 \frac{\partial}{\partial s_2} + \mathbf{u}_3 \frac{\partial}{\partial s_3}, \tag{160}$$

which display the intrinsic nature of the operator $\mathbf{V}$ and which, indeed, could have been anticipated from the results of Section 6.7.

In particular, there follows

$$\nabla u_1 = \frac{\mathbf{u}_1}{h_1}, \qquad \nabla u_2 = \frac{\mathbf{u}_2}{h_2}, \qquad \nabla u_3 = \frac{\mathbf{u}_3}{h_3}, \tag{161}$$

for the gradients of the independent variables u_1, u_2, u_3.

From these special results we now obtain two further results which will be useful in the following work. First, since $\nabla \times \nabla u_k = 0$, from (74f), we conclude that *the expressions* $\mathbf{u}_1/h_1$, $\mathbf{u}_2/h_2$, and $\mathbf{u}_3/h_3$ *have zero curl*,

$$\nabla \times \frac{\mathbf{u}_1}{h_1} = \nabla \times \frac{\mathbf{u}_2}{h_2} = \nabla \times \frac{\mathbf{u}_3}{h_3} = 0. \tag{162}$$

Second, since $\mathbf{u}_1 = \mathbf{u}_2 \times \mathbf{u}_3$, there follows

$$\frac{\mathbf{u}_1}{h_2 h_3} = \frac{\mathbf{u}_2}{h_2} \times \frac{\mathbf{u}_3}{h_3} = (\nabla u_2) \times (\nabla u_3),$$

and two analogous expressions are similarly obtained. Reference to Equation (74i) then shows that *the expressions* $\mathbf{u}_1/(h_2 h_3)$, $\mathbf{u}_2/(h_3 h_1)$, and $\mathbf{u}_3/(h_1 h_2)$ *have zero divergence*,

$$\nabla \cdot \frac{\mathbf{u}_1}{h_2 h_3} = \nabla \cdot \frac{\mathbf{u}_2}{h_3 h_1} = \nabla \cdot \frac{\mathbf{u}_3}{h_1 h_2} = 0. \tag{163}$$

It should be kept in mind that while the vectors $\mathbf{u}_1$, $\mathbf{u}_2$, and $\mathbf{u}_3$ are *of constant length* unity, the *directions* of these vectors will in general change with position in space, and hence their divergence and curl will not, in general, vanish.

To find an expression for the divergence of a vector $\mathbf{F}$ in u_1, u_2, u_3 coordinates,

$$\mathbf{F} = F_1 \mathbf{u}_1 + F_2 \mathbf{u}_2 + F_3 \mathbf{u}_3, \tag{164}$$

we first write

$$\nabla \cdot \mathbf{F} = \nabla \cdot (F_1 \mathbf{u}_1) + \nabla \cdot (F_2 \mathbf{u}_2) + \nabla \cdot (F_3 \mathbf{u}_3).$$

If we now write the first term in the form

$$\nabla \cdot (F_1 \mathbf{u}_1) = \nabla \cdot \left[(h_2 h_3 F_1) \left(\frac{\mathbf{u}_1}{h_2 h_3} \right) \right]$$

and notice that, according to Equation (163), the second factor in the brackets has zero divergence, we obtain from (74a) the result

$$\nabla \cdot (F_1 \mathbf{u}_1) = \frac{\mathbf{u}_1}{h_2 h_3} \cdot \nabla(h_2 h_3 F_1) = \frac{1}{h_1 h_2 h_3} \frac{\partial}{\partial u_1} (h_2 h_3 F_1).$$

Treating the other terms in a similar way, we obtain finally

$$\nabla \cdot \mathbf{F} = \frac{1}{h_1 h_2 h_3} \left[\frac{\partial}{\partial u_1} (h_2 h_3 F_1) + \frac{\partial}{\partial u_2} (h_3 h_1 F_2) + \frac{\partial}{\partial u_3} (h_1 h_2 F_3) \right]. \tag{165}$$

In particular, by combining Equations (158) and (165) we find the expression for the Laplacian operator,

$$\nabla^2 = \nabla \cdot \nabla = \frac{1}{h_1 h_2 h_3} \left[\frac{\partial}{\partial u_1} \left(\frac{h_2 h_3}{h_1} \frac{\partial}{\partial u_1} \right) \right.$$
$$\left. + \frac{\partial}{\partial u_2} \left(\frac{h_3 h_1}{h_2} \frac{\partial}{\partial u_2} \right) + \frac{\partial}{\partial u_3} \left(\frac{h_1 h_2}{h_3} \frac{\partial}{\partial u_3} \right) \right]. \tag{166}$$

In a similar way, to find an expression for the curl of $\mathbf{F}$,

$$\mathbf{V} \times \mathbf{F} = \mathbf{V} \times (F_1 \mathbf{u}_1) + \mathbf{V} \times (F_2 \mathbf{u}_2) + \mathbf{V} \times (F_3 \mathbf{u}_3),$$

we write the first term in the form

$$\mathbf{V} \times (F_1 \mathbf{u}_1) = \mathbf{V} \times \left[(h_1 F_1)\left(\frac{\mathbf{u}_1}{h_1}\right) \right]$$

so that, according to Equation (162), the second factor in the brackets has zero curl. Then, from (74b), we obtain the result

$$\mathbf{V} \times (F_1 \mathbf{u}_1) = -\left(\frac{\mathbf{u}_1}{h_1} \times \mathbf{V}\right)(h_1 F_1)$$

$$= -\frac{\mathbf{u}_3}{h_1 h_2} \frac{\partial}{\partial u_2}(h_1 F_1) + \frac{\mathbf{u}_2}{h_3 h_1} \frac{\partial}{\partial u_3}(h_1 F_1)$$

$$= \frac{1}{h_1 h_2 h_3}\left(h_2 \mathbf{u}_2 \frac{\partial}{\partial u_3} - h_3 \mathbf{u}_3 \frac{\partial}{\partial u_2}\right)(h_1 F_1).$$

If the other terms are treated in a similar way, the result can be written in the form of a determinant

$$\mathbf{V} \times \mathbf{F} = \frac{1}{h_1 h_2 h_3} \begin{vmatrix} h_1 \mathbf{u}_1 & h_2 \mathbf{u}_2 & h_3 \mathbf{u}_3 \\ \dfrac{\partial}{\partial u_1} & \dfrac{\partial}{\partial u_2} & \dfrac{\partial}{\partial u_3} \\ h_1 F_1 & h_2 F_2 & h_3 F_3 \end{vmatrix}. \tag{167}$$

For the rectangular coordinates $(u_1, u_2, u_3) \equiv (x, y, z)$ we have $h_1 = h_2 = h_3 = 1$, and all the preceding results are seen to reduce to the forms originally given in that case.

6.18. Special Coordinate Systems. In *circular cylindrical coordinates* (r, θ, z) we have†

$$x = r \cos \theta, \qquad y = r \sin \theta, \qquad z = z, \tag{168a}$$

where $r \geqq 0$ and $0 \leqq \theta < 2\pi$. Since r, θ, and z *in that order* form a right-handed system, we may take

$$(r, \theta, z) \equiv (u_1, u_2, u_3).$$

The position vector $\mathbf{r}$ has the form

$$\mathbf{r} = \mathbf{i} r \cos \theta + \mathbf{j} r \sin \theta + \mathbf{k} z.$$

Thus we have

$$\mathbf{U}_r = \frac{\partial \mathbf{r}}{\partial r} = \mathbf{i} \cos \theta + \mathbf{j} \sin \theta, \qquad \mathbf{U}_\theta = \frac{\partial \mathbf{r}}{\partial \theta} = -\mathbf{i} r \sin \theta + \mathbf{j} r \cos \theta, \qquad \mathbf{U}_z = \mathbf{k}.$$

From Equation (146), there follows

$$h_r = 1, \qquad h_\theta = r, \qquad h_z = 1, \tag{168b}$$

and hence also

---

†We notice that the coordinate r in this system is distance from the z axis. It must not be confused with $|\mathbf{r}|$, where $\mathbf{r}$ is the position vector and $|\mathbf{r}|$ is distance from the origin.

$$\mathbf{u}_r = \mathbf{i} \cos \theta + \mathbf{j} \sin \theta, \qquad \mathbf{u}_\theta = -\mathbf{i} \sin \theta + \mathbf{j} \cos \theta, \qquad \mathbf{u}_z = \mathbf{k}. \qquad (168c)$$

Equations (150), (151), and (153) give

$$ds = \sqrt{dr^2 + r^2 \, d\theta^2 + dz^2},$$

$$d\tau = r \, d\theta \, dr \, dz, \qquad\qquad (168d)$$

$$d\sigma_r = r \, d\theta \, dz, \qquad d\sigma_\theta = dr \, dz, \qquad d\sigma_z = r \, d\theta \, dr.$$

Finally, Equations (157), (165), (166), and (167) give the results

$$\nabla f = \mathbf{u}_r \frac{\partial f}{\partial r} + \mathbf{u}_\theta \frac{1}{r} \frac{\partial f}{\partial \theta} + \mathbf{u}_z \frac{\partial f}{\partial z},$$

$$\nabla \cdot \mathbf{F} = \frac{1}{r} \frac{\partial}{\partial r}(rF_r) + \frac{1}{r} \frac{\partial F_\theta}{\partial \theta} + \frac{\partial F_z}{\partial z},$$

$$\nabla^2 f = \frac{1}{r} \frac{\partial}{\partial r}\left(r \frac{\partial f}{\partial r}\right) + \frac{1}{r^2} \frac{\partial^2 f}{\partial \theta^2} + \frac{\partial^2 f}{\partial z^2}, \qquad (168e)$$

$$\nabla \times \mathbf{F} = \frac{1}{r} \begin{vmatrix} \mathbf{u}_r & r\mathbf{u}_\theta & \mathbf{u}_z \\ \dfrac{\partial}{\partial r} & \dfrac{\partial}{\partial \theta} & \dfrac{\partial}{\partial z} \\ F_r & rF_\theta & F_z \end{vmatrix}.$$

In *spherical coordinates* (r, φ, θ), where r is distance *from the origin*, θ is the polar angle measured from the xz plane, and φ is the "cone angle" measured from the z axis (Figure 6.15), there follows†

$$x = r \sin \varphi \cos \theta, \qquad y = r \sin \varphi \sin \theta, \qquad z = r \cos \varphi, \qquad (169a)$$

where $r \geq 0, 0 \leq \varphi \leq \pi, 0 \leq \theta < 2\pi$, and the following results are obtained:

$$h_r = 1, \qquad h_\varphi = r, \qquad h_\theta = r \sin \varphi, \qquad\qquad (169b)$$

$$\mathbf{u}_r = \mathbf{i} \sin \varphi \cos \theta + \mathbf{j} \sin \varphi \sin \theta + \mathbf{k} \cos \varphi,$$

$$\mathbf{u}_\varphi = \mathbf{i} \cos \varphi \cos \theta + \mathbf{j} \cos \varphi \sin \theta - \mathbf{k} \sin \varphi, \qquad (169c)$$

$$\mathbf{u}_\theta = -\mathbf{i} \sin \theta + \mathbf{j} \cos \theta,$$

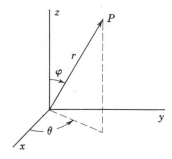

Figure 6.15

†The angles φ and θ are defined in different ways by different writers. The present convention is consistent with that used for circular cylindrical coordinates in the plane $z = 0$.

$$ds = \sqrt{dr^2 + r^2\, d\varphi^2 + r^2 \sin^2 \varphi\, d\theta^2},$$

$$d\tau = r^2 \sin \varphi\, dr\, d\varphi\, d\theta, \tag{169d}$$

$$d\sigma_r = r^2 \sin \varphi\, d\varphi\, d\theta, \quad d\sigma_\varphi = r \sin \varphi\, dr\, d\theta, \quad d\sigma_\theta = r\, dr\, d\varphi,$$

and also

$$\nabla f = \mathbf{u}_r \frac{\partial f}{\partial r} + \frac{\mathbf{u}_\varphi}{r} \frac{\partial f}{\partial \varphi} + \frac{\mathbf{u}_\theta}{r \sin \varphi} \frac{\partial f}{\partial \theta},$$

$$\nabla \cdot \mathbf{F} = \frac{1}{r^2} \frac{\partial}{\partial r}(r^2 F_r) + \frac{1}{r \sin \varphi} \frac{\partial}{\partial \varphi}(F_\varphi \sin \varphi) + \frac{1}{r \sin \varphi} \frac{\partial F_\theta}{\partial \theta},$$

$$\nabla^2 f = \frac{1}{r^2} \frac{\partial}{\partial r}\left(r^2 \frac{\partial f}{\partial r}\right) + \frac{1}{r^2 \sin \varphi} \frac{\partial}{\partial \varphi}\left(\sin \varphi \frac{\partial f}{\partial \varphi}\right) + \frac{1}{r^2 \sin^2 \varphi} \frac{\partial^2 f}{\partial \theta^2}, \tag{169e}$$

$$\nabla \times \mathbf{F} = \frac{1}{r^2 \sin \varphi}
\begin{vmatrix}
\mathbf{u}_r & r\mathbf{u}_\varphi & r \sin \varphi\, \mathbf{u}_\theta \\
\dfrac{\partial}{\partial r} & \dfrac{\partial}{\partial \varphi} & \dfrac{\partial}{\partial \theta} \\
F_r & rF_\varphi & r \sin \varphi\, F_\theta
\end{vmatrix}.$$

6.19. Application to Two-Dimensional Incompressible Fluid Flow. In certain types of fluid flow it is possible to choose a system of orthogonal curvilinear coordinates in such a way that the flow depends only upon two of the coordinates and is independent of the third. The simplest example is that in which flow is parallel to a plane, say the xy plane, and is such that its properties everywhere are then specified by its properties on the xy plane. In the more general case, we assume that the flow is independent of u_3 and is then specified by its properties on any surface for which u_3 is constant. We suppose also that the coordinate system is such that the scale factors h_1, h_2, and h_3 are independent of u_3.

If the flow is divergenceless and irrotational, that is, if there are no distributions of sources or sinks or of vortices, about which the fluid tends to rotate, and if also the fluid is assumed to be incompressible, we have seen that the velocity vector $\mathbf{V}$ is the gradient of a function φ, called the *velocity potential*, and that φ satisfies Laplace's equation.

Since in the present case $\mathbf{V}$ has no component in the u_3 direction, we may write

$$\mathbf{V} = V_1 \mathbf{u}_1 + V_2 \mathbf{u}_2, \tag{170}$$

where V_1 and V_2 are the components of $\mathbf{V}$ in the directions of $\mathbf{u}_1$ and $\mathbf{u}_2$, respectively. Since the velocity is to be independent of u_3, the relation $\mathbf{V} = \nabla \varphi$ becomes, by virtue of (158),

$$\mathbf{V} = \frac{\mathbf{u}_1}{h_1} \frac{\partial \varphi}{\partial u_1} + \frac{\mathbf{u}_2}{h_2} \frac{\partial \varphi}{\partial u_2}, \tag{171}$$

and hence we have

$$V_1 = \frac{1}{h_1} \frac{\partial \varphi}{\partial u_1}, \quad V_2 = \frac{1}{h_2} \frac{\partial \varphi}{\partial u_2}. \tag{172}$$

Also, the equation $\nabla^2\varphi = 0$ becomes

$$\frac{\partial}{\partial u_1}\left(\frac{h_2 h_3}{h_1}\frac{\partial\varphi}{\partial u_1}\right) + \frac{\partial}{\partial u_2}\left(\frac{h_3 h_1}{h_2}\frac{\partial\varphi}{\partial u_2}\right) = 0. \tag{173}$$

When the solution φ of the partial differential equation (173), satisfying appropriate conditions on the boundary of the region considered, has been obtained, the *equipotential lines* in the u_3 surfaces are curves given by equations of the form

$$\varphi(u_1, u_2) = c_1 \tag{174}$$

as c_1 takes on successive constant values. The velocity vectors, and hence the corresponding lines of flow or *streamlines*, are normal to these curves. If we express the streamlines by equations of the form

$$\psi(u_1, u_2) = c_2, \tag{175}$$

then, since we must have

$$\nabla\varphi \cdot \nabla\psi = 0, \tag{176}$$

there follows

$$\left(\frac{1}{h_1}\frac{\partial\varphi}{\partial u_1}\right)\left(\frac{1}{h_1}\frac{\partial\psi}{\partial u_1}\right) + \left(\frac{1}{h_2}\frac{\partial\varphi}{\partial u_2}\right)\left(\frac{1}{h_2}\frac{\partial\psi}{\partial u_2}\right) = 0,$$

or

$$\frac{\dfrac{\partial\psi}{\partial u_2}}{\dfrac{h_2}{h_1}\dfrac{\partial\varphi}{\partial u_1}} = -\frac{\dfrac{\partial\psi}{\partial u_1}}{\dfrac{h_1}{h_2}\dfrac{\partial\varphi}{\partial u_2}}.$$

If we denote the common value of these two ratios by μ, where μ is an unknown function of u_1 and u_2, there follows

$$\frac{\partial\psi}{\partial u_1} = -\mu\frac{h_1}{h_2}\frac{\partial\varphi}{\partial u_2}, \qquad \frac{\partial\psi}{\partial u_2} = \mu\frac{h_2}{h_1}\frac{\partial\varphi}{\partial u_1}. \tag{177}$$

To determine μ, we make use of the equation

$$\frac{\partial^2\psi}{\partial u_2\,\partial u_1} = \frac{\partial^2\psi}{\partial u_1\,\partial u_2},$$

assuming appropriate continuity, and so obtain from (177)

$$\frac{\partial}{\partial u_1}\left(\frac{\mu h_2}{h_1}\frac{\partial\varphi}{\partial u_1}\right) + \frac{\partial}{\partial u_2}\left(\frac{\mu h_1}{h_2}\frac{\partial\varphi}{\partial u_2}\right) = 0.$$

If this equation is compared with (173), one obtains the result $\mu = ch_3$, where c is an arbitrary constant. As will be seen, it is convenient to take $c = 1$, so that[†]

$$\mu = h_3. \tag{178}$$

[†]Equation (178) defines a *particular* solution of the equation determining μ. The most general solution is $\mu = h_3 f(\psi)$, where f is an arbitrary function of the expression ψ being determined. The more general solution leads only to the obvious fact that any function of the ψ determined by Equation (179) will also be constant along the streamlines.

Thus, from Equations (172) and (178) we obtain the relationship

$$d\psi = -\frac{h_3 h_1}{h_2}\frac{\partial\varphi}{\partial u_2}\,du_1 + \frac{h_2 h_3}{h_1}\frac{\partial\varphi}{\partial u_1}\,du_2, \tag{179}$$

from which ψ can be determined by integration when φ is known. By making use of (172), Equation (179) can also be written in the form

$$d\psi = h_3(-h_1 V_2\,du_1 + h_2 V_1\,du_2). \tag{180}$$

By using Equations (160), (171), and (172), we obtain the relation

$$\nabla\psi = -\frac{h_3}{h_2}\frac{\partial\varphi}{\partial u_2}\mathbf{u}_1 + \frac{h_3}{h_1}\frac{\partial\varphi}{\partial u_1}\mathbf{u}_2 = h_3(-V_2\mathbf{u}_1 + V_1\mathbf{u}_2), \tag{181}$$

which can also be written in the form

$$\nabla\psi = h_3\mathbf{u}_3 \times \mathbf{V}. \tag{182}$$

Hence there follows also

$$\nabla^2\psi = \nabla\cdot(h_3\mathbf{u}_3\times\mathbf{V}) = \mathbf{V}\cdot\nabla\times(h_3\mathbf{u}_3) - h_3\mathbf{u}_3\cdot\nabla\times\mathbf{V}$$

$$= \mathbf{V}\cdot\nabla\times\left(h_3^2\frac{\mathbf{u}_3}{h_3}\right) = \mathbf{V}\cdot\left(\nabla h_3^2\times\frac{\mathbf{u}_3}{h_3}\right) = \frac{\nabla h_3^2}{h_3^2}\cdot h_3\mathbf{u}_3\times\mathbf{V}$$

or

$$\nabla^2\psi = (\nabla\log h_3^2)\cdot\nabla\psi, \tag{183}$$

when use is made of Equations (74c), (74b), and (162), and of the fact that $\nabla\times\mathbf{V} = 0$ in consequence of the assumed irrotational nature of the flow.

Thus we conclude that if h_3 is constant then the function ψ in the flow under consideration also satisfies Laplace's equation, and hence may be considered as the velocity potential corresponding to a second flow having the curves $\varphi = c_1$ as streamlines. The two flows so related are said to be *conjugate*.

If h_3 is not constant, the function ψ does not satisfy Laplace's equation, and hence no conjugate (nondivergent) flow exists.

In any case, the function ψ is known as the *stream function* of the flow for which φ is the velocity potential. We notice from (179) that the expression for ψ involves an arbitrary additive constant (as well as the assigned multiplicative constant c). The additive constant can be chosen so that along a particular reference streamline we have $\psi = 0$.

In order to establish another useful property of the stream function ψ, we consider a curve C in a u_3 surface, joining two points P_0 and P_1. Then the rate of mass flow, in the direction of the normal $\mathbf{n}$, through a "rectangular element" based on a linear element $d\mathbf{r}$ of this curve, and extending in the u_3 direction a distance corresponding to a *unit* increment in u_3, is given by

$$df = \rho\mathbf{V}\cdot\mathbf{n}\,h_3\,ds = \rho\mathbf{V}\cdot\left(\frac{d\mathbf{r}}{ds}\times\mathbf{u}_3\right)h_3\,ds = \rho h_3\mathbf{u}_3\times\mathbf{V}\cdot d\mathbf{r}.$$

If use is made of Equation (182), this relation takes the form

$$df = \rho\,\nabla\psi\cdot d\mathbf{r} = \rho\,d\psi.$$

Thus, since here ρ is constant, we obtain by integration along C from P_0 to P_1 the result

$$\Delta_C f = \rho \, \Delta_C \psi. \tag{184}$$

where Δ_C indicates *change along* C, so that *the difference between the values of ψ at two points in a u_3 surface is numerically equal to the rate of mass flow of a fluid with unit density "across any curve in that surface which joins these points."*

Example. We verify that the function

$$\varphi = x^2 - y^2$$

is a solution of Laplace's equation in the xy plane,

$$\nabla^2 \varphi = \frac{\partial^2 \varphi}{\partial x^2} + \frac{\partial^2 \varphi}{\partial y^2} = 0,$$

and hence is the potential function of a flow of the type considered. In this case we have

$$u_1 = x, \qquad u_2 = y, \qquad h_1 = h_2 = h_3 = 1.$$

The velocity vector $\mathbf{V} = \nabla\varphi$ becomes

$$\mathbf{V} = 2x\mathbf{i} - 2y\mathbf{j},$$

and the equipotential curves in the xy plane are the hyperbolas

$$x^2 - y^2 = c_1.$$

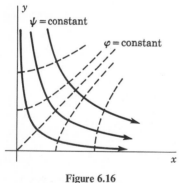

Figure 6.16

The stream function ψ is determined from Equation (179),

$$d\psi = 2y \, dx + 2x \, dy = 2d(xy),$$

and hence the streamlines are the hyperbolas

$$\psi = 2xy = c_2.$$

Since, in particular, the lines $x = 0$, $y = 0$ are streamlines, the flow in the first quadrant is a plane flow around a right-angled corner (Figure 6.16). Also, since $\psi = 2xy$ also satisfies Laplace's equation, ψ can be considered as the velocity potential in a conjugate flow where the curves $x^2 - y^2 = c_1$ are streamlines, and the function $\varphi = x^2 - y^2$ is the stream function. ∎

6.20. Compressible Ideal Fluid Flow. To further illustrate the use of vector methods, we now treat briefly the basic equations governing the flow of non-viscous (ideal) compressible fluids.

We consider first an element of fluid mass dm in the form of a rectangular parallelepiped with edges dx, dy, dz parallel to the coordinate axes at the time t and moving with velocity $\mathbf{V}$ at that instant. If we denote the pressure by p, the differential force exerted by the pressure distributions on the two faces parallel to the yz plane is readily found to be $-[i(\partial p/\partial x)\,dx]\,dy\,dz$. Considering the forces acting on the other four faces similarly, we then find that the resultant differential force due to fluid pressure on the element dm is given by $-\nabla p\,dx\,dy\,dz$ or, equivalently, by $-(\nabla p/\rho)\,dm$. The same expression can be shown (by use of the divergence theorem) to be valid for the differential force due to pressure acting on an element of volume $(dm)/\rho$ of any shape.

As an element moves in space, its shape and volume generally will change. However, its *mass dm* must be conserved. For an element of constant mass dm the rate of change of momentum is given by $(d\mathbf{V}/dt)\,dm$. Hence, if external forces are omitted, Newton's second law of motion leads to the equation

$$\rho\frac{d\mathbf{V}}{dt} + \nabla p = 0. \tag{185}$$

The velocity $\mathbf{V}$ is a function of position x, y, z and time t, whereas the position coordinates x, y, z depend not only upon time t but also upon the initial values x_0, y_0, z_0 at a reference time, say $t = 0$. Thus the time derivative $d\mathbf{V}/dt$ in (185) actually is not a total derivative in the strict sense, but is calculated only for fixed initial coordinates. That is, the time differentiation follows a given particle in its motion. For this reason this derivative is frequently termed the "particle," "substantial," or "material" derivative of $\mathbf{V}$, and is often indicated by the special notation $D\mathbf{V}/Dt$.

For fixed initial coordinates, $\mathbf{V}$ depends directly on x, y, z and t, and x, y, and z themselves depend upon the time t, in such a way that $dx/dt = V_x$, and so on. Hence we may write the substantial derivative of $\mathbf{V}$ in the expanded form (see Section 7.1),

$$\frac{d\mathbf{V}}{dt} = \frac{\partial\mathbf{V}}{\partial x}\frac{dx}{dt} + \frac{\partial\mathbf{V}}{\partial y}\frac{dy}{dt} + \frac{\partial\mathbf{V}}{\partial z}\frac{dz}{dt} + \frac{\partial\mathbf{V}}{\partial t}$$

$$= V_x\frac{\partial\mathbf{V}}{\partial x} + V_y\frac{\partial\mathbf{V}}{\partial y} + V_z\frac{\partial\mathbf{V}}{\partial z} + \frac{\partial\mathbf{V}}{\partial t}$$

or
$$\frac{d\mathbf{V}}{dt} = (\mathbf{V}\cdot\nabla)\mathbf{V} + \frac{\partial\mathbf{V}}{\partial t}. \tag{186}$$

In terms of differential operators we may write also

$$\frac{d}{dt} = \mathbf{V}\cdot\nabla + \frac{\partial}{\partial t} = V_x\frac{\partial}{\partial x} + V_y\frac{\partial}{\partial y} + V_z\frac{\partial}{\partial z} + \frac{\partial}{\partial t}. \tag{187}$$

Thus Equation (185) can be written in the form

$$\rho\left(\mathbf{V}\cdot\nabla\mathbf{V} + \frac{\partial\mathbf{V}}{\partial t}\right) + \nabla p = 0 \tag{188}$$

and is equivalent to three scalar equations, the first of which is of the form

$$\rho\left(V_x\frac{\partial V_x}{\partial x} + V_y\frac{\partial V_x}{\partial y} + V_z\frac{\partial V_x}{\partial z} + \frac{\partial V_x}{\partial t}\right) + \frac{\partial p}{\partial x} = 0.$$

These equations of motion are known as *Euler's equations*.

In addition to the vector equation (188), the condition of conservation of mass must hold, so that Equation (112),

$$\rho \mathbf{V} \cdot \mathbf{V} = -\left(\mathbf{V} \cdot \nabla \rho + \frac{\partial \rho}{\partial t}\right) = -\frac{d\rho}{dt}, \tag{189}$$

must be satisfied. This equation is known as the *equation of continuity*.

Finally, to complete the system of basic equations, we may take the effect of compressibility into account by assuming a suitable relationship between pressure p and density ρ. (For an incompressible fluid, ρ is constant.) Assuming such a relationship, we may then write

$$\nabla p = \mathbf{i}\frac{\partial p}{\partial x} + \mathbf{j}\frac{\partial p}{\partial y} + \mathbf{k}\frac{\partial p}{\partial z} = \frac{dp}{d\rho}\left(\mathbf{i}\frac{\partial \rho}{\partial x} + \mathbf{j}\frac{\partial \rho}{\partial y} + \mathbf{k}\frac{\partial \rho}{\partial z}\right) = \frac{dp}{d\rho}\nabla\rho.$$

The quantity

$$\sqrt{\frac{dp}{d\rho}} = V_s \tag{190}$$

is known as the *local sonic velocity*, and is seen to vary with ρ from point to point in the fluid, and with time.

With these results Euler's equations in the vector form (188) become

$$\rho \mathbf{V} \cdot \nabla \mathbf{V} + \rho \frac{\partial \mathbf{V}}{\partial t} + V_s^2 \nabla \rho = \mathbf{0}. \tag{191}$$

Equations (189) and (191) are sufficient to determine ρ and $\mathbf{V}$ if suitable boundary conditions and/or initial conditions are prescribed.

In many problems the flow is nearly uniform. That is, the vector $\mathbf{V}$ and the scalar functions ρ and p are nearly constant. In such cases, Equations (189) and (191) can be *linearized* by first writing

$$\mathbf{V} = \mathbf{U} + \mathbf{u}, \qquad \rho = \rho_0 + \delta, \qquad V_s = V_{s0} + v_s, \tag{192}$$

where $\mathbf{U}$, ρ_0, and V_{s0} are the *constant values* corresponding to the uniform flow and the quantities $\mathbf{u}$, δ, and v_s are considered as small deviations. With this notation Equations (189) and (191) become

$$(\rho_0 + \delta)\mathbf{V} \cdot \mathbf{u} + (\mathbf{U} + \mathbf{u}) \cdot \nabla \delta + \frac{\partial \delta}{\partial t} = 0,$$

$$(\rho_0 + \delta)(\mathbf{U} + \mathbf{u}) \cdot \nabla\mathbf{u} + (\rho_0 + \delta)\frac{\partial \mathbf{u}}{\partial t} + (V_{s0} + v_s)^2 \nabla\delta = \mathbf{0}.$$

If products of small deviations are assumed to be relatively negligible, these equations are reduced to the linear forms

$$\rho_0 \mathbf{V} \cdot \mathbf{u} + \left(\mathbf{U} \cdot \mathbf{V} + \frac{\partial}{\partial t}\right)\delta = 0, \tag{193}$$

$$\left(\mathbf{U} \cdot \mathbf{V} + \frac{\partial}{\partial t}\right)\mathbf{u} + \frac{V_{s0}^2}{\rho_0}\nabla\delta = \mathbf{0}. \tag{194}$$

For an *irrotational* flow (see Section 6.15), the velocity $\mathbf{V}$, and hence also $\mathbf{u}$, can be expressed as the gradient of a scalar function. Hence, if we write

$$\mathbf{u} = \nabla\varphi, \tag{195}$$

Equation (194) becomes

$$\mathbf{V}\left[\delta + \frac{\rho_0}{V_{s0}^2}\left(\mathbf{U} \cdot \mathbf{V} + \frac{\partial}{\partial t}\right)\varphi\right] = 0,$$

from which the density deviation δ is expressed in terms of the *velocity deviation potential* φ in the form

$$\delta = -\frac{\rho_0}{V_{s0}^2}\left(\mathbf{U} \cdot \mathbf{V} + \frac{\partial}{\partial t}\right)\varphi, \tag{196}$$

except for an additive function of time only, which can be considered as incorporated in φ. It should be noticed that all the preceding relations involve only the invariant operator $\mathbf{V}$, insofar as space differentiation is concerned, and hence may be readily expressed in terms of any convenient coodinate system.

We now suppose that the uniform flow is parallel to the x axis, and so write

$$\mathbf{U} = U\mathbf{i}, \tag{197}$$

in which case Equation (196) becomes

$$\delta = -\frac{\rho_0}{V_{s0}^2}\left(U\frac{\partial\varphi}{\partial x} + \frac{\partial\varphi}{\partial t}\right). \tag{198}$$

If Equations (195), (197), and (198) are introduced into (193), we obtain the equation

$$\mathbf{V}^2\varphi - \frac{1}{V_{s0}^2}\left(U\frac{\partial}{\partial x} + \frac{\partial}{\partial t}\right)^2\varphi = 0, \tag{199}$$

which must be satisfied by the potential function φ. For an *incompressible* fluid $V_s = \infty$, and the equation reduces to Laplace's equation, in accordance with the results of Section 6.15.

In the special case of *steady flow in two dimensions*, where the flow is parallel to the xy plane and the velocity and density do not vary with time, Equation (194) reduces to the form

$$(1 - M^2)\frac{\partial^2\varphi}{\partial x^2} + \frac{\partial^2\varphi}{\partial y^2} = 0, \tag{200}$$

where M is the so-called local *Mach number*,

$$M = \frac{U}{V_{s0}}, \tag{201}$$

and is the ratio of the speed of the uniform flow to the sonic velocity corresponding to this flow. A flow is said to be *subsonic* when $M < 1$ and *supersonic* when $M > 1$. It is important to notice that the coefficient of $\partial^2\varphi/\partial x^2$ in Equation (200) changes sign when the sonic velocity is passed.

In the special case of *nonsteady one-dimensional flow*, when the velocities in the y and z directions are assumed to be negligible, Equation (199) becomes

$$(U^2 - V_{s0}^2)\frac{\partial^2\varphi}{\partial x^2} + 2U\frac{\partial^2\varphi}{\partial x\,\partial t} + \frac{\partial^2\varphi}{\partial t^2} = 0. \tag{202}$$

If the above developments are still assumed to be satisfactorily approximate when *small deviations* are measured from a *state of rest*, so that $U = 0$, we may replace **u** by **V** and rewrite (198) and (199) in the form

$$\delta = p - p_0 = -\frac{p_0}{V_{s0}^2}\frac{\partial\varphi}{\partial t} \tag{203}$$

and

$$\nabla^2\varphi = \frac{1}{V_{s0}^2}\frac{\partial^2\varphi}{\partial t^2}, \tag{204}$$

where

$$\mathbf{V} = \nabla\varphi. \tag{205}$$

Here V_{s0}^2 is essentially an effective "mean value" of $dp/d\rho$. Thus we have also, approximately,

$$p - p_0 = V_{s0}^2(\rho - \rho_0)$$

so that the pressure change due to the disturbing flow is given approximately by

$$p - p_0 = -\rho_0\frac{\partial\varphi}{\partial t}. \tag{206}$$

To conclude this section, we obtain an explicit exact integral of Euler's equation, when external forces are absent, in the case of irrotational flow ($\nabla \times \mathbf{V} = 0$). From the identity

$$\nabla(\mathbf{V}\cdot\mathbf{V}) = 2\mathbf{V}\cdot\nabla\mathbf{V} + 2\mathbf{V}\times(\nabla\times\mathbf{V})$$

[see Equation (74e)], we here obtain the result

$$\mathbf{V}\cdot\nabla\mathbf{V} = \nabla(\tfrac{1}{2}V^2),$$

where $V = |\mathbf{V}| = \sqrt{\mathbf{V}\cdot\mathbf{V}}$. Thus Euler's equation (188) can be written in the form

$$\nabla\left(\frac{1}{2}V^2\right) + \frac{\partial\mathbf{V}}{\partial t} + \frac{\nabla p}{\rho} = \mathbf{0}. \tag{207}$$

If we now write

$$\mathbf{V} = \nabla\varphi \tag{208}$$

and *define* a function P by the relation

$$\nabla P = \frac{1}{\rho}\nabla p \quad\text{or}\quad P = \int\frac{dp}{\rho}, \tag{209}$$

Equation (207) takes the form

$$\nabla\left(\frac{1}{2}V^2\right) + \frac{\partial}{\partial t}\nabla\varphi + \nabla P = \mathbf{0}$$

or

$$\nabla\left(\frac{1}{2}V^2 + \frac{\partial\varphi}{\partial t} + P\right) = \mathbf{0}. \tag{210}$$

Hence the quantity $\frac{1}{2}V^2 + (\partial\varphi/\partial t) + P$ must be independent of the space coordinates and, accordingly, a function of time only,

$$\frac{1}{2}V^2 + \frac{\partial\varphi}{\partial t} + P = f(t). \tag{211}$$

This is one form of *Bernoulli's equation.* For *steady flow* (in which the flow is independent of time), this equation becomes

$$\tfrac{1}{2}V^2 + P = \text{constant.} \tag{212}$$

In particular, for an *incompressible fluid* ($\rho = \text{constant}$) we may take $P = p/\rho$, in accordance with (209), and hence (212) can be written in the form

$$\tfrac{1}{2}\rho V^2 + p = \text{constant} \tag{213}$$

in this case.

In illustration, for the example at the end of Section 6.19 involving flow around a right-angled corner, Equation (213) determines the fluid pressure p in the form

$$p = p_c - 2\rho(x^2 + y^2),$$

where p_c is the pressure at the corner.

REFERENCES

1. BRAND, L., *Vector Analysis*, John Wiley & Sons, Inc., New York, 1957.

2. KELLOGG, O. D., *Foundations of Potential Theory*, Dover Publications, Inc., New York, 1953.

3. PHILLIPS, H. B., *Vector Analysis*, John Wiley & Sons, Inc., New York, 1933.

4. STRUIK, D. J., *Lectures on Classical Differential Geometry*, Addison-Wesley Publishing Company, Inc., Reading, Mass., 1950.

PROBLEMS

Section 6.1

1. Find the length and direction cosines of the vector **a** from the point $(1, -1, 3)$ to the midpoint of the line segment from the origin to the point $(6, -6, 4)$.

2. The vectors **a** and **b** extend from the origin O to the points A and B. Determine the vector **c** which extends from O to the point C which divides the line segment from A to B in the ratio $m:n$.

3. (a) If θ denotes the angle between the vectors **a** and **b**, use a theorem of elementary geometry to show that

$$|\mathbf{a} + \mathbf{b}|^2 = |\mathbf{a}|^2 + |\mathbf{b}|^2 + 2\,|\mathbf{a}|\,|\mathbf{b}|\cos\theta.$$

(b) If $\mathbf{a} = a_x\mathbf{i} + a_y\mathbf{j} + a_z\mathbf{k}$ and $\mathbf{b} = b_x\mathbf{i} + b_y\mathbf{j} + b_z\mathbf{k}$, use the preceding result to show that

$$\cos\theta = \frac{a_x b_x + a_y b_y + a_z b_z}{|\mathbf{a}|\,|\mathbf{b}|}.$$

4. Prove that $|\mathbf{a} + \mathbf{b}| \leqq |\mathbf{a}| + |\mathbf{b}|$ and $|\mathbf{a} + \mathbf{b}| \geqq |\mathbf{a}| - |\mathbf{b}|$.

5. (a) Show that a straight line with direction angles α, β, γ can be specified by the equations

$$\frac{x - x_0}{\cos \alpha} = \frac{y - y_0}{\cos \beta} = \frac{z - z_0}{\cos \gamma},$$

where (x_0, y_0, z_0) is a point on the line.

(b) Show that the above equations can also be written in the parametric form

$$x = x_0 + t \cos \alpha, \qquad y = y_0 + t \cos \beta, \qquad z = z_0 + t \cos \gamma,$$

where t is a parameter.

(Notice that in either set of equations $\cos \alpha$, $\cos \beta$, and $\cos \gamma$ can be replaced by *direction ratios* A, B, and C which are proportional to them.)

6. The equations of a straight line are of the form

$$\frac{x - 1}{1} = \frac{y + 2}{2} = \frac{z}{-2}.$$

(a) Show that the line includes the point $(1, -2, 0)$ and determine its direction cosines.

(b) Determine a unit vector in the direction of the line.

Section 6.2

7. Prove geometrically, from the definition of the scalar product, that the distributive law $\mathbf{a} \cdot (\mathbf{b} + \mathbf{c}) = \mathbf{a} \cdot \mathbf{b} + \mathbf{a} \cdot \mathbf{c}$ is valid.

8. The vectors $\mathbf{a}$ and $\mathbf{b}$ are defined as follows:

$$\mathbf{a} = 3\mathbf{i} - 4\mathbf{k}, \qquad \mathbf{b} = 2\mathbf{i} - 2\mathbf{j} + \mathbf{k}.$$

(a) Find the scalar projection of $\mathbf{a}$ on $\mathbf{b}$.

(b) Find the angle between the positive directions of the vectors.

9. Find the magnitude of the scalar component of the force vector

$$\mathbf{F} = \mathbf{i} + 2\mathbf{j} + 2\mathbf{k}$$

in the direction of the straight line with equations $x = y = 2z$.

10. A plane is determined by a point $P_0(x_0, y_0, z_0)$ on it and by a vector

$$\mathbf{N} = A\mathbf{i} + B\mathbf{j} + C\mathbf{k}$$

normal to it. Show that the requirement that the vector from P_0 to a point $P(x, y, z)$ in the plane be perpendicular to $\mathbf{N}$ determines the equation of the plane in the form

$$A(x - x_0) + B(y - y_0) + C(z - z_0) = 0.$$

11. (a) Show that the vector $A\mathbf{i} + B\mathbf{j} + C\mathbf{k}$ is normal to the plane

$$Ax + By + Cz = D.$$

(b) Prove that the shortest distance from the point $P_0(x_0, y_0, z_0)$ to the plane $Ax + By + Cz = D$ is given by

$$d = \frac{|Ax_0 + By_0 + Cz_0 - D|}{\sqrt{A^2 + B^2 + C^2}}.$$

[Let $P_1(x_1, y_1, z_1)$ be any point on the plane and determine the projection of the vector from P_0 to P_1 on the normal to the plane.]

12. Find the angle between the planes $2x - y + 2z = 1$ and $x - y = 2$.

Section 6.3

13. (a) Determine a unit vector perpendicular to the plane of the vectors $\mathbf{a} = \mathbf{i} + 3\mathbf{j} - \mathbf{k}$, $\mathbf{b} = 2\mathbf{i} + \mathbf{j} + \mathbf{k}$.

(b) Find the area of the triangle of which these two vectors form coterminous sides.

14. (a) Determine a unit vector normal to the plane determined by the points $(0, 0, 0)$, $(1, 1, 1)$, and $(2, 1, 3)$.

(b) Find the area of the triangle with vertices at the points defined in part (a).

15. A rigid body rotates with angular velocity ω about the line $x = y = z$. Find the speed of a particle at the point $(1, 2, 2)$.

16. Find the scalar moment of the force $\mathbf{F} = \mathbf{i} - 3\mathbf{j} + 2\mathbf{k}$, acting at the point $(1, 2, 1)$, about the z axis.

17. Show that the shortest distance from a point P_0 to the line joining the points P_1 and P_2 is given by

$$d = \frac{|\mathbf{v}_1 \times \mathbf{v}_2|}{|\mathbf{v}_3|},$$

where $\mathbf{v}_1 = \overrightarrow{P_0 P_1}$, $\mathbf{v}_2 = \overrightarrow{P_0 P_2}$, and $\mathbf{v}_3 = \overrightarrow{P_1 P_2}$. (Notice that $|\mathbf{v}_1 \times \mathbf{v}_2|$ is twice the area of the triangle $P_0 P_1 P_2$, and that the desired distance is the altitude of that triangle normal to the base $P_1 P_2$.)

18. Show that the shortest distance between the lines AB and CD is the projection of $\overrightarrow{AC}$ (or of the vector joining *any* point of AB with any point of CD) on the vector $\overrightarrow{AB} \times \overrightarrow{CD}$.

Section 6.4

19. From the last result of Section 6.3, deduce that, if $\mathbf{r}$ denotes the vector from a point O to the point of application of a force $\mathbf{F}$, then the scalar moment of $\mathbf{F}$ about an axis OA is given by $M_{OA} = (\mathbf{r}\, \mathbf{F}\, \mathbf{u})$, where $\mathbf{u}$ is a unit vector in the direction of OA.

20. Find the volume of the tetrahedron with vertices at the points $(0, 0, 0)$, $(1, 1, 1)$, $(2, 1, 1)$, and $(1, 2, 1)$.

21. Show that the assertion

$$(\mathbf{a}\,\mathbf{b}\,\mathbf{c})(\mathbf{d}\,\mathbf{e}\,\mathbf{f}) = \begin{vmatrix} \mathbf{a}\cdot\mathbf{d} & \mathbf{a}\cdot\mathbf{e} & \mathbf{a}\cdot\mathbf{f} \\ \mathbf{b}\cdot\mathbf{d} & \mathbf{b}\cdot\mathbf{e} & \mathbf{b}\cdot\mathbf{f} \\ \mathbf{c}\cdot\mathbf{d} & \mathbf{c}\cdot\mathbf{e} & \mathbf{c}\cdot\mathbf{f} \end{vmatrix}$$

can be interpreted as a statement of the fact that the determinant of the product of two 3×3 matrices is the product of their determinants.

22. Given that

$$(\mathbf{a} \times \mathbf{b}) \times \mathbf{c} = \lambda([\mathbf{a} \cdot \mathbf{c}]\mathbf{b} - (\mathbf{b} \cdot \mathbf{c})\mathbf{a}],$$

for some scalar λ [see Equation (33)], determine λ as follows: Let $\mathbf{u}$ be a unit vector parallel to $\mathbf{a}$, $\mathbf{v}$ a second unit vector perpendicular to $\mathbf{a}$ and such that $\mathbf{b}$ is in the plane of u and v, and $w = u \times v$. Substitute the relations $a = \alpha_1 u$, $b = \beta_1 u + \beta_2 v$, and $c = \gamma_1 u + \gamma_2 v + \gamma_3 w$ into the given relation and show that $\lambda = 1$.

23. Deduce Equation (34b) from (34a).

24. Determine a unit vector in the plane of the vectors $i + j$ and $j + k$ and perpendicular to the vector $i + j + k$.

25. If u is a unit vector, prove that $u \times (a \times u)$ is the vector projection of a on a plane perpendicular to u.

26. Prove that $a \times (b \times c) + b \times (c \times a) + c \times (a \times b) = 0$.

27. (a) Prove that

$$(a \times b) \times (c \times d) = \begin{cases} (a\ b\ d)c - (a\ b\ c)d, \\ (a\ c\ d)b - (b\ c\ d)a. \end{cases}$$

(To obtain the first form, write temporarily $u = a \times b$.)

(b) Show that this vector, if not a zero vector, is in the direction of the intersection of a plane including the vectors a and b with one including c and d.

Section 6.5

28. If F is a function of t, find the derivative of

$$F \cdot \frac{dF}{dt} \times \frac{d^2F}{dt^2}.$$

29. At time t, the vector from the origin to a moving point is

$$r = a \cos \omega t + b \sin \omega t,$$

where a, b, and ω are constants.

(a) Find the velocity $v = dr/dt$ and prove that $r \times v$ is constant, so that the curve traced out lies in a plane.

(b) Show that the acceleration is directed toward the origin and is proportional to the distance from the origin.

30. Let r represent the vector from a fixed origin O to a moving particle of mass m, subject to a force F.

(a) If H denotes the moment of the momentum vector $mv = m(dr/dt)$ about O, prove that

$$\frac{dH}{dt} = m\frac{d}{dt}(r \times v) = r \times F = M,$$

where M is the moment of the force F about O.

(b) If the force F always passes through the fixed point O, show that

$$\frac{d}{dt}(r \times v) = 0.$$

and hence deduce that $r \times v = h$, where h is a constant vector. Deduce also that the motion is in a plane.

(c) Show that $|r \times v|$ is twice the rate $dA/dt = \frac{1}{2}r^2\, d\theta/dt$ at which area A is swept over by the vector r, and hence deduce that, when a mass at P is subject to a "central" force, which always passes through a fixed point O, the vector $\overrightarrow{OP}$ moves in a plane and sweeps over equal areas in equal times.

31. If **u** is a unit vector originating at a fixed point O, and rotating about a fixed vector ω through O, with angular velocity of constant magnitude ω, show that

$$\frac{d\mathbf{u}}{dt} = \omega \times \mathbf{u}.$$

32. Let $\mathbf{r} = x\mathbf{i} + y\mathbf{j} + z\mathbf{k}$ represent the position vector from a fixed origin O to a point P, and suppose that the xyz axis system is rotating about a fixed vector ω through O, with angular velocity of constant magnitude ω.

   (a) By calculating $d\mathbf{r}/dt$, and noticing that $d\mathbf{i}/dt = \omega \times \mathbf{i}$, and so forth, obtain the velocity vector in the form

$$\mathbf{v} = \mathbf{v}_0 + \omega \times \mathbf{r},$$

where the vector

$$\mathbf{v}_0 = \frac{dx}{dt}\mathbf{i} + \frac{dy}{dt}\mathbf{j} + \frac{dz}{dt}\mathbf{k}$$

is the velocity vector which would be obtained if the axes were fixed.

   (b) Obtain the acceleration vector in the form

$$\mathbf{a} = \mathbf{a}_0 + 2\omega \times \mathbf{v}_0 + \omega \times (\omega \times \mathbf{r}),$$

where $\mathbf{v}_0$ is defined in part (a), and where

$$\mathbf{a}_0 = \frac{d^2x}{dt^2}\mathbf{i} + \frac{d^2y}{dt^2}\mathbf{j} + \frac{d^2z}{dt^2}\mathbf{k}.$$

33. If the system of Problem 32 is rotating with angular velocity of constant magnitude ω about the z axis, the z axis being fixed, show that the equations of motion for a point mass m are of the form

$$m\left(\frac{d^2x}{dt^2} - 2\omega\frac{dy}{dt} - \omega^2x\right) = F_x,$$

$$m\left(\frac{d^2y}{dt^2} + 2\omega\frac{dx}{dt} - \omega^2y\right) = F_y,$$

$$m\frac{d^2z}{dt^2} = F_z,$$

where F_x, F_y, and F_z are the components of the external force along the respective rotating axes. [Notice that the mass hence behaves as though the axes were fixed, with an additional force $2m\omega(dy/dt) + m\omega^2x$ acting in the positive x direction and an additional force $-2m\omega(dx/dt) + m\omega^2y$ acting in the positive y direction.]

Section 6.6

34. (a) If θ is polar angle, show that the vectors

$$\mathbf{u}_1 = \mathbf{i}\cos\theta + \mathbf{j}\sin\theta, \qquad \mathbf{u}_2 = -\mathbf{i}\sin\theta + \mathbf{j}\cos\theta$$

are perpendicular unit vectors in the radial and circumferential directions, respectively, in the xy plane, and that

$$\frac{d\mathbf{u}_1}{d\theta} = \mathbf{u}_2, \qquad \frac{d\mathbf{u}_2}{d\theta} = -\mathbf{u}_1.$$

   (b) For points on a plane curve in polar coordinates, the position vector is of the form $\mathbf{r} = r\mathbf{u}_1$. By differentiation with respect to time t, obtain expressions for the vectors of velocity and acceleration of a point moving along the curve and show that the radial and circumferential components are of the form

$$v_r = \dot{r}, \qquad\qquad v_\theta = r\dot{\theta},$$

$$a_r = \ddot{r} - r\dot{\theta}^2, \qquad a_\theta = r\ddot{\theta} + 2\dot{r}\dot{\theta} = \frac{1}{r}\frac{d}{dt}(r^2\dot{\theta}),$$

where a dot denotes time differentiation. [Notice, for example, that $\dot{\mathbf{u}}_1 = (d\mathbf{u}_1/d\theta)\dot{\theta}$.]

35. (a) Verify that the parametric equations

$$x = a\cos t, \qquad y = a\sin t, \qquad z = ct$$

specify a right circular helix in space.

   (b) Show that for this curve there follows

$$ds = \sqrt{a^2 + c^2}\, dt,$$

where s is arc length.

   (c) Determine the unit tangent, principal normal, and binormal vectors, and show that the radii of curvature and torsion are given by

$$\rho = \frac{a^2 + c^2}{a}, \qquad \tau = \frac{a^2 + c^2}{c}.$$

   (d) Show that the osculating, normal, and rectifying planes at the point $(a, 0, 2\pi c)$ are specified by the respective equations

$$az - cy = 2\pi ca, \qquad ay = 2\pi c^2, \qquad x = a.$$

In Problems 36–39, the position vector to a curve C is assumed to be expressed in the form $\mathbf{r} = x(t)\mathbf{i} + y(t)\mathbf{j} + z(t)\mathbf{k}$, *and a prime is used to denote differentiation with respect to the parameter t, whereas s denotes arc length along C.*

36. Establish the relations

   (a) $s' = |\mathbf{r}'| = \sqrt{\mathbf{r}' \cdot \mathbf{r}'}$,   (b) $\mathbf{u} = \dfrac{\mathbf{r}'}{|\mathbf{r}'|} = \dfrac{\mathbf{r}'}{\sqrt{\mathbf{r}' \cdot \mathbf{r}'}}$.

37. (a) Show that

$$\frac{1}{\rho}\mathbf{n} = \frac{d\mathbf{u}}{ds} = \frac{\mathbf{u}'}{s'} = \frac{(\mathbf{r}' \cdot \mathbf{r}')\mathbf{r}'' - (\mathbf{r}' \cdot \mathbf{r}'')\mathbf{r}'}{|\mathbf{r}'|^4} = \frac{(\mathbf{r}' \times \mathbf{r}'') \times \mathbf{r}'}{|\mathbf{r}'|^4}.$$

   (b) Use the identity of Lagrange [Equation (36)] to obtain the result

$$|(\mathbf{r}' \times \mathbf{r}'') \times \mathbf{r}'| = |\mathbf{r}' \times \mathbf{r}''||\mathbf{r}'|.$$

   (c) Deduce the following results:

$$\frac{1}{\rho} = \frac{|\mathbf{r}' \times \mathbf{r}''|}{|\mathbf{r}'|^3}, \qquad \mathbf{n} = \frac{(\mathbf{r}' \times \mathbf{r}'') \times \mathbf{r}'}{|\mathbf{r}' \times \mathbf{r}''||\mathbf{r}'|}.$$

38. Use Equation (52) and the results of Problems 36(b) and 37(c) to obtain the result

$$\mathbf{b} = \frac{\mathbf{r}' \times \mathbf{r}''}{|\mathbf{r}' \times \mathbf{r}''|}.$$

39. (a) Verify that

$$\mathbf{u} \times \frac{d\mathbf{u}}{ds} \cdot \frac{d^2\mathbf{u}}{ds^2} = \left(\frac{\mathbf{r}'}{s'}\right) \times \left(\frac{s'\mathbf{r}'' - s''\mathbf{r}'}{s'^3}\right) \cdot \left(\frac{\mathbf{r}''' + \cdots}{s'^3}\right) = \frac{(\mathbf{r}'\,\mathbf{r}''\,\mathbf{r}''')}{s'^6},$$

where the omitted terms in the third factor of the second expression involve $\mathbf{r}'$ and $\mathbf{r}''$ linearly, and hence do not affect the value of the product.

(b) Using the relation (55a),

$$\frac{1}{\tau} = \rho^2 \left(\mathbf{u} \times \frac{d\mathbf{u}}{ds} \cdot \frac{d^2\mathbf{u}}{ds^2} \right),$$

and the results of Problems 37(c) and 39(a), obtain the result

$$\frac{1}{\tau} = \frac{(\mathbf{r}' \, \mathbf{r}'' \, \mathbf{r}''')}{|\mathbf{r}' \times \mathbf{r}''|^2}.$$

40. Determine $\mathbf{u}$, $\mathbf{n}$, $\mathbf{b}$, ρ, and τ for the right circular helix

$$x = a \cos t, \qquad y = a \sin t, \qquad z = ct$$

by using preceding results, and compare with the results of Problem 35. (The formulas obtained in Problems 36–39 are particularly useful when t cannot be simply expressed in terms of the arc length s.)

41. Use appropriate results of Problems 37 and 38 to show that the curve for which $\mathbf{r} = x(t)\mathbf{i} + y(t)\mathbf{j} + z(t)\mathbf{k}$ is a straight line if $\mathbf{r}' \times \mathbf{r}'' = \mathbf{0}$ and is a plane curve if $\mathbf{r}' \times \mathbf{r}'' \cdot \mathbf{r}''' = 0$.

42. Determine the curvature and torsion of the twisted cubic $x = t$, $y = t^2$, $z = t^3$ at the point $(1, 1, 1)$.

43. *Bending of a rod.* Suppose that a thin rod of uniform circular cross section is bent and twisted in such a way that its axis coincides with a space curve C. Let the intensity of the applied load per unit distance along C have components p_u, p_n, and p_b along the tangent, principal normal, and binormal, respectively, of the deformed rod and write $\mathbf{p} = p_u\mathbf{u} + p_n\mathbf{n} + p_b\mathbf{b}$. Similarly, write $\mathbf{m} = m_u\mathbf{u} + m_n\mathbf{n} + m_b\mathbf{b}$ for the vector whose components are intensities of distributed applied couples along the deformed rod. Let the influence of that part of the rod beyond a section at distance s from one end upon the remaining part be resolved into a force vector $\mathbf{F} = T\mathbf{u} + Q\mathbf{n} + R\mathbf{b}$ and a moment vector $\mathbf{M} = H\mathbf{u} + L\mathbf{n} + M\mathbf{b}$, so that T is tension, Q and R shear forces, H a twisting moment, and L and M bending moments.

(a) By considering static equilibrium of a section between s and $s + \Delta s$, and proceeding to the limit as $\Delta s \rightarrow 0$, show that the conditions

$$\frac{d\mathbf{F}}{ds} + \mathbf{p} = \mathbf{0}, \qquad \frac{d\mathbf{M}}{ds} + \mathbf{u} \times \mathbf{F} + \mathbf{m} = \mathbf{0}$$

must be satisfied.

(b) Show that these two vector equations imply the following six equations of equilibrium:

$$\frac{dT}{ds} - \frac{1}{\rho}Q + p_u = 0, \qquad \frac{dQ}{ds} + \frac{1}{\rho}T - \frac{1}{\tau}R + p_n = 0, \qquad \frac{dR}{ds} + \frac{1}{\tau}Q + p_b = 0,$$

$$\frac{dH}{ds} - \frac{1}{\rho}L + m_u = 0, \qquad \frac{dL}{ds} + \frac{1}{\rho}H - \frac{1}{\tau}M - R + m_n = 0,$$

$$\frac{dM}{ds} + \frac{1}{\tau}L + Q + m_b = 0.$$

[Notice that this set comprises six equations in the eight unknown quantities T, Q, R, H, L, M, ρ, and τ. For an *elastic* rod, these equations are conventionally supplemented by the two equations

$$M = A\left(\frac{1}{\rho} - \frac{1}{\rho_0}\right), \qquad H = B\left(\frac{1}{\tau} - \frac{1}{\tau_0}\right),$$

where ρ_0 and τ_0 are the radii of curvature and torsion of the undeformed rod, and A and B are bending and torsional stiffnesses associated with the material and cross section of the rod.]

(c) Suppose that the rod is originally straight and untwisted, so that $1/\rho_0$ and $1/\tau_0 = 0$, and is subjected only to a transverse load distribution of intensity p_n (and to end constraints). If, in addition, small deformations are assumed, so that $1/\rho$ and $1/\tau$ are assumed to be negligible in the *equilibrium* equations for the deformed rod, show that the equations of part (b) are satisfied by $T = R = H = L = 0$ if M and Q satisfy the equations

$$\frac{dQ}{ds} + p_n = 0, \qquad \frac{dM}{ds} + Q = 0.$$

[With the supplementary equation $M = EI/\rho$, where EI is the bending stiffness, these are the basic equations of the elementary theory of small deflections of laterally loaded, originally straight rods or beams. (See Section 5.3, where $S \equiv -Q$ and where s is replaced by its projection x along the undeformed axis.)]

Section 6.7

44. The temperature at any point in space is given by

$$T = xy + yz + zx.$$

(a) Find the direction cosines of the direction in which the temperature changes most rapidly with distance from the point $(1, 1, 1)$, and determine the maximum rate of change.

(b) Find the derivative of T in the direction of the vector $3\mathbf{i} - 4\mathbf{k}$ at the point $(1, 1, 1)$.

45. If r and θ are polar coordinates in the xy plane, determine grad r and grad θ.

46. Determine the direction cosines of grad $\varphi(x, y, z)$.

47. Determine a unit vector normal to the surface

$$x^3 - xyz + z^3 = 1$$

at the point $(1, 1, 1)$.

Section 6.8

48. If $\mathbf{V} = \text{grad } \varphi$, where $\varphi = xyz$, determine div $\mathbf{V}$ and curl $\mathbf{V}$.

49. Show that

$$\mathbf{\nabla} \cdot (x\mathbf{v}) = x\mathbf{\nabla} \cdot \mathbf{v} + \mathbf{i} \cdot \mathbf{v}$$

and

$$\mathbf{\nabla} \times (x\mathbf{v}) = x\mathbf{\nabla} \times \mathbf{v} + \mathbf{i} \times \mathbf{v}.$$

50. Prove that $\mathbf{u} \times \mathbf{\nabla} \cdot \mathbf{v} = \mathbf{u} \cdot \mathbf{\nabla} \times \mathbf{v}$.

51. Prove that $(\mathbf{u} \cdot \mathbf{\nabla})\varphi = \mathbf{u} \cdot (\mathbf{\nabla}\varphi)$.

52. (a) Determine the derivative of the function $x^2 + y^2 + z^2$ in the direction specified by the direction cosines l, m, and n.

(b) Determine the derivative of the vector $x^2\mathbf{i} + y^2\mathbf{j} + z^2\mathbf{k}$ in the direction specified by the direction cosines l, m, and n.

53. Prove that $\nabla \times \mathbf{F}$ is not necessarily perpendicular to $\mathbf{F}$, by giving a suitable example.

Section 6.9

54. Prove that $\nabla \cdot (\nabla \varphi) = (\nabla \cdot \nabla)\varphi \equiv \nabla^2 \varphi$.

55. (a) Show that

$$\nabla(\mathbf{v} \cdot \mathbf{v}) = 2\mathbf{v} \cdot \nabla\mathbf{v} + 2\mathbf{v} \times (\nabla \times \mathbf{v}),$$

and deduce that the relation

$$\mathbf{v} \cdot \nabla\mathbf{v} = \tfrac{1}{2}\nabla v^2$$

is true when $\nabla \times \mathbf{v} = 0$ or, more generally, when $\nabla \times \mathbf{v}$ is parallel to $\mathbf{v}$.
    (b) Verify the general result of part (a) when $\mathbf{v} = xy\mathbf{i} + y^2\mathbf{j}$.

56. Show that $\nabla^2 u = 0$ if $\nabla \times \mathbf{u} = \nabla\varphi$ and $\nabla \cdot \mathbf{u} = 0$, where φ is a scalar function.

57. If $\mathbf{v} = \varphi_1 \nabla \varphi_2$, prove that $\nabla \times \mathbf{v}$ is perpendicular to $\mathbf{v}$.

58. If there exists a family of surfaces $\varphi(x, y, z) = $ constant such that $\mathbf{v}(x, y, z)$ is perpendicular at every point to the surface $\varphi = $ constant which passes through that point, show that $\mathbf{v} \cdot (\nabla \times \mathbf{v}) = 0$. (Use the result of Problem 57.)

59. If $\mathbf{u}$ and $\mathbf{v}$ have zero divergence and $\mathbf{w}$ has zero curl, show that

$$\nabla \cdot [(\mathbf{u} \times \mathbf{v}) \times \mathbf{w}] = \mathbf{w} \cdot [(\mathbf{v} \cdot \nabla)\mathbf{u} - (\mathbf{u} \cdot \nabla)\mathbf{v}].$$

Section 6.10, 6.11

60. For each of the following vector functions, determine whether the equation $\nabla\varphi = \mathbf{F}$ possesses a solution, and determine that solution if it exists:
    (a) $\mathbf{F} = 2xyz^3\mathbf{i} - (x^2z^3 + 2y)\mathbf{j} + 3x^2yz^2\mathbf{k}$,
    (b) $\mathbf{F} = 2xy\mathbf{i} + (x^2 + 2yz)\mathbf{j} + (y^2 + 1)\mathbf{k}$.

61. For each vector function $\mathbf{F}$ defined in Problem 60, determine the value of the integral $\int_C \mathbf{F} \cdot d\mathbf{r}$ from the origin to the point $(1, 1, 1)$ along the curve specified by the simultaneous equations $y = x^2$, $z = x^3$.

62. For each vector function defined in Problem 60, determine the value of the integral $\oint_C \mathbf{F} \cdot d\mathbf{r}$ around the unit circle with center at the origin, in the xy plane.

63. (a) If $P\,dx + Q\,dy + R\,dz = d\varphi$ in a region $\Re$ including a point (x_0, y_0, z_0), show that the result of integrating $d\varphi$ along straight line segments from (x_0, y_0, z_0) to (x, y_0, z_0) to (x, y, z_0) to (x, y, z) is

$$\varphi(x, y, z) = \int_{x_0}^{x} P(t, y_0, z_0)\,dt + \int_{y_0}^{y} Q(x, t, z_0)\,dt + \int_{z_0}^{z} R(x, y, t)\,dt + \text{constant},$$

if those segments remain in $\Re$, where x is held constant in the second integral and both x and y are held constant in the third one.
    (b) If φ is defined by the expression obtained in part (a), and if the first partial derivatives of P, Q, and R are continuous and such that the conditions (88a, b, c) are satisfied in $\Re$, show that

$$\frac{\partial \varphi}{\partial x} = P, \qquad \frac{\partial \varphi}{\partial y} = Q, \qquad \frac{\partial \varphi}{\partial z} = R,$$

so that $d\varphi = P\, dx + Q\, dy + R\, dz$ in $\mathfrak{R}$. [Notice, for example, that then

$$\varphi_x(x, y, z) = P(x, y_0, z_0) + \int_{y_0}^{y} Q_x(x, t, z_0)\, dt + \int_{z_0}^{z} R_x(x, y, t)\, dt$$

and that $Q_x(x, t, z_0) = P_t(x, t, z_0)$ and $R_x(x, y, t) = P_t(x, y, t)$.]

64. Show that the use of the formula obtained in Problem 63, with $(x_0, y_0, z_0) = (0, 0, 0)$, for the determination of φ such that $d\varphi = y^2\, dx + 2(xy + z)\, dy + 2y\, dz$, leads to the result

$$\varphi(x, y, z) = \int_0^x 0\, dt + \int_0^y 2xt\, dt + \int_0^z 2y\, dt + c = xy^2 + 2yz + c.$$

(Compare the determination in the text, page 291.)

65. Evaluate the integral $\int_C h\, ds$, where s is arc length along C, in each of the following cases:

(a) $h = x^2 + y^2 + z^2$; C is the helical arc $x = \cos\theta$, $y = \sin\theta$, $z = \theta/(2\pi)$ between $(0, 1, 0)$ and $(0, 1, 1)$.

(b) $h = x + y + z$; C is the circle $x^2 + y^2 = 1$, $z = 1$.

(c) $h = xyz$; C is the arc of the twisted cubic $x = t$, $y = 3t^2$, $z = 6t^3$ between $(0, 0, 0)$ and $(1, 3, 6)$.

Section 6.12

66. (a) If S is a portion of a surface specified by an equation of the form $z = f(x, y)$, show that the unit normal vector **n** which points in the positive z direction from S is given by

$$\mathbf{n} = \frac{-\mathbf{i}(\partial f/\partial x) - \mathbf{j}(\partial f/\partial y) + \mathbf{k}}{\sqrt{1 + (\partial f/\partial x)^2 + (\partial f/\partial y)^2}}.$$

(b) Deduce that, in this case, if $\mathbf{F} = P\mathbf{i} + Q\mathbf{j} + R\mathbf{k}$, there follows

$$\iint_S \mathbf{F} \cdot d\boldsymbol{\sigma} = \iint_D \left(R - P\frac{\partial f}{\partial x} - Q\frac{\partial f}{\partial y} \right) dx\, dy,$$

where D is the projection of S on the xy plane, and where z is to be replaced by $f(x, y)$ in the expressions for P, Q, and R.

67. Evaluate the surface integral of the vector $\mathbf{F} = x\mathbf{i} + y\mathbf{j} + z\mathbf{k}$ over that portion of the surface $z = xy + 1$ which covers the square $0 \le x \le 1$, $0 \le y \le 1$ in the xy plane.

68. Evaluate the surface integral of the vector $\mathbf{F} = x\mathbf{i} + y\mathbf{j} + z\mathbf{k}$ over that portion of the paraboloid $z = x^2 + y^2$ which is inside the cylinder $x^2 + y^2 = 1$.

69. Evaluate the surface integral of the vector $\mathbf{F} = x\mathbf{i} + y\mathbf{j} + z\mathbf{k}$ over the closed surface of the cube bounded by the planes $x = \pm 1$, $y = \pm 1$, $z = \pm 1$.

70. Evaluate the surface integral of $\mathbf{F} = yz\mathbf{i} + xz\mathbf{j} + xy\mathbf{k}$ over the closed boundary of the region bounded below by $z = x^2 + y^2$ and above by $z = 1$.

71. Evaluate the integral $\iint_S h\, d\sigma$ in each of the following cases:

(a) $h = x^2$; S is the portion of the plane $x + y + z = 1$ inside the cylinder $x^2 + y^2 = 1$.

(b) $h = xy + \sqrt{z}$; S is the portion of the cylinder $z = x^2/2$ for which $0 \le x \le 1$ and $0 \le y \le 1$.

Section 6.13

72. Evaluate the integral of Problem 69 by using the divergence theorem.

73. Determine the value of the surface integral $\iint_S \mathbf{F} \cdot \mathbf{n}\, d\sigma$ in each of the following cases, by use of the divergence theorem:

(a) $\mathbf{F} = x\mathbf{i} + y\mathbf{j} + z\mathbf{k}$; S is the closed spherical surface $x^2 + y^2 + z^2 = 1$.

(b) $\mathbf{F} = xy\mathbf{i} + xz\mathbf{j} + (1 - z - yz)\mathbf{k}$; S is the closed surface composed of the portion of the paraboloid $z = 1 - x^2 - y^2$ for which $z \geq 0$ and the circular disk $x^2 + y^2 \leq 1$, $z = 0$.

(c) $\mathbf{F} = xy\mathbf{i} + xz\mathbf{j} + (1 - z - yz)\mathbf{k}$; S is the portion of the paraboloid $z = 1 - x^2 - y^2$ for which $z \geq 0$.

(d) $\mathbf{F} = x^2\mathbf{i} - (1 + 2x)\mathbf{j} + z\mathbf{k}$; S is the lateral surface of that portion of the cylinder $x^2 + y^2 = 1$ for which $0 \leq z \leq 1$.

74. If $\mathbf{r}$ is the position vector $x\mathbf{i} + y\mathbf{j} + z\mathbf{k}$, show that

(a) $\displaystyle\oiint_S \mathbf{r} \cdot \mathbf{n}\, d\sigma = 3V,$ (b) $\displaystyle\oiint_S x\mathbf{r} \cdot \mathbf{n}\, d\sigma = 4V\bar{x},$

where V is the volume enclosed by S and $\bar{x}$ is the x coordinate of its center of gravity.

75. By using the physical argument of Section 6.13, show that, if $\mathfrak{R}$ is the region inside a closed surface S and outside a closed surface Σ, then the divergence theorem takes the form

$$\iiint_{\mathfrak{R}} \nabla \cdot \mathbf{V}\, d\tau = \oiint_S \mathbf{V} \cdot \mathbf{n}\, d\sigma + \oiint_\Sigma \mathbf{V} \cdot \mathbf{n}\, d\sigma,$$

where $\mathbf{n}$ points outward from $\mathfrak{R}$ along both S and Σ, so that $\mathbf{n}$ points into the deleted region enclosed by Σ in the last integral.

76. *Analytical derivation of the divergence theorem.* Suppose that the closed boundary S of a region $\mathfrak{R}$ is cut by lines in the z direction in not more than two points. Denote the upper part of S by $S+$ and the lower part by $S-$, and denote the projection of either part on the xy plane by D.

(a) Show that

$$\iiint_{\mathfrak{R}} \frac{\partial R}{\partial z} dx\, dy\, dz = \iint_D R_{S+} dx\, dy - \iint_D R_{S-} dx\, dy,$$

where R_{S+} and R_{S-} denote the value of $R(x, y, z)$ on $S+$ and on $S-$, respectively.

(b) Noticing that

$$dx\, dy = d\sigma \cos \gamma \qquad \text{on } S+$$

and $\qquad\qquad\qquad\; dx\, dy = -d\sigma \cos \gamma \qquad \text{on } S-,$

where $\cos \gamma$ is the z direction cosine of the outward normal, deduce the relation

$$\iiint_{\mathfrak{R}} \frac{\partial R}{\partial z} dx\, dy\, dz = \oiint_S R \cos \gamma\, d\sigma$$

in this case. [Corresponding results in the x and y directions establish the divergence theorem (117) analytically when $\mathfrak{R}$ is a convex region. Proofs for other regions essentially depend upon subdividing them into convex regions or taking limits of results corresponding to such subdivisions.

77. Show that the product $d\sigma \cos \alpha$ in the right-hand member of the divergence theorem (117) can be replaced by $\pm dy\, dz$, where the sign chosen at a point of S is to be the sign of $\cos \alpha$ at that point, and similarly that we may write $d\sigma \cos \beta = \pm dx\,dz$ and $d\sigma \cos \gamma = \pm dx\,dy$. (Notice that proper choice of each sign may differ from one part of S to another.)

78. (a) If $\mathbf{V}$ is a two-dimensional vector in the xy plane, show that

$$\iint_D \nabla \cdot \mathbf{V}\,dx\,dy = \oint_C \mathbf{V} \cdot \mathbf{n}\,ds,$$

where D is a simply connected region in the xy plane with a simple closed boundary C, and $\mathbf{n}$ is the unit outward normal vector along the curve C. (Apply the divergence theorem to the vector $\mathbf{V}$ over a three-dimensional region which consists of the interior of a right cylinder of unit height having the region D as its lower base.)

(b) If D is the region inside a simple closed curve C but outside a simple closed interior curve Γ in the xy plane, show that there follows

$$\iint_D \nabla \cdot \mathbf{V}\,dx\,dy = \oint_C \mathbf{V} \cdot \mathbf{n}\,ds + \oint_\Gamma \mathbf{V} \cdot \mathbf{n}\,ds,$$

when $\mathbf{n}$ points outward from D along both C and Γ, so that $\mathbf{n}$ points into the deleted region enclosed by Γ in the last integral.

79. Let $\mathfrak{R}$ be a region bounded by a closed surface S.

(a) Establish the "gradient theorem,"

$$\iiint_\mathfrak{R} \nabla\varphi\,d\tau = \oiint_S \varphi\,d\boldsymbol{\sigma},$$

in rectangular coordinates by applying the divergence theorem (117) to each component.

(b) In a similar way, establish the "curl theorem,"

$$\iiint_\mathfrak{R} \nabla \times \mathbf{F}\,d\tau = \oiint_S d\boldsymbol{\sigma} \times \mathbf{F}.$$

(c) Show that

$$\iiint_\mathfrak{R} (\varphi\nabla \cdot \mathbf{F} + \mathbf{F} \cdot \nabla\varphi)\,d\tau = \oiint_S \varphi\mathbf{F} \cdot d\boldsymbol{\sigma},$$

by applying (116) to the vector $\mathbf{V} = \varphi\mathbf{F}$.

Section 6.14

80. Show that Equations (123), (126), and (127) imply the relations

$$\iint_D (\varphi_1 \nabla^2\varphi_2 - \varphi_2\nabla^2\varphi_1)\,dx\,dy = \oint_C \left(\varphi_1 \frac{\partial \varphi_2}{\partial n} - \varphi_2 \frac{\partial \varphi_1}{\partial n} \right) ds,$$

$$\iint_D [\varphi\nabla^2\varphi + (\nabla\varphi)^2]\,dx\,dy = \oint_C \varphi \frac{\partial \varphi}{\partial n}\,ds,$$

$$\iint_D \nabla^2\varphi\,dx\,dy = \oint_C \frac{\partial \varphi}{\partial n}\,ds,$$

where $\varphi = \varphi(x, y)$ and $\nabla^2 = (\partial^2/\partial x^2) + (\partial^2/\partial y^2)$, and where D is the region in the xy plane bounded by C.

81. Verify the validity of the second equation of Problem 80 when $\varphi = x$ and D is the circular disk of radius a with center at the origin. Use polar coordinates in the right-hand member and notice that, on the boundary C, there follows

$$\frac{\partial x}{\partial n} = \frac{\partial x}{\partial r} = \cos\theta.$$

82. Suppose that $\nabla^2\varphi = 0$ everywhere in a region $\Re$ bounded by a closed surface S. Establish the following results in that case:

(a) $\displaystyle\oint\!\!\!\oint_S \frac{\partial\varphi}{\partial n}\, d\sigma = 0$, (b) $\displaystyle\oint\!\!\!\oint_S \varphi\frac{\partial\varphi}{\partial n}\, d\sigma = \int\!\!\!\int\!\!\!\int_\Re (\nabla\varphi)^2\, d\tau$.

83. By writing $\varphi_1 = \varphi$ and taking $\varphi_2 = \nabla^2\varphi$ in Equation (123), obtain the relation

$$\int\!\!\!\int\!\!\!\int_\Re [\varphi\nabla^4\varphi - (\nabla^2\varphi)^2]\, d\tau = \oint\!\!\!\oint_S \left[\varphi\frac{\partial\,\nabla^2\varphi}{\partial n} - (\nabla^2\varphi)\frac{\partial\phi}{\partial n}\right] d\sigma.$$

84. Derive the following generalizations of Green's theorems (122) and (123) from the divergence theorem, assuming that p is continuously differentiable:

(a) $\displaystyle\int\!\!\!\int\!\!\!\int_\Re [\varphi_1\nabla\cdot p\nabla\varphi_2 + p\nabla\varphi_1\cdot\nabla\varphi_2]\, d\tau = \oint\!\!\!\oint_S p\varphi_1\frac{\partial\varphi_2}{\partial n}\, d\sigma$,

(b) $\displaystyle\int\!\!\!\int\!\!\!\int_R [\varphi_1\nabla\cdot p\nabla\varphi_2 - \varphi_2\nabla\cdot p\nabla\varphi_1]\, d\tau = \oint\!\!\!\oint_S p\left(\varphi_1\frac{\partial\varphi_2}{\partial n} - \varphi_2\frac{\partial\varphi_1}{\partial n}\right) d\sigma$.

[Write $\mathbf{V} = p\varphi_1\nabla\varphi_2$ (in place of $\varphi_1\nabla\varphi_2$) in the divergence theorem.]

Section 6.15

85. Suppose that φ satisfies Laplace's equation everywhere in a region $\Re$ bounded by a closed surface S.

(a) If $\partial\varphi/\partial n$ vanishes everywhere on S, deduce from Problem 82(b) that φ must have a constant value in $\Re$.

(b) If φ vanishes everywhere on S, deduce from Problem 82(b) that φ must vanish everywhere in $\Re$.

86. Suppose that φ satisfies the equation

$$\nabla^4\varphi \equiv \frac{\partial^4\varphi}{\partial x^4} + 2\frac{\partial^4\varphi}{\partial x^2\,\partial y^2} + \frac{\partial^4\varphi}{\partial y^4} = 0,$$

known as the *bi-Laplacian equation*, everywhere in a region $\Re$ bounded by a closed surface S, and that both φ and $\partial\varphi/\partial n$ vanish everywhere on S. Deduce from Problem 83 that φ then must also satisfy Laplace's equation everywhere in $\Re$ and hence, by virtue of the result of Problem 85(b), must vanish everywhere in $\Re$.

Section 6.16

87. If S is a closed surface in a region $\Re$ where the vector $\mathbf{V}$ is continuously differentiable, show that

$$\oint\!\!\!\oint_S \mathbf{n}\cdot\nabla\times\mathbf{V}\, d\sigma = 0.$$

88. Verify the truth of Stokes's theorem, as given in Equation (133), in the case when $\mathbf{V} = y\mathbf{i} + 2x\mathbf{j} + z\mathbf{k}$, if C is the circle $x^2 + y^2 = 1$ (or $x = \cos t$, $y = \sin t$) in the xy plane, and S is the plane area bounded by C.

89. Use Stokes's theorem to determine the value of the integral $\iint_S \mathbf{n} \cdot \boldsymbol{\nabla} \times \mathbf{V} \, d\sigma$ over the part of the unit sphere $x^2 + y^2 + z^2 = 1$ above the xy plane, when $\mathbf{V} = y\mathbf{i}$.

90. Use Stokes's theorem to determine the value of the integral

$$\oint_C [(1 + y)z \, dx + (1 + z)x \, dy + (1 + x)y \, dz]$$

with each of the following definitions of C:

(a) The circle $x = \cos\theta$, $y = \sin\theta$, $z = 1$, oriented in the direction of increasing θ.

(b) The triangle with vertices at $P_1(1, 0, 0)$, $P_2(0, 1, 0)$, and $P_3(0, 0, 1)$, oriented from P_1 to P_2.

(c) A closed curve in the plane $x - 2y + z = 1$.

91. Show that the formulas

$$\oint_C x \, dy = -\oint_C y \, dx = A,$$

and infinitely many others, can be used in place of Equation (140) for the purpose of determining the plane area bounded by a simple closed curve.

92. Show that

(a) $\frac{1}{2}\oint_C x^2 \, dy = -\oint_C xy \, dy = \frac{1}{3}\oint_C (x^2 \, dy - xy \, dx) = A\bar{x}$,

(b) $\frac{1}{3}\oint_C x^3 \, dy = -\oint_C x^2 y \, dy = \frac{1}{4}\oint_C (x^3 \, dy - x^2 y \, dx) = I_y$,

where A is the area in the xy plane bounded by a simple closed curve C, $(\bar{x}, \bar{y})$ is its center of gravity, and I_y its moment of inertia about the y axis.

93. For any simple closed circuit C bounding an open surface S in a region $\mathfrak{R}$, the electric intensity vector $\mathbf{E}$ and the magnetic intensity vector $\mathbf{H}$ satisfy the relations

$$\oint_C \mathbf{E} \cdot d\mathbf{r} = -\alpha\frac{\partial}{\partial t}\iint_S \mathbf{H} \cdot d\boldsymbol{\sigma}, \qquad \oint_C \mathbf{H} \cdot d\mathbf{r} = \beta\frac{\partial}{\partial t}\iint_S \mathbf{E} \cdot d\boldsymbol{\sigma},$$

where α and β are certain constants. Use Stokes's theorem to transform these relations to the form

$$\iint_S \left(\boldsymbol{\nabla} \times \mathbf{E} + \alpha\frac{\partial \mathbf{H}}{\partial t}\right) \cdot d\boldsymbol{\sigma} = 0, \qquad \iint_S \left(\boldsymbol{\nabla} \times \mathbf{H} - \beta\frac{\partial \mathbf{E}}{\partial t}\right) \cdot d\boldsymbol{\sigma} = 0.$$

From the arbitrariness of S, deduce *Maxwell's equations* in the form

$$\boldsymbol{\nabla} \times \mathbf{E} = -\alpha\frac{\partial \mathbf{H}}{\partial t}, \qquad \boldsymbol{\nabla} \times \mathbf{H} = \beta\frac{\partial \mathbf{E}}{\partial t}.$$

94. In a region free of electric and magnetic charges, it is true that $\boldsymbol{\nabla} \cdot \mathbf{E} = 0$ and $\boldsymbol{\nabla} \cdot \mathbf{H} = 0$. By eliminating $\mathbf{E}$ and $\mathbf{H}$ successively between the equations of Problem 93, and using this fact, deduce that $\mathbf{E}$ and $\mathbf{H}$ both satisfy the *vector wave equation*

$$\nabla^2\mathbf{V} = \alpha\beta\frac{\partial^2\mathbf{V}}{\partial t^2}$$

in such a region.

95. Let S be an open surface bounded by a simple closed curve C.

(a) Establish the relation

$$\iint_S d\boldsymbol{\sigma} \times \nabla\varphi = \oint_C \varphi \, d\mathbf{r}$$

in rectangular coordinates by applying Stokes's theorem (138) to each component.

(b) Show that

$$\oint_C \varphi_1 \nabla\varphi_2 \cdot d\mathbf{r} = -\oint_C \varphi_2 \nabla\varphi_1 \cdot d\mathbf{r},$$

by applying (133) to the vector $\mathbf{V} = \nabla(\varphi_1\varphi_2)$.

(c) Show that

$$\iint_S (\nabla\varphi \times \mathbf{F} + \varphi\nabla \times \mathbf{F}) \cdot d\boldsymbol{\sigma} = \oint_C \varphi\mathbf{F} \cdot d\mathbf{r},$$

by applying (133) to the vector $\mathbf{V} = \varphi\mathbf{F}$.

Section 6.17

96. *Elliptical cylindrical coordinates* may be defined by the equations

$$x = a \cosh u \cos v, \qquad y = a \sinh u \sin v, \qquad z = z,$$

where $u \geq 0$ and $0 \leq v < 2\pi$.

(a) Show that this system of coordinates (u, v, z) is orthogonal [by verifying that Equation (149) is satisfied].

(b) Show that in the xy plane a curve $u = $ constant is an ellipse with semi-axes $a \cosh u$, in the x direction, and $a \sinh u$, in the y direction; also that a curve $v = $ constant is *half of one branch* of an hyperbola with semi-axes $a \cos v$ and $a \sin v$. In particular, show that the locus $u = 0$ degenerates into the segment $(-a, a)$ of the x axis, while the loci $v = 0$ and $v = \pi$ are respectively the positive and negative exteriors of this segment; also that the loci $v = \pi/2$ and $v = 3\pi/2$ are respectively the positive and negative portions of the y axis. Sketch and label in a single diagram the curves $u = 0$, 1 and $v = 0$, $\pi/4$, $\pi/2$, $3\pi/4$, $5\pi/4$, $3\pi/2$, and $7\pi/4$.

97. For the coordinates of Problem 96 derive the relations analogous to those of Equations (168b–e) for circular cylindrical coordinates. In particular, verify that

$$h_u = h_v = a\sqrt{\cosh^2 u - \cos^2 v}, \qquad h_z = 1,$$

$$\mathbf{u}_1 = \frac{\mathbf{i}\sinh u \cos v + \mathbf{j}\cosh u \sin v}{\sqrt{\cosh^2 u - \cos^2 v}},$$

$$\mathbf{u}_2 = \frac{-\mathbf{i}\cosh u \sin v + \mathbf{j}\sinh u \cos v}{\sqrt{\cosh^2 u - \cos^2 v}},$$

$$\nabla f = \frac{1}{a\sqrt{\cosh^2 u - \cos^2 v}}\left(\mathbf{u}_1\frac{\partial f}{\partial u} + \mathbf{u}_2\frac{\partial f}{\partial v}\right) + \mathbf{k}\frac{\partial f}{\partial z},$$

$$\nabla^2 f = \frac{1}{a^2\,(\cosh^2 u - \cos^2 v)}\left(\frac{\partial^2 f}{\partial u^2} + \frac{\partial^2 f}{\partial v^2}\right) + \frac{\partial^2 f}{\partial z^2}.$$

Show also that *for large values of u* there follows

$$\mathbf{u}_1 \sim \mathbf{i} \cos v + \mathbf{j} \sin v, \qquad \mathbf{u}_2 \sim -\mathbf{i} \sin v + \mathbf{j} \cos v,$$

and $$\mathbf{u}_1 \cos v - \mathbf{u}_2 \sin v \sim \mathbf{i} \qquad \mathbf{u}_1 \sin v + \mathbf{u}_2 \cos v \sim \mathbf{j}.$$

98. *Parabolic cylindrical coordinates* may be defined by the equations

$$x = \tfrac{1}{2}(u^2 - v^2), \qquad y = uv, \qquad z = z,$$

where $-\infty < u < \infty$ and $v \geq 0$.

(a) Show that this system is orthogonal.

(b) Show that in the xy plane a curve $v = $ constant is a parabola symmetrical about the x axis and opening to the right, while a curve $u = $ constant is *one half* of a similar parabola opening to the left. In particular, show that the locus $v = 0$ is the positive x axis, while the locus $u = 0$ is the negative x axis, and that the positive y axis is given by $u = v$ and the negative y axis by $u = -v$. Sketch and label in a single diagram the curves $u = 0, \pm 1$ and $v = 0, 1$.

(c) Perform the calculations necessary to show that the Laplacian is of the form

$$\nabla^2 f = \frac{1}{u^2 + v^2}\left(\frac{\partial^2 f}{\partial u^2} + \frac{\partial^2 f}{\partial v^2}\right) + \frac{\partial^2 f}{\partial z^2}.$$

99. *Paraboloidal coordinates* correspond to a system in which the plane configuration of Problem 98 is rotated about the axis of symmetry of the two sets of parabolas. The axis of rotation is then conventionally taken as the z axis.

(a) In Problem 98 replace x by z and y by $\sqrt{x^2 + y^2}$, the distance from the z axis; then write $y = x \tan \theta$, so that θ is the circumferential angle, and show that the coordinate transformation becomes

$$x = uv \cos \theta, \qquad y = uv \sin \theta, \qquad z = \tfrac{1}{2}(u^2 - v^2),$$

where now $u \geq 0$ and $v \geq 0$.

(b) Obtain the Laplacian in the form

$$\nabla^2 f = \frac{1}{u^2 + v^2}\left[\frac{1}{u}\frac{\partial}{\partial u}\left(u\frac{\partial f}{\partial u}\right) + \frac{1}{v}\frac{\partial}{\partial v}\left(v\frac{\partial f}{\partial v}\right)\right] + \frac{1}{u^2 v^2}\frac{\partial^2 f}{\partial \theta^2}$$

100. In a *translation* and *rotation* of axes, in which the origin in the new $x'y'z'$ plane is taken at the point (a, b, c) in the xyz plane, and in which the directions of the x', y', and z' axes are specified by the direction cosines (l_1, m_1, n_1), (l_2, m_2, n_2), and (l_3, m_3, n_3) relative to the original axes, the transformation of coordinates is of the form

$$x = a + l_1 x' + l_2 y' + l_3 z',$$
$$y = b + m_1 x' + m_2 y' + m_3 z',$$
$$z = c + n_1 x' + n_2 y' + n_3 z'.$$

Show that Laplace's equation is of the form

$$\frac{\partial^2 f}{\partial x'^2} + \frac{\partial^2 f}{\partial y'^2} + \frac{\partial^2 f}{\partial z'^2} = 0$$

in terms of the new variables.

101. Suppose that the coordinates u_1 and u_2 are related to x and y by an equation of the form

$$x + iy = F(u_1 + iu_2),$$

where $i^2 = -1$, so that x is the real part of $F(u_1 + iu_2)$ and y the imaginary part, and that u_1, u_2, and z are chosen as curvilinear coordinates in space.

(a) Show that, if F is a differentiable function of the argument $u_1 + iu_2$, there follows

$$\frac{\partial x}{\partial u_1} + i\frac{\partial y}{\partial u_1} = F'(u_1 + iu_2), \qquad \frac{\partial x}{\partial u_2} + i\frac{\partial y}{\partial u_2} = iF'(u_1 + iu_2),$$

where a prime denotes differentiation with respect to the complete argument $u_1 + iu_2$, and hence that

$$\sqrt{\left(\frac{\partial x}{\partial u_1}\right)^2 + \left(\frac{\partial y}{\partial u_1}\right)^2} = \sqrt{\left(\frac{\partial x}{\partial u_2}\right)^2 + \left(\frac{\partial y}{\partial u_2}\right)^2} = |F'(u_1 + iu_2)|.$$

(b) Deduce also that

$$\frac{\partial x}{\partial u_1} + i\frac{\partial y}{\partial u_1} = -i\frac{\partial x}{\partial u_2} + \frac{\partial y}{\partial u_2},$$

and hence, by equating real and imaginary parts, that

$$\frac{\partial x}{\partial u_1} = \frac{\partial y}{\partial u_2}, \qquad \frac{\partial y}{\partial u_1} = -\frac{\partial x}{\partial u_2}.$$

(c) With the notation of Section 6.17, deduce that the vectors $\mathbf{U}_1$, $\mathbf{U}_2$, and $\mathbf{U}_3$ are mutually orthogonal, and that

$$h_1 = h_2 = |F'(u_1 + iu_2)|, \qquad h_3 = 1.$$

Hence show that Laplace's equation is of the form

$$\frac{\partial^2 f}{\partial u_1^2} + \frac{\partial^2 f}{\partial u_2^2} + h^2\frac{\partial^2 f}{\partial z^2} = 0,$$

where $h = |F'(u_1 + iu_2)|$.

102. (a) If x and y are related to u_1 and u_2 by the equation

$$x + iy = \tfrac{1}{2}(u_1 + iu_2)^2,$$

so that $x = \tfrac{1}{2}(u_1^2 - u_2^2)$ and $y = u_1 u_2$, use the result of Problem 101 to obtain Laplace's equation in $u_1 u_2 z$ coordinates in the form

$$\frac{\partial^2 f}{\partial u_1^2} + \frac{\partial^2 f}{\partial u_2^2} + (u_1^2 + u_2^2)\frac{\partial^2 f}{\partial z^2} = 0.$$

(b) Obtain the same result by using the formulas of Section 6.17. (See also Problem 98.)

Section 6.18

103. If $\mathbf{u}_r$ and $\mathbf{u}_\theta$ are the unit tangent vectors in the r and θ directions, in circular cylindrical coordinates, show that

$$\mathbf{i} = \mathbf{u}_r\cos\theta - \mathbf{u}_\theta\sin\theta, \qquad \mathbf{j} = \mathbf{u}_r\sin\theta + \mathbf{u}_\theta\cos\theta.$$

104. If r, θ, and z are circular cylindrical coordinates, evaluate the following quantities:
(a) $\nabla\theta$,   (b) ∇r^n,   (c) $\nabla\times\mathbf{u}_\theta$,
(d) $\nabla\cdot[r^{n-1}(\mathbf{u}_r\sin n\theta + \mathbf{u}_\theta\cos n\theta)]$,   (e) $\nabla^2(r^2\cos\theta)$,
(f) $\nabla^2(r^n\cos n\theta)$.

105. If $\mathbf{u}_r$, $\mathbf{u}_\varphi$, and $\mathbf{u}_\theta$ are the unit vectors tangent to the coordinates curves in spherical coordinates, show that

$$\mathbf{i} = (\mathbf{u}_r \sin\varphi + \mathbf{u}_\varphi \cos\varphi)\cos\theta - \mathbf{u}_\theta \sin\theta,$$
$$\mathbf{j} = (\mathbf{u}_r \sin\varphi + \mathbf{u}_\varphi \cos\varphi)\sin\theta + \mathbf{u}_\theta \cos\theta,$$
$$\mathbf{k} = \mathbf{u}_r \cos\varphi - \mathbf{u}_\varphi \sin\varphi.$$

106. If r, φ, and θ are spherical coordinates, evaluate the following quantities:

(a) $\nabla\varphi$, (b) $\nabla\theta$,

(c) $\nabla \cdot [\mathbf{u}_r \cot\varphi - 2\mathbf{u}_\varphi]$, (d) $\nabla^2\left[\left(r + \dfrac{1}{r^2}\right)\cos\varphi\right]$.

107. Show that the unit tangent vectors in spherical coordinates satisfy the following relations:

$$\frac{\partial \mathbf{u}_r}{\partial \varphi} = \mathbf{u}_\varphi, \qquad \frac{\partial \mathbf{u}_r}{\partial \theta} = \mathbf{u}_\theta \sin\varphi, \qquad \frac{\partial \mathbf{u}_\varphi}{\partial \varphi} = -\mathbf{u}_r, \qquad \frac{\partial \mathbf{u}_\varphi}{\partial \theta} = \mathbf{u}_\theta \cos\varphi,$$

$$\frac{\partial \mathbf{u}_\theta}{\partial \varphi} = 0, \qquad \frac{\partial \mathbf{u}_\theta}{\partial \theta} = -\mathbf{u}_r \sin\varphi - \mathbf{u}_\varphi \cos\varphi.$$

108. By writing the position vector in the form $\mathbf{r} = r\mathbf{u}_r$, and using the results of Problem 107, obtain expressions for components of acceleration along the coordinate curves in spherical coordinates as follows:

$$a_r = \ddot{r} - r\dot{\varphi}^2 - r\dot{\theta}^2 \sin^2\varphi,$$
$$a_\varphi = 2\dot{r}\dot{\varphi} + r\ddot{\varphi} - r\dot{\theta}^2 \sin\varphi \cos\varphi,$$
$$a_\theta = 2\dot{r}\dot{\theta}\sin\varphi + r\ddot{\theta}\sin\varphi + 2r\dot{\varphi}\dot{\theta}\cos\varphi.$$

109. Prove that

$$\nabla^2[r^n P_n(\cos\varphi)] = 0,$$

where r and φ are spherical coordinates and P_n is the Legendre polynomial of order n. [See Equation (169) of Chapter 4.]

110. (a) Evaluate the surface integral

$$\iint_S \mathbf{F} \cdot d\boldsymbol{\sigma},$$

where $\mathbf{F} = x\mathbf{i} - y\mathbf{j} + z\mathbf{k}$ and where S is the lateral surface of the cylinder $x^2 + y^2 = 1$ between the planes $z = 0$ and $z = 1$, using right circular cylindrical coordinates.

(b) Check the result by use of the divergence theorem.

111. (a) Evaluate the surface integral of $\mathbf{F} = x\mathbf{i} - y\mathbf{j} + z\mathbf{k}$ over the closed surface of the sphere $x^2 + y^2 + z^2 = 1$, using spherical coordinates.

(b) Check the result by use of the divergence theorem.

Section 6.19

112. Suppose that a flow of an ideal incompressible fluid is free of distributions of sources and sinks, and of vortices, and that it takes place parallel to the xy plane.

(a) Show that the velocity potential $\varphi(x, y)$ and the stream function $\psi(x, y)$ are such that

$$V_x = \frac{\partial\varphi}{\partial x} = \frac{\partial\psi}{\partial y}, \qquad V_y = \frac{\partial\varphi}{\partial y} = -\frac{\partial\psi}{\partial x},$$

and that φ and ψ satisfy Laplace's equation.

(b) Show that

$$\varphi(x, y) = \int^{(x,y)} (V_x\, dx + V_y\, dy), \qquad \psi(x, y) = \int^{(x,y)} (-V_y\, dx + V_x\, dy),$$

where the line integrals are each evaluated along an arbitrary path from a fixed point to the variable point (x, y).

113. (a) Verify that the function $\varphi(x, y) = x^3 - 3xy^2$ is a velocity potential function.

(b) Determine the velocity vector $\mathbf{V}$, and its magnitude V.

(c) Determine the stream function $\psi(x, y)$, subject to the condition $\psi(0, 0) = 0$, and obtain the equation of the streamlines.

(d) If the uniform density of the fluid is ρ, determine the rate of flow across an arc joining the points $(1, 1)$ and $(2, 2)$ (that is, through a cylindrical surface of unit height having this arc as its base).

114. (a) If elliptical cylindrical coordinates (Problems 96 and 97) are used to describe the flow of an ideal fluid parallel to the xy plane and independent of z, show that the velocity potential satisfies the equation

$$\frac{\partial^2\varphi}{\partial u^2} + \frac{\partial^2\varphi}{\partial v^2} = 0.$$

(b) Show that the corresponding stream function ψ satisfies an equation of the same form and that ψ may be related to φ by the equation

$$d\psi = -\frac{\partial\varphi}{\partial v}\, du + \frac{\partial\varphi}{\partial u}\, dv.$$

115. (a) Show that

$$\varphi = C \cosh u \sin v$$

is a permissible potential function in Problem 114 and that the corresponding stream function is then of the form

$$\psi = -C \sinh u \cos v,$$

if an arbitrary additive constant is discarded.

(b) Show that the streamline $\psi = 0$ consists of the combination of the loci $u = 0$ and both $v = \pi/2$ and $v = 3\pi/2$, and hence deduce from continuity considerations that neighboring streamlines follow near the y axis toward the x axis, thence around a perpendicular barrier (plate) of breadth $2a$ back toward the y axis and onward nearly parallel to the y axis.

(c) Show that the flow velocity is given by

$$\mathbf{V} = \frac{C}{a\sqrt{\cosh^2 u - \cos^2 v}} (\mathbf{u}_1 \sinh u \sin v + \mathbf{u}_2 \cosh u \cos v),$$

and that at large distances from the barrier $(u \to \infty)$ the velocity vector tends to the constant value

$$\mathbf{V} \sim \frac{C}{a}(\mathbf{u}_1 \sin v + \mathbf{u}_2 \cos v) \sim \frac{C}{a}\mathbf{j} \qquad (u \longrightarrow \infty).$$

Deduce that the potential

$$\varphi = aV_0 \cosh u \sin v$$

corresponds to a flow, around a plate of breadth $2a$, which tends to a uniform flow, with velocity V_0 at right angles to the plate, at large distances from it.

(d) Show that at a point (u, v, z) the flow velocity is of magnitude

$$V = V_0\sqrt{\frac{\cosh^2 u - \sin^2 v}{\cosh^2 u - \cos^2 v}}.$$

In particular, verify that at points on the plate $(u = 0)$ there follows

$$V = V_0\frac{|x|}{\sqrt{a^2 - x^2}} \qquad (|x| < a),$$

while for points on the x axis outside the plate $(v = 0, \pi)$

$$V = V_0\frac{|x|}{\sqrt{x^2 - a^2}} \qquad (|x| > a),$$

and for points on the y axis $(v = \pi/2, 3\pi/2)$

$$V = V_0\frac{|y|}{\sqrt{a^2 + y^2}}.$$

116. Suppose that a source-free and vortex-free flow of an incompressible fluid possesses axial symmetry about the z axis, so that its velocity can be expressed in the form

$$\mathbf{V} = V_z\mathbf{k} + V_r\mathbf{u}_r,$$

in terms of circular cylindrical coordinates.

(a) With $(u_1, u_2, u_3) = (z, r, \theta)$, show that $h_1 = h_2 = 1$ and $h_3 = r$, and that

$$V_z = \frac{\partial\varphi}{\partial z}, \qquad V_r = \frac{\partial\varphi}{\partial r},$$

where $\varphi(z, r)$ satisfies the equation

$$\frac{1}{r}\frac{\partial}{\partial r}\left(r\frac{\partial\varphi}{\partial r}\right) + \frac{\partial^2\varphi}{\partial z^2} = 0.$$

(b) Show that the stream function ψ is determined by the relation

$$d\psi = r(-V_r\,dz + V_z\,dr)$$

and verify that the right-hand member is indeed an exact differential.

(c) Show that the rate of flow through the surface obtained by rotating about the z axis an arc C joining two points P_0 and P_1 in a plane $\theta = $ constant is given by

$$2\pi\rho\int_C d\psi = 2\pi\rho[\psi(P_1) - \psi(P_0)].$$

117. (a) Verify that the function $\varphi(z, r) = 2z^2 - r^2$ is a potential function corresponding to the type of flow considered in Problem 116.

(b) Determine the velocity vector $\mathbf{V}$, and its magnitude V.

(c) Determine the stream function ψ, and show that the streamlines are the curves in the planes $\theta = $ constant for which $r^2 z = $ constant.

(d) Show that the rate of flow through the surface obtained by rotating about the z axis an arc joining P_0 and P_1 in a plane $\theta = $ constant is given by $4\pi\rho(r_1^2 z_1 - r_0^2 z_0)$.

Section 6.20

118. If, in the developments of Section 6.20, a *body force* $\mathbf{F}$ per unit mass is assumed to be active, show that $\rho\mathbf{F}$ must be added to the right-hand members of Equations (185) and (188).

119. If a fluid is *viscous*, the force on a surface element $d\sigma$, due to internal friction, is usually assumed to have a component in any direction equal to μ times the product of $d\sigma$ and the derivative, normal to $d\sigma$, of the velocity component in that direction, where μ is a constant known as the *coefficient of viscosity*.

(a) Show that the viscous force in the x direction on an element $d\sigma$ then is of the magnitude

$$\mu\mathbf{n}\cdot\nabla V_x\,d\sigma,$$

and that the net viscous force, in the x direction, on the closed boundary S of a region $\mathfrak{R}$ is of the magnitude

$$\mu\oiint_S \mathbf{n}\cdot\nabla V_x\,d\sigma = \mu\iiint_{\mathfrak{R}} \nabla^2 V_x\,d\tau.$$

Hence deduce that the viscous force on an element $d\tau$ is equivalent to a body force with component $\mu\nabla^2 V_x\,d\tau$ in the x direction, and analogous components in the y and z directions, and hence to a body-force vector $\mu\nabla^2\mathbf{V}\,d\tau$.

(b) Deduce that effects of viscosity then may be taken into account by adding the term $\mu\nabla^2\mathbf{V}$ to the right-hand members of (185) and (188).

120. If a body force $\mathbf{F}$ per unit mass is conservative, and hence can be written in the form

$$\mathbf{F} = -\nabla U,$$

where U is a potential energy function, use the result of Problem 118 to show that Equation (211) is replaced by the equation

$$\frac{1}{2}V^2 + \frac{\partial\varphi}{\partial t} + P + U = f(t),$$

when $\mathbf{F}$ is present, but viscosity effects are neglected.

121. If the Mach number M is greater than unity, show that any expression of the form

$$\varphi(x,y) = f(x + \alpha t) + g(x - \alpha t) \qquad (\alpha = \sqrt{M^2 - 1})$$

satisfies Equation (200), where f and g are any twice-differentiable functions.

122. Verify that the expression

$$\varphi(x,t) = F[x - (U + V_{S0})t] + G[x - (U - V_{S0})t]$$

satisfies Equation (202), where F and G are any twice-differentiable functions.

7

Topics in Higher-Dimensional Calculus

7.1. Partial Differentiation. Chain Rules. In this section we review and discuss certain notations and relations involving partial derivatives which will be needed in the sequel.

The more general case may be illustrated here by considering a function f of three variables x, y, and z,

$$f = f(x, y, z). \tag{1}$$

If y and z are held constant and only x is allowed to vary, the partial derivative with respect to x is denoted by $\partial f/\partial x$ and is defined as the limit

$$\frac{\partial f(x, y, z)}{\partial x} = \lim_{\Delta x \to 0} \frac{f(x + \Delta x, y, z) - f(x, y, z)}{\Delta x}. \tag{2}$$

Similarly we define the functions $\partial f/\partial y$ and $\partial f/\partial z$. In all cases two of the three variables explicitly appearing in the definition of f are held constant, and f is differentiated with respect to the third variable. The total differential of f is defined by the equation

$$df = \frac{\partial f}{\partial x} dx + \frac{\partial f}{\partial y} dy + \frac{\partial f}{\partial z} dz, \tag{3}$$

whether or not x, y, and z are independent of each other, provided only that the partial derivatives involved are continuous. Several types of dependence among x, y, and z are now considered. In each of the formulas to be obtained, the continuity of all derivatives appearing in the right-hand member is to be assumed.

(1) If x, y, and z are all functions of a single variable, say t, then the *dependent variable f* may also be considered as truly a function of the one *independent*

342

variable t, and we may conveniently speak of x, y, and z as *intermediate variables*. Since only one independent variable is present, df/dt has a meaning and it can be shown, by appropriate limiting processes, that

$$\frac{df}{dt} = \frac{\partial f}{\partial x}\frac{dx}{dt} + \frac{\partial f}{\partial y}\frac{dy}{dt} + \frac{\partial f}{\partial z}\frac{dz}{dt}. \tag{4}$$

This result is formally obtained by dividing the expression for df by dt. We notice that df/dt is the sum of three terms, each of which represents the contribution of the change in t through the corresponding change in one of the *intermediate* variables.

(2) More generally, the intermediate variables x, y, and z may be functions of two (or more) independent variables, say s and t. Then if we consider f as a function of s and t, we may investigate the *partial* derivative of f with respect to t when s is held constant. If we denote this function by the notation $(df/dt)_s$, then Equation (4) must be modified to read

$$\left(\frac{\partial f}{\partial t}\right)_s = \frac{\partial f}{\partial x}\left(\frac{\partial x}{\partial t}\right)_s + \frac{\partial f}{\partial y}\left(\frac{\partial y}{\partial t}\right)_s + \frac{\partial f}{\partial z}\left(\frac{\partial z}{\partial t}\right)_s. \tag{5a}$$

The derivatives with respect to t are now not total but partial. The subscript indicates the variable held constant. With this convention we should perhaps also write $(\partial f/\partial x)_{y,z}$ in place of $\partial f/\partial x$, and so on, in (4) and (5a). However, *we will follow the convention that $\partial f/\partial x$, without subscripts, indicates the result of differentiating f with respect to the explicitly appearing variable x, holding all other explicitly appearing variables* (here y and z) *constant*. Frequently, we also omit the subscript s in (5a) and write merely

$$\frac{\partial f}{\partial t} = \frac{\partial f}{\partial x}\frac{\partial x}{\partial t} + \frac{\partial f}{\partial y}\frac{\partial y}{\partial t} + \frac{\partial f}{\partial z}\frac{\partial z}{\partial t}, \tag{5b}$$

if it is clear from the context that s and t are to be associated with each other as the independent variables, with x, y, and z as the (explicit) intermediate variables. Alternatively, we could write $F(s, t)$ for the result of replacing x, y, and z by their equivalents in $f(x, y, z)$, so that

$$f[x(s, t), y(s, t), z(s, t)] = F(s, t),$$

in accordance with which (5a, b) could be written in the form

$$\frac{\partial F}{\partial t} = \frac{\partial f}{\partial x}\frac{\partial x}{\partial t} + \frac{\partial f}{\partial y}\frac{\partial y}{\partial t} + \frac{\partial f}{\partial z}\frac{\partial z}{\partial t}, \tag{5c}$$

without any possible ambiguity. Whereas this is the most elegant way of proceeding in such situations, it is often inconvenient in practice to use two different symbols (here f and F) to represent the same physical or geometrical quantity.

In the preceding cases the independent variable t does not appear explicitly in f, and its changes are reflected in f only through the intermediate changes in x, y, and z. However, it may be convenient to take an explicitly appearing variable as an independent variable. We again distinguish two cases.

(3) If we suppose that y and z are functions of x, then f is a function of the one independent variable x, and y and z are intermediate. Also, identifying t with x in Equation (4), we obtain

$$\frac{df}{dx} = \frac{\partial f}{\partial x} + \frac{\partial f}{\partial y}\frac{dy}{dx} + \frac{\partial f}{\partial z}\frac{dz}{dx}. \tag{6}$$

The term $\partial f/\partial x$ in (6) is obtained, as before, by holding the other two explicit variables (y and z) constant, and it represents the contribution of the explicit variation of x. The other terms add the contributions of the intermediate variations in y and z.

(4) If we suppose that x and y are independent but that z is a function of both x and y, then f can be considered as depending upon x and y *directly* and also *intermediately* through z. Also, identifying t and s with x and y in (5a), we obtain

$$\left(\frac{\partial f}{\partial x}\right)_y = \frac{\partial f}{\partial x} + \frac{\partial f}{\partial z}\left(\frac{\partial z}{\partial x}\right)_y. \tag{7a}$$

Here the notation is rather treacherous. On the left-hand side of (7a) we think of f as being actually expressed in terms of x and y, the variable z having been replaced by its equivalent in terms of these variables. Then we imagine that y is held constant in the x differentiation. On the right-hand side f is expressed in its original form, in terms of x, y, and z. The first term on the right *is again calculated with the other explicit variables y and z held constant*, and it represents the contribution due to the explicit variation of x. The other term adds the contribution of the only intermediate variable z. Since z depends only on x and y, the last subscript may be omitted without confusion. However, the subscript on the left is clearly essential. Thus we may write Equation (7a) in the form

$$\left(\frac{\partial f}{\partial x}\right)_y = \frac{\partial f}{\partial x} + \frac{\partial f}{\partial z}\frac{\partial z}{\partial x}. \tag{7b}$$

These formulas are useful when we deal with a function f in abstract terms. If f is given as a specific function and the dependencies are specifically stated, such formulas usually are not needed. In illustration we consider the function

$$f(x, y, z) = x^2 + xz + 2y^2.$$

If we consider x, y, and z as functions of t, and possibly other independent variables, we merely differentiate term by term and obtain

$$\frac{\partial f}{\partial t} = 2x\frac{\partial x}{\partial t} + x\frac{\partial z}{\partial t} + z\frac{\partial x}{\partial t} + 4y\frac{\partial y}{\partial t}$$

$$= (2x + z)\frac{\partial x}{\partial t} + 4y\frac{\partial y}{\partial t} + x\frac{\partial z}{\partial t}.$$

This result is the same as that given by Equation (5b). If we consider x as independent and assume that y and z are given by other equations as functions of x, we again differentiate term by term and obtain

$$\frac{df}{dx} = 2x + x\frac{dz}{dx} + z + 4y\frac{dy}{dx}$$

$$= 2x + z + 4y\frac{dy}{dx} + x\frac{dz}{dx},$$

in accordance with the result of using Equation (6). If z is given by another equation in terms of x and y, and if x and y are independent, then we obtain directly, holding y constant,

$$\left(\frac{\partial f}{\partial x}\right)_y = 2x + z + x\frac{\partial z}{\partial x},$$

in accordance with the result of using (7b). The term $2x + z$ is equivalent to $\partial f/\partial x$, where y and z are held constant. The term $x\,\partial z/\partial x$ corrects for the fact that here z cannot *actually* be held constant but must vary also with x.

As a further example, suppose that we have the relation

$$x^2 + xz + 2y^2 = 0.$$

We again denote the function of x, y, and z on the left by f,

$$f = x^2 + xz + 2y^2.$$

We may consider the given relation as determining, say, z in terms of x and y, both of which may then be taken as independent. Then, holding y constant and differentiating with respect to x, we obtain

$$2x + z + x\frac{\partial z}{\partial x} = 0.$$

Hence we must have

$$\frac{\partial z}{\partial x} = -\frac{2x + z}{x}.$$

We may also arrive at this result by considering the left-hand member of the given equation as a function of the independent variables x and y and the intermediate variable z. Then, since f constantly satisfies the equation

$$f(x, y, z) = 0, \tag{8}$$

the partial derivative of f with respect to either *independent* variable must vanish. But Equation (7b) then gives

$$\left(\frac{\partial f}{\partial x}\right)_y = \frac{\partial f}{\partial x} + \frac{\partial f}{\partial z}\frac{\partial z}{\partial x} = 0.$$

This equation states that the contributions of the explicit variation of x and the intermediate variation of z must cancel. In order that this be so, we must then have

$$\frac{\partial z}{\partial x} = -\frac{\partial f/\partial x}{\partial f/\partial z}, \tag{9}$$

in accordance with the result obtained above.†

---

†Equation (9) tends to illustrate the dangers associated with routine symbolic manipulations, since a formal (but unjustified) inversion and cancellation in the right-hand member might suggest that the prefixed sign is incorrect.

Partial derivatives of higher order, of a function $f(x, y, z)$, are calculated by successive differentiation. Thus we write, for example,

$$\frac{\partial^2 f}{\partial y \, \partial x} = \frac{\partial}{\partial y} \frac{\partial f}{\partial x},$$

and so forth. In this connection, we review the important fact that *the crossed partial derivatives are equal*,

$$\frac{\partial^2 f}{\partial y \, \partial x} = \frac{\partial^2 f}{\partial x \, \partial y}, \tag{10}$$

that is, the order of differentiation is immaterial, if the derivatives involved are continuous. This statement is true for derivatives of any order if they are continuous, but it may not be true otherwise.

For the purpose of obtaining analytical formulas for higher-order partial derivatives, it is often convenient to use *operational* notation (see also Problems 3, 4, and 5). As an example, we suppose that f is a function of x, y, and z, with x and y independent and z a function of x and y, so that Equation (7) applies. Here, in order to avoid the complexities involved in unambiguously generalizing the notation $(\partial f / \partial x)_y$ to higher-order derivatives, it is particularly desirable to introduce the special notation

$$f[x, y, z(x, y)] = F(x, y), \tag{11}$$

so that $F(x, y)$ is the result of replacing z by its equivalent in terms of x and y in the expression for $f(x, y, z)$. The formulas (7a, b) then can be rewritten in the form

$$\frac{\partial F}{\partial x} = \frac{\partial f}{\partial x} + \frac{\partial f}{\partial z} \frac{\partial z}{\partial x} = \left(\frac{\partial}{\partial x} + \frac{\partial z}{\partial x} \frac{\partial}{\partial z} \right) f, \tag{12}$$

from which there follows, by iteration,

$$\begin{aligned}
\frac{\partial^2 F}{\partial x^2} &= \left(\frac{\partial}{\partial x} + \frac{\partial z}{\partial x} \frac{\partial}{\partial z} \right) \left(\frac{\partial f}{\partial x} + \frac{\partial z}{\partial x} \frac{\partial f}{\partial z} \right) \\
&= \left(\frac{\partial^2 f}{\partial x^2} + \frac{\partial z}{\partial x} \frac{\partial^2 f}{\partial x \, \partial z} \right) + \frac{\partial z}{\partial x} \left(\frac{\partial^2 f}{\partial x \, \partial z} + \frac{\partial z}{\partial x} \frac{\partial^2 f}{\partial z^2} \right) + \left[\left(\frac{\partial}{\partial x} + \frac{\partial z}{\partial x} \frac{\partial}{\partial z} \right) \frac{\partial z}{\partial x} \right] \frac{\partial f}{\partial z} \\
&= \frac{\partial^2 f}{\partial x^2} + 2 \frac{\partial z}{\partial x} \frac{\partial^2 f}{\partial x \, \partial z} + \left(\frac{\partial z}{\partial x} \right)^2 \frac{\partial^2 f}{\partial z^2} + \frac{\partial^2 z}{\partial x^2} \frac{\partial f}{\partial z}.
\end{aligned} \tag{13}$$

Here, as before, the convention is that, in each of the indicated partial derivatives, all variables explicitly involved in the function being differentiated are held constant except the one with respect to which the differentiation is being effected. Thus, since $F = F(x, y)$ and $z = z(x, y)$, there follows

$$\frac{\partial F}{\partial x} \equiv \left(\frac{\partial F}{\partial x} \right)_y \quad \text{and} \quad \frac{\partial z}{\partial x} \equiv \left(\frac{\partial z}{\partial x} \right)_y .$$

On the other hand, since $f = f(x, y, z)$, there follows

$$\frac{\partial f}{\partial x} \equiv \left(\frac{\partial f}{\partial x} \right)_{y, z} .$$

7.2. Implicit Functions. Jacobian Determinants. An equation of the form

$$f(x, y, z, \dots) = 0, \tag{14}$$

involving any finite number of variables, where f possesses continuous first partial derivatives, can be considered as determining *one* of the variables, say z, as a function of the remaining variables, say

$$z = \varphi(x, y, \dots), \tag{15}$$

in *some* region about any point where Equation (14) is satisfied and where the partial derivative of f with respect to that variable exists and is not zero,

$$\frac{\partial f}{\partial z} \neq 0. \tag{16}$$

In such a case we say that Equation (14) defines z as an *implicit* function of the other variables, in the neighborhood of that point. If we consider all the other variables as independent, we can determine the partial derivative of z with respect to any one of them, without solving *explicitly* for z, by differentiating (14) partially with respect to that variable. Thus, to determine $\partial z/\partial x$, we obtain from (14)

$$\frac{\partial f}{\partial x} + \frac{\partial f}{\partial z}\frac{\partial z}{\partial x} = 0, \qquad \frac{\partial z}{\partial x} = -\frac{\partial f/\partial x}{\partial f/\partial z}, \tag{17}$$

the denominator differing from zero by virtue of Equation (16).

If $n + k$ variables are related by n equations, it is *usually* possible to consider n of the variables as functions of the remaining k variables. However, this is not *always* possible.

As an illustration, suppose that x, y, u, and v are related by two equations of the form

$$\begin{aligned} f(x, y, u, v) &= 0, \\ g(x, y, u, v) &= 0. \end{aligned} \tag{18}$$

If these equations determine u and v as differentiable functions of the variables x and y, we may differentiate the system with respect to x and y, considering these two variables to be independent, and so obtain the four relations

$$\frac{\partial f}{\partial x} + \frac{\partial f}{\partial u}\frac{\partial u}{\partial x} + \frac{\partial f}{\partial v}\frac{\partial v}{\partial x} = 0,$$
$$\frac{\partial g}{\partial x} + \frac{\partial g}{\partial u}\frac{\partial u}{\partial x} + \frac{\partial g}{\partial v}\frac{\partial v}{\partial x} = 0, \tag{19}$$

$$\frac{\partial f}{\partial y} + \frac{\partial f}{\partial u}\frac{\partial u}{\partial y} + \frac{\partial f}{\partial v}\frac{\partial v}{\partial y} = 0,$$
$$\frac{\partial g}{\partial y} + \frac{\partial g}{\partial u}\frac{\partial u}{\partial y} + \frac{\partial g}{\partial v}\frac{\partial v}{\partial y} = 0. \tag{20}$$

For brevity, we use the conventional subscript notation for partial derivatives so that, for example, u_x is written for $\partial u/\partial x$. Then, if (19) is solved for $\partial u/\partial x$

and $\partial v/\partial x$ and (20) is solved for $\partial u/\partial y$ and $\partial v/\partial y$, the expressions for these partial derivatives can be written in terms of determinants as follows:

$$u_x = - \frac{\begin{vmatrix} f_x & f_v \\ g_x & g_v \end{vmatrix}}{\begin{vmatrix} f_u & f_v \\ g_u & g_v \end{vmatrix}}, \qquad v_x = - \frac{\begin{vmatrix} f_u & f_x \\ g_u & g_x \end{vmatrix}}{\begin{vmatrix} f_u & f_v \\ g_u & g_v \end{vmatrix}},$$

$$u_y = - \frac{\begin{vmatrix} f_y & f_v \\ g_y & g_v \end{vmatrix}}{\begin{vmatrix} f_u & f_v \\ g_u & g_v \end{vmatrix}}, \qquad v_y = - \frac{\begin{vmatrix} f_u & f_y \\ g_u & g_y \end{vmatrix}}{\begin{vmatrix} f_u & f_v \\ g_u & g_v \end{vmatrix}}. \tag{21}$$

It must be assumed, however, that the common denominator in (21) does not vanish, that is, that

$$\begin{vmatrix} \dfrac{\partial f}{\partial u} & \dfrac{\partial f}{\partial v} \\ \dfrac{\partial g}{\partial u} & \dfrac{\partial g}{\partial v} \end{vmatrix} \neq 0. \tag{22}$$

Unless (22) is satisfied, the desired partial derivatives cannot exist uniquely, so that u and v cannot be differentiable functions of x and y. However, if (18) and (22) *are* satisfied at a point, and if the first partial derivatives of f and g are continuous at and near that point, it can be shown that Equations (18) determine u and v as implicit functions of x and y in *some* region including that point, with partial derivatives given by (21).

The determinant in (22) is known as *the Jacobian of f and g with respect to u and v*, and the notation

$$\frac{\partial(f, g)}{\partial(u, v)} \equiv \begin{vmatrix} \dfrac{\partial f}{\partial u} & \dfrac{\partial f}{\partial v} \\ \dfrac{\partial g}{\partial u} & \dfrac{\partial g}{\partial v} \end{vmatrix} = \frac{\partial f}{\partial u}\frac{\partial g}{\partial v} - \frac{\partial f}{\partial v}\frac{\partial g}{\partial u} \tag{23}$$

is frequently used. In a similar way we write, for example,

$$\frac{\partial(f, g, h)}{\partial(u, v, w)} = \begin{vmatrix} \dfrac{\partial f}{\partial u} & \dfrac{\partial f}{\partial v} & \dfrac{\partial f}{\partial w} \\ \dfrac{\partial g}{\partial u} & \dfrac{\partial g}{\partial v} & \dfrac{\partial g}{\partial w} \\ \dfrac{\partial h}{\partial u} & \dfrac{\partial h}{\partial v} & \dfrac{\partial h}{\partial w} \end{vmatrix} \tag{24}$$

and proceed in the same way to define the Jacobian of any n functions with respect to n variables.

In this notation, if

$$\frac{\partial(f, g)}{\partial(u, v)} \neq 0, \tag{25}$$

the first equation of (21) becomes, for example,

$$\frac{\partial u}{\partial x} = -\frac{\dfrac{\partial(f, g)}{\partial(x, v)}}{\dfrac{\partial(f, g)}{\partial(u, v)}}. \tag{26}$$

More generally, if $n + k$ variables are related by n equations of the form $f_1 = 0, f_2 = 0, \ldots, f_n = 0$, where the functions f_k each have continuous first partial derivatives, then any set of n variables may be considered as functions of the remaining k variables, in *some* neighborhood of a point where the n equations are satisfied, *if the Jacobian of the f's with respect to the n dependent variables is not zero at that point.*

Example. We consider the system

$$x + y + z = 0, \tag{27}$$
$$x^2 + y^2 + z^2 + 2xz - 1 = 0.$$

To investigate whether x and y can be considered as functions of z, we denote the left-hand members by f and g, respectively, and calculate the Jacobian

$$\frac{\partial(f, g)}{\partial(x, y)} = \begin{vmatrix} 1 & 1 \\ 2x + 2z & 2y \end{vmatrix} = -2(x + z - y). \tag{28}$$

Thus, except on the surface $x + z - y = 0$, x and y can be considered as functions of z. That is, z can be taken as the independent variable. When $y = x + z$, the equations become $2(x + z) = 0$ and $2(x + z)^2 = 1$ and are hence incompatible. To investigate whether x and z can be taken as the dependent variables, we calculate the Jacobian

$$\frac{\partial(f, g)}{\partial(x, z)} = \begin{vmatrix} 1 & 1 \\ 2x + 2z & 2x + 2z \end{vmatrix} = 0. \tag{29}$$

Since this determinant is identically zero, we see that x and z cannot be taken as the dependent variables. It is readily verified *directly* that the system (27) cannot be solved for x and z in terms of y. This situation follows from the fact that both equations involve only y and the *combination* $x + z$, and hence cannot be solved for x and z separately. ∎

By direct expansion we can verify that if u and v are functions of r and s, and also r and s are functions of x and y, then the relevant Jacobians satisfy the equation

$$\frac{\partial(u, v)}{\partial(r, s)} \frac{\partial(r, s)}{\partial(x, y)} = \frac{\partial(u, v)}{\partial(x, y)}. \tag{30}$$

As a special case of this result, we find that if u and v are functions of x and y, and conversely, then

$$\frac{\partial(u, v)}{\partial(x, y)} \frac{\partial(x, y)}{\partial(u, v)} = 1. \tag{31}$$

Analogous identities hold for Jacobians of any order. Thus Jacobians behave in certain ways like derivatives, as is suggested by the notation used.

The Jacobian notation is useful in many other applications. In this connection, the identity

$$(\nabla u) \cdot (\nabla v) \times (\nabla w) = \begin{vmatrix} \dfrac{\partial u}{\partial x} & \dfrac{\partial u}{\partial y} & \dfrac{\partial u}{\partial z} \\[2mm] \dfrac{\partial v}{\partial x} & \dfrac{\partial v}{\partial y} & \dfrac{\partial v}{\partial z} \\[2mm] \dfrac{\partial w}{\partial x} & \dfrac{\partial w}{\partial y} & \dfrac{\partial w}{\partial z} \end{vmatrix} = \frac{\partial(u, v, w)}{\partial(x, y, z)} \tag{32}$$

may be noted. From this relation we conclude that if $\partial(u, v, w)/\partial(x, y, z) = 0$ at a point, then the surfaces $u = c_1$, $v = c_2$, $w = c_3$ which pass through this point have coplanar normals.

7.3. Functional Dependence. The general solutions of certain types of partial differential equations, to be dealt with in Chapter 8, are of the form

$$z = f[u_1(x, y)] + g[u_2(x, y)], \tag{33}$$

where u_1 and u_2 are *independent* particular solutions and f and g are arbitrary functions of these expressions. If u_2 were a function of u_1, both terms would then be functions merely of u_1, and (33) would not be the required general solution. Thus in the expression

$$z = f(x + y) + g(x^2 + 2xy + y^2 + 1) \equiv f(u_1) + g(u_2)$$

we have $u_2 = u_1^2 + 1$ and hence both terms are functions of the same combination $u_1 = x + y$. It is thus important in more involved cases to have a criterion for determining whether one function $u_1(x, y)$ is a function of a second function $u_2(x, y)$. If such a functional relationship does exist, the two functions are said to be *functionally dependent*.

Suppose that such a relationship *does* exist, so that, for some F, not *identically* zero, it is true that

$$F(u_1, u_2) \equiv 0, \tag{34}$$

where u_1 and u_2 are functions of the independent variables x and y. Then, if we calculate the partial derivatives of (34), we obtain

$$\frac{\partial F}{\partial u_1} \frac{\partial u_1}{\partial x} + \frac{\partial F}{\partial u_2} \frac{\partial u_2}{\partial x} \equiv 0,$$

$$\frac{\partial F}{\partial u_1} \frac{\partial u_1}{\partial y} + \frac{\partial F}{\partial u_2} \frac{\partial u_2}{\partial y} \equiv 0. \tag{35}$$

These two linear equations in the quantities $\partial F/\partial u_1$ and $\partial F/\partial u_2$ can have nontrivial solutions only if the determinant of their coefficients vanishes. But this determinant is precisely the Jacobian of u_1 and u_2 with respect to x and y. Hence *if $u_1(x, y)$ and $u_2(x, y)$ are functionally dependent in a region, their Jacobian must vanish identically in that region,*

$$\frac{\partial(u_1, u_2)}{\partial(x, y)} \equiv 0. \tag{36}$$

Conversely, it can be shown that if the partial derivatives are continuous and if the Jacobian vanishes identically in a region, the two functions are functionally dependent in that region.

Completely analogous statements apply in the more general case of n functions of n variables. Thus if the functions $u_1, u_2, \ldots, u_n$, are functions of n variables, and if their partial derivatives are continuous, then the functions are functionally dependent, that is, there exists a nontrivial F such that

$$F(u_1, u_2, \ldots, u_n) \equiv 0$$

in a region, if and only if the Jacobian of these functions with respect to the n variables is identically zero in that region.

Example. For the two linear functions

$$u_1 = ax + by + c,$$
$$u_2 = dx + ey + f,$$

the Jacobian is

$$\frac{\partial(u_1, u_2)}{\partial(x, y)} = \begin{vmatrix} a & b \\ d & e \end{vmatrix} = ae - bd.$$

Thus u_1 and u_2 are functionally independent unless $ae = bd$. If $ae = bd$, the functional relationship

$$eu_1 - bu_2 = ec - bf$$

exists between u_1 and u_2. ∎

When there are fewer functions than variables, *several* relations of form (36) must hold. For example, in the case of two functions of three variables, $u_1(x, y, z)$ and $u_2(x, y, z)$, the assumption

$$F(u_1, u_2) \equiv 0 \tag{37}$$

leads to the *three* equations

$$\frac{\partial F}{\partial u_1} \frac{\partial u_1}{\partial x} + \frac{\partial F}{\partial u_2} \frac{\partial u_2}{\partial x} \equiv 0,$$

$$\frac{\partial F}{\partial u_1} \frac{\partial u_1}{\partial y} + \frac{\partial F}{\partial u_2} \frac{\partial u_2}{\partial y} \equiv 0,$$

$$\frac{\partial F}{\partial u_1} \frac{\partial u_1}{\partial z} + \frac{\partial F}{\partial u_2} \frac{\partial u_2}{\partial z} \equiv 0,$$

from which, by considering the equations in pairs, we may deduce that the *three* conditions

$$\frac{\partial(u_1, u_2)}{\partial(x, y)} \equiv 0, \qquad \frac{\partial(u_1, u_2)}{\partial(y, z)} \equiv 0, \qquad \frac{\partial(u_1, u_2)}{\partial(z, x)} \equiv 0 \tag{38}$$

must be satisfied, by the same argument used in deriving (36). Conversely, if the partial derivatives are continuous and if the three conditions of (38) are satisfied identically in a region, then the functions $u_1(x, y, z)$ and $u_2(x, y, z)$ are functionally dependent in that region.

The generalization to m functions of n variables, when $m < n$, is straightforward. When $m > n$, the m functions are *always* functionally dependent.

7.4. Jacobians and Curvilinear Coordinates. Change of Variables in Integrals.

If the equations

$$x = x(u_1, u_2, u_3), \qquad y = y(u_1, u_2, u_3), \qquad z = z(u_1, u_2, u_3) \qquad (39)$$

are interpreted as defining curvilinear coordinates $u_1, u_2,$ and u_3 in space, and if we write

$$\mathbf{U}_k = \mathbf{i}\frac{\partial x}{\partial u_k} + \mathbf{j}\frac{\partial y}{\partial u_k} + \mathbf{k}\frac{\partial z}{\partial u_k} \qquad (k = 1, 2, 3), \qquad (40)$$

then, as has been shown in Section 6.17, the vectors $\mathbf{U}_1, \mathbf{U}_2,$ and $\mathbf{U}_3$ are vectors tangent to the three coordinate curves at any point, with lengths given by $ds_1/du_1, ds_2/du_2,$ and $ds_3/du_3,$ where $s_1, s_2,$ and s_3 represent arc length along the coordinate curves. Then [compare Equation (151), Section 6.17] the element of volume in the new coordinate system, whether or not the system is orthogonal, is seen to be given by

$$d\tau = (\mathbf{U}_1 \cdot \mathbf{U}_2 \times \mathbf{U}_3)\, du_1\, du_2\, du_3,$$

if the coordinates are so ordered that the right-hand member is positive. But from (40) we obtain

$$\mathbf{U}_1 \cdot \mathbf{U}_2 \times \mathbf{U}_3 = \begin{vmatrix} \dfrac{\partial x}{\partial u_1} & \dfrac{\partial y}{\partial u_1} & \dfrac{\partial z}{\partial u_1} \\[2mm] \dfrac{\partial x}{\partial u_2} & \dfrac{\partial y}{\partial u_2} & \dfrac{\partial z}{\partial u_2} \\[2mm] \dfrac{\partial x}{\partial u_3} & \dfrac{\partial y}{\partial u_3} & \dfrac{\partial z}{\partial u_3} \end{vmatrix} = \frac{\partial(x, y, z)}{\partial(u_1, u_2, u_3)},$$

since the determinant is unchanged if rows and columns are interchanged. Thus we may write

$$d\tau = \left| \frac{\partial(x, y, z)}{\partial(u_1, u_2, u_3)} \right| du_1\, du_2\, du_3. \qquad (41)$$

It is seen that the requirement that $\mathbf{U}_1 \cdot \mathbf{U}_2 \times \mathbf{U}_3$ be different from zero is necessary in order that (39) be solvable for $u_1, u_2,$ and u_3. In the special case of orthogonal coordinates, $\mathbf{U}_1 \cdot \mathbf{U}_2 \times \mathbf{U}_3$ has the value $h_1 h_2 h_3$, with the notation of Section 6.17.

Accordingly, we have the change-of-variables formula

$$\iiint_{\mathcal{R}} w(x, y, z)\, dx\, dy\, dz = \iiint_{\mathcal{R}^*} W(u_1, u_2, u_3) \left| \frac{\partial(x, y, z)}{\partial(u_1, u_2, u_3)} \right| du_1\, du_2\, du_3,$$

$$(42)$$

where

$$W(u_1, u_2, u_3) = w[x(u_1, u_2, u_3), y(u_1, u_2, u_3), z(u_1, u_2, u_3)]$$

and where $\mathcal{R}^*$ is the $u_1 u_2 u_3$ region into which (39) transforms the xyz region $\mathcal{R}$.

Here it is assumed that the Jacobian $\partial(x, y, z)/\partial(u_1, u_2, u_3)$ is continuous and nonzero in $\mathfrak{R}^*$.

In a similar way, the equations

$$x = x(u_1, u_2), \qquad y = y(u_1, u_2) \tag{43}$$

can be interpreted as defining curvilinear coordinates u_1 and u_2 in the xy plane. The vectors

$$\mathbf{U}_1 = \mathbf{i}\frac{\partial x}{\partial u_1} + \mathbf{j}\frac{\partial y}{\partial u_1}, \qquad \mathbf{U}_2 = \mathbf{i}\frac{\partial x}{\partial u_2} + \mathbf{j}\frac{\partial y}{\partial u_2} \tag{44}$$

are then tangent to the coordinate curves, with lengths ds_1/du_1 and ds_2/du_2.

The vector element of plane area is then given by

$$d\mathbf{A} = (\mathbf{U}_1 \times \mathbf{U}_2)\, du_1\, du_2 = \begin{vmatrix} \mathbf{i} & \mathbf{j} & \mathbf{k} \\ \dfrac{\partial x}{\partial u_1} & \dfrac{\partial y}{\partial u_1} & 0 \\ \dfrac{\partial x}{\partial u_2} & \dfrac{\partial y}{\partial u_2} & 0 \end{vmatrix} du_1\, du_2,$$

and this relation gives the result

$$dA = |d\mathbf{A}| = \left| \frac{\partial(x, y)}{\partial(u_1, u_2)} \right| du_1\, du_2. \tag{45}$$

Accordingly, we have

$$\iint_D w(x, y)\, dx\, dy = \iint_{D^*} W(u_1, u_2) \left| \frac{\partial(x, y)}{\partial(u_1, u_2)} \right| du_1\, du_2, \tag{46}$$

where $W(u_1, u_2) = w[x(u_1, u_2), y(u_1, u_2)]$ and where (43) transforms D into D^*, if $\partial(x, y)/\partial(u_1, u_2)$ is continuous and nonzero in D^*.

Example. We consider the coordinates u and φ defined by the equations

$$x = au\cos\varphi, \qquad y = bu\sin\varphi \qquad (u \geq 0, 0 \leq \varphi < 2\pi). \tag{47}$$

The curves $u = \text{constant}$ are the ellipses

$$\frac{x^2}{(au)^2} + \frac{y^2}{(bu)^2} = 1 \qquad \cdot$$

with semi-axes au and bu, whereas a curve $\varphi = \text{constant}$ is the portion of the straight line $y = (b/a)x\tan\varphi$ in the quadrant determined by φ. The element of area in $u\varphi$ coordinates is given by (45),

$$dA = \left| \frac{\partial(x, y)}{\partial(u, \varphi)} \right| du\, d\varphi = abu\, du\, d\varphi. \tag{48}$$

Thus, for example, any integral of the form $\iint_D w(x, y)\, dx\, dy$, where the integration is carried out over the interior of the ellipse

$$\frac{x^2}{a^2} + \frac{y^2}{b^2} = 1,$$

corresponding to $u = 1$, can be written in the form

$$\iint_D w(x, y)\, dx\, dy = ab \int_0^{2\pi} \int_0^1 w(au\cos\varphi, bu\sin\varphi)\, u\, du\, d\varphi. \tag{49}$$

In particular, to calculate the moment of inertia I_x of the area about the x axis, we write

$$w(x, y) = y^2 = b^2 u^2 \sin^2 \varphi,$$

and Equation (49) gives

$$I_x = ab^3 \int_0^{2\pi} \int_0^1 u^3 \sin^2 \varphi \, du \, d\varphi = \frac{\pi ab^3}{4}. \quad \blacksquare$$

Formulas permitting the determination of area and arc length on *surfaces* are derived in Problems 25 and 26.

7.5. Taylor Series. Functions of two or more variables often can be expanded in power series which generalize the familiar one-dimensional expansions. The more general situation may be illustrated here by a consideration of the two-variable case.

For this purpose, we begin by defining a function $F(t)$, such that

$$f(x + ht, y + kt) = F(t), \tag{50}$$

where x, y, h, and k are temporarily to be held fixed and, in any case, are to be independent of t. Then, if $F(t)$ has a continuous Nth derivative in some interval about $t = 0$, we may write

$$F(t) = \sum_{n=0}^{N-1} \frac{F^{(n)}(0)}{n!} t^n + \frac{F^{(N)}(\tau)}{N!} t^N, \tag{51}$$

for *some* value of τ between 0 and t, by virtue of Equations (10) and (11) of Chapter 4.

Now, since

$$\frac{d}{dt} F(t) = h \frac{\partial f(x + ht, y + kt)}{\partial x} + k \frac{\partial f(x + ht, y + kt)}{\partial y}$$

$$= \left(h \frac{\partial}{\partial x} + k \frac{\partial}{\partial y} \right) f(x + ht, y + kt),$$

there follows also

$$\frac{d^n}{dt^n} F(t) = \left(h \frac{\partial}{\partial x} + k \frac{\partial}{\partial y} \right)^n f(x + ht, y + kt) \quad (n = 0, 1, \dots).$$

Hence we have the results

$$F^{(n)}(0) = \left(h \frac{\partial}{\partial x} + k \frac{\partial}{\partial y} \right)^n f(x, y) \tag{52}$$

and

$$F^{(N)}(\tau) = \left(h \frac{\partial}{\partial x} + k \frac{\partial}{\partial y} \right)^N f(x + \tau h, y + \tau k). \tag{53}$$

If we introduce Equations (51), (52), and (53) into (50), and specialize the result by taking $t = 1$, we thus obtain the form

$$f(x + h, y + k) = \sum_{n=0}^{N-1} \frac{1}{n!} \left(h \frac{\partial}{\partial x} + k \frac{\partial}{\partial y} \right)^n f(x, y) + R_N, \tag{54}$$

where R_N, the "remainder after N terms," is given by

$$R_N = \frac{1}{N!}\left(h\frac{\partial}{\partial x} + k\frac{\partial}{\partial y}\right)^N f(x + \tau h, y + \tau k) \qquad (0 < \tau < 1), \qquad (55)$$

for *some* τ between 0 and 1.

For example, when $N = 3$ this result becomes

$$f(x + h, y + k) = f(x, y) + [hf_x(x, y) + kf_y(x, y)]$$
$$+ \frac{1}{2!}[h^2 f_{xx}(x, y) + 2hk f_{xy}(x, y) + k^2 f_{yy}(x, y)] + R_3, \qquad (56)$$

where

$$R_3 = \frac{1}{3!}[h^3 f_{xxx} + 3h^2 k f_{xxy} + 3hk^2 f_{xyy} + k^3 f_{yyy}]_{(x+\tau h, y+\tau k)}. \qquad (57)$$

It is seen that the contents of the brackets in R_3 are evaluated at *some* point on the straight line segment between the point (x, y) and the point $(x + h, y + k)$.

More generally, Equation (54) represents an expansion of $f(x + h, y + k)$ in powers of h and k through the $(N - 1)$th, with an error term, and is one form of the two-dimensional *Taylor formula*. Although the form given is perhaps the most compact one, a form which more closely resembles the most familiar one-dimensional form can be obtained from Equation (54) by first replacing (x, y) by (x_0, y_0) and then replacing h and k by $x - x_0$ and $y - y_0$, respectively. When $N = 2$, for example, there then follows

$$f(x, y) = f(x_0, y_0) + (x - x_0)f_x(x_0, y_0) + (y - y_0)f_y(x_0, y_0) + R_2 \qquad (58)$$

with

$$R_2 = \frac{1}{2!}[(x - x_0)^2 f_{xx}(\xi, \eta) + 2(x - x_0)(y - y_0)f_{xy}(\xi, \eta)$$
$$+ (y - y_0)^2 f_{yy}(\xi, \eta)], \qquad (59)$$

where the point (ξ, η) is somewhere on the line segment joining the points (x_0, y_0) and (x, y).

When $f(x, y)$ is sufficiently well behaved, the remainder R_N tends to zero for sufficiently small values of the increments, yielding a (generally infinite) power series of the form

$$f(x, y) = f(x_0, y_0) + [(x - x_0)f_x(x_0, y_0) + (y - y_0)f_y(x_0, y_0)]$$
$$+ \frac{1}{2!}[(x - x_0)^2 f_{xx}(x_0, y_0) + 2(x - x_0)(y - y_0)f_{xy}(x_0, y_0)$$
$$+ (y - y_0)^2 f_{yy}(x_0, y_0)] + \cdots, \qquad (60)$$

which converges when $|x - x_0|$ and $|y - y_0|$ are sufficiently small. Within its region of convergence, the series can be differentiated or integrated term by term and the result will converge to the derivative or integral of f inside the same region.

It can be proved that the expansion (60) is *unique*, in the sense that if an expansion of the form

$$f(x, y) = a_0 + b_1(x - x_0) + b_2(y - y_0) + c_1(x - x_0)^2 + \cdots,$$

which converges to f near (x_0, y_0), can be obtained by *any* method, it is necessarily the same as that defined by (60). For the elementary functions, alternative methods which are preferable to the use of (60) are usually evident.

7.6. Maxima and Minima. The developments of the preceding section are helpful in studying maxima and minima of functions of several variables. We again restrict attention here to the two-dimensional case.

If we write

$$\Delta f(x, y) = f(x + h, y + k) - f(x, y) \tag{61}$$

for the increment of f, corresponding to the increments h and k in x and y, respectively, we say that f has a *relative minimum* at $P(x_0, y_0)$ if $\Delta f(x_0, y_0) \geqq 0$ for all sufficiently small permissible increments h and k, and that f has a *relative maximum* at P if instead $\Delta f(x_0, y_0) \leqq 0$ for all such increments in x and y.

If the point P is an interior point of a region in which f, $\partial f/\partial x$, and $\partial f/\partial y$ exist, Equation (56) shows that *a necessary condition that f assume a relative maximum or a relative minimum at (x_0, y_0) is that*

$$f_x = f_y = 0 \qquad at \; (x_0, y_0). \tag{62}$$

For, when h and k are sufficiently small, the *sign* of $\Delta f(x_0, y_0)$ will be the same as the sign of $hf_x(x_0, y_0) + kf_y(x_0, y_0)$ when this quantity is not zero, and clearly the sign of this quantity will change as the signs of h and/or k change unless Equation (62) holds.

Suppose now that the condition of (62) is satisfied at a certain point P. Then, from (56), there follows

$$\text{sign } [\Delta f(x_0, y_0)] = \text{sign } [h^2 f_{xx}(x_0, y_0) + 2hk f_{xy}(x_0, y_0) + k^2 f_{yy}(x_0, y_0)] \tag{63}$$

when h and k are sufficiently small, unless the bracketed quantity on the right is zero. That quantity is a quadratic expression in h and k, of the form $Ah^2 + 2Bhk + Ck^2$. When the discriminant $B^2 - AC$ is *positive*, and *only* in that case, there will be two distinct values of the ratio k/h for which the expression is zero, the expression having one sign for intermediate values of k/h and the opposite sign for all other values. Hence *a necessary condition that f have either a relative maximum or a relative minimum at $P(x_0, y_0)$ is that*

$$\delta \equiv f_{xx}f_{yy} - f_{xy}^2 \geqq 0 \qquad at \; (x_0, y_0). \tag{64}$$

If $\delta < 0$ at a point $P(x_0, y_0)$ where Equation (62) is satisfied, then Δf is positive for some increments in x and y and negative for others, and the point P is said to be a *saddle point* or a *minimax*.

If $\delta = 0$ at a point $P(x_0, y_0)$, then the bracketed espression on the right in (63) is a perfect square, of the form $(\alpha h - \beta k)^2$, and hence either is identically zero or is zero along a line

$$\frac{k}{h} \equiv \frac{y - y_0}{x - x_0} = \frac{\alpha}{\beta}$$

passing through P. Thus in this case the sign of $\Delta f(x_0, y_0)$ may not be completely determined near (x_0, y_0) by the quadratic terms when (62) is satisfied, and the terms involving higher-order partial derivatives must be considered.

If $\delta > 0$ at (x_0, y_0), then clearly f_{xx} and f_{yy} must be either both positive or both negative at that point. Since $\Delta f(x_0, y_0)$ is of constant sign in either case, when h and k are sufficiently small, it follows from (63) that the former case corresponds to a relative minimum ($\Delta \geqq 0$) and the latter to a relative maximum ($\Delta \leqq 0$).

Thus, in summary, *if $f_x = 0$ and $f_y = 0$ at a point P, then at that point f has*
(a) *a relative maximum if $f_{xx} < 0, f_{xx}f_{yy} > f_{xy}^2$ at P,*
(b) *a relative minimum if $f_{xx} > 0, f_{xx}f_{yy} > f_{xy}^2$ at P,*
(c) *a saddle point if $f_{xx}f_{yy} < f_{xy}^2$ at P.*
When $f_{xx}f_{yy} = f_{xy}^2$ at P, further investigation is necessary.

Cases (a), (b), and (c) are illustrated by the functions $1 - x^2 - y^2$, $x^2 + y^2$, and xy, respectively, at $(0, 0)$. For the functions $1 - x^2 y^2$, $x^2 y^2$, and $x^3 y^2$ the point $(0, 0)$ is exceptional, since for each not only $f_x = f_y = 0$ but also $f_{xx} = f_{xy} = f_{yy} = 0$ at the point. However, it is obvious, by inspection, that these functions have a maximum, a minimum, and a saddle point, respectively, at the origin.

Very frequently, in practice, the use of the preceding criteria is too involved to be feasible, and a direct study of the behavior of $\Delta f(x_0, y_0)$ for small h and k may be necessary. Often physical or geometrical considerations make such investigations unnecessary.

It should be noted that a relative maximum or minimum may also be attained at a point where f_x and/or f_y *fail to exist* and that, when attention is restricted to a region $\Re$ with a finite boundary, it may happen that an extreme value is taken on at a *boundary point*, at which f_x and f_y may or may not exist and may or may not differ from zero. In order to locate an *absolute* maximum or minimum (that is, the largest or smallest value taken on) in a finite region, it is necessary to explore all these possibilities.

7.7. Constraints and Lagrange Multipliers. Situations also may occur in which a function f, to be maximized or minimized, depends upon variables which are not independent, but are interrelated by one or more *constraint* conditions. The more general situation may be illustrated by the problem of maximizing or minimizing a function $f(x, y, z)$,

$$f(x, y, z) = \text{relative max or min}, \tag{65}$$

subject to two constraints of the form

$$g(x, y, z) = 0, \tag{66a}$$

$$h(x, y, z) = 0, \tag{66b}$$

where g and h are not functionally dependent, so that the constraints are neither equivalent nor incompatible. We suppose that the functions f, g, and h have first partial derivatives everywhere in a region which includes the desired point.

An obvious procedure consists of using Equations (66a, b) to eliminate two of the variables from f, leading to a problem of maximizing or minimizing a function of only one variable, without constraints. However, this elimination, by analytical methods, will be feasible only if the functions g and h are of relatively simple form.

Alternatively, we may notice that f can have an extreme value at $P(x_0, y_0, z_0)$ only if the linear terms in the Taylor expansion of Δf about P are zero. This condition can be written in the form

$$f_x \, dx + f_y \, dy + f_z \, dz = 0 \qquad at \ (x_0, y_0, z_0). \tag{67}$$

Here, however, the increments dx, dy, and dz are not independent, so that here we *cannot* conclude that f_x, f_y, and f_z must vanish separately at P. Further, since g and h are each constant, the differential of each must be zero, so that we must have

$$g_x \, dx + g_y \, dy + g_z \, dz = 0 \qquad at \ (x_0, y_0, z_0), \tag{68a}$$

$$h_x \, dx + h_y \, dy + h_z \, dz = 0 \qquad at \ (x_0, y_0, z_0). \tag{68b}$$

Now Equations (68a, b) are *linear* in dx, dy, and dz. They can be solved uniquely for two of those differentials in terms of the third when and only when g and h are functionally independent (see Section 7.3). Thus two of the differentials (say dx and dy) can be eliminated from (67), leaving an equation of the form $F(x, y, z) \, dz = 0$, in which the one remaining differential *can* be arbitrarily assigned. The condition $F(x, y, z) = 0$, together with the conditions (66a, b), at (x_0, y_0, z_0), constitute three equations in the three unknowns x_0, y_0, and z_0.

A third alternative, of frequent usefulness, is based on the observation that one may multiply the equal members of (68a) and of (68b) by any constants λ_1 and λ_2, respectively, and add the results to (67) to yield the requirement

$$(f_x + \lambda_1 g_x + \lambda_2 h_x) \, dx + (f_y + \lambda_1 g_y + \lambda_2 h_y) \, dy + (f_z + \lambda_1 g_z + \lambda_2 h_z) \, dz = 0$$

at (x_0, y_0, z_0) for *any* values of λ_1 and λ_2. Now it is possible to determine λ_1 and λ_2 so that the coefficients of two of the differentials are zero. For if this were not so it would follow that

$$\begin{vmatrix} g_x & h_x \\ g_y & h_y \end{vmatrix} = \begin{vmatrix} g_y & h_y \\ g_z & h_z \end{vmatrix} = \begin{vmatrix} g_z & h_z \\ g_x & h_x \end{vmatrix} = 0,$$

and hence g and h would be functionally dependent, so that the two constraints would be either equivalent or inconsistent. If we imagine that λ_1 and λ_2 have been so determined, then the remaining differential can be arbitrarily assigned, so that *its* coefficient *also* must vanish. Hence we obtain the three equations

$$\left. \begin{array}{l} f_x + \lambda_1 g_x + \lambda_2 h_x = 0 \\ f_y + \lambda_1 g_y + \lambda_2 h_y = 0 \\ f_z + \lambda_1 g_z + \lambda_2 h_z = 0 \end{array} \right\} \qquad at \ (x_0, y_0, z_0), \tag{69}$$

which, together with the conditions

$$g = 0, \quad h = 0 \qquad at \ (x_0, y_0, z_0), \tag{70}$$

comprise five equations in the five unknown quantities x_0, y_0, z_0, λ_1, and λ_2.

The parameters λ_1 and λ_2 are called *Lagrange multipliers*. If they were eliminated from (69), the result would be the relation $F = 0$ obtained by eliminating dx, dy, and dz from (67) and (68a, b). However, the new system of five equations may be preferable because of the additional flexibility which it affords. For example, it may be more convenient to solve (69) for x, y, and z in terms of λ_1 and λ_2 and to introduce the results into (70) for the determination of λ_1 and λ_2.

The equations (69) are easily remembered if one notices that they are the necessary conditions $\varphi_x = 0, \varphi_y = 0, \varphi_z = 0$ that the "auxiliary function"

$$\varphi = f + \lambda_1 g + \lambda_2 h \tag{71}$$

attain a relative maximum or minimum at (x_0, y_0, z_0) when *no* constraints are imposed.

More generally, if f were to be maximized subject to n independent constraints $g_1 = 0, g_2 = 0, \ldots, g_n = 0$, one would form the auxiliary function

$$\varphi = f + \lambda_1 g_1 + \lambda_2 g_2 + \cdots + \lambda_n g_n, \tag{72}$$

where $\lambda_1, \ldots, \lambda_n$ are unknown constants, and write down the necessary conditions for rendering φ a relative maximum or minimum with no constraints.

Example. As a very simple illustration, we seek the point on the plane

$$Ax + By + Cz = D$$

which is nearest the origin. Thus we are to minimize $x^2 + y^2 + z^2$, subject to the single constraint $Ax + By + Cz - D = 0$. With

$$\varphi = (x^2 + y^2 + z^2) + \lambda(Ax + By + Cz - D),$$

the conditions $\varphi_x = \varphi_y = \varphi_z = 0$ at (x_0, y_0, z_0) become

$$2x_0 + \lambda A = 0, \qquad 2y_0 + \lambda B = 0, \qquad 2z_0 + \lambda C = 0,$$

from which there follows

$$x_0 = -\tfrac{1}{2}\lambda A, \qquad y_0 = -\tfrac{1}{2}\lambda B, \qquad z_0 = -\tfrac{1}{2}\lambda C$$

and substitution into the constraint condition determines λ,

$$-\tfrac{1}{2}\lambda(A^2 + B^2 + C^2) = D.$$

Thus, finally, the coordinates of the desired point are found to be

$$x_0 = \frac{AD}{A^2 + B^2 + C^2}, \quad y_0 = \frac{BD}{A^2 + B^2 + C^2}, \quad z_0 = \frac{CD}{A^2 + B^2 + C^2}.$$

The corresponding minimum distance from the origin is

$$\sqrt{x_0^2 + y_0^2 + z_0^2} = \frac{|D|}{\sqrt{A^2 + B^2 + C^2}}.$$

7.8. Calculus of Variations. An important class of problems involves the determination of one or more *functions*, subject to certain conditions, so as to maximize or minimize a certain *definite integral*, whose integrand depends upon the unknown function or functions and/or certain of their derivatives.

For example, to find the equation $y = u(x)$ of the curve along which the distance from $(0, 0)$ to $(1, 1)$ in the xy plane is least, we would seek $u(x)$ such that

$$I \equiv \int_0^1 \sqrt{1 + u'^2}\, dx = \min$$

with $u(0) = 0, \qquad u(1) = 1.$

This section presents a brief treatment of some of the simpler aspects of such problems.

We consider first the case when we are to attempt to maximize or minimize an integral of the form

$$I = \int_a^b F(x, u, u')\, dx, \tag{73}$$

subject to the conditions

$$u(a) = A, \qquad u(b) = B, \tag{74}$$

where a, b, A, and B are given constants. We suppose that F has continuous second-order derivatives with respect to its three arguments and require that the unknown function $u(x)$ possess two derivatives everywhere in (a, b). To fix ideas, we suppose that I is to be *maximized*.

We thus visualize a competition, to which only functions which have two derivatives in (a, b) and which take on the prescribed end values are admissible. The problem is that of selecting, from *all* admissible competing functions, the function (or functions) for which I is largest.

Under the assumption that there *is* indeed a function $u(x)$ having this property, we next consider a one-parameter family of admissible functions which includes $u(x)$, namely, the set of all functions of the form

$$u(x) + \epsilon\eta(x)$$

where $\eta(x)$ is any arbitrarily chosen twice-differentiable function which *vanishes* at the end points of the interval (a, b),

$$\eta(a) = \eta(b) = 0, \tag{75}$$

and where ϵ is a parameter which is constant for any one function in the set but which varies from one function to another. The increment $\epsilon\eta(x)$, representing the difference between the varied function and the actual solution function, is often called a *variation* of $u(x)$.

If the result of replacing $u(x)$ by $u(x) + \epsilon\eta(x)$ in I is denoted by $I(\epsilon)$,

$$I(\epsilon) = \int_a^b F(x, u + \epsilon\eta, u' + \epsilon\eta')\, dx, \tag{76}$$

it then follows that $I(\epsilon)$ takes on its maximum value when $\epsilon = 0$, that is, when the variation of u is zero. Hence it must follow that

$$\frac{dI(\epsilon)}{d\epsilon} = 0 \qquad when \ \epsilon = 0. \tag{77}$$

The assumed continuity of the partial derivatives of F with respect to its three arguments implies the continuity of $dF/d\epsilon$, so that we may differentiate $I(\epsilon)$ under the integral sign (see Section 7.9) to obtain

$$\frac{dI(\epsilon)}{d\epsilon} = \int_a^b \left[\frac{\partial F(x, u + \epsilon\eta, u' + \epsilon\eta')}{\partial(u + \epsilon\eta)} \eta + \frac{\partial F(x, u + \epsilon\eta, u' + \epsilon\eta')}{\partial(u' + \epsilon\eta')} \eta' \right] dx.$$

Hence, by setting $\epsilon = 0$, we obtain an expression for the condition (77) in the form

$$I'(0) = \int_a^b \left[\frac{\partial F}{\partial u} \eta(x) + \frac{\partial F}{\partial u'} \eta'(x) \right] dx = 0. \tag{78}$$

Here we write $F \equiv F(x, u, u')$, noticing that the partial derivatives $\partial F/\partial u$ and $\partial F/\partial u'$ have been formed with x, u, and u' treated as independent variables.

The next step consists of transforming the integral of the second product in (78) by an integration by parts, to give

$$\int_a^b \frac{\partial F}{\partial u'} \eta'(x) \, dx = \left[\frac{\partial F}{\partial u'} \eta(x) \right]_a^b - \int_a^b \frac{d}{dx} \left(\frac{\partial F}{\partial u'} \right) \eta(x) \, dx$$

$$= - \int_a^b \frac{d}{dx} \left(\frac{\partial F}{\partial u'} \right) \eta(x) \, dx,$$

in consequence of (75). Hence Equation (78) becomes

$$\int_a^b \left[\frac{d}{dx} \left(\frac{\partial F}{\partial u'} \right) - \frac{\partial F}{\partial u} \right] \eta(x) \, dx = 0. \tag{79}$$

It is possible to prove rigorously that, since (79) is true for *any* function $\eta(x)$ which is twice differentiable in (a, b) and zero at the ends of that interval, consequently *the coefficient of $\eta(x)$ in the integrand must be zero everywhere in (a, b)*, so that the condition

$$\frac{d}{dx} \left(\frac{\partial F}{\partial u'} \right) - \frac{\partial F}{\partial u} = 0 \tag{80}$$

must be satisfied. This is the so-called *Euler equation* associated with the problem of maximizing (or minimizing) the integral (73), subject to (74).

If we recall that F, and hence also its partial derivatives, may depend upon x both directly and indirectly, through the intermediate variables $u(x)$ and $u'(x)$, we deduce from the chain rule that

$$\frac{dM}{dx} = \frac{\partial M}{\partial x} + \frac{\partial M}{\partial u} \frac{du}{dx} + \frac{\partial M}{\partial u'} \frac{d^2u}{dx^2}, \tag{81}$$

where the function M may be identified with F or with one of its partial derivatives. Thus, in particular, we can use (81) with $M = \partial F/\partial u'$ to write the Euler equation (80) in the expanded form

$$F_{u'u'}\frac{d^2u}{dx^2} + F_{uu'}\frac{du}{dx} + (F_{xu'} - F_u) = 0. \tag{82}$$

Although the original form (80) usually is more convenient in practice, the expanded form shows that, except in the special cases when $F_{u'u'} \equiv \partial^2 F/\partial u'^2$ is zero, the equation is in fact a differential equation of second order in u, subject to the two boundary conditions $u(a) = A$ and $u(b) = B$.

Since the coefficients in Equation (82) may depend not only on x but also on u and du/dx, the equation is not necessarily *linear*. However, it involves the *highest*-order derivative d^2u/dx^2 in a linear way and hence (as is customary) it may be described as a *quasi-linear* equation in those cases when it is not in fact linear.

It may be noticed that from (80) there follows

$$\frac{\partial F}{\partial u'} = \text{constant} \qquad when\ F = F(x, u'), \tag{83a}$$

so that, when F does not involve u explicitly, the *first-order* equation $\partial F/\partial u' =$ constant comprises a "first integral" of the Euler equation. In addition, it is shown in Problem 45 that (80) also implies the relation

$$\frac{\partial F}{\partial u'}u' - F = \text{constant} \qquad when\ F = F(u, u'), \tag{83b}$$

so that a first integral is also available when F does not involve x explicitly (see also Problem 47).

It is of some importance to notice that we have *not* shown that (80) *has* a solution satisfying (74) or, if it has such a solution, that this solution *does* indeed maximize or minimize I. We have indicated only that (80) is a *necessary* condition, which must be satisfied by u if u is to qualify. The formulation of *sufficient* conditions, which ensure that a function $u(x)$ so obtained truly maximizes or minimizes I (or makes it a *relative* maximum or minimum in some sense), is much more difficult.

Not infrequently, in practice, one can be certain in advance that an admissible maximizing (or minimizing) function *exists*. In such a case that function necessarily will be obtained as the solution of (80) which satisfies (74), or will be one such solution if there are several. Solutions of (80) are often called *extremals* of the variational problem, whether or not they satisfy (74) and maximize or minimize I. A solution of (80) which also satisfies the prescribed end conditions (74) is said to make the integral (73) *stationary*, whether or not the corresponding *stationary value* of the integral is a maximum or minimum.

In the case of the example cited at the beginning of this section, where $F = (1 + u'^2)^{1/2}$, the fact that F depends only upon u' shows that the relevant Euler equation (82) reduces to the form $u'' = 0$, so that (as was to be expected) u must be a linear function, $u = c_1x + c_2$. The given end conditions yield $c_1 = 1$ and $c_2 = 0$ and hence $u(x) = x$.

Example 1. We seek to minimize the integral

$$I = \int_0^{\pi/2} \left[\left(\frac{dy}{dt}\right)^2 - y^2 + 2ty \right] dt$$

with
$$y(0) = 0, \qquad y(\pi/2) = 0.$$

The Euler equation (80), with u and x replaced by y and t, respectively, becomes

$$\frac{d}{dt}\left(2\frac{dy}{dt}\right) - (-2y + 2t) = 0 \quad \text{or} \quad \frac{d^2y}{dt^2} + y = t,$$

from which there follows $y = c_1 \cos t + c_2 \sin t + t$. The end conditions then give $c_1 = 0$, $c_2 = -\pi/2$, and hence

$$y = t - \frac{\pi}{2} \sin t,$$

in correspondence with which

$$I_{\min} = -\frac{\pi}{2}\left(1 - \frac{\pi^2}{12}\right). \qquad \blacksquare$$

Generalizations, in which more dependent and/or independent variables are involved or which involve other modifications, as well as formulations of sufficiency conditions, may be found in the literature.

Two such generalizations, which are particularly straightforward, may be described here:

(a) If (73) is replaced by the integral

$$I = \int_a^b F(x; u_1, \ldots, u_n; u_1', \ldots, u_n') \, dx, \tag{84}$$

where values of the n independent unknown functions $u_1(x), \ldots, u_n(x)$ are each given at the end points $x = a$ and $x = b$, we obtain an Euler equation similar to (80) in correspondence with each u_r,

$$\frac{d}{dx}\left(\frac{\partial F}{\partial u_r'}\right) - \frac{\partial F}{\partial u_r} = 0 \qquad (r = 1, 2, \ldots, n). \tag{85}$$

Example 2. The Euler equations associated with the integral

$$\int_a^b (u_1'^2 + u_2'^2 - 2u_1u_2 + 2xu_1) \, dx$$

are obtained by use of (85) in the form

$$\frac{d}{dx}(2u_1') - (-2u_2 + 2x) = 0 \quad \text{and} \quad \frac{d}{dx}(2u_2') - (-2u_1) = 0$$

or
$$u_1'' + u_2 = x \quad \text{and} \quad u_2'' + u_1 = 0. \qquad \blacksquare$$

(b) Suppose that we are to maximize or minimize (73),

$$\int_a^b F(x, u, u') \, dx = \text{max or min}, \tag{86}$$

where $u(x)$ is to satisfy the prescribed end conditions

$$u(a) = A, \qquad u(b) = B, \tag{87}$$

as before, but that also a *constraint* condition is imposed in the form

$$\int_a^b G(x, u, u') \, dx = K, \tag{88}$$

where K is a prescribed constant. In this case, the appropriate Euler equation is found to be the result of replacing F in (80) by the auxiliary function

$$H = F + \lambda G, \tag{89}$$

where λ is an unknown constant. This constant, which is of the nature of a Lagrange multiplier (Section 7.7), thus generally will appear in the Euler equation and in its solution, and is to be determined together with the two constants of integration in such a way that the *three* conditions of (87) and (88) are satisfied.

Example 3. To minimize the integral

$$\int_0^1 y'^2 \, dx,$$

subject to the end conditions $y(0) = 0$ and $y(1) = 0$ and also to the constraint

$$\int_0^1 y \, dx = 1,$$

we write $H = y'^2 + \lambda y$, in correspondence with which the Euler equation is

$$2y'' - \lambda = 0.$$

Hence y must be of the form $y = \frac{1}{4}\lambda x^2 + c_1 x + c_2$. The end conditions and the constraint condition yield $c_1 = 6$, $c_2 = 0$, and $\lambda = -24$, and hence there follows

$$y = 6x(1 - x). \qquad \blacksquare$$

7.9. Differentiation of Integrals Involving a Parameter. Rather frequently it is necessary to deal with a function $\varphi(x)$ defined by an integral of the form

$$\varphi(x) = \int_{A(x)}^{B(x)} f(x, t) \, dt, \tag{90}$$

where f is such that the integration cannot be effected analytically. In particular, an expression for the derivative $\varphi'(x)$ often is required.

If the limits A and B are finite *constants*, differentiation with respect to x under the integral sign can be justified for all x in an interval (a, b) when f and *the resultant integrand $\partial f/\partial x$ are continuous for $a \leqq x \leqq b$ and $A \leqq t \leqq B$.*

More generally, when the limits are not constant we can think of φ as a function of x directly and also indirectly, through the intermediate variables A and B, and hence write $\varphi = \varphi(x, A, B)$. It then follows as an application of (6) that

$$\frac{d\varphi}{dx} = \frac{\partial\varphi}{\partial x} + \frac{\partial\varphi}{\partial B}\frac{dB}{dx} + \frac{\partial\varphi}{\partial A}\frac{dA}{dx}, \tag{91}$$

if the derivatives on the right are continuous, where $\partial\varphi/\partial x$ is to be calculated by treating A and B as constants, and hence by merely differentiating with respect to x under the integral sign (when f and $\partial f/\partial x$ are continuous). To evaluate the other partial derivatives of φ in (91), let $F(x, t)$ be a function such that

$$f(x, t) = \frac{\partial F(x, t)}{\partial t}.$$

There then follows

$$\varphi(x, A, B) = \int_A^B \frac{\partial F}{\partial t} \, dt = F(x, B) - F(x, A)$$

and hence, when x is held constant as A and B are imagined to vary, there follows

$$\frac{\partial \varphi}{\partial B} = \frac{\partial F(x, B)}{\partial B} = f(x, B), \qquad \frac{\partial \varphi}{\partial A} = -\frac{\partial F(x, A)}{\partial A} = -f(x, A).$$

By introducing these results into Equation (91), we thus obtain the useful formula

$$\frac{d}{dx} \int_A^B f(x, t) \, dt = \int_A^B \frac{\partial f(x, t)}{\partial x} \, dt + f(x, B) \frac{dB}{dx} - f(x, A) \frac{dA}{dx}, \qquad (92)$$

which is valid *for all values of x in an interval (a, b) when f and $\partial f/\partial x$ are continuous for $a \leq x \leq b$ and $A \leq t \leq B$, and also $A'(x)$ and $B'(x)$ are continuous in (a, b).* This formula is often known as *Leibnitz's rule.*

Example 1. If

$$y(x) = \int_a^x h(t) \sin(x - t) \, dt, \qquad (93)$$

then the repeated use of Equation (92) yields

$$y'(x) = \int_a^x h(t) \cos(x - t) \, dt$$

and

$$y''(x) = -\int_a^x h(t) \sin(x - t) \, dt + h(x). \qquad (94)$$

Hence it follows that $y(x)$ satisfies the differential equation

$$y''(x) + y(x) = h(x). \qquad (95)$$

In fact, by setting $x = a$ in the expressions for y and dy/dx, we find that (93) defines that solution of (95) which satisfies the initial conditions

$$y(a) = 0, \qquad y'(a) = 0. \qquad \blacksquare$$

In the case of a function defined by an *improper* integral,

$$\varphi(x) = \int_{A(x)}^{\infty} f(x, t) \, dt, \qquad (96)$$

where the range of integration is infinite, the situation is somewhat more complicated. Here, assuming that the integral (96) in fact converges, it is known that the formula

$$\frac{d}{dx} \int_{A(x)}^{\infty} f(x, t) \, dt = \int_A^{\infty} \frac{\partial f(x, t)}{\partial x} \, dt - f(x, A) \frac{dA}{dx} \qquad (97)$$

is valid *for x in (a, b) when f and ∂f/∂x are continuous for all t ≧ A and a ≦ x ≦ b, and A'(x) is continuous in (a, b), and when also there exists a function M(t), independent of x, such that*

$$\left| \frac{\partial f(x, t)}{\partial x} \right| \leq M(t)$$

for all t ≧ A and a ≦ x ≦ b, and such that the integral

$$\int_A^\infty M(t) \, dt$$

converges. The situations in which $A = -\infty$ in (96) are included as special cases.

Example 2. We consider the evaluation of the integral

$$\varphi(x) = \int_0^\infty e^{-t^2} \cos(2tx) \, dt. \tag{98}$$

Formally, Equation (97) gives

$$\frac{d\varphi}{dx} = -2 \int_0^\infty t e^{-t^2} \sin(2tx) \, dt, \tag{99}$$

and the validity of the use of that formula is established for all real x when we notice that $|t e^{-t^2} \sin(2tx)| \leq t e^{-t^2}$ for all real x and that the integral $\int_0^\infty t e^{-t^2} dt$ converges.

Further, if Equation (99) is integrated by parts, there follows

$$\frac{d\varphi}{dx} = [e^{-t^2} \sin(2tx)]_{t=0}^{t=\infty} - 2x \int_0^\infty e^{-t^2} \cos(2tx) \, dt$$

$$= -2x \int_0^\infty e^{-t^2} \cos(2tx) \, dt.$$

Hence φ satisfies the differential equation $(d\varphi/dx) + 2x\varphi = 0$, and therefore is of the form $\varphi(x) = c e^{-x^2}$. But when $x = 0$ Equation (98) gives

$$\varphi(0) = \int_0^\infty e^{-t^2} dt = \frac{1}{2}\sqrt{\pi},$$

by Equation (58) of Chapter 2, so that the constant c is determined and there follows

$$\int_0^\infty e^{-t^2} \cos(2tx) \, dt = \frac{1}{2}\sqrt{\pi} e^{-x^2}. \tag{100}$$

If we introduce the change in variables $t = au$, where a is real and positive, and write $x = b/(2a)$, this result takes the useful form

$$\int_0^\infty e^{-a^2 u^2} \cos bu \, du = \frac{\sqrt{\pi}}{2a} e^{-b^2/4a^2} \qquad (a > 0). \tag{101}$$

∎

It should be noted that the stated conditions are *sufficient*, but by no means *necessary*, for the applicability of (97).

Example 3. For the integral

$$\varphi(x) = \int_0^\infty \frac{e^{-xt}}{\sqrt{t}} \, dt \qquad (x > 0), \tag{102}$$

the conditions stated in connection with (97) are strongly violated. *Formal* use of that relation would give

$$\varphi'(x) = -\int_0^\infty e^{-xt}\sqrt{t} \, dt. \tag{103}$$

If the substitution $xt = u^2$ is made before the differentiation, there follows

$$\varphi(x) = \frac{2}{\sqrt{x}} \int_0^\infty e^{-u^2} \, du = \sqrt{\frac{\pi}{x}}, \tag{104}$$

and hence the correct formula is

$$\varphi'(x) = -\frac{1}{2x}\sqrt{\frac{\pi}{x}}. \tag{105}$$

The same substitution in (103) yields a result equivalent to (105) (see Problem 57) and hence shows that differentiation under the integral sign was indeed permissible, in spite of the unpleasant behavior of the integrand. ■

Example 4. For an integral of the form

$$\varphi(x) = \int_0^\infty \frac{e^{-xt}}{\sqrt{t}} g(t) \, dt \qquad (x > 0), \tag{106}$$

the substitution $t = v^2$ leads to the equivalent form

$$\varphi(x) = 2\int_0^\infty e^{-xv^2} g(v^2) \, dv \qquad (x > 0), \tag{107}$$

for which differentiation under the integral sign may be more easily justified. ■

Problem 56 illustrates the existence of situations in which the interchange of t integration and x differentiation *cannot* be justified.

7.10. Newton's Iterative Method. The use of the procedure known as Newton's method (or as the Newton–Raphson method) in obtaining successive approximations to roots of algebraic and transcendental equations in one variable is introduced in courses in elementary calculus. In this section we rederive the basic equation and indicate the role of Jacobian determinants in the generalization to the solution of simultaneous equations in several variables.

Suppose that we have obtained a first approximation to a certain root $x = \alpha$ of an equation of the form $f(x) = 0$ and require a more nearly accurate value. If we denote the first approximation by x_0, we then attempt to determine h so that $f(x_0 + h) = 0$. But for small values of h we have the Taylor series expansion

$$f(x_0 + h) = f(x_0) + hf'(x_0) + \frac{h^2}{2}f''(x_0) + \cdots. \tag{108}$$

If the initial approximation is sufficiently accurate and if the higher derivatives of f are not excessively large at x_0, we may neglect terms involving higher powers of h in the equation $f(x_0 + h) = 0$ and hence obtain for h the approximation

$$h_0 = -\frac{f(x_0)}{f'(x_0)}. \tag{109}$$

The next approximation for x is then taken as $x_1 = x_0 + h_0$, and the process is repeated. Geometrically, this procedure consists of approximating the curve representing $y = f(x)$ by its tangent line at $x = x_0$, and of determining the intersection of the tangent line with the x axis (see Figure 7.1). In unfavorable cases the process may not converge.

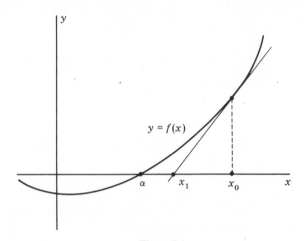

Figure 7.1

If the curve representing $y = f(x)$ is concave away from the x axis at the point for which $x = x_0$, and if the curve has no turning points or inflections in the closed interval between x_0 and α (as, for example, in Figure 7.1), it is obvious geometrically (and easily proved analytically) that x_1 will fall between x_0 and α, x_2 between x_1 and α, and so forth, and that convergence of the sequence of approximations $x_0, x_1, \ldots, x_k, \ldots$ to α then is guaranteed. The specified geometrical conditions correspond to the mathematical requirements that $f(x_0)f''(x_0) > 0$ and that $f'(x)$ and $f''(x)$ each be of constant sign in the closed interval between x_0 and α, and are known as the *Fourier conditions*. When they are not satisfied, convergence to α (or to *another* root) may or may not follow.

It can be proved (see Problem 59) that, when $f''(x)$ is continuous, the error e_k in the kth approximation x_k satisfies the equation

$$e_k = -\frac{f''(t_{k-1})}{2f'(x_{k-1})}e_{k-1}^2, \tag{110}$$

where t_{k-1} is *some* number between x_{k-1} and the true root α.

Example 1. To illustrate Newton's method, we seek the real root of the equation

$$f(x) = x^3 + 3x - 1 = 0.$$

Since $f(0) = -1$ and $f(1) = 3$, and since $f'(x) > 0$ for all x, there is one and only one real root and it lies in $(0, 1)$. Further, a consideration of the form of f suggests the starting value $x_0 = \frac{1}{3}$, for which it is also found that the desirable condition $ff'' > 0$ is satisfied. Since also $f''(x) > 0$ for all $x > 0$, the Fourier conditions are satisfied and convergence is guaranteed. The formula

$$h_k = -\frac{f(x_k)}{f'(x_k)} = -\frac{x_k^3 + 3x_k - 1}{3(x_k^2 + 1)}$$

then yields $h_0 = -\frac{1}{90}$ when $x_0 = \frac{1}{3}$, so that $x_1 = 0.3333 - 0.0111 = 0.3222$ when four significant figures are retained. Next, with $k = 1$, there follows $h_1 = -0.00001465$, to four significant figures, and hence $x_2 = 0.32218535$, to eight places.

Equation (110) here is of the form

$$e_k = -\frac{t_{k-1}}{1 + x_{k-1}^2}e_{k-1}^2$$

and, since here t_{k-1} certainly is in $(0, 1)$ when x_{k-1} is in that interval, it follows that certainly $|e_k| < e_{k-1}^2$ in the present case. If we estimate e_1 as approximately $x_2 - x_1 \approx 2 \times 10^{-5}$, we deduce that probably $|e_2| < 4 \times 10^{-10}$, indicating that the eight digits in x_2 are correct and, indeed, that a ninth digit might have been properly retained in its calculation. ■

Now suppose that a certain solution of the *simultaneous* equations

$$f(x, y) = 0, \qquad g(x, y) = 0 \tag{111}$$

is required, and that a reasonably accurate initial approximation (x_0, y_0) has been obtained by some method. We next attempt to determine values of h and k such that the equations

$$f(x_0 + h, y_0 + k) = 0, \qquad g(x_0 + h, y_0 + k) = 0$$

are simultaneously satisfied. If the left-hand members are expanded in Taylor series about the initial point and if only linear terms are retained, these equations become

$$\begin{aligned}
f_0 + h f_{x0} + k f_{y0} = 0, \\
g_0 + h g_{x0} + k g_{y0} = 0,
\end{aligned} \tag{112}$$

where the zero subscripts indicate that the functions involved are evaluated at the point (x_0, y_0). Thus approximate corrections h_0 and k_0 are given by the solution of these equations, in the form

$$h_0 = -\frac{\begin{vmatrix} f_0 & f_{y0} \\ g_0 & g_{y0} \end{vmatrix}}{\dfrac{\partial(f, g)_0}{\partial(x, y)}}, \qquad k_0 = -\frac{\begin{vmatrix} f_{x0} & f_0 \\ g_{x0} & g_0 \end{vmatrix}}{\dfrac{\partial(f, g)_0}{\partial(x, y)}}. \tag{113}$$

It is seen that the success of this method depends in part upon the magnitude of the Jacobian determinant in the neighborhood of the desired solution.

To see the geometrical significance of this dependence, we recall that the vectors

$$\nabla f = f_x \mathbf{i} + f_y \mathbf{j}, \qquad \nabla g = g_x \mathbf{i} + g_y \mathbf{j}$$

are normal to the curves representing $f = 0$ and $g = 0$ in the xy plane. Further, we obtain the result

$$(\nabla f) \times (\nabla g) = \mathbf{k}(f_x g_y - f_y g_x) = \mathbf{k}\frac{\partial(f, g)}{\partial(x, y)}, \tag{114}$$

and it follows from the definition of the cross product (Section 6.3) that the magnitude of the Jacobian is proportional to the magnitude of the sine of the angle between the normals (and hence also between the tangents) to the curves at points of intersection. Thus small values of the Jacobian may be expected to correspond to cases in which the intersecting curves have nearly equal slope at the intersection, in which cases slow convergence (or divergence) of the iterative process is to be anticipated.

Example 2. To illustrate this procedure in a favorable case, we consider the simultaneous equations

$$f \equiv x^3 - y^2 - 3x + y + 2 = 0, \qquad g \equiv x^2 + y^2 - 4 = 0.$$

By rough graphical methods it is found that the circle $g = 0$ intersects the cubic curve $f = 0$ at a point with *approximate* coordinates $(1.5, 1.5)$. With the choice $x_0 = y_0 = 1.5$, Equations (113) give $h_0 = -0.080$ and $k_0 = -0.087$, leading to the first improved coordinates $x_1 = 1.420$, $y_1 = 1.413$. The exact values are $x = y = \sqrt{2} = 1.414. \ldots$
∎

The methods of this section clearly can be generalized to the solution of n equations in n variables.

REFERENCES

1. APOSTOL, T. M., *Mathematical Analysis*, Addison-Wesley Publishing Company, Inc., Reading, Mass., 1957.

2. BLISS, G. A., *Calculus of Variations*, First Carus Mathematical Monograph, Open Court Publishing Company, La Salle, Ill., 1925.

3. BUCK, R. C., *Advanced Calculus*, McGraw-Hill Book Company, Inc., New York, 1956.

4. FRANKLIN, P., *A Treatise on Advanced Calculus*, John Wiley & Sons, Inc., New York, 1940.

5. HARDY, G. H., *A Course of Pure Mathematics*, Cambridge University Press, New York, 1959.

6. KAPLAN, W., *Advanced Calculus*, Addison-Wesley Publishing Company, Inc., Reading, Mass., 1952.

7. TAYLOR, A. E., *Advanced Calculus*, Ginn and Company, Boston, 1955.

8. WEINSTOCK, R., *Calculus of Variations: With Applications to Physics and Engineering*, McGraw-Hill Book Company, Inc., New York, 1952.

9. WIDDER, D. V., *Advanced Calculus*, 2nd ed., Prentice-Hall. Inc., Englewood Cliffs, N.J., 1961.

PROBLEMS

Section 7.1

1. If $x = r \cos \theta$ and $y = r \sin \theta$, determine expressions for each of the following and express each result as a function of r and θ:

(a) $\left(\dfrac{\partial x}{\partial r}\right)_{\theta}$,   (b) $\left(\dfrac{\partial x}{\partial \theta}\right)_{r}$,   (c) $\left(\dfrac{\partial y}{\partial r}\right)_{\theta}$,   (d) $\left(\dfrac{\partial y}{\partial \theta}\right)_{r}$,

(e) $\left(\dfrac{\partial r}{\partial x}\right)_{y}$,   (f) $\left(\dfrac{\partial r}{\partial y}\right)_{x}$,   (g) $\left(\dfrac{\partial \theta}{\partial x}\right)_{y}$,   (h) $\left(\dfrac{\partial \theta}{\partial y}\right)_{x}$.

2. If $x = r \cos \theta$ and $y = r \sin \theta$, express each of the following as a function of r and θ:

(a) $\left(\dfrac{\partial y}{\partial x}\right)_{r}$,   (b) $\left(\dfrac{\partial x}{\partial y}\right)_{r}$,   (c) $\left(\dfrac{\partial y}{\partial x}\right)_{\theta}$,   (d) $\left(\dfrac{\partial x}{\partial y}\right)_{\theta}$,

(e) $\left(\dfrac{\partial r}{\partial \theta}\right)_{x}$,   (f) $\left(\dfrac{\partial \theta}{\partial r}\right)_{x}$,   (g) $\left(\dfrac{\partial r}{\partial \theta}\right)_{y}$,   (h) $\left(\dfrac{\partial \theta}{\partial r}\right)_{y}$.

3. If s and t are functions of x and y, say $s = f(x, y)$ and $t = g(x, y)$, show that

$$\frac{\partial F}{\partial x} = \left(s_x \frac{\partial}{\partial s} + t_x \frac{\partial}{\partial t}\right)F, \qquad \frac{\partial F}{\partial y} = \left(s_y \frac{\partial}{\partial s} + t_y \frac{\partial}{\partial t}\right)F.$$

4. With the notation of Problem 3, show that

$$\frac{\partial^2 F}{\partial x\, \partial y} = \left(s_x \frac{\partial}{\partial s} + t_x \frac{\partial}{\partial t}\right)\left(s_y \frac{\partial F}{\partial s} + t_y \frac{\partial F}{\partial t}\right)$$

$$= s_x s_y \frac{\partial^2 F}{\partial s^2} + (s_x t_y + s_y t_x)\frac{\partial^2 F}{\partial s\, \partial t} + t_x t_y \frac{\partial^2 F}{\partial t^2}$$

$$+ \left[\left(s_x \frac{\partial}{\partial s} + t_x \frac{\partial}{\partial t}\right)s_y\right]\frac{\partial F}{\partial s} + \left[\left(s_x \frac{\partial}{\partial s} + t_x \frac{\partial}{\partial t}\right)t_y\right]\frac{\partial F}{\partial t},$$

and hence obtain the result

$$\frac{\partial^2 F}{\partial x\, \partial y} = s_x s_y \frac{\partial^2 F}{\partial s^2} + (s_x t_y + s_y t_x)\frac{\partial^2 F}{\partial s\, \partial t} + t_x t_y \frac{\partial^2 F}{\partial t^2} + s_{xy}\frac{\partial F}{\partial s} + t_{xy}\frac{\partial F}{\partial t}.$$

[Notice that the variables (x, y) and (s, t) may be interchanged throughout.]

5. (a) By identifying x and y in the result of Problem 4, obtain the relation

$$\frac{\partial^2 F}{\partial x^2} = s_x^2 \frac{\partial^2 F}{\partial s^2} + 2s_x t_x \frac{\partial^2 F}{\partial s\, \partial t} + t_x^2 \frac{\partial^2 F}{\partial t^2} + s_{xx}\frac{\partial F}{\partial s} + t_{xx}\frac{\partial F}{\partial t},$$

as well as an analogous expression for $\partial^2 F/\partial y^2$.

(b) If $t_y = s_x$ and $t_x = -s_y$, show that

$$\nabla^2 F \equiv \frac{\partial^2 F}{\partial x^2} + \frac{\partial^2 F}{\partial y^2} = (s_x^2 + s_y^2)\left(\frac{\partial^2 F}{\partial s^2} + \frac{\partial^2 F}{\partial t^2}\right).$$

6. If $f(p, v, T) = 0$, show that

$$\left(\frac{\partial v}{\partial T}\right)_p = -\frac{f_T}{f_v}, \qquad \left(\frac{\partial p}{\partial v}\right)_T = -\frac{f_v}{f_p}, \qquad \left(\frac{\partial p}{\partial T}\right)_v = -\frac{f_T}{f_p},$$

and deduce that

$$\left(\frac{\partial p}{\partial T}\right)_v = -\left(\frac{\partial v}{\partial T}\right)_p\left(\frac{\partial p}{\partial v}\right)_T.$$

7. If $E = f(p, T)$ and $T = g(p, v)$, show that

(a) $\left(\dfrac{\partial E}{\partial v}\right)_p = f_T g_v \equiv \left(\dfrac{\partial E}{\partial T}\right)_p\left(\dfrac{\partial T}{\partial v}\right)_p,$

(b) $\left(\dfrac{\partial E}{\partial p}\right)_v = f_p + f_T g_p \equiv \left(\dfrac{\partial E}{\partial p}\right)_T + \left(\dfrac{\partial E}{\partial T}\right)_p\left(\dfrac{\partial T}{\partial p}\right)_v.$

Section 7.2

8. If u and v are functions of r and s, and also r and s are functions of x and y, prove that

$$\frac{\partial(u, v)}{\partial(r, s)} \frac{\partial(r, s)}{\partial(x, y)} = \frac{\partial(u, v)}{\partial(x, y)}.$$

9. The variables x and y are expressed in terms of the variables u and v by the equations

$$x = F(u, v), \qquad y = G(u, v).$$

(a) By writing

$$f(x, y, u, v) = x - F(u, v), \qquad g(x, y, u, v) = y - G(u, v)$$

and thus reducing the given equations to a special case of Equations (18), show that the statement following Equation (22) implies that u and v may be considered as functions of x and y if the Jacobian $J = \partial(F, G)/\partial(u, v)$ does not vanish, and that this condition is equivalent to the requirement

$$J = \frac{\partial(x, y)}{\partial(u, v)} \neq 0$$

when x and y are considered as functions of u and v.

(b) Assuming that $J \neq 0$, show that Equations (21) then become

$$u_x = \frac{y_v}{J}, \quad u_y = -\frac{x_v}{J}; \quad v_x = -\frac{y_u}{J}, \quad v_y = \frac{x_u}{J}.$$

(c) Obtain the same results directly by differentiating the original equations partially with respect to x and with respect to y, in each case holding the second variable constant, to obtain the relations

$$1 = F_u u_x + F_v v_x, \qquad 0 = F_u u_y + F_v v_y,$$
$$0 = G_u u_x + G_v v_x, \qquad 1 = G_u u_y + G_v v_y,$$

and by solving these equations for the desired quantities.

10. The five variables x, y, z, u, and v are related by three equations of the form

$$f(x, u, v) = 0 \qquad g(y, u, v) = 0, \qquad h(z, u, v) = 0.$$

(a) State conditions under which the first two equations determine u and v as functions of x and y and the third equation determines z as a function of u and v, and hence then as a function of x and y.

(b) Considering z, u, and v as functions of the independent variables x and y, differentiate the three equations partially with respect to x, holding y constant, and hence show that u_x, v_x, and z_x satisfy the equations

$$f_u u_x + f_v v_x = -f_x,$$
$$g_u u_x + g_v v_x = 0,$$
$$h_z z_x + h_u u_x + h_v v_x = 0.$$

(c) By solving these equations for z_x (by determinants), obtain the result

$$z_x = \frac{f_x}{h_z} \cdot \frac{\partial(h, g)/\partial(u, v)}{\partial(f, g)/\partial(u, v)}.$$

(d) Show also, by symmetry or otherwise, that

$$z_y = -\frac{g_y}{h_z} \cdot \frac{\partial(h, f)/\partial(u, v)}{\partial(f, g)/\partial(u, v)}.$$

(e) Verify these results in the case when

$$2x^2 - u + v = 0, \qquad 2y^3 - u - v = 0, \qquad z + u - v^2 = 0$$

by first obtaining z explicitly as a function of x and y and differentiating, and also using the results of parts (c) and (d), to determine z_x and z_y.

11. Show that the equations

$$x = F(u, v), \qquad y = G(u, v), \qquad z = H(u, v)$$

determine z as a function of x and y such that

$$z_x = \frac{\partial(z, y)/\partial(u, v)}{\partial(x, y)/\partial(u, v)}, \qquad z_y = -\frac{\partial(z, x)/\partial(u, v)}{\partial(x, y)/\partial(u, v)},$$

if $\partial(x, y)/\partial(u, v) \neq 0$. (Write $f = x - F$, $g = y - G$, $h = z - H$ in Problem 10.)

12. If $f(x, y, z) = 0$ and $g(x, y, z) = 0$, we may in general consider any two of the variables as functions of the third. Show that

$$\frac{dx}{\partial(f, g)/\partial(y, z)} = \frac{dy}{\partial(f, g)/\partial(z, x)} = \frac{dz}{\partial(f, g)/\partial(x, y)}$$

if no denominator vanishes.

13. Show that the tangent line, at a point (x_0, y_0, z_0), to the curve of intersection of the surfaces $f(x, y, z) = 0$ and $g(x, y, z) = 0$ is specified by the equations

$$\frac{x - x_0}{[\partial(f, g)/\partial(y, z)]_0} = \frac{y - y_0}{[\partial(f, g)/\partial(z, x)]_0} = \frac{z - z_0}{[\partial(f, g)/\partial(x, y)]_0}.$$

Section 7.3

14. Prove that the functions

$$u_1 = \frac{x + y}{x - y}, \qquad u_2 = \frac{xy}{(x - y)^2}$$

are functionally dependent.

15. Prove that the functions $u_1 = y + z$, $u_2 = x + 2z^2$, and $u_3 = x - 4yz - 2y^2$ are functionally dependent.

16. Determine whether the functions

$$u_1 = \frac{x - y}{x + z} \quad \text{and} \quad u_2 = \frac{x + z}{y + z}$$

are functionally dependent.

17. Determine whether the functions

$$u_1 = \frac{yz - x}{x}, \qquad u_2 = \frac{xyz - y^2z^2 + x^2}{xyz + y^2z^2}$$

are functionally dependent.

Section 7.4

18. The coordinates u, φ, and θ are defined by the equations

$$x = au \sin \varphi \cos \theta, \qquad y = bu \sin \varphi \sin \theta, \qquad z = cu \cos \varphi,$$

where $u \geq 0$, $0 \leq \varphi \leq \pi$, and $0 \leq \theta < 2\pi$.

(a) Show that the surfaces $u = $ constant are the ellipsoids

$$\left(\frac{x}{au}\right)^2 + \left(\frac{y}{bu}\right)^2 + \left(\frac{z}{cu}\right)^2 = 1$$

whereas the surfaces $\varphi = $ constant and $\theta = $ constant are elliptical cones and planes, respectively.

(b) Show that the element of volume is of the form

$$d\tau = abcu^2 \sin \varphi \, du \, d\varphi \, d\theta.$$

19. Use the result of Problem 18 to show that the z coordinate of the center of gravity of that half of the ellipsoid

$$\frac{x^2}{a^2} + \frac{y^2}{b^2} + \frac{z^2}{c^2} = 1$$

which lies above the xy plane is given by

$$V\bar{z} = abc^2 \int_0^1 u^3 \, du \int_0^{\pi/2} \sin \varphi \cos \varphi \, d\varphi \int_0^{2\pi} d\theta = \frac{\pi}{4} abc^2,$$

where

$$V = abc \int_0^1 u^2 \, du \int_0^{\pi/2} \sin \varphi \, d\varphi \int_0^{2\pi} d\theta = \frac{2\pi}{3} abc,$$

and hence has the value $\bar{z} = 3c/8$.

20. To generalize Problem 18, let

$$x = f(\varphi, \theta), \qquad y = g(\varphi, \theta), \qquad z = h(\varphi, \theta)$$

define a simple closed surface S which surrounds the origin, with $0 \leq \varphi \leq \pi$ and $0 \leq \theta < 2\pi$, and introduce the coordinates

$$x = uf(\varphi, \theta), \qquad y = ug(\varphi, \theta), \qquad z = uh(\varphi, \theta),$$

where $0 \leq u \leq 1$, $0 \leq \varphi \leq \pi$, and $0 \leq \theta < 2\pi$.

(a) Show that the element of volume is

$$d\tau = |Q(\varphi, \theta)|\, u^2\, du\, d\varphi\, d\theta,$$

where

$$Q(\varphi, \theta) = \begin{vmatrix} f & g & h \\ \dfrac{\partial f}{\partial \varphi} & \dfrac{\partial g}{\partial \varphi} & \dfrac{\partial h}{\partial \varphi} \\ \dfrac{\partial f}{\partial \theta} & \dfrac{\partial g}{\partial \theta} & \dfrac{\partial h}{\partial \theta} \end{vmatrix}.$$

(b) Show that

$$\iiint_S w(x, y, z)\, dx\, dy\, dz = \int_0^{2\pi} \int_0^{\pi} \int_0^1 W(u, \varphi, \theta)\, |Q(\varphi, \theta)|\, u^2\, du\, d\varphi\, d\theta,$$

where

$$W(u, \varphi, \theta) = w[uf(\varphi, \theta), ug(\varphi, \theta), uh(\varphi, \theta)],$$

and, in particular, that the volume of S is

$$V = \frac{1}{3} \int_0^{2\pi} \int_0^{\pi} |Q(\varphi, \theta)|\, d\varphi\, d\theta.$$

(Notice that only in special cases are φ and θ the cone angle and the polar angle of spherical coordinates.)

21. Suppose that a simple closed curve C, surrounding the origin in the xy plane, is specified by the parametric equations

$$x = f(\varphi), \qquad y = g(\varphi), \qquad (0 \leq \varphi < 2\pi).$$

(a) Show that, if coordinates u and φ are defined by the equations

$$x = uf(\varphi), \qquad y = ug(\varphi),$$

then the plane region D enclosed by C is specified by the ranges $0 \leq u \leq 1$ and $0 \leq \varphi < 2\pi$.

(b) Show that the element of area in $u\varphi$ coordinates is given by

$$dA = |Q(\varphi)|\, u\, du\, d\varphi,$$

where $Q(\varphi)$ is the Wronskian (see Section 1.2) of $f(\varphi)$ and $g(\varphi)$,

$$Q(\varphi) \equiv f(\varphi)g'(\varphi) - f'(\varphi)g(\varphi).$$

Deduce that

$$\iint_D w(x, y)\, dx\, dy = \int_0^{2\pi} \int_0^1 w[uf(\varphi), ug(\varphi)]\, |Q(\varphi)|\, u\, du\, d\varphi,$$

and, in particular, that the area of D is

$$A = \frac{1}{2} \int_0^{2\pi} |Q(\varphi)|\, d\varphi.$$

22. (a) Verify that the results of Problem 21 specialize to those of the text Example when $f(\varphi) = a \cos \varphi$ and $g(\varphi) = b \sin \varphi$.

(b) If $b = a$ in part (a), so that C is the circle $x^2 + y^2 = a^2$ and D is its interior, show that the result of Problem 21(b) takes the form

$$\iint_D w(x, y)\, dx\, dy = a^2 \int_0^{2\pi} \int_0^1 w(au \cos \varphi, au \sin \varphi) u\, du\, d\varphi$$

and that it reduces to a more familiar form with the substitutions $\theta = \varphi$ and $r = au$. (See also Problem 23.)

23. If φ can be identified with the polar angle $\theta = \cos^{-1}(x/r)$ in Problem 21, verify that the result of Problem 21 corresponds simply to replacing the polar coordinates (r, θ) by new coordinates (u, φ) such that

$$\theta = \varphi, \qquad r = up(\theta),$$

with

$$(\theta) = p(\theta)\cos\theta, \quad g(\theta) = p(\theta)\sin\theta, \quad p(\theta) = \sqrt{[f(\theta)]^2 + [g(\theta)]^2},$$

where $r = p(\theta)$ on C. [Show that here the Wronskian of f and g is $Q(\theta) = [p(\theta)]^2$ and that the integral

$$\int_0^{2\pi}\int_0^{p(\theta)} w(r\cos\theta, r\sin\theta)r\,dr\,d\theta$$

takes the form provided by Problem 21(b) with the change of variables $\theta = \varphi, r = up(\theta)$. Also specialize to the case when $f(\varphi) = a\cos\varphi$ and $g(\varphi) = a\sin\varphi$. See also Problem 22(b).]

24. If D is the plane region bounded by the curve

$$x = a(2\cos\varphi - \cos 2\varphi), \qquad y = a(2\sin\varphi - \sin 2\varphi) \qquad (0 \le \varphi < 2\pi),$$

use the result of Problem 21(b) to show that

$$\iint_D w(x, y)\,dx\,dy = 6a^2 \int_0^{2\pi}\int_0^1 W(u, \varphi)(1 - \cos\varphi)u\,du\,d\varphi,$$

where

$$W(u, \varphi) = w[au(2\cos\varphi - \cos 2\varphi), au(2\sin\varphi - \sin 2\varphi)].$$

In particular, show that the enclosed area A is $6\pi a^2$. (Here φ clearly is *not* identifiable with the polar angle θ, in contrast with the cases considered in Problem 23. The curve is an *epicycloid*, traced out by a point on the circumference of a circle of radius a as that circle rolls around a fixed circle with the same radius.)

25. (a) If a surface S is defined by the equations

$$x = x(u_1, u_2), \qquad y = y(u_1, u_2), \qquad z = z(u_1, u_2),$$

show that the directed element of *surface area* on that surface is given by

$$d\boldsymbol{\sigma} = \mathbf{U}_1 \times \mathbf{U}_2 \, du_1 \, du_2,$$

where $\mathbf{U}_1 = \partial\mathbf{r}/\partial u_1$ and $\mathbf{U}_2 = \partial\mathbf{r}/\partial u_2$, and where $\mathbf{r}$ is the position vector $x\mathbf{i} + y\mathbf{j} + z\mathbf{k}$.

    (b) Use Lagrange's identity [Equation (36) of Section 6.4] to show that the scalar element of surface area is of the form

$$d\sigma = \sqrt{EG - F^2}\,du_1\,du_2$$

where

$$E = \frac{\partial\mathbf{r}}{\partial u_1} \cdot \frac{\partial\mathbf{r}}{\partial u_1}, \qquad F = \frac{\partial\mathbf{r}}{\partial u_1} \cdot \frac{\partial\mathbf{r}}{\partial u_2}, \qquad G = \frac{\partial\mathbf{r}}{\partial u_2} \cdot \frac{\partial\mathbf{r}}{\partial u_2}.$$

26. Show that the element of *arc length* ds along any curve C on the surface considered in Problem 25 can be obtained, from the fact that

$$\frac{d\mathbf{r}}{ds} = \frac{\partial\mathbf{r}}{\partial u_1}\frac{du_1}{ds} + \frac{\partial\mathbf{r}}{\partial u_2}\frac{du_2}{ds}$$

is a unit vector tangent to the curve C, in the form

$$ds = \sqrt{E\, du_1^2 + 2F\, du_1\, du_2 + G\, du_2^2},$$

where E, F, and G are defined in Problem 25.

27. Specialize the results of Problems 25 and 26 to the case when the surface S is specified by the equations

$$x = u_1, \qquad y = u_2, \qquad z = u_1 u_2,$$

showing that

$$d\sigma = \sqrt{1 + u_1^2 + u_2^2}\, du_1\, du_2$$

and

$$ds = \sqrt{(1 + u_2^2)\, du_1^2 + 2u_1 u_2\, du_1\, du_2 + (1 + u_1^2)\, du_2^2}$$

on S. Also show that S is the hyperbolic paraboloid

$$z = xy,$$

and rederive the expression for $d\sigma$ on S by use of Equation (104) of Chapter 6 and that for ds on S by starting with

$$ds = \sqrt{dx^2 + dy^2 + dz^2}.$$

(Notice that here x and y are *not* orthogonal coordinates on S.)

28. If S is the ellipsoid

$$x = a \sin u_1 \cos u_2, \qquad y = b \sin u_1 \sin u_2, \qquad z = c \cos u_1,$$

where $0 \leq u_1 \leq \pi$ and $0 \leq u_2 < 2\pi$, show that

$$E = a^2 \cos^2 u_1 \cos^2 u_2 + b^2 \cos^2 u_1 \sin^2 u_2 + c^2 \sin^2 u_1,$$

$$F = (b^2 - a^2) \sin u_1 \cos u_1 \sin u_2 \cos u_2,$$

$$G = a^2 \sin^2 u_1 \sin^2 u_2 + b^2 \sin^2 u_1 \cos^2 u_2,$$

with the notation of Problems 25 and 26, and hence that

$$d\sigma = \sin u_1 \sqrt{c^2 \sin^2 u_1 (a^2 \sin^2 u_2 + b^2 \cos^2 u_2) + a^2 b^2 \cos^2 u_1}\, du_1\, du_2$$

and

$$ds = \sqrt{E\, du_1^2 + 2F\, du_1\, du_2 + G\, du_2^2}$$

on S. In particular, when $a = b$ show that u_1 and u_2 are orthogonal coordinates on S, with $u_2 = \theta$, where θ is the polar angle, and that then

$$d\sigma = a\sqrt{c^2 \sin^2 u_1 + a^2 \cos^2 u_1}\, \sin u_1\, du_1\, du_2,$$

$$ds = \sqrt{(a^2 \cos^2 u_1 + c^2 \sin^2 u_1)\, du_1^2 + a^2 \sin^2 u_1\, du_2^2},$$

whereas when $a = b = c$ the ellipsoid becomes a sphere and the coordinates become the spherical coordinates $u_1 = \varphi$, $u_2 = \theta$, with $r = a$.

Section 7.5

29. Obtain the terms of degree two or less in the Taylor expansion

$$e^{x+y} = a_0 + b_1 x + b_2 y + c_1 x^2 + c_2 xy + c_3 y^2 + \cdots$$

(a) by direct use of the two-dimensional Taylor formula,

(b) by first expanding e^{x+y} in a series of powers of $x + y$,

(c) by multiplying together appropriate expansions of e^x and e^y.

30. Obtain the terms of degree two or less in the Taylor expansion

$$\frac{1}{1 + x - y} = a_0 + b_1(x - 1) + b_2(y - 1) + c_1(x - 1)^2$$
$$+ c_2(x - 1)(y - 1) + c_3(y - 1)^2 + \cdots$$

   (a) by direct use of the two-dimensional Taylor formula,
   (b) by any shorter method.

31. For a certain function $f(x, y)$, it is known that

$$f(0, 0) = 1, \qquad f_x(0, 0) = 0.25, \qquad f_y(0, 0) = 0.50$$

and that $|f_{xx}| \leq 0.15, |f_{xy}| \leq 0.05$, and $|f_{yy}| \leq 0.05$ everywhere along that segment of the line $y = x$ which joins $(0, 0)$ and $(0.1, 0.1)$. Show that

$$1.0735 \leq f(0.1, 0.1) \leq 1.0765.$$

32. Obtain a three-decimal place approximation to the value of

$$\int_0^1 \int_0^1 e^{-x^2 y^2} \, dx \, dy$$

by use of Taylor series.

Section 7.6

33. Locate and identify all relative maxima and minima and saddle points (if such exist) of the following functions:

   (a) $x^2 + 2bxy + y^2$,   (b) $x^2 - xy + y^2 - 2x + y$,
   (c) $x^2 y^2 - x^2 - y^2$,   (d) $x^3 y^3 - 3x - 3y$.

34. Locate and identify all points at which absolute maxima or minima are assumed by the following functions:

   (a) $x^{2/3} + y^{2/3}$,   (b) $(x^2 + y^2 - 2x + 2y + 2)^{1/2}$,
   (c) $(y^2 - x^2)^{1/2}$,   (d) $(2x - 2y - x^2 - y^2)^{1/2}$.

35. Locate and classify all maxima and minima and determine any saddle points for the following functions:

   (a) $2x^4 - x^2 y^2 + 2y^4$,   (b) $x^4 - 2x^2 y^2 + y^4$,
   (c) $x^4 - 4x^2 y^2 + y^4$,   (d) $x^4 + y^4$.

36. Find the absolute maximum and minimum of the function $x^2 - y^2 - 2x$ in the region $x^2 + y^2 \leq 1$. (Write $x = \cos \theta, y = \sin \theta$ on the boundary.)

Section 7.7

37. Determine when, if ever, $x^2 + y^2$ takes on a minimum or a maximum value under the constraint $xy = 1$.

38. Determine when, if ever, xy takes on a maximum or a minimum value under the constraint $x^2 + y^2 = 2$.

39. Minimize $x^2 + 4y^2 + 16z^2$ under each of the following constraints:

   (a) $xyz = 1$,   (b) $xy = 1$,   (c) $x = 1$.

40. Determine the points on the ellipse, defined by the intersection of the surfaces $x + y = 1$ and $x^2 + 2y^2 + z^2 = 1$, which are nearest to and furthest from the origin.

Section 7.8

41. Obtain the Euler equation associated with maximizing or minimizing $\int_a^b F \, dx$ in the following cases:

(a) $F = p(x)y'^2 - q(x)y^2 + 2f(x)y$,

(b) $F = A(x)y'^2 + 2B(x)yy' + C(x)y^2 + 2D(x)y' + 2E(x)y$.

42. Obtain the two Euler equations associated with maximizing or minimizing $\int_a^b F \, dx$ in the following cases:

(a) $F = A(x)u_1'^2 + 2B(x)u_1' u_2' + C(x)u_2'^2$,

(b) $F = a_{11}(x)u_1'^2 + 2a_{12}(x)u_1' u_2' + a_{22}(x)u_2'^2$
$$+ b_{11}(x)u_1^2 + 2b_{12}(x)u_1 u_2 + b_{22}(x)u_2^2.$$

43. *Geodesics* on a surface are the extremals of the problem of minimizing the distance, along that surface, between two points on the surface. Determine the geodesics on the right circular cylinder $x^2 + y^2 = a^2$, by writing $x = a \cos \theta$ and $y = a \sin \theta$ and proceeding by each of the following methods:

(a) Assume $z = u(\theta)$, so that $ds = \sqrt{a^2 + u'^2} \, d\theta$,

(b) Assume $\theta = v(z)$, so that $ds = \sqrt{a^2 v'^2 + 1} \, dz$,

(c) Assume $z = f(t)$ and $\theta = g(t)$, where t is a parameter, so that

$$ds = \sqrt{f'^2 + a^2 g'^2} \, dt.$$

[Notice that in (a) we assume that z can be expressed as a single-valued differentiable function of θ, whereas in (b) the roles of the variables are reversed. In (c) neither assumption is made.]

44. Determine the geodesics on the sphere $x^2 + y^2 + z^2 = a^2$, using the spherical coordinates defined by Equation (169) of Chapter 6 and assuming that the required equation can be written in the form $\theta = u(\varphi)$. (Take $\cot \varphi$ as a new variable in evaluating a relevant integral.)

45. *Transformation of the Euler equation.*

(a) By multiplying the equal members of the Euler equation (80) by $u' \equiv du/dx$, transforming the first resultant left-hand member by using the relation $UV' = (UV)' - U'V$, and referring to (81) with $M = F$, show that *the Euler equation* (80) *implies the equation*

$$\frac{d}{dx}\left(\frac{\partial F}{\partial u'} u' - F\right) + \frac{\partial F}{\partial x} = 0.$$

[Apart from the fact that this equation admits the generally extraneous solution $u = $ constant, the derivation shows that it is *equivalent* to (80).]

(b) Deduce that *if F does not depend explicitly upon x, then a first integral of the Euler equation is*

$$\frac{\partial F}{\partial u'} u' - F = \text{constant},$$

with the understanding that the solution $u = $ constant may not be admissible. (The use of this result may or may not be advantageous in a specific case, as is illustrated by Problems 46 and 47.)

46. (a) Use the result of Problem 45 to obtain the extremals associated with the integral

$$\int_a^b \sqrt{u^2 + u'^2}\, dx$$

in the form

$$u(x) = c_1 \sec (x - c_2).$$

(b) Obtain the Euler equation (80) when $F = \sqrt{u^2 + u'^2}$ and compare the labor required for its solution with that needed in part (a).

[Note that the solution $u = c$, discarded in part (a), in fact does not satisfy the Euler equation unless $c = 0$, in which case it is a specialization of the basic solution.]

47. Compare the determination of the extremals associated with the integral

$$\int_a^b (u^2 + u'^2)\, dx$$

by the method of Problem 45 with that based on (80). (In general, if the Euler equation is a *linear* equation for which the solution is known, the "short-cut" method of Problem 45 is not to be recommended.)

48. (a) Determine the minimum and maximum values of the integral

$$\int_0^1 yy'\, dx,$$

subject to the end conditions $y(0) = 0$, $y(1) = 1$. Also account for the degenerate form of the Euler equation in this case.

(b) Show that, of all the twice-differentiable functions such that $y(-1) = -1$, $y(1) = 1$, there is no one for which the integral

$$\int_{-1}^1 x^2 y'^2\, dx$$

takes on either a minimum value or a maximum value.

49. Show that the Euler equation corresponding to minimizing the integral

$$I \equiv \int_a^b [p(x)y'^2 - q(x)y^2]\, dx,$$

subject to the end conditions $y(a) = 0$, $y(b) = 0$ and to the constraint

$$\int_a^b r(x)y^2\, dx = 1,$$

is

$$\frac{d}{dx}\left(p \frac{dy}{dx}\right) + (q + \lambda r)y = 0,$$

when the Lagrange multiplier is denoted by $-\lambda$. [Notice that here λ is a characteristic number associated with a Sturm–Liouville problem (see Section 5.6).]

50. Find the minimum value of $\int_0^1 y'^2\, dx$, subject to the conditions

$$y(0) = y(1) = 0 \quad \text{and} \quad \int_0^1 y^2\, dx = 1.$$

51. If the length of a curvilinear arc joining the points $(0, 0)$ and $(1, 0)$ is prescribed as l, and if the area bounded by this arc and the x axis in the upper half-plane is to be maximized, show that the arc must be circular, under the assumption that we may write $y = u(x)$ along the arc. Show also that this assumption is contradicted unless $l \leq \pi/2$, and that the additional assumption that $u'(x)$ exist at the ends requires in fact that $l < \pi/2$.

52. To avoid the restriction $l < \pi/2$ in Problem 51, use Equation (140) of Chapter 6 to write

$$A = \frac{1}{2} \int_0^1 (xy' - yx') \, dt$$

where

$$x(0) = 0, \qquad x(1) = 1, \qquad y(0) = 0, \qquad y(1) = 1$$

and where

$$\int_0^1 \sqrt{x'^2 + y^{2'}} \, dt = l.$$

By introducing a Lagrange multiplier λ, integrating the two Euler equations once, and then eliminating λ between the results, show again that the arc must be circular, but now with no restriction on l. [Omit the determination of $x(t)$ and $y(t)$.]

53. Determine $y(x)$ and $z(x)$ such that

$$y(0) = z(0) = 0, \qquad y(\pi) = z(\pi) = 2$$

and such that the integral

$$I = \int_0^\pi (y'^2 + z'^2 + 2y'z) \, dx$$

is stationary, and also determine the corresponding stationary value of I.

Section 7.9

54. If

$$I(a) = \int_0^1 \frac{x^a - x^b}{\log x} \, dx,$$

where $a > b > -1$, show that

$$I'(a) = \int_0^1 x^a \, dx = \frac{1}{a+1},$$

and hence, noticing that $I(b) = 0$, obtain the result

$$\int_0^1 \frac{x^a - x^b}{\log x} \, dx = \int_b^a \frac{du}{u+1} = \log \frac{a+1}{b+1} \qquad (a > b > -1).$$

55. (a) If

$$I(a) = \int_0^\infty \frac{e^{-ax} \sin x}{x} \, dx,$$

where $a > 0$, show that $I'(a) = -1/(a^2 + 1)$. Hence, noticing that $I(\infty) = 0$, obtain the result

$$\int_0^\infty \frac{e^{-ax} \sin x}{x} \, dx = \frac{\pi}{2} - \tan^{-1} a \qquad (a > 0).$$

(b) By considering the limit of this result as $a \rightarrow 0$, and making the justifiable assumption that the limit can be taken under the integral sign, deduce also that

$$\int_0^\infty \frac{\sin x}{x}\, dx = \frac{\pi}{2}.$$

56. (a) By replacing x by ax in the result of Problem 55(b), obtain the result

$$I(a) \equiv \int_0^\infty \frac{\sin ax}{x}\, dx = \frac{\pi}{2} \qquad (a > 0).$$

(b) Show that, although hence $I'(a) = 0$ when $a > 0$, the derivative *cannot* be determined by differentiating under the integral sign.

(c) Show that $I(0) = 0$ and also that $I(-a) = -I(a)$, so that $I(a)$ must be given by $-\pi/2$ when a is negative.

57. Given the evaluation

$$\int_{-\infty}^\infty e^{-x^2}\, dx = \sqrt{\pi}.$$

[see Equation (58) of Chapter 2], deduce the formula

$$\int_{-\infty}^\infty x^2 e^{-x^2}\, dx = \frac{1}{2}\sqrt{\pi}$$

by replacing x by ux in the original integral, dividing by u, differentiating with respect to u, and finally setting $u = 1$.

58. Obtain a differential equation, together with appropriate initial conditions, satisfied by the function

$$y(x) = \frac{1}{2!}\int_a^x (x - t)^2 h(t)\, dt.$$

Section 7.10

59. (a) If x_k is the kth approximation afforded by Newton's method to a zero α of $f(x)$, show that

$$\alpha - x_k = \alpha - x_{k-1} - \frac{f(\alpha) - f(x_{k-1})}{f'(x_{k-1})}.$$

(b) If $f(x)$ has a continuous second derivative in the interval between x_{k-1} and α, show that

$$f(\alpha) - f(x_{k-1}) = (\alpha - x_{k-1})f'(x_{k-1}) + \frac{(\alpha - x_{k-1})^2}{2}f''(t_{k-1}),$$

where t_{k-1} is some number between x_{k-1} and α.

(c) By introducing the result of part (b) into the result of part (a), and writing $e_k = \alpha - x_k$ for the error in x_k, deduce Equation (110).

60. Determine to five places the real root of the equation $x^3 - 2x - 5 = 0$.

61. Determine to three places the smallest positive root of the equation

$$\tan x = x$$

by Newton's method. [*Suggestions:* Notice that $x = 3\pi/2$ is a fair approximation, obtained by sketching the curves $y = \tan x$ and $y = x$ on the same graph and estimating the abscissa of the intersection. Take $f(x) = \sin x - x \cos x$, rather than $f(x) = \tan x - x$, to avoid the infinity of the latter function.]

62. Determine to three places the smallest positive root of the equation

$$\tanh x = \tan x$$

by Newton's method.

63. Determine to two places the smallest characteristic value of μ for the problem

$$\frac{d^2y}{dx^2} + \mu^2 y = 0, \qquad y'(0) = y(0), \quad y'(1) = -y(1).$$

64. Determine to three places the real solution of the simultaneous equations

$$x^2 + y^2 = 1, \qquad xy = x - y$$

in the first quadrant.

65. Determine to three places the real solution of the simultaneous equations

$$4x^3 - 27xy^2 + 25 = 0, \qquad 4x^2 y - 3y^3 - 1 = 0$$

in the first quadrant.

66. The path of a projectile moving in the xy plane is specified by the parametric equations $x = f(t)$, $y = g(t)$, where t is time. It is required to determine the time at which its trajectory will intercept a curve specified by the equation $\varphi(x, y) = 0$. If the approximate time is t_0 and the approximate coordinates of the interception are (x_0, y_0), show that Newton's method yields a new estimate $t = t_0 + \tau$, where

$$\tau = -\frac{\varphi_0 + (f_0 - x_0)\varphi_{x0} + (g_0 - y_0)\varphi_{y0}}{f_0'\varphi_{x0} + g_0'\varphi_{y0}},$$

and where the zero subscripts indicate evaluation for $t = t_0$, $x = x_0$, and $y = y_0$. [Linearize the relations $x_0 + h - f(t_0 + \tau) = 0$, $y_0 + k - g(t_0 + \tau) = 0$, and $\varphi(x_0 + h, y_0 + k) = 0$ and solve for τ.]

8

Partial Differential Equations

8.1. Definitions and Examples. A partial differential equation is said to be *linear* if, when the equation has been rationalized and cleared of fractions, no powers or products of the unknown function or its partial derivatives are present. If it is true only that no powers or products of the partial derivatives *of highest order* are present, the equation is said to be *quasi-linear*. Thus, for example, the equation

$$x\frac{\partial z}{\partial x} + y\frac{\partial z}{\partial y} = z$$

is a linear equation in z, of first order, whereas the equation

$$z\frac{\partial^2 z}{\partial x^2} + \left(\frac{\partial z}{\partial y}\right)^2 = 0$$

is a quasi-linear equation of second order.

By a *solution* of such an equation we mean either an explicit expression for z, of the form $z = \varphi(x, y)$, which reduces the equation to an identity, or a relation $F(x, y, z) = 0$ which determines z implicitly as such a function of x and y. The term *integral surface* is frequently used to denote either the geometrical representation of a solution or the solution itself.

In this work we deal principally with linear equations, although certain procedures to be developed can also be applied to the solution of certain quasi-linear equations.

We recall that in the case of an *ordinary* differential equation of order n, a general solution involves n independent arbitrary constants. In particular, the general solution of a *linear* ordinary differential equation expresses the unknown variable as a linear combination of n independent functions, the n arbitrary constants appearing as the coefficients of the n functions in the linear combination. For *nonlinear* ordinary equations the constants in general appear in a more complicated way in the solution, and there may exist so-called "singular solutions" which do not involve the arbitrary constants and cannot be obtained from the first solution by specializing these constants.

In the case of *partial* differential equations, general solutions are found to involve *arbitrary functions* of *specific functions*. As a simple example, we readily verify that the equation

$$2\frac{\partial z}{\partial x} - \frac{\partial z}{\partial y} = 0 \tag{1}$$

is satisfied by the expression

$$z = f(x + 2y), \tag{2}$$

no matter what functional relationship is indicated by f, so long as f is differentiable. For from Equation (2) we obtain

$$\frac{\partial z}{\partial x} = \frac{df(x + 2y)}{d(x + 2y)}\frac{\partial(x + 2y)}{\partial x} = f'(x + 2y),$$

$$\frac{\partial z}{\partial y} = \frac{df(x + 2y)}{d(x + 2y)}\frac{\partial(x + 2y)}{\partial y} = 2f'(x + 2y) = 2\frac{\partial z}{\partial x},$$

and hence (2) implies (1).

To see that (2) is indeed the *most general* solution of (1), we introduce the new variables

$$t = x + 2y, \qquad s = x. \tag{3}$$

There then follows

$$\frac{\partial z}{\partial x} = \frac{\partial z}{\partial t}\frac{\partial t}{\partial x} + \frac{\partial z}{\partial s}\frac{\partial s}{\partial x} = \frac{\partial z}{\partial t} + \frac{\partial z}{\partial s},$$

$$\frac{\partial z}{\partial y} = \frac{\partial z}{\partial t}\frac{\partial t}{\partial y} + \frac{\partial z}{\partial s}\frac{\partial s}{\partial y} = 2\frac{\partial z}{\partial t},$$

and (1) becomes merely

$$2\frac{\partial z}{\partial s} = 0, \tag{4}$$

when z is considered as a function of s and t. The general solution of this equation is clearly $z = f(t) = f(x + 2y)$, where f is an arbitrary function, in accordance with (2). We see that such expressions as $z = x + 2y + 1$, $z = \sin(x + 2y), z = 4\sqrt{x + 2y} + \cos(x + 2y), \ldots$, are all particular solutions of the partial differential equation (1).

As a further example, we determine a partial differential equation which has the expression

$$z = f(2x + y) + g(x - y) - xy \tag{5}$$

as its general solution. The procedure consists of attempting to obtain, by differentiation, sufficiently many relations to permit elimination of the arbitrary functions. With the notations

$$f' = \frac{df(2x+y)}{d(2x+y)}, \qquad g' = \frac{dg(x-y)}{d(x-y)}, \qquad \cdots,$$

the first and second derivatives of (5) are obtained in the form

$$\frac{\partial z}{\partial x} = 2f' + g' - y, \qquad \frac{\partial z}{\partial y} = f' - g' - x, \qquad (6a)$$

$$\frac{\partial^2 z}{\partial x^2} = 4f'' + g'', \qquad \frac{\partial^2 z}{\partial x\,\partial y} = 2f'' - g'' - 1, \qquad \frac{\partial^2 z}{\partial y^2} = f'' + g''. \qquad (6b)$$

Equations (6b) constitute three relations involving the two arbitrary functions f'' and g''. If two of these relations are used to determine f'' and g'', and if the results are introduced into the third relation, there follows

$$\frac{\partial^2 z}{\partial x^2} - \frac{\partial^2 z}{\partial x\,\partial y} - 2\frac{\partial^2 z}{\partial y^2} = 1. \qquad (7)$$

This is the differential equation of lowest order satisfied by (5), with f and g arbitrary functions.

We may verify that, in fact, Equation (5) defines the *most general* solution of (7) by first noticing that, since $z = -xy$ satisfies (7), the function

$$w = z + xy \qquad (8)$$

is to be the general solution of the equation

$$\frac{\partial^2 w}{\partial x^2} - \frac{\partial^2 w}{\partial x\,\partial y} - 2\frac{\partial^2 w}{\partial y^2} = 0. \qquad (9)$$

Now, if we make the substitutions

$$2x + y = s, \qquad x - y = t \qquad (10)$$

suggested by (5), there follows

$$\frac{\partial}{\partial x} = \frac{\partial s}{\partial x}\frac{\partial}{\partial s} + \frac{\partial t}{\partial x}\frac{\partial}{\partial t} = 2\frac{\partial}{\partial s} + \frac{\partial}{\partial t},$$

$$\frac{\partial}{\partial y} = \frac{\partial s}{\partial y}\frac{\partial}{\partial s} + \frac{\partial t}{\partial y}\frac{\partial}{\partial t} = \frac{\partial}{\partial s} - \frac{\partial}{\partial t},$$

and (9) takes the form

$$\frac{\partial^2 w}{\partial s\,\partial t} = 0 \qquad (11)$$

from which the general solution

$$w = f(s) + g(t) = f(2x+y) + g(x-y) \qquad (12)$$

is obtained by direct integration. The introduction of (12) into (8) then leads indeed to (5).

In a similar way, the expression

$$z = f(ax + by) + g(cx + dy), \qquad (13)$$

where a, b, c, and d are constants such that $ad \neq bc$, is readily shown to be the general solution of a linear partial differential equation of the form

$$A \frac{\partial^2 z}{\partial x^2} + B \frac{\partial^2 z}{\partial x \, \partial y} + C \frac{\partial^2 z}{\partial y^2} = 0, \tag{14}$$

where A, B, and C are constants depending upon a, b, c, and d.

In more involved partial differential equations, the arbitrary functional relations may enter into a general solution in much more complicated ways, and additional solutions may exist which cannot be obtained by specializing the general solution so obtained. Because of the fact that "general" solutions of partial differential equations involve arbitrary functions, the specialization of such solutions to particular forms which satisfy prescribed boundary conditions involves the determination of *functional relations*, rather than merely the determination of *constants*, and is usually not feasible. For this reason, we generally prefer to determine a set of *particular* solutions directly, and to attempt to combine these solutions in such a way that the prescribed boundary conditions are satisfied. The development of such procedures, in cases of linear equations, forms the basis of much of Chapter 9. However, since certain general properties of solutions of certain equations are most readily obtained by studying the general solutions, we show in Sections 8.2 through 8.5 in what way such solutions can be obtained in certain simple cases of frequent occurrence.

8.2. The Quasi-Linear Equation of First Order. The most general quasi-linear partial differential equation of first order can be written in the form

$$P(x, y, z) \frac{\partial z}{\partial x} + Q(x, y, z) \frac{\partial z}{\partial y} = R(x, y, z), \tag{15}$$

where z is the dependent variable and x and y are independent. If (15) is to be truly *linear*, then P and Q must be independent of z and R must be a linear function of z, say $R = R_1 z + R_2$, so that (15) becomes

$$P(x, y) \frac{\partial z}{\partial x} + Q(x, y) \frac{\partial z}{\partial y} = R_1(x, y)z + R_2(x, y) \tag{16}$$

in this special case.

Suppose that the equation

$$u(x, y, z) = c \tag{17}$$

defines a solution (or "integral surface") of (15), in the sense that (17) determines z as a function of x and y which satisfies (15). Then, by partial differentiation we obtain the two results

$$\frac{\partial u}{\partial x} + \frac{\partial u}{\partial z} \frac{\partial z}{\partial x} = 0, \qquad \frac{\partial u}{\partial y} + \frac{\partial u}{\partial z} \frac{\partial z}{\partial y} = 0, \tag{18}$$

where *in the derivatives $\partial u/\partial x$, $\partial u/\partial y$, and $\partial u/\partial z$ the variables x, y, and z are con-*

sidered as independent. Thus we may write

$$\frac{\partial z}{\partial x} = -\frac{\partial u/\partial x}{\partial u/\partial z}, \qquad \frac{\partial z}{\partial y} = -\frac{\partial u/\partial y}{\partial u/\partial z}, \tag{19}$$

assuming, of course, that $\partial u/\partial z \neq 0$. If these expressions are introduced into (15), an equation governing the function u is obtained in the form

$$P\frac{\partial u}{\partial x} + Q\frac{\partial u}{\partial y} + R\frac{\partial u}{\partial z} = 0. \tag{20}$$

This form has the advantage that in it the variables x, y, and z play completely symmetrical roles, and also we are readily led to a geometrical interpretation of the equation. Equation (20) obviously can be written in terms of a dot product,

$$(P\mathbf{i} + Q\mathbf{j} + R\mathbf{k}) \cdot \nabla u = 0. \tag{21}$$

Since ∇u is a vector normal to the surface $u = c$, Equation (21) states that the vector $P\mathbf{i} + Q\mathbf{j} + R\mathbf{k}$ is perpendicular to the normal to that surface at any point on the surface, and hence lies in the tangent plane. Thus we see that *at any point the vector $P\mathbf{i} + Q\mathbf{j} + R\mathbf{k}$ is tangent to a curve in the integral surface $u = c$ which passes through that point.*

Thus the differential equation (15) may be considered as determining at any point in some three-dimensional region a *direction*, specified by the vector $\mathbf{V} = P\mathbf{i} + Q\mathbf{j} + R\mathbf{k}$. If a particle moves from a given initial point in such a way that its direction at any point coincides with the direction of the vector $\mathbf{V}$ at that point, a space curve is traced out. Such curves are called the *characteristic curves* of the differential equation.

In a similar way, the ordinary differential equation of first order, $dy/dx = f(x, y)$, defines an angle $\varphi = \tan^{-1} f(x, y)$ at any point in some region in the xy plane, which the tangent to an integral curve must make with the x axis, and the general solution of the equation is represented by the set of curves having the prescribed direction at any point in the two-dimensional region of definition.

We see next that any surface which is built up from such characteristic curves in space will have the property that the tangent plane at any point contains the vector $\mathbf{V} = P\mathbf{i} + Q\mathbf{j} + R\mathbf{k}$, so that the normal vector at any point of such a surface is perpendicular to $\mathbf{V}$, as is required by (21). That is, *any smooth surface built up from characteristic curves is an integral surface of the differential equation* (15), provided only that its equation is not independent of z.

If $\mathbf{r}$ is the position vector to a point of a characteristic curve, and if s represents arc length along the curve, then the unit tangent vector to the curve at that point is given by

$$\frac{d\mathbf{r}}{ds} = \frac{dx}{ds}\mathbf{i} + \frac{dy}{ds}\mathbf{j} + \frac{dz}{ds}\mathbf{k}.$$

The requirement that this vector have the same direction as the vector

$P\mathbf{i} + Q\mathbf{j} + R\mathbf{k}$ yields the conditions

$$P = \mu\frac{dx}{ds}, \qquad Q = \mu\frac{dy}{ds}, \qquad R = \mu\frac{dz}{ds}, \tag{22}$$

where μ may be a function of x, y, and z. These relations can be written in the differential form

$$\frac{dx}{P} = \frac{dy}{Q} = \frac{dz}{R}, \tag{23}$$

and hence are equivalent to two ordinary differential equations.

Let the solutions of two independent equations which imply (23) be denoted by

$$u_1(x, y, z) = c_1, \qquad u_2(x, y, z) = c_2, \tag{24}$$

where c_1 and c_2 are independent constants. Then these equations represent two families of surfaces, such that a surface of one family intersects a surface of the second family in a characteristic curve. If we consider the one-parameter family of characteristic curves which are the intersections of those pairs of surfaces for which c_1 and c_2 are related by an equation of the general form $F(c_1, c_2) = 0$, we see that the locus of these intersections will determine a surface which is an integral surface, for any differentiable function F.† That is, any surface specified by an equation of the form

$$F[u_1(x, y, z), u_2(x, y, z)] = 0 \tag{25a}$$

will be an integral surface of the partial differential equation (15) provided, again, that F is differentiable and that the left-hand member of (25a) actually depends upon z. In many cases [particularly, if u_1 does not involve z, so that $u_1 = c_1$ is *not* an integral surface of (15)], it is more convenient to write the solution in the alternative form

$$u_2(x, y, z) = f[u_1(x, y, z)]. \tag{25b}$$

The procedure illustrated in Section 8.1 will confirm the fact that (25a) or (25b) does indeed define the *most general* solution of (15), so that the result can be summarized as follows:

The general solution of the equation

$$P\frac{\partial z}{\partial x} + Q\frac{\partial z}{\partial y} = R \tag{26}$$

is of the form

$$F(u_1, u_2) = 0 \quad or \quad u_2 = f(u_1), \tag{27}$$

where $u_1(x, y, z) = c_1$ and $u_2(x, y, z) = c_2$ are solutions of any two independent ordinary differential equations which imply the relationships

$$\frac{dx}{P} = \frac{dy}{Q} = \frac{dz}{R}. \tag{28}$$

---

†In particular, $u_1 = c_1$ itself is an integral surface of (15) unless u_1 is independent of z, and the same statement applies to the surface $u_2 = c_2$.

The intersection of any two of the surfaces $u_1 = c_1$ and $u_2 = c_2$ is a characteristic curve whose tangent at any point has the direction ratios (P, Q, R).

We notice that, since (28) is a consequence of (22), in case P, Q, or R is identically zero we must take dx, dy, or dz, respectively, equal to zero. Thus, the relations $dx/P = dy/Q = dz/0$ imply the two equations $Q\,dx = P\,dy$ and $dz = 0$.

We may now establish the validity of the general result by an argument which is independent of geometrical considerations. Let $u_1(x, y, z) = c_1$ and $u_2(x, y, z) = c_2$ be solutions of two independent equations which imply (23). Then since (23) implies (22), we have

$$0 = du_1 = \frac{\partial u_1}{\partial x}\,dx + \frac{\partial u_1}{\partial y}\,dy + \frac{\partial u_1}{\partial z}\,dz = \left(P\frac{\partial u_1}{\partial x} + Q\frac{\partial u_1}{\partial y} + R\frac{\partial u_1}{\partial z}\right)\frac{ds}{\mu}.$$

Hence $u = u_1$ and (similarly) $u = u_2$ are solutions of (20). If follows also that, if $u_1 = c_1$ and $u_2 = c_2$ determine z as functions of x and y, these functions are solutions of (15). We next replace the independent variables x, y, and z in (20) by the new variables $r = u_1(x, y, z)$, $s = u_2(x, y, z)$, and $t = \varphi(x, y, z)$, where φ is *any* function which is independent of u_1 and u_2 and which does *not* satisfy (20). There then follows

$$\frac{\partial u}{\partial x} = \frac{\partial u}{\partial r}\frac{\partial r}{\partial x} + \frac{\partial u}{\partial s}\frac{\partial s}{\partial x} + \frac{\partial u}{\partial t}\frac{\partial t}{\partial x} = \frac{\partial u}{\partial r}\frac{\partial u_1}{\partial x} + \frac{\partial u}{\partial s}\frac{\partial u_2}{\partial x} + \frac{\partial u}{\partial t}\frac{\partial \varphi}{\partial x},$$

together with similar expressions for $\partial u/\partial y$ and $\partial u/\partial z$. With these substitutions (20) takes the form

$$\left(P\frac{\partial u_1}{\partial x} + Q\frac{\partial u_1}{\partial y} + R\frac{\partial u_1}{\partial z}\right)\frac{\partial u}{\partial r} + \left(P\frac{\partial u_2}{\partial x} + Q\frac{\partial u_2}{\partial y} + R\frac{\partial u_2}{\partial z}\right)\frac{\partial u}{\partial s}$$
$$+ \left(P\frac{\partial \varphi}{\partial x} + Q\frac{\partial \varphi}{\partial y} + R\frac{\partial \varphi}{\partial z}\right)\frac{\partial u}{\partial t} = 0,$$

where u is now considered as a function of the independent variables $r = u_1$, $s = u_2$, and $t = \varphi$. But since u_1 and u_2 satisfy (20), the coefficients of $\partial u/\partial r$ and $\partial u/\partial s$ vanish, and since φ does *not* satisfy (20), *Equation (20) is thus equivalent to the equation $\partial u/\partial t = 0$.* Hence the *most general* solution of (20) is $u = F(r, s) = F(u_1, u_2)$, where F is an arbitrary differentiable function. Thus, finally, the most general solution of (15) is $u = c$ or $F(u_1, u_2) = c$, where c can be incorporated into F, and hence can be replaced by zero, as was to be shown.

Example 1. We consider Equation (1),

$$2\frac{\partial z}{\partial x} - \frac{\partial z}{\partial y} = 0, \tag{29}$$

and deduce the solution (2) by the method derived above. In this case (28) becomes

$$\frac{dx}{2} = \frac{dy}{-1} = \frac{dz}{0},$$

or
$$dx + 2\,dy = 0, \qquad dz = 0.$$

The solutions of these equations are

$$x + 2y = c_1, \qquad z = c_2,$$

and hence, with $u_1 = x + 2y$, $u_2 = z$, Equation (27) gives the general solution of (29) in the alternative forms

$$F(x + 2y, z) = 0 \quad \text{or} \quad z = f(x + 2y), \tag{30}$$

in accordance with (2). In this case the surfaces $u_1 = c_1$ and $u_2 = c_2$ are planes, and the characteristic curves are the straight lines parallel to the xy plane, given by the intersections of the planes $x + 2y = c_1$ with the planes $z = c_2$. The direction cosines of these lines are proportional to the coefficients P, Q, R in (29), that is, to the values $(2, -1, 0)$. In this case the characteristic "curves" coincide with their tangents. Any surface built up from straight lines of the type noted is of the form (30), for *some* choice of f or F, and is an integral surface of (29) unless it is a plane $x + 2y = c$. It should be noticed that any solution z of (29) is constant along any line $x + 2y = $ constant in the xy plane. ∎

To generalize the result of Example 1, we may verify that the general solution of the equation

$$a\frac{\partial z}{\partial x} + b\frac{\partial z}{\partial y} = 0, \tag{31}$$

where a and b are constants, defines all *cylindrical surfaces* with elements parallel to the direction $(a, b, 0)$, and is of the form

$$z = f(bx - ay). \tag{32}$$

It should be noticed that this solution can also be expressed in other related forms such as

$$z = F(ay - bx), \qquad z = g\left(\frac{y}{b} - \frac{x}{a}\right), \qquad z = G\left(y - \frac{b}{a}x\right),$$

and so on, if exceptions when a or b vanish are taken into account.

Example 2. For the equation

$$x\frac{\partial z}{\partial x} + y\frac{\partial z}{\partial y} = z \tag{33}$$

the associated equations become

$$\frac{dx}{x} = \frac{dy}{y} = \frac{dz}{z},$$

and solutions of two of them are obtained in the form

$$\frac{y}{x} = c_1, \qquad \frac{z}{x} = c_2.$$

The general solution of (33) then can be written in the form

$$F\left(\frac{y}{x}, \frac{z}{x}\right) = 0 \quad \text{or} \quad z = xf\left(\frac{y}{x}\right) \tag{34}$$

or in several other equivalent forms, such as

$$z = yg\left(\frac{y}{x}\right), \qquad z = xF\left(\frac{x}{y}\right), \qquad z = yG\left(\frac{x}{y}\right).$$

Since the characteristic curves are straight lines with the direction ratios (x, y, z), it follows that any surface (34) contains the straight lines from the origin to points on the surface, and hence is a *conical surface* with vertex at the origin. ■

8.3. Special Devices. Initial Conditions. Although the *geometric interpretation* of the general solution of an equation of type (26) usually is readily obtained, the determination of explicit solutions of two of the associated equations (28) may present difficulties. In the *linear* case,

$$P(x, y)\frac{\partial z}{\partial x} + Q(x, y)\frac{\partial z}{\partial y} = R_1(x, y)z + R_2(x, y), \tag{35}$$

since P and Q are independent of z, the equation

$$\frac{dy}{dx} = \frac{Q}{P}$$

involves only x and y. If it can be solved in the form $u_1(x, y) = c_1$, then this result may be used to express, say, y in terms of x and c_1. Then the equation

$$\frac{dz}{dx} = \frac{R_1 z + R_2}{P}$$

becomes a linear ordinary differential equation in z.† The solution of this equation can be written in the form

$$u_2 \equiv z\alpha_1(x, c_1) + \alpha_2(x, c_1) = c_2.$$

If c_1 is then replaced by its equivalent in terms of x and y, the result is of the general form

$$u_2 \equiv z\beta_1(x, y) + \beta_2(x, y) = c_2.$$

The general solution of the partial differential equation then becomes $u_2 = f(u_1)$ or

$$z = \frac{1}{\beta_1(x, y)} f[u_1(x, y)] - \frac{\beta_2(x, y)}{\beta_1(x, y)}.$$

Thus, *the general solution of a linear equation of first order can be put in the form*

$$z = s_1(x, y)f[s_2(x, y)] + s_3(x, y), \tag{36}$$

where s_1, s_2, *and* s_3 *are specific functions and* f *is an arbitary function.*

Example 1. In the case of the equation

$$y\frac{\partial z}{\partial x} + x\frac{\partial z}{\partial y} = z - 1, \tag{37}$$

---

†It is proper to hold c_1 constant in solving this equation, since the condition $u_1(x, y) = c_1$ is to be combined with its solution to define a characteristic curve.

Equations (28) become

$$\frac{dx}{y} = \frac{dy}{x} = \frac{dz}{z-1}.$$

Equality of the first two members gives

$$y\,dy - x\,dx = 0, \qquad y^2 - x^2 = c_1.$$

Equality of the first and third members then gives

$$\frac{dz}{z-1} = \frac{dx}{\sqrt{x^2 + c_1}}; \qquad \frac{z-1}{x + \sqrt{x^2 + c_1}} = c_2 \quad \text{or} \quad \frac{z-1}{x+y} = c_2.$$

Thus, with $u_1 = y^2 - x^2$ and $u_2 = (z-1)/(x+y)$, the general solution of (37) becomes $u_2 = f(u_1)$ or

$$z = (x+y)f(y^2 - x^2) + 1. \tag{38}$$

∎

In other cases, if one integrable equation can be obtained by suitably rearranging (28), a similar procedure can be followed. In particular, since the equal ratios in (28) are also equal to the ratio

$$\frac{k_1\,dx + k_2\,dy + k_3\,dz}{k_1 P + k_2 Q + k_3 R}, \tag{39}$$

where $k_1, k_2,$ and k_3 are entirely arbitrary, any one of the given ratios can be replaced by a new ratio of this form. If $k_1, k_2,$ and k_3 can be chosen in such a way that the denominator of the new ratio vanishes and the numerator is an exact differential du_1, then the corresponding requisite vanishing of the numerator leads to the fact that one integral is $u_1 = c_1$.

Example 2. As a classic example, we consider the equation

$$(mz - ny)\frac{\partial z}{\partial x} + (nx - lz)\frac{\partial z}{\partial y} = ly - mx. \tag{40}$$

The associated equations are

$$\frac{dx}{mz - ny} = \frac{dy}{nx - lz} = \frac{dz}{ly - mx},$$

and no pairs are immediately integrable. We next determine $k_1, k_2,$ and k_3 so that the ratio (39) has a zero denominator and hence also a vanishing numerator, and so write

$$k_1(mz - ny) + k_2(nx - lz) + k_3(ly - mx) = 0.$$

If we require the coefficients of $l, m,$ and n to vanish independently, we obtain

$$k_1 = x, \qquad k_2 = y, \qquad k_3 = z,$$

and if we require the coefficients of $x, y,$ and z to vanish independently, we obtain

$$k_1 = l, \qquad k_2 = m, \qquad k_3 = n.$$

With these particular choices of the k's, we obtain the two additional equivalent ratios

$$\frac{x\,dx + y\,dy + z\,dz}{0}, \qquad \frac{l\,dx + m\,dy + n\,dz}{0}.$$

Since the numerators happen to be the exact differentials $d(x^2 + y^2 + z^2)/2$ and $d(lx + my + nz)$, their vanishing leads immediately to the integrals

$$u_1 \equiv x^2 + y^2 + z^2 = c_1, \qquad u_2 \equiv lx + my + nz = c_2,$$

and hence the general solution of (40) can be written in the form

$$F(x^2 + y^2 + z^2, lx + my + nz) = 0. \tag{41}$$

The characteristic curves in this case are the circles in space determined as the intersections of the spheres $u_1 = c_1$ and the planes $u_2 = c_2$.   ∎

In the case of an *ordinary* differential equation of first order, the arbitrary constant can, in general, be determined so that the integral curve passes through a specified *point* in the xy plane. In a similar way, the arbitrary function in the general solution of a first-order *partial* differential equation can be determined, in general, so that the integral surface includes a specified *curve* in xyz space. If the equation of the curve is given by the pair of equations

$$\varphi_1(x, y, z) = 0, \qquad \varphi_2(x, y, z) = 0, \tag{42}$$

that is, as the intersection of the two surfaces $\varphi_1 = 0$ and $\varphi_2 = 0$, and if the solutions of two of the associated equations (28) are written in the form

$$u_1(x, y, z) = c_1, \qquad u_2(x, y, z) = c_2, \tag{43}$$

the determination of the function F in the general solution (27) is equivalent to the determination of a functional relationship between c_1 and c_2 in (43) such that (42) and (43) are compatible. Thus, if the elimination of x, y, and z from the four equations involved in (42) and (43) leads to an equation of the form $F(c_1, c_2) = 0$, the required integral surface is given by $F(u_1, u_2) = 0$.

Example 3. We require the solution of (37) which passes through the curve

$$z = x^2 + y + 1, \qquad y = 2x. \tag{44}$$

First method. We eliminate x, y, and z between these equations and the equations $u_1 = c_1, u_2 = c_2$,

$$y^2 - x^2 = c_1, \qquad \frac{z-1}{x+y} = c_2.$$

The result of the elimination is

$$c_2 = \frac{\sqrt{c_1}}{3\sqrt{3}} + \frac{2}{3},$$

and hence the desired solution is

$$z = \frac{x+y}{3}\left(\frac{\sqrt{y^2 - x^2}}{\sqrt{3}} + 2\right) + 1. \tag{45}$$

Second method. If we take x as the independent variable along the curve, we have, from (44), $y = 2x$, $z = (x + 1)^2$, and the introduction of these results directly into the general solution (38) gives

$$x^2 + 2x = 3xf(3x^2), \qquad f(3x^2) = \frac{x}{3} + \frac{2}{3}.$$

If we write $u = 3x^2$, this equation becomes

$$f(u) = \frac{\sqrt{u}}{3\sqrt{3}} + \frac{2}{3}$$

and so determines the function f in (38) in accordance with the result given by (45). *Third method.* In an equivalent procedure, which may be a bit longer, but in which the steps are more easily interpreted geometrically, we first determine the characteristic curve passing through the point $P_0(x_0, y_0, z_0)$, in the form

$$y^2 - x^2 = y_0^2 - x_0^2, \qquad \frac{z-1}{x+y} = \frac{z_0 - 1}{x_0 + y_0},$$

assuming that $x_0 + y_0 \neq 0$, then require that P_0 lie on the specified curve (44), so that

$$z_0 = x_0^2 + y_0 + 1, \qquad y_0 = 2x_0,$$

and hence

$$y_0 = 2x_0, \qquad z_0 = (x_0 + 1)^2.$$

Thus the characteristic curve passing through a point P_0 on C is defined by the simultaneous equations

$$y^2 - x^2 = 3x_0^2, \qquad z - 1 = \frac{x_0 + 2}{3}(x + y).$$

The required surface is the locus traced out by this curve as x_0 varies through all real values, and is obtained in the form (45) by eliminating x_0. ■

It should be noticed that if a unique integral surface is to be determined, *the prescribed curve cannot be taken as a characteristic curve, since,* in general, *infinitely many integral surfaces include a specified characteristic curve.* Further, in general there exist exceptional curves through which *no* integral surface passes (see Section 8.7).

For example, in the case of (37), if the prescribed curve C projects onto the straight line $y = x$ in the plane $z = 0$, then along that curve (38) becomes

$$z = 2xf(0) + 1.$$

Consequently, unless also $z = Ax + 1$ along C, where A is a constant, there is *no* integral surface passing through that curve. If C is defined by the requirement that $z = Ax + 1$ when $y = x$, the above condition becomes

$$A = 2f(0)$$

and is satisfied by *any* function f for which $f(0) = A/2$, so that for any such f which is differentiable the surface (38) passes through the curve C. In this last case, we verify that the prescribed curve is indeed a characteristic curve ($u_1 = 0$, $u_2 = A/2$).

The procedures given above can be extended to the solution of analogous equations in which there are more than two independent variables. Thus, for example, if x, y, and t are independent variables, the general solution of the quasi-linear equation

$$P\frac{\partial z}{\partial x} + Q\frac{\partial z}{\partial y} + R\frac{\partial z}{\partial t} = S \tag{46}$$

is of the form

$$F(u_1, u_2, u_3) = 0, \tag{47}$$

where $u_1(x, y, z, t) = c_1$, $u_2(x, y, z, t) = c_2$, and $u_3(x, y, z, t) = c_3$ are independent solutions of three of the associated ordinary equations

$$\frac{dx}{P} = \frac{dy}{Q} = \frac{dt}{R} = \frac{dz}{S}. \tag{48}$$

8.4. Linear and Quasi-Linear Equations of Second Order. The general *quasi-linear* equation of second order, involving two independent variables x and y, is of the form

$$a\frac{\partial^2 z}{\partial x^2} + b\frac{\partial^2 z}{\partial x\,\partial y} + c\frac{\partial^2 z}{\partial y^2} + F = 0, \tag{49}$$

where a, b, c, and F may depend upon x, y, z, $\partial z/\partial x$, and $\partial z/\partial y$. In particular, when a, b, and c are independent of z, $\partial z/\partial x$, and $\partial z/\partial y$, while F is a linear function of those quantities, the equation is *linear* in z. Such an equation thus is of the form

$$a\frac{\partial^2 z}{\partial x^2} + b\frac{\partial^2 z}{\partial x\,\partial y} + c\frac{\partial^2 z}{\partial y^2} + d\frac{\partial z}{\partial x} + e\frac{\partial z}{\partial y} + fz = g, \tag{50}$$

where the functions $a, \ldots, g$ depend only upon x and y.

It happens that the terms involving *second* derivatives are of principal significance. In fact, a very important role is played by the sign of the *discriminant* $b^2 - 4ac$. From analogy with the terminology associated with conic sections, we say that (49) or (50) is of *hyperbolic* type when $b^2 > 4ac$, of *elliptic* type when $b^2 < 4ac$, and of *parabolic* type in the intermediate case, when $b^2 = 4ac$. The importance of this classification will be indicated in later sections with reference to *linear* equations, to which our attention will be restricted in most of what follows.

In the linear case, it is clear that, as in the case of *ordinary* differential equations, the most general solution of (50) consists of the sum of any *particular solution* of (50) and the most general solution of the equation obtained by replacing the right-hand member of (50) by zero. This second solution is sometimes called the *complementary solution* of (50).

It is seen also that any linear combination of two complementary solutions will also be a complementary solution. However, such a combination cannot be expected to be the most general complementary solution, since the general solution should involve arbitrary *functions*. It might be expected that the most general solution of (50) would be a generalization of the form (36),

$$z = s_1(x, y)f[s_2(x, y)] + s_3(x, y)g[s_4(x, y)] + s_5(x, y), \tag{51}$$

where $s_1, \ldots, s_5$ are specific functions and f and g are arbitrary functions. However, this condition exists only in special cases. That is, *the most general complementary solution of* (50) *cannot always be written as a sum of terms involving arbitrary functions.* (For an example establishing this assertion, see Problem 22.)

In Section 8.5 we consider an important special class of linear equations in which the general solution is of simple form, and in Section 8.6 we illustrate other special types of linear equations.

8.5. Special Linear Equations of Second Order, with Constant Coefficients. We here consider equations of the very special form

$$a\frac{\partial^2 z}{\partial x^2} + b\frac{\partial^2 z}{\partial x\,\partial y} + c\frac{\partial^2 z}{\partial y^2} = 0, \tag{52}$$

where a, b, and c are *constants.*† To obtain the general solution of (52), we assume a solution of the form

$$z = f(y + mx), \tag{53}$$

where f is an arbitrary twice-differentiable function and m is a constant, and attempt to determine values of m for which (53) is a solution of (52). By differentiation, we obtain from (53) the expressions

$$\frac{\partial^2 z}{\partial x^2} = m^2 f''(y + mx), \quad \frac{\partial^2 z}{\partial x\,\partial y} = mf''(y + mx), \quad \frac{\partial^2 z}{\partial y^2} = f''(y + mx), \tag{54}$$

and the introduction of (54) into (52) gives the condition

$$am^2 + bm + c = 0 \tag{55}$$

which must be satisfied by m.

In general, (55) will determine two distinct values of m, say m_1 and m_2. Then, because of the linearity of (52), it follows that any expression of the form

$$z = f(y + m_1 x) + g(y + m_2 x) \tag{56}$$

is a solution of (52). Further, if the new variables $s = y + m_1 x$ and $t = y + m_2 x$ are introduced, it is readily verified that (52) takes the form $\partial^2 z/(\partial s\,\partial t) = 0$, from which it follows that (56) is indeed the *most general* solution of (52).

Example 1. For the equation

$$\frac{\partial^2 z}{\partial x^2} - 3\frac{\partial^2 z}{\partial x\,\partial y} + 2\frac{\partial^2 z}{\partial y^2} = 0, \tag{57}$$

the determinant equation becomes

$$m^2 - 3m + 2 = 0, \quad m_1 = 1, \quad m_2 = 2,$$

---

†Equations of this general type, in which all terms involve derivatives of the same order, are often called "homogeneous equations." However, we will continue to use this designation only as it is defined in Section 5.1.

and the general solution is

$$z = f(x + y) + g(2x + y).\qquad(58)$$

∎

If $a = 0$ and $b \neq 0$, only one solution of Equation (55) is obtained. However, in this case it is clear by inspection that the second term in the general solution is then an arbitrary function of x alone. If $a = b = 0$, the general solution is clearly $z = f(x) + yg(x)$. In these and other cases, the alternative assumption of a solution of the form $z = f(x + ny)$ may be convenient.

The remaining exceptional case is that in which (55) is a perfect square, so that the two roots are equal. The second term in the general solution can be found by a method similar to that used in analogous cases in Chapter 1. If m_1 and m_2 are distinct roots of (55), then the expression

$$\frac{h(y + m_2 x) - h(y + m_1 x)}{m_2 - m_1}$$

is clearly a solution of (52). As $m_2 \rightarrow m_1$, this solution approaches the limiting value

$$\left[\frac{\partial}{\partial m} h(y + mx)\right]_{m=m_1} = xh'(y + m_1 x).$$

Thus, if we write $g = h'$, *the general solution in the case of equal roots can be taken in the form*

$$z = f(y + m_1 x) + xg(y + m_1 x).\qquad(59)$$

Example 2. The general solution of

$$\frac{\partial^2 z}{\partial x^2} - 2\frac{\partial^2 z}{\partial x\, \partial y} + \frac{\partial^2 z}{\partial y^2} = 0$$

is found in this way to be

$$z = f(x + y) + xg(x + y).\qquad\blacksquare$$

It may be noticed that since we can write

$$xg(y + m_1 x) \equiv \frac{1}{m_1}[(y + m_1 x)g(y + m_1 x) - yg(y + m_1 x)]$$

if $m_1 \neq 0$, the second solution can equally well be taken in the form $yg(y + m_1 x)$ unless $m_1 = 0$.

It is seen that the solutions of (55) will both be real if $b^2 > 4ac$, so that (52) is *hyperbolic*, and will be conjugate complex if $b^2 < 4ac$, so that (52) is *elliptic*. The intermediate case, when $b^2 = 4ac$, so that the equation is of *parabolic* type, is that in which (55) is a perfect square.

Thus Laplace's equation

$$\frac{\partial^2 z}{\partial x^2} + \frac{\partial^2 z}{\partial y^2} = 0 \tag{60a}$$

is *elliptic*, with general solution of the form

$$z = f(x + iy) + g(x - iy), \tag{60b}$$

the equation

$$\frac{\partial^2 z}{\partial x^2} - \frac{\partial^2 z}{\partial y^2} = 0 \tag{61a}$$

is *hyperbolic*, with general solution

$$z = f(x + y) + g(x - y), \tag{61b}$$

and the equation

$$\frac{\partial^2 z}{\partial x^2} - 2\frac{\partial^2 z}{\partial x\, \partial y} + \frac{\partial^2 z}{\partial y^2} = 0 \tag{62a}$$

is *parabolic*, with general solution

$$z = f(x + y) + xg(x + y). \tag{62b}$$

A procedure completely analogous to that given above can be applied in obtaining the general solution of a linear equation of any order N, in which each term is a constant multiple of a derivative of order N. In such a case, the assumption of a solution of form (53) leads to an equation analogous to (55), but of Nth degree in m. If N distinct roots exist, the general solution is of a form similar to (56) but involving N independent terms. If a root m_1 is repeated r times, the part of the solution corresponding to the r equal roots is readily shown to be of the form

$$f_1(y + m_1 x) + x f_2(y + m_1 x) + \cdots + x^{r-1} f_r(y + m_1 x). \tag{63}$$

Occasionally it is feasible to solve a problem, governed by a partial differential equation together with appropriate side conditions, by obtaining the general solution of the equation and determining the relevant arbitrary functions by imposing the side conditions, in analogy with the usual procedure relating to *ordinary* differential equations.

Perhaps the best-known situation of this type is that in which the solution $\varphi(x, t)$ of the "one-dimensional wave equation"

$$\frac{\partial^2 \varphi}{\partial t^2} = c^2 \frac{\partial^2 \varphi}{\partial x^2} \tag{64}$$

is required, for all real values of x and t, subject to the conditions

$$\varphi(x, 0) = F(x), \qquad \varphi_t(x, 0) = G(x), \tag{65}$$

prescribed for *all* real x when $t = 0$. Here c is a positive real constant. This problem is often called the *Cauchy problem* for Equation (64).

If the general solution of (64) is written in the form

$$\varphi(x, t) = f(x + ct) + g(x - ct), \tag{66}$$

the conditions (65) require that f and g satisfy the conditions

$$f(x) + g(x) = F(x), \qquad c[f'(x) - g'(x)] = G(x),$$

for all values of x. The elimination of $g(x)$ yields

$$f'(x) = \frac{1}{2}F'(x) + \frac{1}{2c}G(x), \qquad f(x) = \frac{1}{2}F(x) + \frac{1}{2c}\int_0^x G(\xi)\,d\xi + C,$$

where C is an arbitrary constant. The substitution of this result and the corresponding expression for $g(x)$ into (66) gives the desired solution

$$\varphi(x, t) = \frac{1}{2}[F(x + ct) + F(x - ct)] + \frac{1}{2c}\int_{x-ct}^{x+ct} G(\xi)\,d\xi. \tag{67}$$

This solution is associated with the name of *d'Alembert*. A physical interpretation of the problem and its solution is included in Section 9.12, and a generalization is obtained in Problem 37.

Analogous solutions of certain other problems governed by (64) are obtained in Problems 33 and 35.

8.6. Other Linear Equations. As a simple example of other solvable linear equations of second order, we consider the equation

$$\frac{\partial^2 z}{\partial x^2} - z = 0, \tag{68}$$

where x and y are the independent variables. Since y is not involved explicitly, we may integrate (68) as though it were an ordinary equation, holding y constant in the process, and hence replacing the two arbitrary constants of integration by arbitrary functions of y. Thus the general solution of (68) is of the form

$$z = e^x f(y) + e^{-x} g(y). \tag{69}$$

However, even though the coefficients in the general second-order linear equation (50) be constants, it is not possible in *all* such cases to obtain general solutions of similar forms. The following procedure is frequently useful.

From the analogy with the ordinary equations, we are led to expect that *exponential* solutions of the equation

$$a\frac{\partial^2 z}{\partial x^2} + b\frac{\partial^2 z}{\partial x\,\partial y} + c\frac{\partial^2 z}{\partial y^2} + d\frac{\partial z}{\partial x} + e\frac{\partial z}{\partial y} + fz = 0, \tag{70}$$

with constant coefficients, may be of importance. If we assume a solution of the form

$$z = Ae^{\alpha x + \beta y}, \tag{71}$$

where A, α, and β are unspecified constants, the introduction of (71) into (70) gives the condition

$$a\alpha^2 + b\alpha\beta + c\beta^2 + d\alpha + e\beta + f = 0 \tag{72}$$

to be satisfied by α and β. This equation will, in general, determine β as either of two functions of α, say

$$\beta = \varphi_1(\alpha), \qquad \beta = \varphi_2(\alpha). \tag{73}$$

It then follows that any expression of either of the forms

$$Ae^{\alpha x + \varphi_1(\alpha)y}, \qquad Be^{\alpha x + \varphi_2(\alpha)y}$$

will satisfy (70) for arbitrary values of A, B, and α, and hence will be a *particular* solution of (70). The same is true for any linear combination or suitably convergent infinite series of such solutions, or for an integral superposition of the form

$$\int_{\alpha_1}^{\alpha_2} [A(\alpha)e^{\alpha x + \varphi_1(\alpha)y} + B(\alpha)e^{\alpha x + \varphi_2(\alpha)y}] \, d\alpha$$

when $A(\alpha)$ and $B(\alpha)$ are sufficiently respectable.

As is next illustrated, this procedure leads to a precise form of the *general* solution if (73) expresses β as linear functions of α, as is the case when (70) reduces to (52) and, more generally, when the condition

$$ae^2 + fb^2 + cd^2 = bde + 4acf \tag{74}$$

is satisfied. The algebraic equation (72), associated with (70), represents a conic section in the $\alpha\beta$ plane, and is a hyperbola, ellipse, or parabola, according as $b^2 - 4ac > 0$, < 0, or $= 0$, respectively. In the special cases just mentioned, the conic sections degenerate into two straight lines.

Example 1. We consider the equation

$$\frac{\partial^2 z}{\partial x^2} - \frac{\partial^2 z}{\partial y^2} - \frac{\partial z}{\partial x} + \frac{\partial z}{\partial y} = 0, \tag{75}$$

for which the condition (74) is satisfied. The assumption (71) leads to the equation

$$\alpha^2 - \beta^2 - \alpha + \beta = (\alpha - \beta)(\alpha + \beta - 1) = 0$$

and hence (71) is a solution of (75) if

$$\beta = \alpha \quad \text{or} \quad \beta = -\alpha + 1.$$

Thus, by superposition, any expression of the form

$$z = \sum_{\alpha} A(\alpha)e^{\alpha(x+y)} + \sum_{\alpha} B(\alpha)e^{y + \alpha(x-y)} \tag{76}$$

is also a solution, the notation indicating that α may take on arbitrary values in the summation. But since the coefficients A and B are also arbitrary, the first term may be expected to represent an arbitrary respectable function of $x + y$, and the second term to represent the product of e^y and an arbitrary respectable function of $x - y$. Thus we may suspect that the general solution can be written in the form

$$z = f(x + y) + e^y g(x - y), \tag{77}$$

where f and g are twice differentiable, and this can be verified by direct substitution. ∎

The sort of simplification achieved in this example clearly is always possible when the left-hand member of (72) contains distinct linear factors, and the special case of a repeated linear factor is easily dealt with by use of a method similar to that which led to (59).

In other cases, whereas this procedure usually does not yield the *general* solution, it may provide a useful set of *particular* solutions which (as will be seen in Chapter 9) is sufficiently extensive for the treatment of a class of specific problems.

Example 2. For the two-dimensional *Helmholtz equation*,

$$\frac{\partial^2 z}{\partial x^2} + \frac{\partial^2 z}{\partial y^2} + k^2 z = 0, \tag{78}$$

the assumption (71) leads to the equation

$$\alpha^2 + \beta^2 + k^2 = 0,$$

and hence

$$\beta = \pm i\sqrt{\alpha^2 + k^2}.$$

Any expression of the form

$$A(\alpha)e^{\alpha x + i\sqrt{\alpha^2 + k^2}\,y} + B(\alpha)e^{\alpha x - i\sqrt{\alpha^2 + k^2}\,y}$$

is then a solution of (78), for any choice of A, B, and α. If we write

$$A + B = C, \qquad i(A - B) = D,$$

this expression takes the real form

$$e^{\alpha x}(C \cos \sqrt{\alpha^2 + k^2}\,y + D \sin \sqrt{\alpha^2 + k^2}\,y).$$

Any combination of expressions of this sort, as α, C, and D take on arbitrary values, will also satisfy (78). Finally, any expression of the form

$$z = \int_{-\infty}^{\infty} [C(\alpha) \cos \sqrt{\alpha^2 + k^2}\,y + D(\alpha) \sin \sqrt{\alpha^2 + k^2}\,y]e^{\alpha x}\, d\alpha, \tag{79}$$

for which the indicated integral is suitably convergent, will be a particular solution of (78). (See also Problems 38 and 39.) ∎

When a partial differential equation has *variable* coefficients, it is rarely possible to obtain its general solution. However if, for example, the equation can be written in the form

$$a\frac{\partial^2(pz)}{\partial x^2} + b\frac{\partial^2(pz)}{\partial x\,\partial y} + c\frac{\partial^2(pz)}{\partial y^2} = 0, \tag{80}$$

where a, b, and c are constants and p is a function of x and/or y, the general solution clearly expresses z as the product of $1/p$ and a combination of arbitrary functions. (See Problem 42.)

Example 3. We find that the equation

$$\frac{\partial^2 \varphi}{\partial r^2} + \frac{4}{r}\frac{\partial \varphi}{\partial r} + \frac{2}{r^2}\varphi = \frac{1}{\alpha^2}\frac{\partial^2 \varphi}{\partial t^2} \tag{81}$$

also can be written in the form

$$\frac{\partial^2(r^2\varphi)}{\partial r^2} - \frac{1}{\alpha^2}\frac{\partial^2(r^2\varphi)}{\partial t^2} = 0,$$

so that its general solution is

$$\varphi = \frac{1}{r^2}[f(r + \alpha t) + g(r - \alpha t)] \tag{82}$$

when $\alpha \neq 0.$ ∎

8.7. Characteristics of Linear First-Order Equations. In Section 8.2 we have defined a *characteristic curve* of the first-order linear equation

$$P\frac{\partial z}{\partial x} + Q\frac{\partial z}{\partial y} = R_1 z + R_2, \tag{83}$$

where P, Q, R_1, and R_2 are functions of x and y only, as a curve in space whose tangent at any point has the direction ratios $(P, Q, R_1 z + R_2)$. Any such curve is the intersection of two surfaces of the form

$$u_1(x, y) = c_1, \qquad u_2(x, y, z) = c_2, \tag{84}$$

where $u_1(x, y) = c_1$ is an integral of the ordinary equation

$$\frac{dx}{P} = \frac{dy}{Q} \tag{85}$$

and $u_2(x, y, z) = c_2$ is an independent integral of one of the associated equations

$$\frac{dx}{P} = \frac{dy}{Q} = \frac{dz}{R_1 z + R_2}. \tag{86}$$

Thus any characteristic curve of (83) is the intersection of the *cylinder*

$$u_1(x, y) = c_1, \tag{87}$$

with elements parallel to the z axis, and a second surface of the form $u_2(x, y, z) = c_2$. It follows that any characteristic curve can be specified by first choosing a particular cylinder of the family (87), and so satisfying (85), and then prescribing z on this cylinder in such a way that another independent equation obtained from (86) is satisfied. We shall find it convenient to speak of any cylinder obtained by specializing c_1 in (87) as a *characteristic cylinder*. Also, we refer to the intersection of this cylinder with the xy plane as a *characteristic base curve*. Such a curve is clearly the *projection* of a characteristic space curve onto the xy plane.

Example 1. In the case of the equation

$$y\frac{\partial z}{\partial x} - x\frac{\partial z}{\partial y} = y, \tag{88}$$

two associated equations can be written in the form

$$\frac{dx}{y} = -\frac{dy}{x}, \qquad dz = dx,$$

with solutions

$$u_1 = x^2 + y^2 = c_1, \qquad u_2 = z - x = c_2.$$

Thus the characteristic curves are ellipses in space determined as the intersections of the right circular cylinders $x^2 + y^2 = c_1$ and the planes $z - x = c_2$. The cylinders $x^2 + y^2 = c_1$ are the characteristic cylinders, and the circles $x^2 + y^2 = c_1$ in the xy plane are the characteristic base curves, which are the projections of the characteristic ellipses onto the xy plane. Along this base curve we can express x and y in terms of a single variable λ by writing

$$x = \sqrt{c_1} \cos \lambda, \tag{89a}$$

$$y = \sqrt{c_1} \sin \lambda. \tag{89b}$$

If along this curve we define z by an equation of the form

$$z = x + c_2 = \sqrt{c_1} \cos \lambda + c_2, \tag{89c}$$

the corresponding locus in space is a characteristic curve. Each characteristic cylinder is seen to include infinitely many characteristic curves. ∎

We now reconsider, from a somewhat different point of view, the question discussed in Section 8.3, as to whether an equation of form (83) has as a particular solution an integral surface which includes an arbitrarily prescribed curve in space.

Along any curve in space, the coordinates x, y, and z can be considered as functions of a single variable, say λ. Thus the curve can be specified by three equations of the form

$$C: \qquad x = x(\lambda), \qquad y = y(\lambda), \qquad z = z(\lambda). \tag{90}$$

The variable λ may be taken as arc length along the curve, for example, or it may be identified with one of the coordinates x or y. The projection C_0 of this curve onto the xy plane is given by

$$C_0: \qquad x = x(\lambda), \qquad y = y(\lambda) \tag{91}$$

and $z = 0$. Thus (90) can be considered as specifying z along the curve C_0 in the xy plane in terms of a parameter λ which varies along C_0.

Suppose then that z is prescribed along an arbitrary curve C_0 in the xy plane, in accordance with (90), and that $x(\lambda)$, $y(\lambda)$ have continuous first derivatives and $z(\lambda)$ continuous derivatives of all orders with respect to λ. We inquire whether (83) has a solution expressing z as a function of x and y everywhere in a region including C_0 in such a way that z takes on the prescribed values on C_0. This query is clearly equivalent to asking whether (83) possesses an integral surface which includes the space curve (90).

Let us assume that such a solution does exist and that the solution $z = f(x, y)$ can be expanded in a series of powers of x and y, about a point (x_0, y_0) on C_0, of the form

$$z = f(x_0, y_0) + \frac{\partial f(x_0, y_0)}{\partial x}(x - x_0) + \frac{\partial f(x_0, y_0)}{\partial y}(y - y_0) + \cdots$$

$$= (z)_{(x_0, y_0)} + \left(\frac{\partial z}{\partial x}\right)_{(x_0, y_0)}(x - x_0) + \left(\frac{\partial z}{\partial y}\right)_{(x_0, y_0)}(y - y_0) + \cdots. \tag{92}$$

Then the solution is determined in the neighborhood of a point on C_0 if we know all its partial derivatives at that point.

It may be recalled that in Chapter 3 we made use of such ideas to calculate the series solution of an ordinary differential equation of the form

$$\frac{dy}{dx} = F(x, y)$$

for which y is prescribed as y_0 when $x = x_0$. The differential equation then gives dy/dx when $x = x_0$, and by successive differentiation all higher derivatives are generally obtainable. We then have the required solution

$$y(x) = y(x_0) + y'(x_0)(x - x_0) + \frac{y''(x_0)}{2}(x - x_0)^2 + \cdots$$

if suitable convergence is assumed.

On the curve C_0, the coefficients P, Q, R_1, and R_2, as well as z itself, are now known functions of λ, and hence (83) gives one equation involving the values of the two first partial derivatives of z at points on C_0,

$$P(\lambda)\frac{\partial z}{\partial x} + Q(\lambda)\frac{\partial z}{\partial y} = R_1(\lambda)z(\lambda) + R_2(\lambda), \tag{93}$$

where, for example, $P(\lambda)$ has been written for $P[x(\lambda), y(\lambda)]$. To obtain a second equation, we notice that, since z is known on C_0, its derivative $dz/d\lambda$ along C_0 is also known, and it must satisfy the equation

$$\frac{dz(\lambda)}{d\lambda} = \frac{\partial z}{\partial x}\frac{dx}{d\lambda} + \frac{dz}{\partial y}\frac{dy}{d\lambda}.$$

Thus a second equation complementing (93) is of the form

$$\frac{dx}{d\lambda}\frac{\partial z}{\partial x} + \frac{dy}{d\lambda}\frac{\partial z}{\partial y} = \frac{dz}{d\lambda}. \tag{94}$$

Equations (93) and (94) can be solved uniquely for the unknown quantities $\partial z/\partial x$ and $\partial z/\partial y$ everywhere on C_0 if and only if the determinant of their coefficients is never zero,

$$\begin{vmatrix} P(\lambda) & Q(\lambda) \\ \dfrac{dx}{d\lambda} & \dfrac{dy}{d\lambda} \end{vmatrix} = P(\lambda)\frac{dy}{d\lambda} - Q(\lambda)\frac{dx}{d\lambda} \neq 0. \tag{95}$$

But this condition is equivalent to the restriction

$$\frac{dx(\lambda)}{P(\lambda)} \neq \frac{dy(\lambda)}{Q(\lambda)}, \tag{96}$$

and *hence (95) requires that the curve C_0 along which z is prescribed never be tangent to a characteristic base curve.* If (95) is satisfied at all points on C_0, the values of $\partial z/\partial x$ and $\partial z/\partial y$ then can be calculated by (93) and (94) at all points on C_0.

To calculate the second-order partial derivatives of z on C_0, we may differentiate (83) with respect to x, and so obtain the equation

$$P\frac{\partial^2 z}{\partial x^2} + Q\frac{\partial^2 z}{\partial x\,\partial y} = R_1\frac{\partial z}{\partial x} + \frac{\partial R_1}{\partial x}z + \frac{\partial R_2}{\partial x} - \frac{\partial P}{\partial x}\frac{\partial z}{\partial x} - \frac{\partial Q}{\partial x}\frac{\partial z}{\partial y}, \tag{97}$$

where all quantities but $\partial^2 z/\partial x^2$ and $\partial^2 z/(\partial x \, \partial y)$ are now known functions of λ on C_0. A second equation involving these unknown quantities is obtained by differentiating the known function $\partial z/\partial x$ along C_0, with respect to λ,

$$\frac{dx}{d\lambda}\frac{\partial^2 z}{\partial x^2} + \frac{dy}{d\lambda}\frac{\partial^2 z}{\partial x \, \partial y} = \frac{d}{d\lambda}\left(\frac{\partial z}{\partial x}\right). \tag{98}$$

Equations (97) and (98) determine $\partial^2 z/\partial x^2$ and $\partial^2 z/(\partial x \, \partial y)$ uniquely in terms of known functions if (95) is true, and the remaining derivative $\partial^2 z/\partial y^2$ then can be calculated either from the equation obtained by differentiating $\partial z/\partial y$ with respect to λ or from that obtained by differentiating (83) with respect to y. This process can be continued indefinitely, to determine the values of all partial derivatives of z on C_0 in such a case and, assuming convergence of the series (92), it follows that *the function z is determined uniquely for values of x and y in a region about C_0 by its prescribed values on C_0 if C_0 is never tangent to a characteristic base curve.* This assertion is equivalent to the statement that *if C is never tangent to a characteristic cylinder of* (83), *there exists a unique integral surface of* (83) *which includes C.*

If, however, (95) is violated for *all* λ on C_0, so that C is *on* a characteristic cylinder, then (93) and (94) are *incompatible* unless they are *equivalent*, that is, unless

$$\frac{dx(\lambda)}{P(\lambda)} = \frac{dy(\lambda)}{Q(\lambda)} = \frac{dz(\lambda)}{R_2(\lambda) + z R_1(\lambda)}. \tag{99}$$

In this case one of the two equations (93) and (94) implies the other, and hence only one restriction on the two derivatives is present. Thus the partial derivatives $\partial z/\partial x$ and $\partial z/\partial y$ *are not determined uniquely* in this case but may be chosen in infinitely many ways. But this result is to be expected, since if (99) is satisfied, C is then a *characteristic curve*, which does indeed lie on infinitely many integral surfaces. Thus it follows that *if C lies on a characteristic cylinder of* (83), *there is no integral surface of* (83) *which includes C unless C is a characteristic curve, in which case there exist infinitely many such surfaces.* It follows also that *if C_0 is a characteristic base curve, then z cannot be prescribed arbitrarily on it but must be taken in such a way that C is a characteristic curve. In such a case the value of z at a point near C_0 is not determined by the values of z on C_0.*

Thus, in the case of Equation (88), there exists a unique integral surface including any space curve whose projection onto the xy plane is not a circle with center at the origin. The only curves whose projections are of this type and which lie on integral surfaces are the ellipses which are characteristic curves, and these curves each lie on infinitely many integral surfaces.

When (95) is violated only for certain isolated values of λ, so that C_0 is *tangent* to a characteristic base curve at each corresponding point, the relevant solution $z(x, y)$ may behave exceptionally at each such point of tangency.

We next show that if the equation (87) of the characteristic cylinders of the differential equation

$$P\frac{\partial z}{\partial x} + Q\frac{\partial z}{\partial y} = R_1 z + R_2 \tag{100}$$

is known, the equation can be reduced to a normal ("canonical") form by a suitable change in variables. If $P \equiv 0$, Equation (100) already is in the desired form. Otherwise, we take as a new independent variable the expression

$$\varphi = u_1(x, y), \tag{101}$$

where $u_1(x, y) = c_1$ is a solution of (85), and retain x as the second independent variable, since then x and u_1 assuredly are functionally independent. There then follows

$$\left(\frac{\partial z}{\partial x}\right)_y = \left(\frac{\partial z}{\partial x}\right)_\varphi + \left(\frac{\partial z}{\partial \varphi}\right)_x \frac{\partial u_1}{\partial x},$$

$$\left(\frac{\partial z}{\partial y}\right)_x = \left(\frac{\partial z}{\partial \varphi}\right)_x \frac{\partial u_1}{\partial y},$$

and hence, with z now considered as a function of x and φ, Equation (100) becomes

$$P\frac{\partial z}{\partial x} + \left(P\frac{\partial u_1}{\partial x} + Q\frac{\partial u_1}{\partial y}\right)\frac{\partial z}{\partial \varphi} = R_1 z + R_2, \tag{102}$$

where P, Q, R_1, and R_2 now are to be expressed as functions of x and φ. But since $u_1 = c_1$ is an integral of (85), the function u_1 satisfies the partial differential equation

$$P\frac{\partial u_1}{\partial x} + Q\frac{\partial u_1}{\partial y} = 0.$$

Hence the coefficient of $\partial z/\partial \varphi$ in (102) vanishes, and the equation takes the form

$$P\frac{\partial z}{\partial x} = R_1 z + R_2, \tag{103}$$

when z is considered as a function of x and φ.†

Example 2. In the case of Equation (88),

$$y\frac{\partial z}{\partial x} - x\frac{\partial z}{\partial y} = y, \tag{104}$$

we write

$$\varphi = x^2 + y^2, \qquad y = \sqrt{\varphi - x^2}.$$

Then (103) shows that (88) can be put into the simpler form

$$\sqrt{\varphi - x^2}\frac{\partial z}{\partial x} = \sqrt{\varphi - x^2},$$

with z considered as a function of x and φ. Since $\sqrt{\varphi - x^2}$ does not involve z, it may be canceled, and the equation becomes

$$\frac{\partial z}{\partial x} = 1, \tag{105}$$

---

†The need for such a qualifying phrase (particularly here, but also elsewhere) would be avoided by using a new symbol Z for the dependent variable when it is considered as a function of the new independent variables, so that $Z(x, \varphi) = Z[x, u_1(x, y)] = z(x, y)$. (See also Section 7.1.)

with the obvious general solution

$$z = x + f(\varphi) \equiv x + f(x^2 + y^2).$$

This result is equivalent to the form $u_2 = f(u_1)$, with the notation following (88), as is to be expected.   ◼

In the more general case of a quasi-linear equation

$$P(x, y, z)\frac{\partial z}{\partial x} + Q(x, y, z)\frac{\partial z}{\partial y} = R(x, y, z), \tag{106}$$

an integral $u_1 = c_1$ of the equation

$$\frac{dx}{P} = \frac{dy}{Q} \tag{107}$$

generally cannot be obtained without reference to another equation obtained from the relations

$$\frac{dx}{P} = \frac{dy}{Q} = \frac{dz}{R}, \tag{108}$$

and, in any case, u_1 generally will not be independent of z. Thus here the concept of a *characteristic cylinder* generally is no longer meaningful. However, it still remains true that, if the condition (107) is never satisfied on a specified curve C, then there is a unique integral surface of (106) which contains C. If (107) is satisfied everywhere on C, then there is *no* integral surface containing C unless (108) is satisfied on C, in which case C is a characteristic curve and there are *infinitely many* integral surfaces containing C.

8.8. Characteristics of Linear Second-Order Equations. We next investigate in a similar way the problem of determining solutions of linear second-order partial differential equations satisfying appropriate initial conditions. The most general linear second-order equation is of the form

$$a\frac{\partial^2 z}{\partial x^2} + b\frac{\partial^2 z}{\partial x\,\partial y} + c\frac{\partial^2 z}{\partial y^2} + d\frac{\partial z}{\partial x} + e\frac{\partial z}{\partial y} + fz = g, \tag{109}$$

where the coefficients and the right-hand member may be functions of x and y. In the case of *ordinary* equations of second order, the initial conditions prescribe a point in the plane, through which the integral curve is to pass, and also prescribe the slope of the integral curve at that point. This is equivalent to prescribing the values of the dependent variable y and its derivative dy/dx corresponding to a given value of the independent variable x. In the case of a *partial* differential equation, the analogous initial conditions prescribe a curve C in space which is to lie in the integral surface, and also prescribe the orientation of the tangent plane to the integral surface along that curve. These conditions are equivalent to conditions which prescribe the values of z and its two partial derivatives $\partial z/\partial x$ and $\partial z/\partial y$ along the projection C_0 of the curve C onto the xy plane. However, the values of z, $\partial z/\partial x$, and $\partial z/\partial y$ cannot be prescribed in a completely independent way if z is to be differentiable along the curve C_0, since if λ is a parameter specifying position along C_0, we must have

$$\frac{dz}{d\lambda} = \frac{\partial z}{\partial x}\frac{dx}{d\lambda} + \frac{\partial z}{\partial y}\frac{dy}{d\lambda}. \tag{110}$$

Thus, for example, if C_0 is the x axis and if z is prescribed as $f(x)$ along C_0, then the derivative $\partial z/\partial x$ *along* C_0 cannot also be prescribed but must be given as $f'(x)$. However, the derivative $\partial z/\partial y$ *normal to* C_0 can be independently prescribed. This limitation is in accordance with (110), which here becomes

$$f'(x) = \frac{\partial z}{\partial x} \cdot 1 + \frac{\partial z}{\partial y} \cdot 0$$

if we identify λ with the distance x along C_0.

A curve C in space, together with values of $\partial z/\partial x$ and $\partial z/\partial y$ prescribed along C in such a way that (110) is satisfied, is called a *strip*, and Equation (110) is referred to as the *strip condition*. If it is recalled [see Equation (103), Chapter 6] that the tangent plane to a surface $z = z(x, y)$ has as its normal a vector with direction ratios $(\partial z/\partial x, \partial z/\partial y, -1)$, we can think of the prescribed values of $\partial z/\partial x$ and $\partial z/\partial y$ at points along C as determining the normal direction to differential elements of surface area at these points on the required integral surface. The strip condition (110) requires that the elements join together in a regular way (Figure 8.1).

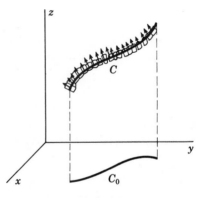

Figure 8.1

Suppose now that a curve C_0 in the xy plane is given by the equations

$$C_0: \qquad x = x(\lambda), \qquad y = y(\lambda), \tag{111}$$

where $x(\lambda)$ and $y(\lambda)$ have continuous derivatives, and that z, $\partial z/\partial x$, and $\partial z/\partial y$ are prescribed as functions of λ having derivatives of all orders at all points of C_0, in such a way that (110) is satisfied. For brevity, we introduce the conventional abbreviations

$$p = \frac{\partial z}{\partial x}, \quad q = \frac{\partial z}{\partial y}, \quad r = \frac{\partial^2 z}{\partial x^2}, \quad s = \frac{\partial^2 z}{\partial x\,\partial y}, \quad t = \frac{\partial^2 z}{\partial y^2}. \tag{112}$$

Then the differential equation (109) becomes

$$ar + bs + ct + dp + eq + fz = g, \tag{113}$$

and the strip condition takes the form

$$\frac{dz}{d\lambda} = p\frac{dx}{d\lambda} + q\frac{dy}{d\lambda}. \tag{114}$$

The prescribed conditions along C_0 are then

$$z = z(\lambda), \qquad p = p(\lambda), \qquad q = q(\lambda), \tag{115}$$

where $z(\lambda)$, $p(\lambda)$, and $q(\lambda)$ satisfy (114).

As in Section 8.7, we now attempt to calculate the values of the higher derivatives of z along C_0 in terms of the prescribed values. If all such derivatives can be determined and if z can be expanded in a power series in x and y about points on C_0, then an integral surface satisfying the prescribed conditions is determined for values of x and y in a region including C_0, and the value of z at a point near C_0 is determined in terms of the known values on C_0 by use of this series.

The first problem, then, is to attempt to determine the values of the second derivatives r, s, and t for points on C_0. One condition involving these unknown quantities is given by (113),

$$ar + bs + ct = g - (dp + eq + fz). \tag{116}$$

If we notice that, along C_0, there follows also

$$\frac{d}{d\lambda}\left(\frac{\partial z}{\partial x}\right) = \frac{\partial}{\partial x}\left(\frac{\partial z}{\partial x}\right)\frac{\partial x}{\partial \lambda} + \frac{\partial}{\partial y}\left(\frac{\partial z}{\partial x}\right)\frac{\partial y}{\partial \lambda},$$

$$\frac{d}{d\lambda}\left(\frac{\partial z}{\partial y}\right) = \frac{\partial}{\partial x}\left(\frac{\partial z}{\partial y}\right)\frac{dx}{dy} + \frac{\partial}{\partial y}\left(\frac{\partial z}{\partial y}\right)\frac{dy}{d\lambda},$$

we thus obtain two additional conditions involving r, s, and t, of the form

$$\frac{dx}{d\lambda}r + \frac{dy}{d\lambda}s = \frac{dp}{d\lambda}, \tag{117}$$

$$\frac{dx}{d\lambda}s + \frac{dy}{d\lambda}t = \frac{dq}{d\lambda}. \tag{118}$$

Equations (116), (117), and (118) determine r, s, and t uniquely if and only if the determinant of their coefficients is not zero,

$$\begin{vmatrix} a & b & c \\ \dfrac{dx}{d\lambda} & \dfrac{dy}{d\lambda} & 0 \\ 0 & \dfrac{dx}{d\lambda} & \dfrac{dy}{d\lambda} \end{vmatrix} = a\left(\frac{dy}{d\lambda}\right)^2 - b\frac{dx}{d\lambda}\frac{dy}{d\lambda} + c\left(\frac{dx}{d\lambda}\right)^2 \neq 0. \tag{119}$$

If (119) is satisfied for all points on C_0, then (116), (117), and (118) determine r, s, and t everywhere along C_0. If we proceed by a method similar to that used in Section 8.7, we then find that all higher partial derivatives of z also can be calculated at points on C_0 if only (119) is satisfied. In such a case we may say that the strip consisting of the curve C in space and the associated values of $\partial z/\partial x$ and $\partial z/\partial y$ along C, satisfying (110), is a *proper strip*, in the sense that *there exists a unique integral surface of* (109) *which includes this strip.*

Now suppose that at all points of C_0 the determinant in (119) *is* zero,

$$a\left(\frac{dy}{d\lambda}\right)^2 - b\frac{dx}{d\lambda}\frac{dy}{d\lambda} + c\left(\frac{dx}{d\lambda}\right)^2 = 0. \tag{120}$$

Then (116), (117), and (118) cannot be solved for s unless the numerator determinant in the formal solution by Cramer's rule also vanishes,

$$\begin{vmatrix} a & g - (dp + eq + fz) & c \\ \dfrac{dx}{d\lambda} & \dfrac{dp}{d\lambda} & 0 \\ 0 & \dfrac{dq}{d\lambda} & \dfrac{dy}{d\lambda} \end{vmatrix} = 0,$$

or $\qquad a\dfrac{dp}{d\lambda}\dfrac{dy}{d\lambda} + c\dfrac{dq}{d\lambda}\dfrac{dx}{d\lambda} + (dp + eq + fz - g)\dfrac{dx}{d\lambda}\dfrac{dy}{d\lambda} = 0. \tag{121}$

In this case generally two of the three equations imply the other one,† and hence only two restrictions on the three unknown quantities r, s, and t are present. Then, if (120) and (121) are *both* satisfied, Equations (116) to (118) serve merely to express two of the second derivatives in terms of the third (or to determine only two of those derivatives) and the remaining one is yet to be determined (or partially determined). Finally, if (120) is satisfied but the left-hand member of (121) is *not* zero, *no solution exists.* Equation (121), or its replacement when it is nonrestrictive, is known as the *compatibility condition.*

If (120) is satisfied at all points on C_0, then at all such points the slope dy/dx of the curve in the xy plane must satisfy the equation

$$a\left(\frac{dy}{dx}\right)^2 - b\frac{dy}{dx} + c = 0. \tag{122}$$

That is, we must have

$$\frac{dy}{dx} = \frac{b \pm \sqrt{b^2 - 4ac}}{2a}, \tag{123}$$

and hence *the quantity* $b^2 - 4ac$ *must be nonnegative.* Suppose that $b^2 - 4ac$ is *positive* and a is not zero. Then (123) determines two families of curves in the xy plane, say

$$\varphi(x, y) = c_1, \qquad \psi(x, y) = c_2. \tag{124}$$

That is, if (120) is satisfied, the curve C_0 must be a member of one of these families. In su:h a case *no integral surface including the prescribed strip exists unless along the curve C* (which projects into C_0) *z and its partial derivatives satisfy a compatibility condition.* When that condition is also satisfied, at least one integral surface including the prescribed strip exists, and the strip is called a *characteristic strip.* The curve C in space which bears the strip is called a

---

†Exceptional cases where both a and $dx/d\lambda$ or both c and $dy/d\lambda$ vanish on C_0 must be treated separately. In each of these two situations, (121) is *identically* satisfied and a different condition relating p and q on C_0 is obtained. (See Problem 53 for an example.)

characteristic curve, and its projection C_0 in the xy plane, which must be a member of one of the families of (124), is called a *characteristic base curve*. Equations (124) determine two families of cylinders in space, with generators parallel to the z axis, which may be called the *characteristic cylinders*. Some writers refer to the *projection* C_0 itself as a "characteristic" of the differential equation.

If $a = 0$, Equation (120) can be solved for dx/dy and two families of characteristic cylinders again are obtained unless also $c = 0$. If $a = c = 0$, the cylinders reduce to the planes $x = c_1$ and $y = c_2$. If $b^2 = 4ac$, so that (109) is of parabolic type, the two families coincide.

From Equation (123) we see that an *elliptic linear equation of second order has no real characteristic curves*, since for such an equation the discriminant $b^2 - 4ac$ is negative. Thus an integral surface of an elliptic equation can always be determined so as to include a prescribed regular strip in space. Equivalently, if z and its first partial derivatives are prescribed in a sufficiently regular way along any curve in the xy plane, the solution of an elliptic equation which takes on these values is determined uniquely in the neighborhood of this curve.

The situation in the case of a hyperbolic or parabolic equation is, however, quite different. In these cases real characteristics exist, and if a strip is built up along a curve C which lies on a characteristic cylinder, there will be *no* integral surface including this strip *unless the strip is a characteristic strip*. In this case, when the differential equation is *hyperbolic* it is found that there then are infinitely many integral surfaces including the strip, and the solution in the neighborhood of the strip is *not* determined uniquely by the prescribed information *on* the strip. However, when the equation is of *parabolic* type, a characteristic strip generally determines a unique integral surface. (Problem 53 provides an example.)†

Thus, for the hyperbolic equation

$$\frac{\partial^2 z}{\partial x^2} - \frac{\partial^2 z}{\partial y^2} = 0, \tag{125}$$

Equation (122) becomes

$$\left(\frac{dy}{dx}\right)^2 - 1 = 0, \qquad \frac{dy}{dx} = \pm 1,$$

and hence the characteristic base curves in the xy plane are the straight lines

$$x + y = c_1, \qquad x - y = c_2. \tag{126}$$

Along any such line the values of $z, p = \partial z/\partial x$, and $q = \partial z/\partial y$ cannot be independently prescribed in an arbitrary way. The compatibility condition (121),

$$\frac{dp}{d\lambda}\frac{dy}{d\lambda} - \frac{dq}{d\lambda}\frac{dx}{d\lambda} = 0, \tag{127a}$$

---

†Proofs of these assertions are somewhat involved, but can be based on the reducibility of (109) to one of the standard forms (132) and (134) in those cases. In the parabolic case, the uninteresting situations in which $E = 0$ in (134) are exceptional.

as well as the usual strip condition (110),

$$\frac{dz}{d\lambda} = p\frac{dx}{d\lambda} + q\frac{dy}{d\lambda}, \tag{127b}$$

must be satisfied.

If we identify λ with arc length, say s, then along a line of the first family $x + y = c_1$ we have

$$\frac{dx}{ds} = -\frac{dy}{ds} = \frac{\sqrt{2}}{2}$$

and hence Equations (127a, b) lead to the requirements

$$p + q = \text{constant}, \quad p - q = \sqrt{2}\frac{dz}{ds} \quad (\textit{along } x + y = c_1). \tag{128a}$$

Similarly, along a line of the second family $x - y = c_2$, we have the relations

$$\frac{dx}{ds} = \frac{dy}{ds} = \frac{\sqrt{2}}{2},$$

and hence (127a, b) give

$$p - q = \text{constant}, \quad p + q = \sqrt{2}\frac{dz}{ds} \quad (\textit{along } x - y = c_2). \tag{128b}$$

A comparison of these conditions shows that the equation of compatibility requires that, at all points of a line of one family, the derivative of z in the direction of the intersecting lines of the other family must be constant. Since here the two families of characteristic lines are perpendicular to each other, *the derivative of z normal to a characteristic line must be constant along that line* in this particular example.

The general solution of (125) is of the form

$$z = f(x + y) + g(x - y). \tag{129}$$

We may notice that each function in (129) is constant along one of the characteristic lines. If we write (126) in the form $\varphi(x, y) = c_1, \psi(x, y) = c_2$, the general solution (129) becomes

$$z = f(\varphi) + g(\psi). \tag{130}$$

Also, if φ and ψ are taken as new independent variables, Equation (125) becomes

$$\frac{\partial^2 z}{\partial\varphi\,\partial\psi} = 0. \tag{131}$$

Although it usually is not possible to express the general solution of a specific equation of type (109) in a form similar to (130), still it is not difficult to show that if the expressions φ and ψ of (124) are taken as new independent variables, in the case where (109) is *hyperbolic*, then (109) is transformed to the so-called *normal* (or *canonical*) *form*

$$\frac{\partial^2 z}{\partial\varphi\,\partial\psi} = D\frac{\partial z}{\partial\varphi} + E\frac{\partial z}{\partial\psi} + Fz + G, \tag{132}$$

where D, E, F, and G are functions of φ and ψ.

By choosing suitable new *real* independent variables α and β, any *elliptic* equation of type (109) can be put in the normal form

$$\frac{\partial^2 z}{\partial \alpha^2} + \frac{\partial^2 z}{\partial \beta^2} = D\frac{\partial z}{\partial \alpha} + E\frac{\partial z}{\partial \beta} + Fz + G, \qquad (133)$$

and any *parabolic* equation of this type can be put in the form

$$\frac{\partial^2 z}{\partial \alpha^2} = D\frac{\partial z}{\partial \alpha} + E\frac{\partial z}{\partial \beta} + Fz + G. \qquad (134)$$

It may be noticed that the two families of characteristic base curves for (132) comprise the curves $\varphi = $ constant and $\psi = $ constant, and that the one family for (134) comprises the curves $\beta = $ constant.

In the more general case of a quasi-linear equation of second order,

$$a\frac{\partial^2 z}{\partial x^2} + b\frac{\partial^2 z}{\partial x\,\partial y} + c\frac{\partial^2 z}{\partial y^2} + F = 0, \qquad (135)$$

the curves in the xy plane which satisfy (120) again are exceptional, and are often called the *characteristics* of the differential equation. However, here the situation is considerably complicated by the fact that, since the coefficients a, b, and c may depend upon z, $\partial z/\partial x$, and $\partial z/\partial y$, the characteristics of the equation (in fact, their very *reality*) may depend upon the *solution*, which in turn also depends upon the side conditions (initial and/or boundary conditions) that supplement the differential equation in the complete formulation of a problem. Characteristic cylinders and related entities generally do not exist.

8.9. Singular Curves on Integral Surfaces. Let S be a particular integral surface corresponding to the linear first-order equation

$$P\frac{\partial z}{\partial x} + Q\frac{\partial z}{\partial y} = R_1 z + R_2, \qquad (136)$$

where P, Q, R_1, and R_2 are continuous functions of x and y. Then the existence of the partial derivatives $\partial z/\partial x$ and $\partial z/\partial y$ in (136) implies that z itself be continuous; that is, the surface S must be a continuous surface. Hence the equal members of (136) must both be continuous functions of x and y. This condition, however, does not exclude the possibility that the two terms on the left may be *each* discontinuous, so long as their *sum* is continuous. In particular, there may be a curve C on S along which the surface possesses a "corner" or "edge," as is indicated in Figure 8.2. We denote by C_0 the projection of C onto the xy plane, and consider z, $\partial z/\partial x$, and $\partial z/\partial y$ as functions of x and y. Then, while z is an everywhere continuous function of x and y, and while the derivative of z *in the direction of C_0* is continuous, we suppose that the derivative of z *normal*

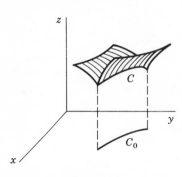

Figure 8.2

to C_0 has a finite jump as the curve C_0 is crossed in the xy plane (Figure 8.2). Thus here the partial derivatives $\partial z/\partial x$ and $\partial z/\partial y$ will exist (in general) only as "one-sided" derivatives on C_0, that is, as limits in which C_0 is approached from one side or the other. Surfaces of this general nature are of frequent interest in the study of physical phenomena where abrupt changes of some sort may occur. We now investigate more explicitly the possibility of their existence, as integral surfaces of an equation of type (136).

Let the curve C_0 be specified by the parametric equations

$$x = x(\lambda), \qquad y = y(\lambda), \tag{137}$$

where λ is a convenient variable indicating position on C_0. Then λ may also be considered as representing position in the direction of C_0 along parallel curves in the immediate neighborhood of C_0. In particular, the derivative of z in the direction of C_0 at points on such curves is given by

$$\frac{dz}{d\lambda} = \frac{\partial z}{\partial x}\frac{dx}{d\lambda} + \frac{\partial z}{\partial y}\frac{dy}{d\lambda} = p\frac{dx}{d\lambda} + q\frac{dy}{d\lambda}. \tag{138}$$

Now consider any point P on the curve C_0. Since the left-hand member of (136) must be continuous across C_0 at this point, we must have

$$P\,\Delta p + Q\,\Delta q = 0, \tag{139}$$

where Δp and Δq are the values of the jumps associated with the partial derivatives $p = \partial z/\partial x$ and $q = \partial z/\partial y$ across C_0 at P. Since the derivative of z in the direction of C_0 is also assumed to be continuous across C_0, the right-hand member of (138) also must be continuous, and we obtain a second equation

$$\frac{dx}{d\lambda}\Delta p + \frac{dy}{d\lambda}\Delta q = 0 \tag{140}$$

relating the jumps in p and q across C_0 at P. Equations (139) and (140) are compatible (with Δp and Δq not both zero) only if the coefficients of Δp and Δq are proportional, that is, if

$$\frac{dx}{P} = \frac{dy}{Q} \tag{141}$$

along C_0. Thus C_0 must lie on a characteristic cylinder of (136), and hence, since we have assumed that C lies on an integral surface of (136), we conclude that the "singular curve" C *must be a characteristic curve of* (136). That is, corners of the type considered can exist on an integral surface of (136) only along characteristic curves.

In the case of the linear second-order equation

$$a\frac{\partial^2 z}{\partial x^2} + b\frac{\partial^2 z}{\partial x\,\partial y} + c\frac{\partial^2 z}{\partial y^2} = g - \left(d\frac{\partial z}{\partial x} + e\frac{\partial z}{\partial y} + fz\right), \tag{142}$$

the existence of the second derivatives presupposes continuity of z and its first partial derivatives, and hence also continuity of the equal members of (142) if the coefficients and the function g are continuous. However, the possibility that the *separate terms* on the left may be individually discontinuous is not excluded. Let C be a curve in an integral surface S, with projection C_0 in the xy plane, and denote position along C_0 by the parameter λ. We investigate the possibility that z, p, and q be continuous on S, and that the derivatives of p and q in the *direction of C_0*,

$$\frac{dp}{d\lambda} = \frac{\partial p}{\partial x}\frac{dx}{d\lambda} + \frac{\partial p}{\partial y}\frac{dy}{d\lambda} = r\frac{dx}{d\lambda} + s\frac{dy}{d\lambda}, \tag{143}$$

$$\frac{dq}{d\lambda} = \frac{\partial q}{\partial x}\frac{dx}{d\lambda} + \frac{\partial q}{\partial y}\frac{dy}{d\lambda} = s\frac{dx}{d\lambda} + t\frac{dy}{d\lambda}, \tag{144}$$

be continuous across C_0, but that the derivatives of p and q *normal to C_0* be discontinuous across C_0. Here the separate second partial derivatives may exist along C_0 only as one-sided derivatives, as C_0 is approached from one side or the other.

The continuity of the right member of (142) and of the left members of (143) and (144) implies continuity of their equivalents, and hence we obtain the three equations

$$a\ \Delta r + b\ \Delta s + c\ \Delta t = 0,$$

$$\frac{dx}{d\lambda}\ \Delta r + \frac{dy}{d\lambda}\ \Delta s \qquad\qquad = 0, \tag{145}$$

$$\frac{dx}{d\lambda}\ \Delta s + \frac{dy}{d\lambda}\ \Delta t = 0,$$

relating the jumps in the three partial derivatives r, s, and t across C_0. These equations are consistent (with at least one nonzero jump) only if the determinant of the coefficients vanishes. But this condition is the same as that of Equation (120), and hence the curve C_0 must lie on a characteristic cylinder of (142). Since C lies on an integral surface and projects into C_0, it follows that C *must be a characteristic curve of* (142). In particular, since an equation of *elliptic* type has no real characteristic curves, an integral surface of such an equation cannot contain a "singular curve" of the type required.

Results of this type are of importance, for example, in the study of two-dimensional *compressible* fluid flow around rigid bodies [see Equation (200), Section 6.20]. Here it is found that for low velocities the problem is governed by a partial differential equation of elliptic type, whereas if sufficiently high velocities are attained the governing equation becomes hyperbolic, and discontinuous phenomena may then be present.

Situations also arise in which the *first* partial derivatives of the "solution" of a problem governed by a second-order hyperbolic differential equation necessarily exhibit jumps along certain characteristic base curves, or in which the "solution" itself has this property (see Section 9.15). Consideration of the physical (or other) basis of the mathematical formulation then must determine whether the "solution" is to be acceptable.

8.10. Remarks on Linear Second-Order Initial-Value Problems. In this section we summarize briefly certain preceding results and present additional facts bearing on their significance.

For the general linear second-order equation of the form

$$a\frac{\partial^2 z}{\partial x^2} + b\frac{\partial^2 z}{\partial x\, \partial y} + c\frac{\partial^2 z}{\partial y^2} + d\frac{\partial z}{\partial x} + e\frac{\partial z}{\partial y} + fz = g, \qquad (146)$$

we define the *characteristic base curves*, in the xy plane, as those curves for which

$$a(dy)^2 - b(dx)(dy) + c(dx)^2 = 0. \qquad (147)$$

In the *elliptic* case, for which $b^2 < 4ac$, no real characteristic base curves can exist. In the *hyperbolic* case, for which $b^2 > 4ac$, two distinct sets of such curves are obtained, whereas in the *parabolic* case, $b^2 = 4ac$, the two sets become coincident.

In the *elliptic* case, if along any curve c in the xy plane z and its derivative in the direction *normal* to that curve are prescribed as functions of position along c in such a way that these prescribed values are *regular* everywhere along c, then a solution to the resulting initial-value problem is determined for values of x and y in *some* neighborhood of c. However, here two additional facts are of importance. *First*, unless the prescribed values are regular along c (in particular, unless derivatives of all orders exist at all points), *no* solution to the defined problem can exist. *Second*, even though a solution exist, in many cases it will exist only in a restricted neighborhood of c and will not be valid over the whole xy plane.

In the *hyperbolic* and *parabolic* cases, if c is *not* a characteristic base curve, the initial-value problem again has a solution. However, here the prescribed initial values need not be regular. The higher derivatives of the prescribed values may have finite jumps at certain points on c. These jumps are then found to be propagated along those characteristic base curves which pass through the relevant points on the initial curve, and the solution obtained is valid throughout the xy plane. If c *is* a characteristic base curve, then the initial-value problem does *not* have a solution unless the initial values of z and its derivative normal to c satisfy a certain compatibility condition, in which case infinitely many solutions then exist in the hyperbolic case and a single solution is determined in the parabolic case, in general.

Certain of these facts are illustrated in Problems 53 and 60–62 at the end of this chapter.

8.11. The Characteristics of a Particular Quasi-Linear Problem. To illustrate the nature of characteristics in other types of problems, we here consider the simultaneous equations

$$\frac{\partial u}{\partial x} + \frac{\partial v}{\partial x}\cos 2v + \frac{\partial v}{\partial y}\sin 2v = 0, \qquad (148a)$$

$$\frac{\partial u}{\partial y} + \frac{\partial v}{\partial x}\sin 2v - \frac{\partial v}{\partial y}\cos 2v = 0. \qquad (148b)$$

These equations, which are quasi-linear in u and v, are of basic importance in certain two-dimensional problems in the mathematical theory of plasticity. We suppose that u and v are prescribed at all points along a curve C_0 in the xy plane. If λ specifies position along C_0, we may define C_0 by the equations

$$C_0: \qquad x = x(\lambda), \qquad y = y(\lambda). \tag{149}$$

For points on C_0 the dependent functions u and v are then to be considered as given functions of λ.

We now ask whether there exists a curve C_0, or a set of such curves, with the property that u and v are not uniquely determined for points near C_0 by their prescribed values *along* C_0. It is assumed here that these prescribed values are continuously differentiable along C_0.

Since u and v are known along C_0, as functions of λ, their derivatives along C_0 are calculable, and must satisfy the equations

$$\frac{du(\lambda)}{d\lambda} = \frac{\partial u}{\partial x}\frac{dx}{d\lambda} + \frac{\partial u}{\partial y}\frac{dy}{d\lambda}, \tag{150a}$$

$$\frac{dv(\lambda)}{d\lambda} = \frac{\partial v}{\partial x}\frac{dx}{d\lambda} + \frac{\partial v}{\partial y}\frac{dy}{d\lambda}. \tag{150b}$$

Equations (148a, b) and (150a, b), constituting four linear equations in the four first partial derivatives $\partial u/\partial x$, $\partial u/\partial y$, $\partial v/\partial x$, and $\partial v/\partial y$, may be rewritten in the form

$$u_x \qquad\qquad + v_x \cos 2v + v_y \sin 2v = 0, \tag{151a}$$

$$u_y \; + v_x \sin 2v - v_y \cos 2v = 0, \tag{151b}$$

$$u_x x' + u_y y' \qquad\qquad\qquad = u', \tag{151c}$$

$$v_x x' \qquad + v_y y' \quad = v', \tag{151d}$$

where a prime denotes λ differentiation. A unique solution is assured *unless* the determinant of coefficients vanishes, that is, unless

$$\begin{vmatrix} 1 & 0 & \cos 2v & \sin 2v \\ 0 & 1 & \sin 2v & -\cos 2v \\ x' & y' & 0 & 0 \\ 0 & 0 & x' & y' \end{vmatrix} = 0. \tag{152}$$

By expansion, this equation can be written in the form

$$y'^2 \sin 2v + 2x'y' \cos 2v - x'^2 \sin 2v = 0,$$

from which we obtain the two alternatives

$$y' = x' \tan v, \tag{153a}$$

$$y' = -x' \cot v. \tag{153b}$$

Thus two systems of curves are determined in the plane, for one of which $dy/dx = \tan v$, and for the second $dy/dx = -\cot v$. The two systems are seen to be orthogonal, and the slope of each curve is seen to depend upon the value of the dependent function v at the point under consideration.

Along any such curve, Equations (151a–d) cannot have a unique solution. In particular, the set cannot be solved for u_x and hence *no* solution exists unless the numerator determinant of Cramer's rule also vanishes, that is, unless

$$\begin{vmatrix} 0 & 0 & \cos 2v & \sin 2v \\ 0 & 1 & \sin 2v & -\cos 2v \\ u' & y' & 0 & 0 \\ v' & 0 & x' & y' \end{vmatrix} = 0,$$

or, expanding, unless

$$u' = v'\left(\cos 2v - \frac{x'}{y'}\sin 2v\right). \tag{154}$$

Along the curves (153a) we then obtain the requirement $u' = -v'$, or

$$\frac{dy}{dx} = \tan v: \qquad u + v = \text{constant}; \tag{155a}$$

and along the curves (153b) there follows $u' = v'$, or

$$\frac{dy}{dx} = -\cot v: \qquad u - v = \text{constant}. \tag{155b}$$

If either (155a) or (155b) is satisfied, u_x may be taken *arbitrarily*, and it is readily verified that Equations (151a–c) then determine u_y, v_x, and v_y in terms of u_x, u', and v'. Thus in these cases *infinitely many* solutions exist, all corresponding to the same prescribed values of u and v along C_0.

The curves *in the xy plane* for which (155a) or (155b) is true may be called the *characteristics* of the simultaneous equations (148a, b).

We next show that the equations (148a, b) are simplified if we take as new *dependent variables* the combinations $u + v$ and $u - v$ suggested by (155a, b), say

$$\xi = u + v, \qquad \eta = u - v. \tag{156}$$

The variable ξ is then constant along the first set of characteristics, whereas η is constant along curves in the second set. Then since we also have

$$u = \tfrac{1}{2}(\xi + \eta), \qquad v = \tfrac{1}{2}(\xi - \eta), \tag{157}$$

Equations (148a, b) take the form

$$(\xi_x + \eta_x) + (\xi_x - \eta_x)\cos 2v + (\xi_y - \eta_y)\sin 2v = 0,$$
$$(\xi_y + \eta_y) + (\xi_x - \eta_x)\sin 2v - (\xi_y - \eta_y)\cos 2v = 0,$$

or, after a rearrangement,

$$(\eta_x - \eta_y \cot v) + \cot^2 v(\xi_x + \xi_y \tan v) = 0,$$
$$-(\eta_x - \eta_y \cot v) + \qquad (\xi_x + \xi_y \tan v) = 0. \tag{158}$$

These equations imply the simplified relations

$$\xi_x + \xi_y \tan v = 0,$$
$$\eta_x - \eta_y \cot v = 0, \tag{159}$$

where $v = \frac{1}{2}(\xi - \eta)$. Equations (159) are still nonlinear, but they are more tractable than the equivalent equations (148) (see Problem 64), and form the basis for further treatment of certain problems in plasticity. Physically, the characteristics in this case are the so-called *shear lines* (lines of maximum shearing stress) in a plastic problem of plane strain for an incompressible material.†

REFERENCES

1. BATEMAN, H., *Partial Differential Equations of Mathematical Physics*, Cambridge University Press, New York, 1959.

2. BERG, P. W., and J. L. MCGREGOR, *Elementary Partial Differential Equations*, Holden-Day, Inc., San Francisco, 1966.

3. COURANT, R., and D. HILBERT, *Methods of Mathematical Physics*, Vol. 1, John Wiley & Sons, Inc. (Interscience Division), New York, 1953.

4. HOPF, L., *Introduction to the Differential Equations of Physics*, Dover Publications, Inc., New York, 1948.

5. JEFFREYS, H., and B. S. JEFFREYS, *Methods of Mathematical Physics*, 3rd ed., Cambridge University Press, New York, 1956.

6. MORSE, P. M., and H. FESHBACH, *Methods of Theoretical Physics*, 2 pts., McGraw-Hill Book Company, Inc., New York, 1953.

7. PRAGER, W., *Discontinuous Solutions in the Theory of Plasticity*, Courant Anniversary Volume, pp. 289–300, John Wiley & Sons, Inc. (Interscience Division), New York, 1948.

8. SOMMERFELD, A., *Partial Differential Equations in Physics*, Academic Press, Inc., New York, 1949.

9. WEBSTER, A. G., *Partial Differential Equations of Mathematical Physics*, 2nd ed., Dover Publications, Inc., New York, 1956.

10. WEINBERGER, H. F., *A First Course in Partial Differential Equations*, Xerox College Publishing, Lexington, Mass., 1965.

PROBLEMS

Section 8.1

1. Find the differential equation of lowest order which possesses each of the following solutions, with f and g arbitrary functions:

(a) $z = (x - y)f(x + y)$,

(b) $z = f(ax + by) + g(cx + dy)$ $(ad - bc \neq 0)$,

(c) $z = f(ax + by) + xg(ax + by)$.

†See Reference 7.

2. (a) Obtain the partial differential equation of first order satisfied by $z = f(\psi)$, where ψ is a given function of x and y and f is an arbitrary function, in the form

$$\psi_y \frac{\partial z}{\partial x} - \psi_x \frac{\partial z}{\partial y} = 0.$$

(b) With the new independent variables s and t, where $t = \psi(x, y)$ and s is any independent function of x and y, show that, when z is considered as a function of s and t, the differential equation of part (a) takes the form

$$(s_x \psi_y - s_y \psi_x) \frac{\partial z}{\partial s} = 0.$$

Hence deduce that the *most general* solution of the equation is of the form $z = f(\psi)$, where f is arbitrary.

3. Noticing that $z = \psi$ is a solution of the equation considered in Problem 2(a), deduce that, if one solution of the equation

$$P(x, y) \frac{\partial z}{\partial x} + Q(x, y) \frac{\partial z}{\partial y} = 0$$

is of the form $z = \psi(x, y)$, then the most general solution is $z = f(\psi)$, where f is arbitrary.

4. Obtain the partial differential equation of first order satisfied by $z = \varphi f(\psi)$, where φ and ψ are given functions of x and y and f is an arbitrary function, in the form

$$\varphi \left(\psi_y \frac{\partial z}{\partial x} - \psi_x \frac{\partial z}{\partial y} \right) = (\varphi_x \psi_y - \varphi_y \psi_x) z.$$

Section 8.2

5. Determine the general solution of each of the following equations (with a, b, and c constant), writing each solution in a form solved for z:

(a) $a \dfrac{\partial z}{\partial x} + b \dfrac{\partial z}{\partial y} = c,$ (b) $a \dfrac{\partial z}{\partial x} + b \dfrac{\partial z}{\partial y} = cz,$

(c) $y \dfrac{\partial z}{\partial x} - x \dfrac{\partial z}{\partial y} = 0,$ (d) $\dfrac{\partial z}{\partial x} + \dfrac{\partial z}{\partial y} + 2xz = 0,$

(e) $x \dfrac{\partial z}{\partial x} - y \dfrac{\partial z}{\partial y} = z,$ (f) $x^2 \dfrac{\partial z}{\partial x} + y^2 \dfrac{\partial z}{\partial y} = z^2.$

6. (a) Show that if a particular solution of the equation

$$a \frac{\partial z}{\partial x} + b \frac{\partial z}{\partial y} = f_1(x) + f_2(y),$$

where a and b are constant, is assumed in the form $z_p = \varphi_1(x) + \varphi_2(y)$, there follows

$$z_p = \frac{1}{a} \int f_1(x) \, dx + \frac{1}{b} \int f_2(y) \, dy \qquad (ab \neq 0),$$

so that the general solution is of the form

$$z = f(bx - ay) + z_p.$$

(b) Illustrate the results of part (a) in the case of the equation

$$\frac{\partial z}{\partial x} - 2 \frac{\partial z}{\partial y} = 2x - e^y + 1.$$

7. Use the results of Problem 9 of Chapter 7, with an appropriate change in notation, to show that if the dependent and independent variables are interchanged in a pair of simultaneous *quasi-linear* differential equations in u and v, each of the form

$$Au_x + Bu_y + Cv_x + Dv_y = 0,$$

where A, B, C, and D depend only on u and v, a pair of equivalent *linear* equations in x and y is obtained, each of the form

$$Ay_v - Bx_v - Cy_u + Dx_u = 0.$$

Section 8.3

8. (a) Show that the characteristic curves of the equation

$$\frac{\partial z}{\partial x} + \frac{\partial z}{\partial y} = 1$$

are straight lines parallel to the vector $\mathbf{i} + \mathbf{j} + \mathbf{k}$, and hence that the characteristic curve passing through a point (x_0, y_0, z_0) is specified by the equations

$$x - x_0 = y - y_0 = z - z_0.$$

(b) Determine the characteristic curve which passes through the point $(0, y_0, z_0)$ in the yz plane, and show that those characteristic curves which pass through points on the curve $z = y^2$ in the plane $x = 0$ are specified by the equations $y - x = y_0$, $z - x = y_0^2$. Thus deduce that the surface traced out by these curves is the parabolic cylinder

$$z = x + (y - x)^2.$$

(c) Verify directly that the surface $z = x + (y - x)^2$ is an integral surface of the equation $(\partial z/\partial x) + (\partial z/\partial y) = 1$ which includes the curve $z = y^2$ in the yz plane.

9. (a) Obtain the general solution of the differential equation of Problem 8(a), in the form

$$z = x + f(y - x).$$

(b) Determine the function f in such a way that this solution is consistent with the equations $z = y^2$, $x = 0$. Hence rederive the result of Problem 8(c).

10. (a) Show that the solution of the differential equation of Problem 8(a), for which $z = \varphi(x)$ along the line $y = 2x$ in the xy plane, is of the form

$$z = 2x - y + \varphi(y - x).$$

(b) If z is prescribed as $\varphi(x)$ along the line $y = x$ in the xy plane, show that *no solution* exists unless $\varphi(x)$ is prescribed in the form $\varphi(x) = x + k$, where k is a constant. In this last case, show that $z = x + f(y - x)$ is a solution for *any* differentiable f such that $f(0) = k$, so that *infinitely many solutions* then exist.

(c) Show that the projections of the characteristic curves onto the xy plane are the lines $y = x + c$, and that if z is prescribed as $\varphi(x)$ along any such curve there is no solution unless $\varphi(x) = x + k$, in which case infinitely many solutions exist.

11. Find the solution of the equation

$$\frac{\partial z}{\partial x} = \frac{\partial z}{\partial y}$$

for which $z = (t + 1)^4$ when $x = t^2 + 1$ and $y = 2t$.

12. (a–f) For each equation of Problem 5, determine the integral surface which includes the straight line $x = y = z$, if such a surface exists.

13. (a–f) For each equation of Problem 5, determine the solution for which $z = x^2$ along the straight line $y = 2x$ in the xy plane, if such a solution exists.

14. Determine the general solution of each of the following equations, writing each solution in a form solved for z:

(a) $(x + y)\left(\dfrac{\partial z}{\partial x} + \dfrac{\partial z}{\partial y}\right) = z - 1$,   (b) $\dfrac{\partial z}{\partial x} = xy$,

(c) $xz\dfrac{\partial z}{\partial x} + yz\dfrac{\partial z}{\partial y} = xy$,   (d) $\dfrac{\partial z}{\partial x} + \dfrac{\partial z}{\partial y} = 6xyz$.

15. Suppose that the solution of the equation

$$\frac{\partial z}{\partial x} + \frac{\partial z}{\partial y} = 1$$

is to be such that $z = \varphi(x)$ when $y = mx + b$.

(a) Determine that solution in the form

$$z = \varphi\left(\frac{y - x - b}{m - 1}\right) - \frac{y - mx - b}{m - 1}$$

when $m \neq 1$.

(b) Describe the situation when $m = 1$.

16. (a) Obtain the general solution of the equation

$$\frac{\partial z}{\partial x} + \frac{\partial z}{\partial y} + \frac{\partial z}{\partial t} = 1$$

in the form

$$F(y - x, t - x, z - x) = 0.$$

(b) If that solution for which $z = \varphi(x, y)$ when $x + y = t$ is required, show that there must follow

$$F[y - x, y, \varphi(x, y) - x] = 0$$

and hence

$$F[u, v, \varphi(v - u, v) + u - v] = 0.$$

Thus deduce that if we write $y - x = u$, $t - x = v$, we must have also $z - x = \varphi(v - u, v) + u - v$, and hence obtain the required solution in the form

$$z = \varphi(t - y, t - x) + x + y - t.$$

17. (a) Obtain the general solution of the equation

$$x\frac{\partial z}{\partial x} + y\frac{\partial z}{\partial y} + t\frac{\partial z}{\partial t} = 1.$$

(b) Determine that solution for which

$$z = \varphi(x, y) \quad \text{when} \quad x^2 + y = t$$

in the form

$$z = \frac{x^2}{t - y}\varphi\left[\frac{t - y}{x}, \frac{y(t - y)}{x^2}\right].$$

Section 8.4

18. Suppose that $z = u_1(x, y)$, $z = u_2(x, y)$, ... each reduce the left-hand member of (50) to zero and that $z = p(x, y)$ reduces that member to $g(x, y)$.

(a) Show that, assuming appropriate convergence,

$$z = \sum_{k=1}^{\infty} c_k u_k(x, y)$$

satisfies (50) with g replaced by zero, when the c's are constants.

(b) Again assuming appropriate convergence, show that

$$z = p(x, y) + \sum_{k=1}^{\infty} c_k u_k(x, y)$$

is a solution of (50).

19. (a) Show that the result of setting $z = uv$ in (50) is of the form

$$v\left(a\frac{\partial^2 u}{\partial x^2} + b\frac{\partial^2 u}{\partial x \partial y} + c\frac{\partial^2 u}{\partial y^2}\right) + Fu = g,$$

if v satisfies both of the first-order equations

$$2a\frac{\partial v}{\partial x} + b\frac{\partial v}{\partial y} + dv = 0, \qquad b\frac{\partial v}{\partial x} + 2c\frac{\partial v}{\partial y} + ev = 0.$$

(b) When the coefficients a, b, c, d, and e are *constants*, show that constants α and β can be determined in such a way that $v = e^{\alpha x + \beta y}$ satisfies the two equations obtained in part (a) provided only that $b^2 \neq 4ac$, so that the equation is not parabolic. Verify also that, in this case, there follows

$$F = [f - (a\alpha^2 + b\alpha\beta + c\beta^2)]v.$$

(c) Illustrate the preceding transformation in the case of the equation

$$\frac{\partial^2 z}{\partial x \partial y} + d\frac{\partial z}{\partial x} + e\frac{\partial z}{\partial y} + fz = g.$$

(Notice that if $f = 0$ the elimination of the terms involving the first partial derivatives generally is at the expense of introducing a term involving the function itself.)

20. Suppose that g is replaced by $\partial z/\partial t$ in (50), where t is a third independent variable, so that z is to satisfy the equation

$$a\frac{\partial^2 z}{\partial x^2} + b\frac{\partial^2 z}{\partial x \partial y} + c\frac{\partial^2 z}{\partial y^2} + d\frac{\partial z}{\partial x} + e\frac{\partial z}{\partial y} + fz = \frac{\partial z}{\partial t}.$$

When a, b, c, d, e, and f are *constants*, show that constants α, β, and γ can be determined in such a way that the substitution

$$z = ue^{\alpha x + \beta y + \gamma t}$$

requires that u satisfy the equation

$$a\frac{\partial^2 u}{\partial x^2} + b\frac{\partial^2 u}{\partial x \partial y} + c\frac{\partial^2 u}{\partial y^2} = \frac{\partial u}{\partial t},$$

provided only that $b^2 \neq 4ac$.

21. (a) Show that the equation

$$\frac{\partial^2 z}{\partial x^2} + 2x\frac{\partial^2 z}{\partial x\,\partial y} + (1 - y^2)\frac{\partial^2 z}{\partial y^2} = 0$$

is elliptic for values of x and y in the region $x^2 + y^2 < 1$, parabolic on the boundary, and hyperbolic outside the region.

(b) Show that the quasi-linear equation

$$x\frac{\partial^2 z}{\partial x^2} + z\frac{\partial^2 z}{\partial x\,\partial y} + y\frac{\partial^2 z}{\partial y^2} = 0$$

is elliptic when $z^2 < 4xy$, parabolic when $z^2 = 4xy$, and hyperbolic otherwise. (Notice that the nature of a quasi-linear equation thus may depend not only upon position in the xy plane but also upon the solution z, which in turn depends upon the boundary conditions.)

22. (a) Show that the assumption that the equation

$$\frac{\partial^2 z}{\partial x^2} = \frac{\partial z}{\partial y}$$

possesses a solution of the form $z = \varphi(x, y)f[\psi(x, y)]$, where φ and ψ are specific functions and f is arbitrary, leads to the requirement

$$\varphi\psi_x^2 f''(\psi) + (2\varphi_x\psi_x + \varphi\psi_{xx} - \varphi\psi_y)f'(\psi) + (\varphi_{xx} - \varphi_y)f(\psi) = 0,$$

for arbitrary f, and hence to the three requirements

$$\varphi\psi_x^2 = 0, \qquad 2\varphi_x\psi_x + \varphi\psi_{xx} - \varphi\psi_y = 0, \qquad \varphi_{xx} - \varphi_y = 0.$$

(b) Show that these conditions imply that either $\varphi \equiv 0$ or $\psi =$ constant, so that there exists no solution of the assumed form which actually depends upon an arbitrary function f.

Section 8.5

23. Obtain the general solution of each of the following equations:

(a) $\dfrac{\partial^2 z}{\partial x^2} - 2\dfrac{\partial^2 z}{\partial x\,\partial y} - 3\dfrac{\partial^2 z}{\partial y^2} = 0,$

(b) $\dfrac{\partial^2 \varphi}{\partial x^2} - 2\dfrac{\partial^2 \varphi}{\partial x\,\partial y} + 2\dfrac{\partial^2 \varphi}{\partial y^2} = 0,$

(c) $\dfrac{\partial^2 \varphi}{\partial x^2} - \dfrac{\partial^2 \varphi}{\partial x\,\partial y} = 0,$

(d) $\dfrac{\partial^2}{\partial r^2}(rw) = \dfrac{r}{c^2}\dfrac{\partial^2 w}{\partial t^2},$

(e) $(U^2 - V^2)\dfrac{\partial^2 \varphi}{\partial x^2} + 2U\dfrac{\partial^2 \varphi}{\partial x\,\partial t} + \dfrac{\partial^2 \varphi}{\partial t^2} = 0,$

(f) $\dfrac{\partial^4 \varphi}{\partial x^4} - \dfrac{\partial^4 \varphi}{\partial y^4} = 0,$

(g) $\dfrac{\partial^4 z}{\partial x^4} - 2\dfrac{\partial^4 z}{\partial x^2\,\partial y^2} + \dfrac{\partial^4 z}{\partial y^4} = 0,$

(h) $\nabla^4 \varphi \equiv \dfrac{\partial^4 \varphi}{\partial x^4} + 2\dfrac{\partial^4 \varphi}{\partial x^2\,\partial y^2} + \dfrac{\partial^4 \varphi}{\partial y^4} = 0.$

24. Obtain the general solution of the simultaneous equations

$$\frac{\partial u}{\partial x} = -c^2 \frac{\partial v}{\partial y}, \qquad \frac{\partial v}{\partial x} = -k^2 \frac{\partial u}{\partial y}$$

in the form

$$u = f\left(x + \frac{y}{kc}\right) + g\left(x - \frac{y}{kc}\right),$$

$$v = \frac{k}{c}\left[-f\left(x + \frac{y}{kc}\right) + g\left(x - \frac{y}{kc}\right)\right].$$

(First eliminate v and solve the resultant equation for u; then use the original equations to determine v. Notice, for example, that

$$\frac{\partial f}{\partial x} = kc \frac{\partial f}{\partial y}.$$

An arbitrary constant so introduced can be absorbed into the definitions of f and g.)

25. (a) Prove that the general solution of Laplace's equation, in the form

$$\frac{\partial^2 \varphi}{\partial x^2} + \frac{\partial^2 \varphi}{\partial y^2} = 0,$$

can be written in the form

$$\varphi(x, y) = f(x + iy) + g(x - iy),$$

where f and g are twice-differentiable functions of the complex conjugate arguments $x + iy$ and $x - iy$.

(b) Deduce that the real and imaginary parts of both $f(x + iy)$ and $g(x - iy)$ satisfy Laplace's equation.

(c) Use the fact that the real and imaginary parts of a twice-differentiable function $g(x - iy)$ are respectively the real part and the negative of the imaginary part of the twice-differentiable conjugate function $\bar{g}(x + iy)$ to deduce that any real solution of Laplace's equation is the real or imaginary part of a twice-differentiable function of the complex variable $x + iy$, and conversely.

26. (a) Obtain as the real and imaginary parts of the function $(x + iy)^n$, for $n = 0, 1, 2, 3,$ and 4, the functions 1, x, y, $x^2 - y^2$, $2xy$, $x^3 - 3xy^2$, $3x^2y - y^3$, $x^4 - 6x^2y^2 + y^4$, and $4x^3y - 4xy^3$, and verify directly that they each satisfy Laplace's equation.

(b) Show that the functions $r^n \cos n\theta$ and $r^n \sin n\theta$, where n is integral, are the real and imaginary parts of $(x + iy)^n$ when x and y are expressed in polar coordinates, and verify directly that they each satisfy Laplace's equation. [See Equation (26) of Section 1.5 and Equation (168e) of Section 6.18.]

27. In each of the following cases, first obtain a particular solution of the equation as a function of one variable only, and then obtain the general solution by adding this solution to the general complementary solution:

(a) $\dfrac{\partial^2 z}{\partial x^2} - \dfrac{\partial^2 z}{\partial y^2} = x,$ (b) $\dfrac{\partial^2 z}{\partial x^2} - 3\dfrac{\partial^2 z}{\partial x \, \partial y} + 2\dfrac{\partial^2 z}{\partial y^2} = \cos y.$

28. Suppose that the right-hand member of a linear partial differential equation is a homogeneous polynomial of degree k in x and y (each term being of degree k), whereas the left-hand member is homogeneous of order n in z (each term involving an nth derivative) with constant coefficients. Show that the assumption of a particular solution in the form of a homogeneous polynomial of degree $n + k$ leads to $k + 1$ linear equations in $n + k + 1$ unknown coefficients. (It can be shown that this set always possesses a solution and, indeed, that a certain set of at least n of the coefficients can be assigned arbitrarily, in a convenient way.)

29. Use the procedure outlined in Problem 28 (and/or Problem 27) to obtain any particular solution of each of the following equations:

(a) $\dfrac{\partial z}{\partial x} + \dfrac{\partial z}{\partial y} = x^2 y + xy^2,$ (b) $\dfrac{\partial^2 z}{\partial x^2} + \dfrac{\partial^2 z}{\partial y^2} = x^2 + xy,$

(c) $\dfrac{\partial^2 z}{\partial x^2} - \dfrac{\partial^2 z}{\partial y^2} = xy + x,$ (d) $\dfrac{\partial^2 z}{\partial x^2} - 2\dfrac{\partial^2 z}{\partial x\,\partial y} + \dfrac{\partial^2 z}{\partial y^2} = x^2 + y.$

[In parts (c) and (d), consider the two terms on the right separately and use superposition.]

30. (a) If $am^2 + bm + c = a(m - m_1)(m - m_2)$, show that the equation

$$a\frac{\partial^2 z}{\partial x^2} + b\frac{\partial^2 z}{\partial x\,\partial y} + c\frac{\partial^2 z}{\partial y^2} = f(x, y),$$

where a, b, and c are constants, can be written in the operational form

$$a\left(\frac{\partial}{\partial x} - m_1\frac{\partial}{\partial y}\right)\left(\frac{\partial z}{\partial x} - m_2\frac{\partial z}{\partial y}\right) = f(x, y)$$

and deduce the general solution when $f(x, y) = 0$ by the methods of Section 8.2.

(b) Show that a particular solution of the general equation of part (a) can be obtained as a particular solution of the equation

$$\frac{\partial z}{\partial x} - m_2\frac{\partial z}{\partial y} = z_1(x, y),$$

where z_1 is a particular solution of the equation

$$a\left(\frac{\partial z_1}{\partial x} - m_1\frac{\partial z_1}{\partial y}\right) = f(x, y).$$

31. Use the method of Problem 30 to obtain the general solution of the equation

$$\frac{\partial^2 z}{\partial x^2} - \frac{\partial^2 z}{\partial y^2} = (x^2 - y^2)\sin xy.$$

32. Use Equation (67) to find the solution of the equation

$$\frac{\partial^2 \varphi}{\partial t^2} = c^2 \frac{\partial^2 \varphi}{\partial x^2}$$

for which $\varphi = x^2$ and $\partial\varphi/\partial t = \cos x$ for all values of x when $t = 0$.

33. It is required to find the solution of the equation

$$\frac{\partial^2 \varphi}{\partial t^2} = c^2 \frac{\partial^2 \varphi}{\partial x^2}$$

which satisfies the conditions

$$\varphi(0, t) = 0, \quad \varphi(l, t) = 0 \qquad \text{(for all values of } t\text{)}$$

along the boundary of the strip $0 \leq x \leq l$ in an xt plane, and the conditions

$$\varphi(x, 0) = F(x), \qquad \frac{\partial \varphi(x, 0)}{\partial t} = G(x) \qquad \text{(when } 0 < x < l\text{)}$$

along the line segment $t = 0$ in that strip.

(a) Taking the general solution of the equation in the form

$$\varphi = f(x + ct) + g(x - ct),$$

show that the conditions along the boundaries $x = 0$ and $x = l$ lead to the relations

$$f(u) + g(-u) = 0 \qquad \text{(writing } u = ct\text{)},$$
$$f(u + 2l) = f(u) \qquad \text{(writing } u = ct - l\text{)},$$

for all values of u, while the conditions along the segment $t = 0$ for $0 < x < l$ become

$$\left. \begin{array}{l} f(u) + g(u) = F(u) \\ c[f'(u) - g'(u)] = G(u) \end{array} \right\} \qquad \text{(when } 0 < u < l\text{)}.$$

(b) Deduce that, if we write $H'(u) = G(u)$, then the last two conditions of part (a) give

$$f(u) = \frac{1}{2} F(u) + \frac{1}{2c} H(u),$$

$$g(u) = \frac{1}{2} F(u) - \frac{1}{2c} H(u),$$

when $0 < u < l$, and that also, *for all values of u, f(u)* must be a *periodic function,* of period $2l$, and the condition

$$f(u) = -g(-u)$$

must be satisfied.

(c) Show that these conditions are all satisfied if we take

$$f(u) = \frac{1}{2} \mathcal{F}(u) + \frac{1}{2c} \mathcal{H}(u),$$

$$g(u) = \frac{1}{2} \mathcal{F}(u) - \frac{1}{2c} \mathcal{H}(u),$$

for all values of u, where $\mathcal{F}(u)$ *and* $\mathcal{G}(u)$ *are odd period functions of period 2l, agreeing with* $F(u)$ *and* $G(u)$, *respectively, when* $0 < u < l$, and where $\mathcal{H}'(u) = \mathcal{G}(u)$. Thus, with this notation, obtain the required solution in the form

$$\varphi(x, t) = \frac{1}{2} [\mathcal{F}(x + ct) + \mathcal{F}(x - ct)] + \frac{1}{2c} \int_{x-ct}^{x+ct} \mathcal{G}(\xi) \, d\xi.$$

34. To illustrate the solution of Problem 33, notice that if $F(x) = \sin \pi x / l$ and $G(x) = 0$, there follows $\mathcal{F} = F$ and $\mathcal{G} = G = 0$, and obtain the solution in the form

$$\varphi = \frac{1}{2} \left(\sin \pi \frac{x + ct}{l} + \sin \pi \frac{x - ct}{l} \right) = \sin \frac{\pi x}{l} \cos \frac{c\pi t}{l}.$$

35. By using the method of Problem 33, obtain the solution of the problem

$$\frac{\partial^2 \varphi}{\partial t^2} = c^2 \frac{\partial^2 \varphi}{\partial x^2},$$

$$\varphi(x, 0) = F(x), \qquad \frac{\partial \varphi(x, 0)}{\partial t} = G(x) \qquad \text{(when } x > 0\text{)},$$

$$\varphi(0, t) = 0$$

in the half-plane $x > 0$ in the form

$$\varphi(x, t) = \frac{1}{2}[\mathfrak{F}(x + ct) + \mathfrak{F}(x - ct)] + \frac{1}{2c} \int_{x-ct}^{x+ct} \mathfrak{G}(\xi) \, d\xi,$$

where here $\mathfrak{F}(u)$ and $\mathfrak{G}(u)$ are odd functions of u agreeing with $F(u)$ and $G(u)$, respectively, when $u > 0$.

36. Suppose that a function $u(x, y)$ satisfies the equation

$$\frac{\partial^2 u}{\partial x^2} - \frac{\partial^2 u}{\partial y^2} = q(x, y)$$

in a region D which includes the triangle with vertices at the points $P(x_0, y_0)$, $P_1(x_0 - y_0, 0)$, and $P_2(x_0 + y_0, 0)$ in Figure 8.3.

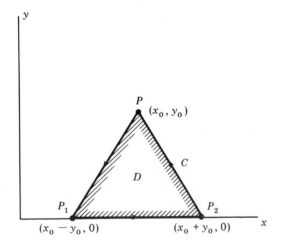

Figure 8.3

(a) By integrating the equal members of the equation over D and using Stokes's theorem in the plane [Equation (139) of Chapter 6], show that

$$\oint_C \left(\frac{\partial u}{\partial x} \, dy + \frac{\partial u}{\partial y} \, dx \right) = \int \int_D q(x, y) \, dx \, dy,$$

where C is the boundary of D traversed in the counterclockwise direction.

(b) By noticing that $dy = -dx$ along $P_2 P$ and $dy = dx$ along PP_1, show that the integrand in the line integral is $-du$ along $P_2 P$ and $+du$ along PP_1, and hence deduce that

$$u(x_0, y_0) = \frac{1}{2}[u(x_0 - y_0, 0) + u(x_0 + y_0, 0)] + \frac{1}{2} \int_{x_0 - y_0}^{x_0 + y_0} u_y(x, 0) \, dx$$

$$- \frac{1}{2} \int_0^{y_0} \int_{x_0 - y_0 + y}^{x_0 + y_0 - y} q(x, y) \, dx \, dy.$$

37. *A nonhomogeneous Cauchy problem.*

(a) Obtain the solution of the problem

$$c^2 \frac{\partial^2 \varphi}{\partial x^2} - \frac{\partial^2 \varphi}{\partial t^2} = h(x, t),$$

$$\varphi(x, 0) = f(x), \qquad \varphi_t(x, 0) = g(x)$$

when $-\infty < x < \infty$, $0 \le t < \infty$, by making appropriate changes in notation in Problem 36, in the form

$$\varphi(x, t) = \frac{1}{2}[f(x + ct) + f(x - ct)] + \frac{1}{2c} \int_{x-ct}^{x+ct} g(\xi)\, d\xi$$

$$- \frac{1}{2c} \int_0^t \int_{x-c(t-\tau)}^{x+c(t-\tau)} h(\xi, \tau)\, d\xi\, d\tau.$$

[Notice that this result reduces to (67) when $h = 0$.]

(b) Use the result of part (a) to obtain the solution of the specified problem when $f(x) = x^2$, $g(x) = 1$, and $h(x, t) = 1$, and verify its correctness.

Section 8.6

38. (a) By assuming a solution of the Helmholtz equation

$$\frac{\partial^2 z}{\partial x^2} + \frac{\partial^2 z}{\partial y^2} + k^2 z = 0$$

in the form $z = e^{\alpha x + \beta y}$, and satisfying the resultant requirement $\alpha^2 + \beta^2 + k^2 = 0$ by writing $\alpha = ik \cos \varphi$ and $\beta = ik \sin \varphi$, where φ is an arbitrary real parameter, obtain particular solutions in the forms

$$\cos[k(x \cos \varphi + y \sin \varphi)], \qquad \sin[k(x \cos \varphi + y \sin \varphi)].$$

(b) Deduce that the real or imaginary part of any expression of the form

$$U(x, y) = \int_{\varphi_1}^{\varphi_2} A(\varphi) e^{ik(x \cos \varphi + y \sin \varphi)}\, d\varphi$$

is a formal solution, for an arbitrarily chosen function $A(\varphi)$ and arbitrary constant limits φ_1 and φ_2.

39. By changing to polar coordinates, deduce from the results of Problem 38 that the real and imaginary parts of any expression of the form

$$V(r, \theta) = \int_{\varphi_1}^{\varphi_2} A(\varphi) e^{ikr \cos(\varphi - \theta)}\, d\varphi$$

formally satisfy the Helmholtz equation in polar coordinates,

$$\frac{\partial^2 z}{\partial r^2} + \frac{1}{r} \frac{\partial z}{\partial r} + \frac{1}{r^2} \frac{\partial^2 z}{\partial \theta^2} + k^2 z = 0.$$

40. By assuming a solution of the heat flow equation

$$\frac{\partial^2 z}{\partial x^2} + \frac{\partial^2 z}{\partial y^2} = k^2 \frac{\partial z}{\partial t}$$

in the form $z = e^{\alpha x + \beta y + \gamma t}$, and satisfying the resultant requirement by setting

$\alpha = ik\rho \cos \varphi$, $\beta = ik\rho \sin \varphi$, $\gamma = -\rho^2$, deduce particular solutions in the forms

$$e^{-\rho^2 t} \cos [k\rho(x \cos \varphi + y \sin \varphi)], \qquad e^{-\rho^2 t} \sin [k\rho(x \cos \varphi + y \sin \varphi)]$$

or as the real or imaginary parts of an expression of the form

$$U(x, y, t) = \iint_{\mathfrak{R}} A(\rho, \varphi) e^{-\rho^2 t + ik\rho(x \cos \varphi + y \sin \varphi)} \, d\rho \, d\varphi.$$

41. By changing to polar coordinates in Problem 40, deduce particular solutions of the equation

$$\frac{\partial^2 z}{\partial r^2} + \frac{1}{r} \frac{\partial z}{\partial r} + \frac{1}{r^2} \frac{\partial^2 z}{\partial \theta^2} = k^2 \frac{\partial z}{\partial t}$$

in the forms

$$V_1(r, \theta, t) = \iint_{\mathfrak{R}} A(\rho, \varphi) e^{-\rho^2 t} \cos [k\rho r \cos (\varphi - \theta)] \, d\rho \, d\varphi$$

and

$$V_2(r, \theta, t) = \iint_{\mathfrak{R}} B(\rho, \varphi) e^{-\rho^2 t} \sin [k\rho r \cos(\varphi - \theta)] \, d\rho \, d\varphi.$$

42. Suppose that a function $p(x, y)$ exists such that the equation

$$a \frac{\partial^2 z}{\partial x^2} + b \frac{\partial^2 z}{\partial x \, \partial y} + c \frac{\partial^2 z}{\partial y^2} + d \frac{\partial z}{\partial x} + e \frac{\partial z}{\partial y} + fz = g$$

can be written in the form

$$\frac{1}{p} \left[a \frac{\partial^2 (pz)}{\partial x^2} + b \frac{\partial^2 (pz)}{\partial x \, \partial y} + c \frac{\partial^2 (pz)}{\partial y^2} \right] = g.$$

(a) Show that p must satisfy the three equations

$$2a \frac{p_x}{p} + b \frac{p_y}{p} = d, \qquad b \frac{p_x}{p} + 2c \frac{p_y}{p} = e,$$

and

$$a p_{xx} + b p_{xy} + c p_{yy} = fp.$$

(b) Deduce that, with the assumption $b^2 \neq 4ac$ and the abbreviations

$$M = \frac{be - 2cd}{b^2 - 4ac}, \qquad N = \frac{bd - 2ac}{b^2 - 4ac},$$

the function p is determined by the equations

$$\frac{\partial}{\partial x} \log p = M, \qquad \frac{\partial}{\partial y} \log p = N,$$

and that M and N must satisfy the two "compatibility conditions"

$$\frac{\partial M}{\partial y} = \frac{\partial N}{\partial x}$$

and

$$a \frac{\partial M}{\partial x} + \frac{1}{2} b \left(\frac{\partial M}{\partial y} + \frac{\partial N}{\partial x} \right) + c \frac{\partial N}{\partial y} + \frac{1}{2} (dM + eN) = f.$$

43. Consider the application of the results of Problem 42 in the following cases:

(a) For the equation

$$\frac{\partial^2 z}{\partial x^2} - \frac{1}{\alpha^2} \frac{\partial^2 z}{\partial y^2} + \frac{4}{x} \frac{\partial z}{\partial x} + \frac{2}{x^2} z = 0,$$

show that $p = x^2$ and deduce the result of text Example 3.

(b) For Equation (75), show that

$$p = e^{-(x+y)/2},$$

deduce that the general solution of (75) is

$$z = e^{(x+y)/2}[F(x+y) + G(x-y)],$$

and show that this result is equivalent to (77).

(c) When M and N are both constant, show that the only restrictive compatibility equation reduces to the condition (74). [Notice that part (b) is a special case.]

(d) For the equation

$$\frac{\partial^2 z}{\partial x^2} - \frac{\partial^2 z}{\partial y^2} + 4x \frac{\partial z}{\partial x} = 0,$$

show that p does not exist.

44. Show that the general solution of the equation

$$\frac{\partial^2 \varphi}{\partial x^2} + \frac{\partial^2 \varphi}{\partial y^2} - \frac{2}{x}(Ax + m)\frac{\partial \varphi}{\partial x} - \frac{2}{y}(By + n)\frac{\partial \varphi}{\partial y}$$

$$+ \left[\frac{m(m+1)}{x^2} + \frac{n(n+1)}{y^2} + \frac{2Am}{x} + \frac{2Bn}{y} + A^2 + B^2 \right] \varphi = 0$$

can be expressed in terms of arbitrary functions when A, B, m, and n are constants, and obtain that solution. (See Problem 42.)

Section 8.7

45. The function $z(x, y)$ is required to satisfy the differential equation

$$\frac{\partial z}{\partial x} - \frac{\partial z}{\partial y} = z,$$

and to take on the value $z(\lambda) = \sin 2\lambda$ along the straight line C_0 specified by the equations $x = y = \lambda$ in the xy plane.

(a) By using equations corresponding to (93) and (94), show that

$$\frac{\partial z}{\partial x} = \cos 2\lambda + \frac{1}{2}\sin 2\lambda, \qquad \frac{\partial z}{\partial y} = \cos 2\lambda - \frac{1}{2}\sin 2\lambda$$

at points of C_0.

(b) By using equations corresponding to (97) and (98), show that

$$\frac{\partial^2 z}{\partial x^2} = \cos 2\lambda - \frac{3}{4}\sin 2\lambda, \qquad \frac{\partial^2 z}{\partial x\, \partial y} = -\frac{5}{4}\sin 2\lambda$$

at points of C_0.

(c) Use the result of differentiating $\partial z/\partial y$ along C_0 to show that

$$\frac{\partial^2 z}{\partial y^2} = -\cos 2\lambda - \frac{3}{4}\sin 2\lambda$$

at points of C_0.

(d) Check the result of part (c) by using the result of differentiating the governing differential equation with respect to y.

46. (a) Show that the Taylor expansion of the solution of the problem considered in Problem 45, in the neighborhood of the point $(0, 0)$, can be written in the form

$$z(x, y) = \frac{1}{1!}(x + y) + \frac{1}{2!}(x^2 - y^2) + \cdots,$$

in consequence of the results of Problem 45.

(b) Obtain the solution in the closed form

$$z = e^{(x-y)/2} \sin(x + y).$$

(c) Verify the correctness of the result of part (a) by obtaining the leading terms in the expansion of the closed form in a Taylor series about $(0, 0)$.

47. Verify directly that the procedure of Problem 45 fails if the curve C_0 is taken instead to be the characteristic base curve $x = -y = \lambda$.

48. (a) If Problem 45 is modified in such a way that C_0 is taken as the characteristic base curve $x = -y = \lambda$ and z is prescribed as $z(\lambda) = e^\lambda$ along C_0, verify directly that $\partial z/\partial x$ and $\partial z/\partial y$ then can be determined in infinitely many ways.

(b) Show that the curve $x = \lambda$, $y = -\lambda$, $z = e^\lambda$ is a characteristic curve of the differential equation.

49. (a) If Problem 45 is modified in such a way that C_0 is taken as the circle $x = \cos \lambda$, $y = \sin \lambda$, and z is prescribed as $z(\lambda) = f(\lambda)$ along C_0, show that $\partial z/\partial x$ and $\partial z/\partial y$ can be determined uniquely at all points along C_0 except the points

$$(\sqrt{2}/2, \sqrt{2}/2) \quad \text{and} \quad (-\sqrt{2}/2, -\sqrt{2}/2).$$

(b) Show that C_0 is tangent to characteristic base curves at the exceptional points of part (a).

50. Consider the quasi-linear equation

$$\frac{\partial z}{\partial x} + z\frac{\partial z}{\partial y} = 1.$$

(a) Obtain the general solution in the form

$$F\left((z - x, y + \frac{x^2}{2} - xz\right) = 0.$$

(b) Show that the integral surface which contains the line on which $z = x/2$ when $y = x$ has the equation

$$z = \frac{4x - 2y - x^2}{2(2 - x)} \qquad (x \neq 2).$$

(c) Show that the integral surface which contains the line on which $z = 2$ when $y = x$ is

$$z = 1 + \sqrt{1 + 2(y - x)}.$$

(d) Show that there are *two* integral surfaces,

$$z = 1 \pm \sqrt{2(y - x)},$$

which contain the line on which $z = 1$ when $y = x$.

(e) Show that the curve on which $z = x + A$ when $y = x^2/2 + Ax + B$ lies on infinitely many integral surfaces.

[Notice that the sufficient condition stated in the text, for the existence of a unique solution, here becomes the requirement that $z \neq dy/dx$ everywhere along C, that this requirement is violated when $x = 2$ in part (b) and for all x in parts (d) and (e), and that the consequences of the violation in those cases are quite dissimilar.]

Section 8.8

51. The solution of Laplace's equation

$$\frac{\partial^2 z}{\partial x^2} + \frac{\partial^2 z}{\partial y^2} = 0$$

is required in the neighborhood of the line $C_0: x = y = \lambda$, subject to the requirement that $z = 0$ along C_0 and that $\partial z/\partial x \equiv p = \lambda$ along C_0.

(a) Show that the strip condition requires that $\partial z/\partial y \equiv q = -\lambda$ along C_0.

(b) With the notation of Equation (112), verify that Equations (116) to (118) take the form $r + t = 0$, $r + s = 1$, $s + t = -1$, so that there must follow $r = 1$, $s = 0$, and $t = -1$ along C_0.

(c) Deduce that, at the origin $(0, 0)$, there follows

$$z = 0, \qquad \frac{\partial z}{\partial x} = 0, \qquad \frac{\partial z}{\partial y} = 0,$$

$$\frac{\partial^2 z}{\partial x^2} = 1, \qquad \frac{\partial^2 z}{\partial x \, \partial y} = 0, \quad \text{and} \quad \frac{\partial^2 z}{\partial y^2} = -1,$$

and hence show that the Taylor series expansion of z near the origin must be of the form

$$z(x, y) = 0 + \frac{1}{1!}(0x + 0y) + \frac{1}{2!}(x^2 + 2 \cdot 0xy - y^2) + \cdots = \frac{1}{2}(x^2 - y^2) + \cdots.$$

(d) Verify the correctness of the result of part (c) by showing that the function $z = \frac{1}{2}(x^2 - y^2)$ in fact satisfies the conditions of the problem (so that the remaining terms in the expansion all vanish).

52. (a) Show that the characteristic base curves of the equation

$$\frac{\partial^2 z}{\partial x^2} + 3 \frac{\partial^2 z}{\partial x \, \partial y} + 2 \frac{\partial^2 z}{\partial y^2} - \frac{\partial z}{\partial x} = 0$$

are the lines $y = x + c_1$ and $y = 2x + c_2$.

(b) Show that, along the characteristic base curve $C_0: x = y = \lambda$, the strip condition takes the form $z' = p + q$ (where $z' \equiv dz/d\lambda$) and the compatibility condition becomes $p' + 2q' = p$.

(c) Verify directly that, if z, p, and q are prescribed along C_0 in such a way that these two conditions are satisfied, then Equations (116) to (118) permit any one of the second derivatives, say s, to be chosen arbitrarily along C_0, after which there follows $r = p' - s$ and $t = q' - s$ along C_0.

[In the next step, consistency of the equations governing z_{xxx}, z_{xxy}, and z_{xyy} on C_0 provides the additional requirement $r' + 2s' = r$ which, when combined with the first result of part (c), finally determines s along C_0 as any solution of the equation $s' + s = p'' - p'$, and hence as a function depending upon an *arbitrary constant*.]

53. Consider the initial-value problem

$$\frac{\partial^2 z}{\partial x^2} = \frac{\partial z}{\partial y},$$

$$z(x, 0) = F(x), \qquad z_x(x, 0) = G(x), \qquad z_y(x, 0) = H(x),$$

assuming that F, G, and H have derivatives of all orders.

(a) Show that the differential equation is parabolic, that the line $y = 0$ is a characteristic base curve, and that here Equation (121) is nonrestrictive (so that a new compatibility condition is to be expected).

(b) Show that the strip condition (110) requires that

$$G = F',$$

obtain the additional conditions

$$r = q, \qquad r = p', \qquad s = q'$$

when $y = 0$, from (116) to (118) [where, for example, $p' \equiv dp(x, 0)/dx$], and deduce the compatibility condition

$$p' = q,$$

replacing (121). Thus deduce that F, G, and H must be such that

$$H = G' = F''$$

and hence also

$$r = F'', \qquad s = F''',$$

whereas $t \equiv z_{yy}$ is as yet undetermined on the line $y = 0$.

(c) Obtain the additional conditions

$$z_{xxx} = s, \qquad z_{xxy} = t$$

and

$$z_{xxx} = r', \qquad z_{xxy} = s', \qquad z_{xyy} = t'$$

when $y = 0$, and deduce that

$$t = s' = F^{iv}$$

and also

$$z_{xxx} = F''', \qquad z_{xxy} = F^{iv}, \qquad z_{xyy} = F^{v}$$

when $y = 0$, whereas $z_{yyy}(x, 0)$ is not yet determined.

(d) Thus obtain (formally) the expansion

$$z(x, y) = F(x_0) + (x - x_0)F'(x_0) + yF''(x_0) + \frac{1}{2!}[(x - x_0)^2 F''(x_0)$$

$$+ 2(x - x_0)yF'''(x_0) + y^2 F^{iv}(x_0)] + \cdots$$

and also the expansion

$$z(x, y) = F(x) + \frac{y}{1!}F''(x) + \frac{y^2}{2!}F^{iv}(x) + \cdots.$$

54. (a) Show that, if $\varphi = y - x$ and $\psi = y - 2x$ are taken as new independent variables in Problem 52, then there follows

$$\frac{\partial}{\partial x} = -\left(\frac{\partial}{\partial \varphi} + 2\frac{\partial}{\partial \varphi}\right), \qquad \frac{\partial}{\partial y} = \frac{\partial}{\partial \varphi} + \frac{\partial}{\partial \psi}$$

and the differential equation takes the normal form

$$\frac{\partial^2 z}{\partial \varphi \, \partial \psi} = \frac{\partial z}{\partial \varphi} + 2 \frac{\partial z}{\partial \psi}.$$

(b) Determine constants α and β such that the change of variables

$$z = u e^{\alpha \varphi + \beta \psi}$$

transforms the normal form obtained in part (a) to an alternative normal form

$$\frac{\partial^2 u}{\partial \varphi \, \partial \psi} = \gamma u,$$

where γ is a constant.

55. (a) By writing $r = x + ct$ and $s = x - ct$, transform the equation

$$\frac{\partial^2 \varphi}{\partial x^2} - \frac{1}{c^2} \frac{\partial^2 \varphi}{\partial t^2} = h(x, t)$$

to the equation

$$\frac{\partial^2 \varphi}{\partial r \, \partial s} = \frac{1}{4} h\left(\frac{r+s}{2}, \frac{r-s}{2c}\right).$$

(b) Use the result of part (a) to obtain the general solution of the original equation when $h(x, t) = 1$ and also when $h(x, t) = \cos(x + ct)$.

Section 8.9

56. Show that the solution of the equation

$$\frac{\partial z}{\partial x} + \frac{\partial z}{\partial y} = 0,$$

for which $z = |x|$ along the x axis, is of the form $z = |y - x|$. (Notice that the "corner" at the origin is propagated along the characteristic base curve $y = x$ which passes through the origin.)

57. Show that, if z satisfies the equation

$$\frac{\partial^2 z}{\partial x^2} - \frac{\partial^2 z}{\partial y^2} = 0,$$

and if $\partial^2 z / \partial x^2$ has a jump of one unit across the line C_0: $y = x$ in the xy plane, then $\partial^2 z / (\partial x \, \partial y)$ must have a jump of -1 and $\partial^2 z / \partial y^2$ a jump of $+1$ across C_0, if $\partial z / \partial x$ and $\partial z / \partial y$ are to be continuously differentiable along C_0.

58. If $z = \frac{1}{2}(x + y)^2$ when $y \geq x$ and $z = 2xy$ when $y \leq x$, show that z, $\partial z / \partial x$, and $\partial z / \partial y$ are continuous on the line $y = x$ and that z satisfies the equation

$$\frac{\partial^2 z}{\partial x^2} - \frac{\partial^2 z}{\partial y^2} = 0,$$

and also verify that z possesses the properties specified in Problem 57.

59. If z satisfies Laplace's equation

$$\frac{\partial^2 z}{\partial x^2} + \frac{\partial z^2}{\partial y^2} = 0,$$

show that the assumption that $\partial^2 z/\partial^2 x$ has a jump across the line $C_0 : y = x$ in the xy plane leads to a contradiction if $\partial z/\partial x$ and $\partial z/\partial y$ are to be continuously differentiable along C_0.

Section 8.10

60. (a) By replacing ct by iy, where $i^2 = -1$, in Equation (67), show that *formally* we obtain a solution of Laplace's equation

$$\frac{\partial^2 z}{\partial x^2} + \frac{\partial^2 z}{\partial y^2} = 0$$

for which $z(x, 0) = F(x)$ and $\partial z(x, 0)/\partial y = G(x)$ in the form

$$z = \frac{1}{2}[F(x + iy) + F(x - iy)] + \frac{1}{2i} \int_{x-iy}^{x+iy} G(\xi)\, d\xi.$$

(b) Show that if we specify $F(x) = x$, $G(x) = e^x$, then an expression

$$z = x + e^x \sin y$$

is obtained, and verify that this expression satisfies Laplace's equation everywhere and also satisfies the conditions imposed along the line $y = 0$.

(c) Notice that the formal solution is *meaningless* if, for example, we specify that $F(x)$ be zero along the negative x axis and unity along the positive x axis. Show also that if

$$F(x) = \frac{1}{1 + x^2}, \qquad G(x) = 0$$

the formal solution becomes

$$z = \frac{1 + x^2 - y^2}{(1 + x^2 - y^2)^2 + 4x^2 y^2},$$

and that this expression is not defined at the points $(0, \pm 1)$. In particular, show that this formal solution becomes infinite as either of these points is approached along the y axis, so that the differential equation is not satisfied at this point.

[In general, the *initial-value* problem for Laplace's equation (or *any elliptic* equation) has no solution unless the prescribed values are sufficiently regular. Also, even though a solution be determined by the above method, it may be valid only *near the curve* along which the function and its normal derivative are prescribed. A deeper insight into the problem is afforded by the theory of analytic functions of a complex variable (Chapter 10).]

61. Suppose that $z(x, y)$ is to satisfy the equation

$$\frac{\partial^2 z}{\partial x^2} - \frac{\partial^2 z}{\partial y^2} = 0$$

and the conditions

$$z(x, 0) = f(x), \qquad z_y(x, 0) = g(x)$$

along the entire x axis.

(a) By appropriately modifying the derivation of (67), show that

$$z(x, y) = \frac{1}{2}[f(x + y) + f(x - y)] + \frac{1}{2} \int_{x-y}^{x+y} g(\xi)\, d\xi.$$

(b) In the special case when

$$f(x) = \begin{cases} \frac{1}{2}x^2 & (x \geq 0), \\ 0 & (x \leq 0), \end{cases}$$

and $g(x) = 0$, show that $z = x^2 + y^2$ when $x \geq |y|$, $z = \frac{1}{2}(x + y)^2$ when $y \geq |x|$, $z = \frac{1}{2}(x - y)^2$ when $y \leq |x|$, and $z = 0$ when $x \leq |y|$. Also verify that z, z_x, and z_y are continuous everywhere but that the jump in $f''(x)$ propagates into jumps in z_{xx}, z_{xy} and z_{yy} along the lines $y = \pm x$.

62. Writing the general solution of the equation

$$\frac{\partial^2 z}{\partial x \, \partial y} = 0$$

in the form $z = f(x) + g(y)$, show that the initial conditions

$$z(x, 0) = F(x), \qquad p(x, 0) \equiv \frac{\partial z(x, 0)}{\partial x} = F'(x), \qquad q(x, 0) \equiv \frac{\partial z(x, 0)}{\partial y} = G(x),$$

imposed along the infinite line $y = 0$, lead to the conditions

$$f(x) + g(0) = F(x), \qquad g'(0) = G(x).$$

Hence deduce that the problem has no solution unless $G(x)$ is prescribed as a constant, say $G(x) = C$, in which case the problem has infinitely many solutions of the form

$$z = F(x) + g(y) - g(0),$$

where $g(y)$ is any (differentiable) function of y for which $g'(0) = C$. (Notice that the line $y = 0$ in the xy plane is a characteristic base curve in this case.)

Section 8.11

63. Show by the methods of Section 8.11 that the simultaneous equations

$$\frac{\partial U}{\partial x} + \frac{\partial V}{\partial y} = 0,$$

$$2 \frac{\partial U}{\partial x} + \left(\frac{\partial U}{\partial y} + \frac{\partial V}{\partial x} \right) \tan 2V = 0.$$

possess the same characteristics [Equations (153a, b)] as the problem of that section.

64. Use the results of Problem 7 to show that if the Jacobian $\partial(\xi, \eta)/\partial(x, y)$ is not zero, the dependent and independent variables in Equations (159) may be interchanged to give the equations

$$y_\eta - x_\eta \tan v = 0, \qquad y_\xi + x_\xi \cot v = 0,$$

where $v = \frac{1}{2}(\xi - \eta)$. [Notice that these equations are *linear* in x and y, whereas Equations (159) are *quasi-linear* in ξ and η, since v depends upon ξ and η.]

9

Solutions of Partial Differential
Equations of Mathematical Physics

9.1. Introduction. Many linear problems in mathematical physics involve
the solution of an equation obtained by suitably specializing the form

$$\mathbf{V}^2\varphi = \lambda\frac{\partial^2\varphi}{\partial t^2} + \mu\frac{\partial\varphi}{\partial t} + h, \tag{1}$$

where h is a specified function of position and λ and μ are certain specified
physical constants. Here the operator $\mathbf{V}^2$ is the Laplacian operator in the space
of one, two, or three dimensions under consideration and is of the form

$$\mathbf{V}^2 = \frac{\partial^2}{\partial x^2} + \frac{\partial^2}{\partial y^2} + \frac{\partial^2}{\partial z^2} \tag{2}$$

in rectangular coordinates of three-space. The unknown function φ is then, in
general, a function of the position coordinates (x, y, z) and the time coordinate t.

In particular, *Laplace's equation*,

$$\mathbf{V}^2\varphi = 0, \tag{3}$$

is satisfied, for example, by the velocity potential in an ideal incompressible
fluid without vorticity or continuously distributed sources and sinks, by gravitational potential in free space, electrostatic potential in the steady flow of
electric currents in solid conductors, and by the steady-state temperature distribution in solids.

Also, *Poisson's equation*,

$$\mathbf{V}^2\varphi = h, \tag{4}$$

is satisfied, for example, by the velocity potential of an incompressible, irrotational, ideal fluid with continuously distributed sources or sinks, by steady-
state temperature distributions due to distributed heat sources, and by a "stress

function" involved in the elastic torsion of prismatic bars, with a suitably prescribed function h.

The so-called *wave equation*,

$$\mathbf{V}^2\varphi = \frac{1}{c^2}\frac{\partial^2\varphi}{\partial t^2},$$
(5)

arises in the study of propagation of waves with velocity c, independent of the wave length. In particular, it is satisfied by the components of the electric or magnetic vector in electromagnetic theory, by suitably chosen components of displacements in the theory of elastic vibrations, and by the velocity potential in the theory of sound (acoustics) for a perfect gas.

The equation of heat conduction,

$$\mathbf{V}^2\varphi = \frac{1}{\alpha^2}\frac{\partial\varphi}{\partial t},$$
(6)

is satisfied, for example, by the temperature at a point of a homogeneous body and by the concentration of a diffused substance in the theory of diffusion, with a suitably prescribed constant α.

The *telegraph equation*,

$$\frac{\partial^2\varphi}{\partial x^2} = \lambda\frac{\partial^2\varphi}{\partial t^2} + \mu\frac{\partial\varphi}{\partial t},$$
(7)

which is a one-dimensional specialization of Equation (1), is satisfied by the potential in a telegraph cable, where $\lambda = LC$ and $\mu = RC$, if leakage is neglected (L is inductance, C capacity, and R resistance per unit length).

Differential equations of higher order, involving the operator $\mathbf{V}^2$, also are rather frequently encountered. In particular, the *bi-Laplacian equation* in two dimensions,

$$\mathbf{V}^4\varphi = \mathbf{V}^2\mathbf{V}^2\varphi = \frac{\partial^4\varphi}{\partial x^4} + 2\frac{\partial^4\varphi}{\partial x^2\partial y^2} + \frac{\partial^4\varphi}{\partial y^4} = 0,$$
(8)

is involved in many two-dimensional problems of the theory of elasticity.

The solution of a given problem must satisfy the proper differential equation, together with suitably prescribed *boundary conditions* and/or *initial conditions*, the nature of these conditions depending upon the problem. However, the requirement that the solution be *single-valued* in the relevant region $\mathcal{R}$ may substitute in part for boundary conditions when that region is not simply connected. In addition, it is to be assumed always (unless the contrary is stated) that the solution is to be *bounded* in every finite subregion of $\mathcal{R}$.

The process of attempting to obtain the *most general* solution of the governing differential equation and then specializing it so as to satisfy the given conditions (as in d'Alembert's solution of the problem at the end of Section 8.5) is rarely feasible. Instead, one often seeks to determine a suitable set of *particular* solutions, each of which satisfies certain of the conditions, and then attempts to combine a finite or infinite number of such solutions in such a way that the combination satisfies all the prescribed conditions. Here one must rely upon knowledge that the problem does not in fact possess more than one solution.

The particular solutions frequently are obtained by a certain method of *separation of variables*. It may be mentioned that, although this method is restricted in a mathematical sense to a comparatively narrow range of differential equations, fortunately this range includes a large number of those equations which arise in practice.

Methods which are somewhat more direct, but which again are applicable only to specific classes of problems, are considered in Sections 9.15 and 9.16 and in Section 11.8.

It should be emphasized that most problems arising in practice, which involve partial differential equations, are in fact not amenable to exact analytical solution and either must be dealt with by approximate numerical methods, or must be simulated by problems which can be treated analytically.

Before illustrating the treatment of some analytically solvable problems, we first consider the mathematical formulation of the study of heat flow in space, as a basis for some of these problems.

9.2. Heat Flow. We consider the flow of heat in a region in space such that the temperature T at a point (x, y, z) may depend upon the time t. For any region $\mathfrak{R}$ bounded by a closed surface S, the rate at which heat flows *outward* from $\mathfrak{R}$ through a surface element $d\sigma$ with unit outward normal $\mathbf{n}$ is given by

$$-dQ_1 = -K\frac{\partial T}{\partial n}\,d\sigma = -K(\nabla T) \cdot \mathbf{n}\,d\sigma,$$

where K is the thermal conductivity of the material. (The negative sign clearly corresponds to the fact that heat flow is in the direction of *decreasing* temperature.) Thus the net rate of heat flow *into* $\mathfrak{R}$ is given by

$$Q_1 = +\oiint_S K(\nabla T) \cdot \mathbf{n}\,d\sigma. \tag{9}$$

However, the rate at which heat is absorbed by a *volume* element $d\tau$ is given by

$$dQ_2 = s\rho\frac{\partial T}{\partial t}\,d\tau,$$

where s is the specific heat and ρ the mass density. Hence, if there are no sources or sinks in $\mathfrak{R}$, the rate of heat flow into $\mathfrak{R}$ is also given by

$$Q_2 = \iiint_{\mathfrak{R}} s\rho\frac{\partial T}{\partial t}\,d\tau. \tag{10}$$

Before equating Q_1 and Q_2, we first transform Equation (9) to a volume integral by making use of the divergence theorem (Section 6.13), and so obtain

$$Q_1 = \iiint_{\mathfrak{R}} K\nabla^2 T\,d\tau, \tag{11}$$

442 Solutions of Partial Differential Equations of Mathematical Physics

when K does not vary with position.† The requirement $Q_1 = Q_2$ then becomes

$$\int\int\int_\mathfrak{R} \left(K\mathbf{V}^2 T - s\rho \frac{\partial T}{\partial t} \right) d\tau = 0; \tag{12}$$

but since this result must be true for any region $\mathfrak{R}$ not containing heat sources or sinks, the integrand must vanish. Hence T must satisfy the equation

$$\mathbf{V}^2 T = \frac{1}{\alpha^2} \frac{\partial T}{\partial t}, \tag{13}$$

where we have written

$$\alpha^2 = \frac{K}{s\rho}. \tag{14}$$

The quantity α^2 is known as the *thermal diffusivity* of the material.

In particular, in the *steady-state* condition when the temperature at a point does not vary with time, the temperature satisfies *Laplace's equation*,

$$\mathbf{V}^2 T = 0. \tag{15}$$

We now consider two basic problems in the theory of steady-state heat flow. (Generalizations are considered in Problems 11, 12, and 13.) First, it would be expected intuitively that in the steady state the temperatures at points inside a given region $\mathfrak{R}$ would be uniquely determined if the temperature were prescribed along the bounding surface S. Second, one might prescribe the rate of (steady) heat flow outward per unit of area,

$$-\frac{dQ}{d\sigma} = -K\frac{\partial T}{\partial n},$$

at all points of the boundary S and require the temperature at internal points. Let T represent a function which satisfies Equation (15) and one of these boundary conditions. Then Equations (126) and (127) of Section 6.14 give the useful results

$$\int\int\int_\mathfrak{R} (\mathbf{V}T)^2 \, d\tau = \oint\!\!\!\oint_S T\frac{\partial T}{\partial n} \, d\sigma \tag{16}$$

and

$$\oint\!\!\!\oint_S \frac{\partial T}{\partial n} \, d\sigma = 0. \tag{17}$$

Equation (17) is readily interpreted as requiring that the net flow through the closed boundary S must be zero, as must obviously be the case in steady-state flow without sources or sinks inside $\mathfrak{R}$. Thus $\partial T/\partial n$ cannot be specified in a perfectly arbitrary way on S, but its mean value on S must be zero.

Assume now that two solutions T_1 and T_2 exist, both satisfying Laplace's Equation (15) and both taking on the same prescribed values as a point on the boundary S is approached from the interior of $\mathfrak{R}$. Then clearly the difference

---

†If K is a function of position, $K\mathbf{V}^2 T$ must be replaced by $\mathbf{V} \cdot K\mathbf{V}T$ in Equations (11) and (12), and (13) is replaced by $\mathbf{V} \cdot K\mathbf{V}T = s\rho \, \partial T/\partial t$.

$T_2 - T_1$ also satisfies (15), and hence may be substituted for T in (16). But since $T_2 - T_1$ takes on the value *zero* at all points of S, the right-hand side of the resultant equation vanishes, and there follows

$$\iiint_{\mathfrak{R}} [\nabla(T_2 - T_1)]^2 \, d\tau = 0. \tag{18}$$

Now since the integrand is nonnegative, it must itself vanish everywhere in $\mathfrak{R}$, and hence we have

$$\nabla(T_2 - T_1) = \mathbf{0}, \qquad T_2 - T_1 = c = \text{constant}. \tag{19}$$

Finally, since $T_2 - T_1$ vanishes on S, there follows $c = 0$ and consequently $T_2 = T_1$. That is, *there can be only one solution of Laplace's equation valid in $\mathfrak{R}$ and taking on prescribed values on the boundary S.* In a similar way we find that *two solutions of Laplace's equation valid in $\mathfrak{R}$ and having the same specified value of the normal derivative on the boundary S can differ at most by a constant,* noticing that here $T_2 - T_1$ is not necessarily zero on S.

The two problems considered are known respectively as the *Dirichlet* and *Neumann problems.* The solution to the Dirichlet problem, where the function itself is prescribed on the boundary, is *unique,* whereas the solution to the Neumann problem, where the normal derivative is prescribed on the boundary, is determined only to within an additive constant. These statements clearly apply to the solution of any problem governed by Laplace's equation, regardless of any physical interpretation of the unknown function.

Although other types of boundary-value problems involving Laplace's equation may occur, the two discussed here are of most frequent occurrence.

Next we consider the solution of several simple problems of the Dirichlet type. Although, *to fix ideas,* we choose to identify the quantity to be determined with steady-state temperature, the results may be equally well interpreted in terms of many other physical quantities which also satisfy Laplace's equation.

9.3. Steady-State Temperature Distribution in a Rectangular Plate. Suppose that the three edges $x = 0$, $x = l$, and $y = 0$ of a thin rectangular plate are maintained at zero temperature,

$$T(0, y) = 0, \tag{20a}$$

$$T(l, y) = 0, \tag{20b}$$

$$T(x, 0) = 0, \tag{20c}$$

and that the fourth edge $y = d$ is maintained at a temperature distribution $f(x)$,

$$T(x, d) = f(x), \tag{21}$$

until steady-state conditions are realized (Figure 9.1). The temperature distribution throughout the plate is required. Thus we

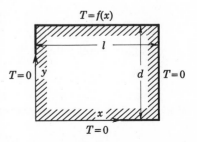

Figure 9.1

must determine that solution of Laplace's equation in two dimensions,

$$\frac{\partial^2 T}{\partial x^2} + \frac{\partial^2 T}{\partial y^2} = 0, \tag{22}$$

which takes on the prescribed boundary values (20) and (21).

The method of *separation of variables* consists of seeking particular "product solutions" of (22) in the form

$$T_p(x, y) = X(x)Y(y), \tag{23}$$

where X is a function of x alone and Y is a function of y alone. If (23) is introduced into (22), there follows

$$\frac{d^2 X}{dx^2} Y + X \frac{d^2 Y}{dy^2} = 0,$$

or, after *separating the variables*,

$$-\frac{1}{X} \frac{d^2 X}{dx^2} = \frac{1}{Y} \frac{d^2 Y}{dy^2}. \tag{24}$$

Since, by hypothesis, the left-hand member of (24) is independent of y and the equivalent right-hand member is independent of x, it follows that both sides must be independent of both x and y, and hence must be equal to a constant. If we call this arbitrary constant k^2, there follows

$$\frac{d^2 X}{dx^2} + k^2 X = 0, \tag{25a}$$

$$\frac{d^2 Y}{dy^2} - k^2 Y = 0. \tag{25b}$$

Thus we see that the product (23) will satisfy (22) if X and Y are solutions of (25a, b), regardless of the value of k. Because of the *linearity* and *homogeneity* of (22), it follows that any linear combination of such solutions, corresponding to different values of k, will also satisfy (22).

We next notice that three of the boundary conditions also are homogeneous. Thus, if *each* of the particular product solutions is required to satisfy (20a, b, c), any linear combination will also satisfy the same conditions. Equations (20a, b) will be satisfied if

$$X(0) = 0, \tag{26a}$$

$$X(l) = 0, \tag{26b}$$

whereas (20c) implies the condition

$$Y(0) = 0. \tag{27}$$

Equations (25a) and (26) constitute a previously considered Sturm–Liouville problem for which the characteristic values are

$$k = k_n = \frac{n\pi}{l} \qquad (n = 1, 2, 3, \dots), \tag{28}$$

and the corresponding solutions (characteristic functions) are of the form

$$X = X_n = A_n \sin \frac{n\pi x}{l}. \qquad (29)$$

Corresponding to (28), the solution of (25b) which satisfies (27) is of the form

$$Y = Y_n = B_n \sinh \frac{n\pi y}{l}. \qquad (30)$$

Thus it follows that any particular solution of the form

$$T_n = a_n \sin \frac{n\pi x}{l} \sinh \frac{n\pi y}{l} \qquad (n = 1, 2, 3, \ldots), \qquad (31)$$

where we have written $a_n = A_n B_n$, satisfies Equation (22) and the three boundary conditions (20a, b, c). The same is true for any series of the form

$$T = \sum_{n=1}^{\infty} a_n \sin \frac{n\pi x}{l} \sinh \frac{n\pi y}{l} \qquad (32)$$

if suitable convergence is assumed. It remains, then, to attempt to determine the coefficients a_n in (32) in such a way that the remaining condition (21) is satisfied, so that

$$f(x) = \sum_{n=1}^{\infty} \left(a_n \sinh \frac{n\pi d}{l} \right) \sin \frac{n\pi x}{l} \qquad (0 < x < l). \qquad (33)$$

But, from the theory of Fourier sine series, the coefficients $a_n \sinh (n\pi d/l)$ in this series must be of the form

$$a_n \sinh \frac{n\pi d}{l} = \frac{2}{l} \int_0^l f(x) \sin \frac{n\pi x}{l} \, dx,$$

and hence, writing $c_n = a_n \sinh (n\pi d/l)$, the required solution (32) takes the form

$$T(x, y) = \sum_{n=1}^{\infty} c_n \sin \frac{n\pi x}{l} \frac{\sinh (n\pi y/l)}{\sinh (n\pi d/l)}, \qquad (34)$$

where

$$c_n = \frac{2}{l} \int_0^l f(x) \sin \frac{n\pi x}{l} \, dx, \qquad (35)$$

assuming appropriate convergence.

If $f(x)$ is, in fact, representable by a convergent Fourier series of the form (33) in $(0, l)$, it can be established that the series (34), subject to (35), truly converges to the solution of the stated problem inside the rectangle of definition. This situation exists, in particular, when $f(x)$ is piecewise differentiable in $(0, l)$.

It is clear that the solution of the more general problem where T is pre-scribed arbitrarily along all four edges can be obtained by superimposing four solutions analogous to the one obtained here, each corresponding to a problem in which zero temperatures are prescribed along three of the four edges, al-though more convenient alternative procedures frequently suggest themselves in such cases. (See Problems 15, 22, and 23.)

The fact of basic importance is that permissible values of the "separation constant" k^2 were determined by the characteristic-value problem arising from the presence of *homogeneous* boundary conditions along the *two* edges $x =$ constant. The additional fact that a homogeneous condition also was imposed along one of the other boundaries afforded a simplification, but was not necessary to the success of the method (see Problem 15).

By virtue of (23) and (25), any expression of the form

$$T_p = (c_1 \cos kx + c_2 \sin kx)(c_3 \cosh ky + c_4 \sinh ky) \qquad (36a)$$

is a particular solution of (22) for arbitrary values of k. It is readily verified that, if the equal members of (24) were set equal to $-k^2$, rather than $+k^2$, the signs in (25a, b) would be reversed and particular solutions of the form

$$T_p = (c_5 \cosh kx + c_6 \sinh kx)(c_7 \cos ky + c_8 \sin ky) \qquad (36b)$$

would be obtained. Finally, if k^2 is replaced by zero in (25a, b), the solutions of the resultant equations lead to the further particular product solutions

$$T_p = (c_9 x + c_{10})(c_{11} y + c_{12}). \qquad (36c)$$

Clearly, exponential forms could be used in (36a, b) in place of the hyperbolic functions, if this procedure were desirable. However, *only the product solutions listed in* (31) *satisfy the homogeneous boundary conditions* (20).

The choice of the sign of the separation constant associated with (24) was motivated by the knowledge that solutions of type (36a) would lead to a Fourier series expansion in the x direction along the edge $y = d$ where the nonhomogeneous condition is prescribed. However, it should be noted that the alternative process of equating the two members of (24) to $-k^2$ (rather than $+k^2$) would not be *incorrect* (if nonreal values of k are admitted) since it would lead to the Sturm–Liouville problem

$$\frac{d^2 X}{dx^2} - k^2 X = 0, \qquad X(0) = X(l) = 0$$

for which the characteristic values of k^2 are *negative* and such that $-k^2 = n^2\pi^2/l^2$. The characteristic values of k then would be imaginary, of the form $k_n = in\pi/l$, but the same *product solutions* (31) would be obtained. The use of $+k^2$ to denote the separation constant here is preferable only because, with $\lambda = k^2$, the Sturm–Liouville problem determining λ then is *proper* (see Section 5.6), and the permissible values of k accordingly then turn out to be real.

9.4. Steady-State Temperature Distribution in a Circular Annulus. Suppose that the temperature distributions along the inner and outer arcs of a circular annulus are maintained as $f_1(\theta)$ and $f_2(\theta)$, respectively,

$$T(r_1, \theta) = f_1(\theta), \qquad (37a)$$

$$T(r_2, \theta) = f_2(\theta), \qquad (37b)$$

until steady-state conditions are realized (Figure 9.2). The temperatures at internal points of the annulus are required.

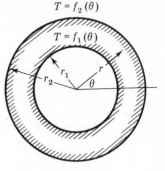

Figure 9.2

If polar coordinates are used, Laplace's equation in the plane becomes (see Section 6.18)

$$\nabla^2 T = \frac{\partial^2 T}{\partial r^2} + \frac{1}{r}\frac{\partial T}{\partial r} + \frac{1}{r^2}\frac{\partial^2 T}{\partial \theta^2} = 0. \qquad (38)$$

Assuming a particular product solution of the form

$$T_p(r, \theta) = R(r)\Theta(\theta), \qquad (39)$$

we find that Equation (38) becomes

$$R''\Theta + \frac{1}{r}R'\Theta + \frac{1}{r^2}R\Theta'' = 0,$$

where a prime denotes differentiation with respect to the argument. By separating the variables, there follows

$$\frac{1}{R}(r^2 R'' + rR') = -\frac{\Theta''}{\Theta} = k^2, \qquad (40)$$

where k^2 is the separation constant. This condition implies the two ordinary equations

$$r^2 R'' + rR' - k^2 R = 0, \qquad (41)$$

$$\Theta'' + k^2\Theta = 0. \qquad (42)$$

The sign of the separation constant was chosen in such a way that, with real values of k, sines and cosines (rather than exponential functions) will be introduced in the θ direction where expansions along the circles $r = r_1$ and $r = r_2$ presumably will be required.

Equation (41) is an equidimensional equation (see Section 1.6) with general solution

$$\begin{aligned} R &= A_k r^k + B_k r^{-k} & (k \neq 0), \\ R &= A_0 + B_0 \log r & (k = 0), \end{aligned} \qquad (43)$$

whereas (42) has the solution

$$\begin{aligned} \Theta &= C_k \cos k\theta + D_k \sin k\theta & (k \neq 0), \\ \Theta &= C_0 + D_0\theta & (k = 0). \end{aligned} \qquad (44)$$

Thus any expression of the form

$$T = a_0 + b_0 \log r + (c_0 + d_0 \log r)\theta$$
$$+ \sum_k [(a_k r^k + b_k r^{-k}) \cos k\theta + (c_k r^k + d_k r^{-k}) \sin k\theta], \qquad (45)$$

where k takes on arbitrary nonzero values, is a solution of Equation (38).

In order that T be *single-valued* in the annulus, we must take

$$c_0 = d_0 = 0. \tag{46}$$

Also, for the same reason, the trigonometric functions must possess a common period 2π. In the absence of boundaries $\theta = $ constant, *this requirement*, in the present case, *serves to determine the permissible values of the separation constant*,

$$k = n \qquad (n = 1, 2, 3, \ldots). \tag{47}$$

Hence we are led to assume the solution of the present problem in the form

$$T = (a_0 + b_0 \log r) + \sum_{n=1}^{\infty} [(a_n r^n + b_n r^{-n}) \cos n\theta$$
$$+ (c_n r^n + d_n r^{-n}) \sin n\theta] \qquad (r_1 \leqq r \leqq r_2). \tag{48}$$

The boundary conditions (37a, b) then take the form

$$f_1(\theta) = (a_0 + b_0 \log r_1) + \sum_{n=1}^{\infty} [(a_n r_1^n + b_n r_1^{-n}) \cos n\theta + (c_n r_1^n + d_n r_1^{-n}) \sin n\theta],$$
$$\tag{49}$$

$$f_2(\theta) = (a_0 + b_0 \log r_2) + \sum_{n=1}^{\infty} [(a_n r_2^n + b_n r_2^{-n}) \cos n\theta + (c_n r_2^n + d_n r_2^{-n}) \sin n\theta],$$

so that, according to the theory of Fourier series, the constants a_0 and b_0 are determined by the equations

$$a_0 + b_0 \log r_1 = \frac{1}{2\pi} \int_0^{2\pi} f_1(\theta)\, d\theta, \tag{50a}$$

$$a_0 + b_0 \log r_2 = \frac{1}{2\pi} \int_0^{2\pi} f_2(\theta)\, d\theta, \tag{50b}$$

the constants a_n and b_n ($n > 0$) by the equations

$$a_n r_1^n + b_n r_1^{-n} = \frac{1}{\pi} \int_0^{2\pi} f_1(\theta) \cos n\theta\, d\theta, \tag{50c}$$

$$a_n r_2^n + b_n r_2^{-n} = \frac{1}{\pi} \int_0^{2\pi} f_2(\theta) \cos n\theta\, d\theta, \tag{50d}$$

and the constants c_n and d_n by the equations

$$c_n r_1^n + d_n r_1^{-n} = \frac{1}{\pi} \int_0^{2\pi} f_1(\theta) \sin n\theta\, d\theta, \tag{50e}$$

$$c_n r_2^n + d_n r_2^{-n} = \frac{1}{\pi} \int_0^{2\pi} f_2(\theta) \sin n\theta\, d\theta. \tag{50f}$$

Two limiting cases are of particular interest. First, in the limiting case $r_1 = 0$, the region considered becomes *the interior of the circle $r = r_2$*. In order that the left-hand members of (50a, c, e) remain finite as $r_1 \to 0$ we must take

$$b_0 = b_n = d_n = 0 \qquad (n = 1, 2, 3, \ldots). \tag{51}$$

If we also write A_n for $a_n r_2^n$, C_n for $c_n r_2^n$, a for r_2, and $f(\theta)$ for $f_2(\theta)$, we deduce that *the solution of the problem*

$$\nabla^2 T = 0 \qquad (r \leq a),$$
$$T(a, \theta) = f(\theta) \tag{52}$$

is of the form

$$T = A_0 + \sum_{n=1}^{\infty} \left(\frac{r}{a}\right)^n (A_n \cos n\theta + C_n \sin n\theta) \qquad (r \leq a), \tag{53}$$

where

$$A_0 = \frac{1}{2\pi} \int_0^{2\pi} f(\theta)\, d\theta,$$

$$A_n = \frac{1}{\pi} \int_0^{2\pi} f(\theta) \cos n\theta\, d\theta \qquad (n = 1, 2, 3, \ldots), \tag{54}$$

$$C_n = \frac{1}{\pi} \int_0^{2\pi} f(\theta) \sin n\theta\, d\theta \qquad (n = 1, 2, 3, \ldots),$$

from (50b, d, f).

In this case $f_1(\theta)$ is merely the constant value, say T_0, of the temperature at the center of the circle $r = a$, so that (50c, e) reduce to the trivial identity $0 = 0$. Also, Equation (50a) becomes

$$A_0 = \frac{1}{2\pi} \int_0^{2\pi} T_0\, d\theta = T_0. \tag{55}$$

Since this result must be compatible with the first equation of (54), which states that A_0 is the mean value of the temperature distribution along the circle $r = a$, we conclude that *the temperature at the center of the circle is the mean of the temperature distribution along the boundary of the circle*. This result also follows directly from (53) when $r = 0$.

In the second limiting case $r_2 \to \infty$, the region considered becomes *the exterior of the circle $r = r_1$*. In this case we must take

$$b_0 = a_n = c_n = 0 \qquad (n = 1, 2, 3, \ldots) \tag{56}$$

in (50b, d, f). Equations (50d, f) then become, in the limit,

$$r_2 \to \infty: \qquad 0 = \int_0^{2\pi} f_2(\theta) \cos n\theta\, d\theta = \int_0^{2\pi} f_2(\theta) \sin n\theta\, d\theta \qquad (n = 1, 2, 3, \ldots), \tag{57}$$

where $f_2(\theta)$ is the prescribed temperature distribution on the circle with increasing radius r_2. These relations can be true only if $f_2(\theta)$ is constant in the limit. Thus the temperature must approach a uniform value, say T_∞, as $r \to \infty$. Equations (50a, b) become

$$A_0 \equiv a_0 = \frac{1}{2\pi} \int_0^{2\pi} f_1(\theta)\, d\theta = T_\infty, \tag{58}$$

showing that T_∞ is the mean of the prescribed temperature distribution on the inner boundary, and Equations (50d, f) reduce to $0 = 0$.

If we also write B_n for $r_1^{-n}b_n$, D_n for $r_1^{-n}d_n$, a for r_1, and $f(\theta)$ for $f_1(\theta)$, we deduce that *the solution of the problem*

$$\nabla^2 T = 0 \qquad (r \geq a),$$
$$T(a, \theta) = f(\theta) \tag{59}$$

is of the form

$$T = A_0 + \sum_{n=1}^{\infty} \left(\frac{a}{r}\right)^n (B_n \cos n\theta + D_n \sin n\theta) \qquad (r \geq a), \tag{60}$$

where

$$A_0 = \frac{1}{2\pi} \int_0^{2\pi} f(\theta)\, d\theta,$$

$$B_n = \frac{1}{\pi} \int_0^{2\pi} f(\theta) \cos n\theta\, d\theta \qquad (n = 1, 2, 3, \ldots), \tag{61}$$

$$D_n = \frac{1}{\pi} \int_0^{2\pi} f(\theta) \sin n\theta\, d\theta \qquad (n = 1, 2, 3, \ldots).$$

9.5. Poisson's Integral. We now show that, if the values given by (54) are introduced into (53), then the resultant series can be summed, and the solution can be expressed as an integral. For this purpose, we first replace the dummy variable θ in (54) by φ, to distinguish this variable from the current variable θ in (53). The introduction of (54) into (53) then leads to the relation

$$T(r, \theta) = \frac{1}{2\pi} \int_0^{2\pi} f(\varphi)\, d\varphi + \frac{1}{\pi} \sum_{n=1}^{\infty} \left(\frac{r}{a}\right)^n \left[\cos n\theta \int_0^{2\pi} f(\varphi) \cos n\varphi\, d\varphi \right.$$

$$\left. + \sin n\theta \int_0^{2\pi} f(\varphi) \sin n\varphi\, d\varphi \right]$$

$$= \frac{1}{\pi} \int_0^{2\pi} f(\varphi) \left[\frac{1}{2} + \sum_{n=1}^{\infty} \left(\frac{r}{a}\right)^n (\cos n\theta \cos n\varphi + \sin n\theta \sin n\varphi)\right] d\varphi,$$

or

$$T(r, \theta) = \frac{1}{\pi} \int_0^{2\pi} f(\varphi) \left[\frac{1}{2} + \sum_{n=1}^{\infty} \left(\frac{r}{a}\right)^n \cos n(\varphi - \theta)\right] d\varphi \qquad (r < a), \tag{62}$$

assuming the legitimacy of the interchange of summation and integration in the first form. But since†

$$\cos n(\varphi - \theta) = \text{Re}\,\{e^{in(\varphi-\theta)}\},$$

---

†"Re" means "real part of" and "Im" means "imaginary part of" in the sense that $f = \text{Re}\,f + i\,\text{Im}\,f$, so that $\text{Im}\,f$ is *real* (see Section 10.1).

there follows

$$\sum_{n=1}^{\infty} \left(\frac{r}{a}\right)^n \cos n(\varphi - \theta) = \mathrm{Re}\left\{\sum_{n=1}^{\infty}\left[\frac{r}{a}e^{i(\varphi-\theta)}\right]^n\right\}$$

$$= \mathrm{Re}\left\{\frac{(r/a)e^{i(\varphi-\theta)}}{1 - (r/a)e^{i(\varphi-\theta)}}\right\} \qquad (r < a)$$

$$= \mathrm{Re}\left\{\frac{(r/a)[e^{i(\varphi-\theta)} - (r/a)]}{[1 - (r/a)e^{i(\varphi-\theta)}][1 - (r/a)e^{-i(\varphi-\theta)}]}\right\}$$

$$= \frac{(r/a)\cos(\varphi-\theta) - (r^2/a^2)}{1 - 2(r/a)\cos(\varphi-\theta) + (r^2/a^2)} \qquad (r < a).$$

Thus the quantity in braces in (62) becomes

$$\frac{1}{2} + \sum_{n=1}^{\infty}\left(\frac{r}{a}\right)^n \cos n(\varphi - \theta) = \frac{1}{2}\frac{a^2 - r^2}{a^2 - 2ar\cos(\varphi-\theta) + r^2} \qquad (r < a), \tag{63}$$

and, noticing that

$$f(\theta) = T(a, \theta),$$

we find that Equation (62) takes the form

$$T(r, \theta) = \frac{1}{2\pi}\int_0^{2\pi} \frac{a^2 - r^2}{a^2 - 2ar\cos(\theta - \varphi) + r^2} T(a, \varphi)\, d\varphi \qquad (r < a). \tag{64}$$

This remarkable integral formula, due to *Poisson*, expresses the value of a harmonic function T at all points in a circular disk of radius a in terms of the values of T on the *boundary* of the disk.

It can also be derived by methods which do not presume the validity of (53), or of the interchange of summation and integration which led to (62), and it holds, in particular, when $T(a, \theta) = f(\theta)$ is piecewise *continuous* in $(0, 2\pi)$. The validity of the *series* representation (53) can be guaranteed only when somewhat more stringent conditions are imposed, such as the requirement that $f(\theta)$ be piecewise *differentiable*. The corresponding integral formula for the *exterior* problem is treated in Problem 35.

9.6. Axisymmetrical Temperature Distribution in a Solid Sphere. On the surface of a solid sphere, let the temperature distribution be prescribed in such a way that it is symmetrical with respect to a diameter. The temperature distribution inside the sphere is to be determined. Using spherical coordinates (Figure 9.3), with the z axis coinciding with the axis of symmetry, and noticing that the temperature T is independent of θ, we recall that Laplace's equation takes the form (see

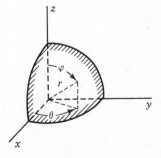

Figure 9.3

Section 6.18)

$$\frac{\partial}{\partial r}\left(r^2 \frac{\partial T}{\partial r}\right) + \frac{1}{\sin \varphi}\frac{\partial}{\partial \varphi}\left(\sin \varphi \frac{\partial T}{\partial \varphi}\right) = 0. \tag{65}$$

The boundary condition on the surface $r = a$ is then

$$T(a, \varphi) = f(\varphi). \tag{66}$$

If a product solution is assumed,

$$T_p(r, \varphi) = R(r)\Phi(\varphi), \tag{67}$$

Equation (65) is separable into the form

$$\frac{1}{R}(r^2 R')' = -\frac{1}{\Phi \sin \varphi}(\Phi' \sin \varphi)' = k^2, \tag{68}$$

where k^2 is a separation constant. This condition is equivalent to the equations

$$(r^2 R')' - k^2 R = r^2 R'' + 2rR' - k^2 R = 0 \tag{69}$$

and

$$\frac{1}{\sin \varphi}\frac{d}{d\varphi}\left(\sin \varphi \frac{d\Phi}{d\varphi}\right) + k^2 \Phi = 0. \tag{70}$$

Equation (70) is brought into accordance with Equation (169), Section 4.12, if we write

$$k^2 = n(n + 1). \tag{71}$$

The general solution of (70) is then of the form

$$\Phi = A_n P_n(\cos \varphi) + B_n Q_n(\cos \varphi), \tag{72}$$

where P_n and Q_n are Legendre functions. In order that the solution be finite on the z axis $\varphi = 0, \pi$, *we must restrict n to integral values* and take

$$B_n = 0. \tag{73}$$

This follows from the fact that the Legendre *polynomials* (for which n is an integer) are the only Legendre functions of type (72) which are finite at both $\varphi = 0$ and $\varphi = \pi$ (see Section 4.12). With the notation of (71), the general solution of (69) is obtained in the form

$$R = C_n r^n + D_n r^{-n-1}. \tag{74}$$

To avoid infinite temperature at the center of the sphere ($r = 0$), we must set

$$D_n = 0. \tag{75}$$

Hence we are led to assume the desired temperature distribution in the form

$$T(r, \varphi) = \sum_{n=0}^{\infty} c_n \left(\frac{r}{a}\right)^n P_n(\cos \varphi) \qquad (0 < \varphi < \pi), \tag{76}$$

where we have written $A_n C_n = c_n/a^n$. The boundary condition (66) requires that the constants c_n be determined so that the representation

$$f(\varphi) = \sum_{n=0}^{\infty} c_n P_n(\cos \varphi) \qquad (0 < \varphi < \pi) \tag{77}$$

is valid. If we introduce the new variable

$$\mu = \cos \varphi, \tag{78}$$

the function $f(\varphi)$ becomes a new function of μ, say $F(\mu)$, such that

$$F(\mu) = f(\cos^{-1} \mu), \tag{79}$$

and (77) becomes

$$F(\mu) = \sum_{n=0}^{\infty} c_n P_n(\mu) \qquad (-1 < \mu < 1). \tag{80}$$

The results of Section 5.14 [Equation (213)] then given the result

$$c_n = \frac{2n+1}{2} \int_{-1}^{1} F(\mu) P_n(\mu) \, d\mu, \tag{81}$$

or alternatively, returning to the original variable φ and using the relation $F(\cos \varphi) = f(\varphi)$,

$$c_n = \frac{2n+1}{2} \int_{0}^{\pi} f(\varphi) P_n(\cos \varphi) \sin \varphi \, d\varphi. \tag{82}$$

If the first n derivatives of $F(\mu)$ with respect to its argument are continuous for $-1 \leq \mu \leq 1$, the second alternative form given by Equation (213), Section 5.13,

$$c_n = \frac{2n+1}{2^{n+1} n!} \int_{-1}^{1} (1 - \mu^2)^n \frac{d^n F(\mu)}{d\mu^n} \, d\mu \tag{83}$$

may in certain cases be more useful for the determination of c_n.

The general Dirichlet problem for a sphere, when axial symmetry is not presumed, is considered in Problem 47. In addition, certain such problems for a *right circular cylinder* are treated in Problems 39, 40, 41, and 43.

9.7. Temperature Distribution in a Rectangular Parallelepiped. Let the temperatures of five faces of a rectangular parallelepiped be maintained at zero,

$$T(0, y, z) = T(l_1, y, z) = T(x, 0, z) = T(x, l_2, z) = T(x, y, 0) = 0 \tag{84}$$

and suppose that the sixth face is maintained at a prescribed temperature distribution

$$T(x, y, d) = f(x, y) \tag{85}$$

until steady-state conditions are attained (Figure 9.4). We investigate the resultant distribution of temperature in the interior.

If we assume a product solution of the relevant equation

$$\frac{\partial^2 T}{\partial x^2} + \frac{\partial^2 T}{\partial y^2} + \frac{\partial^2 T}{\partial z^2} = 0 \tag{86}$$

in the form

$$T_p = X(x)Y(y)Z(z), \tag{87}$$

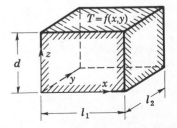

Figure 9.4

the equation may be separated in the form

$$-\frac{X''}{X} = \frac{Y''}{Y} + \frac{Z''}{Z} = k_1^2, \tag{88}$$

the separation here depending upon the fact that the first member is independent of both y and z and the second equal member is independent of x. Hence we must have

$$X'' + k_1^2 X = 0 \tag{89}$$

and, after a second separation,

$$-\frac{Y''}{Y} = \frac{Z''}{Z} - k_1^2 = k_2^2. \tag{90}$$

Thus Y and Z are determined by the equations

$$Y'' + k_2^2 Y = 0, \tag{91}$$

$$Z'' - (k_1^2 + k_2^2)Z = 0. \tag{92}$$

The *homogeneous* boundary conditions (84) are satisfied by the product solution if the factors satisfy the conditions

$$X(0) = X(l_1) = 0, \tag{93}$$

$$Y(0) = Y(l_2) = 0, \tag{94}$$

$$Z(0) = 0. \tag{95}$$

We thus obtain from Equations (89), (91), (93), and (94)

$$k_1 = \frac{m\pi}{l_1} \qquad (m = 1, 2, 3, \ldots), \tag{96}$$

$$X = X_m = A_m \sin \frac{m\pi x}{l_1}, \tag{97}$$

$$k_2 = \frac{n\pi}{l_2} \qquad (n = 1, 2, 3, \ldots), \tag{98}$$

$$Y = Y_n = B_n \sin \frac{n\pi y}{l_2}. \tag{99}$$

If we write further

$$k_1^2 + k_2^2 = \pi^2 \left(\frac{m^2}{l_1^2} + \frac{n^2}{l_2^2}\right) \equiv k_{mn}^2$$

or

$$k_{mn} = \pi \sqrt{\frac{m^2}{l_1^2} + \frac{n^2}{l_2^2}}, \tag{100}$$

the solution of (92) satisfying (95) becomes

$$Z_{mn} = C_{mn} \sinh k_{mn} z. \tag{101}$$

Thus, writing $a_{mn} = A_m B_n C_{mn}$, we are led to assume the desired solution in the form

$$T(x, y, z) = \sum_{m=1}^{\infty} \sum_{n=1}^{\infty} a_{mn} \sin \frac{m\pi x}{l_1} \sin \frac{n\pi y}{l_2} \sinh k_{mn} z. \tag{102}$$

This expression formally satisfies (86), as well as conditions (84), for arbitrary values of the coefficients a_{mn}. It remains, then, to determine these coefficients in such a way that the remaining condition (85) is satisfied. If we introduce the abbreviation

$$c_{mn} = a_{mn} \sinh k_{mn}d, \tag{103}$$

this condition takes the form

$$f(x, y) = \sum_{m=1}^{\infty} \sum_{n=1}^{\infty} c_{mn} \sin \frac{m\pi x}{l_1} \sin \frac{n\pi y}{l_2} \qquad (0 < x < l_1, 0 < y < l_2). \tag{104}$$

Thus the coefficients c_{mn} are the coefficients of the *double Fourier sine-series expansion* of $f(x, y)$ over the indicated rectangle.

These coefficients are readily determined by a simple extension of the methods used in earlier work. If both sides of (104) are multiplied by

$$\sin (p\pi x/l_1) \sin (q\pi y/l_2),$$

where p and q are arbitrary positive integers, and if the results are integrated over the rectangle, there follows

$$\int_0^{l_1} \int_0^{l_2} f(x, y) \sin \frac{p\pi x}{l_1} \sin \frac{q\pi y}{l_2} \, dy \, dx$$

$$= \sum_{m=1}^{\infty} \sum_{n=1}^{\infty} c_{mn} \int_0^{l_1} \int_0^{l_2} \sin \frac{p\pi x}{l_1} \sin \frac{q\pi y}{l_2} \sin \frac{m\pi x}{l_1} \sin \frac{n\pi y}{l_2} \, dy \, dx. \tag{105}$$

The double integral on the right can be written as the product

$$\left(\int_0^{l_1} \sin \frac{p\pi x}{l_1} \sin \frac{m\pi x}{l_1} \, dx \right) \left(\int_0^{l_2} \sin \frac{q\pi y}{l_2} \sin \frac{n\pi y}{l_2} \, dy \right),$$

and hence, by virtue of Equations (140) and (141), Section 5.10, this product *vanishes unless $p = m$ and $q = n$*, in which case it has the value

$$\frac{l_1}{2} \frac{l_2}{2} = \frac{l_1 l_2}{4}.$$

Thus the double series in the right-hand member of (105) reduces to a single term, for which $m = p$ and $n = q$, and there follows

$$c_{mn} = \frac{4}{l_1 l_2} \int_0^{l_1} \int_0^{l_2} f(x, y) \sin \frac{m\pi x}{l_1} \sin \frac{n\pi y}{l_2} \, dy \, dx. \tag{106}$$

With these values of c_{mn}, and with the notation of (103), the solution (102) becomes

$$T(x, y, z) = \sum_{m=1}^{\infty} \sum_{n=1}^{\infty} c_{mn} \sin \frac{m\pi x}{l_1} \sin \frac{n\pi y}{l_2} \frac{\sinh k_{mn}z}{\sinh k_{mn}d}, \tag{107}$$

where k_{mn} is defined by (100).

It is useful to notice that an equivalent approach to the present problem consists of first seeking a product solution of (86) in the form

$$T_p = F(x, y)Z(z), \tag{108}$$

where the z factor is treated in a distinct way since only along a boundary $z =$ constant is a *nonhomogeneous* condition imposed, and deducing the requirement

$$-\frac{\partial^2 F/\partial x^2 + \partial^2 F/\partial y^2}{F} = \frac{Z''}{Z} = k^2, \tag{109}$$

where k^2 is the separation constant. Thus F is to be a nontrivial solution of the two-dimensional characteristic-value problem

$$\frac{\partial^2 F}{\partial x^2} + \frac{\partial^2 F}{\partial y^2} + \lambda F = 0,$$
$$F(0, y) = F(l_1, y) = 0, \tag{110}$$
$$F(x, 0) = F(x, l_2) = 0,$$

involving the *Helmholtz equation*, and Z is to be such that

$$\frac{d^2 Z}{dz^2} - \lambda Z = 0, \tag{111}$$
$$Z(0) = 0.$$

The functions

$$\varphi_{mn}(x, y) = \sin\frac{m\pi x}{l_1} \sin\frac{n\pi y}{l_2}, \tag{112}$$

where m and n are positive integers, accordingly are characteristic functions of the problem (110), in correspondence with the characteristic numbers

$$\lambda_{mn} = k_{mn}^2 = \pi^2\left(\frac{m^2}{l_1^2} + \frac{n^2}{l_2^2}\right). \tag{113}$$

These functions are seen to have the two-dimensional orthogonality property

$$\iint_D \varphi_{mn}(x, y)\varphi_{pq}(x, y)\, dx\, dy = 0 \qquad \text{if } (p, q) \neq (m, n), \tag{114}$$

where D is the rectangular region $(0 \leq x \leq l_1, 0 \leq y \leq l_2)$ associated with the characteristic-value problem (110). (See also Problem 45.)

With $Z_{mn}(z)$ correspondingly of the form (101), superposition again leads to the assumption (102) and to the ensuing determination of the required solution.

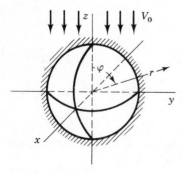

Figure 9.5

9.8. Ideal Fluid Flow about a Sphere. The preceding problems have all been of the Dirichlet type. To illustrate the solution of a Neumann problem, we investigate the effect of the presence of a stationary sphere of radius a in an initially *uniform* flow of an ideal incompressible fluid (Figure 9.5). We introduce spherical coordinates and require that for large values of r the flow be in the negative direction parallel to the z axis ($\varphi = 0, \pi$). The flow then clearly will be independent of the circumferential angle θ. The velocity potential is here denoted by $P(r, \varphi)$. According to the results of Section 6.15, this function satisfies Laplace's equation, which takes the form

$$\frac{\partial}{\partial r}\left(r^2 \frac{\partial P}{\partial r}\right) + \frac{1}{\sin \varphi}\frac{\partial}{\partial \varphi}\left(\sin \varphi \frac{\partial P}{\partial \varphi}\right) = 0. \tag{115}$$

Also, the velocity vector V is determined as the gradient of P,

$$V = \nabla P = u_r \frac{\partial P}{\partial r} + u_\varphi \frac{1}{r}\frac{\partial P}{\partial \varphi} \equiv u_r V_r + u_\varphi V_\varphi. \tag{116}$$

[See Equation (169e) Section 6.18.] At the surface of the sphere ($r = a$) there must be no component of V normal to the sphere; that is, $V_r = \partial P/\partial r$ must vanish,

$$r = a: \qquad \frac{\partial P}{\partial r} = 0. \tag{117}$$

As $r \to \infty$, the velocity vector must approach $-V_0 k$, where V_0 is the undisturbed velocity. Making use of the readily established relation

$$k = u_r \cos \varphi - u_\varphi \sin \varphi, \tag{118}$$

we see that as $r \to \infty$ we must have $V \to -V_0 u_r \cos \varphi + V_0 u_\varphi \sin \varphi$. Referring to (116), it follows that as $r \to \infty$ the radial velocity must satisfy the condition

$$r \to \infty: \qquad \frac{\partial P}{\partial r} \to -V_0 \cos \varphi. \tag{119}$$

According to Section 9.2, the boundary conditions (117) and (119) are sufficient to determine P, except for an irrelevant additive constant.

As was shown in Section 9.6, particular product solutions of (115) may be superimposed to give solutions of the form

$$P = \sum_{n=0}^{\infty} [(A_n r^n + B_n r^{-n-1})P_n(\cos \varphi) + (C_n r^n + D_n r^{-n-1})Q_n(\cos \varphi)] \tag{120}$$

for which also

$$\frac{\partial P}{\partial r} = \sum_{n=0}^{\infty} \{[nA_n r^{n-1} - (n+1)B_n r^{-n-2}]P_n(\cos \varphi)$$
$$+ [nC_n r^{n-1} - (n+1)D_n r^{-n-2}]Q_n(\cos \varphi)\}. \tag{121}$$

To avoid infinite velocities along the axis of symmetry, we must take

$$C_n = D_n = 0, \tag{122}$$

and to avoid infinite velocities as $r \to \infty$, in accordance with (119), we must have

$$A_n = 0 \qquad (n = 2, 3, \dots). \tag{123}$$

Finally, to satisfy (117) we then must have

$$nA_n a^{n-1} - (n+1)B_n a^{-n-2} = 0$$

or

$$B_n = \frac{n}{n+1} a^{2n+1} A_n. \tag{124}$$

Thus, with (122), (123), and (124), Equations (120) and (121) become

$$P = A_0 P_0(\cos \varphi) + A_1 \left(r + \frac{a^3}{2r^2} \right) P_1(\cos \varphi) \tag{125}$$

and

$$\frac{\partial P}{\partial r} = A_1 \left(1 - \frac{a^3}{r^3} \right) P_1(\cos \varphi). \tag{126}$$

If we recall [see Equation (171), Section 4.12] that

$$P_0(\cos \varphi) = 1, \qquad P_1(\cos \varphi) = \cos \varphi,$$

it follows that the remaining condition (119) is satisfied if

$$A_1 = -V_0. \tag{127}$$

Hence the velocity potential is of the form

$$P = -V_0 \left(r + \frac{a^3}{2r^2} \right) \cos \varphi + A_0, \tag{128}$$

where A_0 is an irrelevant constant which may be set equal to zero. The velocity vector is of the form

$$\mathbf{V} = -\left[V_0 \left(1 - \frac{a^3}{r^3} \right) \cos \varphi \right] \mathbf{u}_r + \left[V_0 \left(1 + \frac{a^3}{2r^3} \right) \sin \varphi \right] \mathbf{u}_\varphi, \tag{129}$$

by virtue of (116).

To find the streamlines, we make use of Equation (179), Section 6.19, with

$$h_1 = h_r = 1, \qquad h_2 = h_\varphi = r, \qquad h_3 = h_\theta = r \sin \varphi,$$

in accordance with Equation (169b), Section 6.18. The stream function ψ is then determined from the relation

$$d\psi = -\frac{\partial P}{\partial \varphi} \sin \varphi \, dr + \frac{\partial P}{\partial r} r^2 \sin \varphi \, d\varphi$$

$$= -V_0 \left[\left(r + \frac{a^3}{2r^2} \right) \sin^2 \varphi \, dr + \left(r^2 - \frac{a^3}{r} \right) \sin \varphi \cos \varphi \, d\varphi \right].$$

By integration we then obtain

$$\psi = -\frac{V_0}{2} \left(r^2 - \frac{a^3}{r} \right) \sin^2 \varphi + C, \tag{130}$$

where C is an arbitrary constant. The streamlines are thus the traces of the surfaces $\psi = $ constant, or, equivalently,

$$r^2 \left(1 - \frac{a^3}{r^3} \right) \sin^2 \varphi = \text{constant} \tag{131}$$

in the diametral planes $\theta = $ constant.

A considerably shorter but less direct procedure for solving the problem specified by (115), (117), and (119) would consist of *guessing* initially that the solution will be of the form $P(r, \varphi) = R(r) \cos \varphi$, and of then attempting to determine $R(r)$ in such a way that this expression satisfies (115), and such that $R'(a) = 0$ and $R'(\infty) = -V_0$.

9.9. The Wave Equation. Vibration of a Circular Membrane. If we consider equilibrium of an element $(x, x + dx; y, y + dy)$ of a membrane under a distributed normal load of intensity $f(x, y)$, and denote the deflection in the z direction by $w(x, y)$ and the tension by $T(x, y)$, the differential resultant vertical component of tensile force normal to the xy plane is found to be

$$\frac{\partial}{\partial x}\left(T\frac{\partial w}{\partial x}\right) dx\, dy + \frac{\partial}{\partial y}\left(T\frac{\partial w}{\partial y}\right) dx\, dy,$$

if only small slopes and deflections are considered. Since the distributed load on the element is given by $f\, dx\, dy$, the differential equation satisfied by w is of the form

$$\frac{\partial}{\partial x}\left(T\frac{\partial w}{\partial x}\right) + \frac{\partial}{\partial y}\left(T\frac{\partial w}{\partial y}\right) + f = 0 \tag{132a}$$

or

$$T\left(\frac{\partial^2 w}{\partial x^2} + \frac{\partial^2 w}{\partial y^2}\right) + \left(\frac{\partial T}{\partial x}\frac{\partial w}{\partial x} + \frac{\partial T}{\partial y}\frac{\partial w}{\partial y}\right) + f = 0$$

or, in terms of the invariant operator ∇,

$$T\nabla^2 w + (\nabla T) \cdot (\nabla w) + f = 0. \tag{132b}$$

For small deflections of an initially tightly stretched membrane we may assume that T is uniform in the membrane. In addition, if we consider here only free vibration of the membrane, and hence replace f by the inertia loading $-\rho\, \partial^2 w/\partial t^2$, Equation (132b) reduces to the two-dimensional *wave equation*

$$\nabla^2 w = \frac{1}{c^2}\frac{\partial^2 w}{\partial t^2}, \tag{133}$$

where

$$c^2 = \frac{T}{\rho}. \tag{134}$$

In dealing with vibrations of a *circular* membrane, we introduce polar coordinates and write (133) in the form

$$\frac{\partial^2 w}{\partial r^2} + \frac{1}{r}\frac{\partial w}{\partial r} + \frac{1}{r^2}\frac{\partial^2 w}{\partial \theta^2} = \frac{1}{c^2}\frac{\partial^2 w}{\partial t^2}. \tag{135}$$

If a product solution is assumed in the form

$$w_p = R(r)\Theta(\theta)T(t), \tag{136}$$

the usual process of separation leads to equations determining the factors in the form

$$T'' + \omega^2 T = 0, \tag{137a}$$

$$r^2 R'' + rR' + \left(\frac{\omega^2}{c^2}r^2 - k^2\right)R = 0, \tag{137b}$$

$$\Theta'' + k^2\Theta = 0, \tag{137c}$$

where ω^2/c^2 and k^2 are arbitrary separation constants. The general solution of (137b) is of the form [see Equation (130a), Section 4.10]

$$R = AJ_k\left(\frac{\omega r}{c}\right) + BY_k\left(\frac{\omega r}{c}\right), \tag{138}$$

and, if k and ω are not zero, (137c) and (137a) give

$$\Theta = C \cos k\theta + D \sin k\theta, \tag{139}$$

$$T = E \cos \omega t + F \sin \omega t. \tag{140}$$

In order that the deflection w be single-valued, the functions in (139) must be of period 2π, and hence k must be integral,

$$k = m \qquad (m = 0, 1, 2, \ldots). \tag{141}$$

If $k = m = 0$, (139) does not represent the general solution of (137c). However, the missing solution $D\theta$ is not periodic. Next, to avoid infinite deflections at the center ($r = 0$), we must take

$$B = 0. \tag{142}$$

Finally, since w must vanish on the boundary $r = a$,

$$w(a, \theta, t) = 0, \tag{143}$$

we require that *each* of the product solutions w_p, to be superimposed, vanish when $r = a$. By virtue of (142) and (141), this condition requires that ω be a solution of the equation

$$J_m\left(\frac{\omega a}{c}\right) = 0. \tag{144}$$

We notice that the solution $\omega = 0$, when $m > 0$, would reduce the factor $R(r)$ to zero (identically), and hence it need not be considered.

If the nth positive solution of this equation is denoted by ω_{mn},

$$J_m\left(\frac{\omega_{mn} a}{c}\right) = 0 \qquad (n = 1, 2, \ldots), \tag{145}$$

the superposition of permissible product solutions leads to an expression for w in the form

$$w(r, \theta, t) = \sum_{m=0}^{\infty} \sum_{n=1}^{\infty} J_m\left(\frac{\omega_{mn} r}{c}\right)[(a_{mn} \cos m\theta + b_{mn} \sin m\theta) \cos \omega_{mn} t$$
$$+ (c_{mn} \cos m\theta + d_{mn} \sin m\theta) \sin \omega_{mn} t]. \tag{146}$$

In addition to the *boundary condition* (143), we may be given two *initial conditions* which prescribe the deflection w and the velocity $\partial w/\partial t$ as functions of r and θ at an initial time $t = 0$,

$$w(r, \theta, 0) = f_1(r, \theta), \tag{147a}$$

$$\frac{\partial w(r, \theta, 0)}{\partial t} = f_2(r, \theta). \tag{147b}$$

Thus the constants in (146) must be determined so as to satisfy the relations

$$f_1(r, \theta) = \sum_{m=0}^{\infty} \sum_{n=1}^{\infty} (a_{mn} \cos m\theta + b_{mn} \sin m\theta) J_m\left(\frac{\omega_{mn} r}{c}\right), \tag{148a}$$

$$f_2(r, \theta) = \sum_{m=0}^{\infty} \sum_{n=1}^{\infty} \omega_{mn}(c_{mn} \cos m\theta + d_{mn} \sin m\theta) J_m\left(\frac{\omega_{mn} r}{c}\right), \tag{148b}$$

when $0 \leq r \leq a$ and $0 < \theta < 2\pi$. The determination follows closely the procedure used in Section 9.7, if use is made of Equations (161) of Section 5.11 and of the corresponding results in Section 5.13 (see Problem 51).

For simplicity, we consider further only the special case when f_1 is independent of θ and $f_2 = 0$; that is, we assume that initially the membrane is deflected into a *radially symmetrical form* and is *released from rest*. The ensuing motion of interior points is to be determined. In this case only the terms for which $m = 0$ are independent of θ, so that all other terms may be suppressed. The initial velocities of all points will vanish if the c and d coefficients are set equal to zero. Hence, writing ω_n for ω_{0n}, Equation (146) reduces to the form

$$ w = \sum_{n=1}^{\infty} A_n J_0 \left(\frac{\omega_n r}{c} \right) \cos \omega_n t \qquad (0 \leq r \leq a, t \geq 0), \tag{149} $$

where ω_n is the nth (positive) solution of

$$ J_0 \left(\frac{\omega_n a}{c} \right) = 0 \qquad (n = 1, 2, 3, \ldots). \tag{150} $$

If the membrane is initially deflected to the form

$$ w(r, 0) = F(r), \tag{151} $$

where $F(a) = 0$, the coefficients A_n must be so determined that

$$ F(r) = \sum_{n=1}^{\infty} A_n J_0 \left(\frac{\omega_n r}{c} \right) \qquad (0 \leq r \leq a). \tag{152} $$

Reference to Equations (179) and (185a) of Section 5.13, with $\mu_n = \omega_n/c$, $p = 0$, and $l = a$, then leads to the determination

$$ A_n = \frac{2}{a^2 [J_1(\omega_n a/c)]^2} \int_0^a r F(r) J_0 \left(\frac{\omega_n r}{c} \right) dr. \tag{153} $$

It may be noticed that the motion specified by (149) is a superposition of modes having frequencies $\omega_n/2\pi$, where ω_n satisfies (150), the amplitude of vibration in each such mode varying in the radial direction as a Bessel function. The smallest solution of (150) is given by

$$ \frac{\omega_1 a}{c} = 2.405, $$

to three places (see Table 1 of Section 5.13), so that, for example, the fundamental frequency $f_1 = \omega_1/2\pi$ of a circular drumhead is

$$ f_1 = \frac{2.405}{2\pi} \frac{c}{a} = 0.3828 \sqrt{\frac{T}{\rho a^2}}. $$

Conversely, the tension T required to produce a desired fundamental note is given by

$$ T = 6.825 \rho a^2 f_1^2. $$

9.10. The Heat-Flow Equation. Heat Flow in a Rod.

We next consider the one-dimensional problem of heat flow in a homogeneous rod of length l with insulated sides, the temperature depending only on the distance x from one end of the rod and the time t. The heat-flow equation (13), Section 9.2, then becomes

$$\frac{\partial^2 T}{\partial x^2} = \frac{1}{\alpha^2} \frac{\partial T}{\partial t}, \tag{154}$$

where α is a constant defined by (14). It is apparent from (154) that in the steady state, when $\partial T/\partial t = 0$, T is a *linear* function of the distance x.

Initially, the temperature is prescribed along the rod as a function of x, say

$$T(x, 0) = f(x). \tag{155}$$

In particular, if steady-state conditions exist initially, $f(x)$ must be a linear function, of the form

$$f(x) = T_1^{(0)} + (T_2^{(0)} - T_1^{(0)})\frac{x}{l},$$

where $T_1^{(0)}$ and $T_2^{(0)}$ are the initial temperatures of the ends.

At the instant $t = 0$ we suppose that the temperature at the end $x = 0$ is changed to a new value, say T_1, and the temperature at the end $x = l$ is changed to the value T_2, and these constant values are assumed to be maintained thereafter:

$$T(0, t) = T_1, \qquad T(l, t) = T_2 \qquad (t > 0). \tag{156}$$

The temperature distribution throughout the rod is required as a function of x and t.

In a problem of this sort, lacking homogeneous boundary conditions, it is often convenient to express the desired solution as the sum of *two* expressions, the first of which is taken here to be the limiting steady-state distribution (independent of t) after transient effects have become negligible, and the other of which is to represent the transient distribution (which must then approach zero as t increases indefinitely). Thus, if we write

$$T(x, t) = T_S(x) + T_T(x, t), \tag{157}$$

the function $T_S(x)$ must be a linear function of x satisfying (156), and hence is of the form

$$T_S(x) = T_1 + (T_2 - T_1)\frac{x}{l}, \tag{158}$$

and $T_T(x, t)$ is a particular solution of (154). The function T_T must be determined in such a way that it vanishes when $t \to \infty$,

$$T_T(x, \infty) = 0, \tag{159}$$

and so that the sum $T_S + T_T$ satisfies the initial condition (155). Also, since $T_S(x)$ satisfies (156), it follows that T_T must *vanish* at the ends $x = 0$ and $x = l$ *for all positive values of t*,

$$T_T(0, t) = T_T(l, t) = 0 \qquad (t > 0). \tag{160}$$

Thus the *transient* distribution satisfies *homogeneous* end conditions. *It is for this reason that the steady-state distribution was first separated out.*†

---

†A procedure of the same basic type is used in Problem 22 of Section 9.3, with the first ("homogenizing") part of the solution determined by a quite different method.

Product solutions of (154) satisfying (159) and (160) are readily obtained in the form

$$T_{T,p} = a_n \sin \frac{n\pi x}{l} e^{-n^2\pi^2\alpha^2 t/l^2} \qquad (n = 1, 2, 3, \ldots).$$

Thus, by combining (158) and a superposition of solutions of this type, the required function $T(x, t)$ may be assumed in the form

$$T(x, t) = T_1 + (T_2 - T_1)\frac{x}{l} + \sum_{n=1}^{\infty} a_n \sin \frac{n\pi x}{l} e^{-n^2\pi^2\alpha^2 t/l^2}. \qquad (161)$$

We may verify that (161) satisfies the end conditions (156), and that this solution approaches the proper steady-state solution as $t \to \infty$. It remains, then, to determine the coefficients a_n in such a way that the *initial condition* (155) is satisfied, and hence

$$f(x) - T_1 - (T_2 - T_1)\frac{x}{l} = \sum_{n=1}^{\infty} a_n \sin \frac{n\pi x}{l} \qquad (0 < x < l). \qquad (162)$$

The Fourier coefficients a_n in (162) are determined in the usual way,

$$a_n = \frac{2}{l} \int_0^l f(x) \sin \frac{n\pi x}{l} \, dx + \frac{2}{n\pi}(T_2 \cos n\pi - T_1). \qquad (163)$$

If we write

$$\lambda = \frac{l^2}{\pi^2\alpha^2} = \frac{s\rho l^2}{\pi^2 K}, \qquad (164)$$

the parameter λ has the units of time.† Since the exponential factor in (161) is then of the form $e^{-n^2 t/\lambda}$, we see that λ is closely related to the time required for the transient effect to become negligible.

9.11. Duhamel's Superposition Integral. If we think of the problem treated in Section 9.10 as being defined over a semi-infinite strip $(0 \leq x \leq l, 0 \leq t < \infty)$ in a distance–time plane (Figure 9.6), the non-homogeneous conditions along the lines $x = 0$ and $x = l$ prescribe T as a constant along each of these boundaries. The problem was solved by determining a *particular solution* T_S of the governing differential equation which satisfies those conditions, and then using conventional separation-of-variable methods to determine the "correction" $T_T = T - T_S$, which then is to satisfy *homogeneous* conditions along the two boundaries $x = $ constant. Problems 22, 23, 60, and 61 illustrate the use of this proce-

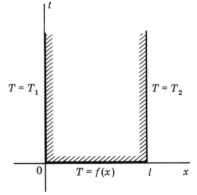

Figure 9.6

---

†In cgs units the relevant quantities are of the following dimensions:

$$[s] = \text{cal}/(\text{g})(^\circ\text{C}), \qquad [\rho] = \text{g}/\text{cm}^3, \qquad [l] = \text{cm},$$
$$[K] = \text{cal}/(\text{cm})(\text{sec})(^\circ\text{C}), \qquad [\lambda] = \text{sec}.$$

dure in other such situations, where a different differential equation may be involved, where a prescribed function need not be constant along a boundary, and where the corresponding particular solution need not be a function of only one of the variables (neither of which need be *time*).

Unless the prescribed functions are of sufficiently simple form, the determination of an appropriate particular solution may present a challenge. In order to illustrate an alternative approach to such situations, when the problem is governed by the one-dimensional heat-flow equation

$$\frac{\partial^2 T}{\partial x^2} = \frac{1}{\alpha^2}\frac{\partial T}{\partial t}, \tag{165}$$

we now consider the case where a rod is initially at zero temperature,

$$T(x, 0) = f(x) \equiv 0, \tag{166}$$

and where when $t > 0$ the end $x = 0$ is kept at zero temperature while the temperature of the end $x = l$ is varied in a prescribed way with time,

$$T(0, t) = 0, \qquad T(l, t) = F(t). \tag{167}$$

As a first step toward the solution of this problem, we solve the problem in the special case when $F(t)$ *is unity*. Denoting this solution by $T = A(x, t)$, we obtain from (162) and (163), with $f(x) = 0$, $T_1 = 0$, $T_2 = 1$, the result

$$A(x, t) = \frac{x}{l} + \frac{2}{\pi}\sum_{n=1}^{\infty}\frac{(-1)^n}{n}\sin\frac{n\pi x}{l}e^{-n^2 t/\lambda} \qquad (t > 0), \tag{168}$$

with the notation of (164).

Next, suppose instead that the temperature at the end $x = l$ is maintained at zero until a certain time $t = \tau_1$, and at that instant is raised to temperature unity and is maintained at unit temperature thereafter. Then it is clear from physical considerations (and possible to show mathematically) that the resulting temperature distribution will be zero everywhere when $t < \tau_1$, and will be given by the result of replacing t by $t - \tau_1$ in (168) when $t > \tau_1$; that is, in this case

$$T = A(x, t - \tau_1) \qquad (t > \tau_1). \tag{169}$$

Here $t - \tau_1$ is time measured from the instant of change.

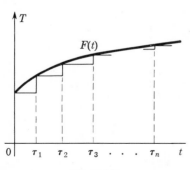

Figure 9.7

Next, suppose instead that the temperature is raised abruptly to the value $F(0)$ when $t = 0$ and held at this value until $t = \tau_1$, then is again abruptly raised by an amount $F(\tau_1) - F(0)$ to the value $F(\tau_1)$ at the time τ_1 and held at that value until $t = \tau_2$, then is abruptly raised by an amount $F(\tau_2) - F(\tau_1)$ at the time τ_2, and so on (Figure 9.7). From the linearity of the problem it is seen that, at the instant following $t = \tau_n$, the temperature distribution corresponding to this *step-function* approximation to $F(t)$ is given by the sum

$$T = F(0)A(x, t) + [F(\tau_1) - F(0)]A(x, t - \tau_1) + [F(\tau_2) - F(\tau_1)]A(x, t - \tau_2)$$
$$+ \cdots + [F(\tau_n) - F(\tau_{n-1})]A(x, t - \tau_n). \quad (170)$$

If we write

$$F(\tau_{k+1}) - F(\tau_k) = \Delta F_k, \qquad \tau_{k+1} - \tau_x = \Delta \tau_k, \quad (171)$$

Equation (170) can be written in the form

$$T = F(0)A(x, t) + \sum_{k=0}^{n-1} A(x, t - \tau_{k+1}) \left(\frac{\Delta F}{\Delta \tau} \right)_k \Delta \tau_k. \quad (172)$$

Finally, as the number n of jumps becomes infinite, in such a way that all jumps and intervals between successive jumps tend to zero, the definition of the integral suggests the limiting form

$$T(x, t) = F(0)A(x, t) + \int_0^t A(x, t - \tau)F'(\tau) \, d\tau, \quad (173)$$

assuming that $F(t)$ is differentiable. This is a version of *Duhamel's principle*, and it gives the desired solution in terms of the basic function $A(x, t)$. An alternative form is obtained by an integration by parts,

$$T(x, t) = F(0)A(x, t) + [A(x, t - \tau)F(\tau)]_{\tau=0}^{\tau=t} - \int_0^t F(\tau) \frac{\partial}{\partial \tau} A(x, t - \tau) \, d\tau,$$

or, since $(\partial/\partial\tau)A(x, t - \tau) = -(\partial/\partial t)A(x, t - \tau)$, we also have

$$T(x, t) = A(x, 0)F(t) + \int_0^t F(\tau) \frac{\partial A(x, t - \tau)}{\partial t} \, d\tau. \quad (174)$$

If we notice further that in the present case $A(x, 0)$ vanishes when $0 \leq x < l$, this result reduces to the form

$$T(x, t) = \int_0^t F(\tau) \frac{\partial A(x, t - \tau)}{\partial t} \, d\tau \qquad (0 \leq x < l). \quad (175)$$

The integral appearing in (173) or (175) is often known as a *superposition integral*.

When (175) is used, the solution of the given problem takes the form

$$T(x, t) = \frac{2}{\pi\lambda} \sum_{n=1}^{\infty} (-1)^{n+1} n \left[\int_0^t F(\tau) e^{n^2\tau/\lambda} \, d\tau \right] e^{-n^2 t/\lambda} \sin \frac{n\pi x}{l}, \quad (176)$$

whereas (173) leads to the equivalent form

$$T(x, t) = \frac{x}{l} F(t) + \frac{2}{\pi} \sum_{n=1}^{\infty} \frac{(-1)^n}{n} \left[F(0) + \int_0^t F'(\tau) e^{n^2\tau/\lambda} \, d\tau \right] e^{-n^2 t/\lambda} \sin \frac{n\pi x}{l}.$$
$$(177)$$

Although the second form appears to be somewhat more complicated than the first, it has the advantage that in many cases the convergence of the infinite series in (177) is more rapid than the convergence of (176).

This follows from the fact that since the term $(x/l)F(t)$ satisfies the prescribed conditions at $x = 0$ and at $x = l$ for all $t > 0$, the series part of (177) represents a function which vanishes when $x = 0$ and when $x = l$ for $t > 0$, and hence it then converges to that function for all x such that $0 \leq x \leq l$ (when F is differentiable). On the other hand, since $T(x, t)$ does *not* vanish when $x = l$ and $t > 0$ unless $F(t) = 0$, the series in (176) does *not* converge to the solution when $x = l$, but only when $0 \leq x < l$. Convergence of that series accordingly will be slow, particularly when x is near l.

Example. Suppose that the temperature at the end $x = l$ is increased uniformly with time from a zero value, at the rate of T_0 degrees per second,

$$F(t) = T_0 t. \tag{178}$$

Then (176) gives

$$T(x, t) = \frac{2\lambda T_0}{\pi} \sum_{n=1}^{\infty} \frac{(-1)^{n+1}}{n^3} \left[\frac{n^2 t}{\lambda} - (1 - e^{-n^2 t/\lambda}) \right] \sin \frac{n\pi x}{l}, \tag{179a}$$

and (177) gives the alternative form

$$T(x, t) = T_0 \frac{x}{l} + \frac{2\lambda T_0}{\pi} \sum_{n=1}^{\infty} (-1)^n \frac{1 - e^{-n^2 t/\lambda}}{n^3} \sin \frac{n\pi x}{l}. \tag{179b}$$

The equivalence of these forms is verified by noticing the validity of the expansion

$$\frac{x}{l} = -\frac{2}{\pi} \sum_{n=1}^{\infty} \frac{(-1)^n}{n} \sin \frac{n\pi x}{l} \qquad (0 \leq x < l).$$

It is clear that the second form (179b) is better adapted to numerical calculation. A third form of the solution is obtained in Problem 61 by a generalization of the method of Section 9.10. ∎

The formulas (173) and (174) continue to apply when the homogeneous conditions $T(x, 0) = 0$ and $T(0, t) = 0$ are replaced by more general homogeneous conditions $\alpha_1 T(x, 0) + \alpha_2 T_t(x, 0) = 0$ and $\beta_1 T(0, t) + \beta_2 T_x(0, t) = 0$, where the α's and β's are constants, with the appropriate modification of $A(x, t)$. Here (175) obviously will continue to apply only if $\alpha_2 = 0$.

However, the same techniques do *not* necessarily apply when the heat-flow equation (165) is replaced by another one. For example, if Laplace's equation is substituted for the heat-flow equation in (165), with t replaced by y, to yield the problem

$$\frac{\partial^2 T}{\partial x^2} + \frac{\partial^2 T}{\partial y^2} = 0,$$

$$T(0, y) = 0, \qquad T(l, y) = F(y), \tag{180}$$

$$T(x, 0) = 0$$

in the strip $(0 \leq x \leq l, \; 0 \leq y < \infty)$, with appropriate restrictions on the behavior of T as $y \to \infty$, and if $T = A(x, y)$ denotes the solution when $F(y) = 1$, then the solution when $F(y) = 0$ if $0 \leq y < \eta_1$ and $F(y) = 1$ if $y > \eta_1$ is *not* given by $T = 0$ if $0 \leq y < \eta_1$ and $T = A(x, y - \eta_1)$ if $y > \eta_1$. Physical (or mathematical) considerations make this fact clear when it is noted that here $T(x, y)$ is the *steady-state* temperature in a long plate, with a portion of one edge permanently maintained at unit temperature and the remainder of the boundary held at zero temperature, and that the temperature inside the plate then certainly will be *positive everywhere*.

An alternative method of dealing with the problem (180) is presented in Section 9.14, and another technique which also can be used to cope with non-homogeneous conditions in more general situations is described in Section 9.17. The formulas (173) and (175) are rederived, by methods to be treated in Sections 9.15 and 9.16, in Problem 101 for the heat-flow problem (165) to (167) and in Problem 102 for a similar problem governed by the *wave* equation.

9.12. Traveling Waves. It was found in Section 8.5 that the general solution of the one-dimensional wave equation

$$\frac{\partial^2 \varphi}{\partial x^2} = \frac{1}{c^2}\frac{\partial^2 \varphi}{\partial t^2} \tag{181}$$

is of the form

$$\varphi = f(x - ct) + g(x + ct), \tag{182}$$

where f and g are arbitrary functions. If we consider the graphical representation of φ as a function of x for varying values of the time t, we see that a solution of the form $\varphi = f(x - ct)$ is represented at the time $t = 0$ by $\varphi = f(x)$, and is represented at any following time t by the same curve moved parallel to itself a distance ct in the positive x direction. That is, *a solution $\varphi = f(x - ct)$ is represented by a curve moving in the positive x direction with velocity c.* Similarly, *a solution $\varphi = g(x + ct)$ is represented by a curve moving in the negative x direction with velocity c.* We may speak of these solutions, in a general sense, as *traveling waves.*

D'Alembert's formula (67), of Section 8.5, in the form

$$\varphi(x, t) = \frac{1}{2}[f(x + ct) + f(x - ct)] + \frac{1}{2c}\int_{x-ct}^{x+ct} g(\xi)\, d\xi, \tag{183}$$

defines that solution of (181) which satisfies the so-called *Cauchy conditions*

$$\varphi(x, 0) = f(x), \qquad \varphi_t(x, 0) = g(x) \tag{184}$$

for all real values of x, where f and g are any twice-differentiable functions. Here, for example, we may interpret φ and φ_t as the lateral deflection and velocity, respectively, of a tightly stretched uniform string of infinite length with $c = \sqrt{T/\rho}$, where T is tension and ρ is linear mass density, and where small slopes and deflections are assumed. (The case of a string of finite length is treated in Problem 33 of Chapter 8 and in Problem 72 at the end of this chapter.)

In many other applications we are principally interested in solutions of the wave equation which are *periodic in time*. Such solutions, for (181), must be combinations of terms of the form $f_1(x) \cos \omega t$ or $f_2(x) \sin \omega t$ or, equivalently, must be the real or imaginary part of a function of the form

$$\varphi = F(x)\,e^{i\omega t}, \tag{185}$$

where $F(x)$ may be complex. By introducing (185) into (181) and canceling the resultant common factor $e^{i\omega t}$, there follows

$$F'' + \frac{\omega^2}{c^2}F = 0. \tag{186}$$

If we write the general solution of (186) in the complex form

$$F = c_1 e^{i\omega x/c} + c_2 e^{-i\omega x/c}, \tag{187}$$

the required solution (185) becomes

$$\varphi = c_1 e^{i\omega(x+ct)/c} + c_2 e^{-i\omega(x-ct)/c}. \tag{188}$$

This solution is of the form of (182). By taking real and imaginary parts, we see that linear combinations of terms of the form

$$\cos\frac{\omega}{c}(x-ct), \qquad \sin\frac{\omega}{c}(x-ct) \tag{189}$$

for arbitrary values of ω are available for the representation of periodic plane waves moving in the *positive* x direction, whereas combinations of terms of the form

$$\cos\frac{\omega}{c}(x+ct), \qquad \sin\frac{\omega}{c}(x+ct) \tag{190}$$

are solutions of (181) representing periodic plane waves moving in the *negative* x direction. In certain problems it may be more convenient to retain such solutions in the complex exponential form of (188).

We next investigate the existence of analogous *spherical wave* solutions of the wave equation. In spherical coordinates, with the solution φ dependent only on the radius r from the origin, the wave equation becomes

$$\frac{1}{r^2}\frac{\partial}{\partial r}\left(r^2\frac{\partial\varphi}{\partial r}\right) = \frac{1}{c^2}\frac{\partial^2\varphi}{\partial t^2} \tag{191}$$

and the general solution can be obtained in the form

$$\varphi = \frac{1}{r}[f(r-ct) + g(r+ct)]. \tag{192}$$

(See Problem 70.) If we assume a solution periodic in time, of the form

$$\varphi = F(r)e^{i\omega t}, \tag{193}$$

substitution of (193) into (191) gives the equation

$$rF'' + 2F' + \frac{\omega^2}{c^2}rF = 0 \tag{194}$$

to be satisfied by the amplitude function F. The general solution of (194) can be expressed in terms of Bessel functions by identifying (194) with Equation (127), Section 4.10. The solution is then given by Equations (128) and (129) of that section in the form

$$F = \frac{A}{\sqrt{r}} J_{1/2}\left(\frac{\omega r}{c}\right) + \frac{B}{\sqrt{r}} J_{-1/2}\left(\frac{\omega r}{c}\right). \tag{195}$$

This result is in turn expressible in terms of elementary functions, by virtue of Equation (115), Section 4.9, in the form

$$F = \sqrt{\frac{2c}{\pi\omega}}\left[A \frac{\sin(\omega r/c)}{r} + B \frac{\cos(\omega r/c)}{r}\right]$$

or, finally, in the complex form

$$F = c_1 \frac{e^{i\omega r/c}}{r} + c_2 \frac{e^{-i\omega r/c}}{r}. \tag{196}$$

Thus Equation (193) becomes

$$\varphi = c_1 \frac{e^{i\omega(r+ct)/c}}{r} + c_2 \frac{e^{-i\omega(r-ct)/c}}{r}. \tag{197}$$

Linear combinations of such functions, or of their real and imaginary parts, are thus solutions of the wave equation which represent periodic spherical waves moving inward toward or outward from the origin, respectively. It may be noticed that the amplitude of the oscillation is inversely proportional to r.

In cylindrical coordinates, with φ dependent only on the distance r from the z axis, the wave equation becomes

$$\frac{1}{r}\frac{\partial}{\partial r}\left(r\frac{\partial\varphi}{\partial r}\right) = \frac{1}{c^2}\frac{\partial^2\varphi}{\partial t^2}. \tag{198}$$

The assumption

$$\varphi = F(r)e^{i\omega t} \tag{199}$$

then leads to the equation

$$F'' + \frac{1}{r}F' + \frac{\omega^2}{c^2}F = 0. \tag{200}$$

Although the general solution can be written in the form

$$F = AJ_0\left(\frac{\omega r}{c}\right) + BY_0\left(\frac{\omega r}{c}\right),$$

we are guided by the preceding developments in writing this solution instead in the *complex* form

$$F = c_1 H_0^{(1)}\left(\frac{\omega r}{c}\right) + c_2 H_0^{(2)}\left(\frac{\omega r}{c}\right), \tag{201}$$

where, as defined in Section 4.8, the Hankel functions are given by

$$H_0^{(1)}\left(\frac{\omega r}{c}\right) = J_0\left(\frac{\omega r}{c}\right) + iY_0\left(\frac{\omega r}{c}\right),$$

$$H_0^{(2)}\left(\frac{\omega r}{c}\right) = J_0\left(\frac{\omega r}{c}\right) - iY_0\left(\frac{\omega r}{c}\right). \tag{202}$$

The solution (199) then becomes

$$\varphi = c_1 e^{i\omega t} H_0^{(1)}\left(\frac{\omega r}{c}\right) + c_2 e^{i\omega t} H_0^{(2)}\left(\frac{\omega r}{c}\right). \tag{203}$$

The usefulness of the definition of the Hankel functions is now seen if we consider the asymptotic expressions for $H_0^{(1)}$ and $H_0^{(2)}$ as given by Equation (106), Section 4.9. By making use of these results, the behavior of the expression (203) for large values of r is found to be given by the asymptotic expression

$$\varphi \sim \sqrt{\frac{2c}{\pi\omega}}\left[(c_1 e^{-i\pi/4})\frac{e^{i\omega(r+ct)/c}}{\sqrt{r}} + (c_2 e^{i\pi/4})\frac{e^{-i\omega(r-ct)/c}}{\sqrt{r}}\right] \quad (r \to \infty). \tag{204}$$

Hence we see that *for large values of r* the real and imaginary parts of $e^{i\omega t} H_0^{(1)}(\omega r/c)$ represent cylindrical waves moving *inward* toward the z axis, whereas the real and imaginary parts of $e^{i\omega t} H_0^{(2)}(\omega r/c)$ represent *outward*-traveling waves. The amplitude of the oscillation here is inversely proportional to $\sqrt{r}$.

More generally, reference to the asymptotic expressions for $H_p^{(1)}$ and $H_p^{(2)}$ shows that the real and imaginary parts of $e^{i\omega t} H_p^{(1)}(kr)$ represent *inward* traveling cylindrical waves for large values of r (for any nonnegative value of p), whereas the real and imaginary parts of $e^{i\omega t} H_p^{(2)}(kr)$ represent *outward* traveling waves.

Although in most problems these complex forms are preferable, the explicit real and imaginary parts may be listed as follows:

$$e^{i\omega t} H_p^{(1)}\left(\frac{\omega r}{c}\right) = \left[J_p\left(\frac{\omega r}{c}\right)\cos \omega t - Y_p\left(\frac{\omega r}{c}\right)\sin \omega t\right]$$
$$+ i\left[J_p\left(\frac{\omega r}{c}\right)\sin \omega t + Y_p\left(\frac{\omega r}{c}\right)\cos \omega t\right], \tag{205a}$$

$$e^{i\omega t} H_p^{(2)}\left(\frac{\omega r}{c}\right) = \left[J_p\left(\frac{\omega r}{c}\right)\cos \omega t + Y_p\left(\frac{\omega r}{c}\right)\sin \omega t\right]$$
$$+ i\left[J_p\left(\frac{\omega r}{c}\right)\sin \omega t - Y_p\left(\frac{\omega r}{c}\right)\cos \omega t\right]. \tag{205b}$$

The preceding special results are useful principally in dealing with problems, governed by the appropriate form of the wave equation, in which the supplementary conditions are such that the solution may be expected to have a time factor of the form $e^{i\omega t}$, where ω is a known real constant, or to be expressible as a linear combination of terms having such time factors. Their use in such cases is particularly effective when a boundary condition (perhaps "at infinity") involves the specification of inward- or outward-traveling waves, as is illustrated in the following section.

9.13. The Pulsating Cylinder. As a simple example of the usefulness of the results obtained in Section 9.12, we consider a problem which is related to certain aerodynamic investigations of nonuniform motion. The lateral boundary of an infinite circular cylinder surrounded by ideal compressible fluid is caused to pulsate radially with circular frequency ω in such a way that the radial velocity of the boundary varies periodically about a zero value, with maximum value V_0 (Figure 9.8) The periodically varying velocities imparted to the points in the surrounding fluid are to be determined.

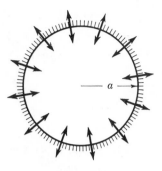

According to hydrodynamic theory [see Equations (204) and (205), Section 6.20], the velocity vector **V** is here the gradient of a velocity potential φ which satisfies the wave equation

$$\nabla^2 \varphi = \frac{1}{c^2} \frac{\partial^2 \varphi}{\partial t^2}, \tag{206}$$

Figure 9.8

where c is the velocity of sound in the fluid. If we introduce cylindrical coordinates, noticing that φ is independent of z and θ, Equation (206) becomes

$$\frac{\partial^2 \varphi}{\partial r^2} + \frac{1}{r} \frac{\partial \varphi}{\partial r} = \frac{1}{c^2} \frac{\partial^2 \varphi}{\partial t^2}. \tag{207}$$

The magnitude V of the (radial) velocity is given by the equation

$$V = \frac{\partial \varphi}{\partial r}. \tag{208}$$

On the boundary the fluid velocity must equal the velocity of the pulsating boundary, say $V_0 \cos \omega t$, whereas for large values of r the velocity must correspond to outward-traveling cylindrical waves (in the absence of a reflecting outer boundary).†

As will be seen, it is convenient to consider φ as the *real part* of a complex function φ_c, and to seek to determine φ_c, since then we are to require that $\partial \varphi_c/\partial r$ reduce to $V_0 e^{i\omega t}$ when $r = a$. Thus φ_c is to satisfy the wave equation and the boundary condition

$$r = a: \qquad \frac{\partial \varphi_c}{\partial r} = V_0 e^{i\omega t}. \tag{209}$$

Accordingly, we are led to write

$$\varphi_c = F(r) e^{i\omega t} \tag{210}$$

and Equation (203) of Section 9.12 shows that a suitable expression for φ_c then is of the form

$$\varphi_c = c_1 e^{i\omega t} H_0^{(1)}\left(\frac{\omega r}{c}\right) + c_2 e^{i\omega t} H_0^{(2)}\left(\frac{\omega r}{c}\right). \tag{211}$$

If we make use of the derivative formula (114), Section 4.9, there then follows

$$\frac{\partial \varphi_c}{\partial r} = -\frac{\omega}{c}\left[c_1 e^{i\omega t} H_1^{(1)}\left(\frac{\omega r}{c}\right) + c_2 e^{i\omega t} H_1^{(2)}\left(\frac{\omega r}{c}\right)\right]. \tag{212}$$

---

†The requirement that the solution of a problem involve only *outward*-traveling waves is sometimes called the Sommerfeld *radiation condition*.

To ensure *outward*-traveling waves for large values of r, we must take

$$c_1 = 0. \tag{213}$$

The remaining constant c_2 is determined by (209) in the form

$$V_0 e^{i\omega t} = -\frac{\omega}{c} c_2 e^{i\omega t} H_1^{(2)}\left(\frac{\omega a}{c}\right),$$

$$c_2 = -\frac{c}{\omega} \frac{V_0}{H_1^{(2)}(\omega a/c)}. \tag{214}$$

With (213) and (214), the required velocity thus is quickly determined in the remarkably compact form

$$V = \operatorname{Re}\left\{\frac{\partial \varphi_c}{\partial r}\right\} = \operatorname{Re}\left\{V_0 e^{i\omega t} \frac{H_1^{(2)}(\omega r/c)}{H_1^{(2)}(\omega a/c)}\right\}, \tag{215}$$

and only elementary algebraic computations remain. For this purpose, we write (215) in the expanded form

$$V = V_0 \operatorname{Re}\left\{(\cos \omega t + i \sin \omega t)\frac{J_1(\omega r/c) - i Y_1(\omega r/c)}{J_1(\omega a/c) - i Y_1(\omega a/c)}\right\}$$

and the usual process of rationalization and separation of real and imaginary parts leads to the explicit form

$$V = \frac{V_0}{J_1^2(\omega a/c) + Y_1^2(\omega a/c)}\left\{J_1\left(\frac{\omega a}{c}\right)\left[J_1\left(\frac{\omega r}{c}\right)\cos \omega t + Y_1\left(\frac{\omega r}{c}\right)\sin \omega t\right]\right.$$
$$\left. - Y_1\left(\frac{\omega a}{c}\right)\left[J_1\left(\frac{\omega r}{c}\right)\sin \omega t - Y_1\left(\frac{\omega r}{c}\right)\cos \omega t\right]\right\}. \tag{216}$$

The amplitude of the velocity oscillation at a given distance r from the axis is given by

$$A(r) = V_0\left[\frac{J_1^2(\omega r/c) + Y_1^2(\omega r/c)}{J_1^2(\omega a/c) + Y_1^2(\omega a/c)}\right]^{1/2}. \tag{217}$$

With this notation, Equation (216) can be written in the form

$$V = A(r)\cos[\omega t - \alpha(r)], \tag{218}$$

where the phase shift α at distance r from the axis is given by

$$\alpha(r) = \tan^{-1}\left[\frac{J_1(\omega a/c)Y_1(\omega r/c) - Y_1(\omega a/c)J_1(\omega r/c)}{J_1(\omega a/c)J_1(\omega r/c) + Y_1(\omega a/c)Y_1(\omega r/c)}\right]. \tag{219}$$

For values of r and a large with respect to c/ω there follows, in particular,

$$A(r) \sim V_0\sqrt{\frac{a}{r}}, \qquad \alpha(r) \sim \frac{\omega}{c}(r - a),$$

if use is made of Equation (105), Section 4.9. Thus, as $r \to \infty$, (218) becomes

$$V \sim V_0\sqrt{\frac{a}{r}}\cos\left[\omega\left(t - \frac{r-a}{c}\right)\right]$$

if $a\omega \gg c$.

We may notice that if the boundary condition had been left in the form $\partial\varphi(a, t)/\partial r = V_0 \cos \omega t$, the assumed solution $\varphi(r, t)$ necessarily would have involved *both* cos ωt *and* sin ωt, and also the imposition of the "outward-wave" condition as $r \rightarrow \infty$ would have been less simply effected.

9.14. Examples of the Use of Fourier Integrals. When the formulation of a problem involves a nonhomogeneous condition specified along a boundary consisting of a line segment of infinite length, and use is made of the method of separation of variables, the superposition of particular solutions generally is accomplished by integration rather than by summation.

As a first illustration, we require the solution of the two-dimensional Laplace equation

$$\frac{\partial^2\varphi}{\partial x^2} + \frac{\partial^2\varphi}{\partial y^2} = 0 \tag{220}$$

valid in the half-plane $y > 0$, taking on prescribed values along the x axis for all values of x,

$$\varphi(x, 0) = f(x), \tag{221}$$

and vanishing at large distance from the origin,

$$\lim_{x^2+y^2\to\infty} \varphi(x, y) = 0, \tag{222}$$

in that half-plane.

Physically, we may interpret this problem (for example) as determining the steady-state temperature distribution in a thin plane sheet of infinite extent on one side of a straight edge, when the temperature is prescribed along that edge.

Product solutions of (220) which vanish as $y \rightarrow \infty$ are readily obtained in the form

$$\varphi_p = e^{-uy}(A \cos ux + B \sin ux) \qquad (u > 0), \tag{223}$$

where u^2 has been written for the arbitrary separation constant and A and B are arbitrary. If we think of A and B as functions of u,

$$A = A(u), \qquad B = B(u), \tag{224}$$

we may superimpose solutions of form (223) for all positive values of u by integration, and so write

$$\varphi(x, y) = \int_0^\infty e^{-uy}[A(u) \cos ux + B(u) \sin ux] \, du. \tag{225}$$

This expression formally satisfies (220) and (222) for arbitrary forms of the functions A and B. The remaining condition (221) will then be satisfied if we determine A and B such that

$$f(x) = \int_0^\infty [A(u) \cos ux + B(u) \sin ux] \, du \qquad (-\infty < x < \infty). \tag{226}$$

But, in accordance with Equations (235) and (236) of Section 5.15, the right-hand member of (226) reduces to the Fourier integral representation of $f(x)$

if we take

$$A(u) = \frac{1}{\pi} \int_{-\infty}^{\infty} f(\xi) \cos u\xi \, d\xi, \tag{227a}$$

$$B(u) = \frac{1}{\pi} \int_{-\infty}^{\infty} f(\xi) \sin u\xi \, d\xi. \tag{227b}$$

With these results, Equation (225) takes the form

$$\varphi(x, y) = \frac{1}{\pi} \int_0^{\infty} e^{-uy} \left\{ \left[\int_{-\infty}^{\infty} f(\xi) \cos u\xi \, d\xi \right] \cos ux \right. $$
$$\left. + \left[\int_{-\infty}^{\infty} f(\xi) \sin u\xi \, d\xi \right] \sin ux \right\} du \tag{228}$$

or, equivalently,

$$\varphi(x, y) = \frac{1}{\pi} \int_0^{\infty} \left[\int_{-\infty}^{\infty} e^{-uy} f(\xi) \cos u(\xi - x) \, d\xi \right] du. \tag{229}$$

We remark that, when the integral $\int_{-\infty}^{\infty} |f(x)| \, dx$ does not exist, the Fourier integral representation (226) usually is not valid. Still, in certain cases when (226) is not valid, if the u integration is carried out before the ξ integration, Equation (229) may nevertheless lead to a specific expression for φ, which, however, may not satisfy the condition (222). The validity of the *formal* solution so obtained should be checked by direct calculation in such cases. Physical considerations will generally indicate whether violation of the condition (222) is permissible in exceptional cases.

When the order of integration is reversed, the solution (229) takes the form

$$\varphi(x, y) = \frac{1}{\pi} \int_{-\infty}^{\infty} \left[\int_0^{\infty} e^{-uy} \cos u(\xi - x) \, du \right] f(\xi) \, d\xi$$

or, after an evaluation of the inner integral,

$$\varphi(x, y) = \frac{1}{\pi} \int_{-\infty}^{\infty} \frac{yf(\xi) \, d\xi}{y^2 + (\xi - x)^2}. \tag{230}$$

It can be proved that this integral converges and defines a solution of Laplace's equation for $y > 0$ when $f(x)$ is bounded and piecewise continuous for all real x.

In the special case when

$$f(x) = \begin{cases} 0 & (x < 0), \\ 1 & (x > 0), \end{cases} \tag{231}$$

there follows, from (230),

$$\varphi(x, y) = \frac{1}{\pi} \int_0^{\infty} \frac{y}{y^2 + (\xi - x)^2} d\xi$$
$$= \frac{1}{\pi} \left(\frac{\pi}{2} + \tan^{-1} \frac{x}{y} \right)$$

or

$$\varphi(x, y) = 1 - \frac{\theta}{\pi}, \tag{232}$$

where θ is the polar angle. In this case $\int_{-\infty}^{\infty} |f(x)|\, dx$ does not exist and, further, *the right-hand members of* (227) *do not exist!* However, (232) is easily shown to satisfy (220) and to reduce to (231) when $y = 0$. Since (222) is violated by the prescribed value of φ on the x axis ($\theta = 0$), it is not surprising that it is also violated by the solution (232) for all values of θ except $\theta = \pi$. Here (232) is the only solution of (220) which reduces to (231) on the x axis and is *bounded* as $y \longrightarrow \infty$.†

Other one-dimensional problems of similar type are solved in an analogous way. As a second example, we consider the problem of one-dimensional heat flow in a rod of infinite length with insulated sides (see Section 9.10), and so deal with the differential equation

$$\frac{\partial^2 T}{\partial x^2} = \frac{1}{\alpha^2} \frac{\partial T}{\partial t}. \tag{233}$$

We assume that an initial distribution of temperature is prescribed along the rod,

$$T(x, 0) = f(x), \tag{234}$$

and that at the instant $t = 0$ the temperature at the end $x = 0$ is changed to zero and maintained at zero thereafter,

$$T(0, t) = 0 \qquad (t > 0). \tag{235}$$

The temperature at a point x at time t is required.

Bounded product solutions of (233) satisfying (235) are of the form

$$T_p = Ae^{-u^2\alpha^2 t} \sin ux,$$

where u is arbitrary. Considering A as a function of u, we may superimpose such solutions for all positive values of u by writing

$$T(x, t) = \int_0^\infty A(u) e^{-u^2\alpha^2 t} \sin ux\, du \qquad (x > 0). \tag{236}$$

The condition (234) then requires that $A(u)$ be determined so that

$$T(x, 0) = f(x) = \int_0^\infty A(u) \sin ux\, du \qquad (x > 0). \tag{237}$$

But, according to Equation (232) of Section 5.14, we have the Fourier sine-integral representation

$$f(x) = \frac{2}{\pi} \int_0^\infty \left[\int_0^\infty f(\xi) \sin u\xi\, d\xi \right] \sin ux\, du \qquad (x > 0),$$

---

†As this example illustrates, the result of a formal manipulation may be correct even in cases when the steps in that manipulation cannot be justified (or are clearly incorrect). If the validity of the *final result* can be verified, there is no need to seek an alternative derivation (except perhaps on esthetic grounds). Accordingly, in a case such as the present one, if the solution were required only when $f(x)$ is defined by (231), it would be appropriate to replace that specified function by a literal one, to solve the resultant problem formally, and finally to replace the literal function $f(x)$ by the specific function in question only after certain questionable steps have been taken, *provided that the validity of the final result is verified directly.*

if $f(x)$ is sufficiently well behaved, and hence the right-hand member of (237) then reduces to $f(x)$ if we take

$$A(u) = \frac{2}{\pi} \int_0^\infty f(\xi) \sin u\xi \, d\xi. \tag{238}$$

Thus the solution (236) becomes

$$T(x, t) = \frac{2}{\pi} \int_0^\infty \int_0^\infty e^{-u^2\alpha^2 t} f(\xi) \sin u\xi \sin ux \, d\xi \, du. \tag{239}$$

This form can be simplified if we carry out the u integration first. Thus we obtain

$$T(x, t) = \frac{2}{\pi} \int_0^\infty f(\xi) \left(\int_0^\infty e^{-u^2\alpha^2 t} \sin u\xi \sin ux \, du \right) d\xi. \tag{240}$$

The inner integral is first simplified by writing

$$\sin u\xi \sin ux = \tfrac{1}{2}[\cos u(\xi - x) -- \cos u(\xi + x)].$$

Then if use is made of the known result†

$$\int_0^\infty e^{-a^2 x^2} \cos bx \, dx = \frac{\sqrt{\pi}}{2a} e^{-b^2/4a^2} \qquad (a > 0), \tag{241}$$

Equation (240) is reduced to the form

$$T(x, t) = \frac{1}{2\alpha\sqrt{\pi t}} \int_0^\infty f(\xi)[e^{-(\xi-x)^2/4\alpha^2 t} - e^{-(\xi+x)^2/4\alpha^2 t}] \, d\xi, \tag{242}$$

which can be shown to be a valid solution for any bounded and piecewise continuous f, in spite of the fact that (236), with (238), is valid only under much stronger restrictions on f.

When the initial temperature is uniform,

$$T(x, 0) = f(x) = T_0, \tag{243}$$

Equation (242) leads, after obvious substitutions, to the solution

$$T(x, t) = \frac{T_0}{\sqrt{\pi}} \left(\int_{-x/2\alpha\sqrt{t}}^\infty e^{-u^2} \, du - \int_{x/2\alpha\sqrt{t}}^\infty e^{-u^2} \, du \right)$$

$$= \frac{T_0}{\sqrt{\pi}} \int_{-x/2\alpha\sqrt{t}}^{x/2\alpha\sqrt{t}} e^{-u^2} \, du = \frac{2T_0}{\sqrt{\pi}} \int_0^{x/2\alpha\sqrt{t}} e^{-u^2} \, du.$$

This result takes the form

$$T(x, t) = T_0 \operatorname{erf} \frac{x}{2\alpha\sqrt{t}} \tag{244}$$

in terms of the so-called *error function*

$$\operatorname{erf} x = \frac{2}{\sqrt{\pi}} \int_0^x e^{-u^2} \, du, \tag{245}$$

†See Equation (101) of Chapter 7.

which is of frequent occurrence in the study of heat flow and in other fields, and which is a tabulated function. The multiplicative constant in (245) is so chosen that

$$\operatorname{erf}(\infty) = 1. \tag{246}$$

If we integrate both sides of (241) with respect to the parameter b, from $b = 0$ to $b = \beta$, we obtain for future reference the relation

$$\int_0^\infty e^{-a^2 x^2} \frac{\sin \beta x}{x} \, dx = \frac{\sqrt{\pi}}{2a} \int_0^\beta e^{-b^2/4a^2} \, db$$

$$= \frac{\pi}{2} \operatorname{erf} \frac{\beta}{2a}, \tag{247}$$

which is frequently useful in dealing with problems which involve the error function. It should be noted that the result of omitting the factor $2/\sqrt{\pi}$ in (245) is often denoted by Erf x, and that other usages (some of which are inconsistent with the present one) sometimes appear in the literature.

The preceding problems can be solved alternatively by use of Fourier (or Fourier sine) *transforms* (see Problems 79 to 81), which have the additional useful property that they are also applicable to either problem when a "forcing term" h is inserted in the right-hand member of the relevant differential equation, and to the second problem when the homogeneous boundary condition $T(0, t) = 0$ is replaced by a nonhomogeneous one, such as $T(0, t) = F(t)$.

9.15. Laplace Transform Methods. When a problem governed by a partial differential equation with constant coefficients is specified for all positive values of the time t, with the initial condition or conditions prescribed when $t = 0$ and with no conditions prescribed for any *other* fixed value of t, the use of the Laplace transform may be convenient. (Clearly, t may be replaced by a different variable, which may or may not represent *time*.)

Here we may note, for example, that the Laplace transform $\bar{f}(x, s)$ of a function $f(x, t)$ is defined as

$$\bar{f}(x, s) = \int_0^\infty e^{-st} f(x, t) \, dt, \tag{248}$$

where x is held fixed in the integration, in accordance with the definition in Chapter 2. We see that the transform of the x derivative of f is the x derivative of the transform $\bar{f}$,

$$\mathcal{L} \left\{ \frac{\partial f(x, t)}{\partial x} \right\} = \int_0^\infty e^{-st} \frac{\partial f(x, t)}{\partial x} \, dt = \frac{\partial \bar{f}(x, s)}{\partial x}, \tag{249}$$

and that the same statement applies to x derivatives of higher order. However, as regards t differentiation, the results of Chapter 2 can be applied to give

$$\mathcal{L} \left\{ \frac{\partial f(x, t)}{\partial t} \right\} = s \bar{f}(x, s) - f(x, 0), \tag{250a}$$

$$\mathcal{L} \left\{ \frac{\partial^2 f(x, t)}{\partial t^2} \right\} = s^2 \bar{f}(x, s) - s f(x, 0) - \frac{\partial f(x, 0)}{\partial t}, \tag{250b}$$

and so forth [see formulas (T3), (T4), and (T5), page 67]. In a similar way, all the results of Chapter 2 are again applicable here, insofar as time variation is concerned.

In applications to the solution of problems governed by partial differential equations, it is frequently necessary to make use of the fact that, if $f(x, t)$ is *bounded* as a function of x for all relevant values of x when t has a fixed positive value, then the transform $\bar{f}(x, s)$ has the same property when s is fixed. In addition, the transform must tend to zero as $s \longrightarrow +\infty$, so that

$$\lim_{s \to +\infty} \bar{f}(x, s) = 0 \tag{251}$$

for all such values of x.

As a simple example of the usefulness of the Laplace transform, we consider the solution of the wave equation

$$\frac{\partial^2 \varphi}{\partial x^2} = \frac{1}{c^2} \frac{\partial^2 \varphi}{\partial t^2} \tag{252}$$

in the first quadrant $(0 \leqq x < \infty, 0 \leqq t < \infty)$ of the distance–time plane (Figure 9.9), subject to the boundary condition

$$\varphi(0, t) = 0 \tag{253}$$

and to the initial conditions

$$\varphi(x, 0) = 0, \tag{254a}$$

$$\frac{\partial \varphi(x, 0)}{\partial t} = 1. \tag{254b}$$

If we take the transform (with respect to t) of (252) and use (254a, b), we obtain the differential equation

$$\frac{\partial^2 \bar{\varphi}}{\partial x^2} - \frac{s^2}{c^2} \bar{\varphi} = -\frac{1}{c^2} \tag{255}$$

to be satisfied by $\bar{\varphi}(x, s)$. In addition, from (253) it follows that the condition

$$\bar{\varphi}(0, s) = 0 \tag{256}$$

must be satisfied.

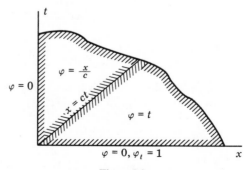

Figure 9.9

Whereas (255) is a *partial* differential equation, the fact that only x differentiation is involved permits it to be solved as though it were an *ordinary* differential equation, with s held constant in the process, and with the arbitrary "constants of integration" accordingly replaced by arbitrary functions of s. This is the principal advantage gained by use of the Laplace transform technique.

The general solution of (255) is thus obtained in the form

$$\bar{\varphi}(x, s) = A(s)e^{-sx/c} + B(s)e^{sx/c} + \frac{1}{s^2}. \tag{257}$$

The condition (256) and the requirement that the transform be bounded as $x \to +\infty$ for $s > 0$ then give†

$$A(s) = -\frac{1}{s^2}, \qquad B(s) = 0, \tag{258}$$

so that the transform of the required solution is

$$\bar{\varphi}(x, s) = \frac{1}{s^2} - \frac{1}{s^2}e^{-sx/c}. \tag{259}$$

The inverse transform of the right-hand member is identifiable by direct use of (T35) and (T9), pages 67 and 68, and we thus obtain the result

$$\varphi(x, t) = t - \begin{cases} 0 & \left(t \leqq \dfrac{x}{c}\right), \\ t - \dfrac{x}{c} & \left(t \geqq \dfrac{x}{c}\right), \end{cases}$$

or, equivalently,

$$\varphi(x, t) = \begin{cases} t & \left(t \leqq \dfrac{x}{c}\right), \\ \dfrac{x}{c} & \left(t \geqq \dfrac{x}{c}\right). \end{cases} \tag{260}$$

This example was chosen not only to illustrate the efficiency of the use of the Laplace transform in a favorable situation, but also to provide an example of an occurrence mentioned at the end of Section 8.9. Specifically, if we examine the "solution" obtained here, we notice that, whereas φ is continuous, the two first partial derivatives are discontinuous at all points on the line $x = ct$ in the first quadrant of the xt plane. Hence, strictly speaking, the *second* partial derivatives do not exist on that line (even as one-sided derivatives), and the differential equation accordingly is not satisfied along it. The prescribed boundary and initial conditions *are* indeed satisfied.

The fault is not with the Laplace transform procedure (which would have yielded a proper solution had one existed) but with the formulation of the

†Here it is implied that $\varphi(x, t)$ is required to be bounded as $x \to +\infty$ for any fixed positive value of t. In fact, this requirement here is irrelevant (see Problem 35 of Chapter 8), and the requisite vanishing of $B(s)$ can be shown to follow from the fact that $B(s)e^{sx/c}$ cannot be a Laplace transform as $x \to +\infty$ unless it vanishes identically.

problem. It can be shown that the only continuous function which satisfies the specified side conditions and which satisfies the differential equation everywhere except along the line $x = ct$ is the one obtained. Whether it is in fact acceptable as the required description of a certain entity (physical or otherwise) depends upon the nature of that entity. If it is not acceptable, then it must be concluded that the mathematical simulation [(252) to (254)] of the phenomenon in question is at fault (and probably is oversimplified, by linearization or otherwise).

It should be noticed that the use of Laplace transforms permits the solution of certain problems governed by nonhomogeneous partial differential equations. In illustration, we consider the determination of $\varphi(x, t)$ satisfying the equation

$$\frac{\partial^2 \varphi}{\partial x^2} = \frac{1}{c^2} \frac{\partial^2 \varphi}{\partial t^2} - \cos \omega t \qquad (261)$$

and the homogeneous side conditions

$$\varphi(0, t) = 0, \qquad (262a)$$

$$\varphi(x, 0) = 0, \qquad (262b)$$

$$\frac{\partial \varphi(x, 0)}{\partial t} = 0, \qquad (262c)$$

when $0 \leq x < \infty$ and $0 \leq t < \infty$. The associated problem governing the transform $\bar{\varphi}(x, s)$ is then

$$\frac{\partial^2 \bar{\varphi}}{\partial x^2} - \frac{s^2}{c^2} \bar{\varphi} = -\frac{s}{s^2 + \omega^2} \qquad (263a)$$

$$\bar{\varphi}(0, s) = 0, \qquad (263b)$$

from which there follows easily

$$\bar{\varphi}(x, s) = (1 - e^{-sx/c}) \frac{c^2}{s(s^2 + \omega^2)} \qquad (264)$$

and hence

$$\varphi(x, t) = \begin{cases} \dfrac{c^2}{\omega^2}(1 - \cos \omega t) & \left(t \leq \dfrac{x}{c}\right), \\[2ex] \dfrac{c^2}{\omega^2}\left[\cos \omega\left(t - \dfrac{x}{c}\right) - \cos \omega t\right] & \left(t \geq \dfrac{x}{c}\right). \end{cases} \qquad (265)$$

Here φ, φ_x, and φ_t are continuous when $x = ct$.

In less contrived situations, the determination of the *inverse transform* may present a greater challenge. Methods to be derived in Sections 11.2 and 11.3 can be used (to supplement the use of tables) in many cases. The example treated in the following section exploits an expansion process which is of rather frequent usefulness.

9.16. Application of the Laplace Transform to the Telegraph Equations for a Long Line. In dealing with the flow of electricity in a long insulated cable, the potential v (volts) and the current i (amperes) are found to be related approximately by the simultaneous equations

$$\frac{\partial v}{\partial x} = -L\frac{\partial i}{\partial t}, \tag{266a}$$

$$\frac{\partial i}{\partial x} = -C\frac{\partial v}{\partial t}, \tag{266b}$$

where x is distance (miles) from one end of the cable, if the effects of leakage and resistance are neglected. Here L is inductance per unit length of cable (henries/mile), and C is the capacitance to ground per unit length (farads/mile). Both L and C are assumed to be *constants*. If i is eliminated between (266a) and (266b), we find that v satisfies the wave equation

$$\frac{\partial^2 v}{\partial x^2} = LC\frac{\partial^2 v}{\partial t^2}, \tag{267}$$

and a completely analogous equation is satisfied by i.

Initially, the line is here considered to be dead,

$$v(x, 0) = \frac{\partial v(x, 0)}{\partial t} = 0. \tag{268}$$

Then, beginning at the instant $t = 0$, a voltage is impressed at the end $x = 0$, and thereafter that end is maintained at a prescribed potential which may vary with time,

$$v(0, t) = f(t) \qquad (t > 0). \tag{269}$$

If the other end $(x = l)$ is *open*, then $i = 0$ when $x = l$. With the use of (266a), this condition becomes

$$\frac{\partial v(l, t)}{\partial x} = 0 \qquad (t > 0), \tag{270a}$$

whereas if the end $x = l$ is *grounded*, then $v = 0$ when $x = l$,

$$v(l, t) = 0 \qquad (t > 0). \tag{270b}$$

If we take the Laplace transform of Equation (267), with respect to t, there follows

$$\frac{\partial^2 \bar{v}(x, s)}{\partial x^2} = LC\left[s^2 \bar{v}(x, s) - sv(x, 0) - \frac{\partial v(x, 0)}{\partial t}\right] \tag{271}$$

or, taking into account the initial conditions (268),

$$\frac{\partial^2 \bar{v}(x, s)}{\partial x^2} - LCs^2 \bar{v}(x, s) = 0. \tag{272}$$

It will be convenient to write

$$c = \frac{1}{\sqrt{LC}}, \tag{273}$$

so that (272) becomes

$$\frac{\partial^2 \bar{v}}{\partial x^2} - \frac{s^2}{c^2}\bar{v} = 0, \tag{274}$$

with the general solution

$$\bar{v}(x, s) = A(s)e^{-sx/c} + B(s)e^{sx/c}. \tag{275}$$

We consider here only the case where the line is *open* at the end $x = l$, so that (270a) and (269) are to be satisfied. If we take the transforms of these two equations we find that $\bar{v}(x, s)$ then must satisfy the end conditions

$$\bar{v}(0, s) = \bar{f}(s), \tag{276a}$$

$$\frac{\partial \bar{v}(l, s)}{\partial x} = 0, \tag{276b}$$

where $\bar{f}(s)$ is the transform of $f(t)$. Thus the functions $A(s)$ and $B(s)$ in (275) are determined by the conditions

$$A(s) + B(s) = \bar{f}(s), \tag{277a}$$

$$-A(s)e^{-sl/c} + B(s)e^{sl/c} = 0, \tag{277b}$$

and hence are of the form

$$A(s) = \frac{\bar{f}(s)}{1 + e^{-2sl/c}}, \tag{278a}$$

$$B(s) = \frac{\bar{f}(s)}{1 + e^{+2sl/c}}. \tag{278b}$$

In the limiting case of an *infinite line*,

$$l = \infty, \tag{279}$$

there follows

$$A(s) = \bar{f}(s), \qquad B(s) = 0, \tag{280}$$

and (275) becomes

$$\bar{v}(x, s) = \bar{f}(s)e^{-sx/c}. \tag{281}$$

This is the transform of the desired solution, and the use of (T9), page 67, determines $v(x, t)$ in the form

$$v(x, t) = \begin{cases} f\left(t - \dfrac{x}{c}\right) & \text{when } t > \dfrac{x}{c} \text{ or } x < ct, \\ 0 & \text{when } t < \dfrac{x}{c} \text{ or } x > ct. \end{cases} \tag{282}$$

Thus we see that at a given time t the effect of introducing the voltage at $x = 0$ is present only at distances not greater than $ct = t/\sqrt{LC}$ miles from that end. That is, a voltage wave is propagated along the line with velocity $c = 1/\sqrt{LC}$, in a manner specified by Equation (282). For example, if $v(0, t) = f(t)$ varies periodically with time as is indicated in Figure 9.10(a), the voltage along the line at time t is as indicated in Figure 9.10(b). In general, we may verify that the representation of $v(x, t)$ as a function of x at time t is obtained by reflecting the curve for $v(0, t)$ about the axis $t = 0$, replacing the t scale by an x/c scale, and then translating the reflected curve to the right through t units.

In the more general case of a *finite line*, the introduction of (278) into (275) gives the result

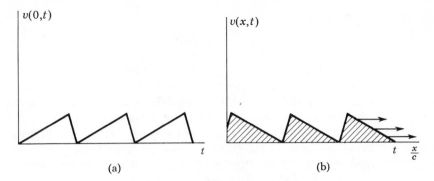

Figure 9.10

$$\bar{v}(x, s) = \bar{f}(s)\frac{\cosh\,[s(l-x)/c]}{\cosh\,(sl/c)}. \tag{283}$$

To determine the solution having this transform, we first rewrite (283) in the form

$$\bar{v}(x, s) = \bar{f}(s)\frac{e^{-sx/c} + e^{-s(2l-x)/c}}{1 + e^{-2sl/c}},$$

then expand $1/(1+e^{-2sl/c})$ in a series of ascending powers of $e^{-2sl/c}$, and so obtain

$$\bar{v}(x, s) = \bar{f}(s)[e^{-sx/c} + e^{-s(2l-x)/c} - e^{-s(2l+x)/c} - e^{-s(4l-x)/c} + e^{-s(4l+x)/c} + \cdots]. \tag{284}$$

Again making use of formula (T9), page 67, we obtain $v(x, t)$ in the form

$$
v(x, t) =
\begin{cases}
f\!\left(t - \dfrac{x}{c}\right) & \text{when } t > \dfrac{x}{c} \\
0 & \text{when } t < \dfrac{x}{c}
\end{cases}
$$

$$
+
\begin{cases}
f\!\left(t - 2T + \dfrac{x}{c}\right) & \text{when } t > 2T - \dfrac{x}{c} \\
0 & \text{when } t < 2T - \dfrac{x}{c}
\end{cases}
$$

$$
-
\begin{cases}
f\!\left(t - 2T - \dfrac{x}{c}\right) & \text{when } t > 2T + \dfrac{x}{c} \\
0 & \text{when } t < 2T + \dfrac{x}{c}
\end{cases}
$$

$$
-
\begin{cases}
f\!\left(t - 4T + \dfrac{x}{c}\right) & \text{when } t > 4T - \dfrac{x}{c} \\
0 & \text{when } t < 4T - \dfrac{x}{c}
\end{cases}
$$

$$
+
\begin{cases}
f\!\left(t - 4T - \dfrac{x}{c}\right) & \text{when } t > 4T + \dfrac{x}{c} \\
0 & \text{when } t < 4T + \dfrac{x}{c}
\end{cases}
+ \cdots, \tag{285}
$$

where we have written

$$T = \frac{l}{c} = \sqrt{LC}\, l, \tag{286}$$

so that T is the time required for a wave to travel the length of the line.

Noticing that $0 \leq x/c \leq T$, we see that in the first part of the propagation $(0 \leq t \leq T)$ only the first brace in (285) may differ from zero, and hence in this part of the process the wave decribed above is traveling toward the end $x = l$, reaching it at the time $t = T$. Then when $T \leq t \leq 2T$, only the first *two* braces in (285) are not zero. In this second interval the second brace represents a *reflected* wave returning from the open end of the line without change in sign. The actual voltage in this time interval thus consists of the superposition of an outward- and an inward-traveling wave. When $t = 2T$, the reflected wave reaches the closed end $(x = 0)$ and, according to the third brace in (285), is again reflected, but this time *with reversal of sign*. This process is continued indefinitely, reflections *without* sign change occurring at the open end and reflections *with* change of sign occurring at the closed end. The various inward and outward waves so generated may combine in various ways, for various periodic impressed end voltages, the nature of the superposition depending, in particular, upon the relationship between this period and the time interval T.

It may be noted that, with an abbreviation such as

$$f^+(t) = \begin{cases} f(t) & (t > 0), \\ 0 & (t < 0), \end{cases} \tag{287}$$

Equation (285) can be rewritten in the more compact form

$$v(x, t) = f^+\left(t - \frac{x}{c}\right) + f^+\left(t - 2T + \frac{x}{c}\right) - f^+\left(t - 2T - \frac{x}{c}\right)$$
$$- f^+\left(t - 4T + \frac{x}{c}\right) + \cdots. \tag{285'}$$

9.17. Nonhomogeneous Conditions. The Method of Variation of Parameters. In each of the problems considered in this chapter, a certain linear partial differential equation is to be satisfied in a certain region $\mathfrak{R}$ and the solution is to satisfy a *boundary condition* along each boundary which serves to limit $\mathfrak{R}$. If *time* also is involved, these boundary conditions may be time-dependent and the problem may also specify one or more *initial conditions* to be satisfied by the unknown function φ throughout $\mathfrak{R}$ at an initial time.

The method of *separation of variables* requires for its success that $\mathfrak{R}$ be such that the relevant differential equation is *separable* (or becomes separable when a nonhomogeneous "forcing term" is suppressed) in terms of a coordinate system for which one of the coordinates is constant along each of the boundaries which make up the border of $\mathfrak{R}$. In addition, however, when the differential equation is homogeneous and time is not involved, nonhomogeneous boundary conditions can be imposed only along one pair of boundaries (or, in limiting cases, along the single boundary) on which one of the variables is constant. When the equation either is nonhomogeneous† or is time-dependent, all the boundary conditions must be homogeneous; combined nonhomogeneity and time dependence of the differential equation generally are not acceptable. In the favorable situations noted, the presence of pairs of homogeneous boundary conditions leads to a set of *characteristic-value problems,* for which the permissible values of the separation constant or constants are the characteristic numbers. (In some cases, requirements of boundedness or periodicity, which also are homogeneous requirements, replace one or both of a pair of homogeneous boundary conditions.) The representation assumed for the required solution φ then comprises a superposition (by summation and/or integration) of product terms containing these characteristic functions.

Special devices also permit the treatment of certain other types of problems by such methods. In particular, an excess of nonhomogeneity often can be conquered by the process of superimposing the solutions of two or more related problems, for each of which no such excess occurs. This process, however, may be tedious and may provide the final solution in a cumbersome form. Alternatively, it may be possible to determine (by inspection or otherwise) a function u which satisfies the *excess* nonhomogeneous requirement (or requirements) as well as the homogeneous ones. The *difference* $\varphi - u$ then is to satisfy an acceptable set of conditions. (As was noted in Section 9.11, this procedure is used, for example, in Section 9.10, as well as in Problems 22, 23, 60, and 61.) However, except in relatively simple cases, this process of removing one sort of nonhomogeneity, *without introducing another*, usually is not feasible.

Other methods, which also are available in these and in certain less tractable situations, include the Duhamel integral superposition technique which is illustrated in Section 9.11 (and which can cope with time-dependent boundary conditions in certain cases), the use of the Laplace transform (which also serves this purpose, and further permits the differential equation to be nonhomogeneous), and the use of other transforms such as those of Fourier, Hankel, and Mellin (which can deal with combined nonhomogeneities in appropriate space regions).

In the remainder of this section we describe and illustrate a rather powerful technique, somewhat analogous to that treated in Section 1.9 (in connection with *ordinary* differential equations), which incorporates many of the features of the methods just mentioned, and which is capable of dealing with a variety of reasonably troublesome problems analytically, in a compact and uniform way.

As a first illustration of the method, we seek the solution of the nonhomogeneous heat-flow equation

$$\alpha^2 \frac{\partial^2 T}{\partial x^2} - \frac{\partial T}{\partial t} = h(x, t), \tag{288}$$

†See Problems 45 and 46.

subject to the homogeneous boundary conditions

$$T(0, t) = 0, \qquad T(l, t) = 0 \tag{289}$$

when $t > 0$, and to the initial condition

$$T(x, 0) = f(x). \tag{290}$$

In the absence of the "forcing function" $h(x, t)$, the method of separation of variables would lead to an assumed expression for T, satisfying (288) with $h = 0$ as well as (289), in the form

$$T(x, t) = \sum_{n=1}^{\infty} a_n e^{-n^2 \pi^2 \alpha^2 t / l^2} \sin \frac{n\pi x}{l}, \tag{291}$$

where the a's are constants to be determined in such a way that (290) also is satisfied.

The *method of variation of parameters* here consists of replacing the constants a_n by unknown functions $C_n(t)$, and attempting to determine these functions in such a way that *both* (288) and (290) are satisfied. Since the exponential function of t can be absorbed into $C_n(t)$, we thus are led to the assumption

$$T(x, t) = \sum_{n=1}^{\infty} C_n(t) \sin \frac{n\pi x}{l}. \tag{292}$$

Since $T(x, t)$ satisfies the end conditions (289), the consequence of the requirement that it also satisfy (288) can be obtained by inserting (292) into (288) and differentiating term by term (see Section 5.12). Whereas this procedure is to be recommended for the purpose of minimizing computation, when it is permissible, there are other situations (as will be seen) in which its use is improper and leads to false conclusions. Hence we here illustrate an alternative process of more general applicability.

For the purpose of determining $C_n(t)$, we next deduce from (292) that

$$C_n(t) = \frac{2}{l} \int_0^l T(x, t) \sin \frac{n\pi x}{l} \, dx, \tag{293}$$

noticing that $T(x, t)$ must be a *continuously* differentiable function of x for $0 \leq x \leq l$ when $t > 0$, since T satisfies (288). Accordingly, there follows

$$C_n'(t) = \frac{2}{l} \int_0^l \frac{\partial T}{\partial t} \sin \frac{n\pi x}{l} \, dx$$

and hence, by using (288) to replace T_t by $\alpha^2 T_{xx} - h$, we obtain the relation

$$C_n'(t) = \frac{2\alpha^2}{l} \int_0^l \frac{\partial^2 T}{\partial x^2} \sin \frac{n\pi x}{l} \, dx - \frac{2}{l} \int_0^l h(x, t) \sin \frac{n\pi x}{l} \, dx. \tag{294}$$

If the first term on the right is integrated by parts twice, with account taken of (289), the result takes the form

$$C_n'(t) + \frac{n^2}{\lambda} C_n(t) = -\frac{2}{l} \int_0^l h(x, t) \sin \frac{n\pi x}{l} \, dx, \tag{295}$$

with the previously used abbreviation

$$\lambda = \frac{l^2}{\pi^2 \alpha^2}. \tag{296}$$

It should be verified that the same equation is obtained (*with less effort*) by introducing (292) into (288), differentiating term by term, and then identifying the coefficients in the resultant sine-series representation.

Thus we have a linear ordinary differential equation determining $C_n(t)$, where $h(x, t)$ is known. The associated constant of integration is determined by the requirement that (292) satisfy (290) when $t = 0$, so that

$$\sum_{n=1}^{\infty} C_n(0) \sin \frac{n\pi x}{l} = f(x) \qquad (0 < x < l).$$

Hence we must have

$$C_n(0) = a_n, \tag{297}$$

where

$$a_n = \frac{2}{l} \int_0^l f(x) \sin \frac{n\pi x}{l} \, dx. \tag{298}$$

The solution of (295) which satisfies (297) is obtained in the form

$$C_n(t) = a_n e^{-n^2 t/\lambda} - \frac{2}{l} \int_0^t e^{-n^2(t-\tau)/\lambda} \left[\int_0^l h(x, \tau) \sin \frac{n\pi x}{l} \, dx \right] d\tau \tag{299}$$

(see Problem 110), and the introduction of (299) into (292) yields the required solution.

We consider next the generalized problem

$$\alpha^2 \frac{\partial^2 T}{\partial x^2} - \frac{\partial T}{\partial t} = h(x, t), \tag{300}$$

$$T(0, t) = G(t), \qquad T(l, t) = F(t), \tag{301}$$

$$T(x, 0) = f(x), \tag{302}$$

with $0 \leq x \leq l$ and $t \geq 0$, where both the differential equation and the pair of boundary conditions may be nonhomogeneous. One method of dealing with this problem consists of again assuming the solution in the form (292) and of determining an appropriately modified differential equation for $C_n(t)$. However, since a better method is available in this case, we consider that method first and then investigate the former method.

The proposed method consists of first homogenizing the boundary conditions (301) by subtracting from T *any* (respectable) function $u(x, t)$ which satifies (301). Here $u(x, t)$ need not reduce the left-hand member of (300) to zero, since failure to do so merely will require the correction to satisfy a differential equation with a modified right-hand member, and we now can deal with such a modification. With the choice

$$u(x, t) = \frac{x}{l} F(t) + \left(1 - \frac{x}{l}\right) G(t), \tag{303}$$

which is convenient since it is linear in x, we then write

$$T(x, t) = u(x, t) + v(x, t) \tag{304}$$

and find that the *correction* $v(x, t)$ must solve the new problem

$$\alpha^2 \frac{\partial^2 v}{\partial x^2} - \frac{\partial v}{\partial t} = h^*(x, t),$$

$$v(0, t) = 0, \qquad v(l, t) = 0, \tag{305}$$

$$v(x, 0) = f^*(x),$$

where

$$h^*(x, t) = h(x, t) + \frac{x}{l} F'(t) + \left(1 - \frac{x}{l}\right) G'(t) \tag{306}$$

and

$$f^*(x) = f(x) - \frac{x}{l} F(0) - \left(1 - \frac{x}{l}\right) G(0), \tag{307}$$

under the assumption that $F(t)$ and $G(t)$ are differentiable.

Hence we see that the correction $v(x, t)$ can be obtained in the form

$$v(x, t) = \sum_{n=1}^{\infty} C_n^*(t) \sin \frac{n\pi x}{l}, \tag{308}$$

where $C_n^*(t)$ is defined by the right-hand member of (299) with h replaced by h^* and with a_n determined from f^*.

In particular, if we take

$$h(x, t) = G(t) = f(x) = 0, \tag{309}$$

and so deal with the problem previously treated in Section 9.11, we obtain the solution in the form

$$T(x, t) = \frac{x}{l} F(t) + \sum_{n=1}^{\infty} C_n^*(t) \sin \frac{n\pi x}{l}, \tag{310}$$

where

$$C_n^*(t) = e^{-n^2 t/\lambda} \left\{ a_n - \frac{2}{l} \int_0^t e^{n^2 \tau/\lambda} \left[\int_0^l \frac{x}{l} F'(\tau) \sin \frac{n\pi x}{l} \, dx \right] d\tau \right\} \tag{311}$$

and where

$$a_n = -\frac{2}{l} F(0) \int_0^l \frac{x}{l} \sin \frac{n\pi x}{l} \, dx. \tag{312}$$

When use is made of the evaluation

$$\int_0^l \frac{x}{l} \sin \frac{n\pi x}{l} \, dx = \frac{(-1)^{n+1} l}{n\pi}, \tag{313}$$

we may verify that the solution just obtained is identified with the form given in Equation (177) of Section 9.11, and obtained there by use of the Duhamel superposition process.

The alternative procedure suggested above for the treatment of (300) to (302) has the disadvantage that the series solution which it produces converges less rapidly than does the series (308). However, it has the advantages that it does not presume the differentiability of $F(t)$ and $G(t)$ and, perhaps more importantly, that it can be used in other cases when it is not feasible to determine a homogenizing function analogous to $u(x, t)$. For this reason, we next outline its use in the present example.

Accordingly, we assume the solution of (300) to (302) in the form

$$T(x, t) = \sum_{n=1}^{\infty} K_n(t) \sin \frac{n\pi x}{l}, \tag{314}$$

and deduce that

$$K_n(t) = \frac{2}{l} \int_0^l T(x, t) \sin \frac{n\pi x}{l} \, dx \tag{315}$$

and that

$$K'_n(t) = \frac{2\alpha^2}{l} \int_0^l \frac{\partial^2 T}{\partial x^2} \sin \frac{n\pi x}{l} \, dx - \frac{2}{l} \int_0^l h(x, t) \sin \frac{n\pi x}{l} \, dx, \tag{316}$$

as before. But now when the first term on the right in (316) is integrated twice by parts, and use is made of the nonhomogeneous conditions (301), we obtain the differential equation

$$K'_n(t) + \frac{n^2}{\lambda} K_n(t) = \frac{2n}{\pi\lambda}[G(t) + (-1)^{n+1}F(t)] - \frac{2}{l} \int_0^l h(x, t) \sin \frac{n\pi x}{l} \, dx, \tag{317}$$

in place of (295), with the initial condition

$$K_n(0) = \frac{2}{l} \int_0^l f(x) \sin \frac{n\pi x}{l} \, dx. \tag{318}$$

The introduction of the function $K_n(t)$, so determined, into (314) then yields the required solution.

In particular, in the special case (309) it is found that the solution obtained in this way is identified with the form given by Equation (176) of Section 9.11.

The comments made in Section 9.11 apply also in the more general case considered here, in that whereas the series (308) will converge to $v(x, t)$ for all x such that $0 \le x \le l$ with $t > 0$ (when F, G, and f are respectable), the same is not true of the series (314) unless $F(t) = G(t) = 0$. Since each term of that series vanishes when $x = 0$ and when $x = l$ for all t, and since $T(0, t)$ and $T(l, t)$ do *not* vanish unless $F(t)$ and $G(t)$ do so, the series (314) will converge to the required solution only when $0 < x < l$ if $t > 0$.† Since T is *known* when $x = 0$ and when $x = l$, this fact would be of no consequence if it were not a corollary that the *rate* of convergence of (314) will be inferior to that of (308) when $0 < x < l$ unless $F(t) = G(t) = 0$.

The preceding examples illustrate the use of the method of variation of parameters in a rather extensive class of linear problems, specifically those in

---

†In this connection, it is easily verified that if expressions for T_{xx} and T_t are obtained by *formally* differentiating (314) term by term, and are introduced into (300), the requirement that the resultant equation be satisfied yields an *incorrect* differential equation for $K_n(t)$ unless $F(t) = G(t) = 0$. This is a consequence of the fact that term-by-term differentiation of (314) is not justifiable if the represented function does not vanish when $x = 0$ and when $x = l$ (Section 5.12).

which a certain nonhomogeneity is troublesome, but in which the method of separation of variables would succeed if that particular nonhomogeneity were absent. The procedure essentially consists of first *imagining* the troublesome nonhomogeneity to be absent and writing down the characteristic-function representation (series or integral) which would be assumed for the solution in that case. The next step replaces the unknown constants of combination by unknown functions of the variable or variables not appearing in the characteristic functions (or, when superposition is by integration, replaces unknown functions of a parameter by unknown functions depending also upon those variables), and deduces differential equations determining these functions by methods analogous to those used in the preceding examples. Several additional examples are treated in Problems 111–116.

9.18. Formulation of Problems. From the preceding examples it is apparent that the *number* and *nature* of the conditions to be imposed along a physical boundary or at a given time depend upon the type of partial differential equation which governs the problem.

Thus, for example, in the case of Laplace's equation in two dimensions, $\varphi_{xx} + \varphi_{yy} = 0$, we have seen that the solution φ is determined everywhere inside a region $\mathfrak{R}$ if the *boundary values* of φ are prescribed along the closed boundary C of $\mathfrak{R}$. The same can be shown to be true also for other linear *elliptic* equations, that is, for any equation $a\varphi_{xx} + b\varphi_{xy} + c\varphi_{yy} + \cdots = 0$, where $b^2 < 4ac$. However, this statement is *not* true in general for *hyperbolic* equations, where $b^2 > 4ac$. For example, we readily verify that the expression $\varphi = \sin k\pi x \sin k\pi y$ satisfies the equation $\varphi_{xx} - \varphi_{yy} = 0$, as well as the requirement that φ vanish along the closed boundary of the rectangle $(0 \leq x \leq 1, 0 \leq y \leq 1)$ *for any integral value of* k. Thus for this equation this particular boundary-value problem has infinitely many solutions, whereas for Laplace's equation the only solution is that for which $\varphi = 0$ everywhere inside the rectangle.

In the case of the particular *initial-value* problem (the Cauchy problem) for which along the entire x axis the function φ *and* its normal derivative φ_y are prescribed, say

$$\varphi(x, 0) = f(x), \qquad \frac{\partial \varphi(x, 0)}{\partial y} = g(x),$$

and a solution valid (say) for *all positive values of* y is required, it can be shown that, for Laplace's equation (see Problem 60 of Chapter 8) and other *elliptic* equations, the solution exists only if $f(x)$ and $g(x)$ satisfy very stringent restrictions. In particular, it is necessary that *all derivatives* of $f(x)$ and $g(x)$ exist for *all real values of* x, but even this condition is by no means sufficient to guarantee the existence of a solution valid for all positive values of y. On the other hand, unless the x axis is a characteristic base curve, this problem possesses a proper solution in the case of linear *hyperbolic* equations when $f(x)$ and $g(x)$ satisfy only very mild conditions.

These examples are typical of the general linear case. Thus it may be expected that, in general, *elliptic* equations are associated with *boundary*-value problems, whereas *hyperbolic* equations are associated with *initial*-value problems. *Parabolic* equations, where the coefficient discriminant $b^2 - 4ac$ vanishes, are intermediate in nature. We next list certain general types of problems which commonly arise in connection with such equations, many of which have been illustrated in the present chapter. The list is not intended to be exhaustive.

Typical problems associated with *elliptic* equations may be illustrated by considering Laplace's equation in the two-dimensional form

$$\frac{\partial^2 T}{\partial x^2} + \frac{\partial^2 T}{\partial y^2} = 0, \tag{319}$$

where (for example) T may represent steady-state temperature. Along the closed boundary of a region we may prescribe T (the temperature) *or* the normal derivative $\partial T/\partial n$ (a quantity proportional to the rate of heat flow through the boundary). In the second case the temperature is determined only within an arbitrary additive constant. Alternatively, a condition of the more general type

$$aT + b\frac{\partial T}{\partial n} = c$$

may be prescribed along the boundary (as, for example, in the case of heat radiation from the boundary according to Newton's law of cooling). In case the region involved is not simply connected, and in certain other cases, it may be necessary to add requirements of single-valuedness or periodicity. The region in which the solution is to be valid may be the interior or the exterior of the closed boundary. However, in case the region extends to infinity, restrictions concerning desirable or permissible behavior of the solution at large distances from the origin generally must be added. In such cases it is frequently convenient to imagine that the conditions "at infinity" are prescribed along a circle (or an arc of a circle) of infinite radius, forming the outer boundary (or the remainder of a boundary which extends to infinity) of the infinite region involved (Figure 9.11).

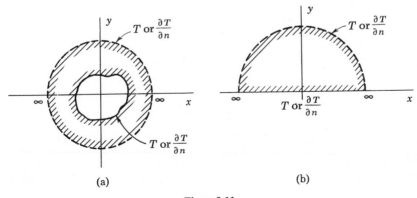

(a)   (b)

Figure 9.11

Typical problems associated with *hyperbolic* equations may be illustrated in the case of the one-dimensional wave equation

$$\frac{\partial^2 w}{\partial x^2} = \frac{1}{c^2}\frac{\partial^2 w}{\partial t^2}, \tag{320}$$

where (for example) w may represent the deflection of a vibrating string (x is distance along the string, t time). For a string so long that end conditions are irrelevant, the prescribed *initial values*

$$w(x, 0) = f(x), \qquad \frac{\partial w(x, 0)}{\partial t} = v(x)$$

of the function w (deflection) *and* w_t (velocity) at the time $t = 0$ determine the deflection for all values of x at all following times ($t > 0$). No additional limiting conditions can be prescribed as $t \to \infty$ or as $|x| \to \infty$. However, for a string of finite length l with both ends fixed (at $x = 0$ and $x = l$) these conditions are prescribed only over the interval $0 < x < l$, and the end conditions

$$w(0, t) = 0, \qquad w(l, t) = 0$$

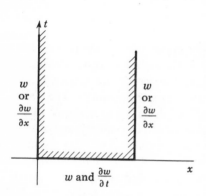

w or $\dfrac{\partial w}{\partial x}$

w or $\dfrac{\partial w}{\partial x}$

w and $\dfrac{\partial w}{\partial t}$

Figure 9.12

must also be satisfied, for all time such that $t > 0$. In this case, if we represent x and t as rectangular coordinates in an artificial xt plane, we see that the solution is required in the *semi-infinite strip* ($0 < x < l$, $0 < t < \infty$) (Figure 9.12). Along the boundaries $x = $ constant, the function w is prescribed, whereas along the "time boundary" $t = 0$ the quantities w *and* $\partial w/\partial t$ are prescribed [for those values of x in the interval $(0 < x < l)$]. With this interpretation, we may think of $\partial w/\partial t$ as the "normal derivative" of w along the "time boundary."

It can be shown (see Problem 33 of Chapter 8) that the solution of this problem, when $0 < x < l$, is the same as that for an *infinitely long* string extending along the entire x axis, for which at the time $t = 0$ the deflection and velocity are prescribed for *all* values of x as odd periodic functions of x, of period $2l$, agreeing with $f(x)$ and $v(x)$ in the interval $0 < x < l$. That is, the effect of the added semi-infinite portions of the string is then the same as the effect of the physical restraints at the ends. Thus, in a sense, this problem can be considered as purely an initial-value problem.

The prescribed value of w along a boundary (end) $x = $ constant may vary with time; also, an end condition of the more general form $aw + bw_x = c$ may be substituted, as for example in the case of an elastic end support. In many problems in which w is prescribed as varying periodically with time at one boundary, a periodic response is eventually propagated throughout the region, and certain transient effects introduced at the beginning of the motion become negligible at sufficiently later times. In such cases it may be that only.the limiting periodic response is of interest. Here we may suppose that the motion was initiated at $t = -\infty$; then only end conditions need be prescribed, and if a periodic solution is obtained it may be considered in this sense as valid for all time. (We remark that, if the nature of the propagation and the character of the transient effects are of interest, the use of the Laplace transform is particularly advantageous, as is illustrated by a comparison of Problems 75 and 99 at the end of this chapter.)

Considerations of this general nature (but obviously with entirely different physical interpretations) apply equally well to the formulation of problems governed by a hyperbolic equation analogous to the one-dimensional wave equation but in which the *time* variable t is replaced by a second *space* variable y. For example (see Section 6.20), in the linearized theory of steady two-dimensional flow of a nonviscous compressible fluid, if the flow is nearly a uniform flow in the x direction, the deviation in velocity from uniformity is the gradient of a potential φ which satisfies the equation

$$(M^2 - 1)\frac{\partial^2 \varphi}{\partial x^2} - \frac{\partial^2 \varphi}{\partial y^2} = 0, \tag{321}$$

where M is the ratio of the uniform flow velocity to the velocity of sound in the fluid. When this ratio is less than unity, this equation is elliptic and, in fact, is reducible to Laplace's equation by an obvious change in variables. The associated problem in this case then normally is a boundary-value problem. However, when the flow is supersonic ($M > 1$), the equation is hyperbolic, and conditions of entirely different type must be prescribed. (When $M \approx 1$, the equation is not valid, since then the linearization which leads to this equation is not permissible.) Suitable conditions in certain problems of this sort may be obtained by replacing the time variable t by y in the discussion of the preceding paragraph. In such cases the semi-infinite strip inside which the solution is to be obtained may extend to infinity in either the x or the y direction. In particular, certain basic problems involve the region consisting of the entire half-plane $y > 0$. The nature of prescribed conditions which make a problem determinate is suggested by the consideration that the half plane may be taken as the limit of either of the semi-infinite strips indicated in Figure 9.13, as the dashed-line boundaries are moved parallel to themselves indefinitely far from the origin.

In the first limiting case, where the half-plane is considered as the limit of a strip extending in the y direction, the problem tends toward the usual initial-value problem [Figure 9.14(a)], where φ and its normal derivative are *both* prescribed along the entire x axis. Since this problem is completely determined by the conditions along the boundary $y = 0$, it appears that the "end conditions" along the sides $x = $ constant must be omitted in the limit. That is, generally there are no additional conditions to be prescribed "at infinity" in this limiting case. However, in the second limiting case [Figure 9.14(b)], where

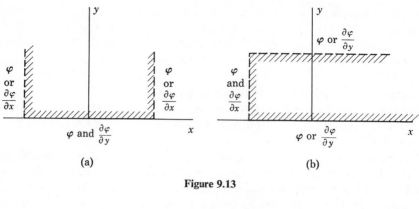

Figure 9.13

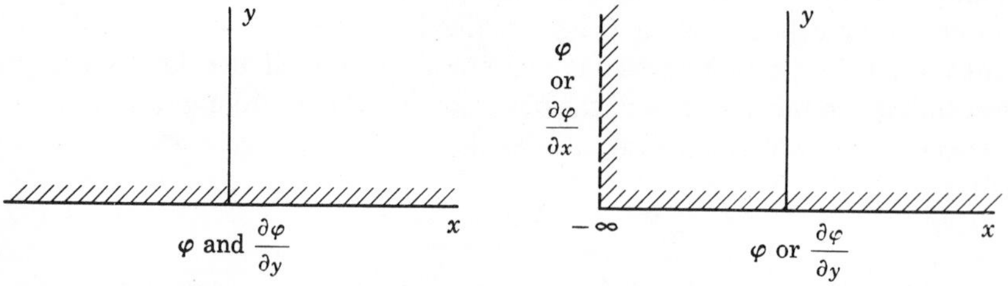

Figure 9.14

the region is considered as the limit of a strip extending in the x direction, the prescribed conditions along the dashed-line boundaries cannot be completely lost in the limit, since the single condition prescribed along the x axis in this case is not sufficient to determine a unique solution. Although the end condition prescribed along the upper boundary $y =$ constant *is* lost in the limit, as in the preceding case, the same is *not* true for the initial-value conditions prescribed along the boundary $x = x_0$ as $x_0 \longrightarrow -\infty$. However, the number of conditions prescribable along this line is, in general, reduced from two to one in the limit. In such a problem nothing can be prescribed with reference to the behavior of the solution as $x \longrightarrow +\infty$. Obviously, the limiting boundary could be taken instead as the line $x = x_0$ as $x_0 \longrightarrow +\infty$, in which case no conditions could be prescribed as $x \longrightarrow -\infty$. A problem of the type just considered is treated in the following section.

Typical problems associated with *parabolic* equations in practice may be illustrated in the case of the equation

$$\frac{\partial^2 T}{\partial x^2} = \frac{1}{\alpha^2}\frac{\partial T}{\partial t}, \tag{322}$$

which (for example) governs one-dimensional heat flow in the x direction. When the region is infinite in extent $(-\infty < x < +\infty)$, a prescribed distribution

$T(x, 0)$ at the time $t = 0$ determines T everywhere at all following times, $t > 0$. When the region is of finite extent $(0 < x < l)$, as for example in the case of a thin rod of length l with insulated sides (Figure 9.15), the distribution $T(x, 0)$ may be prescribed in this interval when $t = 0$, and in addition at each end we may prescribe either the temperature T *or* the rate of heat flow through that end (KT_x); *or* we may prescribe a condition of the more general form $aT + bT_x = c$ at either end, as, for example, in the case of Newtonian cooling by radiation.† As a further alternative, the temperature distribution $T(x, 0)$ may not

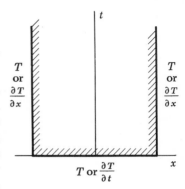

Figure 9.15

be given explicitly but may be determined by specified values of T at the boundaries of the region, under the assumption that initially (at the time $t = 0$) steady-state conditions prevail. In still another type of problem, a periodically varying source of heat may be present at one boundary, leading to a temperature distribution throughout the region which varies periodically with time and tends to zero at large distances from the boundary (see Problem 63).

The problems described in this section are typical of those commonly occurring in many diverse fields, where only two independent variables are present. We have chosen to identify the dependent variable with, say, temperature or deflection of a string for the purpose of fixing ideas and making intuitively plausible the suitability of the formulation. The above discussion can be extended readily, in most cases, to problems involving a greater number of space variables, the region ℛ becoming three-dimensional for Laplace's equation, and the x interval becoming a two- or three-dimensional region in space for the wave and heat-flow equations and for their analogies. In the latter cases it may happen that an initial condition (at the time $t = 0$) itself consists of satisfying a partial differential equation (independent of time) together with associated boundary conditions.

9.19. Supersonic Flow of Ideal Compressible Fluid Past an Obstacle. As a further illustration of physical problems involving hyperbolic differential equations, we consider the determination of the two-dimensional flow of a nonvis-

†The fact that for (322) one cannot prescribe *both* T and $\partial T/\partial t$ when $t = 0$ is associated with the fact that the line $t = 0$ in the xt plane is a "characteristic base curve" for (322) (see Section 8.8). Further, we may notice that if (322) is satisfied when $t = 0$, and if $T(x, 0) = f(x)$, then there must follow also $T_t(x, 0) = \alpha^2 T_{xx}(x, 0) = \alpha^2 f''(x)$, so that $T_t(x, 0)$ is in fact directly determined if $T(x, 0)$ is prescribed as a twice-differentiable function. (Compare Problem 53 of Chapter 8.) Since (322) involves only the *first* time derivative, it is sometimes said to be "of first order" in t.

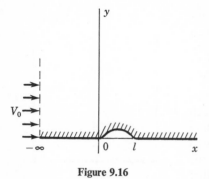

Figure 9.16

cous compressible fluid past an obstacle in the form of a small protuberance from an otherwise straight wall. The flow is pre-scribed as having a uniform velocity V_0 parallel to the wall at large distances upstream from the obstacle (Figure 9.16).

The flow velocity vector at any point can be considered as the sum of the uni-form velocity vector $V_0\mathbf{i}$ and a small devia-tion $\mathbf{u}$. According to Equations (195) and (200) of Section 6.20, for an *irrotational* flow the deviation $\mathbf{u}$ is the gradient of a "potential function" φ,

$$\mathbf{u} = \nabla\varphi, \tag{323}$$

where, for *supersonic* flow, φ satisfies the equation

$$(M^2 - 1)\frac{\partial^2\varphi}{\partial x^2} - \frac{\partial^2\varphi}{\partial y^2} = 0 \qquad (M > 1). \tag{324}$$

Here M, the *Mach number* of the flow, is the ratio of V_0 to the local velocity of sound in the fluid.

The velocity of flow at any point can thus be expressed in the form

$$\mathbf{V} = V_0\mathbf{i} + \nabla\varphi. \tag{325}$$

Along the boundary comprising the wall and protuberance, the normal com-ponent of $\mathbf{V}$ must vanish. If we take the equation of this boundary in the form

$$F(x, y) = 0, \tag{326}$$

the normal vector is proportional to ∇F, and hence this condition becomes

$$[(V_0\mathbf{i} + \nabla\varphi) \cdot \nabla F]_{\text{boundary}} = 0. \tag{327}$$

By writing (326) now in the explicit form

$$y = H(x) \tag{328}$$

and setting $F(x, y) = y - H(x)$, we find that this condition becomes

$$\left[\left\{\mathbf{i}\left(V_0 + \frac{\partial\varphi}{\partial x}\right) + \mathbf{j}\frac{\partial\varphi}{\partial y}\right\} \cdot \{-\mathbf{i}H'(x) + \mathbf{j}\}\right]_{\text{boundary}} = 0$$

or

$$\left[-\left(V_0 + \frac{\partial\varphi}{\partial x}\right)H'(x) + \frac{\partial\varphi}{\partial y}\right]_{\text{boundary}} = 0. \tag{329}$$

The quantity $\partial\varphi/\partial x$ is the x component, u_x, of the velocity deviation; and hence, in accordance with the linearizing approximations leading to (324), $\partial\varphi/\partial x$ is to be neglected with respect to V_0 in (329). Finally, noticing that $H(x)$ is zero except along the protuberance, say

$$H(x) = 0 \qquad unless \ 0 < x < l, \tag{330}$$

and assuming that $H(x)$ is *small* in any case, we may further simplify the analysis by satisfying (329) along the *projection* of the protuberance on the x axis. Thus we replace (329) by the *linearized* boundary condition

$$\frac{\partial \varphi(x, 0)}{\partial y} = V_0 H'(x). \tag{331}$$

Along any line $x = x_0$, as $x_0 \rightarrow -\infty$ the velocity must tend toward uniformity. In particular, the x component of the velocity deviation must tend to zero,

$$\lim_{x_0 \rightarrow -\infty} \frac{\partial \varphi(x_0, y)}{\partial x} = 0. \tag{332}$$

Thus *we require the solution of* (324) *which satisfies* (331) *and* (332) *in the upper half-plane* $y \geqq 0$. This problem is seen to be a particular case of that indicated in Figure 9.14(b).

If we write

$$\alpha = \sqrt{M^2 - 1}, \tag{333}$$

the *general* solution of (324) can be written in the form

$$\varphi(x, y) = f(x - \alpha y) + g(x + \alpha y), \tag{334}$$

where f and g are arbitrary functions (see Section 8.5). The condition (331) then becomes

$$-\alpha f'(x) + \alpha g'(x) = V_0 H'(x),$$

from which there follows

$$f(x) = g(x) - \frac{V_0}{\alpha} H(x) + C, \tag{335}$$

where C is an arbitrary constant. If we use (335) to eliminate f from (334), there follows

$$\varphi(x, y) = g(x + \alpha y) + g(x - \alpha y) - \frac{V_0}{\alpha} H(x - \alpha y) + C, \tag{336}$$

and also

$$\frac{\partial \varphi(x, y)}{\partial x} = g'(x + \alpha y) + g'(x - \alpha y) - \frac{V_0}{\alpha} H'(x - \alpha y). \tag{337}$$

Now let $y \rightarrow +\infty$ along any "characteristic line" $x + \alpha y = c$. Along any of these lines we have, from (337),

$$\frac{\partial \varphi}{\partial x} = g'(c) + g'(c - 2\alpha y) - \frac{V_0}{\alpha} H'(c - 2\alpha y).$$

But since (see Figure 9.17) any such line intersects the boundary $x = x_0 \rightarrow -\infty$ as $y \rightarrow +\infty$, it follows that $\partial \varphi / \partial x$ must then tend to zero along each such line as $y \rightarrow +\infty$, in accordance with (332). Hence we must have, in the limit,

$$g'(c) + g'(-\infty) - \frac{V_0}{\alpha} H'(-\infty) = 0,$$

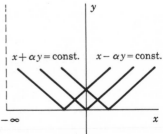

Figure 9.17

or, since $H(x) = 0$ when $x < 0$ and hence $H'(-\infty) = 0$, we must have

$$g'(c) = -g'(-\infty). \qquad (338)$$

But since this result must be true for *all positive values of c,* we then conclude that $g'(c)$ must be a fixed constant independent of c and hence also that $g'(c) = g'(-\infty)$. This statement contradicts (338) unless $g'(c) = 0$, so that we conclude that *the function g is a constant.*

Thus, finally, the deviation potential (336) becomes merely

$$\varphi(x, y) = -\frac{V_0}{\alpha} H(x - \alpha y) + A, \qquad (339)$$

where A is an irrelevant constant. The velocity components are then given by

$$V_x = V_0 + \frac{\partial \varphi}{\partial x} = V_0 \left[1 - \frac{1}{\alpha} H'(x - \alpha y)\right],$$
$$V_y = \frac{\partial \varphi}{\partial y} = V_0 H'(x - \alpha y). \qquad (340)$$

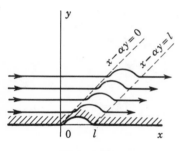

Figure 9.18

From these results we conclude that *the velocity components are constant along any line* $x - \alpha y = constant$. Hence, in particular, the flow is modified from uniformity only in the diagonal strip between the two lines $x - \alpha y = 0$ and $x - \alpha y = l$. The streamlines are as indicated in Figure 9.18.

It is important to notice that any irregularity at a point P of the x axis would propagate a corresponding irregularity in the flow along the particular "characteristic line" $x - \alpha y = c$ which passes through the point P (see Section 8.10).

REFERENCES

1. References at end of Chapters 5 and 8.

2. Carslaw, H. S., and J. C. Jaeger, *Conduction of Heat in Solids,* 2nd ed., Oxford University Press, Inc., New York, 1959.

3. Courant, R., and K. O. Friedrichs, *Supersonic Fluid Flow and Shock Waves,* John Wiley & Sons, Inc. (Interscience Division), New York, 1948.

4. Kellogg, O. D., *Foundations of Potential Theory,* Dover Publications, Inc., New York, 1953.

PROBLEMS

Section 9.1

1. Verify that the equation

$$\nabla^2\varphi + 2a\frac{\partial\varphi}{\partial x} + 2b\frac{\partial\varphi}{\partial y} = 0$$

is transformed into the "modified Helmholtz equation"

$$\nabla^2 U = (a^2 + b^2)U$$

by the substitution

$$\varphi = e^{-(ax+by)}U,$$

if a and b are constant. (See also Problem 19 of Chapter 8.)

2. Verify that the equation

$$\nabla^2\varphi + 2a\frac{\partial\varphi}{\partial x} = \frac{1}{\alpha^2}\frac{\partial\varphi}{\partial t}$$

is transformed into the heat-flow equation

$$\nabla^2 U = \frac{1}{\alpha^2}\frac{\partial U}{\partial t}$$

by the substitution

$$\varphi = e^{-(ax+a^2\alpha^2 t)}U,$$

if a and α are constant. (See also Problem 20 of Chapter 8.)

3. Verify that the equation

$$\nabla^2\varphi + b\varphi = \frac{1}{\alpha^2}\frac{\partial\varphi}{\partial t}$$

is transformed into the heat-flow equation

$$\nabla^2 U = \frac{1}{\alpha^2}\frac{\partial U}{\partial t}$$

by the substitution

$$\varphi = e^{b\alpha^2 t}U,$$

if b and α are constant. (See also Problem 20 of Chapter 8.)

4. If $U(x, y, z)$ satisfies the equation $\nabla^2\varphi = 0$, show also that the function $(ax + by + cz)U(x, y, z)$ satisfies the equation $\nabla^4\varphi = 0$ for any constant values of a, b, and c.

5. If $U(x, y, z)$ satisfies Laplace's equation, show that $\partial U/\partial x$, $\partial U/\partial y$, and $\partial U/\partial z$ are also solutions.

6. If $V(r, \theta, z)$ satisfies Laplace's equation (in circular cylindrical coordinates), show that $\partial V/\partial\theta$ and $\partial V/\partial z$ are also solutions but that $\partial V/\partial r$ generally is *not* a solution.

7. By assuming a solution of Laplace's equation

$$\nabla^2\varphi = 0$$

in the form

$$\varphi(x, y, z) = f(px + qy + z),$$

where p and q are constants and f is an arbitrary twice-differentiable function of its argument, deduce that $\varphi = f(ix \cos u + iy \sin u + z)$ is such a solution and, more generally, show that

$$\varphi(x, y, z) = \int_0^{2\pi} F(ix \cos u + iy \sin u + z, u) \, du$$

is a solution provided that F is such that differentiation under the integral sign is permitted. (This is a "general" solution of Laplace's equation in the sense that any solution which can be expanded in a three-dimensional Taylor series at each point of a region can be expressed in the form given in that region.)

8. By proceeding as in Problem 7, obtain a "general" solution of the wave equation

$$\nabla^2 \varphi = \frac{1}{c^2} \frac{\partial^2 \varphi}{\partial t^2}$$

in the form

$$\varphi(x, y, z, t) = \int_0^{2\pi} \int_0^{2\pi} F(x \cos u \sin v + y \sin u \sin v + z \cos v + ct, u, v) \, du \, dv.$$

Section 9.2

9. Suppose that heat is flowing in a uniform rod of cross section a and perimeter p, and that it is assumed that the temperature T does not vary over a cross section, and hence is a function only of time t and distance x measured along the rod. Assume also that heat escapes from the lateral boundary by radiation, in such a way that the rate of heat loss per unit of area is $\mu K(T - T_0)$, where K is the conductivity of the rod material, T_0 the temperature of the surrounding medium, and μ is a constant.

(a) By considering differential thermal equilibrium in an element $(x, x + dx)$ of the rod, show that there must follow

$$\frac{\partial}{\partial x}\left(K \frac{\partial T}{\partial x}\right) a \, dx - \mu K(T - T_0)p \, dx = s\rho a \frac{\partial T}{\partial t} \, dx,$$

and hence deduce that T then must satisfy the equation

$$\frac{\partial}{\partial x}\left(K \frac{\partial T}{\partial x}\right) = s\rho \frac{\partial T}{\partial t} + \mu K \frac{p}{a}(T - T_0).$$

(b) For a rod of circular cross section, of diameter d and of uniform conductivity K, show that $T(x, t)$ must satisfy the equation

$$\frac{\partial^2 T}{\partial x^2} = \frac{1}{\alpha^2} \frac{\partial T}{\partial t} + \frac{4\mu}{d}(T - T_0),$$

where $\alpha^2 = K/s\rho$.

(c) Verify that, if T_0 is assumed to remain constant, the substitution

$$T(x, t) = T_0 + U(x, t)e^{-4\mu\alpha^2 t/d}$$

leads to the normal heat-flow equation

$$\frac{\partial^2 U}{\partial x^2} = \frac{1}{\alpha^2} \frac{\partial U}{\partial t}.$$

10. Suppose that the *orthogonal coordinates* u_1, u_2, and u_3 of Section 6.17 are used to specify position in a heat-flow problem, and that the conductivity K of the medium is constant.

(a) Show that the heat-flow equation (13) becomes

$$\frac{\partial}{\partial u_1}\left(\frac{h_2 h_3}{h_1}\frac{\partial T}{\partial u_1}\right) + \frac{\partial}{\partial u_2}\left(\frac{h_3 h_1}{h_2}\frac{\partial T}{\partial u_2}\right) + \frac{\partial}{\partial u_3}\left(\frac{h_1 h_2}{h_3}\frac{\partial T}{\partial u_3}\right) = \frac{h_1 h_2 h_3}{\alpha^2}\frac{\partial T}{\partial t}.$$

(b) If **F** is the *flux vector*, representing the rate of heat flow per unit area normal to **F**, and K is the thermal conductivity, show that

$$\mathbf{F} = -K\nabla T = -K\left(\frac{\mathbf{u}_1}{h_1}\frac{\partial T}{\partial u_1} + \frac{\mathbf{u}_2}{h_2}\frac{\partial T}{\partial u_2} + \frac{\mathbf{u}_3}{h_3}\frac{\partial T}{\partial u_3}\right).$$

Also show that the *streamlines* (to which **F** is tangent) are determined by the equations

$$\frac{h_1^2\,du_1}{\partial T/\partial u_1} = \frac{h_2^2\,du_2}{\partial T/\partial u_2} = \frac{h_3^2\,du_3}{\partial T/\partial u_3}.$$

(c) In the case when the temperature and flow are independent of u_3, and when h_1, h_2, and h_3 are independent of u_3, show that the *stream function* $\psi(u_1, u_2)$, determined such that

$$\frac{\partial\psi}{\partial u_1} = -K\frac{h_3 h_1}{h_2}\frac{\partial T}{\partial u_2}, \qquad \frac{\partial\psi}{\partial u_2} = +K\frac{h_3 h_2}{h_1}\frac{\partial T}{\partial u_1},$$

where K is the thermal conductivity, has the property that the streamlines in a surface $u_3 = $ constant are given by $\psi = $ constant.

(d) Show also that the function ψ of part (c) has the additional property that the total rate of heat flow through a surface based on the arc of any curve in a u_3 surface joining points P_1 and P_2, and extending through a *unit* increment of u_3, is given by

$$\int_{P_1}^{P_2}\mathbf{F}\cdot\mathbf{n}h_3 ds = \int_{P_1}^{P_2} h_3\mathbf{F}\cdot d\mathbf{r}\times\mathbf{u}_3 = -\int_{P_1}^{P_2} d\psi = \psi(P_1) - \psi(P_2),$$

and that, when also $h_3 = $ constant, the function ψ also satisfies Laplace's equation.

(See also Section 6.19. Notice that sign differences correspond to the fact that the velocity potential in *fluid* flow has been so defined that flow is from lower to higher potential, whereas temperature is so defined that *heat* flows in the direction of *decreasing* temperature.)

11. Suppose that, in a problem of steady-state heat flow, the temperature T is prescribed as zero over a portion of the boundary S of a region $\mathcal{R}$, the normal derivative $\partial T/\partial n$ is zero over another part of S, and finally a condition of the form $(\partial T/\partial n) + \mu T = 0$ is prescribed over the remainder of S, say S', where μ is a *positive* constant or function of position on S'.

(a) Show that Equation (16) takes the form

$$\iiint_{\mathcal{R}} (\nabla T)^2\,d\tau = -\iint_{S'} \mu T^2\,d\sigma,$$

and deduce that T then must vanish throughout $\mathcal{R}$.

(b) Use this result to show that *the solution of Laplace's equation $\nabla^2 T = 0$ is uniquely determined in a region $\mathcal{R}$ if either T, $\partial T/\partial n$, or $(\partial T/\partial n) + \mu T$ ($\mu > 0$) is prescribed at each point of the closed boundary, except in the case when $\partial T/\partial n$ is prescribed* *over the complete boundary* (*in which case an arbitrary additive constant is present in the solution*).

12. Suppose that the heat-flow equation

$$\nabla^2 T = \frac{1}{\alpha^2} \frac{\partial T}{\partial t}$$

is to be satisfied throughout a region $\Re$ bounded by a closed surface S, for all positive values of time t.

(a) Show formally that Equation (16) then is replaced by the more general form

$$\int \int \int_{\Re} (\nabla T)^2 \, d\tau = -\frac{1}{2\alpha^2} \frac{\partial}{\partial t} \int \int \int_{\Re} T^2 \, d\tau + \oiint_S T \frac{\partial T}{\partial n} \, d\sigma.$$

(b) Suppose that either T or $\partial T / \partial n$ is maintained as zero at all points of S, for all time $t > 0$, and that T is initially zero throughout $\Re$ when $t = 0$. Show that there must follow

$$\int_0^t \left[\int \int \int_{\Re} (\nabla T)^2 \, d\tau \right] dt = -\frac{1}{2\alpha^2} \int \int \int_{\Re} T^2 \, d\tau$$

for all $t > 0$. Hence deduce that T must vanish throughout $\Re$ for all $t > 0$.

(c) Use the result of part (b) to show that the solution of the heat flow equation is uniquely determined for $t > 0$ if T is prescribed in $\Re$ when $t = 0$ and either T or $\partial T / \partial n$ is prescribed on the boundary S for all $t > 0$.

(A rigorous treatment requires for uniqueness that the solution T also be *bounded* in $\Re$ for all $t \geq 0$.)

13. Generalize the result of Problem 12 to the case when either T, $\partial T / \partial n$, or the combination $(\partial T / \partial n) + \mu T$ $(\mu > 0)$ is prescribed at each point of the boundary S for $t > 0$ and T is prescribed throughout $\Re$ when $t = 0$. (See also Problem 11.)

Section 9.3

14. The temperature T is maintained at $0°$ along three edges of a square plate of length 100 cm, and the fourth edge is maintained at $100°$ until steady-state conditions prevail. (Small areas near two corners must be considered as excluded.)

(a) Find an expression for the temperature T at any point (x, y) in the plate, using the notation of Section 9.3.

(b) Calculate the approximate value of the temperature at the center of the plate.

15. Determine the solution of Laplace's equation in the rectangle $0 \leq x \leq l, 0 \leq y \leq d$ which satisfies the conditions

$$T(0, y) = T(l, y) = 0, \qquad T(x, 0) = g(x), \qquad T(x, d) = f(x).$$

[*Suggestion:* Notice that a convenient form of a particular solution of the equation, satisfying the homogeneous conditions, is

$$T_n = \left[a_n \sinh \frac{n\pi y}{l} + b_n \sinh \frac{n\pi(d - y)}{l} \right] \sin \frac{n\pi x}{l},$$

where n is a positive integer.]

16. Suppose that the plate of Section 9.3 is of infinite extent in the y direction, on one side of the boundary $y = 0$, so that it occupies the semi-infinite strip $0 \leq x \leq l$, $0 \leq y < \infty$.

   (a) If the temperature is to vanish on the lateral boundaries $x = 0$ and $x = l$, is to tend to zero as $y \to \infty$, and is to reduce to $f(x)$ along the edge $y = 0$, obtain the temperature distribution in the form

$$T(x, y) = \sum_{n=1}^{\infty} c_n e^{-n\pi y/l} \sin \frac{n\pi x}{l}, \qquad c_n = \frac{2}{l} \int_0^l f(x) \sin \frac{n\pi x}{l} \, dx.$$

   (b) Obtain the same result formally from Equations (34) and (35), by first replacing y by $d - y$ and then considering the limit as $d \to \infty$.

17. Let Problem 16 be modified in such a way that the *rate of heat flow*, per unit distance, into the plate through points of the boundary $y = 0$ is prescribed as $g(x)$, where $g(x)$ may be measured in calories per second per centimeter length along the boundary $y = 0$.

   (a) Show that the condition along the line $y = 0$ then is of the form

$$-Kh \frac{\partial T(x, 0)}{\partial y} = g(x),$$

where h is the thickness of the plate.

   (b) Obtain the solution in the form

$$T(x, y) = \sum_{n=1}^{\infty} c_n e^{-n\pi y/l} \sin \frac{n\pi x}{l}, \qquad c_n = \frac{2}{n\pi Kh} \int_0^l g(x) \sin \frac{n\pi x}{l} \, dx.$$

18. Suppose that $g(x)$ is prescribed in the form

$$g(x) = \begin{cases} 0 & (0 < x < x_0 - \epsilon), \\ \dfrac{1}{2\epsilon} & (x_0 - \epsilon < x < x_0 + \epsilon), \\ 0 & (x_0 + \epsilon < x < l). \end{cases}$$

   (a) Show that the coefficients in Problem 17(b) are then given by

$$c_n = \frac{2l}{n^2 \pi^2 Kh\epsilon} \sin \frac{n\pi\epsilon}{l} \sin \frac{n\pi x_0}{l}.$$

   (b) By considering the limit of c_n as $\epsilon \to 0$, so that the function $g(x)$ tends to the unit singularity function $\delta(x - x_0)$, obtain the temperature distribution due to a permanent *unit heat source*, supplying one calorie of heat per second at the boundary point $x = x_0$, in the form

$$T(x, y) = \frac{2}{\pi Kh} \sum_{n=1}^{\infty} \frac{1}{n} \sin \frac{n\pi x_0}{l} \sin \frac{n\pi x}{l} e^{-n\pi y/l}.$$

[Notice that this result also can be obtained directly from the result of Problem 17, by making use of the fact that

$$\int_a^b \delta(x - x_0)\varphi(x) \, dx = \varphi(x_0) \qquad (a < x_0 < b),$$

for any continuous function φ.]

(c) If the solution of part (b) is denoted by $V(x, y; x_0)$, show that the solution of Problem 17 for any prescribed $g(x)$ can be written in the form

$$T(x, y) = \int_0^l V(x, y; x_0)g(x_0)\, dx_0.$$

Interpret this relation.

19. (a) If $f(x)$ is identified with $\delta(x - x_0)$ in Problem 16, show that the solution of that problem becomes

$$T(x, y) = \frac{2}{l} \sum_{n=1}^{\infty} \sin\frac{n\pi x_0}{l} \sin\frac{n\pi x}{l} e^{-n\pi y/l}.$$

(b) If this solution is denoted by $U(x, y; x_0)$, show that the solution of Problem 16 for any prescribed $f(x)$ can be written in the form

$$T(x, y) = \int_0^l U(x, y; x_0)f(x_0)\, dx_0.$$

(c) Verify that, with the notations of Problems 18(c) and 19(b), there follows

$$U(x, y; x_0) = -Kh\frac{\partial V(x, y; x_0)}{\partial y},$$

and hence that $U(x, y; x_0)$ is the distribution due to a permanent *heat doublet* of strength Kh located at $x = x_0$, the doublet being formed by the confluence of a heat source at $(x_0, 0)$ and a heat sink at (x_0, ϵ), each of strength Kh/ϵ, as $\epsilon \longrightarrow 0$.

20. If Problem 16 is modified in such a way that the lateral edges $(x = 0$ and $x = l)$ of the strip are *insulated* [so that $\partial T/\partial n = \pm(\partial T/\partial x)$ vanishes along those lines], obtain the temperature distribution in the strip in the form

$$T(x, y) = c_0 + \sum_{n=1}^{\infty} c_n e^{-n\pi y/l} \cos\frac{n\pi x}{l},$$

where

$$c_0 = \frac{1}{l}\int_0^l f(x)\, dx, \qquad c_n = \frac{2}{l}\int_0^l f(x) \cos\frac{n\pi x}{l}\, dx \qquad (n = 1, 2, \ldots).$$

21. By imposing the condition $\nabla^2\varphi = 0$ on an expression of the form

$$\varphi = A_0 + A_1 x + A_2 y + A_3 x^2 + A_4 xy + A_5 y^2 + A_6 x^3 + \cdots,$$

deduce that *polynomial solutions of Laplace's equation* are specializations of the expression

$$\varphi = a_0 + a_1 x + a_2 y + a_3 xy + a_4(x^2 - y^2) + a_5(x^3 - 3xy^2)$$
$$+ a_6(y^3 - 3x^2 y) + \cdots.$$

(See also Problem 26 of Chapter 8.)

22. Find the solution $T(x, y)$ of Laplace's equation in the semi-infinite strip $0 \leq x \leq l$, $0 \leq y < \infty$, which does not grow exponentially with y, and is such that

$$T(0, y) = \alpha_1 + \beta_1 y, \qquad T(l, y) = \alpha_2 + \beta_2 y, \qquad T(x, 0) = f(x).$$

[*Suggestion*: Write $T = p(x, y) + u(x, y)$, where $p(x, y)$ is a *polynomial* solution of Laplace's equation, satisfying the conditions prescribed along $x = 0$ and $x = l$, and obtained from the result of Problem 21 in the form

$$p = \alpha_1 + (\alpha_2 - \alpha_1)\frac{x}{l} + \beta_1 y + (\beta_2 - \beta_1)\frac{xy}{l},$$

and where the correction $u(x, y)$ accordingly also satisfies Laplace's equation, but is subject to *homogeneous* conditions $u(0, y) = 0$, $u(l, y) = 0$ along $x = 0$ and $x = l$, together with the initial condition $u(x, 0) = f(x) - \alpha_1 - (\alpha_2 - \alpha_1)(x/l)$.]

23. Find the solution of Laplace's equation, in the rectangle $0 \leq x \leq l$, $0 \leq y \leq d$, which satisfies the conditions

$$T(0, y) = 0, \qquad T(l, y) = T_0 + \frac{cy}{d},$$

$$T(x, 0) = 0, \qquad T(x, d) = f(x).$$

(See Problems 22 and 15.)

24. Find the solution of the equation

$$\frac{\partial^2 T}{\partial x^2} + \frac{\partial^2 T}{\partial y^2} = \sin \omega x \qquad (\omega > 0),$$

in the strip $0 \leq x \leq l$, $0 \leq y < \infty$, which satisfies the conditions

$$T(x, 0) = T(0, y) = T(l, y) = 0$$

and the requirement that T be bounded as $y \longrightarrow \infty$ in the strip. Consider separately the case in which $\omega = k\pi/l$, where k is an integer.

25. Show that the equation

$$a \frac{\partial^2 z}{\partial x^2} + b \frac{\partial^2 z}{\partial x \, \partial y} + c \frac{\partial^2 z}{\partial y^2} + d \frac{\partial z}{\partial x} + e \frac{\partial z}{\partial y} + fz = 0$$

is separable if and only if it can be written in one of the three forms

$$H(y)\left[A(x) \frac{\partial^2 z}{\partial x^2} + D(x) \frac{\partial z}{\partial x} \right] + G(x)\left[C(y) \frac{\partial^2 z}{\partial y^2} + E(y) \frac{\partial z}{\partial y} \right]$$
$$+ [H(y)F_1(x) + G(x)F_2(y)]z = 0,$$

$$H(y)\left[A(x) \frac{\partial^2 z}{\partial x^2} + F_1(x)z \right] + G(x) \frac{\partial^2 z}{\partial x \, \partial y} + [H(y)D_1(x) + G(x)D_2(y)] \frac{\partial z}{\partial x} = 0,$$

$$G(x)\left[C(y) \frac{\partial^2 z}{\partial y^2} + F_2(y)z \right] + H(y) \frac{\partial^2 z}{\partial x \, \partial y} + [H(y)E_1(x) + G(x)E_2(y)] \frac{\partial z}{\partial y} = 0,$$

and, in each case, give the ordinary differential equations which must be satisfied by factors in a product solution $z = X(x)Y(y)$. (The first form is the most important one in practice.)

Section 9.4

26. Show that the necessity of determining constants in Equation (48) by solving successive pairs of equations can be avoided by writing (48) in the form

$$T = a_0 \log \frac{r}{r_1} + b_0 \log \frac{r}{r_2} + \sum_{n=1}^{\infty} \left\{ a_n\left[\left(\frac{r}{r_1}\right)^n - \left(\frac{r_1}{r}\right)^n \right] + b_n\left[\left(\frac{r}{r_2}\right)^n - \left(\frac{r_2}{r}\right)^n \right] \right\} \cos n\theta$$

$$+ \sum_{n=1}^{\infty} \left\{ c_n\left[\left(\frac{r}{r_1}\right)^n - \left(\frac{r_1}{r}\right)^n \right] + d_n\left[\left(\frac{r}{r_2}\right)^n - \left(\frac{r_2}{r}\right)^n \right] \right\} \sin n\theta.$$

(Compare Problem 15.)

27. Along the inner boundary of a circular annulus of radii 10 cm and 20 cm the temperature is maintained as $T(10, \theta) = 15 \cos \theta$, and along the outer boundary the distribution $T(20, \theta) = 30 \sin \theta$ is maintained. Find an expression for the steady-state temperature at an arbitrary point (r, θ) in the annulus.

28. Along the circumference of the unit circle $r = 1$ a solution of Laplace's equation is required to take on the value unity when $0 < \theta < \pi$ and the value zero when $\pi < \theta < 2\pi$.
    (a) Determine an expression for T valid when $r < 1$.
    (b) Determine a corresponding expression valid when $r > 1$.

29. (a) Determine the steady-state temperature at points of the sector $0 \leq \theta \leq \alpha$, $0 \leq r \leq a$ of a circular plate if the temperature is maintained at zero along the straight edges and at a prescribed distribution $T(a, \theta) = f(\theta)$ when $0 < \theta < \alpha$, along the curved edge.
    (b) In the special case when $f(\theta) = T_0 = $ constant, show that the temperature at interior points is given by

$$T(r, \theta) = \frac{4T_0}{\pi} \sum_{n \text{ odd}} \frac{1}{n} \left(\frac{r}{a}\right)^{n\pi/\alpha} \sin \frac{n\pi\theta}{\alpha}.$$

30. (a) Show that the solution of Problem 29(b) tends to the form

$$T(r, \theta) = \frac{4T_0}{\pi} \sum_{n \text{ odd}} \frac{1}{n} \left(\frac{r}{a}\right)^{n/2} \sin \frac{n\theta}{2}$$

as the opening angle α of the sector tends to 2π, so that the sector fans out into the interior of the complete circle with a *cut* along the radius which coincides with the positive x axis.
    (b) Noticing that $\theta \longrightarrow 0$ as the cut is approached from above, whereas $\theta \longrightarrow 2\pi$ as the cut is approached from below, show that the interior temperature distribution of part (a) is *continuous* across the cut.
    (c) Show that the derivative of the temperature in the positive y direction tends to

$$\left[\frac{1}{r} \frac{\partial T}{\partial \theta}\right]_{\theta=0} = \frac{2T_0}{\pi a} \sum_{n \text{ odd}} \left(\frac{r}{a}\right)^{n/2-1} = \frac{2T_0}{\pi} \frac{(r/a)^{-1/2}}{a - r} \qquad (0 < r < a)$$

as the cut is approached from above, whereas it tends to the *negative* of that quantity as the cut is approached from below. [Since $-(\partial T/\partial y)$ is proportional to the rate of heat flow in the positive y direction, this means merely that heat must be continually drawn off from the plate at all points of the cut in order that the desired temperature distribution may be maintained.]

31. The function $\varphi(r, \theta)$ is required subject to the conditions that φ satisfy Laplace's equation inside the circle $r = a$ and at all finite points outside that circle, that φ be continuous across the circle $r = a$, that $\partial\varphi/\partial r$ decrease abruptly by a prescribed function $F(\theta)$ as r increases through a, and that $\partial\varphi/\partial r$ approach zero as $r \longrightarrow \infty$.
    (a) Show that φ may be assumed in the form

$$\varphi(r, \theta) = A_0 + \begin{cases} \sum_{n=1}^{\infty} \left(\frac{r}{a}\right)^n (A_n \cos n\theta + B_n \sin n\theta) & (0 \leq r \leq a) \\[2mm] C \log \frac{a}{r} + \sum_{n=1}^{\infty} \left(\frac{a}{r}\right)^n (A_n \cos n\theta + B_n \sin n\theta) & (a \leq r < \infty). \end{cases}$$

(b) Show that the additive constant A_0 is arbitrary, and that the remaining coefficients are given by

$$C = \frac{a}{2\pi} \int_0^{2\pi} F(\theta)\, d\theta, \qquad A_n = \frac{a}{2n\pi} \int_0^{2\pi} F(\theta) \cos n\theta\, d\theta,$$

$$B_n = \frac{a}{2n\pi} \int_0^{2\pi} F(\theta) \sin n\theta\, d\theta \qquad (n = 1, 2, \ldots).$$

32. Suppose that the outer boundary $r = r_2$ of an annular plate is insulated, and that the temperature is prescribed as $f(\theta)$ along the inner boundary $r = r_1$.

(a) Show that the expression (48) must be specialized to the form

$$T = a_0 + \sum_{n=1}^{\infty} (r^n + r_2^{2n} r^{-n})(a_n \cos n\theta + c_n \sin n\theta)$$

when $r_1 \leq r \leq r_2$.

(b) By writing $A_n = r_1^n a_n$ and $C_n = r_1^n c_n$, and introducing the abbreviation $\beta = r_2/r_1$, obtain the desired temperature distribution in the annulus in the form

$$T = A_0 + \sum_{n=1}^{\infty} \left[\left(\frac{r}{r_1}\right)^n + \beta^{2n}\left(\frac{r_1}{r}\right)^n \right] (A_n \cos n\theta + C_n \sin n\theta),$$

where

$$A_0 = \frac{1}{2\pi} \int_0^{2\pi} f(\theta)\, d\theta, \quad A_n = \frac{\int_0^{2\pi} f(\theta) \cos n\theta\, d\theta}{\pi(1 + \beta^{2n})}, \quad C_n = \frac{\int_0^{2\pi} f(\theta) \sin n\theta\, d\theta}{\pi(1 + \beta^{2n})}.$$

Section 9.5

33. By making use of the relation

$$\int_0^{2\pi} \frac{d\varphi}{1 - A \cos \varphi} = \frac{2\pi}{\sqrt{1 - A^2}} \qquad (|A| < 1),$$

obtain the result

$$\frac{1}{2\pi} \int_0^{2\pi} \frac{a^2 - r^2}{a^2 - 2ar \cos (\theta - \varphi) + r^2}\, d\varphi = 1 \qquad (|r| < a),$$

and thus verify directly the validity of the Poisson integral formula (64) in the case when $T(a, \theta) = T_0 = $ constant. (Notice that the integrand is periodic in φ.)

34. Use the Poisson integral formula (64) and the result of Problem 33 to show that, when $0 \leq r \leq a$, there follows

$$T(r, \theta) \leq \frac{1}{2\pi} \int_0^{2\pi} \frac{a^2 - r^2}{a^2 - 2ar \cos (\theta - \varphi) + r^2} [T(a, \varphi)]_{\max}\, d\varphi = [T(a, \theta)]_{\max},$$

and hence, by considering also the corresponding inequality for $-T(r, \theta)$, deduce that, *if T satisfies Laplace's equation inside a circle, then T cannot take on its maximum or minimum value inside the boundary unless it is constant throughout the circle.* (The same result can be established for an arbitrary finite region.)

35. Show that when $r > a$, the *negative* of the right-hand side of Equation (64) represents the harmonic function $T(r, \theta)$ at all points *outside* the circle $r = a$ in terms of $T(a, \theta)$. [Compare (53) and (60).]

Section 9.6

36. (a) The temperature T on the surface of a sphere of radius a is maintained at $T = T_0(1 - \cos \varphi)$, where φ is the cone angle. Find the steady-state temperature at an arbitrary point inside the sphere.

(b) If, instead, the temperature is maintained at a constant value T_0 over the upper hemisphere $(0 \leq \varphi < \pi/2)$ and at zero over the remainder of the surface, determine the first three terms in the series solution of Section 9.6.

37. If the temperature of a spherical surface $r = a$ is maintained at $T(a, \varphi) = f(\varphi)$, where φ is the cone angle, and if the temperature tends to zero as $r \longrightarrow \infty$, show that the temperature distribution *outside* the sphere is given by

$$T(r, \varphi) = \sum_{n=0}^{\infty} c_n \left(\frac{a}{r}\right)^{n+1} P_n(\cos \varphi),$$

where c_n is defined by Equation (82).

38. If a spherical surface $r = a$ is maintained at constant temperature T_0, show that the temperature at all internal points is also T_0, whereas the temperature at points outside the sphere is given by $T_0 a/r$ $(r \geq a)$.

39. Suppose that the steady-state temperature T in a solid right circular cylinder possesses axial symmetry, and hence is of the form $T = T(r, z)$, where r is distance from the z axis.

(a) Show that T then must satisfy the equation

$$\frac{1}{r} \frac{\partial}{\partial r}\left(r \frac{\partial T}{\partial r}\right) + \frac{\partial^2 T}{\partial z^2} = 0$$

inside the cylinder and that, if a product solution is assumed in the form $T(r, z) = R(r)Z(z)$, then R and Z must satisfy equations of the form

$$r \frac{d^2 R}{dr^2} + \frac{dR}{dr} + \alpha r R = 0, \qquad \frac{d^2 Z}{dz^2} - \alpha Z = 0,$$

where α is an arbitrary constant.

(b) If only solutions which remain finite along the z axis are admissible, show that the choice $\alpha = k^2$, where k is real and positive, leads to exponential or hyperbolic functions of z and to the function $J_0(kr)$, whereas the choice $\alpha = -k^2$ leads to circular functions of z and to the function $I_0(kr)$. Finally, show that the choice $\alpha = 0$ leads to $R = c_1$ and $Z = d_1 z + d_2$.

40. (a) Suppose that a solid right circular cylinder of radius a is of infinite extent on one side of the plane face $z = 0$, and that the temperature is maintained at zero along the lateral boundary, whereas the temperature distribution over the face $z = 0$ is prescribed as $T(r, 0) = f(r)$. Use results of Problem 39 to show that the steady-state temperature at interior points is given by

$$T(r, z) = \sum_{n=1}^{\infty} A_n e^{-k_n z} J_0(k_n r),$$

where k_n is the nth positive root of the equation $J_0(ka) = 0$, and where

$$A_n = \frac{2}{a^2 [J_1(k_n a)]^2} \int_0^a r f(r) J_0(k_n r) \, dr.$$

(b) If the temperature over the face $z = 0$ is maintained at a constant value T_0, show that the resultant interior distribution is given by

$$T(r, z) = \frac{2T_0}{a} \sum_{n=1}^{\infty} e^{-k_n z} \frac{J_0(k_n r)}{k_n J_1(k_n a)}.$$

41. Suppose that the faces $z = 0$ and $z = l$ of a solid right circular cylinder are maintained at temperature zero, and that the temperature distribution along the lateral boundary $r = a$ is dependent only on z, and is prescribed in the form $T(a, z) = f(z)$. Use results of Problem 39 to obtain the resultant steady-state temperature distribution inside the cylinder in the form

$$T(r, z) = \sum_{n=1}^{\infty} A_n \frac{I_0(n\pi r/l)}{I_0(n\pi a/l)} \sin \frac{n\pi z}{l},$$

where

$$A_n = \frac{2}{l} \int_0^l f(z) \sin \frac{n\pi z}{l} \, dz.$$

Section 9.7

42. Suppose that a column with rectangular cross section can be considered to be of infinite height on one side of its base $0 \leq x \leq l_1$, $0 \leq y \leq l_2$ in the plane $z = 0$. If the lateral boundaries are maintained at zero temperature and the temperature distribution over the base is prescribed as $T(x, y, 0) = f(x, y)$, obtain the internal temperature distribution in the form

$$T(x, y, z) = \sum_{m=1}^{\infty} \sum_{n=1}^{\infty} c_{mn} e^{-k_{mn} z} \sin \frac{m\pi x}{l_1} \sin \frac{n\pi y}{l_2},$$

where k_{mn} and c_{mn} are defined by Equations (100) and (106).

43. (a) If the column considered in Problem 42 is of *circular* cross section, of radius a, and if the temperature over the base is prescribed as $T(r, \theta, 0) = f(r, \theta)$, show that the internal temperature distribution is given by

$$T(r, \theta, z) = \sum_{n=1}^{\infty} a_{0n} e^{-k_{0n} z} J_0(k_{0n} r) + \sum_{m=1}^{\infty} \sum_{n=1}^{\infty} e^{-k_{mn} z} J_m(k_{mn} r)(a_{mn} \cos m\theta + b_{mn} \sin m\theta),$$

where k_{mn} is the nth positive root of the equation $J_m(ka) = 0$, and where a_{mn} and b_{mn} are to be determined in such a way that

$$\sum_{n=1}^{\infty} \left[a_{0n} J_0(k_{0n} r) + \sum_{m=1}^{\infty} (a_{mn} \cos m\theta + b_{mn} \sin m\theta) J_m(k_{mn} r) \right] = f(r, \theta)$$

when $0 \leq r < a$ and $0 < \theta < 2\pi$.

(b) By multiplying the equal members of the preceding equation by

$$J_p(k_{pq} r) \sin p\theta \, r \, dr \, d\theta,$$

where p and q are any positive integers, and integrating the result over the base of the column, deduce formally that

$$b_{pq} \frac{\pi a^2}{2} [J_{p+1}(k_{pq} a)]^2 = \int_0^{2\pi} \int_0^a J_p(k_{pq} r) f(r, \theta) \sin p\theta \, r \, dr \, d\theta.$$

(The remaining coefficients can be determined in a similar way.)

44. *Multidimensional characteristic functions.*

(a) Consider the characteristic-value problem in which λ is to be determined such that the differential equation

$$\nabla^2 F + \lambda F = 0 \qquad in \ \Re$$

admits a nontrivial solution which vanishes on the boundary of the region $\Re$. If λ_1 and λ_2 are two distinct characteristic numbers and Φ_1 and Φ_2 are corresponding characteristic functions, show that

$$(\lambda_2 - \lambda_1) \iiint_\Re \Phi_1 \Phi_2 \, d\tau = \iiint_\Re (\Phi_2 \nabla^2 \Phi_1 - \Phi_1 \nabla^2 \Phi_2) \, d\tau.$$

By transforming the right-hand member to an integral over the boundary of $\Re$ [see Equation (123) of Chapter 6], deduce that

$$\iiint_\Re \Phi_1 \Phi_2 \, d\tau = 0 \qquad if \ \lambda_2 \neq \lambda_1.$$

Show also that the orthogonality property still follows if, at each boundary point, the generating characteristic-value problem requires that either $F = 0$ or $\partial F/\partial n = 0$ or $F + \alpha(\partial F/\partial n) = 0$. (Notice that the preceding results apply equally to a two-dimensional problem, in which $\Re$ is a portion of a plane or surface and its boundary is a curve.)

(b) Generalize the results of part (a) to the case when the governing equation is

$$\nabla \cdot (p\nabla F) + \lambda r F = 0,$$

where p is continuously differentiable in $\Re$ and r is continuous, showing that

$$\iiint_\Re r \Phi_1 \Phi_2 \, d\tau = 0 \qquad if \ \lambda_2 \neq \lambda_1.$$

(See Problem 84 of Chapter 6.)

45. *Poisson's equation.* Suppose that the function φ is required such that

$$\nabla^2 \varphi = h \qquad in \ \Re,$$

where φ is to satisfy a *homogeneous* condition, of the form $\varphi = 0$ or $\partial \varphi/\partial n = 0$ or $\varphi + \alpha(\partial \varphi/\partial n) = 0$, on each part of the boundary of $\Re$.

(a) If the associated characteristic-value problem

$$\nabla^2 F + \lambda F = 0 \qquad in \ \Re,$$

subject to the same boundary conditions as those in part (a), has the characteristic numbers $\lambda_1, \lambda_2, \ldots$ and corresponding characteristic functions $\Phi_1, \Phi_2, \ldots$, so that $\nabla^2 \Phi_k + \lambda_k \Phi_k = 0$, use the results of Problem 44(a) to show formally that the solution of the original problem is of the form

$$\varphi = -\sum_{k=1}^\infty \frac{A_k}{\lambda_k} \Phi_k,$$

provided that $\lambda = 0$ *is not a characteristic number*, where A_k is the coefficient of Φ_k in the representation of h in $\Re$,

$$h = \sum_{k=1}^\infty A_k \Phi_k.$$

Show also that

$$A_k = \frac{\int_\Re \Phi_k h \, d\tau}{\int_\Re \Phi_k^2 \, d\tau},$$

where $\int_\Re (\cdots) \, d\tau$ here denotes integration over the two- or three-dimensional region $\Re$. [In a two-dimensional problem, it often is desirable to label the characteristic numbers and functions with a *double* subscript (say, as λ_{mn} and Φ_{mn}) in which case the sums on k are replaced by double sums on m and n. Similarly, in a three-dimensional problem three indices may be desirable.]

(b) Specialize the results of part (a) to obtain the solution of Poisson's equation in a *rectangle*,

$$\frac{\partial^2 \varphi}{\partial x^2} + \frac{\partial^2 \varphi}{\partial y^2} = h(x, y) \qquad (0 \leq x \leq l_1, \, 0 \leq y \leq l_2),$$

with $\varphi = 0$ along the complete boundary, in the form

$$\varphi(x, y) = -\sum_{m=1}^{\infty} \sum_{n=1}^{\infty} \frac{A_{mn}}{\pi^2(m^2/l_1^2 + n^2/l_2^2)} \sin \frac{m\pi x}{l_1} \sin \frac{n\pi y}{l_2},$$

where

$$A_{mn} = \frac{4}{l_1 l_2} \int_0^{l_1} \int_0^{l_2} h(x, y) \sin \frac{m\pi x}{l_1} \sin \frac{n\pi y}{l_2} \, dy \, dx,$$

(Notice that, if φ were prescribed otherwise on the boundary, the required solution could be obtained by first solving that problem with h replaced by zero, and then adding the above expression to that solution.)

46. Let the Laplace operator ∇^2 be replaced by the Helmholtz operator $\nabla^2 + \Lambda$ in the differential equation for φ in Problem 45, so that φ is to satisfy the equation

$$\nabla^2 \varphi + \Lambda \varphi = h \qquad in \ \Re,$$

with the boundary conditions specified as before.

(a) Show that then

$$\varphi = \sum_{k=1}^{\infty} \frac{A_k}{\Lambda - \lambda_k} \Phi_k,$$

with the notation of Problem 45(a), provided that Λ is not identified with one of the characteristic values of λ.

(b) If $\Lambda = \lambda_r$, show that there is no solution unless $A_r = 0$, so that

$$\int_\Re \Phi_r h \, d\tau = 0,$$

in which case the coefficient of Φ_r in the expression for φ is arbitrary, the other coefficients being determined as before.

(c) Specialize the preceding results to the rectangle of Problem 45(b).
(Compare the results of Section 5.8.)

47. *The general Dirichlet problem for a sphere.* It is required to determine the solution of Laplace's equation, $\nabla^2 T = 0$, inside the sphere $r = a$, such that $T = f$ on the sphere surface, where f is prescribed.

(a) Using spherical coordinates (r, φ, θ), show that a product $T_p = R(r)F(\varphi, \theta)$ satisfies Laplace's equation if and only if

$$\frac{d^2R}{dr^2} + 2r\frac{dR}{dr} - \lambda R = 0$$

and

$$\mathbf{V}_s^2 F + \lambda F = 0,$$

where $\mathbf{V}_s^2$ is the two-dimensional Laplacian operator on the unit sphere $r = 1$,

$$\mathbf{V}_s^2 F = \frac{1}{\sin \varphi}\frac{\partial}{\partial \varphi}\left(\frac{\partial F}{\partial \varphi}\sin \varphi\right) + \frac{1}{\sin^2 \varphi}\frac{\partial^2 F}{\partial \theta^2},$$

and λ is a separation constant.

(b) Show that product solutions of the F equation which are periodic in θ must be of the form $\Phi(\varphi)\Theta(\theta)$, where

$$\frac{d^2\Theta}{d\theta^2} + m^2\Theta = 0$$

and

$$\frac{d^2\Phi}{d\varphi^2} + \frac{d\Phi}{d\varphi}\cot \varphi + (\lambda - m^2\csc^2 \varphi)\Phi = 0,$$

with m an integer (which can be taken to be nonnegative) and that such a product will be finite at the poles of the sphere ($\varphi = 0, \pi$) if and only if $\lambda = n(n + 1)$, where n is an integer, and the φ factor is a multiple of $P_n^m(\cos \varphi)$. (Use relevant results stated in Section 4.13.) Hence deduce that, for each nonnegative integer n, $\lambda = \lambda_n = n(n + 1)$ is a characteristic number with a $(2n + 1)$-parameter characteristic function

$$Y_n(\varphi, \theta) = a_{n0} P_n(\cos \varphi) + \sum_{m=1}^{n} (a_{nm}\cos m\theta + b_{nm}\sin m\theta) P_n^m (\cos \varphi).$$

(Any such expression is called a *surface spherical harmonic* of degree n.)

(c) Verify that

$$(\lambda_p - \lambda_q) \oiint_S Y_p Y_q \, d\sigma = \int_0^\pi \int_0^{2\pi} \left\{\frac{\partial}{\partial \varphi}\left[\left(Y_p\frac{\partial Y_q}{\partial \varphi} - Y_q\frac{\partial Y_p}{\partial \varphi}\right)\sin \varphi\right]\right.$$
$$\left. + \frac{1}{\sin \varphi}\frac{\partial}{\partial \theta}\left(Y_p\frac{\partial Y_q}{\partial \theta} - Y_q\frac{\partial Y_p}{\partial \theta}\right)\right\} d\theta \, d\varphi$$
$$= 0 \quad (p \neq q),$$

where S is the unit sphere $r = 1$ and, accordingly, $d\sigma = \sin \varphi \, d\theta \, d\varphi$, and deduce that the product functions

$$P_n(\cos \varphi), \qquad P_n^m(\cos \varphi)\cos m\theta, \qquad P_n^m(\cos \varphi)\sin m\theta$$

($n = 0, 1, 2, \ldots$; $m = 0, 1, 2, \ldots$) comprise an orthogonal set on the surface of that sphere.

(d) Assuming that the function $f(\varphi, \theta)$ can be represented by a series of the spherical harmonics on the surface of the unit sphere, and that the series can be integrated term by term over that surface, and making use of the known formula

$$\int_0^\pi [P_n^m(\cos \varphi)]^2 \sin \varphi \, d\varphi = \frac{2(n + m)!}{(2n + 1)(n - m)!},$$

deduce that

$$f(\varphi, \theta) = \sum_{n=0}^{\infty} Y_n(\varphi, \theta)$$

$$= \sum_{n=0}^{\infty} a_{n0} P_n(\cos \varphi) + \sum_{n=0}^{\infty} \sum_{m=1}^{\infty} (a_{nm} \cos m\theta + b_{nm} \sin m\theta) P_n^m(\cos \varphi)$$

for $0 \leqq \varphi \leqq \pi$, $0 \leqq \theta < 2\pi$, where

$$a_{n0} = \frac{2n + 1}{4\pi} \oiint_S f(\varphi, \theta) P_n(\cos \varphi) \, d\sigma,$$

$$a_{nm} = \frac{2n + 1}{2\pi} \frac{(n - m)!}{(n + m)!} \oiint_S f(\varphi, \theta) P_n^m(\cos \varphi) \cos m\theta \, d\sigma,$$

$$b_{nm} = \frac{2n + 1}{2\pi} \frac{(n - m)!}{(n + m)!} \oiint_S f(\varphi, \theta) P_n^m(\cos \varphi) \sin m\theta \, d\sigma.$$

(This representation is valid, in particular, when f and its first partial derivatives are continuous functions of φ and θ on the sphere S, but these conditions are by no means necessary.)

(e) From the preceding results, deduce that the solution of Laplace's equation $\nabla^2 T = 0$ in the sphere $r = a$, which tends to $f(\varphi, \theta)$ as $r \longrightarrow a$ (and is finite at $r = 0$), is of the form

$$T(r, \varphi, \theta) = \sum_{n=0}^{\infty} \left(\frac{r}{a}\right)^n Y_n(\varphi, \theta),$$

with the notation of part (d). (The terms superimposed in this representation are sometimes called *solid spherical harmonics*, or simply *spherical harmonics*.)

Section 9.8

48. Let a right circular cylinder of radius a be placed in an initially uniform flow of an ideal incompressible fluid, in such a way that the axis of the cylinder coincides with the z axis and the flow tends to a uniform flow with velocity V_0 in the x direction at large distances from the cylinder. Assume that the cylinder can be considered to be of infinite length.

(a) If polar coordinates are used in the xy plane of flow, show that the velocity vector $\mathbf{V}$ is related to the velocity potential $\varphi(r, \theta)$ by the equation

$$\mathbf{V} = \mathbf{u}_r \frac{\partial \varphi}{\partial r} + \mathbf{u}_\theta \frac{1}{r} \frac{\partial \varphi}{\partial \theta},$$

and that φ must satisfy the equation

$$\frac{1}{r} \frac{\partial}{\partial r} \left(r \frac{\partial \varphi}{\partial r}\right) + \frac{1}{r^2} \frac{\partial^2 \varphi}{\partial \theta^2} = 0$$

and the boundary conditions

$$\frac{\partial \varphi(a, \theta)}{\partial r} = 0, \qquad \frac{\partial \varphi(\infty, \theta)}{\partial r} = V_0 \cos \theta,$$

in addition to the requirement that $V_r \equiv \partial \varphi / \partial r$ be a single-valued function of θ.

(b) By imposing these restrictions on an expression of form (45), obtain the velocity potential in the form

$$\varphi(r, \theta) = c_0 + k\theta + V_0\left(r + \frac{a^2}{r}\right)\cos\theta,$$

where c_0 and k are arbitrary constants, and show that the velocity vector then is given by

$$\mathbf{V} = \left[V_0\left(1 - \frac{a^2}{r^2}\right)\cos\theta\right]\mathbf{u}_r - \left[V_0\left(1 + \frac{a^2}{r^2}\right)\sin\theta\right]\mathbf{u}_\theta + \frac{k}{r}\mathbf{u}_\theta.$$

(c) Verify that V_r vanishes when $r = a$ and that $\mathbf{V}$ tends to $V_0\mathbf{i}$ as $r \to \infty$, regardless of the value of the constant k.

49. (a) Show that the stream function $\psi(r, \theta)$ corresponding to the flow obtained in Problem 48(b) can be expressed in the form

$$\psi(r, \theta) = d_0 - k\log r + V_0\left(r - \frac{a^2}{r}\right)\sin\theta,$$

where d_0 is an arbitrary constant, and hence that the streamlines are defined by the equation

$$V_0 y\left(1 - \frac{a^2}{x^2 + y^2}\right) - \frac{k}{2}\log(x^2 + y^2) = \text{constant},$$

in rectangular coordinates.

(b) Show that the flow is symmetrical about the x axis when $k = 0$ and sketch typical streamlines in this case.

(c) Show that the added velocity potential $k\theta$ (which is not single-valued) corresponds to a circulatory flow about the cylinder, in which the velocity $\mathbf{V}_c$ is circumferential, with magnitude k/r at distance r from the center of the cylinder section, and that the circulation $\oint_C \mathbf{V}_c \cdot d\mathbf{r}$ is given by $2\pi k$, around any circle $r = \text{constant}$.

(Notice that the flow is not determined unless the circulation is prescribed, *in addition to* the values of V_r at $r = a$ and of $\mathbf{V}$ as $r \to \infty$. Notice also that the flow is indeed vortex-free, since the point $r = 0$ about which the fluid tends to rotate is not a point in the fluid itself.)

50. The "stagnation points" of a flow are the points at which the flow velocity is zero.

(a) Show that when no circulation is present the stagnation points in the flow of Problem 48 are the points $(a, 0)$ and (a, π) at the intersections of the x axis and the boundary of the cylinder section.

(b) With the abbreviation $\alpha = k/(2V_0a)$, show that when circulation is present the stagnation points are $(a, \sin^{-1}\alpha)$ and $(a, \pi - \sin^{-1}\alpha)$ when $|\alpha| < 1$. Show also that when $|\alpha| \geq 1$ there is only one stagnation point in the flow, and that that point is at a distance $a(|\alpha| + \sqrt{\alpha^2 - 1})$ from the center of the cylinder section, on the positive or negative y axis, according to whether α is positive or negative.

Section 9.9

51. By multiplying both sides of Equation (148a) by

$$J_p\left(\frac{\omega_{pq}r}{c}\right)\cos p\theta\; r\, dr\, d\theta$$

and integrating over the area of the membrane, obtain the results

$$a_{mn} \frac{\pi a^2}{2} \left[J_{m+1}\left(\frac{\omega_{mn}a}{c}\right) \right]^2 = \int_0^{2\pi} \int_0^a J_m\left(\frac{\omega_{mn}r}{c}\right) f_1(r, \theta) \cos m\theta \; r \; dr \; d\theta \qquad (m \neq 0)$$

and

$$a_{0n} \pi a^2 \left[J_1\left(\frac{\omega_{0n}a}{c}\right) \right]^2 = \int_0^{2\pi} \int_0^a J_0\left(\frac{\omega_{0n}r}{c}\right) f_1(r, \theta) \; r \; dr \; d\theta \qquad (m = 0).$$

(The coefficients b_{mn}, where $m \geq 1$, are determined in a similar way.)

52. For a freely vibrating square membrane of side l, supported along the boundary $x = 0$, $x = l$, $y = 0$, $y = l$, obtain a permissible expression for the deflection $w(x, y, t)$ by methods analogous to those of Section 9.9, in the form

$$w = \sum_{m=1}^{\infty} \sum_{n=1}^{\infty} \sin\frac{m\pi x}{l} \sin\frac{n\pi y}{l} (a_{mn} \cos \omega_{mn}t + b_{mn} \sin \omega_{mn}t),$$

where

$$\omega_{mn} = \pi\sqrt{m^2 + n^2}\sqrt{\frac{T}{\rho l^2}}.$$

53. (a) Suppose that the square membrane of Problem 52 is initially deflected in the form

$$w(x, y, 0) = f(x, y)$$

and is released from rest. Obtain an expression for the ensuing deflection in the form

$$w = \sum_{m=1}^{\infty} \sum_{n=1}^{\infty} a_{mn} \sin\frac{m\pi x}{l} \sin\frac{n\pi y}{l} \cos \omega_{mn}t,$$

where

$$a_{mn} = \frac{4}{l^2} \int_0^l \int_0^l f(x, y) \sin\frac{m\pi x}{l} \sin\frac{n\pi y}{l} \; dx \; dy.$$

(b) Obtain an expression for w in Problem 52 if initially the membrane is undeflected but if the membrane is in motion such that

$$\frac{\partial w(x, y, 0)}{\partial t} = v(x, y).$$

54. (a) From the results of Problems 52 and 53, show that free vibrations of a square membrane are compounded of "natural modes" of the form

$$w_{mn}(x, y, t) = A_{mn} \sin\frac{m\pi x}{l} \sin\frac{n\pi y}{l} \cos (\omega_{mn}t + \alpha_{mn}),$$

with circular frequencies

$$\omega_{mn} = \pi\sqrt{m^2 + n^2}\sqrt{\frac{T}{\rho l^2}} \equiv \sqrt{\frac{m^2 + n^2}{2}}\,\Omega,$$

where $\Omega = \pi\sqrt{(2T)/(\rho l^2)}$ is the fundamental circular frequency, and with corresponding amplitudes

$$\varphi_{mn}(x, y) = A_{mn} \sin\frac{m\pi x}{l} \sin\frac{n\pi y}{l},$$

where m and n are positive integers.

(b) Deduce that in the fundamental natural mode ($\omega = \omega_{11} = \Omega$) the amplitude

$$\varphi = a \sin\frac{\pi x}{l} \sin\frac{\pi y}{l}$$

is zero only along the boundary, so that there are no interior nodal lines.

(c) Show that the second natural circular frequency corresponds to *two* modes, in which either $m = 1$ and $n = 2$ or $m = 2$ and $n = 1$, and is given by $\omega = \omega_{12} = \omega_{21} = \sqrt{10}\ \Omega/2$, and that the most general motion with this circular frequency possesses an amplitude of the form

$$\varphi = a \sin \frac{2\pi x}{l} \sin \frac{\pi y}{l} + b \sin \frac{\pi x}{l} \sin \frac{2\pi y}{l}$$

$$= 2 \sin \frac{\pi x}{l} \sin \frac{\pi y}{l} \left(a \cos \frac{\pi x}{l} + b \cos \frac{\pi y}{l} \right),$$

where a and b are constants. Deduce that there is one interior nodal line, along the curve $a \cos \pi x/l + b \cos \pi y/l = 0$, that this curve always passes through the center of the square, and that in the cases $b = \pm a$ the line is a diagonal of the square.

Section 9.10

55. The temperatures at the ends $x = 0$ and $x = 100$ of a rod 100 cm in length, with insulated sides, are held at $0°$ and $100°$, respectively, until steady-state conditions prevail. Then, at the instant $t = 0$, the temperatures of the two ends are interchanged. Find the resultant temperature distribution as a function of x and t.

56. (a) Obtain permissible product solutions of the heat-flow equation (154) which satisfy the conditions $\partial T(0, t)/\partial x = 0$, $\partial T(l, t)/\partial x = 0$, corresponding to the requirement that there be no heat flow through the boundaries (ends) $x = 0$ and $x = l$.

(b) Use these results to solve the modification of Problem 55 in which at the time $t = 0$ the two ends of the rod are suddenly insulated.

57. Suppose that the rod considered in Section 9.10 is such that heat escapes from the lateral boundary according to Newton's law of cooling, so that $T(x, t)$ satisfies the equation

$$\alpha^2 \frac{\partial^2 T}{\partial x^2} = \frac{\partial T}{\partial t} + \beta(T - T_0),$$

where β is a constant and T_0 is the temperature of the surrounding medium. (See Problem 9.) The initial temperature distribution is $T(x, 0) = f(x)$ and the ends $x = 0$ and $x = l$ are maintained at T_1 and T_2, respectively, when $t > 0$.

(a) Show that the substitution

$$T(x, t) = T_0 + U(x, t)e^{-\beta t}$$

reduces the problem to the following one:

$$\frac{\partial^2 U}{\partial x^2} = \frac{1}{\alpha^2} \frac{\partial U}{\partial t},$$

$$U(x, 0) = f(x) - T_0, \quad U(0, t) = (T_1 - T_0)e^{\beta t}, \quad U(l, t) = (T_2 - T_0)e^{\beta t}.$$

(b) In the important special case when $T_1 = T_2 = T_0$, obtain the solution of the original problem in the form

$$T(x, t) = T_0 + \sum_{n=1}^{\infty} a_n \sin \frac{n\pi x}{l} e^{-[\beta + (n^2 \pi^2 \alpha^2)/l^2]t},$$

where

$$a_n = \frac{2}{l} \int_0^l f(x) \sin \frac{n\pi x}{l} dx - \frac{1 - \cos n\pi}{n\pi} 2T_0.$$

58. A rod of length l, with insulated lateral boundaries, has the end $x = 0$ maintained at $T = T_1$ when $t > 0$, whereas heat escapes through the end $x = l$ according to Newton's law of cooling in the form

$$\left[hl\frac{\partial T}{\partial x} + (T - T_0) \right]_{x=l} = 0,$$

where T_0 is the temperature of the surrounding medium and h is a constant. The initial temperature distribution along the rod is prescribed as $T(x, 0) = f(x)$.

(a) Obtain the steady-state distribution in the form

$$T_S = T_1 - \frac{T_1 - T_0}{1 + h}\frac{x}{l}.$$

(b) Show that the transient distribution can be assumed in the form

$$T_T(x, t) = \sum_{n=1}^{\infty} a_n \sin\frac{k_n x}{l}e^{-k_n^2\alpha^2 t/l^2},$$

where k_n is the nth positive root of the equation $\tan k + hk = 0$.

(c) Obtain the required temperature distribution in the form $T = T_S + T_T$, where a_n is determined by the relation

$$a_n \int_0^l \sin^2\frac{k_n x}{l}\,dx = \int_0^l [f(x) - T_S(x)] \sin\frac{k_n x}{l}\,dx.$$

59. (a) Show that the assumption of a solution of the one-dimensional heat flow equation as a linear function of t, in the form $T = tf(x) + g(x)$, leads to the requirements $f'' = 0$ and $\alpha^2 g'' = f$. Hence obtain the particular solution

$$T_p(x, t) = c_1(x^3 + 6\alpha^2 tx) + c_2(x^2 + 2\alpha^2 t) + c_3 x + c_4,$$

where the c's are arbitrary constants.

(b) Use this result to obtain a particular solution for which $\partial T(0, t)/\partial x = 0$ and $\partial T(l, t)/\partial x = C$ in the form

$$T_p = \frac{C}{2l}(x^2 + 2\alpha^2 t).$$

(c) In a similar way, obtain a particular solution for which $T(0, t) = 0$ and $T(l, t) = T_0 t$ in the form

$$T_p = T_0\left[\frac{x}{l}t - \frac{1}{6\alpha^2 l}x(l^2 - x^2)\right].$$

60. A rod with insulated sides has its end $x = 0$ insulated, whereas heat is introduced into the end $x = l$ at a constant rate, so that $\partial T(l, t)/\partial x = C$, where C is a constant. The initial temperature distribution is prescribed as $T(x, 0) = f(x)$.

(a) Show that no steady state can exist in which T is independent of time, but that [see Problem 59(b)] the temperature distribution $T_p = C(x^2 + 2\alpha^2 t)/2l$ satisfies the end conditions (and also happens to specify a state in which the *rate of heat flow* at any point does not vary with time).

(b) Obtain the desired distribution in the form

$$T(x, t) = \frac{C}{2l}(x^2 + 2\alpha^2 t) + a_0 + \sum_{n=1}^{\infty} a_n \cos\frac{n\pi x}{l}e^{-n^2\pi^2\alpha^2 t/l^2},$$

where

$$a_0 = \frac{1}{l} \int_0^l \left[f(x) - \frac{C}{2l} x^2 \right] dx, \qquad a_n = \frac{2}{l} \int_0^l \left[f(x) - \frac{C}{2l} x^2 \right] \cos \frac{n\pi x}{l} \, dx,$$

when $n = 1, 2, \ldots$.

61. Use the result of Problem 59(c) to obtain the solution of the one-dimensional heat-flow equation for which $T(0, t) = 0$ and $T(l, t) = T_0 t$ when $t > 0$ and $T(x, 0) = f(x)$, in the form

$$T(x, t) = T_0 \left[\frac{x}{l} t - \frac{1}{6\alpha^2 l} x(l^2 - x^2) \right] + \sum_{n=1}^{\infty} a_n \sin \frac{n\pi x}{l} e^{-n^2\pi^2\alpha^2 t/l^2},$$

where
$$a_n = \frac{2}{l} \int_0^l \left[f(x) + \frac{T_0}{6\alpha^2 l} x(l^2 - x^2) \right] \sin \frac{n\pi x}{l} \, dx.$$

62. The boundary of a circular plate of radius a is maintained at a temperature which varies periodically with time, according to the law $T(a, t) = T_0 \cos \omega t$, where T_0 and ω are constant. Determine the steady periodic temperature variation at interior points as follows:

(a) Let $T(r, t)$ be the *real part* of a complex function $U = F(r)e^{i\omega t}$, and show that $F(r)$ then must satisfy the equation

$$\frac{d^2 F}{dr^2} + \frac{1}{r} \frac{dF}{dr} - \frac{i\omega}{\alpha^2} F = 0$$

and the condition $F(a) = T_0$, together with the requirement that $F(r)$ be finite at $r = 0$.

(b) By using the results of Section 4.11, deduce that

$$U(r, t) = T_0 \frac{J_0(i^{3/2} kr)}{J_0(i^{3/2} ka)} e^{i\omega t},$$

where $k = \sqrt{\omega}/\alpha$.

(c) Deduce that

$$T(r, t) = T_0 \, \mathrm{Re} \left\{ \frac{\mathrm{ber}\,(kr) + i\,\mathrm{bei}\,(kr)}{\mathrm{ber}\,(ka) + i\,\mathrm{bei}\,(ka)} (\cos \omega t + i \sin \omega t) \right\}$$

$$= T_0 \, \mathrm{Re} \left\{ \frac{M_0(kr)}{M_0(ka)} e^{i[\theta_0(kr) - \theta_0(ka) + \omega t]} \right\}$$

$$= T_0 \frac{M_0(kr)}{M_0(ka)} \cos [\theta_0(kr) - \theta_0(ka) + \omega t],$$

where

$$M_0(kr) = \sqrt{\mathrm{ber}^2\,(kr) + \mathrm{bei}^2\,(kr)}, \qquad \theta_0(kr) = \tan^{-1} \frac{\mathrm{bei}\,(kr)}{\mathrm{ber}\,(kr)}.$$

(d) Show that the temperature at the center $r = 0$ is given by

$$T(0, t) = \frac{T_0}{M_0(ka)} \cos [\omega t - \theta_0(ka)].$$

63. Assume that a portion of the earth's surface may be considered as plane and that effects of the periodic heating due to the sun are transmitted only in the x direction, perpendicular to the surface. If the resultant temperature on the surface of the earth is taken in the form $T(0, t) = T_0 \cos \omega t$, show that the temperature at depth x is given by

$$T(x, t) = T_0 e^{-\sqrt{\omega/2}\,(x/\alpha)} \cos \left(\omega t - \sqrt{\frac{\omega}{2}} \frac{x}{\alpha} \right)$$

with the notation of Section 9.2. [Consider the boundary value $T_0 \cos \omega t$ as the real part of $T_0 e^{i\omega t}$ and determine the *real part* of an expression $U(x, t) = X(x)e^{i\omega t}$ which satisfies the one-dimensional heat-flow equation, reduces to $T_0 e^{i\omega t}$ when $x = 0$, and vanishes when $x \to \infty$.]

Section 9.11

64. The temperature at the end $x = 0$ of a rod of length l is held at $0°$ while the temperature at the other end is varied periodically according to the law

$$T(l, t) = T_0 \sin \omega t$$

when $t > 0$. Suppose that when $t = 0$ the temperature is zero at all points in the rod.

(a) Find the temperature distribution in the rod. [Use Equation (177).]

(b) Show also that for prescribed oscillation so slow that $(\lambda \omega)^2 \ll 1$, and for times such that $t \gg \lambda$, the approximation

$$T \approx T_0 \left[\frac{x}{l} \sin \omega t + \frac{2\lambda \omega}{\pi} \left(\sum_{n=1}^{\infty} \frac{(-1)^n}{n^3} \sin \frac{n\pi x}{l} \right) \cos \omega t \right]$$

is valid. [By integrating the Fourier sine expansion of x twice, and determining the constants of integration, it can be shown that the series in parentheses represents the function $\pi^3 x(x^2 - l^2)/12l^3$.]

65. Show that Equation (177) can be written in the form

$$T(x, t) = \frac{x}{l} F(t) + \frac{2}{\pi} \sum_{n=1}^{\infty} \frac{(-1)^n}{n} \left[F(t) - \frac{n^2}{\lambda} \int_0^t F(\tau)e^{-n^2(t-\tau)/\lambda} \, d\tau \right] \sin \frac{n\pi x}{l}.$$

66. (a) If $F(t)$ is the unit singularity function $\delta(t)$, show that the temperature distribution given by Equation (176) takes the form

$$T(x, t) = \frac{2}{\pi \lambda} \sum_{n=1}^{\infty} (-1)^{n+1} n e^{-n^2 t/\lambda} \sin \frac{n\pi x}{l} \qquad (t > 0).$$

(b) If this expression is denoted by $U(x, t)$, verify that $U(x, t) = \partial A(x, t)/\partial t$, where A is defined by (168), and that the temperature distribution (176) corresponding to a general function $F(t)$ can be written in the form

$$T(x, t) = \int_0^t U(x, t - \tau)F(\tau) \, d\tau.$$

67. If a rod with insulated sides initially is at a temperature distribution $T(x, 0) = f(x)$, and if the conditions $T(0, t) = 0$ and $T(l, t) = F(t)$ are imposed when $t > 0$, show that the resultant temperature distribution can be expressed as the sum of the right-hand member of Equation (176) or (177) and the right-hand member of (161) with $T_1 = T_2 = 0$.

68. (a) Make use of the formal argument of Section 9.11 to show that the temperature distribution for which $T(x, 0) = 0$ and for which $T(0, t) = 0$ and $\partial T(l, t)/\partial x = F(t)$ when $t > 0$ is again given by Equation (173) or (175) if $A(x, t)$ is the distribution corresponding to $F(t) = 1$.

(b) Show that here

$$A(x, t) = x - \sum_{n \text{ odd}} \frac{8l}{n^2 \pi^2} \sin \frac{n\pi}{2} \sin \frac{n\pi x}{2l} e^{-n^2 \pi^2 \alpha^2 t/4l^2}.$$

Section 9.12

69. (a) Show that if $U = Fe^{i\omega t}$ is a solution of the wave equation (5) in three dimensions, where F is independent of the time t, then F satisfies the *Helmholtz equation*

$$\nabla^2 F + \frac{\omega^2}{c^2} F = 0,$$

where ∇^2 is the three-dimensional Laplace operator.

(b) By seeking separable solutions of the preceding equation, show in particular that the real and imaginary parts of the functions

$$U = e^{i\omega(lx+my+nz+ct)/c} \qquad (l^2 + m^2 + n^2 = 1),$$

$$V = r^{-1/2} J_{\pm(n+1/2)}\left(\frac{\omega r}{c}\right) P_n^m (\cos \varphi) e^{i(m\theta + \omega t)},$$

$$W = e^{i[n\theta + \omega(lz+ct)/c]} H_n^{(1 \text{ or } 2)}\left(m\frac{\omega r}{c}\right) \qquad (l^2 + m^2 = 1)$$

are solutions of the wave equation in rectangular, spherical, and circular cylindrical coordinates, respectively. [Here P_n^m is an associated Legendre function (see Section 4.12). The solutions given in Section 9.12 are special cases.]

70. (a) Verify the identity

$$\frac{1}{r^2}\frac{\partial}{\partial r}\left(r^2 \frac{\partial \varphi}{\partial r}\right) = \frac{1}{r}\frac{\partial^2}{\partial r^2}(r\varphi).$$

(b) Hence show that, if φ depends only upon distance r from the origin (in spherical coordinates) and upon time t, then the wave equation (5) can be written in the form

$$\frac{\partial^2}{\partial r^2}(r\varphi) = \frac{1}{c^2}\frac{\partial^2}{\partial t^2}(r\varphi).$$

(c) From this result obtain the general solution

$$r\varphi = f(r - ct) + g(r + ct).$$

71. Prove that the *standing waves* $\cos (\omega x/c) \cos \omega t$, $\cos (\omega x/c) \sin \omega t$, $\sin (\omega x/c) \cos \omega t$, and $\sin (\omega x/c) \sin \omega t$ each can be expressed as the superposition of two waves traveling with velocity c in opposite directions.

72. The equation governing the lateral displacement $w(x, t)$ in small free vibrations of a tightly stretched string of uniform linear density ρ, under constant tension T, is of the form

$$\frac{\partial^2 w}{\partial x^2} = \frac{1}{c^2}\frac{\partial^2 w}{\partial t^2},$$

where $c^2 = T/\rho$. (See Problem 8 of Chapter 5.)

(a) If the string is fixed at the ends $x = 0$ and $x = l$, and if the string is initially deflected in the form $w(x, 0) = f(x)$ and released from rest, so that also $\partial w(x, 0)/\partial t = 0$, obtain the ensuing deflection in the form

$$w(x, t) = \sum_{n=1}^{\infty} a_n \sin \frac{n\pi x}{l} \cos \frac{n\pi ct}{l},$$

where

$$a_n = \frac{2}{l}\int_0^l f(x) \sin \frac{n\pi x}{l}\, dx.$$

(b) Show that the solution of part (a) can be written in the form

$$w(x, t) = \tfrac{1}{2}[F(x - ct) + F(x + ct)],$$

where
$$F(x) = \sum_{n=1}^{\infty} a_n \sin \frac{n\pi x}{l},$$

and hence where $F(x)$ is an odd periodic function of x, of period $2l$, which coincides with $f(x)$ when $0 < x < l$ and is defined for all values of x. (Compare Problem 33 of Chapter 8.)

73. When resistive forces proportional to the velocity are taken into account, the differential equation of Problem 72 is replaced by the equation

$$c^2 \frac{\partial^2 w}{\partial x^2} = \frac{\partial^2 w}{\partial t^2} + 2\gamma \frac{\partial w}{\partial t},$$

where γ is a constant.

(a) Show that this equation admits solutions of the form

$$w = e^{-\gamma t} e^{ik[x \pm c\sqrt{1 - (\gamma^2/k^2 c^2)} \, t]},$$

where k is an arbitrary nonzero constant, and that, if $\gamma^2 < k^2 c^2$, these solutions represent plane waves which are damped in time, and which move along the x axis with velocity $c' = c\sqrt{1 - (\gamma^2/k^2 c^2)}$.

(b) Show that the solution of Problem 72(a) here takes the form

$$w(x, t) = e^{-\gamma t} \sum_{n=1}^{\infty} a_n \left(\cos \omega_n t + \frac{\gamma}{\omega_n} \sin \omega_n t \right) \sin \frac{n\pi x}{l},$$

where
$$\omega_n = \sqrt{\frac{n^2 \pi^2 c^2}{l^2} - \gamma^2}$$

and where
$$a_n = \frac{2}{l} \int_0^l f(x) \sin \frac{n\pi x}{l} \, dx.$$

(Notice that the presence of resistive forces not only causes time damping, but also decreases the frequency of each component oscillation, and that the higher frequencies are no longer integral multiples of the fundamental frequency. When $\gamma > \pi c/l$, one or more of the ω's is pure imaginary, and the corresponding time functions then become real exponential functions.)

Section 9.13

74. When the problem solved in Section 9.13 is modified in such a way that fluid surrounding the pulsating cylinder is contained in a fixed coaxial cylinder $r = b$ ($b > a$), on which the velocity V must vanish, show that the fluid velocity is given by

$$V = V_0 \, \text{Re} \left[e^{i\omega t} \frac{H_1^{(1)}(\omega b/c) H_1^{(2)}(\omega r/c) - H_1^{(2)}(\omega b/c) H_1^{(1)}(\omega r/c)}{H_1^{(1)}(\omega b/c) H_1^{(2)}(\omega a/c) - H_1^{(2)}(\omega b/c (H_1^{(1)}(\omega a/c)} \right],$$

provided that the denominator does not vanish. (For the purpose of this problem, it is not necessary to proceed beyond this point.)

75. The boundary of a sphere of radius a, surrounded by an ideal compressible fluid, is caused to pulsate radially with circular frequency ω, so that the radial velocity of the boundary is given by $V = V_0 \cos \omega t$. By methods analogous to those of Section 9.13, determine the steady-state periodically varying velocity of surrounding points

in the form

$$V = \frac{V_0 a^2/r^2}{c^2 + a^2\omega^2}\left[(c^2 + ar\omega^2)\cos\omega\left(\frac{r-a}{c} - t\right) + c\omega(r-a)\sin\omega\left(\frac{r-a}{c} - t\right)\right],$$

where c is the velocity of sound in the fluid. [Use spherical coordinates with the velocity potential dependent only on r and t, and notice that a suitable assumption for a complex potential φ_c consists of the second term of Equation (197), with c_2 a *complex* constant to be determined. The transient effects corresponding to the *initiation* of the pulsation are considered in Problem 99.]

76. Let $\varphi(x, t)$ satisfy the one-dimensional wave equation $c^2\varphi_{xx} = \varphi_{tt}$ when $0 \le x \le a$, subject to the end conditions $\varphi(0, t) = \cos\omega t$ and $\varphi(a, t) = 0$. Determine a particular solution φ which varies periodically with time in the alternative forms

$$\varphi = \frac{\sin[\omega(a-x)/c]}{\sin(\omega a/c)}\cos\omega t$$

$$= \frac{1}{2\sin(\omega a/c)}\left[\sin\omega\left(t + \frac{a-x}{c}\right) - \sin\omega\left(t - \frac{a-x}{c}\right)\right],$$

under the assumption that $\sin(\omega a/c) \ne 0$, by each of the following procedures:

(a) Assume $\varphi = f(x)\cos\omega t + g(x)\sin\omega t$ and determine f and g.

(b) Assume φ as a linear combination of those terms in Equations (189) and (190) which reduce to $\cos\omega t$ when $x = 0$.

[Notice that φ could represent a permissible lateral displacement of a string with one end fixed and the other oscillating, or it could represent either velocity potential or velocity of an ideal compressible fluid in a tube, closed at $x = a$, and with an oscillatory valve at $x = 0$. Notice also that a steady oscillatory response of the type assumed cannot exist if ω is such that $\sin(\omega a/c) = 0$, that is, if $\omega = n\pi c/a$ where n is integral, and that additional transient responses satisfying the prescribed end conditions may exist. (See Problem 78.)]

77. When Problem 76 is modified in such a way that the wave equation is to be satisfied when $0 \le x < \infty$ and the condition $\varphi(a, t) = 0$ is replaced by the requirement that there be no inward-traveling waves as $x \to \infty$ (and hence for all $x > 0$), use results of Section 9.12 to write down the steady-state solution. (The determination of transient effects is considered in Problem 95.)

78. Let the initial conditions $\varphi(x, 0) = 0$ and $\partial\varphi(x, 0)/\partial t = 0$ be added to the end conditions of Problem 76.

(a) By using the result of Problem 76, show that the complete solution can be assumed in the form

$$\varphi = \frac{\sin[\omega(a-x)/c]}{\sin(\omega a/c)}\cos\omega t - \sum_{n=1}^{\infty} A_n \sin\frac{n\pi x}{a}\cos\frac{n\pi ct}{a} \qquad (t > 0)$$

if $\omega \ne k\pi c/a$, where the A's are to be determined in such a way that $\varphi(x, 0) = 0$.

(b) Deduce that then

$$A_n = \frac{2n\pi}{n^2\pi^2 - (\omega^2 a^2/c^2)} \qquad \left(\omega \ne \frac{n\pi c}{a}\right)$$

and hence obtain the solution in the form

$$\varphi = \frac{\sin[\omega(a-x)/c]}{\sin(\omega a/c)}\cos\omega t - 2\sum_{n=1}^{\infty}\frac{n\pi}{n^2\pi^2 - (\omega^2 a^2/c^2)}\sin\frac{n\pi x}{a}\cos\frac{n\pi ct}{a}$$

when $t > 0$, if $\omega \ne k\pi c/a$ $(k = 1, 2, \ldots)$.

(If $\omega \longrightarrow k\pi c/a$, the kth term of the sum can be extracted and combined with the particular solution, and the combination can be shown to tend to a function of x and t which involves t as a multiplicative factor, the remaining terms of the sum remaining finite in the limit. This is a case of *resonance*. If small resistive forces were present in the nonresonant case, the portion of the solution represented by the series would be damped out with increasing time. Since such forces always exist in practice, the particular solution can be considered as a *quasi*-steady-state solution, regardless of the initial conditions, in the sense that it is the limit, as resistive forces tend to zero, of the true steady-state solution of the problem in which resistance is present.)

Section 9.14

79. Obtain the solution $\varphi(x, y)$ of Laplace's equation in the half-plane $y \geq 0$, with

$$\varphi(x, 0) = f(x), \qquad \lim_{x^2 + y^2 \to \infty} \varphi(x, y) = 0,$$

by use of the *Fourier transform*, as follows:
   (a) If $\bar{\varphi}(u, y)$ is the Fourier transform of $\varphi(x, y)$ with respect to x, so that

$$\bar{\varphi}(u, y) = \int_{-\infty}^{\infty} e^{-iux}\varphi(x, y)\, dx,$$

show that the transform of $\partial^2\varphi/\partial x^2$ is $-u^2\bar{\varphi}$ and the transform of $\partial^2\varphi/\partial y^2$ is $\partial^2\bar{\varphi}/\partial y^2$ (assuming in the former case that both φ and $\partial\varphi/\partial x$ tend to zero as $|x| \longrightarrow \infty$ and in the latter case that differentiation with respect to y can be effected under the integral sign). Hence show that the result of taking the Fourier transform of the equal members of the equation $\varphi_{xx} + \varphi_{yy} = 0$ is the equation

$$\frac{\partial^2\bar{\varphi}}{\partial y^2} - u^2\bar{\varphi} = 0,$$

from which there follows

$$\bar{\varphi}(u, y) = A(u)e^{-uy} + B(u)e^{uy}.$$

   (b) By requiring that $\bar{\varphi}(u, 0) = \bar{f}(u)$ and using the fact that the Fourier transform cannot be unbounded as $y \longrightarrow \infty$, deduce that

$$\bar{\varphi}(u, y) = \bar{f}(u)e^{-|u|y}.$$

   (c) Show that $e^{-|u|y}$ is the Fourier transform of the function

$$g(x, y) = \frac{1}{\pi} \frac{y}{x^2 + y^2}$$

and use the *convolution* property (Problem 91 of Chapter 5) to deduce that

$$\varphi(x, y) = \int_{-\infty}^{\infty} g(x - \xi, y)f(\xi)\, d\xi$$

$$= \frac{1}{\pi} \int_{-\infty}^{\infty} \frac{yf(\xi)\, d\xi}{(x - \xi)^2 + y^2},$$

in accordance with (230). (See also Problem 81.)

80. Obtain the solution of the equation

$$\alpha^2 T_{xx} = T_t \qquad (0 \leq x < \infty, 0 \leq t < \infty),$$

subject to the conditions

$$T(x, 0) = f(x), \qquad T(0, t) = 0,$$

by use of the *Fourier sine transform*, as follows:

(a) If $T_S(u, t)$ is the sine transform of $T(x, t)$ with respect to x, so that

$$T_S(u, t) = \int_0^\infty T(x, t) \sin ux \, dx,$$

show that the transform of $\partial^2 T/\partial x^2$ is $uT(0, t) - u^2 T_S$, assuming that T and $\partial T/\partial x$ tend to zero as $x \longrightarrow \infty$, and the transform of $\partial T/\partial t$ is $\partial T_S/\partial t$. Hence, noting that here $T(0, t) = 0$, deduce that T_S must satisfy the equation

$$\frac{\partial T_S}{\partial t} + \alpha^2 u^2 T_S = 0$$

and the condition

$$T_S(u, 0) = f_S(u),$$

so that

$$T_S(u, t) = f_S(u)e^{-\alpha^2 u^2 t}.$$

(b) Use (241) to show that $e^{-\alpha^2 u^2 t}$ is the *cosine* transform of the function

$$g(x, t) = \frac{1}{\alpha\sqrt{\pi t}} e^{-x^2/4\alpha^2 t}$$

and use the convolution property of Problem 92(c), Chapter 5, to deduce the solution as given by Equation (242).

81. Rederive the solution of the problem

$$\varphi_{xx} + \varphi_{yy} = 0 \qquad (-\infty < x < \infty, 0 \leqq y < \infty),$$

$$\varphi(x, 0) = f(x), \qquad \lim_{x^2+y^2\to\infty} \varphi(x, y) = 0$$

(see Problem 79) by the following method:

(a) If $\varphi_S(x, v)$ denotes the *Fourier sine transform* of $\varphi(x, y)$ with respect to y, so that

$$\varphi_S(x, v) = \int_0^\infty \varphi(x, y) \sin vy \, dy,$$

show that the transform of $\partial^2\varphi/\partial x^2$ is $\partial^2\varphi_S/\partial x^2$ and the transform of $\partial^2\varphi/\partial y^2$ is $v\varphi(x, 0) - v^2\varphi_S \doteq vf(x) - v^2\varphi_S$, assuming that both φ and $\partial\varphi/\partial y$ tend to zero as $y \longrightarrow \infty$, so that φ_S must satisfy the equation

$$\frac{\partial^2\varphi_S}{\partial x^2} - v^2\varphi_S = -vf(x).$$

(b) In order to avoid the solution of this differential equation, take the *Fourier transform* of the equal members. Thus show that the "mixed" double transform $\bar\varphi_S(u, v)$ must satisfy the equation

$$-u^2\bar\varphi_S(u, v) - v^2\bar\varphi_S(u, v) = -v\bar f(u),$$

and hence is determined in the form

$$\bar\varphi_S(u, v) = \frac{v}{u^2 + v^2}\bar f(u).$$

(c) Show that $v/(u^2 + v^2)$ is the sine transform, with respect to y, of the function

$$g(u, y) = e^{-|u|y}$$

[See Problem 86(c) of Chapter 5] so that

$$\bar{\varphi}(u, y) = e^{-|u|y}\bar{f}(u),$$

in accordance with the result of Problem 79(b), after which the required solution $\varphi(x, y)$ is obtained as before.

[This problem is intended to illustrate procedures which are available in situations where a simpler approach, such as that of Problem 79, cannot be followed. As Problems 80 and 81 illustrate, it may be convenient to take the Fourier *sine* transform with respect to a variable in a direction normal to a line segment along which the unknown *function* is prescribed. The *cosine* transform would be similarly appropriate if the *normal derivative* of the function were prescribed instead on that segment. Notice also that the result of introducing a "forcing term" $h(x, y)$ into the governing differential equation, so that it becomes *Poisson's equation* $\varphi_{xx} + \varphi_{yy} = h$, can be treated in the same way.]

82. (a) If $\varphi(x, y)$ satisfies Laplace's equation in the infinite strip $0 \leqq y \leqq b$, and if $\varphi(x, 0) = f(x)$, $\varphi(x, b) = 0$, and φ is bounded as $|x| \to \infty$, obtain φ in the form

$$\varphi(x, y) = \frac{1}{\pi} \int_0^\infty \left[\int_{-\infty}^\infty \frac{\sinh u(b - y)}{\sinh ub} f(\xi) \cos u(\xi - x)\, d\xi \right] du,$$

when $f(x)$ is sufficiently respectable.

(b) By making use of the formula

$$\int_0^\infty \frac{\sinh px}{\sinh qx} \cos rx\, dx = \frac{\pi}{2q} \frac{\sin (p\pi/q)}{\cos (p\pi/q) + \cosh (r\pi/q)} \qquad (p^2 < q^2),$$

and interchanging the order of integration in the result of part (a), express the solution in the form

$$\varphi(x, y) = \frac{1}{2b} \sin \frac{\pi y}{b} \int_{-\infty}^\infty \frac{f(\xi)}{\cosh [\pi(\xi - x)/b] - \cos (\pi y/b)}\, d\xi.$$

(When f is a *polynomial*, the method used in Problems 21 and 22 is also available, yielding a *series* solution. Here the boundedness condition must be modified.)

83. (a) If $\varphi(x, y)$ satisfies Laplace's equation in the quadrant $x \geqq 0$, $y \geqq 0$, and if φ vanishes along the positive y axis, reduces to $f(x)$ along the positive x axis, and is bounded as $x^2 + y^2 \to \infty$, obtain φ in the form

$$\varphi(x, y) = \frac{2}{\pi} \int_0^\infty \int_0^\infty e^{-uy} f(\xi) \sin ux \sin u\xi\, d\xi\, du,$$

when $f(x)$ is sufficiently respectable.

(b) By formally integrating first with respect to u, transform this expression to the form

$$\varphi(x, y) = \frac{1}{\pi} \int_0^\infty f(\xi) \left[\frac{y}{y^2 + (\xi - x)^2} - \frac{y}{y^2 + (\xi + x)^2} \right] d\xi.$$

(c) Deduce the result of part (b) from Equation (230) by replacing $f(x)$ in (230) by an odd function $f_o(x)$ such that $f_o(x) = f(x)$ when $x > 0$ and $f_o(x) = -f(-x)$ when $x < 0$.

84. (a) If Problem 83 is modified in such a way that $\partial\varphi/\partial x$ rather than φ vanishes along the positive y axis, show that

$$\varphi(x, y) = \frac{2}{\pi} \int_0^\infty \int_0^\infty e^{-uy} f(\xi) \cos ux \cos u\xi \, d\xi \, du$$

$$= \frac{1}{\pi} \int_0^\infty f(\xi) \left[\frac{y}{y^2 + (\xi - x)^2} + \frac{y}{y^2 + (\xi + x)^2} \right] d\xi.$$

(b) Deduce the result of part (a) from Equation (230) by replacing $f(x)$ in (230) by an even function $f_e(x)$ such that $f_e(x) = f(x)$ when $x > 0$ and $f_e(x) = f(-x)$ when $x < 0$.

85. A rod with insulated sides extends from $x = -\infty$ to $x = +\infty$. If the initial temperature distribution is given by $T(x, 0) = f(x)$, where $-\infty < x < \infty$, show that

$$T(x, t) = \frac{1}{\pi} \int_{-\infty}^\infty f(\xi) \left[\int_0^\infty e^{-u^2\alpha^2 t} \cos u(\xi - x) \, du \right] d\xi$$

$$= \frac{1}{2\alpha\sqrt{\pi t}} \int_{-\infty}^\infty f(\xi) e^{-(\xi - x)^2/4\alpha^2 t} \, d\xi.$$

86. A rod of infinite length, with insulated sides, has its end $x = 0$ *insulated*. If the initial temperature distribution is given by $T(x, 0) = f(x)$, where $0 < x < \infty$, show that

$$T(x, t) = \frac{1}{2\alpha\sqrt{\pi t}} \int_0^\infty f(\xi)[e^{-(\xi - x)^2/4\alpha^2 t} + e^{-(\xi + x)^2/4\alpha^2 t}] \, d\xi.$$

87. *Source functions.*

(a) If $f(x)$ is the unit singularity function $\delta(x)$ (so that a heat source of intensity ps is present at $x = 0$ at the instant $t = 0$), show that the solution of Problem 85 is of the form

$$T(x, t) = \frac{1}{2\alpha\sqrt{\pi t}} e^{-x^2/4\alpha^2 t} \equiv S(x, t).$$

(b) Deduce that the temperature in a rod extending over $(-\infty, \infty)$, due to *a unit heat source* at $x = x_0$ at the instant $t = t_0$, is given by

$$T(x, t) = \frac{1}{ps} S(x - x_0, t - t_0).$$

(c) Show that Equation (242) takes the form

$$T(x, t) = \int_0^\infty [S(x - \xi, t) - S(x + \xi, t)] f(\xi) \, d\xi,$$

the solution of Problem 86 takes the form

$$T(x, t) = \int_0^\infty [S(x - \xi, t) + S(x + \xi, t)] f(\xi) \, d\xi,$$

and the solution of Problem 85 becomes

$$T(x, t) = \int_{-\infty}^\infty S(x - \xi, t) f(\xi) \, d\xi.$$

[In *two* dimensions, the corresponding *source function* is

$$S(x, y, t) = \frac{1}{4\alpha^2\pi t} e^{-(x^2+y^2)/4\alpha^2 t}$$

(see Problem 92), whereas the function

$$S(x, y, z, t) = \frac{1}{(4\alpha^2\pi t)^{3/2}}e^{-(x^2+y^2+z^2)/4\alpha^2 t}$$

is the *three*-dimensional generalization.]

88. The end $x = 0$ of a rod with insulated sides is maintained at the temperature $T(0, t) = F(t)$ for all $t > 0$. Initially, all points of the rod are at zero temperature, $T(x, 0) = 0$, where $0 < x < \infty$.

(a) Show that the formal argument of Section 9.11 leads to the solution

$$T(x, t) = \int_0^t F(\tau) \frac{\partial A(x, t - \tau)}{\partial t} d\tau,$$

where $A(x, t)$ is the solution of the problem when $F(t) = 1$.

(b) Use the result of Equation (244) to deduce that

$$A(x, t) = 1 - \operatorname{erf}\left(\frac{x}{2\alpha\sqrt{t}}\right) \equiv 1 - \frac{2}{\sqrt{\pi}} \int_0^{x/2\alpha\sqrt{t}} e^{-u^2} du.$$

(c) From the results of parts (a) and (b), obtain the solution of the stated problem in the form

$$T(x, t) = \frac{x}{2\alpha\sqrt{\pi}} \int_0^t F(\tau)e^{-x^2/4\alpha^2(t-\tau)} \frac{d\tau}{(t - \tau)^{3/2}}.$$

89. (a) If $F(t) = \delta(t)$ in Problem 88, show that

$$T(x, t) = \frac{x}{2\alpha t\sqrt{\pi t}}e^{-x^2/4\alpha^2 t},$$

and that, if this function is denoted by $D(x, t)$, then the solution of Problem 88 takes the form

$$T(x, t) = \int_0^t D(x, t - \tau)F(\tau) d\tau.$$

(b) Verify that

$$D(x, t) = -2\alpha^2 \frac{\partial S(x, t)}{\partial x},$$

where $S(x, t)$ is defined in Problem 87 and is the solution of Problem 85 which corresponds to the initial function $\delta(x)$. [Hence $D/2\alpha^2$ is the solution of Problem 85 corresponding to the initial *unit doublet function* $-\delta'(x)$ at $x = 0$ at the instant $t = 0$.]

90. (a) Verify that $D(x, t)$, as defined in Problem 89, does indeed satisfy the heat-flow equation when $x > 0$ and $t > 0$, and that

$$\lim_{x \to 0+} D(x, t) = 0 \quad (t > 0), \qquad \lim_{t \to 0+} D(x, t) = 0 \quad (x > 0).$$

(b) Deduce that if $\varphi = U(x, t)$ is a solution of the problem

$$\frac{\partial^2\varphi}{\partial x^2} = \frac{1}{\alpha^2} \frac{\partial\varphi}{\partial t} \quad (0 < x < \infty, 0 < t < \infty),$$

$$\lim_{x \to 0+} \varphi(x, t) = F(t) \quad (t > 0), \qquad \lim_{t \to 0+} \varphi(x, t) = f(x) \quad (x > 0),$$

then to $U(x, t)$ can be added any constant multiple of $D(x, t)$, so that the problem as stated does not have a unique solution.

(c) Show that if $x \longrightarrow 0+$ and $t \longrightarrow 0+$ in such a way that $x^2 = kt$, where k is a positive constant, the function $D(x, t)$ becomes infinite. [Thus the added specification that φ be *bounded* when $0 \leq x \leq A$ and $0 \leq t \leq B$, for all positive values of A and B, will eliminate the situation described in part (b). This illustrates the need for the boundedness assumption made in Section 9.1.]

91. At the time $t = 0$, the temperature in unbounded space is dependent only upon radial distance r from the z axis, and is prescribed as $T(r, 0) = f(r)$. The ensuing temperature distribution $T(r, t)$ is required, under the assumption that $f(r)$ behaves satisfactorily for large values of r.

(a) Show that T must satisfy the equation

$$\frac{1}{r}\frac{\partial}{\partial r}\left(r\frac{\partial T}{\partial r}\right) = \frac{1}{\alpha^2}\frac{\partial T}{\partial t}$$

when $t > 0$, and obtain a product solution which remains finite at $r = 0$ and as $r \longrightarrow \infty$ in the form

$$Ae^{-u^2\alpha^2 t}J_0(ur).$$

(b) Deduce that the required distribution is given formally by the expression

$$T(r, t) = \int_0^\infty uA(u)e^{-u^2\alpha^2 t}J_0(ur)\,du,$$

where the factor u is inserted for convenience in the following step, and where $A(u)$ is to be determined such that

$$f(r) = \int_0^\infty uA(u)J_0(ur)\,du \qquad (0 < r < \infty),$$

and hence, by referring to Equations (249a, b) of Section 5.15, obtain the desired solution formally as

$$T(r, t) = \int_0^\infty ue^{-u^2\alpha^2 t}J_0(ur)\left[\int_0^\infty \xi f(\xi)J_0(u\xi)\,d\xi\right]du.$$

(c) By assuming the validity of interchange of order of integration, and making use of the relation

$$\int_0^\infty ue^{-a^2u^2}J_n(bu)J_n(cu)\,du = \frac{1}{2a^2}e^{-(b^2+c^2)/4a^2}I_n\left(\frac{bc}{2a^2}\right),$$

express the formal solution of part (c) in the form

$$T(r, t) = \frac{e^{-r^2/4\alpha^2 t}}{2\alpha^2 t}\int_0^\infty \xi f(\xi)e^{-\xi^2/4\alpha^2 t}I_0\left(\frac{\xi r}{2\alpha^2 t}\right)d\xi \qquad (t > 0).$$

92. (a) Suppose that the initial temperature in unbounded space is zero except throughout an infinitely extended right circular cylinder of radius ϵ, with its axis coinciding with the z axis, and has the constant value T_0 inside that cylinder. Show that the excess heat required per unit length of cylinder is given by

$$2\pi \int_0^\epsilon \rho s T_0 r\,dr = \pi \rho s T_0 \epsilon^2,$$

where ρs is the heat capacity per unit volume. Hence deduce that, if *unit* excess heat is to be present per unit length, there must follow $T_0 = 1/(\pi \rho s \epsilon^2)$, and use Problem 91 to show that the corresponding resultant temperature distribution in space at all following times is given by

$$T(r, t) = \frac{e^{-r^2/4\alpha^2 t}}{2\alpha^2 t} \cdot \frac{1}{2\pi\rho s} \cdot \left[\frac{2}{\epsilon^2} \int_0^\epsilon \xi e^{-\xi^2/4\alpha^2 t} I_0\left(\frac{\xi r}{2\alpha^2 t}\right) d\xi \right] \qquad (t > 0).$$

(b) By applying L'Hospital's rule (or otherwise) and recalling that $I_0(0) = 1$, show that the quantity in brackets tends to *unity* as $\epsilon \longrightarrow 0$, and deduce that the temperature distribution due to an instantaneous line source along the infinite line $r = 0$, emitting one calorie of heat per unit length at the instant $t = 0$, is given by

$$T(r, t) = \frac{e^{-r^2/4a^2 t}}{4\pi\rho s\alpha^2 t} \qquad (t > 0)$$

in unbounded space.

(c) Deduce that if the line source is instead perpendicular to the xy plane at the point (x_0, y_0), and if the impulse takes place at the time t_0, then the temperature at any point in unbounded space at any following time is given by

$$T = \frac{1}{4\pi\rho s\alpha^2(t - t_0)} e^{-[(x-x_0)^2 + (y-y_0)^2]/[4\alpha^2(t-t_0)]} \qquad (t > t_0).$$

(See also Problem 87 for the analogs in one and three dimensions.)

Section 9.15

93. Use the Laplace transform to solve the problem

$$\frac{\partial z}{\partial x} + \frac{\partial z}{\partial t} = z,$$

$$z(0, t) = 1, \qquad z(x, 0) = 1,$$

when $0 \leq x < \infty$ and $0 \leq t < \infty$.

94. Use the Laplace transform to solve the problem

$$\frac{\partial z}{\partial x} + \frac{\partial z}{\partial t} = z,$$

$$z(x, 0) = 1,$$

when $-\infty < x < \infty$ and $0 \leq t < \infty$.

95. Use the Laplace transform to solve the problem†

$$\frac{\partial^2\varphi}{\partial x^2} = \frac{1}{c^2}\frac{\partial^2\varphi}{\partial t^2},$$

$$\varphi(x, 0) = 0, \qquad \varphi_t(x, 0) = 0, \qquad \varphi(0, t) = \cos \omega t,$$

when $0 \leq x < \infty$ and $0 \leq t < \infty$, and verify that as $t \longrightarrow \infty$ the solution tends to that obtained in Problem 77.

96. Use the Laplace transform to solve the problem†

$$\frac{\partial^2\varphi}{\partial x^2} - \frac{\partial^2\varphi}{\partial t^2} = 1,$$

$$\varphi(x, 0) = 1, \qquad \varphi_t(x, 0) = 1, \qquad \varphi(0, t) = 1,$$

when $0 \leq x < \infty$ and $0 \leq t < \infty$.

†Use the fact that $C(s)e^{sx/c}$ cannot be a Laplace transform as $x \longrightarrow \infty$ unless $C(s) = 0$.

97. (a) Derive the formula

$$\mathcal{L}\{f(x + \alpha t)\} = \frac{1}{\alpha} \int_x^{\infty \, \text{sgn} \, \alpha} e^{-s(u-x)/\alpha} f(u) \, du,$$

when $\alpha \neq 0$, where $\infty \, \text{sgn} \, \alpha \equiv +\infty$ when $\alpha > 0$ and $-\infty$ when $\alpha < 0$. (Replace $x + \alpha t$ by u in the definition of the left-hand member.)

(b) Deduce that, if $c > 0$, there follows

$$\mathcal{L}\{f(x - ct)\} = \frac{1}{c} \int_{-\infty}^x e^{s(u-x)/c} f(u) \, du$$

and

$$\mathcal{L}\{f(x + ct)\} = \frac{1}{c} \int_x^{\infty} e^{-s(u-x)/c} f(u) \, du.$$

(c) Also deduce that

$$\mathcal{L}\{e^{\beta t} f(x + \alpha t)\} = \frac{1}{\alpha} \int_x^{\infty \, \text{sgn} \, \alpha} e^{-(s-\beta)(u-x)/\alpha} f(u) \, du$$

when $\alpha \neq 0$. [Use formula (T8), page 67.]

98. Use the Laplace transform and the results of Problem 97 to solve the problem

$$\frac{\partial z}{\partial x} + \frac{\partial z}{\partial t} = z,$$

$$z(x, 0) = \varphi(x),$$

when $-\infty < x < \infty$ and $0 \leq t < \infty$. [Show that

$$\bar{z}(x, s) = \int_{-\infty}^x e^{(s-1)(u-x)} \varphi(u) \, du + A(s) e^{-(s-1)x},$$

where the lower limit of the integral is so chosen that the coefficient of s is never positive in the exponential function occurring in the integrand.]

99. In Problem 75 suppose that until the instant $t = 0$ the sphere and surrounding medium are at rest in complete equilibrium, and that at that instant the pulsation begins and continues indefinitely. Determine the velocity of surrounding points by using the Laplace transform as indicated below.

(a) Make use of the result of Problem 70 [see also Equations (203) to (206), Section 6.20] to obtain the relevant relations involving the velocity potential φ and radial velocity V in the form

$$\frac{\partial^2 (r\varphi)}{\partial r^2} = \frac{1}{c^2} \frac{\partial^2 (r\varphi)}{\partial t^2};$$

$$r = a: \quad V = \frac{\partial \varphi}{\partial r} = V_0 \cos \omega t; \quad r \longrightarrow \infty: \quad \frac{\partial \varphi}{\partial r} \text{ finite};$$

$$t = 0: \quad \varphi = \frac{\partial \varphi}{\partial t} = 0.$$

(b) By calculating the Laplace transforms of these relations, obtain the equations

$$\frac{\partial^2 (r\bar{\varphi})}{\partial r^2} - \frac{s^2}{c^2} (r\bar{\varphi}) = 0; \quad \bar{V} = \frac{\partial \bar{\varphi}}{\partial r};$$

$$r = a: \quad \bar{V} = \frac{V_0 s}{s^2 + \omega^2}; \quad r \longrightarrow \infty: \quad \bar{V} \text{ finite}.$$

(c) Show that $r\bar{\varphi}$ must be of the form $Ae^{-sr/c}$ and hence also

$$\bar{V} = -\frac{A}{cr}\left(s + \frac{c}{r}\right)e^{-sr/c}.$$

(d) Determine A, as a function of s, so that the condition at $r = a$ is satisfied. Thus obtain the result

$$\bar{V} = \bar{f}(s)e^{-[(r-a)/c]s},$$

where

$$\bar{f}(s) = \frac{aV_0}{r}\frac{s\left(s + \dfrac{c}{r}\right)}{(s^2 + \omega^2)\left(s + \dfrac{c}{a}\right)}.$$

(e) Expand $\bar{f}(s)$ in partial fractions and use pairs T9, 12, 15, and 16 of Table 1, Chapter 2, to determine the inverse transform of $\bar{V}$. The solution is of the form

$$V = 0 \qquad \text{when } r - a > ct,$$

$$V = \frac{V_0 a^2/r^2}{c^2 + a^2\omega^2}\left[(c^2 + ar\omega^2)\cos\omega\left(t - \frac{r-a}{c}\right) - c\omega(r - a)\sin\omega\left(t - \frac{r-a}{c}\right)\right.$$

$$\left. + \frac{c^2}{a}(r - a)e^{-\frac{c}{a}\left(t - \frac{r-a}{c}\right)}\right] \qquad \text{when } r - a < ct.$$

(Notice that the disturbance travels radially outward with the speed c and at time t is present only at distances less than ct from the sphere surface. As $t \to \infty$ all the surrounding fluid becomes disturbed, the exponential term tends to zero, and the solution approaches that of Problem 75.)

100. *The convolution and Duhamel superposition.* Suppose that, for a certain problem, the Laplace transform of the required solution $\varphi(x, t)$ is determined in the form

$$\bar{\varphi}(x, s) = \bar{F}(s)\bar{G}(x, s) + \bar{P}(x, s),$$

where $\bar{F}$, $\bar{G}$, and $\bar{P}$ are the transforms of identifiable functions $F(t)$, $G(x, t)$, and $P(x, t)$, respectively.

(a) Show that then

$$\varphi(x, t) = \int_0^t G(x, t - \tau)F(\tau)\, d\tau + P(x, t).$$

(b) Show that if

$$\frac{1}{s}\bar{G}(x, s) = \bar{A}(x, s)$$

is the Laplace transform of $A(x, t)$, and if $F(t)$ is differentiable, then also

$$\varphi(x, t) = F(0)A(x, t) + \int_0^t A(x, t - \tau)F'(\tau)\, d\tau + P(x, t).$$

[Notice that $s\bar{F}(s) = \mathcal{L}\{F'(t)\} + F(0)$.]

(c) Show that $G(x, t)$ is the solution of the relevant problem when $P(x, t) = 0$ and $F(t)$ is the *unit impulse function* ("delta function") $\delta(t)$, whereas $A(x, t)$ is the solution when $P(x, t) = 0$ and $F(t)$ is a *unit step function* such that $F(t) = 0$ when $t < 0$ and $F(t) = 1$ when $t > 0$. [The function $A(x, t)$ is sometimes called the *indicial admittance.*]

101. If $T(x, t)$ is the solution of the problem specified by Equations (165), (166), and (167), show that

$$\bar{T}(x, s) = \bar{F}(s) \frac{\sinh (s^{1/2}x/\alpha)}{\sinh (s^{1/2}l/\alpha)}.$$

Hence deduce from Problem 100 that the function $A(x, t)$ defined by (168) must be the inverse of the transform

$$\bar{A}(x, s) = \frac{1}{s} \frac{\sinh (s^{1/2}x/\alpha)}{\sinh (s^{1/2}l/\alpha)},$$

and also rederive (173) and (175). (See also Problem 108, below, and Problem 7 of Chapter 11.)

102. If $\varphi(x, t)$ is the solution of the problem

$$\frac{\partial^2 \varphi}{\partial x^2} = \frac{1}{c^2} \frac{\partial^2 \varphi}{\partial t^2},$$

$$\varphi(0, t) = 0, \quad \varphi(l, t) = f(t), \quad \varphi(x, 0) = 0, \quad \varphi_t(x, 0) = 0,$$

when $0 \leq x \leq l$ and $0 \leq t < \infty$, show that

$$\bar{\varphi}(x, s) = \bar{f}(s) \frac{\sinh (sx/c)}{\sinh (sl/c)}$$

and deduce that if

$$A(x, t) = \mathcal{L}^{-1}\left\{ \frac{1}{s} \frac{\sinh (sx/c)}{\sinh (sl/c)} \right\}$$

(see Problem 106, below, and Problem 6 of Chapter 11), there follows

$$\varphi(x, t) = \int_0^t \frac{\partial A(x, t - \tau)}{\partial t} f(\tau) \, d\tau$$

and also

$$\varphi(x, t) = f(0)A(x, t) + \int_0^t A(x, t - \tau) f'(\tau) \, d\tau,$$

if $f(t)$ is differentiable. (Thus the Duhamel principle of Section 9.11 also applies here.)

Section 9.16

103. (a) When $f(t) = 1$, show that the voltage given by Equation (285), for any fixed value of x, is given by $v = A(x, t)$, where

$$A(x, t) = \begin{cases} 0, & 0 \leq t < x/c, \\ 1, & x/c < t < (2l - x)/c, \\ 2, & (2l - x)/c < t < (2l + x)/c, \\ 1, & (2l + x)/c < t < (4l - x)/c, \\ 0, & (4l - x)/c < t < 4l/c, \end{cases}$$

when $0 \leq t \leq 4l/c$, and A is periodic, with period $4l/c = 4T$.

(b) At the midpoint, $x = l/2$, show that $v = 0$ for $T/2$ seconds, then $v = 1$ for T seconds, then $v = 2$ for T seconds, then $v = 1$ for T seconds, and then $v = 0$ for the remaining $T/2$ seconds of the period $4T$.

(c) At the end $x = l$, show that $v = 0$ for T seconds, then $v = 2$ for $2T$ seconds, and then $v = 0$ for the remaining T seconds of the period $4T$.

(If *resistance* effects were taken into account, the amplitude of the voltage oscillation about the mean value $v = 1$ would damp out with increasing time.)

104. With the notation of Problem 103, show that the solution (285) can be expressed in the alternative form

$$v(x, t) = f(0)A(x, t) + \int_0^t A(x, t - \tau)f'(\tau)\, d\tau.$$

Also, by making use of this result, present $v(x, t)$ graphically as a function of t for $0 \leq t \leq 4l/c$ when $f(t) = 1$ and when $f(t) = v_0 t$. [Notice that the equivalence

$$\int_0^t A(x, t - \tau)f'(\tau)\, d\tau = \int_0^t A(x, \tau)f'(t - \tau)\, d\tau$$

is useful here.]

105. (a) If the end of the line considered in Section 9.16 is *grounded*, so that $v(l, t) = 0$, show that

$$\bar{v}(x, s) = \bar{f}(s)\frac{\sinh\,[s(l - x)/c]}{\sinh sl/c} = \bar{f}(s)\frac{e^{-sx/c} - e^{-s(2l-x)/c}}{1 - e^{-2sl/c}}$$

$$= \bar{f}(s)[e^{-sx/c} - e^{-s(2l-x)/c} + e^{-s(2l+x)/c} - e^{-s(4l-x)/c} + \cdots].$$

(Notice that reflected waves are reversed at *each* end of the line.)

(b) When $f(t) = 1$, show that, for any fixed value of x, this expression takes a form similar to that given in Problem 103(a), with the successive values 0, 1, 0, 1, 0 replacing the values 0, 1, 2, 1, 0 given in that expression, and that, in fact, the voltage here is of period $2T$.

106. Show that the function $A(x, t)$ in Problem 102 is the solution of Problem 105(b) with x replaced by $l - x$. Also use this fact in presenting $v(x, t)$ graphically as a function of t for $0 \leq t \leq 4l/c$ when $f(t) = 1$ and when $f(t) = v_0 t$. (See note to Problem 104.)

107. Let $T(x, t)$ satisfy the heat-flow equation $\alpha^2 T_{xx} - T_t = 0$ for $t \geq 0$ and for $0 \leq x < \infty$, subject to the initial condition $T(x, 0) = 1$ and to the end condition $T(0, t) = 0$.

(a) Show that the Laplace transform $\bar{T}(x, s)$ then must satisfy the equation $\alpha^2 \bar{T}_{xx} - s\bar{T} = -1$ and the end condition $\bar{T}(0, s) = 0$ and, making use of the fact that $\bar{T}$ must be bounded as $x \to +\infty$ for $s > 0$ [or of the more relevant (but less evident) fact that $C(s) \exp(s^{1/2}x/\alpha)$ cannot be a Laplace transform as $x \to +\infty$ unless it vanishes identically], deduce that

$$\bar{T}(x, s) = \frac{1}{s}(1 - e^{-s^{1/2}x/\alpha}).$$

(b) By noticing that the solution of this problem is in fact given by $T(x, t) = \text{erf}\,[x/(2\alpha\sqrt{t})]$, according to Equation (244), deduce that

$$\frac{1}{s}(1 - e^{-qx}) = \mathcal{L}\left\{\text{erf}\,\frac{x}{2\alpha\sqrt{t}}\right\}$$

and

$$\frac{1}{s}e^{-qx} = \mathcal{L}\left\{1 - \text{erf}\,\frac{x}{2\alpha\sqrt{t}}\right\},$$

with the abbreviation

$$q = \frac{s^{1/2}}{\alpha}.$$

(A *direct* derivation of these results is given in Problem 8 of Chapter 11.)

108. Let $T(x, t)$ satisfy the heat-flow equation $\alpha^2 T_{xx} - T_t = 0$ for $t \geq 0$ and $0 \leq x \leq l$, subject to the initial condition $T(x, 0) = 0$ and to the end conditions $T(0, t) = 0$ and $T(l, t) = 1$ when $t > 0$, so that $T = A(x, t)$, where $A(x, t)$ is the function of which the series (168) is an expansion.

(a) Show that the Laplace transform $\bar{T}(x, s)$ must satisfy the equation

$$\alpha^2 \bar{T}_{xx} - s\bar{T} = 0$$

and the end conditions $\bar{T}(0, s) = 0$, $\bar{T}(l, s) = 1/s$, and deduce that

$$\bar{T}(x, s) = \frac{1}{s}\frac{\sinh qx}{\sinh ql} = \frac{1}{s}e^{-q(l-x)}\frac{1 - e^{-2qx}}{1 - e^{-2ql}},$$

where $q = s^{1/2}/\alpha$.

(b) By expanding $1/(1 - e^{-2ql})$ in a series of ascending powers of e^{-2ql} (when $q > 0$), express this transform in the form

$$\bar{T}(x, s) = \frac{1}{s}e^{-q(l-x)} - \frac{1}{s}e^{-q(l+x)} + \frac{1}{s}e^{-q(3l-x)} - \cdots.$$

(c) By evaluating inverse transforms term by term (as can be justified here), and using the results of Problem 107, deduce that $T = A(x, t)$, where

$$A(x, t) = \left(1 - \mathrm{erf}\,\frac{l - x}{2\alpha\sqrt{t}}\right) - \left(1 - \mathrm{erf}\,\frac{l + x}{2\alpha\sqrt{t}}\right) + \left(1 - \mathrm{erf}\,\frac{3l - x}{2\alpha\sqrt{t}}\right) - \cdots.$$

[Notice that the series form (168) converges rapidly when $\alpha\sqrt{t}/l$ is large. Since erf u approaches unity very rapidly as $u \to +\infty$, the terms in parentheses in the form of part (c) are small when $\alpha\sqrt{t}/l$ is *small*. Hence the latter form is particularly convenient when the former one is not. The abbreviation

$$\mathrm{erfc}\,x = 1 - \mathrm{erf}\,x = \frac{2}{\sqrt{\pi}}\int_x^\infty e^{-u^2}\,du,$$

defining the so-called *complementary error function*, is a conventional one.]

Section 9.17

109. Obtain the differential equation (295) by the alternative method suggested in the text.

110. (a) If $C_n''(t) + \mu_n^2 C_n(t) = h_n(t)$, with $\mu_n \neq 0$, show that

$$C_n(t) = \frac{1}{\mu_n}\int_{t_0}^t \sin \mu_n(t - \tau)h_n(\tau)\,d\tau + A_n \cos \mu_n t + B_n \sin \mu_n t,$$

where A_n and B_n are arbitrary constants. (See Section 1.9.)

(b) If the conditions $C_n(0) = a_n$, $C_n'(0) = b_n$ are prescribed, show that $A_n = a_n$ and $B_n = b_n/\mu_n$.

(Notice that μ_n may be replaced by $i\mu_n$.)

111. For the problem

$$\frac{\partial^2 \varphi}{\partial x^2} + \frac{\partial^2 \varphi}{\partial y^2} = h(x, y),$$

$$\varphi(0, y) = 0, \qquad \varphi(l, y) = 0, \qquad \varphi(x, 0) = f(x),$$

subject to the requirement that $\varphi(x, y)$ be *bounded* as $y \to \infty$, obtain the solution (when it exists) in the form

$$\varphi(x, y) = \sum_{n=1}^{\infty} C_n(y) \sin \frac{n\pi x}{l},$$

where

$$C_n''(y) - \frac{n^2\pi^2}{l^2} C_n(y) = h_n(y)$$

with

$$C_n(0) = a_n,$$

where $C_n(y)$ also is to be *bounded* as $y \longrightarrow \infty$ and where

$$h_n(y) = \frac{2}{l} \int_0^l h(x, y) \sin \frac{n\pi x}{l} dx, \qquad a_n = \frac{2}{l} \int_0^l f(x) \sin \frac{n\pi x}{l} dx.$$

Also determine $C_n(y)$ explicitly when $h(x, y) = 1$ and $f(x) = 1$. [If $h(x, y)$ is unbounded as $y \longrightarrow \infty$, but does not grow exponentially, then the boundedness requirement on φ and on C_n generally must be replaced by a restriction against exponential growth, if a solution is to exist.]

112. For the problem

$$\frac{\partial^2 \varphi}{\partial x^2} + \frac{\partial^2 \varphi}{\partial y^2} = h(x, y),$$

$$\varphi(0, y) = G(y), \qquad \varphi(l, y) = F(y), \qquad \varphi(x, 0) = f(x),$$

with $\varphi(x, y)$ bounded as $y \longrightarrow \infty$, obtain the solution (when it exists) in the form

$$\varphi(x, y) = u(x, y) + v(x, y),$$

where

$$u(x, y) = \frac{x}{l} F(y) + \left(1 - \frac{x}{l}\right) G(y)$$

and where $v(x, y)$ is the solution of Problem 111 when $f(x)$ and $h(x, y)$ are replaced by $f^*(x)$ and $h^*(x, y)$, with

$$f^*(x) = f(x) - \frac{x}{l} F(0) - \left(1 - \frac{x}{l}\right) G(0),$$

$$h^*(x, y) = h(x, y) - \frac{x}{l} F''(y) - \left(1 - \frac{x}{l}\right) G''(y),$$

when F'' and G'' exist. Also determine the solution explicitly when $h(x, y) = 1$, $f(x) = 0$, and $F(y) = G(y) = 1$.

113. Suppose that $C_n(y)$ is to satisfy the equation

$$C_n''(y) - \mu_n^2 C_n(y) = h_n(y) \qquad (\mu_n > 0),$$

the condition $C_n(0) = a_n$, and the requirement that $C_n(y)$ be *bounded* as $y \longrightarrow \infty$ (when h_n is sufficiently respectable).

(a) Show that

$$C_n(y) = \frac{1}{2\mu_n} \int_0^y [e^{\mu_n(y-\eta)} - e^{-\mu_n(y-\eta)}] h_n(\eta) \, d\eta + A_n e^{\mu_n y} + B_n e^{-\mu_n y},$$

for some choice of the constants A_n and B_n. (Compare Problem 110).

(b) For large positive values of y, show that

$$C_n(y) \sim \frac{1}{2\mu_n} e^{\mu_n y} \int_0^\infty e^{-\mu_n \eta} h_n(\eta) \, d\eta + A_n e^{\mu_n y},$$

and hence deduce that

$$A_n = -\frac{1}{2\mu_n} \int_0^\infty e^{-\mu_n\eta} h_n(\eta)\, d\eta, \qquad B_n = a_n - A_n.$$

(c) Show that the result of introducing these values into the expression for $C_n(y)$ can be written in the compact form

$$C_n(y) = \int_0^\infty G(y, \eta) h_n(\eta)\, d\eta + a_n e^{-\mu_n y},$$

where $G(y, \eta)$ is the *Green's function*

$$G(y, \eta) = \begin{cases} -\dfrac{1}{\mu_n} e^{-\mu_n\eta} \sinh \mu_n y & (y \le \eta), \\[2mm] -\dfrac{1}{\mu_n} e^{-\mu_n y} \sinh \mu_n\eta & (\eta \le y). \end{cases}$$

Also use this result to redetermine $C_n(y)$ in Problem 111 when $h(x, y) = 1$ and $f(x) = 1$.

114. For the problem

$$\frac{\partial^2\varphi}{\partial x^2} + \frac{\partial^2\varphi}{\partial y^2} = h(x, y),$$

$$\varphi(x, 0) = f(x),$$

in the half-plane $-\infty < x < \infty,\ 0 \le y < \infty$, with $\varphi(x, y)$ required to be *bounded* as $y \longrightarrow \infty$, obtain the solution in the form

$$\varphi(x, y) = \int_0^\infty [C(u, y) \cos ux + D(u, y) \sin ux]\, du,$$

where

$$C_{yy} - u^2C = \frac{1}{\pi} \int_{-\infty}^\infty h(x, y) \cos ux\, dx, \qquad D_{yy} - u^2D = \frac{1}{\pi} \int_{-\infty}^\infty h(x, y) \sin ux\, dx,$$

with

$$C(u, 0) = \frac{1}{\pi} \int_{-\infty}^\infty f(x) \cos ux\, dx, \qquad D(u, 0) = \frac{1}{\pi} \int_{-\infty}^\infty f(x) \sin ux\, dx,$$

and with C and D to be bounded as $y \longrightarrow \infty$. [Here, in addition to assuming that h behaves appropriately as $y \longrightarrow \infty$, we must assume $f(x)$ to be such that the indicated integrals exist.]

115. For the problem

$$\frac{\partial^2\varphi}{\partial r^2} + \frac{1}{r}\frac{\partial\varphi}{\partial r} = \frac{\partial\varphi}{\partial t} + h(r, t),$$

$$\varphi(0, t) = 0, \qquad \varphi(r, 0) = f(r),$$

when $0 \le r \le a,\ 0 \le t < \infty$, obtain the solution in the form

$$\varphi(r, t) = \sum_{n=1}^\infty C_n(t) J_0\left(\alpha_n \frac{r}{a}\right),$$

where $J_0(\alpha_n) = 0$, with $C_n(t)$ determined by the differential equation

$$C_n'(t) + \frac{\alpha_n^2}{a^2} C_n(t) = \frac{2}{a^2[J_1(\alpha_n)]^2} \int_0^a rh(r, t) J_0\left(\alpha_n \frac{r}{a}\right) dr$$

and the condition

$$C_n(0) = \frac{2}{a^2[J_1(\alpha_n)]^2} \int_0^a rf(r)J_0\left(\alpha_n\frac{r}{a}\right) dr.$$

116. For the problem

$$\frac{\partial^2\varphi}{\partial x^2} + \frac{\partial^2\varphi}{\partial y^2} = \frac{\partial\varphi}{\partial t},$$

$$\varphi(x, 0, t) = f(x, t),$$

$$\varphi(x, \pi, t) = \varphi(0, y, t) = \varphi(\pi, y, t) = 0,$$

$$\varphi(x, y, 0) = 0,$$

when $0 \leq x \leq \pi$, $0 \leq y \leq \pi$, $0 \leq t < \infty$, obtain the solution in the form

$$\varphi(x, y, t) = \sum_{m=1}^{\infty} \sum_{n=1}^{\infty} C_{mn}(t) \sin mx \sin ny,$$

where

$$C'_{mn}(t) + (m^2 + n^2)C_{mn}(t) = \frac{4n}{\pi^2} \int_0^\pi f(x, t) \sin mx \, dx,$$

with

$$C_{mn}(0) = 0.$$

Section 9.19

117. In a one-dimensional irrotational flow of an ideal compressible fluid in the positive x direction, in which the velocity V_x depends upon position x and time t but differs by only a small amount from a constant U, the velocity V_x can be expressed (approximately) in the form

$$V_x = U + \frac{\partial\varphi(x, t)}{\partial x},$$

where φ satisfies the equation

$$(U^2 - V_S^2)\frac{\partial^2\varphi}{\partial x^2} + 2U\frac{\partial^2\varphi}{\partial x\,\partial t} + \frac{\partial^2\varphi}{\partial t^2} = 0.$$

Here V_S is an effective mean value of the sonic velocity in the fluid. The density ρ is then expressible (approximately) in the form

$$\rho = \rho_0\left[1 - \frac{1}{V_S^2}\left(U\frac{\partial\varphi}{\partial x} + \frac{\partial\varphi}{\partial t}\right)\right],$$

where ρ_0 is the constant value associated with a uniform flow. (See Section 6.20).

    (a) Obtain the general solution for $\varphi(x, t)$ in the form

$$\varphi(x, t) = f[x - (V_S + U)t] + g[x + (V_S - U)t],$$

where f and g are arbitrary twice-differentiable functions.

    (b) Show that V_x and ρ then are defined by the equations

$$V_x = U + F[x - (V_S + U)t] + G[x + (V_S - U)t],$$

$$\rho = \rho_0\left[1 + \frac{1}{V_S}\{F[x - (V_S + U)t] - G[x + (V_S - U)t]\}\right],$$

where $F = f'$ and $G = g'$.

118. (a) Show that the functions V_x and ρ in Problem 117 are determined at the point P with coordinate $x = x_1$ at any time $t = t_1$ by *initial* values of V_x and ρ (at the time $t = 0$) at the two points Q_1 and Q_2 with coordinates

$$x = x_1 - (V_S + U)t_1 \quad \text{and} \quad x = x_1 + (V_S - U)t_1.$$

(b) Show that Q_1 and Q_2 are both upstream from P when the flow is *supersonic* $(U > V_S)$, whereas they are on opposite sides of P when the flow is *subsonic*.

(c) Sketch typical pairs of "characteristic lines" $x - (V_S + U)t = \text{constant}$ and $x - (V_S - U)t = \text{constant}$ in the upper half $(t > 0)$ of a fictitious xt plane (compare with Figure 9.17) in both the supersonic and subsonic cases, and show that the values of V_x and ρ at any point $P(x_1, t_1)$ of this fictitious plane are determined by values of V_x and ρ at the points where the two characteristic lines passing through P intersect the x axis.

119. (a) If the density is prescribed as $\rho = [1 + \eta(x)]\rho_0$ when $t = 0$ in Problem 117, where ρ_0 is a constant and where $|\eta(x)| \ll 1$, and where $\eta(x) \to 0$ as $x \to -\infty$, show that there must follow

$$V_x = U + G[x - (V_S + U)t] + G[x + (V_S - U)t] + V_S\eta[x - (V_S + U)t],$$

$$\frac{\rho}{\rho_0} = 1 + \frac{1}{V_S}\{G[x - (V_S + U)t] - G[x + (V_S - U)t]\} + \eta[x - (V_S + U)t],$$

where G is an arbitrary function.

(b) In addition, let it be prescribed that the velocity must tend to the uniform velocity U at an infinite distance upstream (as $x \to -\infty$) for all positive values of t. Show that, at any positive time t, the velocity at any point for which $x + (V_S - U)t = c$, where c is a positive constant, is given by

$$V_x = U + G(c - 2V_St) + G(c) + V_S\eta(c - 2V_St).$$

Noticing that $x \to -\infty$ as $t \to +\infty$, when c is held fixed, *if the flow is subsonic* $(U < V_S)$, show that there then must follow $G(c) = -G(-\infty)$ for all positive values of c, and deduce that the function G must be zero. Hence show that, when the flow is subsonic, the prescribed conditions uniquely determine V_x and ρ in the forms

$$V_x = U + V_S\eta[x - (V_S + U)t], \qquad \rho = \rho_0\{1 + \eta[x - (V_S + U)t]\}$$

when $t > 0$. [Notice that here we have made $t \to +\infty$ along a characteristic line $x + (V_S - U)t = c$ in the xt plane, and that this line intersects the line on which $x = x_0 \to -\infty$ in the upper half-plane $t > 0$ if and only if $U < V_S$.]

(c) Show that the same result would follow in the *supersonic* case if it were prescribed that $V_x \to U$ as $x \to -\infty$ for all *negative* values of t, or that $V_x \to U$ as $x \to +\infty$ for all *positive* time, if $\eta(x) \to 0$ as $x \to +\infty$.

120. Determine V_x and ρ in Problem 117, subject to the conditions that $\rho = \rho_0$ and $V_x = [1 + \epsilon(x)]U$ when $t = 0$, where ρ_0 is a constant and where $|\epsilon(x)| \ll 1$.

10

Functions of a Complex Variable

10.1. Introduction. The Complex Variable. A complex number α is an expression of the form $\alpha = a + ib$, where a and b are real numbers and i is the imaginary unit, satisfying the equation

$$i^2 = -1. \tag{1}$$

We follow convention by speaking of a as the *real part* of α and of b as the *imaginary part* of α, and write

$$\alpha = a + ib: \quad a = \operatorname{Re}(\alpha), \quad b = \operatorname{Im}(\alpha), \tag{2}$$

noting that accordingly the "imaginary part" of a complex number is in fact *real*. A complex number $a + ib$ is said to be *real* when $b = 0$ and to be *imaginary* when $b \neq 0$. In particular, when $b \neq 0$ *and* $a = 0$ the number is said to be *pure imaginary*. In the same way we may define a complex variable $z = x + iy$, where x and y are real variables.

Geometrically, it is convenient to represent a complex quantity $x + iy$ by the point (x, y) in a rectangular coordinate system known as the *complex plane* (Figure 10.1). Thus the points representing *real* numbers are all located on the x axis in this plane, whereas those representing the *pure imaginary* numbers are located on the y axis. For some purposes it is convenient to think of the complex number z as being represented by a *vector* in the complex plane from the origin $(x = y = 0)$ to the point (x, y). We speak of

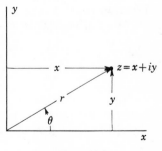

Figure 10.1

the length of this vector as the *absolute value* of z, or as the *modulus* of z, and denote this number by $|z|$,

$$|z| = |x + iy| = \sqrt{x^2 + y^2}. \tag{3}$$

The number $x - iy$ is called the *conjugate* of the number $z = x + iy$ and is denoted by the symbol $\bar{z}$,

$$\bar{z} = x - iy. \tag{4}$$

We say that two complex numbers are *equal* if and only if their real and imaginary parts are respectively equal; that is,

$$x_1 + iy_1 = x_2 + iy_2 \quad \text{implies} \quad x_1 = x_2, \quad y_1 = y_2.$$

In particular, a complex number is *zero* if and only if its real and imaginary parts are *both* zero.

Addition, subtraction, multiplication, and division of complex numbers are defined as being accomplished according to the rules governing real numbers when one writes $i^2 = -1$, in accordance with Equation (1). (See also Problem 1.) Thus, if

$$z_1 = x_1 + iy_1, \qquad z_2 = x_2 + iy_2,$$

there follows

$$z_1 + z_2 = (x_1 + x_2) + i(y_1 + y_2), \tag{5a}$$

$$z_1 - z_2 = (x_1 - x_2) + i(y_1 - y_2), \tag{5b}$$

and

$$z_1 z_2 = (x_1 + iy_1)(x_2 + iy_2) = (x_1 x_2 - y_1 y_2) + i(x_1 y_2 + x_2 y_1). \tag{6}$$

In particular, we notice that

$$z\bar{z} = (x + iy)(x - iy) = x^2 - i^2 y^2 = x^2 + y^2 = |z|^2 = |\bar{z}|^2. \tag{7}$$

That is, *the product of a complex number and its conjugate is a nonnegative real number equal to the square of the absolute value of the complex number.* We then use this fact to deduce that

$$\frac{z_2}{z_1} = \frac{x_2 + iy_2}{x_1 + iy_1} = \frac{(x_2 + iy_2)(x_1 - iy_1)}{x_1^2 + y_1^2} = \left(\frac{x_1 x_2 + y_1 y_2}{x_1^2 + y_1^2}\right) + i\left(\frac{x_1 y_2 - x_2 y_1}{x_1^2 + y_1^2}\right), \tag{8}$$

when $z_1 \neq 0$.

If we introduce polar coordinates (r, θ), such that†

$$x = r\cos\theta, \qquad y = r\sin\theta \qquad (r \geq 0),$$

the complex number z can be written in the *polar form*

$$z = x + iy = r(\cos\theta + i\sin\theta), \tag{9}$$

where r is the modulus of z. It should be noticed that for a given complex number *the angle θ can be taken in infinitely many ways.* With the convention

---

†The restriction $r \geq 0$ (which is imposed throughout this text) is of particular significance here.

that the modulus r be nonnegative, the various possible values of θ differ by integral multiples of 2π unless $z = 0$, in which case θ is arbitrary. Any one of these angles is known as an *argument* or *amplitude* of z, and the abbreviations

$$\theta = \arg z = \operatorname{amp} z = \triangle z$$

are all used.

If we notice that addition or subtraction of complex numbers follows the parallelogram law of vector combination, the truth of the useful inequalities

$$|z_1| - |z_2| \leqq |z_1 + z_2| \leqq |z_1| + |z_2| \quad (10)$$

follows directly from elementary geometrical considerations (Figure 10.2). An analytical proof is supplied in Problem 4.

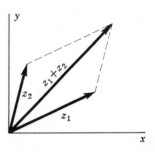

Figure 10.2

10.2. Elementary Functions of a Complex Variable. We now proceed to define functions such as e^z, $\sin z$, $\log z$, and so on, taking care that these definitions reduce to the conventional ones when z becomes the real variable x.†

The simplest such function is the *integral power function*

$$f(z) = z^n, \quad (11)$$

where n is a positive integer or zero. This function naturally is defined recursively by repeated multiplication, according to the law $z^{k+1} = z^k z \ (k = 0, 1, 2, \ldots)$, with $z^0 = 1$, and there follows

$$z^n = (x + iy)^n = r^n(\cos\theta + i\sin\theta)^n \quad (n = 0, 1, 2, \ldots). \quad (12)$$

A *polynomial* in z is then defined as a linear combination of a finite number of these functions, where the constants of combination may be imaginary,

$$f(z) = \sum_{n=0}^{N} A_n z^n. \quad (13)$$

A *rational function* of z is defined as the ratio of two polynomials.

By considering the limit of expressions of form (13) as $N \to \infty$ or, more generally, limits of the form

$$f(z) = \lim_{N \to \infty} \sum_{n=0}^{N} A_n(z - a)^n \equiv \sum_{n=0}^{\infty} A_n(z - a)^n, \quad (14)$$

where a may be real or imaginary, we are led to definitions of functions of a complex variable in terms of *power series*. The convergence of such series may

---

†It should be noticed that it is impossible to "visualize the graphs" of such functions geometrically when z is a complex variable, as one does for functions of a *real* variable, since if (say) $w = \sin z$ one would require four-dimensional "visualization" to relate the real and imaginary parts (say, u and v) of w to the real and imaginary parts (x and y) of z. A method of exhibiting such a relationship geometrically, which uses two planes to supply the necessary four dimensions, is considered in Chapter 11 (Section 11.4).

be investigated by the *ratio test* (see Section 4.1), just as in the case of series of real terms. Thus, if we write

$$L = \lim_{n \to \infty} \left| \frac{A_{n+1}}{A_n} \right| \tag{15}$$

when that limit exists, the series converges when

$$|z - a| < \frac{1}{L}. \tag{16}$$

Geometrically, this restriction is seen to require that z lie inside a circle of radius $R = 1/L$ with center at the point $z = a$ in the complex plane.

Inside this *circle of convergence* the series can be integrated or differentiated term by term, and the resultant series will represent the integral or derivative, respectively, of the represented function.† This fact will be of considerable importance in much of what follows.

The *exponential function* e^z is defined by the power series

$$e^z = \sum_{n=0}^{\infty} \frac{z^n}{n!} = 1 + z + \frac{z^2}{2!} + \frac{z^3}{3!} + \cdots. \tag{17}$$

This definition is acceptable, since the series converges and hence defines a differentiable function of z for *all real or imaginary values* of z, and since this series reduces to the proper one when z is real. If we multiply the series defining e^{z_1} and e^{z_2} together term by term (as is permissible for convergent power series), the resultant series is found to be that defining $e^{z_1+z_2}$ (see Problem 11); that is, the relation

$$e^{z_1} e^{z_2} = e^{z_1+z_2} \tag{18}$$

is true for complex values of z_1 and z_2. Consequently, if n is a positive integer, we have also the relation

$$(e^z)^n = e^{nz} \qquad (n = 1, 2, 3, \ldots) \tag{19}$$

for all complex values of z.

The *circular functions* may be defined, in terms of the already defined exponential functions, by the relations

$$\sin z = \frac{e^{iz} - e^{-iz}}{2i}, \tag{20a}$$

$$\cos z = \frac{e^{iz} + e^{-iz}}{2}, \tag{20b}$$

together with the relations $\tan z = \sin z / \cos z$, and so on. Consequently, we have, from Equations (20) and (17), the corresponding series definitions

$$\sin z = z - \frac{z^3}{3!} + \frac{z^5}{5!} + \cdots = \sum_{n=0}^{\infty} (-1)^n \frac{z^{2n+1}}{(2n+1)!} \tag{21a}$$

---

†This is a consequence of the fact that, if the series (14) converges when $|z - a| < R$, then for any ρ such that $0 < \rho < R$ it is true that the series converges *uniformly* when $|z - a| \leq \rho$, and the same is true of the result of integrating or differentiating the series term by term.

and
$$\cos z = 1 - \frac{z^2}{2!} + \frac{z^4}{4!} + \cdots = \sum_{n=0}^{\infty} (-1)^n \frac{z^{2n}}{(2n)!}, \tag{21b}$$

which reduce to the proper forms when z is real. From these series, or from the definitions (20a, b), it can be shown that the circular functions satisfy the same identities for imaginary values of z as for real values.

Equations (20a, b) imply the important relation

$$e^{iz} = \cos z + i \sin z, \tag{22}$$

which is known as *Euler's formula*.

With this relation Equation (9) takes the form

$$z = x + iy = r(\cos \theta + i \sin \theta) = re^{i\theta}. \tag{23}$$

In consequence of (23) and (19) we then have

$$z^n = r^n(\cos \theta + i \sin \theta)^n = r^n e^{in\theta} \qquad (n = 1, 2, 3, \ldots). \tag{24}$$

But since Equation (22) also implies the relation

$$e^{in\theta} = \cos n\theta + i \sin n\theta,$$

we deduce *DeMoivre's theorem*,

$$(\cos \theta + i \sin \theta)^n = \cos n\theta + i \sin n\theta, \tag{25}$$

and hence may rewrite (24) in the form

$$z^n = r^n(\cos n\theta + i \sin n\theta) \qquad (n = 1, 2, 3, \ldots). \tag{26}$$

Geometrically, Equation (26) shows that *if z has the absolute value r and the angle θ, then z^n has the absolute value r^n and the angle $n\theta$, if n is a positive integer.*

In a similar way we find that if $z_1 = r_1 e^{i\theta_1}$ and $z_2 = r_2 e^{i\theta_2}$, then

$$z_1 z_2 = r_1 r_2 e^{i(\theta_1 + \theta_2)} \tag{27a}$$

and
$$\frac{z_2}{z_1} = \frac{r_2}{r_1} e^{i(\theta_2 - \theta_1)} \qquad (z_1 \neq 0). \tag{27b}$$

That is, if z_1 and z_2 have absolute values r_1 and r_2 and angles θ_1 and θ_2, respectively, then $z_1 z_2$ has the absolute value $r_1 r_2$ and the angle $\theta_1 + \theta_2$, and z_2/z_1 has the absolute value r_2/r_1 and the angle $\theta_2 - \theta_1$.

In particular, we notice that since

$$|e^{i\alpha}| = 1 \qquad (\alpha \text{ real}), \tag{28}$$

the multiplication of any complex number z by a number of the form $e^{i\alpha}$, where α is real, is equivalent to rotating the vector representing the number z through an angle α in the complex plane.

The *hyperbolic functions* are defined, as for functions of a real variable, by the equations

$$\sinh z = \frac{e^z - e^{-z}}{2}, \tag{29a}$$

$$\cosh z = \frac{e^z + e^{-z}}{2}, \tag{29b}$$

and by the equations $\tanh z = \sinh z / \cosh z$, and so on. Consequently, we have also

$$\sinh z = z + \frac{z^3}{3!} + \frac{z^5}{5!} + \cdots = \sum_{n=0}^{\infty} \frac{z^{2n+1}}{(2n+1)!} \tag{30a}$$

and

$$\cosh z = 1 + \frac{z^2}{2!} + \frac{z^4}{4!} + \cdots = \sum_{n=0}^{\infty} \frac{z^{2n}}{(2n)!}. \tag{30b}$$

From these definitions it can be shown that hyperbolic functions of a complex variable satisfy the same identities as the corresponding functions of a real variable. By comparing Equations (20) and (29), we obtain in particular the relations

$$\sinh iz = i \sin z, \qquad \cosh iz = \cos z,$$
$$\sin iz = i \sinh z, \qquad \cos iz = \cosh z, \tag{31}$$

relating the circular and hyperbolic functions.

The results so far obtained permit us to obtain expressions for the functions considered in terms of their real and imaginary parts, including the following ones:

$$e^z = e^{x+iy} = e^x e^{iy} = (e^x \cos y) + i(e^x \sin y),$$
$$\sin z = \sin (x + iy) = \sin x \cos iy + \cos x \sin iy$$
$$= (\sin x \cosh y) + i(\cos x \sinh y),$$
$$\cos z = \cos (x + iy) = \cos x \cos iy - \sin x \sin iy$$
$$= (\cos x \cosh y) - i(\sin x \sinh y), \tag{32}$$
$$\sinh z = \sinh (x + iy) = \sinh x \cosh iy + \cosh x \sinh iy$$
$$= (\sinh x \cos y) + i(\cosh x \sin y),$$
$$\cosh z = \cosh (x + iy) = \cosh x \cosh iy + \sinh x \sinh iy$$
$$= (\cosh x \cos y) + i(\sinh x \sin y).$$

10.3. Other Elementary Functions. We next define the *complex logarithmic function* as the *inverse* of the exponential function. Denoting this function temporarily by $\operatorname{Log} z$, we see that the equation

$$w = \operatorname{Log} z \tag{33}$$

then must be equivalent to

$$z = e^w. \tag{34}$$

To express $\operatorname{Log} z$ in terms of its real and imaginary parts, we write

$$w = u + iv, \tag{35}$$

after which Equation (34) gives

$$z = x + iy = e^{u+iv} = e^u e^{iv} = e^u \cos v + ie^u \sin v.$$

Hence, equating real and imaginary parts, we obtain

$$e^u \cos v = x, \qquad e^u \sin v = y. \tag{36}$$

There then follows

$$e^{2u} = x^2 + y^2 = |z|^2 = r^2, \qquad \cos v = \frac{x}{r} = \cos\theta, \qquad \sin v = \frac{y}{r} = \sin\theta,$$

and hence, if $r \neq 0$,

$$u = \log r = \log|z|, \qquad v = \theta, \tag{37}$$

where r represents, as usual, the absolute value of z and $\log r$ is the ordinary real logarithm, whereas θ is any particular choice of the infinitely many angles (differing by integral multiples of 2π) which may be associated with z. Thus we have obtained the result

$$\text{Log } z = \log|z| + i\theta, \tag{38}$$

for any $z \neq 0$. To emphasize the fact that θ is determinate only within an integral multiple of 2π, we may here denote that particular value of θ which lies in the range $0 \leq \theta < 2\pi$ by θ_P, and may speak of this value as the *principal value* of θ for the logarithm,

$$0 \leq \theta_P < 2\pi. \tag{39}$$

Then any other permissible value of θ is of the form $\theta = \theta_P + 2k\pi$, when k is integral, and Equation (38) becomes

$$\text{Log } z = \log|z| + i(\theta_P + 2k\pi) \qquad (k = 0, \pm 1, \pm 2, \ldots). \tag{40}$$

Thus it follows that the function $\text{Log } z$, defined as the inverse of e^z, is an *infinitely many valued* function. For example, if $z = i$, there follows $|z| = 1$ and $\theta_P = \pi/2$, and hence

$$\text{Log } i = \log 1 + i\left(\frac{\pi}{2} + 2k\pi\right) = \frac{4k+1}{2}\pi i \qquad (k = 0, \pm 1, \pm 2, \ldots). \tag{41}$$

The value corresponding to $k = 0$ may be called the *principal value of the logarithm*.

If, in a particular discussion, z is restricted to *real positive* values, say $z = x$, there follows $|z| = x$ and $\theta_P = 0$, and hence, from (40),

$$\text{Log } x = \log x + 2k\pi i \qquad (x \text{ real and positive}). \tag{42}$$

Thus the complex logarithm of a positive real number may differ from the usual real logarithm by an arbitrary integral multiple of $2\pi i$. In order to conform with conventional usage, henceforth we will identify the complex logarithm $\text{Log } z$ with the real logarithm $\log z$ *when, throughout a given discussion, z is real and positive*, by taking $k = 0$ in (42) in such a case. In the more general case it is also conventional to write $\log z$ in place of $\text{Log } z$, with the understanding that unless z is to take on *only* real positive values, $\log z$ is to be considered as multiply valued. Thus we will write

$$\log z = \log|z| + i(\theta_P + 2k\pi) \qquad (k = 0, \pm 1, \pm 2, \ldots), \tag{43}$$

in place of (40), and avoid resultant contradiction in the case of (42) by taking $k = 0$ when z is a positive real variable.

Suppose now that at a given point z_1 on the positive real axis we choose a particular value of k in (43), say $k = 0$, and hence determine a particular value of $\log z_1$. If a point z moves continuously along a path originating at z_1, the value of $\log z$ then varies continuously from the initial value $\log z_1$. In particular, if z traverses a simple *closed* path surrounding the origin in the positive (counterclockwise) direction and returns toward the initial point, it may be seen that the angle θ increases by an amount approaching 2π; and hence as the circuit is completed the logarithm is increased by $2\pi i$, the *real* part of the logarithm returning to its original value. This statement is true, however, only for a path *enclosing the origin $z = 0$*. If now the point z continues to retrace its first path, the logarithm is now given by a *different (one-valued) function* or by a different "*branch*" of the same (multiple-valued) function. That is, if we write

$$(\log z)_k = \log |z| + i(\theta_P + 2k\pi) \qquad (0 \leqq \theta_P < 2\pi),$$

and if on the first circuit $\log z$ is determined (with $k = 0$) by the branch $(\log z)_0$, then, *if $\log z$ is to vary continuously*, on the second circuit $\log z$ must be determined by the branch $(\log z)_1$, corresponding to $k = 1$. The point $z = 0$, which must be enclosed by the circuit if transition from one branch to another is to be necessary, is known as a *branch point*. We may say that the function $\log z$ has *infinitely many branches*, with a single branch point at $z = 0$.

For any multivalued function, when transition from one branch to another is to be prohibited, so that attention is to be restricted to a single branch as a *single-valued* function, one imagines the complex plane to be "cut" along the line (or curve) along which transition otherwise would take place. The branch in question then will be discontinuous (in general) along the "branch cut" so defined.

Thus, for the function $\log z$, when θ_P is arbitrarily defined as that value of θ for which $0 \leqq \theta_P < 2\pi$, each branch of $\log z$ is discontinuous along the positive real axis. Clearly, in dealing with a problem in which this situation is undesirable, one might instead adopt the convention $-\pi < \theta_P \leqq \pi$, so that the discontinuity along the branch cut occurs along the *negative* real axis, or one could make the transition take place along any other ray, curve, or composite curve which extends from the origin to infinity (without crossing itself).

To fix ideas, we will assume the definition (39) henceforth except when an explicit modified definition is given.

The *generalized power function* $f(z) = z^a$, where a may be real or imaginary, is now defined in terms of the logarithm by the equation

$$z^a = e^{a \log z}, \tag{44}$$

When z is a positive real variable and a is real, this definition is clearly in accord with the usual definition if the logarithm of a positive real number is taken to be real. If a is a positive integer, this definition must be consistent with (24). To see that this is so, we write $a = n$, where n is any integer, and obtain from (44)

$$z^n = e^{n[\log r + i(\theta_P + 2k\pi)]} = e^{n \log r} e^{in\theta_P} e^{2kn\pi i} \qquad (k = 0, \pm 1, \pm 2, \ldots).$$

But since k and n are integers, we have

$$e^{2kn\pi i} = \cos 2kn\pi + i \sin 2kn\pi = 1,$$

and hence there follows, in this case,

$$z^n = r^n e^{in\theta_P} \qquad (n \text{ an integer}). \tag{45}$$

This result is in accordance with (24), since $e^{in\theta} = e^{in(\theta_P + 2k\pi)} = e^{in\theta_P}$ when n is an integer, and if n is a *negative* integer, say $n = -p$, it is identical with the result of the recursive definition $z^{-(p+1)} = z^{-p}/z$ $(p = 0, 1, 2, \ldots)$ when $z \neq 0$, again with $z^0 = 1$.

More generally, if a is a *real rational number*, we can write

$$a = \frac{m}{n},$$

where m and n are integers having no common factor. Then Equation (44) takes the form

$$z^{m/n} = e^{m/n \log r} e^{i(m/n)\theta_P} e^{2k(m/n)\pi i} \qquad (k = 0, \pm1, \pm2, \ldots)$$

or

$$z^{m/n} = [r^{m/n} e^{i(m/n)\theta_P}] e^{2k(m/n)\pi i} \qquad (k = 0, \pm1, \pm2, \ldots).$$

If we remember that m and n are given integers, whereas k is an arbitrary integer, we can easily see that as k takes on any n successive integral values, say $k = 0, 1, 2, \ldots, n-1$, the factor $e^{2k(m/n)\pi i}$ will take on n corresponding different values, but that if k continues to increase through the integers, the n values so obtained are merely repeated periodically. Hence it follows that for any nonzero value of z *the function $z^{m/n}$ has exactly n different values* given by†

$$z^{m/n} = r^{m/n} e^{i(m/n)(\theta_P + 2k\pi)} \qquad (k = 0, 1, 2, \ldots, n-1). \tag{46}$$

It may be seen also that the function $z^{m/n}$ can be considered as having exactly n branches, such that if we again restrict θ_P so that $0 \leq \theta_P < 2\pi$, and if a point traverses a simple closed contour *including the origin* (so that θ changes by 2π), then a continuous variation of $z^{m/n}$ is obtained only through transition from one branch to another. We may verify further that if such a closed contour is traversed exactly n times, starting initially with a certain branch, the transition from the nth branch is back to the initial branch. The point $z = 0$ is again a *branch point*.

As an example, suppose that $m/n = \frac{2}{3}$, and that a point z traverses the unit circle in the positive (counterclockwise) direction, starting at the point $z = 1$. If we arbitrarily start out on the branch $k = 0$ and note that for $z = 1$ there follows $r = 1$, $\theta_P = 0$, the initial value of $z^{2/3}$ at $z = 1$ is given by 1. As the end of a circuit is approached, the angle θ_P approaches 2π, and as $z \to 1$ the power $z^{2/3}$ approaches the value $e^{i(2/3)(2\pi+0)} = e^{(4/3)\pi i}$. In order that $z^{2/3}$ vary continuously as the point $z = 1$ is passed, and hence θ_P drops abruptly to zero and then again

---

†The notation $\sqrt[n]{w}$ is sometimes used to denote the multivalued function $w^{1/n}$ when w is complex. In this text, however, the former notation will not be used at all unless w is *real and positive*, in which case the radical here will denote the *real and positive* nth root. Thus, for example, $3^{1/2} = \pm\sqrt{3}$.

increases, we must now determine $z^{2/3}$ from the *second* branch, for which $k = 1$, since this branch assumes at $\theta_P = 0$ the value approached by the first branch when $\theta_P \longrightarrow 2\pi$. As the end of the *second* circuit is approached and again $z \longrightarrow 1$, the power $z^{2/3}$ approaches the value $e^{i(2/3)(2\pi+2\pi)} = e^{(8/3)\pi i}$; and as this point is passed, the transition to the *third* branch ($k = 2$) must occur. Finally, as the end of the *third* circuit is approached and once more $z \longrightarrow 1$, the power $z^{2/3}$ approaches the value $e^{i(2/3)(2\pi+4\pi)} = e^{4\pi i} = 1$, and hence (for continuity) a transition back to the *first* branch ($k = 0$) must take place. The three values taken on by $z^{2/3}$ at $z = 1$ are 1, $(-1 - i\sqrt{3})/2$, and $(-1 + i\sqrt{3})/2$.

In the general case when a is *complex*, of the form

$$a = a_1 + ia_2,$$

where a_1 and a_2 are real, Equation (44) becomes

$$z^a = e^{(a_1 + ia_2)[\log r + i(\theta_P + 2k\pi)]}$$

$$= e^{[a_1 \log r - a_2(\theta_P + 2k\pi)]}e^{i[a_2 \log r + a_1(\theta_P + 2k\pi)]}$$

$$= r^{a_1}e^{-a_2(\theta_P + 2k\pi)}\{\cos [a_2 \log r + a_1(\theta_P + 2k\pi)]$$

$$+ i \sin [a_2 \log r + a_1(\theta_P + 2k\pi)]\} \qquad (k = 0, \pm 1, \pm 2, \dots). \qquad (47)$$

It is apparent that, in general, the function z^a is infinitely many valued, with $z = 0$ as a branch point. If $a_2 = 0$ and a_1 is rational, then only a finite number of branches exist; in particular, if $a_2 = 0$ and a_1 is integral the function is single-valued and $z = 0$ is *not* a branch point.

The *generalized exponential function* $f(z) = a^z$, where a may be real or imaginary, but $a \neq 0, 1$, is defined similarly by the equation

$$a^z = e^{z \log a}. \qquad (48)$$

If we denote by α_P the principal value of the angle corresponding to a, such that, say, $0 \leq \alpha_P < 2\pi$, there follows

$$a^z = e^{(x+iy)[\log|a| + i(\alpha_P + 2k\pi)]} = e^{[x \log|a| - y(\alpha_P + 2k\pi)]}e^{i[y \log|a| + x(\alpha_P + 2k\pi)]}$$

$$= |a|^x e^{-y(\alpha_P + 2k\pi)}\{\cos [y \log |a| + x(\alpha_P + 2k\pi)]$$

$$+ i \sin [y \log |a| + x(\alpha_P + 2k\pi)]\} \qquad (k = 0, \pm 1, \pm 2, \dots). \qquad (49)$$

Although this function apparently also is infinitely many valued, it is seen that here the ambiguity arises only in the choice of the angle to be associated with the constant a, and not upon the specification of the angular position of the point z in the plane. That is, here there is no possibility of a continuous transition from an expression corresponding to a given value of k to one corresponding to a second value, as the result of motion of a point z around a curve in the complex plane. Thus, in this sense, each choice of k can be considered as determining a *separate function*, rather than a particular *branch* of a single multivalued function. In any specific situation, only one such definition is needed. It is usually convenient to take $k = 0$, and hence write

$$a^z = |a|^x e^{-y\alpha_P}[\cos (y \log |a| + x\alpha_P) + i \sin (y \log |a| + x\alpha_P)]. \qquad (50)$$

We notice in particular that if $a = e$ (and hence $\alpha_P = 0$), the definition (50) reduces to that given in (32). This is the reason for choosing the particular value $k = 0$.

Finally, to conclude the list of elementary functions, we consider the *inverse circular and hyperbolic functions*. In the case of the inverse sine function, the equation

$$w = \sin^{-1} z \qquad (51)$$

implies the equation

$$z = \sin w. \qquad (52)$$

If we make use of the definition (20a), this equation takes the form

$$z = \frac{e^{iw} - e^{-iw}}{2i},$$

or, equivalently,

$$e^{2iw} - 2ize^{iw} - 1 = 0. \qquad (53)$$

Equation (53) is quadratic in e^{iw}, with the solution

$$e^{iw} = iz + (1 - z^2)^{1/2}.$$

When this equation is solved for $w = \sin^{-1} z$, there follows finally

$$\sin^{-1} z = \frac{1}{i} \log [iz + (1 - z^2)^{1/2}]. \qquad (54)$$

It is important to notice that *when $z \neq \pm 1$ the quantity $(1 - z^2)^{1/2}$ has two possible values.* Then corresponding to each such value the logarithm has *infinitely many values*. Hence, for any given value of $z \neq \pm 1$ the function $\sin^{-1} z$ has two infinite sets of values; for $z = \pm 1$, the two sets coincide. This is, of course, already known to be the case when z is real and numerically less than unity so that $\sin^{-1} z$ is real. For example, we have the values $\sin^{-1} \frac{1}{2} = \pi/6 + 2k\pi$ or $5\pi/6 + 2k\pi$, where in either case k may take on arbitrary integral values. In a later section it will be shown that $\sin^{-1} z$ has branch points at $z = \pm 1$.

We may verify that Equation (54) gives the known values for $\sin^{-1} \frac{1}{2}$ by making the calculations

$$\sin^{-1} \frac{1}{2} = \frac{1}{i} \log \left[\frac{i}{2} \pm \frac{\sqrt{3}}{2} \right]$$

$$= \begin{cases} \frac{1}{i} \left[\log 1 + i\left(\frac{\pi}{6} + 2k\pi \right) \right] \\ \frac{1}{i} \left[\log 1 + i\left(\frac{5\pi}{6} + 2k\pi \right) \right] \end{cases}$$

$$= \frac{\pi}{6} + 2k\pi \quad \text{or} \quad \frac{5\pi}{6} + 2k\pi \qquad (k = 0, \pm 1, \pm 2, \ldots).$$

In an entirely analogous way, expressions may be obtained for the other inverse functions, including the following ones:

$$\sin^{-1} z = \frac{1}{i} \log [iz + (1 - z^2)^{1/2}], \tag{55a}$$

$$\cos^{-1} z = \frac{1}{i} \log [z + (z^2 - 1)^{1/2}], \tag{55b}$$

$$\tan^{-1} z = \frac{1}{2i} \log \frac{i - z}{i + z}, \tag{55c}$$

$$\cot^{-1} z = \frac{1}{2i} \log \frac{z + i}{z - i}, \tag{55d}$$

$$\sinh^{-1} z = \log [z + (1 + z^2)^{1/2}], \tag{56a}$$

$$\cosh^{-1} z = \log [z + (z^2 - 1)^{1/2}], \tag{56b}$$

$$\tanh^{-1} z = \frac{1}{2} \log \frac{1 + z}{1 - z}, \tag{56c}$$

$$\coth^{-1} z = \frac{1}{2} \log \frac{z + 1}{z - 1}. \tag{56d}$$

The functions so far considered in this chapter are the basic *elementary functions*. Any linear combination of such functions or any composite function defined in terms of a finite number of such functions is also known as an elementary function.

The *derivative* of a function of a complex variable is defined, as in the real case, by the equation

$$\frac{df(z)}{dz} = f'(z) = \lim_{\Delta z \to 0} \frac{f(z + \Delta z) - f(z)}{\Delta z} \tag{57}$$

when the indicated limit exists. It is readily verified that the derivative formulas established for elementary functions of a real variable are also valid for the corresponding functions of a complex variable, as defined in this section.

Thus, for the function $f(z) = e^{az}$, with a constant, there follows

$$\frac{f(z + \Delta z) - f(z)}{\Delta z} = e^{az} \left(\frac{e^{a\Delta z} - 1}{\Delta z} \right)$$

$$= ae^{az} \left[1 + \frac{a\Delta z}{2!} + \frac{(a\Delta z)^2}{3!} + \cdots \right],$$

for any $\Delta z \neq 0$, and hence $f'(z) = ae^{az}$. Since the other elementary functions are simply related to the exponential function, expressions for their derivatives then are easily derived from this result.

10.4 Analytic Functions of a Complex Variable. A function $f(z)$ is said to be *analytic* in a region $\mathcal{R}$ of the complex plane if $f(z)$ has a finite derivative at each point of $\mathcal{R}$ *and* if $f(z)$ is single-valued in $\mathcal{R}$. It is said to be analytic *at a* *point z* if *z* is an interior point of *some* region where $f(z)$ is analytic.† Thus a function cannot be analytic at a point without also being analytic throughout some circle with center at that point.

It can be shown that, if a function $f(z)$ is analytic at a point *z*, then the derivative $f'(z)$ is *continuous* at *z*. The truth of this theorem (due to *Goursat*, and relatively difficult to prove) is *assumed* here. It then follows (Section 10.7) that, in fact, $f(z)$ has continuous derivatives of *all* orders at the point *z*.

To require that a function $w = f(z)$ have a finite derivative at a point *z* is equivalent to requiring that the limit

$$\frac{dw}{dz} = \lim_{\Delta z \to 0} \frac{f(z + \Delta z) - f(z)}{\Delta z} = \lim_{\Delta z \to 0} \frac{\Delta w}{\Delta z} \qquad (58)$$

exist uniquely as $\Delta z \to 0$ *from any direction* in the complex plane. That is, the value approached by the ratio $\Delta w / \Delta z$ must exist for any direction of approach and must not depend upon the direction.

As an example of a simple function which is not analytic *anywhere*, we consider the relation

$$w = x - iy = \bar{z}. \qquad (59)$$

Here *w* can be considered as a function of $z = x + iy$, since, if *z* is given, the real and imaginary parts (*x* and *y*) of *z* are determined and hence *w* is determined. However, if we examine the ratio

$$\frac{\Delta w}{\Delta z} = \frac{\Delta x - i\Delta y}{\Delta x + i\Delta y}, \qquad (60)$$

we see that if Δz approaches zero along a line parallel to the real (*x*) axis, so that $\Delta y = 0$ throughout the limiting process, the limit of (60) is $+1$, whereas if Δz approaches zero along a line parallel to the imaginary (*y*) axis, so that $\Delta x = 0$ throughout the process, the limit is -1. More generally, if Δz approaches zero along a curve with slope $dy/dx = m$ at the point considered in the complex plane, there follows

$$\lim_{\Delta z \to 0} \frac{\Delta w}{\Delta z} = \lim_{\substack{\Delta x \to 0 \\ \Delta y \to 0}} \frac{1 - i(\Delta y/\Delta x)}{1 + i(\Delta y/\Delta x)} = \frac{1 - mi}{1 + mi} = \frac{(1 - m^2) - 2mi}{1 + m^2},$$

and hence a different limit is approached for each value of *m*. Thus $f(z) = \bar{z}$ does not have a *derivative* anywhere, and hence certainly is not *analytic* anywhere.

Suppose now that the limit (58) *does* exist uniquely, independently of the manner in which $\Delta z \to 0$. We indicate the real and imaginary parts of $w = f(z)$ by *u* and *v*, respectively, where *u* and *v* are certain functions of *x* and *y*, and so write

$$w = u(x, y) + iv(x, y). \qquad (61)$$

---

†A point *P* is an *interior point* of a region $\mathfrak{R}$ if and only if every sufficiently small circle with center at *P* contains *only* points of $\mathfrak{R}$. A *region*, by definition, *must* possess interior points and it also *may* possess *boundary points*.

Then there follows

$$\frac{dw}{dz} = \lim_{\substack{\Delta x \to 0 \\ \Delta y \to 0}} \frac{\Delta u + i\Delta v}{\Delta x + i\Delta y}. \tag{62}$$

For an approach parallel to the real axis we set $\Delta y = 0$ first and so obtain

$$\frac{dw}{dz} = \lim_{\Delta x \to 0} \frac{\Delta u + i\Delta v}{\Delta x} = \lim_{\Delta x \to 0} \left(\frac{\Delta u}{\Delta x} + i\frac{\Delta v}{\Delta x}\right) = \frac{\partial u}{\partial x} + i\frac{\partial v}{\partial x}, \tag{63a}$$

where the partial derivatives are formed with y held constant. For an approach parallel to the imaginary axis, we obtain similarly

$$\frac{dw}{dz} = \lim_{\Delta y \to 0} \frac{\Delta u + i\Delta v}{i\Delta y} = \lim_{\Delta y \to 0} \left(\frac{\Delta v}{\Delta y} - i\frac{\Delta u}{\Delta y}\right) = \frac{\partial v}{\partial y} - i\frac{\partial u}{\partial y}. \tag{63b}$$

But if $f(z)$ has a derivative at the point considered, the two values (63a) and (63b) must be equal,

$$\frac{dw}{dz} = \frac{d(u + iv)}{dz} = f'(z) = \frac{\partial u}{\partial x} + i\frac{\partial v}{\partial x} = \frac{\partial v}{\partial y} - i\frac{\partial u}{\partial y}. \tag{64}$$

Hence the real and imaginary parts of the last two expressions must be respectively equal, so that u and v must satisfy the equations

$$\frac{\partial u}{\partial x} = \frac{\partial v}{\partial y}, \tag{65a}$$

$$\frac{\partial u}{\partial y} = -\frac{\partial v}{\partial x}. \tag{65b}$$

These equations are known as the *Cauchy–Riemann* equations. In order that $w = u(x, y) + iv(x, y)$ be analytic in a region $\mathcal{R}$ of the complex plane, it is thus *necessary* that these equations be satisfied for values of x and y corresponding to all points in that region. Conversely, it is true that, if these equations are satisfied *and* if the four partial derivatives involved are continuous in $\mathcal{R}$, then the derivative exists uniquely at each point in $\mathcal{R}$ for *any* method of approach to the limit.

The proof follows from the fact that, if φ is a real-valued function of x and y with continuous first partial derivatives in the neighborhood of a point, then changes in x, y, and φ are related by the equation

$$\Delta\varphi = \left(\frac{\partial \varphi}{\partial x} + \epsilon_1\right)\Delta x + \left(\frac{\partial \varphi}{\partial y} + \epsilon_2\right)\Delta y,$$

where ϵ_1 and ϵ_2 tend to zero as Δx and Δy both tend to zero. Thus, if $f(z) = u + iv$ is single-valued, and if the real functions u and v have continuous first partial derivatives in the neighborhood of a point P in $\mathcal{R}$, then

$$\Delta f = \left(\frac{\partial u}{\partial x} + \epsilon_1'\right)\Delta x + \left(\frac{\partial u}{\partial y} + \epsilon_2'\right)\Delta y + i\left(\frac{\partial v}{\partial x} + \epsilon_1''\right)\Delta x + i\left(\frac{\partial v}{\partial y} + \epsilon_2''\right)\Delta y,$$

where ϵ_1', ϵ_2', ϵ_1'', and ϵ_2'' all tend to zero with Δx and Δy. If also (65a, b) hold at the point P, this relation can be rewritten in the form

$$\Delta f = \left(\frac{\partial u}{\partial x} + i\frac{\partial v}{\partial x}\right)(\Delta x + i\Delta y) + (\epsilon'_1 + i\epsilon''_1)\Delta x + (\epsilon'_2 + i\epsilon''_2)\Delta y,$$

from which there follows

$$\frac{\Delta f}{\Delta z} - \left(\frac{\partial u}{\partial x} + i\frac{\partial v}{\partial x}\right) = (\epsilon'_1 + i\epsilon''_1)\frac{\Delta x}{\Delta z} + (\epsilon'_2 + i\epsilon''_2)\frac{\Delta y}{\Delta z}.$$

Finally, since $|\Delta x/\Delta z| \leq 1$ and $|\Delta y/\Delta z| \leq 1$, the right-hand members tend to zero as Δx and Δy both tend to zero, and hence *as* $\Delta z \longrightarrow 0$ *in any way*, so that the existence of the derivative

$$\lim_{\Delta z \to 0} \frac{\Delta f}{\Delta z} = \frac{\partial u}{\partial x} + i\frac{\partial v}{\partial x} = \frac{df}{dz}$$

at the point P is established. Since this result is guaranteed for all points in $\mathfrak{R}$, it follows that $f(z)$ is analytic in $\mathfrak{R}$.

Thus a single-valued complex function $w = f(z) = u + iv$, for which the first partial derivatives of u and v are continuous in a region $\mathfrak{R}$, is analytic in $\mathfrak{R}$ *if and only if* the Cauchy–Riemann equations (65a, b) are satisfied. It is clear that the function $w = x - iy$ does not satisfy this condition anywhere.

As a further example, we may verify that for the function

$$w = f(z) = (x - y)^2 + 2i(x + y)$$

the Cauchy–Riemann equations are satisfied only *along the line* $x - y = 1$ in the complex plane. Thus, since $f'(z)$ accordingly exists only on this line, and not throughout any region $\mathfrak{R}$, $f(z)$ *is nowhere analytic*. At all points on the line, direct calculation shows that $f'(z)$ has the constant value $2 + 2i$. (See Problem 22.)

From Equations (65a, b) we obtain the additional relations

$$\frac{\partial^2 u}{\partial x^2} = \frac{\partial^2 v}{\partial x \, \partial y}, \qquad \frac{\partial^2 u}{\partial y^2} = -\frac{\partial^2 v}{\partial y \, \partial x},$$

$$\frac{\partial^2 v}{\partial x^2} = -\frac{\partial^2 u}{\partial x \, \partial y}, \qquad \frac{\partial^2 v}{\partial y^2} = \frac{\partial^2 u}{\partial y \, \partial x}.$$

But if the second partial derivatives of u and v are continuous, the order of differentiation is immaterial and it then follows that u and v satisfy the equations

$$\nabla^2 u = \frac{\partial^2 u}{\partial x^2} + \frac{\partial^2 u}{\partial y^2} = 0, \tag{66a}$$

$$\nabla^2 v = \frac{\partial^2 v}{\partial x^2} + \frac{\partial^2 v}{\partial y^2} = 0. \tag{66b}$$

The previously mentioned fact that an analytic function has derivatives of *all* orders (see Section 10.7) implies, in particular, that the partial derivatives involved *are* indeed continuous. Hence it follows that *the real and imaginary parts of an analytic function are solutions of Laplace's equation.*

The Cauchy–Riemann equations (65a, b) permit the determination of either u or v if the other is known, apart from an arbitrary real additive constant, since, by making use of these equations, we may write

$$du = \frac{\partial u}{\partial x}dx + \frac{\partial u}{\partial y}dy = \frac{\partial v}{\partial y}dx - \frac{\partial v}{\partial x}dy, \tag{67a}$$

$$dv = \frac{\partial v}{\partial x}dx + \frac{\partial v}{\partial y}dy = -\frac{\partial u}{\partial y}dx + \frac{\partial u}{\partial x}dy. \tag{67b}$$

Thus, for example, if we have

$$v = y^2 - x^2,$$

then (67a) gives

$$du = 2y\,dx + 2x\,dy$$

and hence

$$u = 2xy + C,$$

where C is an arbitrary real constant.† It follows that $y^2 - x^2$ is the imaginary part of the function

$$2xy + i(y^2 - x^2) + C = -iz^2 + C.$$

We deduce that if one part of an analytic function is prescribed, the other part, and hence the function itself, is determinate within an arbitrary additive constant. A complication, which will be considered later, is that when the region $\Re$ is not simply connected the determined part may not be single-valued.

Two functions $u(x, y)$ and $v(x, y)$, which are respectively the real and imaginary parts of an analytic function, are known as *harmonic functions*, and v is often called the *harmonic conjugate* of u.

10.5. Line Integrals of Complex Functions. If C is a curve in the complex plane joining the points z_0 and z_1, the line integral of a function $f(z) = u + iv$ along C is defined by the equation

$$\int_C f(z)\,dz = \int_C (u + iv)(dx + i\,dy)$$

$$= \int_C [(u\,dx - v\,dy) + i(v\,dx + u\,dy)]$$

or $$\int_C f(z)\,dz = \int_C (u\,dx - v\,dy) + i\int_C (v\,dx + u\,dy). \tag{68}$$

It is supposed here, and in what follows, that any curve C along which such

† Care should be taken to avoid the errors frequently introduced by the generally improper procedure of determining $u(x, y)$ from $du = P(x, y)\,dx + Q(x, y)\,dy$ by integration, holding y constant in the first term and x constant in the second one. [See Section 1.12 (Example 3) or Section 6.11.]

an integration is to be effected is at least *piecewise smooth* (see Section 6.10). Such a curve is sometimes called a *contour*, although this term also is often used in a less restrictive sense.

The real and imaginary parts of (68) are thus ordinary real line integrals in the plane, of the type considered in Section 6.10. We see that the requirement that the two integrands be exact differentials is precisely the requirement that the Cauchy–Riemann equations (65a, b) be satisfied. The results of Sections 6.10 (or 6.16) and 10.4 then show that *the line integral $\int_C f(z)\,dz$ is independent of the path C joining the end points z_0 and z_1 if C can be enclosed in a simple (simply connected) region $\mathfrak{R}$ in which $f(z)$ is analytic.*

The results cited state that if P_x, P_y, Q_x, and Q_y are continuous and $P_y = Q_x$, in a simple region $\mathfrak{R}$ of the xy plane, then the real integral $\int_C (P\,dx + Q\,dy)$ between two points in $\mathfrak{R}$ is independent of the path C joining those points, provided only that C is piecewise smooth and lies in $\mathfrak{R}$, and it then follows also that $P\,dx + Q\,dy = d\varphi$, where φ is single-valued in $\mathfrak{R}$. The two line integrals in (68) correspond to $P = u$, $Q = -v$ and to $P = v$, $Q = u$, respectively. In addition to using (65), it is at this point that we must exploit the *continuity* of $f'(z)$, and hence of u_x, u_y, v_x, and v_y, in $\mathfrak{R}$.

In such a case the curve need not be prescribed and we may indicate the integral by the notation $\int_{z_0}^{z_1} f(z)\,dz$. It follows also that here $f(z)\,dz$ is the exact differential of a function $F(z)$,

$$f(z)\,dz = dF(z), \tag{69}$$

and that the integral can then be evaluated in the usual way, according to the formula

$$\int_{z_0}^{z_1} f(z)\,dz = F(z_1) - F(z_0), \tag{70}$$

where $F(z)$ is a function whose derivative is $f(z)$.

By considering the case where the end points z_0 and z_1 coincide, we conclude that, if C is any closed curve lying in a simple region $\mathfrak{R}$ where $f(z)$ is analytic, then the line integral of $f(z)$ around C will vanish. If we recall further that, when $f(z)$ is analytic at all points on a closed curve C it is also analytic in some neighborhood (on both sides) of the curve, we deduce finally that *if $f(z)$ is analytic inside and on a closed curve C, then*

$$\oint_C f(z)\,dz = 0. \tag{71}$$

This result is known as *Cauchy's integral theorem.*

In other cases the line integral of a function $f(z)$ around a closed contour may or may not vanish. The *positive direction* around a closed contour is defined, as before, as the direction along which an observer would proceed in keeping the enclosed area to his *left*.

We notice that, if u and v are expressed in polar coordinates, Equation (68) is replaced by the equation

$$\int_C f(z)\,dz = \int_C (u + iv)(e^{i\theta}\,dr + ire^{i\theta}\,d\theta)$$

$$= \int_C [(u\cos\theta - v\sin\theta)\,dr - (v\cos\theta + u\sin\theta)r\,d\theta]$$

$$+ i\int_C [(v\cos\theta + u\sin\theta)\,dr + (u\cos\theta - v\sin\theta)r\,d\theta]. \qquad (72)$$

The requirement that the quantities in brackets be exact differentials can be reduced to the conditions

$$\frac{\partial u}{\partial r} = \frac{1}{r}\frac{\partial v}{\partial \theta},$$
$$\frac{1}{r}\frac{\partial u}{\partial \theta} = -\frac{\partial v}{\partial r}. \qquad (73)$$

These equations are the forms assumed by the Cauchy–Riemann equations (65) in polar coordinates (see also Problem 26).

To illustrate these results, we notice that for the single-valued function

$$f(z) = \frac{1}{z}$$

the derivative $f'(z) = -1/z^2$ exists at all points except $z = 0$. Hence $f(z)$ is analytic in *any* simple region $\mathcal{R}$ *not including the origin* in the complex plane. If we are to evaluate the integral

$$\int_C \frac{dz}{z},$$

where C is, say, the straight line from $z_0 = 1$ to $z_1 = i$, we can avoid the explicit evaluation of a form such as (68) or (72) for this path by choosing a more convenient "equivalent" path joining those points (say, a quadrant of a circle) which, *together with* C, lies in a simple region $\mathcal{R}$ which excludes the origin. That is, C here could be replaced by any curve which also joins z_0 and z_1 *and which lies on the same side of the origin as does* C. Even more conveniently, we could notice that $1/z = d(\log z)/dz$ and write

$$\int_C \frac{dz}{z} = \left[\log z\right]_1^i = \log i - \log 1.$$

But here care must be taken to note that both $\log i$ and $\log 1$ are infinitely ambiguous, and that the correct answer will be assured only when use is made of values of both which correspond to a specific *branch* of $\log z$ which is analytic in a simple region $\mathcal{R}$ which includes the prescribed arc C. For this purpose, the branch for which $-\pi < \theta_P \leq \pi$ will serve,† for example, and with this choice there follows $\log i = \pi i/2$ and $\log 1 = 0$, and hence

$$\int_C \frac{dz}{z} = \frac{\pi}{2} i.$$

For any other permissible branch, both $\log i$ and $\log 1$ would be modified by the addition of the *same* multiple of $2\pi i$, and the same final answer would be obtained.

Here the value of the integral is the change experienced by any branch of $\log z$ in correspondence with a transition from $z_0 = 1$ to $z_1 = i$ along C (or along any equivalent curve, as defined above) provided only that the branch used is *analytic* along C.

Any *closed* curve C not surrounding the origin satisfies the conditions of Cauchy's theorem with $f(z) = 1/z$, and hence for any such curve there follows

$$\oint_C \frac{dz}{z} = 0.$$

However, if C encloses the origin, the integral need not vanish. If C is taken as the unit circle $|z| = 1$, with center at the origin, then on C we may write

$$z = e^{i\theta}, \qquad dz = ie^{i\theta}\, d\theta,$$

and hence for this closed curve we obtain

$$\oint_{C_1} \frac{dz}{z} = \int_0^{2\pi} e^{-i\theta}(ie^{i\theta}\, d\theta) = \int_0^{2\pi} i\, d\theta = 2\pi i, \qquad (74)$$

again without any need for actually splitting the integral into real and imaginary parts, where $\oint_{C_1}$ indicates integration around the positive direction of the unit circle. The value given by (74) is merely the limit of the increase experienced by the imaginary part $i\theta$ of any branch of $\log z$ as z describes a positive circuit about the origin which tends to become closed.

Now consider any *other* simple closed curve C which surrounds the origin and does not intersect C_1. If we make a "crosscut" from C_1 to C, and so determine the simply connected region $\mathfrak{R}$ shown in Figure 10.3, we see that, since $f(z) = 1/z$ is analytic in $\mathfrak{R}$, the integral around the *complete* boundary of $\mathfrak{R}$ must

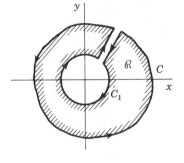

Figure 10.3

---

†This example illustrates the necessity of abandoning the arbitrary definition $0 \leq \theta_P < 2\pi$ when the branch is to be analytic on any part of the positive real axis. In other situations to be encountered, the earlier definition often will be the preferred one.

vanish, by Cauchy's theorem. As the width of the cut is decreased toward zero the integrals along the edges of the cut are taken in opposite directions and hence cancel. Since the part of the complete integration which is carried out along C_1 is taken in the negative direction, there follows in the limit

$$\oint_C \frac{dz}{z} - \oint_{C_1} \frac{dz}{z} = 0 \quad \text{or} \quad \oint_C \frac{dz}{z} = \oint_{C_1} \frac{dz}{z}.$$

It is easy to remove the restriction that C_1 and C not intersect, by choosing a third closed curve C' which encloses both C_1 and C and noticing that then the integrals along C_1 and C are each equal to the integral along C', and hence are equal to each other. A more direct approach, when C and C_1 have a finite number of intersections, consists of considering the sum of the results of integrating around each of the simple regions which lie inside one curve and outside the other (Figure 10.4). By either line of reasoning, we may deduce that

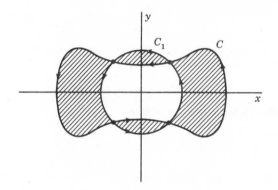

Figure 10.4

for *any* closed curve C enclosing the origin once in the positive direction, we have

$$\oint_C \frac{dz}{z} = 2\pi i. \tag{75}$$

More generally, if $f(z) = z^n$, where n is an integer, there follows

$$\oint_{C_1} z^n \, dz = \int_0^{2\pi} e^{ni\theta}(ie^{i\theta} \, d\theta) = i \int_0^{2\pi} e^{(n+1)i\theta} \, d\theta$$

$$= i \int_0^{2\pi} [\cos(n+1)\theta + i \sin(n+1)\theta] \, d\theta$$

$$= 0 \qquad (n \neq -1), \tag{76}$$

where C_1 again represents the unit circle $|z| = 1$. If n is a *positive integer or zero*, this result is in accordance with Cauchy's theorem, since then $f(z) = z^n$ is analytic for all finite values of z. However, if n is a *negative* integer, z^n is *not* analytic at $z = 0$. Still, Equation (76) states that if $n \neq -1$, the integral $\oint_{C_1} z^n \, dz$ vanishes even in this case. The preceding argument shows that this condition is true also for *any* closed curve C surrounding the origin, and hence, since z^n is analytic except at the origin, we conclude that, when n is an integer,

$$\oint_C z^n \, dz = 0 \qquad (n \neq -1) \tag{77}$$

for *any* closed curve not *passing through* the origin. If $n = -1$, the integral is zero unless C includes the origin, in which case the value is $2\pi i$ if C is simple (and positively oriented).

If we replace z by $z - a$, where a is any complex number, we deduce from (76) and (77) the additional results

$$\oint_C (z - a)^n \, dz = 0 \qquad (n \neq -1), \tag{78a}$$

$$\oint_C \frac{dz}{z - a} = 2\pi i, \tag{78b}$$

where C now is a simple closed curve enclosing the point $z = a$ in the positive direction and n is an integer. These results will be of use in Section 10.12.

Fundamentally, the mathematical definition of the complex line integral $\int_C f(z) \, dz$ along a curve C between the points a and b in the complex plane consists of the limit of the sum

$$\sum_{k=0}^{n} f(z_k) \, \Delta z_k, \qquad \Delta z_k = z_{k+1} - z_k$$

as the number $n + 2$ of the division points $z_0 \, (= a), z_1, \ldots, z_n, z_{n+1} \, (= b)$ along C increases and the chord lengths $|\Delta z_k|$ all tend to zero. This definition can be shown to lead to the evaluation (68), in terms of two real line integrals, whenever $f(z)$ is continuous on C and C is piecewise smooth, whether or not $f(z)$ is *analytic*.

Repeated use of the inequality (10) leads to the inequality

$$\left| \sum_{k=0}^{n} f(z_k) \, \Delta z_k \right| \leq \sum_{k=0}^{n} |f(z_k)| \, |\Delta z_k|,$$

and hence, in the limit as $n \to \infty$, there follows

$$\left| \int_C f(z) \, dz \right| \leq \int_C |f(z)| \, |dz| = \int_C |f(z)| \, |dx + i \, dy| = \int_C |f(z)| \, ds,$$

where s is arc length along C. If now $|f(z)|$ is bounded on C, say

$$|f(z)| \leq M \text{ on } C, \tag{79}$$

the last member of this expression is numerically not greater than

$$\int_C M \, ds = M \int_C ds = ML,$$

where L is the length of the curve C between the specified end points.

Hence we obtain the very useful "*ML* inequality" which states that

$$\left| \int_C f(z)\, dz \right| \le ML, \tag{80}$$

where M is an upper bound for $|f(z)|$ on C, and L is the length of C.

In addition, for future use we generalize preceding considerations by noting that if C_1 and C_2 are two nonintersecting simple closed curves, one enclosing the other, and if $f(z)$ is analytic on C_1 and C_2 and in the region $\Re$ between C_1 and C_2, then by introducing a crosscut as in Figure 10.3 and proceeding as in the development leading to (75), we show that

$$\oint_{C_1} f(z)\, dz = \oint_{C_2} f(z)\, dz.$$

More generally, by also considering situations in which two simple closed curves C_1 and C_2 intersect in two or more points, as in Figure 10.4, with $f(z)$ analytic on C_1 and C_2 and in the regions which lie inside *only one* of the two curves, we deduce that *in dealing with an integral of the form $\oint_{C_1} f(z)\, dz$, where C_1 is a simple closed curve, we may deform C_1 in a continuous way into any other simple closed curve C_2 without changing the value of the integral, so long as $f(z)$ is analytic on C_1 and C_2 and at all points passed over by the curve in the process of the deformation.* In such cases, we say that C_1 and C_2 are *equivalent contours* for the relevant integral.

10.6. Cauchy's Integral Formula. Let C be a simple closed contour inside which and along which $f(z)$ is analytic, and let α be a point inside C. Finally, let C_ϵ be a small circle of radius ϵ, with center at the point α, inside C (Figure 10.5). Then, clearly, the function

$$\frac{f(z)}{z - \alpha}$$

is analytic between C and C_ϵ, and so there follows

$$\oint_C \frac{f(z)}{z - \alpha}\, dz = \oint_{C_\epsilon} \frac{f(z)}{z - \alpha}\, dz \tag{81}$$

for all positive values of ϵ. In particular, Equation (81) must remain true in the limit as $\epsilon \to 0$. But for points on C_ϵ we have

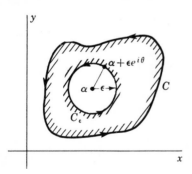

Figure 10.5

$$z = \alpha + \epsilon e^{i\theta}, \qquad dz = i\epsilon e^{i\theta}\, d\theta \qquad (0 \le \theta < 2\pi), \tag{82}$$

and hence there follows

$$\oint_{C_\epsilon} \frac{f(z)}{z - \alpha}\, dz = \int_0^{2\pi} f(\alpha + \epsilon e^{i\theta}) i\, d\theta. \tag{83}$$

In the limit, as $\epsilon \rightarrow 0$, we then obtain from (81) and (83) the result

$$\oint_C \frac{f(z)}{z - \alpha}\, dz = \lim_{\epsilon \to 0} \int_0^{2\pi} f(\alpha + \epsilon e^{i\theta})\, i\, d\theta = 2\pi i f(\alpha), \qquad (84)$$

since $f(z)$ is continuous at $z = \alpha$. Equation (84) expresses the value of $f(z)$ at $z = \alpha$ in terms of the values of $f(z)$ along an enclosing curve C. This result can be written in more convenient form if we interchange z and α in (84),

$$f(z) = \frac{1}{2\pi i} \oint_C \frac{f(\alpha)}{\alpha - z}\, d\alpha, \qquad (85)$$

where C is now a closed contour enclosing the point z. This is *Cauchy's integral formula* for a function $f(z)$ analytic inside and along a simple closed curve C.

If we recall that the real and imaginary parts of $f(z)$ are solutions of Laplace's equation inside C, and hence (see Section 9.2) are determined at points inside C by their values on C, we see that (85) affords exactly this determination, *if both parts are given on C.*

10.7. Taylor Series. We next use this result to show that, if $f(z)$ is analytic at a point $z = a$, then $f(z)$ can be expanded in a Taylor series of powers of $z - a$. For this purpose we first notice that the function $1/(\alpha - z)$ can be expanded in the *geometric series*

$$\frac{1}{\alpha - z} = \frac{1}{(\alpha - a) - (z - a)} = \frac{1}{\alpha - a} \frac{1}{1 - \dfrac{z - a}{\alpha - a}}$$

$$= \frac{1}{\alpha - a} \sum_{n=0}^{\infty} \left(\frac{z - a}{\alpha - a}\right)^n \qquad (|z - a| < |\alpha - a|). \qquad (86)$$

We now take for C any circle, with center at $z = a$, inside which and on which $f(z)$ is analytic (Figure 10.6). Then for points inside C we have

$$|z - a| < |\alpha - a| = \rho,$$

where ρ is the radius of the circle, and hence for such points the power series (86) converges and also can be multiplied by $f(\alpha)$ and integrated around the

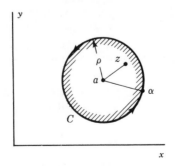

Figure 10.6

curve C term by term. Thus (86) can be introduced into (85) to give the result

$$f(z) = \sum_{n=0}^{\infty} \frac{1}{2\pi i} \left[\oint_C \frac{f(\alpha)\, d\alpha}{(\alpha - a)^{n+1}} \right] (z - a)^n \qquad (|z - a| < \rho),$$

or, equivalently,

$$f(z) = \sum_{n=0}^{\infty} A_n (z - a)^n \qquad (|z - a| < \rho), \tag{87a}$$

where
$$A_n = \frac{1}{2\pi i} \oint_C \frac{f(\alpha)\, d\alpha}{(\alpha - a)^{n+1}}. \tag{87b}$$

Thus, *if $f(z)$ is analytic throughout the circular disk $|z - a| \leq \rho$, it can be expanded in a Taylor series* (87a) *converging inside this disk*, with coefficients given by (87b). Since any Taylor series has derivatives of all orders (which can be obtained by successive term-by-term differentiations) inside its circle of convergence, and since it is now established that $f(z)$ can be represented by such a series when $|z - a| < \rho$, it follows in particular that $f(z)$ indeed has derivatives *of all orders* at any point $z = a$ at which it is analytic.

By differentiating (87a) k times, setting $z = a$ in the result, and using (87b), we obtain the additional formula

$$f^{(k)}(a) = k!\, A_k = \frac{k!}{2\pi i} \oint_C \frac{f(\alpha)\, d\alpha}{(\alpha - a)^{k+1}}$$

or, with a change in notation,

$$f^{(n)}(z) = \frac{n!}{2\pi i} \oint_C \frac{f(\alpha)\, d\alpha}{(\alpha - z)^{n+1}}, \tag{88}$$

where C is now the circle with radius ρ about z, or any equivalent contour

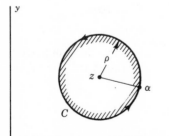

Figure 10.7

(Figure 10.7). We notice that (88) is obtained formally by differentiating (85) n times under the integral sign.

If R is the distance from $z = a$ to the nearest point at which $f(z)$ is not analytic, the series (87a) converges to $f(z)$ when $|z - a| < \rho$ for *every* ρ such that $\rho < R$, and hence indeed the convergence is assured when $|z - a| < R$.

From these results we conclude that *if $f(z)$ is analytic at a point $z = a$* (and hence in some circle around this point), *derivatives of $f(z)$ of all orders exist at that point, and $f(z)$ can be expanded in the Taylor series*

$$f(z) = \sum_{n=0}^{\infty} \frac{f^{(n)}(a)}{n!} (z - a)^n, \tag{89}$$

the circle of convergence of the series coinciding with (or including) the largest circle with center at $z = a$ inside which $f(z)$ is analytic.

Conversely, we see also that, *if $f(z)$ can be represented by a Taylor series in some circle about $z = a$, then $f(z)$ must be analytic at $z = a$,* since then derivatives of all orders exist at and near $z = a$.

Generally, the radius of convergence R_c of the Taylor series (87) will *equal* the distance R from $z = a$ to the nearest point where $f(z)$ ceases to be analytic. However, there exist somewhat unusual situations in which $R_c > R$. For example, suppose that $f(z)$ has the definition

$$f(z) = \begin{cases} 1 & \text{inside } C_0, \\ 0 & \text{outside } C_0, \end{cases} \tag{90}$$

where C_0 is a simple closed curve enclosing the point $z = a$. Then R is the distance from $z = a$ to the nearest point on C_0. But, since $f(a) = 1$ and $f^{(n)}(a) = 0$ for $n \geq 1$, the series representation (87) reduces to the form

$$f(z) = 1 \qquad (|z - a| < R).$$

In this special case, the representation is valid, more generally, *everywhere inside C_0* and, furthermore, the one-term series itself clearly converges for *all* values of z, so that $R_c = \infty$. In fact, the *series* defines a function $f^*(z) = 1$ for all z, which is analytic *everywhere* and which is identical with $f(z)$ inside C_0. Whereas the definition (90) may seem to be somewhat artificial, it may be noted that the more compact definition

$$f(z) = \frac{1}{2\pi i} \oint_{C_0} \frac{d\alpha}{\alpha - z} \tag{90'}$$

is equivalent to it.

More generally, the only situations in which $R_c > R$ are those in which $f(z)$ is so defined that the neighborhood $\mathfrak{R}$ of $z = a$ inside which $f(z)$ is analytic is limited, but in which it is possible to define or redefine $f(z)$ outside that region in such a way that $f(z)$ is analytic in an extended region $\mathfrak{R}^*$ which includes $\mathfrak{R}$. Such a process of extending the region of definition of an analytic function is called *analytic continuation.*†

10.8. Laurent Series. A generalization of the Taylor series expansion of $f(z)$, which may involve both positive and negative integral powers of $z - a$, exists whenever there is a circular *ring* or *annulus* $\mathfrak{R}$, for which $\rho_1 < |z - a| < \rho_2$, such that $f(z)$ is analytic (and hence single-valued) in $\mathfrak{R}$ and on its inner and outer circular boundaries C_1 and C_2 (Figure 10.8).

As a first step toward its derivation, we introduce a crosscut between C_1 and C_2 and apply Cauchy's integral formula to the resultant simply connected region $\mathfrak{R}'$ bounded by the curve C' (Figure 10.9). By considering the limit as the width of the cut tends to zero, we thus deduce that, for any point z in the

---

†In situations less contrived than the one considered here, there generally will be points on the boundary of $\mathfrak{R}$ which cannot be interior points of the extended region $\mathfrak{R}^*$.

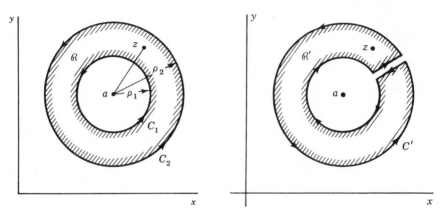

Figure 10.8   Figure 10.9

annulus $\mathfrak{R}$, there follows

$$f(z) = \frac{1}{2\pi i} \oint_{C_2} \frac{f(\alpha)}{\alpha - z}\, d\alpha - \frac{1}{2\pi i} \oint_{C_1} \frac{f(\alpha)}{\alpha - z}\, d\alpha, \tag{91}$$

where both C_1 and C_2 are to be traversed in the positive (counterclockwise) direction.

The *first* integral in (91) now is dealt with exactly as in Section 10.7, by expanding $1/(\alpha - z)$ in powers of $(z - a)/(\alpha - a)$ and integrating term by term, to yield the relation

$$\frac{1}{2\pi i} \oint_{C_2} \frac{f(\alpha)}{\alpha - z}\, d\alpha = \sum_{n=0}^{\infty} A_n (z - a)^n \qquad (|z - a| < \rho_2), \tag{92}$$

where again

$$A_n = \frac{1}{2\pi i} \oint_{C_2} \frac{f(\alpha)\, d\alpha}{(\alpha - a)^{n+1}} \qquad (n = 0, 1, 2, \ldots). \tag{93}$$

An important point here is that the equal members of (92) now generally do *not* also equal $f(z)$, nor does $n!\, A_n$ generally equal $f^{(n)}(a)$, since now $f(z)$ is not necessarily analytic everywhere inside C_2.

In the *second* integral of Equation (91), we instead expand $1/(\alpha - z)$ in powers of $(\alpha - a)/(z - a)$,

$$-\frac{1}{\alpha - z} = \frac{1}{(z - a) - (\alpha - a)} = \frac{1}{z - a} \frac{1}{1 - \dfrac{\alpha - a}{z - a}}$$

$$= \sum_{n=1}^{\infty} \frac{(\alpha - a)^{n-1}}{(z - a)^n}$$

$$= \sum_{n=-\infty}^{-1} \frac{(z - a)^n}{(\alpha - a)^{n+1}} \qquad (|z - a| > \rho_1),$$

so that, after multiplying this convergent power series by $f(\alpha)/(2\pi i)$ and integrating term by term around C_1, there follows

$$-\frac{1}{2\pi i}\oint_{C_1}\frac{f(\alpha)}{\alpha - z}\,d\alpha = \sum_{n=-\infty}^{-1} B_n(z - a)^n \qquad (|z - a| > \rho_1), \qquad (94)$$

where

$$B_n = \frac{1}{2\pi i}\oint_{C_1}\frac{f(\alpha)\,d\alpha}{(\alpha - a)^{n+1}} \qquad (n = -1, -2, \dots). \qquad (95)$$

Now, since Equation (92) is true for z inside C_2 and (94) is true for z outside C_1, *both* are true for z in the ring $\mathfrak{R}$, and hence then can be introduced into (91). Further, since $f(\alpha)/(\alpha - a)^{n+1}$ is an analytic function of α when α is in $\mathfrak{R}$, when n is zero or any positive or negative integer, it follows that, in (93) and (95), C_1 and C_2 can be deformed into *any* simple closed curve C which surrounds $z = a$ and lies in $\mathfrak{R}$.†

Hence, finally, if we write

$$c_n = \frac{1}{2\pi i}\oint_C\frac{f(\alpha)\,d\alpha}{(\alpha - a)^{n+1}} \qquad (n = 0, \pm 1, \pm 2, \dots), \qquad (96)$$

there follows $A_n = c_n$ when $n \geq 0$ and $B_n = c_n$ when $n \leq -1$, and the result of introducing (92) and (95) into (91) thus becomes

$$f(z) = \sum_{n=-\infty}^{\infty} c_n(z - a)^n \qquad (\rho_1 < |z - a| < \rho_2). \qquad (97)$$

This expansion is known as the *Laurent series* expansion of $f(z)$, in the specified annulus $\mathfrak{R}$, and is seen to involve both positive and negative powers of $z - a$, in the general case.

If R_1 is the smallest radius and R_2 the largest radius for which it is true that $f(z)$ is analytic when $R_1 < |z - a| < R_2$, so that also $f(z)$ fails to be analytic somewhere on the circle $|z - a| = R_1$ and somewhere on the circle $|z - a| = R_2$, then since ρ_1 and ρ_2 can be taken to be any radii such that $R_1 < \rho_1 < \rho_2 < R_2$, it follows that the series (97) converges to $f(z)$ when $R_1 < |z - a| < R_2$.

However, if $f(z)$ is analytic everywhere inside the circle $|z - a| = R_2$, (96) shows that $c_n = 0$ when $n < 0$, and (97) then reduces to the *Taylor series* (87), valid when $|z - a| < R_2$. On the other hand, if $f(z)$ is analytic everywhere outside the circle $|z - a| = R_1$, the series involves only a constant term and *negative* integral powers of $z - a$ and can, in fact, be considered as an ordinary power series in $(z - a)^{-1}$, valid when $|z - a| > R_1$. Here (as will be seen in Section 10.10) the function $f(z)$ must, in particular, be analytic "at infinity."

Just as in the case of ordinary power series, it can be proved that, within its annular region of convergence, a Laurent series can be integrated (or differentiated) term by term, and the result will converge to the integral (or derivative) of the function represented by the original series.

---

†It should be noted that this statement generally does *not* apply to the integrals in Equation (91), since $f(\alpha)/(\alpha - z)$ generally is not an analytic function of α when $\alpha = z$.

Further, it is known that the expansion (97) is *unique*, in the sense that if an expansion of the form

$$f(z) = \sum b_n(z - a)^n$$

can be obtained in *any* way, and is valid in the specific annulus

$$R_1 < |z - a| < R_2,$$

then necessarily $b_n = c_n$. (See Problem 56.) This fact frequently permits one to obtain a desired Laurent series by elementary methods, without attempting to evaluate the coefficients by using (96).

Example 1. From the expansion $e^t = \sum_0^\infty t^n/n!$, which is valid for all t, we may deduce other relations, such as

$$\frac{e^z}{z} = \frac{1}{z} + 1 + \frac{z}{2!} + \frac{z^2}{3!} + \cdots + \frac{z^n}{(n + 1)!} + \cdots \qquad (z \neq 0)$$

and

$$e^{1/z} = 1 + \frac{1}{1!\,z} + \frac{1}{2!\,z^2} + \cdots + \frac{1}{n!\,z^n} + \cdots \qquad (z \neq 0).$$

Both these series are Laurent expansions with $a = 0$, $R_1 = 0$, and $R_2 = \infty$. ∎

Example 2. The geometric series expansions

$$\frac{1}{1 - z} = 1 + z + z^2 + \cdots + z^n + \cdots \qquad (|z| < 1)$$

and

$$\frac{1}{1 - z} = \frac{-(1/z)}{1 - (1/z)} = -\frac{1}{z}\left(1 + \frac{1}{z} + \cdots\right)$$

$$= -\frac{1}{z} - \frac{1}{z^2} - \frac{1}{z^3} - \cdots - \frac{1}{z^n} - \cdots \qquad (|z| > 1)$$

have been assumed to be familiar. The first of these expansions is the Taylor series valid inside the circle $|z| = 1$, the second the Laurent series valid outside that circle. ∎

In the general case, if a Laurent expansion of $f(z)$ is required in powers of $z - a$, in the annulus $R_1 < |z - a| < R_2$, we may attempt to express $f(z)$ as the *sum* of two functions $f_1(z)$ and $f_2(z)$, such that $f_1(z)$ is analytic everywhere *outside* the *inner* circle $|z - a| = R_1$ and $f_2(z)$ is analytic everywhere *inside* the *outer* circle $|z - a| = R_2$. The desired representation then can be obtained by expanding $f_2(z)$ in an ordinary power series (Taylor series) in $z - a$, expanding $f_1(z)$ in an ordinary power series in $(z - a)^{-1}$, and adding together the results. Occasionally, it is preferable to write $f(z)$ instead as the *product* $f_1(z)f_2(z)$, where f_1 and f_2 have the properties defined above, and to *multiply* their expansions.

Example 3. The function

$$f(z) = \frac{2}{z(z - 1)(z - 2)}$$

clearly is analytic except when $z = 0$, 1, and 2. To expand $f(z)$ in a Laurent series of powers of z in the annulus $\Re: 1 < |z| < 2$, we may first write

$$f(z) = \frac{1}{z} - \frac{2}{z-1} + \frac{1}{z-2}.$$

Since the first two terms are analytic when $|z| > 1$, whereas the third term is analytic when $|z| < 2$, we may expand the first two terms in powers of $1/z$ and the third in powers of z, to obtain

$$f(z) = \frac{1}{z} - \frac{2/z}{1 - (1/z)} - \frac{1/2}{1 - (z/2)}$$

$$= \frac{1}{z} - \frac{2}{z}\left(1 + \frac{1}{z} + \frac{1}{z^2} + \cdots + \frac{1}{z^n} + \cdots\right)$$

$$-\frac{1}{2}\left(1 + \frac{z}{2} + \frac{z^2}{4} + \cdots + \frac{z^n}{2^n} + \cdots\right).$$

Thus there follows

$$f(z) = \sum_{n=-\infty}^{\infty} c_n z^n \qquad (1 < |z| < 2),$$

where $c_n = -1/2^{n+1}$ when $n \geq 0$, $c_{-1} = -1$, and $c_n = -2$ when $n \leq -2$. The same function also possesses two other Laurent series in powers of z, valid when $0 < |z| < 1$ and when $2 < |z| < \infty$, as well as two, three or four Laurent (or Taylor) series in powers of $z - a$, for any *other* value of a. ∎

10.9. Singularities of Analytic Functions. If *any* region $\Re$ exists such that $f(z)$ or a *branch* of $f(z)$ is analytic in $\Re$, then we speak (somewhat loosely) of $f(z)$ as "an analytic function." Points at which a single-valued function $f(z)$ is *not* analytic are called *singular points* or *singularities* of the function. Also, a point at which a *branch* of a *multivalued* function $f(z)$ is not analytic is called a singular point of $f(z)$. From the definition, we see that any point at which df/dz does not exist is a singular point. Also, when $f(z)$ is multivalued, any point which cannot be an interior point of the region of definition of a *single-valued branch* of $f(z)$ is a singular point.

Points of the latter type are known as *branch points*. At such points the first derivative may or may not exist, but it can be shown that a derivative of *some* order will fail to exist. However, the characteristic feature of a branch point is the property that if a circuit is described around a small simple closed curve enclosing that point, but not enclosing any other singular points, then the value assumed by the function after the circuit differs from the initial value.

In Section 10.3 we have seen that the function $\log z$ and functions of the form z^k, *where k is not an integer*, each have a branch point at $z = 0$. Since $(d/dz)\log z = 1/z$ is analytic elsewhere, this is the *only* singularity of $\log z$ in the finite part of the plane. Similarly, since $(d/dz)z^k = kz^{k-1}$, we see that also z^k has no other finite singularity. Unless the real part of k is smaller than unity, reference to Equation (47) shows that the first derivative of z^k is finite at $z = 0$. However, it is clear that if successive derivatives are calculated, eventually a power of z with a negative real part will be obtained (when k is not a positive integer or zero), and hence the corresponding derivative will not exist at the branch point. It is seen that the functions $\log(z - a)$ and $(z - a)^k$ both have a branch point at $z = a$, with the above restriction on k.

Reference to Section 10.2 and 10.3 then shows that *the elementary functions can have branch points only for those values of z for which a quantity raised to a nonintegral power vanishes, or for which a quantity whose logarithm is taken vanishes or becomes infinite.* It is assumed, for this purpose, that inverse trigonometric and hyperbolic functions are expressed in terms of the logarithm, as in Equations (55) and (56).

Thus, for example, it is to be expected that the function

$$f(z) = (1 - z^2)^{1/2} = (1 + z)^{1/2}(1 - z)^{1/2}$$

has branch points at $z = 1$ and at $z = -1$. This is readily confirmed if we notice, for example, that a circuit about a simple closed curve enclosing $z = 1$, but excluding $z = -1$, will change the second factor but not the first one, and hence will change the product. Similarly, the function $f(z) = \log(z - z^2)$ can be written in the form

$$f(z) = \log[z(1 - z)] = \log z + \log(1 - z),$$

and we may conclude that this function has branch points at $z = 0$ and at $z = 1$. In the cases of the inverse circular and hyperbolic functions, reference to Equations (55) and (56) shows that the functions $\sin^{-1} z$, $\cos^{-1} z$, $\cosh^{-1} z$, $\tanh^{-1} z$, and $\coth^{-1} z$ have branch points at $z = \pm 1$, whereas the functions $\tan^{-1} z$, $\cot^{-1} z$, and $\sinh^{-1} z$ have branch points at $z = \pm i$. The expressions for the first derivatives of these functions show that there are no other singular points (in the finite part of the plane).

It should be noted that the criterion mentioned above applies only to the *elementary* functions, and also that it expresses a necessary but not sufficient condition. For example, the functions $e^{\log z}$ and $(z^{1/2})^2$ clearly do *not* have a branch point at $z = 0$, and expansion in a Taylor series about $z = 0$ shows that the same is true of $\cos(z^{1/2})$.

In working with a multivalued function (possessing one or more branch points) it is usually desirable to specify one particular *branch* of the function and to artificially prevent the possibility of transition from that branch to another branch. Clearly, this result may be accomplished by forbidding curves from surrounding the branch points.

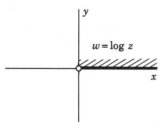

Figure 10.10

In the case of the function $w = \log z$, we have seen that we may imagine the complex plane to be "cut" along the entire positive real axis, so that transition across this part of the axis is impossible (Figure 10.10). In the resultant "cut plane" we then choose an appropriate branch of $\log z$, say the branch such that

$$\log z = \log r + i\theta_p \qquad (0 \leq \theta_p < 2\pi),$$

which is a single-valued function of z, and which is *analytic* everywhere *except on the cut*. The "cut" or "barrier" could equally well be introduced along any other ray or curve extending *from the branch point $z = 0$ to infinity*, with a correspondingly modified definition of the selected branch of the function. Such a modification would be *necessary* if, say, that branch were to be analytic on any part of the positive real axis; a more elaborate modification would be necessary to define a branch which is analytic in the region indicated in Figure 6.10 (page 294).

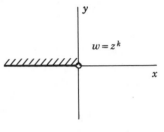

Figure 10.11

For the function $w = z^k$, where k is non-integral, the principal value of θ is rather frequently defined by $-\pi < \theta_p \leq \pi$, so that the cut is taken along the *negative* real axis (Figure 10.11).

For such a function as

$$w = \tan^{-1} z = \frac{1}{2i} \log \frac{i - z}{i + z},$$

with branch points at $z = \pm i$, two cuts may be made along the imaginary axis, from the two branch points outward to infinity, thus avoiding contours which surround *either* of the branch points [Figure 10.12(a)]. In this case it is readily verified that a circuit which encloses *both* branch points does not introduce a transition from branch to branch. Thus, in place of cutting the plane as described, we could merely introduce a *finite* cut *joining the two branch points*, and hence avoid contours which enclose only one of the branch points [Figure 10.12(b)]. The method of cutting to be preferred depends upon the use to which the corresponding branch is to be put. Further facts bearing on the nature of the cuts are brought out in Section 10.10.

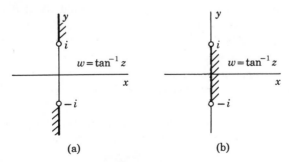

(a)   (b)

Figure 10.12

Thus, in place of dealing with multivalued functions, we may always introduce a suitably cut plane such that only one particular *branch* of the function need be considered, that branch then being single-valued in the cut plane.

If it is necessary to retain the possibility of transition from branch to branch, we may think of constructing a sequence of superimposed planes cut and *joined* along the cuts in such a way that, if a point starts on a given plane, moves around a branch point, and approaches its original position after such a circuit, it moves across a cut from the initial plane onto another superimposed plane corresponding to a second branch of the function. For the function $w = \log z$ an infinite number of such planes would be required, and the resultant configuration would resemble an endless helicoidal surface. For the function $w = z^{1/2}$ only two such planes would be needed, the junctions being so arranged that as a point describes a continuous contour surrounding the origin it moves from the first "sheet" to the second after the first circuit, and returns from the second sheet to the first after a second circuit. Such arrangements are known as *Riemann surfaces*.

In the remainder of this work, however, we will think of the plane as cut in such cases, so that all branch points are removed and the functions considered can be taken to be single-valued. The remaining singular points of a function $f(z)$ are then points at which the derivative df/dz does not exist.

In particular, if $f(z)$ is analytic everywhere throughout some neighborhood of a point $z = a$, say inside the circle $C: |z - a| = R$, except *at* the point $z = a$ itself, then $z = a$ is called an *isolated singular point* of $f(z)$. The results of Section 10.8 show that a Laurent expansion of the form

$$f(z) = \sum_{n=-\infty}^{\infty} c_n(z - a)^n \qquad (0 < |z - a| < R) \tag{98}$$

then is valid.

Now if $f(z)$ were of bounded magnitude as $z \longrightarrow a$, it would necessarily follow that all c's with negative subscripts in (98) would be zero. Thus, everywhere inside the circle C except *at* $z = a$, $f(z)$ would be expressible in the form

$$f(z) = c_0 + c_1(z - a) + c_2(z - a)^2 + \cdots.$$

If it were also true that $f(a) = c_0$, then $f(z)$ would be analytic *everywhere* inside C. That is, $f(z)$ would become analytic at $z = a$ if it were suitably defined (or redefined) there. Such a point is often called a "removable singular point." For example, the functions $f_1(z) = z/z$ and $f_2(z) = (\sin z)/z$ are *undefined* at $z = 0$, but both functions approach 1 as $z \longrightarrow 0$. If we define $f_1(0) = 1$ and $f_2(0) = 1$, then f_1 and f_2 are analytic at $z = 0$ (and elsewhere). If we were to artificially define $f_1(0)$ or $f_2(0)$ to be 0, say, or to refuse to define $f_1(0)$ or $f_2(0)$ at all, then $f_1(z)$ or $f_2(z)$ would have a "removable singularity" at $z = 0$.

Henceforth, we will assume that all such "removable singularities" have indeed been removed by suitable definition.

Apart from such artificialities, it thus follows that $f(z)$ *cannot be of bounded magnitude near an isolated singular point.* To illustrate such points, we may notice that the function $w = 1/z$ has a singular point at $z = 0$, the function $w = 1/(z^2 + 1)$ has singular points at $z = \pm i$, and the function $w = \cot z$ has singular points where $z = k\pi$ $(k = 0, \pm 1, \pm 2, \ldots)$. These singularities are examples of points known as *poles*. Specifically, we say that *if $f(z)$ is not finite at $z = a$ but if for some integer m the product*

$$(z - a)^m f(z)$$

is analytic at $z = a$, then $f(z)$ has a pole at $z = a$.† If m is the smallest integer for which this is so, the pole is said to be *of order m.*

It follows that any *rational function* of the form

$$w = \frac{A_0 + A_1 z + \cdots + A_M z^M}{B_0 + B_1 z + \cdots + B_N z^N} = \frac{N(z)}{D(z)}, \tag{99}$$

where the numerator and denominator are *polynomials* of degree M and N, respectively, without common factors, can have only *poles*, such poles occurring at points where the denominator $D(z)$ vanishes. Thus, if $z - a_1$ is a *simple factor* of the denominator, the product $(z - a_1)w$ is analytic at $z = a_1$ and the point $z = a_1$ is a pole of order one, or a so-called *simple pole.* If $z = a_1$ is an m-fold zero of the denominator, so that $(z - a_1)^m$ is a factor of $D(z)$ but $(z - a_1)^{m+1}$ is not, then $z = a_1$ is a pole of order m.

Example. The function

$$f(z) = \frac{z^3 - 2z + 1}{z^5 + 2z^3 + z} = \frac{z^3 - 2z + 1}{z(z + i)^2(z - i)^2}$$

has a *simple pole* at $z = 0$ and *double poles* at $z = \pm i$ because $zf(z)$ is analytic at $z = 0$, whereas $(z - i)^2 f(z)$ is analytic at $z = i$ and $(z + i)^2 f(z)$ is analytic at $z = -i$. That is, each product listed possesses a derivative *at and near the point noted.* ∎

Functions other than rational functions may also have poles. Thus the function

$$f(z) = \csc z$$

is not finite at the points $z = 0, \pm \pi, \pm 2\pi, \ldots$. Considering first the point $z = 0$, we see that

$$zf(z) = \frac{z}{\sin z}$$

is finite at $z = 0$, since

$$\lim_{z \to 0} \frac{z}{\sin z} = \lim_{z \to 0} \frac{z}{z - z^3/3! + z^5/5! - \cdots} = \lim_{z \to 0} \left(1 - \frac{z^2}{3!} + \cdots\right)^{-1} = 1.$$

†In accordance with the convention just introduced, it is meant here that $(z - a)^m f(z)$ *becomes analytic at $z = a$ when it is defined to take on its limiting value at that point.*

[The same result follows more easily by use of L'Hospital's rule (Problem 49).] Also, $zf(z)$ has a derivative at $z = 0$, since

$$\frac{d}{dz}[zf(z)] = \frac{\sin z - z \cos z}{\sin^2 z} = \frac{(z - z^3/3! + \cdots) - z(1 - z^2/2! + \cdots)}{(z - z^3/3! + \cdots)^2}$$

$$= \frac{z^3/3 + \cdots}{z^2 + \cdots} = \frac{z}{3} + \cdots$$

near $z = 0$, and hence $(d/dz)[zf(z)] = 0$ when $z = 0$. Thus $f(z)$ has a simple pole at $z = 0$. For the purpose of studying a singular point $z = k\pi$ we may make the substitution

$$t = z - k\pi.$$

There then follows

$$(z - k\pi)f(z) = \frac{t}{\sin(t + k\pi)} = (-1)^k \frac{t}{\sin t}.$$

Thus, since $t \to 0$ as $z \to k\pi$, we see that, except possibly for sign, $(z - k\pi)f(z)$ behaves near $z = k\pi$ just as $zf(z)$ behaves near $z = 0$, and hence $f(z)$ has simple poles at all the points $z = k\pi$ (k an integer). From the expression for $f'(z)$,

$$f'(z) = -\frac{\cos z}{\sin^2 z},$$

we see that $f'(z)$ exists at all other points, and hence these points are the only singular points of the function in the finite part of the plane.

From Equation (98) we may deduce that $f(z)$ *has a pole of order m at z = a if and only if f(z) has a Laurent expansion* (98) *in the neighborhood of z = a, with* $c_n = 0$ *when* $n < -m$ *and with* $c_{-m} \neq 0$.

In accordance with the convention adopted at the beginning of this section, we say that "the function

$$f(z) = \frac{\log z}{z + 1}$$

has a simple pole at $z = -1$" when we mean to assert that any *branch* of $f(z)$ has this property (presuming that the relevant branch cut does not pass through the point $z = -1$). However, the function

$$f(z) = \frac{\log z}{z}$$

has a branch point (*not* a simple pole) at $z = 0$, since no branch of $zf(z)$ is analytic at $z = 0$.

All isolated singular points which are not poles (and are not removable) are called *essential singular points*. Thus, when $f(z)$ has an isolated singularity at a point $z = a$, that singularity is *essential* if the derivative df/dz does not exist at $z = a$ and if there is no integer m for which

$$\frac{d}{dz}[(z - a)^m f(z)]$$

exists at $z = a$. It should be noted that a *branch point* is *not* an isolated singularity, since any circle with center at that point must also contain points on *some* branch cut passing through that point, if the branch considered is to be single-valued.

As an example of an essential singular point, we notice that for the function

$$f(z) = e^{1/z}$$

the derivative

$$f'(z) = -\frac{1}{z^2} e^{1/z}$$

exists everywhere except at the origin. Thus the origin is an isolated singular point. For *real* values of z, the function $f(x) = e^{1/x}$ is seen to approach *zero* as $x \to 0$ from *negative* values, and to become *infinite* as $x \to 0$ from positive values. Thus we see that $e^{1/z}$ approaches zero if $z \to 0$ along the negative real axis and that the absolute value of $e^{1/z}$ increases without limit as $z \to 0$ along the positive real axis. Since the same is true for $z^m e^{1/z}$, for *any* value of m, we see that $z^m f(z)$ is not analytic at $z = 0$ for any value of m, and hence the function has an essential singularity at $z = 0$.

To investigate in more detail the behavior of $f(z)$ near the origin, we write $z = re^{i\theta}$ and so express $f(z)$ in the form

$$f(z) = e^{(1/r)e^{-i\theta}} = e^{(1/r)(\cos\theta - i\sin\theta)}$$

$$= e^{(\cos\theta)/r}\left[\cos\left(\frac{\sin\theta}{r}\right) - i\sin\left(\frac{\sin\theta}{r}\right)\right].$$

In particular, we have

$$|f(z)| = |e^{1/z}| = e^{(\cos\theta)/r}.$$

It follows that if $z \to 0$ along any curve for which $\cos\theta > 0$ as $r \to 0$, the absolute value of $f(z)$ increases without limit, whereas if $z \to 0$ along any curve for which $\cos\theta < 0$ as $r \to 0$, then $f(z) \to 0$. Now suppose that $z \to 0$ along a curve $r = c^{-1}\cos\theta$, that is, along a circle of diameter $1/|c|$ tangent to the imaginary axis. For such an approach we have always $|f(z)| = e^c$, and hence $|f(z)|$ can be made to *take on any positive value* as $z \to 0$ along a properly chosen circle of this kind. On such a circle we have also

$$f(z) = e^c[\cos(c\tan\theta) - i\sin(c\tan\theta)].$$

Now as $z \to 0$ along this circle the angle θ tends toward $\pi/2$, and hence the argument of the circular functions increases without limit. Consequently, it is seen that $f(z)$ takes on *all* complex values with absolute value e^c infinitely many times. Thus it follows that in any neighborhood of $z = 0$ the function $e^{1/z}$ takes on all values except zero an infinite number of times![†] Further, the

---

[†]This situation is also evidenced by the fact that the equation $e^{1/z} = \rho e^{i\varphi}$ has infinitely many solutions, of the form

$$z = \frac{1}{\log\rho + i(\varphi_P + 2k\pi)} \qquad (k = 0, \pm 1, \pm 2, \ldots),$$

if $\rho \neq 0$.

exceptional zero value is *approached* as $z \rightarrow 0$ along the negative real axis, or along any curve which passes through the origin along a tangent in the second or third quadrant.

Picard has shown that any analytic function having an isolated essential singularity behaves in this remarkable way near the singularity, all real and imaginary values being *approached* and all values *except one* being actually *taken on infinitely often* in any arbitrarily small region including the singularity.

From Equation (98) we may deduce that $f(z)$ *has an isolated essential singular point at $z = a$ if and only if $f(z)$ has a Laurent expansion in the neighborhood of $z = a$ with infinitely many negative powers of $z - a$.*

In addition to the isolated singularities and branch points, still other types of singular points may exist. For example, in the case of the function

$$f(z) = \frac{1}{2\pi i} \oint_{C_0} \frac{d\alpha}{\alpha - z} = \begin{cases} 1 & \text{inside } C_0, \\ 0 & \text{outside } C_0, \end{cases} \tag{100}$$

mentioned in Section 10.8, all points on the closed curve C_0 are singular points. One might speak of such points as "artificial boundary points" since they comprise a curve across which a transition occurs from one domain, inside which a first analytic function $f_1(z)$ is defined, to another domain in which a second analytic function $f_2(z)$ is defined. In this example there is no difficulty in extending the domain of definition of either $f_1(z)$ or $f_2(z)$ across C_0, so that points on C_0 are not singularities for either of the two analytic functions f_1 and f_2.

However, there exist functions, defined on one side of a curve C, such that their definitions cannot be extended analytically across C. As an example, if $f(z)$ is defined by the series

$$f(z) = \sum_{n=0}^{\infty} z^{2^n} = z + z^2 + z^4 + z^8 + \cdots$$

when $|z| < 1$, it can be shown that the series diverges everywhere on the circle $|z| = 1$ and that $|f(z)|$ is unbounded in the neighborhood of each point on that circle. Thus it is impossible to extend the definition of $f(z)$ in such a way that the result is analytic at any points on the unit circle. (See Problem 66.) Such a curve is called a *natural boundary* for the relevant function, and points on such a curve usually are referred to as (nonisolated) essential singularities.

When branches of a multivalued function are defined, the points on *branch cuts* are also singular points, but are rather artificially so since (except for the *branch points* themselves) they would no longer be exceptional if the branches of the parent function were redefined.

In addition, it may happen that isolated singular points *cluster* about a point $z = a$, in such a way that there are infinitely many such points in *every* circle with center at $z = a$, no matter how small the radius. Then $z = a$ also is a singular point and is included in the category of nonisolated "essential" singularities. The function $f(z) = \tan(1/z)$ has such a singularity at the origin.

10.10. Singularities at Infinity. To study the behavior of a function $f(z)$ for large values of $|z|$, we may make the substitution

$$z = \frac{1}{t}. \tag{101}$$

The function $f(z)$ is then transformed to a new function $g(t)$,

$$g(t) = f\left(\frac{1}{t}\right). \tag{102}$$

Now as a point z moves indefinitely far from the origin *in any direction*, so that $|z| \rightarrow \infty$, the corresponding point t approaches $t = 0$. It is convenient to say that the point $t = 0$ corresponds to a "point at infinity" in the complex plane which may be denoted by $z = z_\infty$, such that this point is approached by any point z receding from the origin indefinitely far in any direction. Thus we "close" the complex plane by adding the point z_∞ "at infinity." The result is often known as the *extended* (or *closed*) *complex plane*. It is then natural to say that $f(z)$ is analytic at z_∞ if $g(t) = f(1/t)$ is analytic at $t = 0$, and similarly that if $g(t)$ has a particular type of singularity at $t = 0$, then $f(z)$ has the same type of singularity at z_∞.

If $g(t) = f(1/t)$ is analytic at $t = 0$, then it can be expanded in a series

$$f\left(\frac{1}{t}\right) = A_0 + A_1 t + A_2 t^2 + \cdots = \sum_{n=0}^{\infty} A_n t^n$$

in some circle about $t = 0$, and conversely. Hence, by replacing t by $1/z$, it follows that $f(z)$ *is analytic at* z_∞ *if and only if it can be represented by a series of the form*

$$f(z) = A_0 + \frac{A_1}{z} + \frac{A_2}{z^2} + \cdots = \sum_{n=0}^{\infty} \frac{A_n}{z^n} \tag{103}$$

for values of z *outside* a sufficiently large circle with center at $z = 0$. In particular, $f(z)$ must approach a limit $f(\infty)$ as $z \rightarrow z_\infty$, that is, as $|z| \rightarrow \infty$ *in any way*, and Equation (103) shows that

$$A_0 = f(\infty). \tag{104}$$

We notice also that if $f(1/t)$ is analytic at $t = 0$, then

$$\frac{d}{dt} f\left(\frac{1}{t}\right) = -\frac{1}{t^2} f'\left(\frac{1}{t}\right)$$

must exist at $t = 0$, and hence $z^2 (df/dz)$ *must exist at* z_∞.

If $f(1/t)$ has a pole of order m at $t = 0$, then $t^m f(1/t)$ can be expanded in a power series in t converging near $t = 0$, and conversely, and hence near $t = 0$ we may write

$$f\left(\frac{1}{t}\right) = \frac{A_0}{t^m} + \frac{A_1}{t^{m-1}} + \cdots + A_m + A_{m+1} t + \cdots = \sum_{n=0}^{\infty} A_n t^{-m+n},$$

where $A_0 \neq 0$. Thus we conclude that $f(z)$ *has a pole of order m at z_∞ if and* *only if we can write*

$$f(z) = A_0 z^m + A_1 z^{m-1} + \cdots + A_m + \frac{A_{m+1}}{z} + \cdots$$

$$= \sum_{n=0}^{\infty} A_n z^{m-n} \qquad (A_0 \neq 0), \tag{105}$$

for values of z outside a sufficiently large circle with center at $z = 0$. In particular, $f(z)/z^m$ must approach a nonzero limit as $z \to z_\infty$, and this limit is identified with the coefficient A_0 in (105),

$$A_0 = \lim_{z \to z_\infty} \frac{1}{z^m} f(z). \tag{106}$$

Also, since $(d/dt)[t^m f(1/t)]$ must exist when $t = 0$, it follows that the expression

$$-z^2 \frac{d}{dz}\left[\frac{f(z)}{z^m}\right] = \frac{m}{z^{m-1}} f(z) - \frac{1}{z^{m-2}} \frac{df(z)}{dz}$$

must tend to a limit as $z \to z_\infty$.

From (105) we conclude that *any polynomial of degree m has a pole of order m at z_∞.* If we consider the general *rational function*

$$f(z) = \frac{a_0 + a_1 z + \cdots + a_N z^N}{b_0 + b_1 z + \cdots \cdots b_M z^M}, \tag{107}$$

with $a_N b_M \neq 0$, we obtain

$$f\left(\frac{1}{t}\right) = \frac{a_0 t^N + a_1 t^{N-1} + \cdots + a_N}{b_0 t^M + b_1 t^{M-1} + \cdots + b_M} t^{M-N}.$$

Thus it follows that $f(1/t)$ is analytic at $t = 0$ if $M \geqq N$ and that $f(1/t)$ has a pole of order $N - M$ at $t = 0$ if $N > M$. The same is thus true for $f(z)$ at z_∞.

Since $f(z)$ clearly is analytic at all finite points except at zeros of the denominator, where poles exist, it follows that *any rational function possesses no singularities other than poles.* If we count a pole of order k as equivalent to k simple poles, we see that since the denominator has M linear factors, the multiplicity of *finite poles* is M. Similarly, the multiplicity of *finite zeros* is N. If $M > N$, the function $f(1/t)$ has a zero of order $M - N$ at $t = 0$, and hence $f(z)$ has a zero of order $M - N$ at z_∞. In this case there are exactly M poles and M zeros (counting multiple poles and zeros separately). Similarly, if $M < N$, there are N poles and N zeros. In any case, *for a rational function the total multiplicity of poles is equal to the total multiplicity of zeros in the extended plane,* that is, when the point z_∞ is taken into account.

Since the expansions of e^z, $\sin z$, $\cos z$, $\sinh z$, and $\cosh z$ involve *infinitely many positive powers* of z, and converge for all finite values of z, reference to the criterion (105) shows that, although these functions have *no finite singularities,* they each have an *essential singularity* at z_∞. In the case of the exponential function this situation may also be seen by noticing that e^z behaves for *large* $|z|$ as $e^{1/z}$ behaves for *small* $|z|$.

For the function $f(z) = \log z$ there follows $f(1/t) = \log (1/t) = -\log t$. Hence it follows that $f(z) = \log z$ has a branch point at z_∞ as well as at $z = 0$. A similar statement applies to $f(z) = z^k$, where k is nonintegral. The "cuts" introduced in Section 10.9 for these functions can be considered as joining the two branch points.

We notice that a *small* circle surrounding the point $t = 0$ corresponds to a *large* circle with center at $z = 0$ in the complex plane. Thus, just as we may consider a "circle of zero radius" as enclosing only the origin, we may in a similar sense consider the exterior of a circle C_∞ "of infinite radius" as consisting only of the point z_∞. Consequently, the exterior of a closed contour C may be considered as composed of the region between C and the infinite circle C_∞ together with an added exterior point z_∞ at infinity.

Also, a *positive* circuit around the *finite* area enclosed by any simple closed contour C can be considered also as a *negative* circuit around the *infinite* area outside C, *including the point z_∞.*

If $f(z)$ is analytic on C and at all *finite* points outside C, it follows that C and C_∞ are equivalent contours, and hence

$$\oint_C f(z)\, dz = \oint_{C_\infty} f(z)\, dz. \tag{108}$$

The right-hand integral may now be considered as taken along a *negative* contour enclosing the exterior of C_∞, that is, enclosing only the immediate neighborhood of the point z_∞. However, *even though $f(z)$ be analytic at z_∞, this integral may not vanish.* It can be shown to vanish if $z^2 f(z)$ is analytic at z_∞ (see Problem 84). This fact leads to the useful result that

$$\oint_C f(z)\, dz = 0 \tag{109}$$

in the case *when $z^2 f(z)$ is analytic on and outside C and at z_∞*, as well as in the case *when $f(z)$ is analytic on and inside C.*

If outside a closed contour C no points, *including the point z_∞*, are *branch points*, and if we think of a closed circuit around C as surrounding the *exterior* of C, we may expect that a function $f(z)$ will return to its initial value after such a closed circuit, whether or not there are finite branch points *inside C*. Thus, for the function

$$f(z) = \tan^{-1} z = \frac{1}{2i} \log \frac{i - z}{i + z}$$

considered in Section 10.9, we obtain

$$f\left(\frac{1}{t}\right) = \frac{1}{2i} \log \frac{t + i}{t - i}.$$

Hence, since $t = 0$ is *not* a branch point for $f(1/t)$, it follows that z_∞ is not a branch point for $f(z)$. In the first method of cutting the plane for this function [Figure 10.12(a)], the cut can be considered as joining the two branch points

$z = \pm i$ along an infinite segment passing through z_∞. However, since z_∞ is not a branch point, closed circuits surrounding z_∞ need not be prohibited, and a finite cut joining the branch points [Figure 10.12(b)] is equally satisfactory. That is, we need only prohibit closed circuits which enclose only *one* of the branch points. It must follow that a closed circuit surrounding both branch points will return $\tan^{-1} z$ to its initial value. As was stated earlier, this statement can be directly verified (see Problem 57).

It may happen that z_∞ is a branch point on certain branches of a multi-valued function but not on other branches, as for the function $\sin^{-1} z + i \log z$ (see Problem 72).

10.11. Significance of Singularities. An important theorem, due to *Liouville*, states that *a function which is analytic at all finite points and at z_∞ must be constant*. To establish this theorem, we consider any two points z_1 and z_2 in the complex plane and apply Cauchy's theorem (Section 10.6) to express $f(z_1)$ and $f(z_2)$ in the form

$$f(z_1) = \frac{1}{2\pi i} \oint_C \frac{f(\alpha)\, d\alpha}{\alpha - z_1}, \qquad f(z_2) = \frac{1}{2\pi i} \oint_C \frac{f(\alpha)\, d\alpha}{\alpha - z_2}, \tag{110}$$

where C is a circle of radius R, with center at the origin, including the points z_1 and z_2. By subtraction we obtain also

$$f(z_2) - f(z_1) = \frac{z_2 - z_1}{2\pi i} \oint_C \frac{f(\alpha)\, d\alpha}{(\alpha - z_1)(\alpha - z_2)}. \tag{111}$$

For points on C we have

$$\alpha = Re^{i\theta}, \qquad |\alpha| = R. \tag{112}$$

Since $f(z)$ is assumed to be analytic everywhere (including z_∞), it must be bounded everywhere. Denoting this bound by K and making use of the inequality (10), we obtain, for points on C,

$$\left| \frac{f(\alpha)}{(\alpha - z_1)(\alpha - z_2)} \right| \leq \frac{K}{(|\alpha| - |z_1|)(|\alpha| - |z_2|)} = \frac{K}{(R - |z_1|)(R - |z_2|)}.$$

Hence, making use of the *ML* inequality (80), we obtain from (111)

$$|f(z_2) - f(z_1)| \leq \frac{|z_2 - z_1|}{2\pi} \cdot \frac{K}{(R - |z_1|)(R - |z_2|)} \cdot 2\pi R$$

$$\leq \frac{|z_2 - z_1|}{R} \cdot \frac{K}{\left(1 - \frac{|z_1|}{R}\right)\left(1 - \frac{|z_2|}{R}\right)}. \tag{113}$$

Since $f(z)$ is analytic everywhere, we are permitted to let the radius R of C increase without limit. But since the right member then tends to zero, it follows that the left-hand member must be zero for any two points z_1 and z_2. Hence $f(z)$ is constant, as was to be shown.

This theorem has many important consequences. In illustration, we have seen that any polynomial of degree m has no singularities except a pole of order m at z_∞. We can now show that, conversely, *a function having no singularities other than a pole of order m at z_∞ is necessarily a polynomial of degree m*. For if $f(z)$ has a pole of order m at z_∞, then for large $|z|$ it must be representable by a series of form (105). Hence $f(z)$ must be expressible as the sum $P_m(z) + g(z)$ of a polynomial

$$P_m(z) = A_0 z^m + A_1 z^{m-1} + \cdots + A_m \tag{114}$$

and a function whose expression, for large $|z|$, is of the form

$$g(z) = \frac{A_{m+1}}{z} + \frac{A_{m+2}}{z^2} + \cdots. \tag{115}$$

But since the function $g(z)$ must then be analytic everywhere, and since it vanishes as $|z| \rightarrow \infty$, it follows from the preceding theorem that $g(z)$ must be zero and hence $f(z)$ must be a polynomial, as stated.

Further, we can show that *if a function is analytic in the extended plane except for a finite number of poles it must be a rational function*; that is, it must be the ratio of two polynomials. For suppose that $f(z)$ has poles of order k_n at the N finite points $z = z_n$ $(n = 1, 2, \ldots, N)$. Then, from the definition of a pole, it follows that the function

$$F(z) = (z - z_1)^{k_1}(z - z_2)^{k_2} \cdots (z - z_N)^{k_N} f(z) \tag{116}$$

is analytic at all finite points. The coefficient of $f(z)$ in (116) is a polynomial of degree K, where

$$K = k_1 + k_2 + \cdots + k_N,$$

and hence it has a pole of order K at z_∞. But since we have supposed that $f(z)$ is either analytic at z_∞ or, at worst, has a pole at z_∞, it follows that $F(z)$ has, at worst, a pole at z_∞. Now since $F(z)$ is analytic at all finite points, the preceding theorem shows that it must be a polynomial. Thus $f(z)$ must be the ratio of this polynomial to the polynomial which multiplies it in (116), as was to be shown.

These examples illustrate the fact that analytic functions are, to a certain extent, essentially *characterized by their singularities*. Those functions which are *single-valued, and analytic at all finite points*, are known as *integral functions* or as *entire functions*. Such a function is either *analytic* also at z_∞, in which case it must be a *constant*; or it may have a *pole* at z_∞, in which case it must be a *polynomial*; or, otherwise, it must have an *essential singularity* at z_∞, in which case it is known as an *integral transcendental function*. The functions e^z, $\sin z$, and $\sinh z$ are examples of such functions. We have seen that if an analytic function is expanded in powers of $z - a$, the circle of convergence extends at least to the nearest singularity. It follows that *the Taylor series representation of any integral function converges for all finite values of z*. Conversely, if a power series converges for all finite values of z, it defines an integral function.

10.12. Residues. Suppose that the analytic function $f(z)$ has a pole of order m at the point $z = a$. Then $(z - a)^m f(z)$ is analytic and hence can be expanded in the Taylor series

$$(z - a)^m f(z) = A_0 + A_1(z - a) + \cdots + A_{m-1}(z - a)^{m-1} + A_m(z - a)^m + \cdots,$$
(117)

where
$$A_k = \frac{1}{k!}\left[\frac{d^k}{dz^k}\{(z - a)^m f(z)\}\right]_{z=a},$$
(118)

in accordance with the results of Section 10.7. This series will converge within any circle about $z = a$ which does not include another singularity. If $z \neq a$, Equation (117) can be rewritten in the form

$$f(z) = \frac{A_0}{(z - a)^m} + \frac{A_1}{(z - a)^{m-1}} + \cdots + \frac{A_{m-1}}{z - a} + A_m + A_{m+1}(z - a) + \cdots.$$
(119)

Now let C_a be any simple closed contour surrounding $z = a$ which lies inside the circle of convergence of (117) and which is such that $f(z)$ is analytic inside and on C_a, except at $z = a$. If we integrate (119) around this contour and review Equations (78a, b), we obtain merely

$$\oint_{C_a} f(z)\, dz = 2\pi i A_{m-1},$$

the only term contributing to the integration being $A_{m-1}/(z - a)$. The same relation then continues to hold when C_a is deformed into any *equivalent* contour, which need not lie inside the circle of convergence of (117).

We call the coefficient A_{m-1} the *residue of $f(z)$ at $z = a$* and denote its value by Res (a), the function $f(z)$ being understood. If a more explicit notation is desirable, we will denote the residue of $f(z)$ at $z = a$ by Res $\{f(z); a\}$. Thus, *if $f(z)$ has a pole of order m at $z = a$, then*

$$\oint_{C_a} f(z)\, dz = 2\pi i \,\text{Res}\,(a)$$
(120)

with
$$\text{Res}\,(a) = \frac{1}{(m - 1)!}\left[\frac{d^{m-1}}{dz^{m-1}}\{(z - a)^m f(z)\}\right]_{z=a},$$
(121)

where C_a is a simple closed contour enclosing $z = a$ but excluding all other singularities of $f(z)$.

We notice that Res (a) is the coefficient of $1/(z - a)$ in the Laurent expansion of $f(z)$, in powers of $z - a$, which is valid *near $z = a$*. The value of the residue can be determined from this fact or can be determined directly from (121).

In the case of a *simple pole* $(m = 1)$, Equation (121) gives

$$m = 1: \qquad \text{Res}\,(a) = [(z - a)f(z)]_{z=a} = \lim_{z \to a}(z - a)f(z).$$
(122)

In this case, if $f(z)$ is expressed as the ratio

$$f(z) = \frac{N(z)}{D(z)},$$ (123)

where $N(a)$ is finite and nonzero, then it follows that $D(z)$ must vanish at $z = a$ in such a way that $D(z)/(z - a)$ approaches a finite limit as $z \to a$. Thus, if we write $(z - a)f(z) = N(z)[(z - a)/D(z)]$, we may use L'Hospital's rule (Problem 49) to evaluate the limit indicated in (122) in the form

$$m = 1: \qquad \text{Res}(a) = \frac{N(a)}{D'(a)}.$$ (124)

When $D(z)$ is a polynomial, $z - a$ must be a factor, and hence (122) is readily evaluated by merely deleting the factor $z - a$ in the denominator and setting $z = a$ in the remaining ratio.

This procedure is applicable, however, *only in the case when $z = a$ is a simple pole*. Otherwise, use should be made of Equation (121) unless the series (119) is easily obtained, in which case the residue is read directly from the series.

Example 1. We consider first the function

$$f(z) = \frac{1}{z^2 + 1} = \frac{1}{(z + i)(z - i)}.$$ (125)

There are simple poles at $z = \pm i$ with residues

$$\text{Res}(i) = \lim_{z \to i}(z - i)f(z) = \left(\frac{1}{z + i}\right)_{z=i} = \frac{1}{2i}$$

and

$$\text{Res}(-i) = \left(\frac{1}{z - i}\right)_{z=-i} = -\frac{1}{2i},$$

from (122). If we use (124), we have, with $N = 1$ and $D = z^2 + 1$,

$$\text{Res}(i) = \left(\frac{1}{2z}\right)_{z=i} = \frac{1}{2i}, \qquad \text{Res}(-i) = \left(\frac{1}{2z}\right)_{z=-i} = -\frac{1}{2i},$$

as before. ∎

Example 2. For the function

$$f(z) = \frac{1}{z^3 - z^2} = \frac{1}{z^2(z - 1)},$$ (126)

there is a double pole at $z = 0$ and a simple pole at $z = 1$. At $z = 1$ we obtain, from (122),

$$\text{Res}(1) = \left(\frac{1}{z^2}\right)_{z=1} = 1.$$

At $z = 0$ we use (121) with $a = 0$, $m = 2$, and find

$$\text{Res}(0) = \left[\frac{d}{dz}\left(\frac{1}{z - 1}\right)\right]_{z=0} = -1.$$ ∎

Example 3. For the function

$$f(z) = \frac{\sin z - z}{z^6},\tag{127}$$

a singularity in the finite part of the plane can occur only at $z = 0$. By expanding the numerator in powers of z, we may write

$$f(z) = \frac{(z - z^3/3! + z^5/5! - z^7/7! + \cdots) - z}{z^6}$$

or

$$f(z) = -\frac{1}{6z^3} + \frac{1}{120z} - \frac{z}{5040} + \cdots.\tag{128}$$

Thus $f(z)$ has a pole of order *three* at $z = 0$. The residue is merely the coefficient of $1/z$ in (128),

$$\operatorname{Res}(0) = \frac{1}{120}.$$

∎

It is easily shown that if $f(z)$ has a pole of order m at $z = a$, the value of the right-hand member of (121) is unchanged when m is replaced by any integer M such that $M > m$. (See Problem 79.) That is, if the order of the pole is *overestimated* when (121) is used, the correct value of the residue will still result. This fortunate state of affairs can be verified in the case of Example 3 by using (121) to calculate Res (0) with $m = 6$, in place of the correct value $m = 3$ which is not readily determined by inspection, and which would lead to a more involved calculation.

If $f(z)$ has an isolated *essential* singularity at $z = a$, but is single-valued in the neighborhood of that point, then reference to (97) shows that (120) again applies, with

$$\operatorname{Res}(a) = c_{-1},\tag{129}$$

where c_{-1} *is the coefficient of* $(z - a)^{-1}$ *in the Laurent expansion of* $f(z)$ *which is valid in the immediate neighborhood of* $z = a$. It has been seen that this statement is true also when $z = a$ is a *pole*, so that, in fact, it applies when $z = a$ is *any* isolated singularity of a single-valued function. However, the alternative formula (121) applies only when $z = a$ is a *pole*.

Example 4. The function

$$f(z) = ze^{1/z}\tag{130}$$

has an isolated essential singularity at $z = 0$, but has no other singularity at a finite point. From the expansion

$$f(z) = z\left(1 + \frac{1}{z} + \frac{1}{2!z^2} + \cdots\right) = z + 1 + \frac{1}{2!z} + \cdots \qquad (z \neq 0),$$

there follows

$$\operatorname{Res}(0) = \tfrac{1}{2}.$$

∎

Suppose now that $f(z)$ is analytic on a simple closed curve C which bounds a finite region inside which $f(z)$ is single-valued and has only isolated singularities, at a finite number of points $z = a_1, a_2, \ldots, a_n$. We enclose these points by small nonintersecting simple closed curves $C_1, C_2, \ldots, C_n$, each of which lies inside C and encloses only one singularity. Then, by introducing a crosscut from each curve C_k to C, a simply connected region $\mathfrak{R}$ is obtained inside which $f(z)$ is analytic (Figure 10.13). Thus the line integral of $f(z)$ around the complete boundary of this region vanishes. Noticing that the integrals along the crosscuts cancel, since $f(z)$ is single-valued, and that

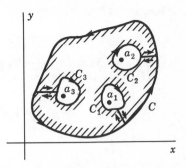

Figure 10.13

the integrations taken around the small contours are in the negative sense, we deduce that

$$\oint_C f(z)\, dz - \left[\oint_{C_1} f(z)\, dz + \oint_{C_2} f(z)\, dz + \cdots + \oint_{C_n} f(z)\, dz \right] = 0. \qquad (131)$$

But since the integral taken in the positive sense around C_k is $2\pi i$ times the residue of $f(z)$ at $z = a_k$, Equation (131) leads to the result

$$\oint_C f(z)\, dz = 2\pi i \sum_{k=1}^n \text{Res}\, (a_k). \qquad (132)$$

Thus, *if $f(z)$ is analytic inside and on a simple closed curve C, except at a finite number of interior isolated singularities, then $\oint_C f(z)\, dz$ is given by $2\pi i$ times the sum of the residues of $f(z)$ at those points.* This result is known as *Cauchy's residue theorem.*

10.13. Evaluation of Real Definite Integrals. The residue theorem is useful in evaluating certain types of real definite integrals. In this section a few examples are presented.

Example 1. Any real integral of the form

$$I = \int_0^{2\pi} R(\sin\theta, \cos\theta)\, d\theta, \qquad (133)$$

where R is a rational function of $\sin\theta$ and $\cos\theta$ which is finite for all (real) values of θ, can be evaluated by residue theory as follows. If we make the substitution

$$z = e^{i\theta}, \qquad dz = ie^{i\theta}\, d\theta \qquad (134)$$

there follows also

$$d\theta = \frac{dz}{iz}, \qquad \sin\theta = \frac{z^2 - 1}{2iz}, \qquad \cos\theta = \frac{z^2 + 1}{2z}. \qquad (135)$$

Thus $R(\sin\theta, \cos\theta)\, d\theta$ takes the form $F(z)\, dz$, where $F(z)$ is a rational function of z. Since z describes a positive circuit around the unit circle C_1 in the complex plane as

θ varies from 0 to 2π, the integral (133) takes the form

$$I = \oint_{C_1} F(z)\, dz = 2\pi i \sum_k \text{Res}\,(a_k),\qquad (136)$$

where the points a_k are the poles of $F(z)$ inside the unit circle.

For example, with (134) and (135) the integral

$$I_1 = \int_0^{2\pi} \frac{d\theta}{A + B\cos\theta}\qquad (A^2 > B^2, A > 0)\qquad (137)$$

takes the form

$$I_1 = \oint_{C_1} F(z)\, dz,\qquad F(z) = \frac{2/i}{Bz^2 + 2Az + B}.$$

The poles of $F(z)$ occur when

$$z = a_1 = -\frac{A}{B} + \frac{\sqrt{A^2 - B^2}}{B}$$

and

$$z = a_2 = -\frac{A}{B} - \frac{\sqrt{A^2 - B^2}}{B}.$$

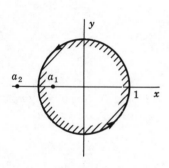

We see that, since $a_1 a_2 = 1$, one pole (that at a_1, since A is assumed to be positive) is inside the unit circle C_1, whereas the second pole is outside C_1 (Figure 10.14). Using (124), we obtain

$$\text{Res}\,(a_1) = \frac{2/i}{2Ba_1 + 2A} = \frac{1}{i\sqrt{A^2 - B^2}}.$$

and hence there follows

$$I_1 = 2\pi i\,\text{Res}\,(a_1) = \frac{2\pi}{\sqrt{A^2 - B^2}}.\qquad (138)$$

Figure 10.14

The restriction $A^2 > B^2$ is necessary in order that the integral (137) exist. ■

Example 2. We next indicate how contour integration may be used to evaluate an integral of the form

$$\int_{-\infty}^{\infty} f(x)\, dx,\qquad (139)$$

where $f(x)$ is a *rational function*, whose denominator is of degree at least two greater than the numerator, and which is finite for all (real) values of x. To illustrate the procedure, we consider the integral

$$I = \int_{-\infty}^{\infty} \frac{x^2}{1 + x^4}\, dx.\qquad (140)$$

We first write

$$f(z) = \frac{z^2}{1 + z^4}\qquad (141)$$

and consider the result of integrating $f(z)$ around the contour indicated in Figure 10.15, consisting of the segment of the real axis extending from $-R$ to R and the

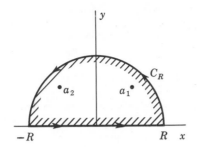

Figure 10.15

semicircle C_R in the upper half-plane. For any value of R there follows

$$\int_{-R}^{R} \frac{x^2}{1+x^4}\,dx + \int_{C_R} f(z)\,dz = 2\pi i \sum_k \text{Res}\,(a_k), \tag{142}$$

where the points a_k are the poles of $f(z)$ inside the contour. As R increases without limit the first integral on the left approaches the required integral. Also, eventually all poles of $f(z)$ in the upper half-plane are absorbed into the contour and hence contribute residues to the right-hand member.

We show next that the integral taken along C_R *tends to zero* as $R \longrightarrow \infty$. On C_R we have $|z| = R$ and hence also, using (10),

$$|f(z)| = \left|\frac{z^2}{1+z^4}\right| \leq \frac{|z|^2}{|z|^4-1} = \frac{R^2}{R^4-1} \equiv M \qquad (R>1).$$

The length of C_R is $L = \pi R$, and hence by Equation (80) there follows

$$\left|\int_{C_R} f(z)\,dz\right| \leq ML = \frac{\pi R^3}{R^4-1} \qquad (R>1). \tag{143}$$

Hence the integral along C_R tends to zero as the radius R increases indefinitely.

Thus, proceeding to the limit as $R \longrightarrow \infty$, we find that Equation (142) gives

$$I = \int_{-\infty}^{\infty} \frac{x^2}{1+x^4}\,dx = 2\pi i \sum_k \text{Res}\,(a_k), \tag{144}$$

where the points a_k are the poles of $f(z)$ in the upper half-plane, namely, the two values of $(-1)^{1/4}$ which have positive imaginary parts,

$$a_1 = e^{\pi i/4}, \qquad a_2 = e^{3\pi i/4}. \tag{145}$$

Also, making use of (124), we obtain

$$\text{Res}\,(a_k) = \left(\frac{z^2}{4z^3}\right)_{z=a_k} = \frac{1}{4a_k}.$$

Thus (144) is evaluated in the form

$$\int_{-\infty}^{\infty} \frac{x^2}{1+x^4}\,dx = 2\pi i\left(\frac{1}{4}e^{-\pi i/4} + \frac{1}{4}e^{-3\pi i/4}\right) = \frac{\pi}{\sqrt{2}}. \tag{146}$$

It should be noticed that the crucial step in this procedure consists of showing that the integral along C_R tends to zero as $R \longrightarrow \infty$. Considering the more general case described in connection with (139), we see that, if the denominator of $f(x)$ is of degree at least two greater than the numerator, then along C_R the maximum value M of $|f(z)|$ is at worst of the order of $1/|z|^2 = 1/R^2$. Since the length of C_R is $L = \pi R$, it follows that ML is at worst of the order of $1/R$ and hence, as in the example, the integral tends to zero as $R \longrightarrow \infty$. Thus in such cases we have

$$\int_{-\infty}^{\infty} f(x)\, dx = 2\pi i \sum_k \mathrm{Res}\,(a_k), \tag{147}$$

where the points a_k are the poles of $f(z)$ in the upper half-plane. It is easily seen that if $f(x)$ is a rational function, the conditions specified are necessary in order that $\int_{-\infty}^{\infty} f(x)\, dx$ exist. Thus (147) is true if $f(x)$ is a rational function and if the integral exists.

We note that if $f(x)$ is an even function, there follows

$$\int_0^{\infty} f(x)\, dx = \frac{1}{2} \int_{-\infty}^{\infty} f(x)\, dx.$$

Thus, for example, Equation (146) also gives

$$\int_0^{\infty} \frac{x^2}{1+x^4}\, dx = \frac{\pi}{2\sqrt{2}}. \tag{148}$$

Example 3. If we attempt to apply the method of Example 2 to the evaluation of an integral of the form

$$\int_{-\infty}^{\infty} f(x) \cos mx\, dx \quad \text{or} \quad \int_{-\infty}^{\infty} f(x) \sin mx\, dx, \tag{149}$$

where $f(x)$ is a rational function of the type described, we encounter difficulty when we attempt to show that the integral of $f(z) \cos mz$ or $f(z) \sin mz$ along C_R in Figure 10.15 tends to zero as $R \longrightarrow \infty$. Thus, if we notice first that on C_R we have $z = Re^{i\theta}$ and hence

$$|e^{imz}| = |e^{imRe^{i\theta}}| = |e^{imR(\cos\theta + i\sin\theta)}| = e^{-mR\sin\theta}, \tag{150a}$$

$$|e^{-imz}| = |e^{-imRe^{i\theta}}| = |e^{-imR(\cos\theta + i\sin\theta)}| = e^{mR\sin\theta}, \tag{150b}$$

it follows that on C_R the functions

$$\cos mz = \frac{e^{imz} + e^{-imz}}{2}, \qquad \sin mz = \frac{e^{imz} - e^{-imz}}{2i} \tag{151}$$

increase exponentially in magnitude as $R \longrightarrow \infty$ because of the presence of the term e^{-imz} when $m > 0$ and because of the term e^{imz} when $m < 0$. Hence, the maximum value of the integrand is unbounded as $R \longrightarrow \infty$, and the integral along C_R cannot be shown to tend to zero.

However, if we notice from (150a) that, when $m > 0$, $|e^{imz}|$ decreases exponentially on C_R as $R \longrightarrow \infty$ in the upper half-plane ($\sin\theta > 0$), we may avoid the difficulty noted by considering the integrals (149) as the real and imaginary parts of the integral

$$\int_{-\infty}^{\infty} e^{imx} f(x)\, dx \qquad (m \geq 0). \tag{152}$$

Then, if the maximum value of $|f(z)|$ on C_R is, at worst, K/R^2, we have, for points on C_R,

$$|e^{imz} f(z)| = e^{-mR\sin\theta} |f(z)| \leq |f(z)| \leq \frac{K}{R^2} \qquad (m \geq 0).$$

Thus the integrand is bounded by $M = K/R^2$ and we have

$$\left| \int_{C_R} e^{imz} f(z)\, dz \right| \leqq ML = \pi \frac{K}{R}$$

so that the integral along C_R again tends to zero as $R \longrightarrow \infty$.

Hence we conclude that if $f(x)$ *is a rational function whose denominator is of degree at least two greater than the numerator, and which is finite for all (real) values of x, then*

$$\int_{-\infty}^{\infty} e^{imx} f(x)\, dx = 2\pi i \sum_k \text{Res}\, \{e^{imz} f(z)\,; a_k\} \qquad (m \geqq 0), \tag{153}$$

where the points a_k are the poles of $f(z)$ in the upper half-plane. The integrals (149) are obtained as the real and imaginary parts of this result.

The theory to be developed in the following section shows in fact that, *if $m > 0$, the degree of the denominator need be only one greater than that of the numerator.*

In illustration, to evaluate the integral

$$\int_{-\infty}^{\infty} \frac{e^{imx}}{a^2 + x^2}\, dx \qquad (m \geqq 0,\ a > 0),$$

where a and m are real, we notice that $f(x) = 1/(a^2 + x^2)$ is a rational function of the required type. The only pole of $f(z) = 1/(a^2 + z^2)$ in the upper half-plane is at $z = ia$. Since the residue of $e^{imz} f(z)$ at $z = ia$ is given by

$$\text{Res}\left\{ \frac{e^{imz}}{a^2 + z^2}\,; ia \right\} = \left(\frac{e^{imz}}{z + ia} \right)_{z=ia} = \frac{1}{2ia} e^{-ma},$$

we obtain the result

$$\int_{-\infty}^{\infty} \frac{e^{imx}}{a^2 + x^2}\, dx = \frac{\pi}{a} e^{-ma} \qquad (m \geqq 0,\ a > 0). \tag{154}$$

By taking real and imaginary parts, we obtain

$$\int_{-\infty}^{\infty} \frac{\cos mx}{a^2 + x^2}\, dx = \frac{\pi}{a} e^{-ma}, \tag{155a}$$

$$\int_{-\infty}^{\infty} \frac{\sin mx}{a^2 + x^2}\, dx = 0, \tag{155b}$$

when $m \geqq 0$, $a > 0$. It is seen from the form of the result that the restriction $m \geqq 0$ can be removed in this case if we replace e^{-ma} by $e^{-|m|a}$. ∎

Example 4. When we are to evaluate an integral of the form

$$\int_{-\infty}^{\infty} f(x) e^{imx}\, dx,$$

where again $f(x)$ is a rational function whose denominator is of higher degree than its numerator and whose zeros are nonreal, but where now $m < 0$, the method of Example 3 may be modified by closing the contour in the *lower* half-plane. In this case, a positive progress along the real axis implies a *negative* (clockwise) circuit about the closed path and there follows accordingly

$$\int_{-\infty}^{\infty} f(x)e^{imx}\,dx = -2\pi i \sum_{k} \operatorname{Res}\{e^{imz}f(z);a_k\} \qquad (m < 0), \tag{156}$$

where the points a_k are the poles of $f(z)$ in the lower half-plane.

Whereas this modification in procedure could be avoided here by merely replacing x by $-x$ in the original integral to yield the equivalent form

$$\int_{-\infty}^{\infty} f(-x)\,e^{-imx}\,dx,$$

where now $-m$ is positive, the use of the modified approach sometimes is more convenient.

In illustration, we consider the evaluation of the real integral

$$I = \int_{0}^{\infty} \frac{\sin tx \sin ax}{x^2 + b^2}\,dx \qquad (a \geqq 0, b > 0), \tag{157}$$

which can be interpreted as the *Fourier sine transform* of $(\sin ax)/(x^2 + b^2)$, expressed as a function of t (see Section 5.15). Since the integrand is an even function of x, we have also

$$I = \frac{1}{2} \int_{-\infty}^{\infty} \frac{\sin tx \sin ax}{x^2 + b^2}\,dx$$

and, since also

$$\sin tx \sin ax = \frac{1}{2}[\cos x(t - a) - \cos x(t + a)]$$

$$= \frac{1}{2} \operatorname{Re}[e^{ix(t-a)} - e^{ix(t+a)}],$$

we can write

$$I = \frac{1}{4} \operatorname{Re}\left[\int_{-\infty}^{\infty} \frac{e^{ix(t-a)}}{x^2 + b^2}\,dx - \int_{-\infty}^{\infty} \frac{e^{ix(t+a)}}{x^2 + b^2}\,dx\right]$$

$$\equiv \frac{1}{4} \operatorname{Re}(I_1 - I_2). \tag{158}$$

Dealing first with I_1, we see that when $t > a$ the formula (153) corresponding to a contour closed in the *upper* half-plane applies and there follows

$$I_1 = 2\pi i\left[\frac{e^{iz(t-a)}}{2z}\right]_{z=ib} = \frac{\pi}{b}e^{-b(t-a)} \qquad (t > a).$$

However, when $t < a$, closure in the *lower* half-plane is needed and hence we use (156) to obtain

$$I_1 = -2\pi i\left[\frac{e^{iz(t-a)}}{2z}\right]_{z=-ib} \doteq \frac{\pi}{b}e^{b(t-a)} \qquad (t < a).$$

When $t = a$, *either* method of closure is permissible, and hence

$$I_1 = 2\pi i\left[\frac{1}{2z}\right]_{z=ib} = -2\pi i\left[\frac{1}{2z}\right]_{z=-ib} = \frac{\pi}{b} \qquad (t = a).$$

These three results can be combined into the form

$$I_1 = \frac{\pi}{b} e^{-b|t-a|}.$$

Consequently, there follows also

$$I_2 = \frac{\pi}{b} e^{-b|t+a|}$$

and hence we have the result

$$I = \int_0^\infty \frac{\sin tx \sin ax}{x^2 + b^2} dx = \frac{\pi}{4b}(e^{-b|t-a|} - e^{-b|t+a|}) \qquad (b > 0), \qquad (159)$$

the condition $a \geq 0$ now being unnecessary since both members of (159) are odd functions of a. ∎

10.14. Theorems on Limiting Contours. In many applications of contour integration it is necessary to evaluate the limit of the result of integrating a function of a complex variable along an arc of a circle as the radius of that circle either increases without limit or tends to zero. In this section we collect and establish certain general results of frequent application. First, however, it is convenient to introduce a useful definition.

If, along a circular arc C_r of radius r, we have $|f(z)| \leq K_r$, where K_r is a bound depending only on r and hence independent of *angular* position on C_r, and if $K_r \to 0$ as $r \to \infty$ (or $r \to 0$), then we will say that $f(z)$ *tends to zero uniformly* on C_r as $r \to \infty$ (or $r \to 0$). Thus, for example, if C_r is a circular arc with center at the origin and $f(z) = z/(z^2 + 1)$, we have

$$|f(z)| = \frac{|z|}{|z^2 + 1|} \leq \frac{|z|}{|z|^2 - 1} = \frac{r}{r^2 - 1} \qquad (r > 1)$$

on C_r, if use is made of the basic inequality (10). Hence, if we then take $K_r = r/(r^2 - 1)$, we conclude that here $f(z)$ tends to zero uniformly on C_r as $r \to \infty$. Also, we may take $K_r = r/(1 - r^2)$ when $r < 1$ to show that the same is true when $r \to 0$.

In particular, any *rational function* (ratio of polynomials) whose denominator is of higher degree than the numerator tends uniformly to zero on any C_r as $r \to \infty$. This follows from the fact that then $|z||f(z)|$ tends to a limit (which may be zero) as $|z| = r \to \infty$, and hence is bounded by some constant k when r is large (say, $r \geq r_0$), so that we may take $K_r = k/r$ (when $r \geq r_0$).

The following theorems now may be stated:

Theorem I. If, on a circular arc C_R with radius R and center at the origin, $zf(z) \to 0$ uniformly as $R \to \infty$, then

$$\lim_{R \to \infty} \int_{C_R} f(z)\, dz = 0.$$

Theorem II. Suppose that, on a circular arc C_R with radius R and center at the origin, $f(z) \to 0$ uniformly as $R \to \infty$. Then:

1. $$\lim_{R \to \infty} \int_{C_R} e^{imz} f(z)\, dz = 0 \qquad (m > 0)$$

if C_R is in the first and/or second quadrants.†

2. $$\lim_{R \to \infty} \int_{C_R} e^{-imz} f(z)\, dz = 0 \qquad (m > 0)$$

if C_R is in the third and/or fourth quadrants.

3. $$\lim_{R \to \infty} \int_{C_R} e^{mz} f(z)\, dz = 0 \qquad (m > 0)$$

if C_R is in the second and/or third quadrants.

4. $$\lim_{R \to \infty} \int_{C_R} e^{-mz} f(z)\, dz = 0 \qquad (m > 0)$$

if C_R is in the first and/or fourth quadrants.

Theorem III. If, on a circular arc C_ρ with radius ρ and center at $z = a$, $(z - a)f(z) \to 0$ uniformly as $\rho \to 0$, then

$$\lim_{\rho \to 0} \int_{C_\rho} f(z)\, dz = 0.$$

Theorem IV. Suppose that $f(z)$ has a simple pole at $z = a$, with residue Res (a). Then, if C_ρ is a circular arc with radius ρ and center at $z = a$, intercepting an angle α at $z = a$, there follows

$$\lim_{\rho \to 0} \int_{C_\rho} f(z)\, dz = \alpha i \, \text{Res}\,(a),$$

where α is positive if the integration is carried out in the counterclockwise direction, and negative otherwise.

The proof of Theorem I follows from the fact that if $|zf(z)| \le K_R$, then $|f(z)| \le K_R/R$. Since the length of C_R is $|\alpha|R$, where α is the subtended angle, Equation (80) gives

$$\left| \int_{C_R} f(z)\, dz \right| \le \frac{K_R}{R} \cdot |\alpha|R = |\alpha|K_R \to 0 \qquad (R \to \infty). \qquad \blacksquare$$

The proof of Theorem II is somewhat more complicated. To prove part 1, we use the relation

$$\left| \int_{C_R} e^{imz} f(z)\, dz \right| \le \int_{C_R} |e^{imz}|\,|f(z)|\,|dz|.$$

†This result is known as *Jordan's lemma*.

But on C_R we have $|dz| = R\,d\theta$, $|f(z)| \leqq K_R$, and $|e^{imz}| = e^{-mR\sin\theta}$, according to (150a). Hence there follows

$$|I_R| \equiv \left| \int_{C_R} e^{imz} f(z)\,dz \right| \leqq RK_R \int_{\theta_0}^{\theta_1} e^{-mR\sin\theta}\,d\theta,$$

where $0 \leqq \theta_0 < \theta_1 \leqq \pi$. Since the last integrand is positive, the right-hand member is not decreased if we take $\theta_0 = 0$ and $\theta_1 = \pi$. Hence we have

$$|I_R| \leqq RK_R \int_0^\pi e^{-mR\sin\theta}\,d\theta = 2RK_R \int_0^{\pi/2} e^{-mR\sin\theta}\,d\theta. \tag{160}$$

This integral cannot be evaluated in terms of elementary functions of R. However, in the range $0 \leqq \theta \leqq \pi/2$ the truth of the relation

$$\sin\theta \geqq \frac{2}{\pi}\theta$$

is easily realized by comparing the graphs of $y = \sin x$ and $y = 2x/\pi$. Thus we have also, from (160),

$$|I_R| \leqq 2RK_R \int_0^{\pi/2} e^{-2mR\theta/\pi}\,d\theta = \frac{\pi}{m} K_R(1 - e^{-mR}), \tag{161}$$

and hence, if $m > 0$, I_R tends to zero with K_R as $R \rightarrow \infty$, as was to be shown.

The other three parts of Theorem II are established by completely analogous methods. ∎

To prove Theorem III we notice that the integrand is not greater than K_p/ρ in absolute value and the length of the path is $|\alpha|\rho$, where α is the subtended angle. Hence the integral tends to zero with K_p. ∎

To establish Theorem IV, we notice that if $f(z)$ has a simple pole at $z = a$ we can write

$$f(z) = \frac{\text{Res}\,(a)}{z - a} + \varphi(z),$$

where $\varphi(z)$ is *analytic*, and hence bounded, in the neighborhood of $z = a$. Hence we have

$$\int_{C_p} f(z)\,dz = \int_{C_p} \frac{\text{Res}\,(a)}{z - a}\,dz + \int_{C_p} \varphi(z)\,dz.$$

On C_p we can write $z = a + \rho e^{i\theta}$, where θ varies from an initial value θ_0 to $\theta_0 + \alpha$. Hence the first integral on the right becomes

$$\text{Res}\,(a) \int_{\theta_0}^{\theta_0+\alpha} \frac{\rho i e^{i\theta}\,d\theta}{\rho e^{i\theta}} = i\,\text{Res}\,(a) \int_{\theta_0}^{\theta_0+\alpha} d\theta = \alpha i\,\text{Res}\,(a).$$

The second integral on the right tends to zero with ρ, in consequence of Theorem III, establishing the desired result. ∎

Several applications of these theorems are presented in the following sections.

10.15. Indented Contours. In many cases the presence of a pole or branch point may lead to the necessity of *indenting* a contour by introducing an arc of a circle of small radius ρ, to avoid integration *through* a singularity. The desired result is then obtained by considering a limit as $\rho \to 0$.

In such cases we frequently encounter a new difficulty which can be explained best by considering what is meant by the Cauchy *principal value* of an improper real integral. To introduce this concept, we consider first the function $y = 1/x$. It is a familiar fact that the integral $\int_{-1}^{2} x^{-1}\, dx$ does not exist in the strict sense because of the strong infinity of the integrand at $x = 0$. We recall that this integral *would* exist, according to the conventional definition of an improper integral, only if the limit

$$\lim_{\delta_1, \delta_2 \to 0} \left(\int_{-1}^{-\delta_1} \frac{dx}{x} + \int_{\delta_2}^{2} \frac{dx}{x} \right)$$

were to exist, and have a unique value, as δ_1 and δ_2 *independently* approach zero through positive values (Figure 10.16). But since this limit is of the form

$$\lim_{\delta_1, \delta_2 \to 0} (\log \delta_1 + \log 2 - \log \delta_2) = \lim_{\delta_1, \delta_2 \to 0} \left(\log 2 - \log \frac{\delta_2}{\delta_1} \right),$$

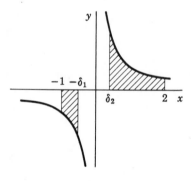

Figure 10.16

it clearly does not exist unless δ_1 and δ_2 tend to zero in such a way that δ_2/δ_1 tends toward a nonzero limit, in which case the value to be assigned to the integral depends upon this limit. However, if δ_1 and δ_2 are taken to be *equal*, so that the gap around the infinity in Figure 10.16 is *symmetrical* about $x = 0$, the limit is seen to be log 2. Incidentally, the same value would be obtained by *formal* substitution in the formula $\int_a^b x^{-1}\, dx = [\log|x|]_a^b$. This limit is defined to be the Cauchy *principal value* of the improper integral, and we write

$$\text{P} \int_{-1}^{2} \frac{dx}{x} = \log 2,$$

the symbol P denoting a principal value.

More generally, if $f(x)$ is not finite at a point $x = A$ inside the interval of integration, we have the definition

$$\text{P} \int_a^b f(x)\, dx = \lim_{\rho \to 0} \left(\int_a^{A-\rho} f(x)\, dx + \int_{A+\rho}^b f(x)\, dx \right), \tag{162}$$

if that limit exists. If the integral exists in the conventional sense, the true value necessarily agrees with the principal value so defined, and the symbol P then may be omitted.

The consideration of principal values of this kind is frequently necessary in the process of evaluating *proper* integrals, as is seen in the following examples. Further, principal values of improper integrals are not infrequently of physical significance in applications.† Clearly, some care should be exercised in dealing with them in such cases.

Example 1. In order to evaluate the integral

$$I = \int_{-\infty}^{\infty} \frac{\sin x}{x} \, dx = 2 \int_{0}^{\infty} \frac{\sin x}{x} \, dx, \qquad (163)$$

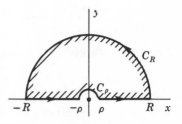

Figure 10.17

we consider the result of integrating the function $F(z) = e^{iz}/z$ around the closed contour of Figure 10.17. As before, we replace the sine by a complex exponential to obtain satisfactory behavior on C_R. In so doing, however, we obtain a function $F(z)$ which then has a *pole* at the origin and hence must introduce the indentation C_ρ. Since $F(z)$ is analytic inside the contour, Cauchy's residue theorem gives

$$\int_{-R}^{-\rho} \frac{e^{ix}}{x} \, dx + \int_{C_\rho} \frac{e^{iz}}{z} \, dz + \int_{\rho}^{R} \frac{e^{ix}}{x} + \int_{C_R} \frac{e^{iz}}{z} \, dz = 0. \qquad (164)$$

Since $|1/z| = 1/R$ on C_R, the fourth integral tends to zero as $R \longrightarrow \infty$, by Theorem II.1 of Section 10.14. Also, since $F(z)$ has a simple pole at $z = 0$, with Res$\,(0) = 1$, Theorem IV states that the second integral tends to $-\pi i$ as $\rho \longrightarrow 0$, the negative sign corresponding to the negative sense of C_ρ. Thus, proceeding to the limit as $\rho \longrightarrow 0$ and $R \longrightarrow \infty$, we have

$$\lim_{\substack{\rho \to 0 \\ R \to \infty}} \left(\int_{-R}^{-\rho} \frac{e^{ix}}{x} \, dx + \int_{\rho}^{R} \frac{e^{ix}}{x} dx \right) - \pi i = 0. \qquad (165)$$

Some care must be taken at this stage, since the integral $\int_{-\infty}^{\infty} x^{-1} e^{ix} \, dx$ does not exist in the strict sense because of the fact that the integrand behaves like $1/x$ near $x = 0$. However, we may notice that the limit in (165) is in fact the Cauchy principal value of this *divergent integral*, and hence (165) takes the form

$$P \int_{-\infty}^{\infty} \frac{e^{ix}}{x} \, dx = \pi i. \qquad (166)$$

By taking imaginary parts of both sides, we obtain the desired result

$$\int_{-\infty}^{\infty} \frac{\sin x}{x} \, dx = \pi, \qquad (167)$$

the symbol P now being omitted since this is a *convergent* integral, the integrand being *finite* at $x = 0$. However, the result of equating *real* parts of (166) should be written

---

†For example, they arise frequently in the aerodynamic theory of airfoils. In technical work, the symbol **P** is often omitted before such integrals. Also, various alternative notations are used.

in the form

$$P \int_{-\infty}^{\infty} \frac{\cos x}{x} \, dx = 0. \tag{168}$$

It may be noted that the principal value in (168) truly is taken in *two* senses, since in (165) we not only have taken the gap $(-\rho, \rho)$ about the origin to be symmetrical, but also have taken the upper and lower cut-off radii to be both equal to R, before proceeding to the limit as the gap closes and the extreme limits become infinite in magnitude. ∎

Example 2. We next consider the integral

$$I = \int_{-\infty}^{\infty} \frac{\cos x}{\pi^2 - 4x^2} \, dx, \tag{169}$$

noticing that since the integrand is finite at $x = \pm\pi/2$, the question of principal values does not arise in the definition of I. To evaluate the integral, we integrate $F(z) = e^{iz}/(\pi^2 - 4z^2)$ around the contour of Figure 10.18, taking into account the poles of

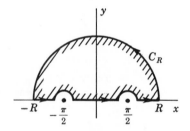

Figure 10.18

$F(z)$ on the real axis. Again making use of the results of Section 10.14, we obtain the result

$$P \int_{-\infty}^{\infty} \frac{e^{ix}}{\pi^2 - 4x^2} \, dx - \pi i \left[\text{Res} \left(-\frac{\pi}{2} \right) + \text{Res} \left(\frac{\pi}{2} \right) \right] = 0,$$

where

$$\pi i \left[\text{Res} \left(-\frac{\pi}{2} \right) + \text{Res} \left(\frac{\pi}{2} \right) \right] = \pi i \left[\frac{e^{-\pi i/2}}{4\pi} - \frac{e^{\pi i/2}}{4\pi} \right] = \frac{1}{2},$$

and hence there follows

$$P \int_{-\infty}^{\infty} \frac{e^{ix}}{\pi^2 - 4x^2} \, dx = \frac{1}{2}. \tag{170}$$

By equating real parts of the two sides of this equation, we obtain the desired result

$$\int_{-\infty}^{\infty} \frac{\cos x}{\pi^2 - 4x^2} \, dx = \frac{1}{2}. \tag{171}$$

∎

10.16. Integrals Involving Branch Points. When contour integration leads to the consideration of a complex function with one or more branch points, care must be taken in selecting an appropriate branch of that function and using that branch consistently throughout the calculation. Two rather typical examples follow and a number of others are considered in Problems 121 to 131.

Example 1. In many cases a real integral can be transformed by contour integration into a more tractable form. To illustrate this procedure and also to consider the treatment of a branch point, we evaluate the integral.

$$I = \int_0^\infty \frac{\cos x}{x^{1-m}}\, dx \qquad (0 < m < 1). \qquad (172)$$

For this purpose we integrate $F(z) = e^{iz}/z^{1-m}$ around the contour of Figure 10.19. The indentation is necessary to avoid the branch point at the origin. We must, of course, choose a branch of the multi-valued function z^{1-m} which is *real* on the positive real axis and analytic in the first quadrant, so that we may define

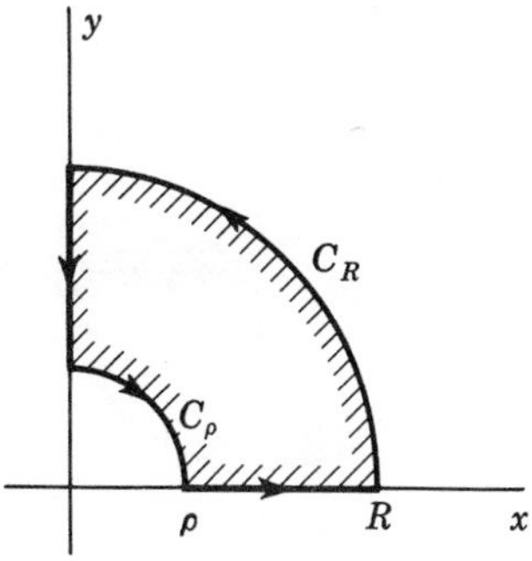

$$z^{1-m} = r^{1-m} e^{i(1-m)\theta} \qquad (-\pi < \theta < \pi)$$

Figure 10.19

when $z = re^{i\theta}$. Cauchy's integral theorem gives

$$\int_\rho^R \frac{e^{ix}}{x^{1-m}}\, dx + \int_{C_R} \frac{e^{iz}}{z^{1-m}}\, dz + \int_R^\rho \frac{e^{-y}}{(iy)^{1-m}}(i\, dy) + \int_{C_\rho} \frac{e^{iz}}{z^{1-m}}\, dz = 0.$$

The integral along C_R vanishes as $R \to \infty$, by Theorem II.1, and that along C_ρ vanishes as $\rho \to 0$, by Theorem III. Hence, proceeding to the limits, we have

$$\int_0^\infty \frac{e^{ix}}{x^{1-m}}\, dx + \int_\infty^0 \frac{e^{-y}}{ie^{-im\pi/2}y^{1-m}} i\, dy = 0$$

or

$$\int_0^\infty \frac{e^{ix}}{x^{1-m}}\, dx = e^{im\pi/2} \int_0^\infty e^{-y}y^{m-1}\, dy \qquad (0 < m < 1). \qquad (173)$$

The integral on the right in (173) may be recognized as that defining $\Gamma(m)$. Thus, by equating real and imaginary parts, we obtain the results

$$\int_0^\infty \frac{\cos x}{x^{1-m}}\, dx = \Gamma(m) \cos \frac{m\pi}{2},$$

$$\int_0^\infty \frac{\sin x}{x^{1-m}}\, dx = \Gamma(m) \sin \frac{m\pi}{2},$$

$$(174)$$

when $0 < m < 1$. We may notice that the right-hand member of (173) is obtained *formally* if we replace x by a new (complex) variable iy in the left-hand member and do not concern ourselves with distinguishing between ∞ and ∞/i (or, more precisely, between the old and new *paths of integration*). The danger of using such a formal substitution without establishing its validity by contour integration (or otherwise) is illustrated by the consideration that *by the same substitution* in (172) we should deduce that $I = i^m \int_0^\infty y^{m-1} \cosh y \, dy$. However, *this integral does not exist!* ■

Example 2. We attempt to evaluate the integral

$$I = \int_0^\infty \frac{x^{m-1}}{x+1} \, dx \qquad (0 < m < 1) \tag{175}$$

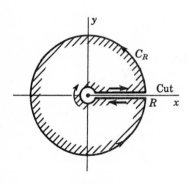

Figure 10.20

by integrating $f(z) = z^{m-1}/(z+1)$ around the contour of Figure 10.20 and proceeding to the limit as the small radius ρ tends to zero and the large radius R tends to infinity. Here we must choose that branch of z^{m-1} for which

$$z^{m-1} = r^{m-1}e^{i(m-1)\theta} \qquad (0 < \theta < 2\pi)$$

when $z = re^{i\theta}$, with the understanding that θ may approach 0 from above and 2π from below. Accordingly, along the upper bank of a cut along the positive real axis we write $z = r$ and along the lower bank we write $z = re^{2\pi i}$, when we calculate z^{m-1}. In the limit, the integrals along the two circles vanish (with the given restrictions on m) and we have

$$\int_0^\infty \frac{r^{m-1}}{r+1} \, dr + \int_\infty^0 \frac{(re^{2\pi i})^{m-1}}{r+1} \, dr = 2\pi i \operatorname{Res}\left\{\frac{z^{m-1}}{z+1}; -1\right\}$$

or, equivalently,

$$(1 - e^{2m\pi i})I = 2\pi i(e^{\pi i})^{m-1} = -2\pi i e^{m\pi i}.$$

Hence we obtain the result

$$\int_0^\infty \frac{x^{m-1}}{x+1} \, dx = \frac{\pi}{\sin m\pi} \qquad (0 < m < 1). \tag{176}$$

(See also Problem 119.) ■

REFERENCES

1. CARRIER, G. F., M. KROOK, and C. E. PEARSON, *Functions of a Complex Variable: Theory and Technique*, McGraw-Hill Book Company, Inc., New York, 1966.

2. CHURCHILL, R. V., *Complex Variables and Applications*, 2nd ed., McGraw-Hill Book Company, Inc., New York, 1960.

3. CURTISS, D. R., *Analytic Functions of a Complex Variable*, Second Carus Mathematical Monograph, Open Court Publishing Company, La Salle, Ill., 1926.

4. FRANKLIN, P., *Functions of Complex Variables*, Prentice-Hall, Inc., Englewood Cliffs, N.J., 1958.

5. KAPLAN, W., *A First Course in Functions of a Complex Variable*, Addison-Wesley Publishing Company, Inc., Reading, Mass., 1953.

6. Knopp, K., *Theory of Functions*, Dover Publications, Inc., New York, 2 pts., 1945, 1947.

7. Titchmarsh, E. C., *The Theory of Functions*, 2nd ed., Oxford University Press, Inc., New York, 1950.

8. Whittaker, E. J., and G. N. Watson, *Modern Analysis*, Cambridge University Press, New York, 1958.

PROBLEMS

Section 10.1

1. Show that, if a complex number $a + ib$ is written as a *number pair* (a, b), then for two such complex numbers the laws of combination take the form

$$(a_1, b_1) + (a_2, b_2) = (a_1 + a_2, b_1 + b_2),$$
$$(a_1, b_1)(a_2, b_2) = (a_1 a_2 - b_1 b_2, a_1 b_2 + a_2 b_1),$$
$$\frac{(a_2, b_2)}{(a_1, b_1)} = \left(\frac{a_1 a_2 + b_1 b_2}{a_1^2 + b_1^2}, \frac{a_1 b_2 - a_2 b_1}{a_1^2 + b_1^2} \right).$$

2. (a) Show that, if the real and imaginary parts of $\alpha \equiv a + ib$ are the components of the vector $\mathbf{v} = a\mathbf{i} + b\mathbf{j}$, then the real and imaginary parts of $\alpha_1 \pm \alpha_2$ are the components of the vector $\mathbf{v}_1 \pm \mathbf{v}_2$, but that no such statement applies to multiplication (with reference to either the dot or the cross product of vectors) or to division.

(b) Show, however, that $\alpha \bar{\alpha} \equiv |\alpha|^2$ is equal to the dot product $\mathbf{v} \cdot \mathbf{v} \equiv |\mathbf{v}|^2$, and that $\alpha_1 \bar{\alpha}_2 + \bar{\alpha}_1 \alpha_2 = 2\mathbf{v}_1 \cdot \mathbf{v}_2$. (Notice that these quantities are real scalar numbers).

3. Establish the following results:
   (a) $\operatorname{Re}(z_1 + z_2) = \operatorname{Re}(z_1) + \operatorname{Re}(z_2)$, but $\operatorname{Re}(z_1 z_2) \neq \operatorname{Re}(z_1) \operatorname{Re}(z_2)$ in general;
   (b) $\operatorname{Im}(z_1 + z_2) = \operatorname{Im}(z_1) + \operatorname{Im}(z_2)$, but $\operatorname{Im}(z_1 z_2) \neq \operatorname{Im}(z_1) \operatorname{Im}(z_2)$ in general;
   (c) $|z_1 z_2| = |z_1| |z_2|$, but $|z_1 + z_2| \neq |z_1| + |z_2|$ in general;
   (d) $\overline{z_1 + z_2} = \bar{z}_1 + \bar{z}_2$ and $\overline{z_1 z_2} = \bar{z}_1 \bar{z}_2$.

4. Establish the following results:
   (a) $z + \bar{z} = 2 \operatorname{Re}(z)$,
   (b) $z - \bar{z} = 2i \operatorname{Im}(z)$,
   (c) $z_1 \bar{z}_2 + \bar{z}_1 z_2 = 2 \operatorname{Re}(z_1 \bar{z}_2) = 2 \operatorname{Re}(\bar{z}_1 z_2)$,
   (d) $\operatorname{Re}(z) \leq |z|$,
   (e) $\operatorname{Im}(z) \leq |z|$,
   (f) $|z_1 \bar{z}_2 + \bar{z}_1 z_2| \leq 2 |z_1 z_2|$,
   (g) $(|z_1| - |z_2|)^2 \leq |z_1 + z_2|^2 \leq (|z_1| + |z_2|)^2$. [Use part (f).]

Section 10.2

5. Express the following quantities in the form $a + ib$, where a and b are real:

   (a) $(1 + i)^3$,   (b) $\dfrac{1 + i}{1 - i}$,   (c) $e^{\pi i/2}$,

   (d) $e^{2 + \pi i/4}$,   (e) $\sin\left(\dfrac{\pi}{4} + 2i\right)$,   (f) $\cosh\left(2 + \dfrac{\pi i}{4}\right)$.

6. Use the definitions (20a, b) to establish the following identities:
(a) $\sin^2 z + \cos^2 z = 1$,
(b) $\sin(z_1 + z_2) = \sin z_1 \cos z_2 + \cos z_1 \sin z_2$,
(c) $\cos 2z = 2\cos^2 z - 1$.

7. Prove that the functions $\sin z$ and $\cos z$ are periodic, with real period 2π, whereas e^z, $\sinh z$, and $\cosh z$ are periodic, with pure imaginary period $2\pi i$. What are the periods of the other circular and hyperbolic functions?

8. Deduce expressions for the derivative of $\sin z$, $\cos z$, $\sinh z$, and $\cosh z$ from the established result $de^{az}/dz = ae^{az}$. Also obtain these expressions by exploiting the permissibility of term-by-term differentiation of the Taylor series representations.

9. Prove that e^z possesses no zeros, that the zeros of $\sin z$ and $\cos z$ all lie on the real axis, and that those of $\sinh z$ and $\cosh z$ all lie on the imaginary axis.

10. If $f(z) = e^{iz}$, show that $\bar{f}(z) = e^{-iz}$, $f(\bar{z}) = e^{i\bar{z}}$, and $\bar{f}(\bar{z}) = \overline{f(z)} = e^{-i\bar{z}}$.

11. Establish (18) by obtaining the relations

$$e^{z_1}e^{z_2} = \sum_{j=0}^{\infty}\sum_{k=0}^{\infty}\frac{z_1^j z_2^k}{j!\,k!} = \sum_{n=0}^{\infty}\left[\sum_{j=0}^{n}\frac{z_1^j z_2^{n-j}}{j!\,(n-j)!}\right],$$

where n was written for $j + k$, and reviewing the binomial expansion of $(z_1 + z_2)^n$.

Section 10.3

12. Show that the nth roots of unity are of the form ω_n^k ($k = 0, 1, \ldots, n-1$), where $\omega_n = \cos(2\pi/n) + i\sin(2\pi/n)$.

13. Determine all possible values of the following quantities in the form $a + ib$, and in each case give also the principal value, assuming the definition (39):
(a) $\log(1 + i)$,   (b) $(i)^{3/4}$,   (c) $(1 + i)^{1/2}$.

14. Express the roots of the equation $z^4 + 2z^2 + 2 = 0$ in the form $a + ib$.

15. Express the function z^π in the form given by Equation (47) and also find the principal value of this function [assuming (39)] when $z = (1 + i)/\sqrt{2}$, in the form $a + ib$.

16. If $f(z) = z^i$ and $g(z) = i^z$, with the convention of Equation (50), distinguish between $f(i)$ and $g(i)$.

17. Derive Equations (55c) and (56a).

18. Determine all possible values of the quantities
(a) $\sin^{-1} 2$,   (b) $\tan^{-1}(2i)$.

19. (a) Verify that, if a is a positive real constant,

$$\coth^{-1}\frac{z}{a} = \frac{1}{2}\log\frac{z+a}{z-a} = \frac{1}{2}\left[\log\left|\frac{z+a}{z-a}\right| + i\arg\left(\frac{z+a}{z-a}\right)\right].$$

(b) Verify that

$$\frac{z+a}{z-a} = \frac{(z+a)(\bar{z}-a)}{(z-a)(\bar{z}-a)} = \frac{|z|^2 - a^2 - 2ai\,\mathrm{Im}\,(z)}{|z|^2 + a^2 - 2a\,\mathrm{Re}\,(z)}.$$

(c) Hence deduce that

$$\coth^{-1}\frac{z}{a} = \frac{1}{4}\log\frac{(x+a)^2+y^2}{(x-a)^2+y^2} + \frac{i}{2}\left[\tan^{-1}\left(\frac{2ay}{a^2-x^2-y^2}\right) + 2k\pi\right]$$

when a is real and positive, where k is any integer.

20. *Bipolar coordinates.*

    (a) If (p_1, ω_1) are polar coordinates relative to the point $(a, 0)$ and (p_2, ω_2) are polar coordinates relative to the point $(-a, 0)$, show that

$$z = a + p_1 e^{i\omega_1} = -a + p_2 e^{i\omega_2}.$$

    (b) Use the result of Problem 19(a) to show that

$$\coth^{-1}\frac{z}{a} = \frac{1}{2}\log\frac{p_2}{p_1} + \frac{i}{2}(\omega_2 - \omega_1),$$

where ω_1 and ω_2 are each defined only within an arbitrary additive integral multiple of 2π.

21. (a) Suppose that the principal values of ω_1 and ω_2 are such that $0 \leq \omega_{1P} < 2\pi$ and $0 \leq \omega_{2P} < 2\pi$ in Problem 20. Verify that, as the segment of the real axis between $-a$ and a is crossed from above, the angle $\omega_2 - \omega_1$ changes abruptly by 2π, whereas no such jump occurs at a crossing outside this segment, so that transition from one branch to another then can and must take place only across the finite segment joining the "branch points" $z = \pm a$.

    (b) Suppose that the definitions $0 \leq \omega_{1P} < 2\pi$ and $-\pi < \omega_{2P} \leq +\pi$ are adopted. Verify that $\omega_2 - \omega_1$ then is continuous across the finite segment joining the branch points, but that it changes abruptly as the real axis is crossed at any point outside this segment.

Section 10.4

22. Let $f(z) = (x - y)^2 + 2i(x + y)$.

    (a) Show that the Cauchy–Riemann equations are satisfied only along the curve $x - y = 1$, and hence deduce that $f(z)$ has a derivative along that curve, but is nowhere analytic.

    (b) Verify directly that $f(z)$ has a derivative along the curve $x - y = 1$ by showing first that

$$\frac{\Delta f}{\Delta z} = \frac{2(x-y)(\Delta x - \Delta y) + (\Delta x - \Delta y)^2 + 2i(\Delta x + \Delta y)}{\Delta x + i\Delta y}$$

in the general case, and that when $x - y = 1$ there follows

$$\frac{\Delta f}{\Delta z} = 2 + 2i + \Delta x - i\Delta y - 2\frac{\Delta x \Delta y}{\Delta z}.$$

    (c) Simplify the expression for $\Delta f/\Delta z$ when $x - y = -1$ as much as possible.

    (d) Verify that

$$f'(z) = \frac{\partial}{\partial x}[(x-y)^2 + 2i(x+y)] = \frac{1}{i}\frac{\partial}{\partial y}[(x-y)^2 + 2i(x+y)]$$

when $f'(z)$ exists, in accordance with Equations (63a, b).

23. (a) If $3x^2y - y^3$ is the real part of an analytic function of z, determine the imaginary part.

(b) Prove that xy^2 cannot be the real part of an analytic function of z.

(c) Determine whether $2xy + i(x^2 - y^2)$ is an analytic function of z.

24. Suppose that $f(z) = u(x, y) + iv(x, y)$ is analytic in a region $\Re$ including part of the real axis.

(a) Show that $f(x) = u(x, 0) + iv(x, 0)$ and hence that

$$f(z) = u(z, 0) + iv(z, 0)$$

when z is in $\Re$. [Hence we then obtain $w = f(z)$ by merely replacing y by zero and x by z in $w = u(x, y) + iv(x, y)$.]

(b) Illustrate the result of part (a) by determining the function f such that $w = f(z)$ when

$$w = \frac{y(x^2 + y^2 - 1) + ix(x^2 + y^2 + 1)}{(x^2 - y^2 + 1)^2 + 4x^2 y^2}.$$

(c) Why is the same result obtained in part (b) when x is replaced by zero and y by $-iz$?

25. Let s represent distance in the counterclockwise direction around a closed curve C in the xy plane.

(a) If, at any point P on C, $\mathbf{t}$ represents the unit tangent vector in this direction, and $\mathbf{n}$ represents the unit outward normal vector, show that

$$\mathbf{t} = \mathbf{i} \cos \varphi + \mathbf{j} \sin \varphi \qquad \text{and} \qquad \mathbf{n} = \mathbf{i} \sin \varphi - \mathbf{j} \cos \varphi,$$

where φ is the slope angle.

(b) If $u(x, y)$ and $v(x, y)$ are the real and imaginary parts of an analytic function of z in a region $\Re$ including C, show that

$$\frac{\partial u}{\partial s} = -\frac{\partial v}{\partial n}, \qquad \frac{\partial u}{\partial n} = \frac{\partial v}{\partial s}$$

at any point of C. [Recall that $\partial/\partial s = \mathbf{t} \cdot \nabla$ and $\partial/\partial n = \mathbf{n} \cdot \nabla$, and use Equations (65a, b).]

26. By applying the result of Problem 25 to a circle $r = $ constant, obtain the Cauchy–Riemann equations in polar coordinates, in the form

$$\frac{\partial u}{\partial r} = \frac{1}{r} \frac{\partial v}{\partial \theta}, \qquad \frac{1}{r} \frac{\partial u}{\partial \theta} = -\frac{\partial v}{\partial r}.$$

27. If $f(z) = u + iv$ is analytic, determine the following:

(a) v when $u = r^{-2} \sin 2\theta \ (r \neq 0)$,

(b) u when $v = r^3(1 - 4 \cos^2 \theta) \sin \theta$.

28. If $f(z) = U(r, \theta) + iV(r, \theta)$ is an analytic function of $z = re^{i\theta}$ in a region including part of the positive real axis, how can f be expressed in terms of U and V? (Compare Problem 24.)

29. If $f(z)$ is analytic in a region $\Re$ and if

$$\text{Re}\,[f(z)] = \log r - r \sin \theta \qquad (r \neq 0)$$

in that region, show that $f(z)$ must be defined as a single-valued *branch* of a multivalued function.

30. Show that the real and imaginary parts of any twice-differentiable function of the form $f(\bar{z})$ satisfy Laplace's equation, but that such a function is nowhere an analytic function of z unless it is a constant. [Compare the values of $\partial f/\partial x$ and $\partial f/\partial(iy)$.]

31. Show that $f(|z|)$ is nowhere an analytic function of z unless it is a constant. (Consider the derivative of f in the θ direction.)

Section 10.5

32. Suppose that $f(z) = u + iv$ is analytic in a simple region $\mathcal{R}$ and that
$$u\,dx - v\,dy = dU, \qquad v\,dx + u\,dy = dV$$
in $\mathcal{R}$ (where U and V are single-valued). If $F(z)$ is the complex function $F = U + iV$, containing an arbitrary additive complex constant, show that $F(z)$ is analytic in $\mathcal{R}$ and that it possesses the properties described by Equations (69) and (70) in $\mathcal{R}$. (Show first that U and V satisfy the Cauchy–Riemann equations in $\mathcal{R}$.)

33. (a) Use the definition (68) to calculate directly the integral $\oint_C z\,dz$, where C is the unit circle $x = \cos t$, $y = \sin t$.

(b) Use the definition (72) to calculate directly the integral $\oint_C \log z\,dz$, where C is the unit circle $r = 1$, taking the principal value of the logarithm.

(c) Obtain the results of parts (a) and (b) by appropriately dealing with the functions $F_1(z) = z^2/2$ and $F_2(z) = z \log z - z$.

34. (a) Show that the value of the integral
$$\int_{-1}^{1} \frac{z+1}{z^2}\,dz$$
is $-2 - \pi i$ if the path is the *upper* half of the circle $r = 1$. [Write $z = e^{i\theta}$, where θ varies from π to 0, or from $(2k+1)\pi$ to $2k\pi$, where k is any integer.]

(b) Show (also by direct integration) that the value is $-2 + \pi i$ if the path is the *lower* half of the same circle.

(c) Obtain the results of parts (a) and (b) by appropriately dealing with the function $F(z) = \log z - z^{-1}$.

35. (a) Evaluate the integral
$$\oint_C \frac{z+1}{z^2}\,dz,$$
where C is the circle $r = 1$, first by using the results of Problem 34(a, b), second by considering the function $\log z - z^{-1}$, and third by using Equations (75) and (77).

(b) Evaluate the integral in part (a) (by any method) when C is the circle $r = a$ $(a > 0)$.

36. Evaluate the integral
$$\oint_C \bar{z}\,dz,$$
when C is the unit circle $r = 1$ and also, more generally, when C is the circle $r = a$ $(a > 0)$.

37. Proceed as in Problem 36 with the integral
$$\oint_C (|z| - e^z \sin z^2)\,dz.$$

38. (a) Prove that the integral

$$\int_{-1}^{2} \frac{dz}{z^2}$$

is independent of the path, so long as that path does not pass through the origin. By integrating along any convenient path (say, around a semicircle and thence along the real axis) show that the value of the integral is $-\frac{3}{2}$.

(b) Show that the *real* integral

$$\int_{-1}^{2} \frac{dx}{x^2}$$

does not exist, but that the value given by formal substitution of limits in the indefinite integral agrees with that obtained in part (a). (Notice that, in spite of this fact, the integrand is never negative!)

39. Show that

$$\oint_C \frac{dz}{z} = 4\pi i$$

when the integration is once around the closed curve C defined by the polar equation

$$r = 2 - \sin^2 \frac{\theta}{4}$$

and explain why the result differs from Equation (75). (*Suggestion:* Sketch the curve.)

40. Let C represent a *semicircle* of radius R, with center at the origin, where $R > 1$, and consider the functions

$$f_1(z) = z^2 - 1, \qquad f_2(z) = \frac{1}{z^2 + 1}.$$

(a) Use Equation (10) to show that on C there follows

$$R^2 - 1 \leqq |f_1(z)| \leqq R^2 + 1, \qquad \frac{1}{R^2 + 1} \leqq |f_2(z)| \leqq \frac{1}{R^2 - 1}.$$

(b) Deduce from (80) that

$$\left| \int_C f_1(z) \, dz \right| \leqq \pi R(R^2 + 1), \qquad \left| \int_C f_2(z) \, dz \right| \leqq \frac{\pi R}{R^2 - 1}.$$

(c) Show also that

$$\left| \int_C f_1(z) f_2(z) \, dz \right| \leqq \pi R \frac{R^2 + 1}{R^2 - 1}.$$

41. (a) Suppose that C_1 and C_2 are two simple closed curves which intersect at exactly two points. If $f(z)$ is analytic on both C_1 and C_2 and also in the two regions which lie inside one of the curves but outside the other, use Cauchy's integral theorem to prove that $\oint_{C_1} f(z) \, dz = \oint_{C_2} f(z) \, dz$.

(b) Generalize the result of part (a) to deduce the "equivalent contour" property stated at the end of Section 10.5 when C_1 and C_2 intersect at a finite number of points.

Section 10.6

42. (a) Use the results of Equations (78a, b) to verify Cauchy's integral formula (85) when $f(z) = z^2$. [Express $f(\alpha)$ in the form $(\alpha - z)^2 + 2z(\alpha - z) + z^2$.]

(b) Verify also the derivative formula (88) in this case.

43. If $F(z) = (z + 6)/(z^2 - 4)$, show that the integral $\oint_C F(z)\,dz$ is 0 if C is the circle $x^2 + y^2 = 1$, is $4\pi i$ if C is the circle $(x - 2)^2 + y^2 = 1$, and is $-2\pi i$ if C is the circle $(x + 2)^2 + y^2 = 1$. [Use Equation (84) in the second and third cases, with the functions $f(z) = (z + 6)/(z + 2)$ and $f(z) = (z + 6)/(z - 2)$, respectively.]

44. *A mean-value theorem.* Let z_0 denote a point in a region $\Re$ where $f(z)$ is analytic, and let C denote any circle, with center at z_0, which lies inside $\Re$. By writing $\alpha = z_0 + ae^{i\varphi}$ in Cauchy's integral formula (85), show that

$$f(z_0) = \frac{1}{2\pi} \int_0^{2\pi} f(z_0 + ae^{i\varphi})\,d\varphi,$$

and deduce that *the value of an analytic function at any point z_0 is the average of its values on any circle, with z_0 as its center, which lies inside the region of analyticity.*

45. *A maximum-modulus theorem.* By applying the inequality (80) to the result of Problem 44, deduce that *the absolute value of an analytic function at a point z_0 cannot exceed the maximum absolute value of that function along any circle, with center at z_0, which lies inside the region of analyticity.*

Section 10.7

46. *Cauchy's inequality.* If $f(z)$ is analytic inside and on a circle C with center a and radius R, and if $|f(z)| \le M$ on C, use Equations (88) and (80) to show that

$$|f^{(n)}(a)| \le \frac{n!\,M}{R^n}.$$

47. *Uniqueness of Taylor series.* Suppose that, by any method, we have obtained a representation

$$f(z) = \sum_{n=0}^{\infty} a_n(z - a)^n$$

which converges to $f(z)$ when $|z - a| < R$. By making use of the fact that such a series can be differentiated term by term any number of times inside the circle of convergence, show that there must follow

$$f^{(n)}(a) = n!\,a_n \qquad (n = 0, 1, 2, \ldots),$$

so that this series necessarily is the Taylor series of $f(z)$ in that circle.

48. Obtain each of the following series expansions by any convenient method:

(a) $\dfrac{\sin z}{z} = 1 - \dfrac{z^2}{3!} + \dfrac{z^4}{5!} - \cdots = \sum_{n=0}^{\infty}(-1)^n\dfrac{z^{2n}}{(2n + 1)!}$ $(|z| < \infty)$,

(b) $\dfrac{\cosh z - 1}{z^2} = \dfrac{1}{2!} + \dfrac{z^2}{4!} + \dfrac{z^4}{6!} + \cdots = \sum_{n=0}^{\infty}\dfrac{z^{2n}}{(2n + 2)!}$ $(|z| < \infty)$,

(c) $\dfrac{e^z}{1 - z} = 1 + 2z + \dfrac{5}{2}z^2 + \dfrac{8}{3}z^3 + \cdots$ $(|z| < 1)$,

(d) $\dfrac{a^2}{z^2} = 1 + 2\dfrac{z + a}{a} + 3\dfrac{(z + a)^2}{a^2} + \cdots$ $(|z + a| < |a|)$.

49. *L'Hospital's rule.* Suppose that $f(z)$ and $g(z)$ are analytic at $z = a$ and that

$$f(a) = f'(a) = \cdots = f^{(k)}(a) = 0$$

and
$$g(a) = g'(a) = \cdots = g^{(k)}(a) = 0$$

but not both $f^{(k+1)}(a)$ and $g^{(k+1)}(a)$ vanish. Prove that

$$\lim_{z \to a} \frac{f(z)}{g(z)} = \frac{f^{(k+1)}(a)}{g^{(k+1)}(a)}$$

when $g^{(k+1)}(a) \neq 0$, and that the limit fails to exist when $g^{(k+1)}(a) = 0$. (*Suggestion:* Consider the Taylor series expansions of f and g near $z = a$.)

50. Use L'Hospital's rule (Problem 49) to evaluate the following limits:

(a) $\displaystyle\lim_{z \to 0} \frac{\sin z}{z}$,

(b) $\displaystyle\lim_{z \to 0} \frac{1 - \cos z}{z^2}$,

(c) $\displaystyle\lim_{z \to \pi i} \frac{\sinh z}{e^z + 1}$,

(d) $\displaystyle\lim_{z \to \pi/2} \left(z - \frac{\pi}{2}\right) \tan z$.

Section 10.8

51. Expand the function $f(z) = 1/(1 - z)$ in each of the following series:
(a) a Taylor series of powers of z for $|z| < 1$;
(b) a Laurent series of powers of z for $|z| > 1$;
(c) a Taylor series of powers of $z + 1$ for $|z + 1| < 2$, by first writing $f(z) = [2 - (z + 1)]^{-1} = \frac{1}{2}[1 - (z + 1)/2]^{-1}$;
(d) a Laurent series of powers of $z + 1$ for $|z + 1| > 2$, by first writing $f(z) = -[1/(z + 1)]/[1 - 2/(z + 1)]$.

52. Expand the function $f(z) = 1/[z(1 - z)]$ in a Laurent (or Taylor) series which converges in each of the following regions:

(a) $0 < |z| < 1$,

(b) $|z| > 1$,

(c) $0 < |z - 1| < 1$,

(d) $|z - 1| > 1$,

(e) $|z + 1| < 1$,

(f) $1 < |z + 1| < 2$,

(g) $|z + 1| > 2$.

53. Determine all Laurent (or Taylor) expansions in powers of z of the function $f(z) = (z - 2)/(z^3 - 1)$, specifying the region of convergence for each series.

54. Without determining the series, specify the region of convergence for the Laurent series of the function $f(z) = 1/(z^4 + 4)$, in powers of $z - 1$, which converges when $z = i$.

55. Determine the first three nonzero terms in the Laurent expansion of $f(z) = \csc z$ which is valid when $0 < |z| < \pi$ by first showing that the expansion must be of the form $f(z) = c_{-1}z^{-1} + c_1 z + c_3 z^3 + \cdots$, then determining c_{-1}, c_1, and c_3 from the condition

$$1 = \left(\frac{c_{-1}}{z} + c_1 z + c_3 z^3 + \cdots\right)\left(z - \frac{1}{6}z^3 + \frac{1}{120}z^5 - \cdots\right).$$

56. *Uniqueness of Laurent series.* Assume that the expansion

$$f(z) = \sum_{n=-\infty}^{\infty} b_n(z - a)^n$$

somehow is known to be valid when $R_1 < |z - a| < R_2$.

(a) Let C be the circle $|z - a| = \rho$, where $R_1 < \rho < R_2$, and deduce the relation

$$\frac{1}{2\pi i} \oint_C \frac{f(\alpha)d\alpha}{(\alpha - a)^{m+1}} = \frac{1}{2\pi i} \oint_C \sum_{n=-\infty}^{\infty} b_n(\alpha - a)^{n-m-1} \, d\alpha,$$

where the left-hand member is the *Laurent* coefficient c_m, by (96).

(b) By making use of the permissibility of integrating term by term around C, and by reviewing Equations (78a, b), show that

$$b_m = c_m \qquad (m = 0, \pm 1, \pm 2, \ldots),$$

so that the expansion must be the Laurent expansion in the specified annulus.

Section 10.9

57. (a) Show that the function $w = A \log (z - a)$ increases by $2\pi i A$ after a simple positive circuit around the point $z = a$, and is unchanged if the circuit does not include that point.

(b) Show that the expression for $w = \tan^{-1} z$ can be written in the form

$$w = \tan^{-1} z = \frac{i}{2} \log (z + i) - \frac{i}{2} \log (z - i) + \frac{\pi}{2}.$$

If the point z describes a simple positive circuit, obtain the increase in w when the circuit includes $z = i$ but excludes $z = -i$, and when the circuit includes $z = -i$ but excludes $z = i$. Thus verify that w is unchanged if the circuit surrounds both branch points.

58. With the bipolar notation of Problem 20, show that

$$\coth^{-1} \frac{z}{a} = \frac{1}{2} \log \frac{p_2}{p_1} + \frac{i}{2}(\omega_2 - \omega_1)$$

defines a single-valued function in the plane cut along the real axis from $-a$ to a if ω_1 and ω_2 are restricted such that $0 \leq \omega_1 < 2\pi$ and $0 \leq \omega_2 < 2\pi$, and a single-valued function in the plane cut by infinite rays from the points $z = \pm a$ along the real axis if $0 \leq \omega_1 < 2\pi$ and $-\pi < \omega_2 \leq +\pi$.

59. (a) With the notation of Problem 20, show that we may write

$$(z^2 - a^2)^{1/2} = (p_1 p_2)^{1/2} e^{i(\omega_1 + \omega_2)/2},$$

and deduce that transition from one branch to the other can take place only if $(\omega_1 + \omega_2)/2$ changes abruptly by an odd integral multiple of π.

(b) Show that, with the restrictions $0 \leq \omega_1 < 2\pi$ and $0 \leq \omega_2 < 2\pi$, the angle $(\omega_1 + \omega_2)/2$ has a jump of π over the finite segment between $-a$ and a, is continuous across the real axis to the left of $-a$, and has a jump of 2π across the real axis to the right of $+a$, if a is real and positive. Deduce that a cut between $-a$ and $+a$ then is necessary and sufficient to make the function so defined single-valued. Show also that the branch so defined takes on values which are real and of the same sign as z when z is real and $z^2 > a^2$.

(c) In a similar way, show that cuts are needed along infinite rays from the branch points $z = \pm a$ if the restrictions $0 \leq \omega_1 < 2\pi$ and $-\pi < \omega_2 \leq +\pi$ are imposed.

60. Prove that the functions

$$f_1(z) = \cos z^{1/2}, \qquad f_2(z) = \frac{\sinh az^{1/2}}{\sinh bz^{1/2}},$$

with $f_2(0) = a/b$, are analytic at $z = 0$.

61. Locate and classify the singularities of the following functions:

(a) $\dfrac{z}{z^2 + 1}$,

(b) $\dfrac{1}{z^3 + 1}$,

(c) $\log (1 + z^2)$,

(d) $(z^2 - 3z + 2)^{2/3}$,

(e) $\tan z$,

(f) $\tan^{-1} (z - 1)$.

62. Show that the function

$$f(z) = \frac{\cosh z - 1}{\sinh z - z}$$

has a simple pole at the origin.

63. Show that the function $f(z) = \csc (1/z)$ has poles at the points $z = 1/(n\pi)$, where n is any integer other than zero, and deduce that $f(z)$ has a nonisolated essential singularity at $z = 0$.

64. Show that the function $w = 1/(1 + z^{1/2})$ has a branch point at $z = 0$, and that if the principal branch of $z^{1/2}$ is chosen, with $-\pi < \theta_P \leq \pi$, there is no other finite singularity, whereas if the second branch is taken, with $\theta = \theta_P + 2\pi$, there is also a pole at $z = 1$. [Notice that we can write $w = (1 - z^{1/2})/(1 - z)$; also that $z^{1/2}$ is $+1$ when $z = 1$ on the principal branch and is -1 when $z = 1$ on the second branch.]

65. *An example of analytic continuation.* Suppose that a function $f(z)$ is defined by a specific series of the form

$$f(z) = A_0 + A_1 z + A_2 z^2 + \cdots$$

when $|z| < \sqrt{2}$ and that $f(z)$ is known to be analytic except at the points $z = \pm\sqrt{2}i$, but that nothing more is known about $f(z)$.

(a) If values of $f(z)$, $f'(z)$, $f''(z)$, ... were calculated from the series when $z = 1$, and were used to determine a new Taylor series in powers of $z - 1$, inside what circle would it converge?

(b) If the value of $f(5)$ were required, show that one additional series expansion (launched, say, from $z = 2.5$) would permit its computation.

(c) If, instead, the value of $f(1.5i)$ were required, how could it be obtained from the result of part (a)?

(A sketch displaying the circular regions of convergence for the successive series may be helpful.)

66. Suppose that $f(z)$ is defined by the series

$$f(z) = \sum_{n=0}^{\infty} z^{2^n} = z + z^2 + z^4 + z^8 + \cdots$$

when $|z| < 1$.

(a) Show that $f(z)$ has a singularity when $z = 1$.

(b) Show that $f(z^2) = f(z) - z$ and hence deduce that $f(z)$ has singularities when $z^2 = 1$.

(c) Show, more generally, that $f(z)$ has singularities at the m points (on the unit circle) for which $z^m = 1$, where m can be assigned any of the values $m = 1, 2, 2^2,$

$2^3, \ldots$. [From this fact it can be deduced that $f(z)$ is not analytic at any point on the circle $|z| = 1$.]

67. Using the convergence theorem stated on page 131, determine inside what real interval an infinite Frobenius series of the form

$$y(x) = \sum_{k=0}^{\infty} A_k x^{k+s},$$

satisfying each of the following differential equations, would converge [excluding the point $x = 0$ when Re $(s) < 0$]:

(a) $(1 - x^2)y'' - 2xy' + p(p + 1)y = 0$,

(b) $x(1 + x^2)y'' + y' + xy = 0$,

(c) $x^2 y'' + xy' + (x^2 - p^2)y = 0$,

(d) $y'' \sin x + y' \cos x + p(p + 1)y \sin x = 0$,

(e) $x(x^2 + 2x + 2)y'' - y' + (x + 1)y = 0$,

(f) $xe^x y'' + y \tan x = 0$.

Section 10.10

68. (a) If $f(z)$ is analytic at z_∞, show that the real and imaginary parts of $f(z)$ must each tend to constant limits (which may be zero) as $\sqrt{x^2 + y^2} \to \infty$ in any way, and that these limits must be independent of the manner in which this limiting process takes place.

(b) If $f(z)$ has a simple pole at z_∞, show that the preceding statement applies instead to the real and imaginary parts of $f'(z)$.

69. (a) Show that $|e^z| = e^x$. Hence deduce that $e^z \to 0$ if $z \to z_\infty$ on any curve along which $x \to -\infty$, that $|e^z| \to \infty$ along a curve for which $x \to +\infty$, and that along a curve with an asymptote parallel to the y axis $|e^z|$ tends to a finite limit whereas e^z does not.

(b) Obtain a corresponding result for each of the functions e^{-z}, e^{iz}, and e^{-iz}.

70. (a) Show that $|\sinh z| = \sqrt{\sinh^2 x + \sin^2 y}$. Hence deduce that $|\sinh z| \to \infty$ as $z \to z_\infty$ in such a way that $|x| \to \infty$, but that, if $z \to z_\infty$ such that $|y| \to \infty$ while x is bounded (for example, along a line parallel to the imaginary axis), $|\sinh z|$ is bounded but does not tend to a limit.

(b) Obtain a corresponding result for each of the functions $\cosh z$, $\sin z$, and $\cos z$.

71. Determine the nature of the point z_∞ for each of the following functions:

(a) z^2,

(b) $\dfrac{z}{z+1}$,

(c) $z \sin \dfrac{1}{z}$,

(d) $(1 + z)^{1/2}$,

(e) $(1 + z^2)^{1/2}$,

(f) $\log (1 + z)$.

72. Show that for the function $w = \sin^{-1} z + i \log z$ [see Equation (55a)] the point z_∞ is a branch point only on those branches for which $|w|$ is not bounded as $z \to z_\infty$.

73. By considering the function $f(z) = 1/z$, show that the integral $\oint_{C_\infty} f(z)\, dz$ may not vanish even though $f(z)$ is analytic at z_∞. (Notice that in this case C_∞ is equivalent to any other contour enclosing the origin.)

Section 10.11

74. Prove that any polynomial of degree N, $f(z) = a_0 + a_1 z + \cdots + a_N z^N$, has at least one zero unless it is constant. [Assume the contrary and apply Liouville's theorem to $F(z) = 1/f(z)$. This result is known as the *fundamental theorem of algebra* and is *assumed* in elementary courses.]

75. (a) If $f(z)$ has a pole of order n_1 at $z = \alpha_1$, show that

$$f(z) = (z - \alpha_1)^{-n_1} g_1(z),$$

where $g_1(z)$ is analytic at $z = \alpha_1$ and at all points where $f(z)$ is analytic.
(b) If $f(z)$ has a zero of order m_1 at $z = \beta_1$, show that

$$f(z) = (z - \beta_1)^{m_1} h_1(z),$$

where $h_1(z)$ is analytic at all points where $f(z)$ is analytic, and $h_1(\beta_1) \neq 0$.
(c) Deduce that *a function $f(z)$ which is analytic in the extended plane except for a finite number of poles is determined except for a multiplicative constant by the position and order of its poles and zeros in the finite part of the plane.*

76. *The argument principle.* Let $f(z)$ be analytic inside and on a simple closed curve C except for a finite number of poles inside C, and suppose that $f(z) \neq 0$ on C.
(a) Make use of the results of Problem 75 to show that

$$\frac{1}{2\pi i} \oint_C \frac{f'(z)}{f(z)} \, dz = M - N,$$

where M is the number of zeros of $f(z)$ inside C and N is the number of poles, poles or zeros of order k being counted k times.
(b) Show that this result also can be expressed in the form

$$M - N = \frac{1}{2\pi} [\Delta_C \arg f(z)],$$

where the bracketed expression denotes the change in the imaginary part of $\log f(z)$ corresponding to a counterclockwise circuit of C.

77. Use Cauchy's inequality (Problem 46) to give an alternative proof of Liouville's theorem. [Show that $f(z) = \sum_0^\infty a_n z^n$ and that $|a_n| \leq M/R^n$, where M is independent of R and where R may be increased without limit.]

Section 10.12

78. Calculate the residues of the following functions at each of the poles in the finite part of the plane:

(a) $\dfrac{e^z}{z^2 + a^2}$,

(b) $\dfrac{1}{z^4 - a^4}$,

(c) $\dfrac{\sin z}{z^2}$,

(d) $\dfrac{\sin z}{z^3}$,

(e) $\dfrac{1 + z^2}{z(z - 1)^2}$,

(f) $\dfrac{1}{(z^2 + a^2)^2}$,

(g) $\dfrac{e^{az}}{2z^2 - 5z + 2}$,

(h) $\dfrac{e^{z-1} - 1}{1 - z^2}$,

(i) $\dfrac{1 - \cos az}{z^9}$,

(j) $\dfrac{\sinh z}{\cosh z - 1}$,

(k) $\dfrac{z}{\sin^2 z}$,

(l) $\dfrac{(1 - \cos z)^2}{z^7}$.

79. If $f(z)$ has a pole of order m at $z = a$, prove that

$$\text{Res}\,(a) = \frac{1}{(M-1)!}\left[\frac{d^{M-1}}{dz^{M-1}}\{(z-a)^M f(z)\}\right]_{z=a}$$

for any positive integer M such that $M \geq m$. [Show that Res (a) is the coefficient of $(z-a)^{M-1}$ in the Taylor series expansion of $(z-a)^M f(z)$ about $z = a$, when $M \geq m$.]

80. (a) If $f(z)$ is that branch of $\log z$ for which $0 \leq \theta_P < 2\pi$, determine the sum of the residues of $f(z)/(z^2+1)$ at its poles.

(b) Proceed as in part (a) when the restriction on θ_P is $-\pi < \theta_P \leq \pi$.

81. (a) If $f(z)$ is that branch of the function $e^{az^{1/2}}$ for which $z^{1/2} = r^{1/2}e^{i\theta_P/2}$ with $0 \leq \theta_P < 2\pi$, determine the sum of the residues of $f(z)/(z^2+1)$ at its poles.

(b) Proceed as in part (a) when $-\pi < \theta_P \leq \pi$.

82. Evaluate the integral

$$\oint_C \frac{dz}{z^2 - 1}$$

when C is the curve sketched in Figure 10.21.

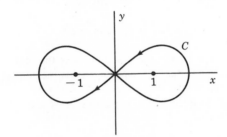

Figure 10.21

83. Show that the substitution $z = 1/t$ transforms the circle $|z| = R$ into the circle $|t| = 1/R$ in such a way that a positive circuit around one circle corresponds to a negative circuit around the other. [Write $z = Re^{i\theta}$, $t = \rho e^{i\varphi}$ and determine ρ, φ in terms of R, θ.]

84. (a) Use the result of Problem 83 to show that

$$\oint_{|z|=R} f(z)\,dz = \oint_{|t|=1/R} f\left(\frac{1}{t}\right)\frac{dt}{t^2}.$$

By letting $R \rightarrow \infty$, deduce that $\oint_{C_\infty} f(z)\,dz$ is given by $2\pi i$ times the residue of $(1/t^2)\,f(1/t)$ at $t = 0$.

(b) Show that this result implies the result of Problem 73.

(c) Use the result of part (a) to show that if $z^2 f(z)$ is analytic at z_∞, then $\oint_{C_\infty} f(z)\,dz = 0$.

85. *The residue at z_∞.* It is conventional to define the residue of $f(z)$ at z_∞, when z_∞ is an isolated singular point, by the equation

$$\oint_{C_\infty} f(z)\,dz = -2\pi i\,\text{Res}\,(z_\infty),$$

the negative sign corresponding to the fact that a positive circuit around C_∞ is described in a negative sense with respect to the *exterior* of the curve.

(a) Use the result of Problem 84 to show that then

$$\text{Res}\{f(z); z_\infty\} = -\text{Res}\left\{\frac{1}{t^2}f\left(\frac{1}{t}\right); 0\right\}.$$

(b) If $f(z)$ has only isolated singularities, use Equation (108) to show that, with the given definition of Res (z_∞), *the sum of the residues of $f(z)$ at all singularites in the finite part of the plane and at z_∞ is zero.* [Notice that $f(z)$ may have a nonzero residue at z_∞ even though $f(z)$ is analytic at z_∞, if $z^2 f(z)$ has a pole at z_∞.]

86. Use the result of Problem 84(a) (or Problem 85) to evaluate the integral

$$\oint_C \frac{a^2 - z^2}{a^2 + z^2}\frac{dz}{z},$$

where C is any simple closed contour enclosing the points $z = 0$, $\pm ia$, and check the result by calculating the residues at those poles.

87. (a) If $f(z)$ can be represented by a Laurent series

$$f(z) = \sum_{n=-\infty}^{\infty} c_n(z - a)^n$$

when $R < |z - a| < \infty$, for some R, show that c_{-1} is the sum of the residues of $f(z)$ at all singularities in the finite part of the plane, so that the number $-c_{-1}$ is the residue of $f(z)$ at z_∞. (See Problem 85.)

(b) By expanding the integrand in increasing powers of $1/z$, and identifying the coefficient of $1/z$, show that

$$\oint_C \frac{a^2 - z^2}{a^2 + z^2}\frac{dz}{z} = -2\pi i$$

if C is any simple closed curve enclosing the points $z = 0$ and $z = \pm ai$. (Compare Problem 86.)

88. Determine the residue of each of the following functions at each singularity:

(a) $e^{1/z}$, (b) e^{1/z^2},

(c) $\cos\dfrac{\pi}{z - \pi}$, (d) $(1 + z^2)e^{1/z}$.

89. Show that for the function

$$f(z) = e^{tz}e^{1/z}$$

it is true that

$$\text{Res}(0) = -\text{Res}(z_\infty) = \frac{1}{\sqrt{t}}I_1(2\sqrt{t}),$$

where I_1 is a modified Bessel function. [Identify the coefficient of z^{-1} in

$$\sum_{j=0}^{\infty}\sum_{k=0}^{\infty}\frac{t^j z^{j-k}}{j!\,k!}$$

and see Equation (95) of Chapter 4.]

Section 10.13

90. Use residue calculus to evaluate the following integrals:

(a) $\displaystyle\int_0^{2\pi} \frac{d\theta}{A + B\sin\theta} = \frac{2\pi}{\sqrt{A^2 - B^2}}$ $(A > |B|)$,

(b) $\displaystyle\int_0^{2\pi} \frac{d\theta}{a^2 + \sin^2\theta} = \int_0^{2\pi} \frac{d\theta}{a^2 + \cos^2\theta} = \frac{\pi}{a\sqrt{a^2 + 1}}$ $(a > 0)$,

(c) $\displaystyle\int_0^{\pi/2} \sin^4\theta \, d\theta = \int_0^{\pi/2} \cos^4\theta \, d\theta = \frac{3\pi}{16}$,

(d) $\displaystyle\int_0^{2\pi} \frac{\sin^2\theta}{5 + 4\cos\theta} \, d\theta = \frac{\pi}{4}$.

91. Use residue calculus to evaluate the following integrals:

(a) $\displaystyle\int_{-\infty}^{\infty} \frac{dx}{(x + b)^2 + a^2} = \frac{\pi}{a}$ $(a > 0)$,

(b) $\displaystyle\int_0^{\infty} \frac{dx}{(x^2 + a^2)(x^2 + b^2)} = \frac{\pi}{2ab(a + b)}$ $(a > 0, b > 0)$,

(c) $\displaystyle\int_0^{\infty} \frac{dx}{x^4 + 4a^4} = \frac{\pi}{8a^3}$ $(a > 0)$,

(d) $\displaystyle\int_0^{\infty} \frac{dx}{(x^2 + a^2)^2} = \frac{\pi}{4a^3}$ $(a > 0)$.

92. Use residue calculus to evaluate the following integrals:

(a) $\displaystyle\int_0^{\infty} \frac{x\sin mx}{a^2 + x^2} \, dx = \frac{\pi}{2}e^{-am}$ $(a > 0, m > 0)$,

(b) $\displaystyle\int_0^{\infty} \frac{\cos mx}{(x^2 + a^2)(x^2 + b^2)} \, dx = \frac{\pi}{2(b^2 - a^2)}\left(\frac{e^{-am}}{a} - \frac{e^{-bm}}{b}\right)$

$(a > 0, b > 0, m \geq 0, b \neq a)$,

(c) $\begin{cases} \displaystyle\int_{-\infty}^{\infty} \frac{\cos mx}{(x + b)^2 + a^2} \, dx = \frac{\pi}{a}e^{-am}\cos bm & (a > 0, m \geq 0), \\[3mm] \displaystyle\int_{-\infty}^{\infty} \frac{\sin mx}{(x + b)^2 + a^2} \, dx = -\frac{\pi}{a}e^{-am}\sin bm & (a > 0, m \geq 0), \end{cases}$

(d) $\displaystyle\int_0^{\infty} \frac{\cos mx}{(x^2 + a^2)^2} \, dx = \frac{\pi}{4a^3}e^{-am}(1 + am)$ $(a > 0, m \geq 0)$,

(e) $\displaystyle\int_0^{\infty} \frac{\cos mx}{x^4 + 4a^4} \, dx = \frac{\pi}{8a^3}e^{-am}(\cos am + \sin am)$ $(a > 0, m \geq 0)$,

(f) $\displaystyle\int_0^{\infty} \frac{x^3\sin mx}{x^4 + 4a^4} \, dx = \frac{\pi}{2}e^{-am}\cos am$ $(a > 0, m > 0)$.

93. (a) By differentiating the equal members in the first result of Problem 92(c) with respect to m and b, deduce the evaluations

$\displaystyle\int_{-\infty}^{\infty} \frac{x\sin mx}{(x + b)^2 + a^2} \, dx = \frac{\pi}{2a}e^{-am}(a\cos bm + b\sin bm)$ $(a > 0, m > 0)$,

$\displaystyle\int_{-\infty}^{\infty} \frac{\cos mx}{[(x + b)^2 + a^2]^2} \, dx = \frac{\pi}{4a^3}(1 + am)e^{-am}\cos bm$ $(a > 0, m \geq 0)$.

(b) Obtain a formula for the integral

$$\int_0^\infty \frac{\cos mx}{(x^4 + 4a^4)^2}\, dx$$

from the result of Problem 92(e).

94. Use residue calculus to show that

$$\int_0^{2\pi} \cos^{2n}\theta\, d\theta = \int_0^{2\pi} \sin^{2n}\theta\, d\theta = \frac{2\pi}{2^{2n}}\binom{2n}{n} = 2\pi \frac{1\cdot 3\cdot 5\cdots (2n-1)}{2\cdot 4\cdot 6\cdots (2n)},$$

where $\binom{2n}{n}$ is the coefficient of x^n in the expansion of $(1+x)^{2n}$.

95. Use residue calculus to show that

$$\int_{-\infty}^\infty \frac{x^2\, dx}{(x^2+1)(x^2-2x\cos\omega+1)} = \frac{\pi}{2|\sin\omega|}$$

if ω is real and $\sin\omega \neq 0$.

96. The Fourier transform of an unknown function $f(t)$ is known to be

$$\bar{f}(u) = \frac{1}{u^2 + a^2} \qquad (a>0).$$

Use residue calculus to determine the function f by use of the formula

$$f(t) = \frac{1}{2\pi}\int_{-\infty}^\infty \frac{e^{itx}}{x^2+a^2}\, dx \qquad (-\infty < t < \infty).$$

(See Section 5.15.)

97. Use residue calculus to show that

$$\int_{-\infty}^\infty \frac{\cos tx \cos cx}{(x+b)^2+a^2}\, dx = \frac{\pi}{2a}\{e^{-a|t-c|}\cos[b(t-c)]+ e^{-a|t+c|}\cos[b(t+c)]\}$$

when $a > 0$.

98. It is required to show that

$$I_1 \equiv \int_0^\infty \frac{e^{-xt}}{x^2+1}\, dx = \int_0^\infty \frac{\sin rt}{r+1}\, dr \equiv I_2 \qquad (t>0).$$

(a) Show that

$$I_1 = \frac{1}{2i}\left(\int_0^\infty \frac{e^{-xt}}{x-i}\, dx - \int_0^\infty \frac{e^{-xt}}{x+i}\, dx\right).$$

(b) By considering the limit of the integral of the function

$$f(z) = \frac{e^{-zt}}{z-i}$$

around the sector $0 \le r \le R$, $-\pi/2 \le \theta \le 0$, as $R \to \infty$, show that

$$\int_0^\infty \frac{e^{-xt}}{x-i}\, dx = \int_0^\infty \frac{e^{irt}}{r+1}\, dr$$

when $t > 0$. (Use Theorem II.4.)

(c) In a similar way, show that

$$\int_0^\infty \frac{e^{-xt}}{x+i} \, dx = \int_0^\infty \frac{e^{-irt}}{r+1} \, dr,$$

when $t > 0$, and hence complete the proof.

99. By writing $u = (r+1)t$ in the definition of I_2, show that the result of Problem 98 can be written, in terms of tabulated functions, in the form

$$\int_0^\infty \frac{e^{-xt}}{x^2+1} \, dx = \int_0^\infty \frac{\sin rt}{r+1} \, dr = \cos t \left[\frac{\pi}{2} - Si(t) \right] + \sin t \, Ci(t),$$

where

$$Si(t) = \int_0^t \frac{\sin u}{u} \, du, \qquad Ci(t) = -\int_t^\infty \frac{\cos u}{u} \, du$$

are the sine-integral and cosine-integral functions.

100. It is required to evaluate

$$\int_{-\infty}^\infty e^{-x^2} \cos 2ax \, dx,$$

by making use of the fact that

$$\int_{-\infty}^\infty e^{-x^2} \, dx = \sqrt{\pi}.$$

Show that, if x is formally replaced by $x + ia$ in the known integral, there follows

$$\int_{-\infty}^\infty e^{-(x+ia)^2} \, dx = e^{a^2} \int_{-\infty}^\infty e^{-x^2} (\cos 2ax - i \sin 2ax) \, dx = \sqrt{\pi},$$

and hence, since $e^{-x^2} \sin 2ax$ is an odd function of x,

$$\int_{-\infty}^\infty e^{-x^2} \cos 2ax \, dx = \sqrt{\pi} e^{-a^2}.$$

(Notice, however, that the validity of such complex substitutions is not established by the familiar rules for real substitutions.)

101. Investigate the validity of the procedure of Problem 100, by considering the integral of $f(z) = e^{-z^2}$ over a closed rectangular path C including the segment of the real axis $y = 0$ from $x = -A$ to $x = +A$ and the segment of the line $y = a$ from $x = +A$ to $x = -A$.

(a) Noticing that $f(z)$ is analytic inside and on C, show that

$$\int_{-A}^A e^{-x^2} \, dx - \int_{-A}^A e^{-(x+ia)^2} \, dx + \int_{S_1} e^{-z^2} \, dz + \int_{S_2} e^{-z^2} \, dz = 0,$$

where S_1 is the line segment from $(A, 0)$ to (A, a) and S_2 is the line segment from $(-A, a)$ to $(-A, 0)$.

(b) Show that, on both S_1 and S_2, there follows

$$|e^{-z^2}| = e^{-(A^2-y^2)} \le e^{-(A^2-a^2)}$$

and, by noticing that the lengths of S_1 and S_2 are equal to a, deduce that the integrals along S_1 and S_2 tend to zero as $A \longrightarrow \infty$ for any fixed value of a. Thus deduce that the relation

$$\int_{-\infty}^{\infty} e^{-x^2}\,dx = \int_{-\infty}^{\infty} e^{-(x+ia)^2}\,dx$$

is valid for any real value of a, and hence that the result obtained formally in Problem 100 is indeed correct.

102. Evaluate the integral

$$\int_{-\infty}^{\infty} \frac{e^{px}}{1 + e^x}\,dx \qquad (0 < p < 1)$$

by the following method:

(a) Show that the result of integrating $f(z) = e^{pz}/(1 + e^z)$ around a closed rectangular contour C, including the real axis from $x = -A$ to $x = +A$ and the line $y = 2\pi$ from $x = +A$ to $x = -A$, can be written in the form

$$\int_{-A}^{A} \frac{e^{px}}{1 + e^x}\,dx - e^{2p\pi i}\int_{-A}^{A} \frac{e^{px}}{1 + e^x}\,dx + \int_{S_1} f(z)\,dz + \int_{S_2} f(z)\,dz$$

$$= 2\pi i \operatorname{Res}\left\{\frac{e^{pz}}{1 + e^z}\,;\, \pi i\right\},$$

where S_1 and S_2 are the closing segments of the rectangle. (Notice that e^z has the period $2\pi i$, and that $f(z)$ has poles at the points $\pi i \pm 2k\pi i$.)

(b) Show that the integrals along S_1 and S_2 tend to zero as $A \longrightarrow \infty$ (if $0 < p < 1$), and hence deduce that

$$\int_{-\infty}^{\infty} \frac{e^{px}}{1 + e^x}\,dx = \left(\frac{2\pi i}{1 - e^{2p\pi i}}\right)(-e^{p\pi i}) = \frac{\pi}{\sin p\pi} \qquad (0 < p < 1).$$

103. Show that the formula obtained in Problem 102 also is valid when p is complex, provided that $0 < \operatorname{Re}(p) < 1$. (Only the integrals along S_1 and S_2 need be reexamined.)

Section 10.14

104. By making the substitution $x = iy$, and noticing that then $y \longrightarrow \infty$ as $x \longrightarrow \infty$, we formally transform the integral

$$\int_0^{\infty} \frac{x}{x^4 + 4}\,dx$$

into the integral

$$-\int_0^{\infty} \frac{y}{y^4 + 4}\,dy,$$

which is the negative of the original integral, and hence are led to the conclusion that the value of the integral is zero. But the true value is found by elementary methods to be $\pi/8$. *What is the fallacy?* [Integrate $f(z) = z/(z^4 + 4)$ around a contour consisting of the portions of the x and y axes for which $0 \le x \le R$ and $0 \le y \le R$, and a quadrant of the circle $|z| = R$, let $R \longrightarrow \infty$, and interpret the result.]

105. Establish part 3 of Theorem II.

106. If $f(z)$ is analytic everywhere on the imaginary axis and has no singularities to the left of that axis except for a finite number of poles at the points $a_1, a_2, \ldots, a_n$, and if $f(z)$ tends to zero uniformly on an arc of the circle $|z| = R$ in the second and third quadrants as $R \longrightarrow \infty$, show that

$$\lim_{R \to \infty} \int_{-iR}^{iR} e^{mz} f(z)\, dz = 2\pi i \sum_{k=1}^{n} \text{Res}\, \{e^{mz} f(z);\, a_k\} \qquad (m > 0),$$

where the integration is carried out along the imaginary axis.

107. Suppose that, on a circular arc C_R with radius R and center at the origin, lying in the sector $-\pi/4 \leqq \theta \leqq +\pi/4$, a function $f(z)$ is such that

$$|f(z)| \leqq RM_R, \qquad \lim_{R \to \infty} M_R = 0,$$

where M_R is independent of θ on C_R. Prove that

$$\lim_{R \to \infty} \int_{C_R} e^{-mz^2} f(z)\, dz = 0 \qquad (m > 0).$$

[If the integral along C_R is denoted by I_R, show that

$$|I_R| \leqq 2R^2 M_R \int_0^{\pi/4} e^{-mR^2 \cos 2\theta}\, d\theta,$$

and set $\theta = (\pi - 2\varphi)/4$.]

108. Use the result of Problem 107 to prove that if $|\beta| \leqq \pi/4$ and if $f(z)$ is analytic throughout the sector bounded by $\theta = 0$, $\theta = \beta$, and $r = R$, except for a finite number of interior poles, and if $|f(z)| \leqq RM_R$ on the curved boundary, where M_R is independent of θ and $M_R \longrightarrow 0$ as $R \longrightarrow \infty$, then

$$\int_0^\infty e^{-mx^2} f(x)\, dx = \int_{(\Gamma)}^\infty e^{-mz^2} f(z)\, dz + 2\pi i \sum_k \text{Res}\, \{e^{-mz^2} f(z);\, a_k\} \qquad (m > 0),$$

where Γ is the radial line $\theta = \beta$, and where the points a_k are the poles of $f(z)$ inside the sector bounded by $\theta = 0$ and $\theta = \beta$. [Notice that hence the formal complex substitution $x = te^{i\beta}$ in the first (real) integral generally would modify the value of the integral if $f(z)$ were not analytic in the relevant sector.]

109. (a) Show that the integrals

$$C \equiv \int_0^\infty \cos t^2\, dt, \qquad S \equiv \int_0^\infty \sin t^2\, dt$$

can be combined in the form

$$C - iS = \int_0^\infty e^{-it^2}\, dt = \int_0^\infty e^{-t^2 e^{i\pi/2}}\, dt = e^{-\pi i/4} \int_{(\Gamma)}^\infty e^{-z^2}\, dz,$$

where Γ is the radial line $\theta = \pi/4$.

(b) From the result of Problem 108, deduce that

$$C - iS = e^{-\pi i/4} \int_0^\infty e^{-x^2}\, dx = e^{-\pi i/4} \frac{\sqrt{\pi}}{2},$$

and hence that

$$\int_0^\infty \cos t^2\, dt = \int_0^\infty \sin t^2\, dt = \frac{1}{2}\sqrt{\frac{\pi}{2}}$$

and also

$$\int_0^\infty \frac{\cos x}{\sqrt{x}}\, dx = \int_0^\infty \frac{\sin x}{\sqrt{x}}\, dx = \sqrt{\frac{\pi}{2}}.$$

Section 10.15

110. By making use of integration around suitably indented contours in the complex plane, evaluate the following integrals:

(a) $\displaystyle\int_{-\infty}^\infty \frac{\sin x}{x(x^2 + a^2)}\, dx \qquad (a > 0),$

(b) $\displaystyle\int_{-\infty}^\infty \frac{\sin x}{x(\pi^2 - x^2)}\, dx.$

111. Show that

$$P\int_{-\infty}^\infty \frac{e^{itx}}{x}\, dx = \begin{cases} \pi i & (t > 0), \\ 0 & (t = 0), \\ -\pi i & (t < 0), \end{cases}$$

and hence also that

$$P\int_{-\infty}^\infty \frac{\cos tx}{x}\, dx = 0$$

and

$$\int_{-\infty}^\infty \frac{\sin tx}{x}\, dx = \begin{cases} \pi & (t > 0), \\ 0 & (t = 0), \\ -\pi & (t < 0). \end{cases}$$

112. The Fourier transform of an unknown function $f(t)$ is known to be

$$\bar{f}(u) = \frac{1 - e^{-iau}}{iu} \qquad (a > 0),$$

and hence (see Section 5.15) there follows

$$f(t) = \frac{1}{2\pi i}\int_{-\infty}^\infty e^{itx}\frac{1 - e^{-iax}}{x}\, dx \qquad (-\infty < t < \infty).$$

Show that

$$f(t) = \begin{cases} 0 & (t > a), \\ \frac{1}{2} & (t = a), \\ 1 & (0 < t < a), \\ \frac{1}{2} & (t = 0), \\ 0 & (t < 0). \end{cases}$$

[Write

$$f(t) = \frac{1}{2\pi i}\left[P\int_{-\infty}^\infty \frac{e^{itx}}{x}\, dx - P\int_{-\infty}^\infty \frac{e^{i(t-a)x}}{x}\, dx \right]$$

and use the result of Problem 111. See also Equations (243) and (246) in Chapter 5.]

113. The Fourier sine transform of an unknown function $f(t)$ is known to be

$$f_S(u) = \frac{1 - \cos au}{u} \qquad (a > 0),$$

and hence (see Section 5.15)

$$f(t) = \frac{2}{\pi} \int_0^\infty \frac{1 - \cos ax}{x} \sin tx \, dx \qquad (t > 0).$$

Show that

$$f(t) = \frac{1}{2\pi} \mathrm{Im} \left[P \int_{-\infty}^\infty \frac{2e^{itx}}{x} \, dx - P \int_{-\infty}^\infty \frac{e^{i(t+a)x}}{x} \, dx - P \int_{-\infty}^\infty \frac{e^{i(t-a)x}}{x} \, dx \right]$$

and consequently that

$$f(t) = \begin{cases} 1 & (0 < t < a), \\ \frac{1}{2} & (t = a), \\ 0 & (t > a). \end{cases}$$

[Use the result of Problem 111. See also Equations (243) and (246) in Chapter 5.]

114. The Fourier cosine transform of an unknown function $f(t)$ is known to be

$$f_C(u) = \frac{\sin au}{u} \qquad (a > 0).$$

By proceeding as in Problem 113, show that $f(t)$ has the same definition here as in that problem. [See also Equations (243) and (246) in Chapter 5.]

115. (a) Suppose that $f(z)$ has a simple pole at a_0 on a simple closed curve C, but is analytic elsewhere inside and on C except for poles at a finite number of interior points $a_1, a_2, \ldots, a_n$. If the contour C is indented at a_0 by a circular arc with center at a_0, show that the limiting form of the integral of $f(z)$ around the indented contour is

$$P \oint_C f(z) \, dz = \pi i \, \mathrm{Res}\,(a_0) + 2\pi i \sum_{k=1}^n \mathrm{Res}\,(a_k)$$

as the radius of the indentation tends to zero, regardless of whether the indentation excludes or includes the point a_0, provided that C is *smooth* at a_0.

(b) What form results in part (a) when C is the boundary of the sector $0 \leq r \leq 1$, $0 \leq \theta \leq \alpha$ and $a_0 = 0$?

116. Obtain the evaluation

$$\int_{-\infty}^\infty \frac{\cos ax - \cos bx}{x^2} \, dx = \pi(b - a).$$

[Notice that $f(z) = (e^{iaz} - e^{ibz})/z^2$ has a simple pole at the origin.] By taking $a = 0$ and $b = 2$, also deduce the formula

$$\int_{-\infty}^\infty \frac{\sin^2 x}{x^2} \, dx = \pi.$$

117. Obtain the evaluation

$$\int_{-\infty}^{\infty} \frac{\sin(x+a)\sin(x-a)}{x^2 - a^2}\, dx = \frac{\pi}{2a} \sin 2a.$$

[Notice that $\sin(x+a)\sin(x-a) = \frac{1}{2}(\cos 2a - \cos 2x)$.]

118. (a) Obtain the evaluation

$$P \int_{-\infty}^{\infty} \frac{e^{px}}{1 - e^x}\, dx = \pi \cot p\pi \qquad (0 < p < 1),$$

by proceeding as in Problem 102, but also introducing indentations at the poles $z = 0$ and $z = 2\pi i$.

(b) Deduce the result

$$\int_{-\infty}^{\infty} \frac{e^{px} - e^{qx}}{1 - e^x}\, dx = \pi(\cot p\pi - \cot q\pi)$$

if $0 < p < 1$ and $0 < q < 1$. (Notice that this integral is convergent, so that principal values are not involved.)

(c) By replacing x by $2x$ and writing $p = (\omega + 1)/2$ in the result of part (a), deduce that

$$P \int_{-\infty}^{\infty} \frac{e^{\omega x}}{\sinh x}\, dx = \pi \tan \frac{\pi\omega}{2} \qquad (-1 < \omega < 1).$$

(d) By replacing x by bx, where $b > 0$, and writing $\omega = a/b$, obtain the result

$$P \int_{-\infty}^{\infty} \frac{e^{ax}}{\sinh bx}\, dx = \frac{\pi}{b} \tan \frac{\pi a}{2b} \qquad (b > |a|).$$

(e) Deduce that

$$\int_{-\infty}^{\infty} \frac{\sinh ax}{\sinh bx}\, dx = \frac{\pi}{b} \tan \frac{\pi a}{2b} \qquad (b > |a|).$$

Section 10.16

119. Use the result of Problem 58, Chapter 2, to evaluate the integral considered in text Example 2 in the alternative form

$$\int_{0}^{\infty} \frac{x^{m-1}}{x + 1}\, dx = \Gamma(m)\Gamma(1 - m) \qquad (0 < m < 1),$$

and hence, by comparing the results of the two calculations, deduce the relation

$$\Gamma(m)\Gamma(1 - m) = \frac{\pi}{\sin m\pi} \qquad (0 < m < 1).$$

[This relation was stated without proof in Equation (59), Chapter 2. Although the present proof is valid only when $0 < m < 1$, the relation is generalized to all non-integral values of m by making use of the recurrence formula for the Gamma function.]

120. Use the method of text Example 2 to show that

$$P \int_{0}^{\infty} \frac{dx}{x^m(1 - x)} = -\pi \cot m\pi \qquad (0 < m < 1).$$

[Indent the upper and lower banks of the cut to exclude the pole at $z = 1$ and deduce that

$$(1 - e^{-2m\pi i})I = \pi i[\text{Res}\,(e^{0i}) + \text{Res}\,(e^{2\pi i})],$$

where I is the required integral.

121. Suppose that $z^m f(z)$ tends uniformly to zero on the circle $|z| = R$ as $R \to \infty$, where $m > 0$, and that $f(z)$ is analytic except for a finite number of poles $a_1, a_2, \cdots,$ a_n, none of which is on the *positive* real axis or at the origin. By proceeding as in text Example 2, show that

$$\int_0^\infty x^{m-1} f(x)\, dx = \frac{\pi e^{-i(m-1)\pi}}{\sin m\pi} \sum_{k=1}^n \text{Res}\,\{z^{m-1} f(z); a_k\},$$

where $z^{m-1} = r^{m-1} e^{i(m-1)\theta}$, with $0 < \theta < 2\pi$, when $z = re^{i\theta}$.

122. Use the result of Problem 121 to obtain the following evaluations:

(a) $\displaystyle\int_0^\infty \frac{x^{m-1}}{x^2 + 1}\, dx = \pi \frac{\sin\,[(1 - m)\pi/2]}{\sin m\pi} = \frac{\pi}{2 \sin m\pi/2}$ $\quad (0 < m < 2),$

(b) $\displaystyle\int_0^\infty \frac{x^{m-1}}{(x + 1)^2}\, dx = \frac{(1 - m)\pi}{\sin m\pi}$ $\quad (0 < m < 2).$

123. (a) Verify that the relation

$$t = \frac{bx + a}{x + 1}, \qquad x = \frac{t - a}{b - t}$$

transforms the finite interval $a \le t \le b$ into the semi-infinite interval $0 \le x < \infty$.

(b) Use the result of part (a) to show that

$$\int_{-1}^1 \left(\frac{1 + t}{1 - t}\right)^{m-1} g(t)\, dt = \int_0^\infty x^{m-1} f(x)\, dx,$$

where

$$f(x) = \frac{2}{(x + 1)^2} g\left(\frac{x - 1}{x + 1}\right),$$

when the integrals exist.

(c) Show similarly that

$$\int_{-1}^1 (1 - t^2)^{m-1} h(t)\, dt = \int_0^\infty x^{m-1} f(x)\, dx,$$

where

$$f(x) = \frac{1}{2}\left(\frac{2}{x + 1}\right)^{2m} h\left(\frac{x - 1}{x + 1}\right),$$

when the integrals exist.

124. Use the results of Problems 122 and 123(b) to deduce the following formulas:

(a) $\displaystyle\int_{-1}^1 \left(\frac{1 + t}{1 - t}\right)^{m-1} dt = \frac{2(1 - m)\pi}{\sin m\pi}$ $\quad (0 < m < 2),$

(b) $\displaystyle\int_{-1}^1 \left(\frac{1 + t}{1 - t}\right)^{m-1} \frac{dt}{t^2 + 1} = \frac{\pi}{2 \sin m\pi/2}$ $\quad (0 < m < 2).$

125. Use the result of Problem 123(c) to obtain the evaluation

$$\int_{-1}^{1} \frac{\sqrt{1 - t^2}}{t^2 + 1} \, dt = \pi(\sqrt{2} - 1).$$

126. Consider the integral

$$\oint_C f(z) \log z \, dz,$$

where C is the contour indicated in Figure 10.20, $f(z)$ is analytic except for a finite number of poles, none of which is on the positive real axis or at the origin, and the branch of $\log z$ for which

$$\log z = \log r + i\theta \qquad (0 < \theta < 2\pi)$$

is to be employed. If $z^{1+c} f(z) \longrightarrow 0$ uniformly on C_R as $R \longrightarrow \infty$, for some positive constant c, show that in the limit as $R \longrightarrow \infty$ and $\rho \longrightarrow 0$ the integral becomes

$$\int_0^\infty f(x) \log x \, dx - \int_0^\infty f(x) (\log x + 2\pi i) \, dx,$$

and hence deduce that

$$\int_0^\infty f(x) \, dx = -\sum_k \text{Res} \{f(z) \log z; a_k\},$$

where the points a_k are the poles of $f(z)$. (Note that $\rho \log \rho \longrightarrow 0$ as $\rho \longrightarrow 0$ and that $R^{-c} \log R \longrightarrow 0$ as $R \longrightarrow \infty$ if $c > 0$. *Care must be taken in using the specified branch of $\log z$ when the residues are evaluated.*)

127. Use the result of Problem 126 to evaluate the following integrals:

(a) $\displaystyle\int_0^\infty \frac{dx}{(x + a)(x^2 + b^2)} = \frac{\pi a + 2b \log (b/a)}{2b(a^2 + b^2)}$ $(a > 0, b > 0)$,

(b) $\displaystyle\int_0^\infty \frac{dx}{x^3 + a^3} = \frac{2\pi\sqrt{3}}{9a^2}$ $(a > 0)$.

128. Proceed as in Problem 126 with the integral

$$\oint_C f(z)(\log z)^2 \, dz,$$

showing that in the limit this integral becomes

$$\int_0^\infty f(x)(\log x)^2 \, dx - \int_0^\infty f(x)(\log x + 2\pi i)^2 \, dx,$$

and hence deducing that

$$\int_0^\infty f(x) \log x \, dx = -\frac{1}{2} \text{Re} \left[\sum_k \text{Res} \{f(z)(\log z)^2 ; a_k\} \right]$$

and

$$\int_0^\infty f(x) \, dx = -\frac{1}{2\pi} \text{Im} \left[\sum_k \text{Res} \{f(z) (\log z)^2 ; a_k\} \right],$$

where the points a_k are the poles of $f(z)$. (The *second* result generally is less convenient than the formula obtained in Problem 126.)

129. Use the first result of Problem 128 to evaluate the following integrals:

(a) $\displaystyle\int_0^\infty \frac{\log x}{x^2 + a^2}\, dx = \frac{\pi}{2a} \log a \qquad (a > 0),$

(b) $\displaystyle\int_0^\infty \frac{\log x}{(x + a)(x + b)}\, dx = \frac{\log (b/a)\log (ab)}{2(b - a)} \qquad (b > a > 0),$

(c) $\displaystyle\int_0^\infty \frac{\log x}{(x + a)^2}\, dx = \frac{\log a}{a} \qquad (a > 0).$

130. Use the first result of Problem 128 to show that

$$\int_0^\infty \frac{\log x}{(x + b)^2 + a^2}\, dx = \frac{\varphi}{a} \log \rho \qquad (a > 0),$$

where $\rho = \sqrt{a^2 + b^2}$ and $\varphi = \cos^{-1}(b/\rho) = \pi/2 - \tan^{-1}(b/a)$. (Write $b + ia = \rho e^{i\varphi}$, where $0 < \varphi < \pi$ since $a > 0$. Then take care when writing $-b \pm ia$ in corresponding forms for the purpose of calculating the residues at those points.)

131. (a) Show that

$$\int_{-1}^1 \log \left(\frac{1 + t}{1 - t}\right) g(t)\, dt = \int_0^\infty f(x) \log x\, dx,$$

where $$f(x) = \frac{2}{(x + 1)^2} g\left(\frac{x - 1}{x + 1}\right).$$

[See Problem 123(a).]

(b) Use the results of part (a) and Problem 129(b) to obtain the evaluation

$$\int_{-1}^1 \log \left(\frac{1 + t}{1 - t}\right) \frac{dt}{1 - ct} = \frac{1}{2c}\left[\log \left(\frac{1 + c}{1 - c}\right)\right]^2 \qquad (|c| < 1).$$

11

Applications of Analytic Function Theory

11.1. Introduction. This chapter relates some of the results obtained in Chapter 10, relative to properties of analytic functions of a complex variable, to considerations of earlier chapters, and is principally focused on methods for dealing with problems governed by partial differential equations.

The first two sections apply contour integration and residue calculus to the determination of inverse Laplace transforms, and hence complement the treatments of Chapters 2 and 9 dealing with the use of Laplace transform methods in the solution of appropriate types of initial-value problems.

In the five following sections the possibility of considering an analytic function of $z = x + iy$ as effecting a mapping of a region in the xy plane onto a region in a new plane, and the distinctive properties of such a mapping, are exploited for the purpose of solving a variety of problems whose formulation involves the Laplacian operator.

The three final sections define the so-called Green's function associated with a two-dimensional region and with a class of problems specified over that region, relates that function to a certain analytic mapping function in important special cases, and illustrates the determination and application of the Green's function in some specific instances.

11.2. Inversion of Laplace Transforms. Whereas Equations (240a, b) of Chapter 5 permit the determination of both the Fourier transform of a given function and the inverse of a given transform, when the relevant functions are sufficiently well behaved, no direct method was given in Chapter 2 for the determination of the inverse Laplace transform. We now use the former relations to derive a formula for this purpose.

Equation (238a) of Chapter 5, which is equivalent to (240a, b) of that chapter, is valid if f is piecewise differentiable and if $\int_{-\infty}^{\infty} |f(t)|\, dt$ exists, but usually is not valid if that integral fails to exist. In order to obtain a relation which is more useful for present purposes, we first suppose that $f(x)$ vanishes for all negative x,

$$f(x) = 0 \qquad \text{when } x < 0, \tag{1}$$

and then replace $f(x)$ by $e^{-cx}f(x)$ in (238a), writing the result in the form

$$e^{-cx}f(x) = \frac{1}{2\pi} \lim_{R \to \infty} \int_{-R}^{R} e^{iux}\left[\int_{0}^{\infty} e^{-iut}e^{-ct}f(t)\, dt \right] du, \tag{2}$$

and correspondingly assuming now that c is a real constant such that

$$\int_{0}^{\infty} e^{-ct} |f(t)|\, dt \quad exists. \tag{3}$$

Hence we deduce the representation

$$f(x) = \frac{1}{2\pi} \lim_{R \to \infty} \int_{-R}^{R} e^{(iu+c)x}\left[\int_{0}^{\infty} e^{-(iu+c)t}f(t)\, dt \right] du, \tag{4}$$

when $x > 0$, of any piecewise differentiable function $f(x)$ for which the real constant c can be chosen so that (3) holds, and with the usual convention that, at a finite jump, $f(x)$ is to be assigned the average of its right- and left-hand limits.

This representation is more general than the Fourier integral representation, for a function $f(x)$ which vanishes when $x < 0$, since it permits $|f(x)|$ to grow exponentially as $x \to \infty$, whereas the latter representation generally requires that $|f(x)|$ tend fairly rapidly to zero.

The inner integral in (4) becomes the *Laplace* transform of $f(t)$ if $iu + c$ is replaced by s. Whereas this replacement is merely a change of notation for the inner integral, it corresponds to a change in the variable of integration for the outer one, leading to the equivalent representation

$$f(x) = \frac{1}{2\pi i} \lim_{R \to \infty} \int_{c-iR}^{c+iR} e^{sx}\left[\int_{0}^{\infty} e^{-st}f(t)\, dt \right] ds. \tag{5}$$

Thus if we here denote the Laplace transform of $f(t)$ by $F(s)$,

$$F(s) \equiv \mathcal{L}\{f(t)\} = \int_{0}^{\infty} e^{-st}f(t)\, dt, \tag{6}$$

then (5) states that

$$f(x) = \frac{1}{2\pi i} \lim_{R \to \infty} \int_{c-iR}^{c+iR} e^{xs}F(s)\, ds,$$

where, since the real variable u increases from $-R$ to R in (4), the complex variable $s = c + iu$ is such that its real part is constantly c while its imaginary part increases from $-R$ to R. If, for convenience, we replace the free variable

x by t and the complex dummy variable s by z in this relation, it takes the form

$$f(t) \equiv \mathcal{L}^{-1}\{F(s)\} = \frac{1}{2\pi i} \lim_{R\to\infty} \int_{c-Ri}^{c+Ri} e^{tz} F(z)\, dz, \qquad (7)$$

where now the integration is along the line $x = c$ in the z plane, indicated in Figure 11.1(a), and where c is assumed to be sufficiently large to ensure the existence of the real integral $\int_0^\infty e^{-ct}|f(t)|\, dt$.

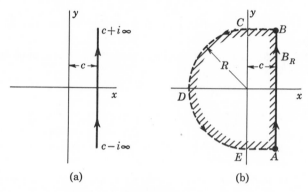

(a)   (b)

Figure 11.1

The right-hand member of (7), specifying the inverse Laplace transform of $F(s)$, is known as the *Bromwich integral*, and often is abbreviated in the form

$$f(t) = \frac{1}{2\pi i} \int_{c-i\infty}^{c+i\infty} e^{tz} F(z)\, dz \equiv \frac{1}{2\pi i} \int_B e^{tz} F(z)\, dz, \qquad (7')$$

where B is the infinite Bromwich line. Generally, it must be evaluated by use of residue calculus, by methods next to be illustrated.

First, we suppose that $F(z)$ *is analytic except for a finite number of poles, all of which lie to the left of the line $x = c$.* In this case we form the closed contour indicated in Figure 11.1(b), consisting of the part B_R of the Bromwich line from $y = -R$ to $y = R$, segments BC and EA parallel to the real axis, and the semicircle CDE. In addition we suppose that $F(z)$ *tends uniformly to zero along the closure $BCDEA$ as $R \to \infty$,* so that $|F(z)| \leq K_R$ on all of $BCDEA$, where K_R depends only on R and where $K_R \to 0$ as $R \to \infty$.

When we consider the integral

$$I \equiv \frac{1}{2\pi i} \int_{ABCDEA} e^{tz} F(z)\, dz,$$

we see that the contribution on AB tends to $f(t)$ as $R \to \infty$, under the assumption that c is also so large that (3) is true. Further, the contribution on CDE tends to zero by virtue of Theorem II.3 of Section 10.14. Along BC we may write $z = x + Ri$, as x varies from c to 0; we then have

$$|e^{tz}| = |e^{t(x+Ri)}| = e^{tx} \leq e^{tc},$$

and since $|F(z)| \leq K_R$ on BC (where $K_R \to 0$ as $R \to \infty$), and also the length of the path BC is constantly c, the ML inequality (80) of Chapter 10 gives

$$\left| \int_{BC} e^{tz} F(z) \, dz \right| \leq ce^{tc} K_R,$$

so that the contribution on BC tends to zero as $R \to \infty$. A similar argument applies to EA.

Finally, since ultimately all poles of $F(z)$ are inside the contour $ABCDEA$, as $R \to \infty$, we deduce that when the preceding conditions are satisfied by $F(z)$, there follows

$$f(t) \equiv \mathcal{L}^{-1}\{F(s)\} = \sum_k \text{Res}\,\{e^{tz}F(z); a_k\}, \tag{8}$$

where the points a_k are the poles of $F(z)$.

Since this function clearly is of exponential order, there assuredly *exists* a value of c for which (3) is true. We have *assumed* that the value of c used in the derivation of (8) is also sufficiently large for this purpose. If this were not the case, then resolving the contradiction by increasing c would not change the result, since no additional residues exist to be introduced into (8). Hence the assumption is irrelevant to the outcome and (8) is valid.

Example. As a simple illustration, we make the determination

$$\mathcal{L}^{-1}\left\{\frac{s}{s^2+1}\right\} = \text{Res}\left\{\frac{ze^{tz}}{z^2+1}; i\right\} + \text{Res}\left\{\frac{ze^{tz}}{z^2+1}; -i\right\}$$

$$= \frac{1}{2}e^{it} + \frac{1}{2}e^{-it}$$

$$= \cos t,$$

in accordance with formula (T16) in the table of transforms, page 67. ∎

11.3. Inversion of Laplace Transforms with Branch Points. The Loop Integral. In order to illustrate the evaluation of the Bromwich integral when $F(z)$ has one or more branch points, we suppose next that $F(z)$ *satisfies the conditions imposed in the preceding section, except that it also possesses a single branch point, at $z = 0$, and has no poles on the negative real axis.*

We then cut the z plane along the negative real axis and form the closed contour of Figure 11.2(a), containing the excursion DOE from the contour of Figure 11.1(b), inward along the upper bank of the cut, around a circle C_ϵ of radius ϵ about the origin, and outward along the lower bank of the cut. Here ϵ is taken sufficiently small to exclude all poles from C_ϵ. Correspondingly, we restrict attention to that branch of the multivalued function $F(z)$ for which the angle θ associated with z is such that $-\pi < \theta < \pi$, so that θ tends to $\pm\pi$ as the upper and lower banks of the cut are approached, and we write $F(re^{i\pi})$ and $F(re^{-i\pi})$ for the limits of $F(re^{i\theta})$ as θ tends to π from smaller values and to $-\pi$ from larger values, respectively.

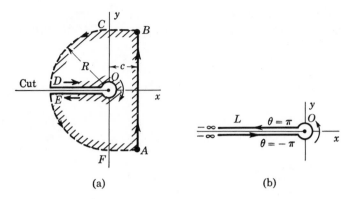

Figure 11.2

Finally, we suppose that $zF(z)$ *tends uniformly to a constant A (which may be zero) on the circle $|z| = \epsilon$, as $\epsilon \to 0$.*

Then, if we consider the integral

$$\frac{1}{2\pi i} \oint_C e^{tz} F(z)\, dz,$$

where C is the closed contour $ABCDOEFA$, and first let $R \to \infty$, so that the contribution on AB again tends to $f(t)$ and the contributions on BCD and EFA again tend to zero, we deduce that

$$f(t) = \sum_k \operatorname{Res}\{e^{tz} F(z); a_k\} + \frac{1}{2\pi i} \int_L e^{tz} F(z)\, dz, \qquad (9)$$

where the "loop integral" is taken along the "loop" L indicated in Figure 11.2(b). Here the orientation of L has been taken as the *negative* of that of DOE, in order that the sign prefixed to the transposed integral in (9) be positive.

On the lower bank of the cut we write $z = re^{-i\pi}$, on the ϵ-circle $z = \epsilon e^{i\theta}$, and on the upper bank of the cut $z = re^{i\pi}$, so that

$$\int_L e^{tz} F(z)\, dz = \int_\infty^\epsilon e^{tre^{-i\pi}} F(re^{-i\pi})(e^{-i\pi}\, dr)$$

$$+ \int_\epsilon^\infty e^{tre^{i\pi}} F(re^{i\pi})(e^{i\pi}\, dr) + i \int_{-\pi}^\pi e^{t\epsilon e^{i\theta}}(\epsilon e^{i\theta}) F(\epsilon e^{i\theta})\, d\theta.$$

Only with reference to the multivalued function $F(z)$ is there need to distinguish between $e^{i\pi}$ and $e^{-i\pi}$ (as indicators of limiting processes involving $e^{i\theta}$ as $\theta \to \pi-$ or $-\pi+$). Hence we can write

$$\int_L e^{tz} F(z)\, dz = \int_\epsilon^\infty e^{-tr}[F(re^{-i\pi}) - F(re^{i\pi})]\, dr + i \int_{-\pi}^\pi e^{t\epsilon e^{i\theta}}(\epsilon e^{i\theta}) F(\epsilon e^{i\theta})\, d\theta, \qquad (10)$$

where ϵ is any sufficiently small positive number.

In the general case, it may happen that *neither* of the two integrals on the right in (10) tends to a finite limit as $\epsilon \to 0$, but only that their *sum* does so. However, if it is true that $zF(z)$ tends uniformly to a certain constant A on C_ϵ

as $\epsilon \rightarrow 0$, we see that the integrand of the θ integral itself tends to A and also that the integral then tends to $2\pi i A$. Accordingly, the r integral also must tend to a limit as $\epsilon \rightarrow 0$ in this case. The introduction of the limiting form of (10) into (9) is then the desired formula

$$f(t) \equiv \mathcal{L}^{-1}\{F(s)\} = \sum_k \text{Res}\,\{e^{tz}F(z); a_k\} + \frac{1}{2\pi i}\int_0^\infty e^{-tr}[F(re^{-i\pi}) - F(re^{i\pi})]\,dr + A, \tag{11}$$

where
$$A = \lim_{z=0} zF(z) \tag{12}$$

and where again the points a_k are the poles of $F(z)$ [or, more specifically, of the *branch* of $F(z)$ under consideration].

Example. We seek the inverse Laplace transform of the function

$$F(s) = s^{-m} \qquad (0 < m < 1). \tag{13}$$

In this case we verify easily that $F(z)$ satisfies all the specified conditions, so that (11) is indeed applicable. Here the inverse transform is given entirely by the loop integral. That is, the infinite Bromwich line and the loop L of Figure 11.2(b) are *equivalent* contours in this case. With $A = 0$, there then follows

$$\mathcal{L}^{-1}\{s^{-m}\} = \frac{1}{2\pi i}\int_0^\infty e^{-tr}[(re^{-i\pi})^{-m} - (re^{i\pi})^{-m}]\,dr$$

$$= \frac{\sin \pi m}{\pi}\int_0^\infty e^{-tr}r^{-m}\,dr \qquad (0 < m < 1). \tag{14}$$

Frequently, in similar situations, it is not possible to proceed beyond this stage, in which the inverse transform is expressed in terms of a certain *real* integral, because of the fact that the integral cannot be expressed in closed form in terms of tabulated functions. (In such cases the integral must be evaluated approximately, in practice, by numerical methods.) Here, however, the substitution $tr = u$ yields the evaluation

$$\int_0^\infty e^{-tr}r^{-m}\,dr = t^{m-1}\int_0^\infty e^{-u}u^{-m}\,du = t^{m-1}\Gamma(1 - m),$$

so that also

$$\mathcal{L}^{-1}\{s^{-m}\} = \frac{\Gamma(1 - m)\sin \pi m}{\pi}t^{m-1} \tag{15}$$

and, indeed, the use of Equation (59) of Chapter 2 simplifies this result to the final form

$$\mathcal{L}^{-1}\{s^{-m}\} = \frac{t^{m-1}}{\Gamma(m)} \equiv \frac{t^{m-1}}{(m - 1)!}, \tag{16}$$

in accordance with formula (T35) in Table 1, page 68.

Whereas we have established this result here only when $0 < m < 1$, it is valid also for all m such that $m \geqq 1$. When $m = 1$, or when m is any other positive integer, the result is very easily derived by the method of Section 11.2, since then $F(z)$ possesses only a *pole*. However, the situation when $m > 1$ and m is nonintegral is one of those noted previously, in which both integrals on the right in (10) diverge as $\epsilon \rightarrow 0$ but their *sum* converges, and the derivation is correspondingly much less simple. ∎

Problems 15 and 19 illustrate the treatment of situations in which the transform function $F(z)$ possesses two or more branch points. In such cases the "loop integral" appears in different forms, as an integral which is equivalent to the Bromwich integral apart, perhaps, from the contributions of certain *poles* of $F(z)$.

11.4. Conformal Mapping. In order to represent geometrically the interpretation of a functional relationship $w = f(z)$, it is conventional to employ two planes, in one of which (the z plane) the real and imaginary parts (x and y) of the independent variable z are plotted as the point (x, y), and in the second of which (the w plane) the real and imaginary parts (u and v) of the dependent variable w are plotted as the point (u, v). In this way a correspondence is set up, in general, between points, curves, and regions in one plane and their images in the other plane. We speak of such a correspondence as a *mapping* between the two planes.

If the function $f(z)$ is *single-valued*, then corresponding to each point z where $f(z)$ is defined there ex;sts one and only one value of $w = u + iv$, and hence one and only one point in the w plane. Otherwise, to a given point z there will, in general, correspond two or more points in the w plane. Usually, if $f(z)$ is multiple-valued, we introduce suitable cuts in the z plane in such a way that a given branch of $f(z)$ is single-valued in the cut plane. Then u and v are single-valued functions of x and y. However, the reverse may not be true, in the sense that two or more values of $z = x + iy$ may correspond to the same point $w = u + iv$. To investigate this possibility, we notice that if we write

$$w = f(z) = f_1(x, y) + if_2(x, y) = u + iv, \qquad (17)$$

then there follows

$$u = f_1(x, y), \qquad v = f_2(x, y). \qquad (18)$$

Thus, to determine x and y in terms of u and v, we must solve Equations (18) for x and y. According to the result of Section 7.3, these equations can be solved uniquely for x and y in some region about any point where (17) holds and where the Jacobian determinant

$$J = \frac{\partial(f_1, f_2)}{\partial(x, y)} = \begin{vmatrix} \dfrac{\partial f_1}{\partial x} & \dfrac{\partial f_1}{\partial y} \\ \dfrac{\partial f_2}{\partial x} & \dfrac{\partial f_2}{\partial y} \end{vmatrix} = \begin{vmatrix} \dfrac{\partial u}{\partial x} & \dfrac{\partial u}{\partial y} \\ \dfrac{\partial v}{\partial x} & \dfrac{\partial v}{\partial y} \end{vmatrix}$$

is not zero. If $f(z)$ is *analytic* at $z_0 = x_0 + iy_0$ (and hence in a region including z_0), the Cauchy–Riemann equations are satisfied at z_0, and hence there then follows, from Equations (64) and (65) of Chapter 10,

$$[J]_{z_0} = \left[\frac{\partial u}{\partial x}\frac{\partial v}{\partial y} - \frac{\partial u}{\partial y}\frac{\partial v}{\partial x} \right]_{z_0} = \left[\left(\frac{\partial u}{\partial x}\right)^2 + \left(\frac{\partial v}{\partial x}\right)^2 \right]_{z_0} = |f'(z_0)|^2. \qquad (19)$$

Thus we conclude that, *if $f(z)$ is analytic at a point z_0 and if $f'(z_0) \neq 0$, then there exists a region including z_0 in the z plane and a region including $w_0 = f(z_0)$*

in the w plane such that the mapping $w = f(z)$ gives a one-to-one correspondence between points in the two regions.

To illustrate such a mapping, we consider the mapping function

$$w = f(z) = z^{1/2}. \tag{20}$$

Since $f(z)$ is double-valued, it follows that to each point $z = x + iy$ except the origin there correspond *two* points in the w plane. However, since the *inverse function*

$$z = F(w) = w^2 \tag{21}$$

is single-valued, we see that to each point $w = u + iv$ there will correspond a *unique* point in the original z plane. To make the mapping one-to-one, we may cut the z plane along the negative real axis, as in Section 10.9, and consider the *principal branch* of $z^{1/2}$, for which $z^{1/2}$ is real and positive when z is real and positive,[†]

$$w = f(z) = z^{1/2} = \sqrt{|z|}\, e^{i\theta_P/2} \qquad (-\pi < \theta_P < \pi). \tag{22}$$

To investigate the nature of the mapping, we may introduce polar coordinates in the two planes by writing

$$z = x + iy = re^{i\theta}, \qquad w = u + iv = \rho e^{i\varphi}. \tag{23}$$

Then (20) becomes $\rho e^{i\varphi} = \sqrt{r}\, e^{i\theta/2}$, from which there follows

$$\rho = \sqrt{r}, \qquad \varphi = \frac{\theta}{2}. \tag{24}$$

We see that as θ varies from $-\pi$ to π, the angle φ varies from $-\pi/2$ to $\pi/2$. Hence the cut z plane is mapped, by the branch of $w = z^{1/2}$ which we have chosen, onto that half of the w plane for which $u > 0$. Equation (24) is convenient for plotting the image in one plane of a given *point* in the other plane, or for mapping curves expressed in polar coordinates.

If (21) is expressed in terms of real and imaginary parts, there follows

$$x + iy = (u + iv)^2 = (u^2 - v^2) + i(2uv),$$

and hence we have the relations

$$x = u^2 - v^2, \qquad y = 2uv. \tag{25}$$

Thus it is seen that any straight line $x = c_1$ is mapped onto that portion of the hyperbola $u^2 - v^2 = c_1$ for which $u > 0$, whereas the straight line $y = c_2$ is mapped onto one branch of the hyperbola $2uv = c_2$. The nature of corresponding regions is indicated in Figure 11.3.

[†]In applications of conformal mapping (and in related considerations) it usually is desirable to define a branch in an *open* region. Thus, with the definition (22), the cut along the negative real axis in the z plane is excluded. In correspondence with the possibility of *approaching* the cut arbitrarily closely from above or below, we continue to speak of the "upper bank" and "lower bank" of the cut.

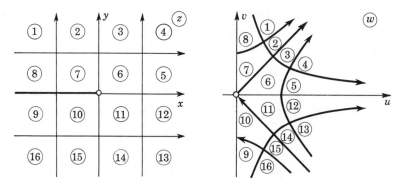

Figure 11.3

If we imagine the z plane to consist of a sort of compressible material and think of the z plane as cut along the negative real axis and then pulled apart along the cut in such a way that the edges of the cut are each rotated through $90°$, so that all the material is compressed into the region to the right of the y axis, then straight lines drawn on the plane may be expected to become distorted into the shapes assumed by the corresponding curves sketched in the w plane.

We notice that the upper and lower banks of the cut in the z plane are mapped onto the positive and negative parts of the v axis, respectively, and that neighboring regions in the z plane are mapped into neighboring regions in the w plane *except for those regions which are separated by the cut*, across which transition is prohibited. Any continuous curve in the z plane *not crossing the cut* maps onto a continuous curve in the w plane.

If neither the function $f(z)$ nor its inverse is multivalued, it may happen that each plane maps as a whole into the other plane. If, as in the present case, $f(z)$ is multivalued and its inverse is single-valued, a branch of $f(z)$ may map the z plane onto only a portion of the w plane. The mapping $w = z^2$ would be described by interchanging the z and w planes in the above sketches. In this case $f(z)$ is single-valued but its inverse is multivalued, and a portion of the z plane maps onto the entire w plane.

Since in the mapping (20) the derivative $f'(z)$ is not zero for any finite value of z, the theorem cited above states that any point in the cut plane (that is, any point not on the cut) can be included in a region which is mapped onto a region in the w plane under a one-to-one correspondence. It is seen, in fact, that *any* region not including points on the cut can be so mapped in this case.

If we indicate the inverse of $w = f(z)$ by $z = F(w)$, we notice that since

$$\frac{dF}{dw} = \frac{dz}{dw} = \frac{1}{dw/dz} = \frac{1}{f'(z)}, \tag{26}$$

the condition $f'(z_0) \neq 0$ ensures that $z = F(w)$ is an analytic function of w when $w = w_0$ if $F(w)$ is single-valued.

If $f(z)$ is analytic at $z = z_0$, then, in particular, to any curve C passing through z_0 in the z plane there corresponds a curve C' passing through the point $w_0 = f(z_0)$ in the w plane. If we consider a second point z_1 on C, and its image w_1 on C', and write

$$\frac{\Delta w}{\Delta z} = \frac{w_1 - w_0}{z_1 - z_0} = \frac{\Delta f}{\Delta z}, \tag{27}$$

we see that this ratio is a complex number whose modulus is the ratio of the lengths of the chords $(w_0 w_1)$ and $(z_0 z_1)$ and whose argument is the angle between the directions of these chords. Thus in the limit as Δz and Δw approach zero, *the limiting argument is the angle between the tangents to C and C' at corresponding points z_0 and w_0, and the limiting modulus represents a local magnification factor in the neighborhood of z_0.* But since for an *analytic function* the ratio (27) tends to $f'(z_0)$, independently of the direction of the chord $(z_0 z_1)$, it follows that if $f(z)$ is analytic at z_0, and if $f'(z_0) \neq 0$, all curves passing through z_0 are mapped onto new curves passing through w_0, all of which are (approximately) rotated through the same angle arg $f'(z_0)$ and magnified in the same ratio $|f'(z_0)|$ in the neighborhood of w_0. Thus, in the neighborhood of a point where $f(z)$ is analytic and $f'(z) \neq 0$, relative angle and shape are preserved in such a mapping, and the mapping is said to be *conformal*. That is, a small closed figure will map onto a similar closed figure with a certain nearly uniform rotation and magnification.

In particular, if $f'(z_0) = k_0 e^{i\alpha_0}$ and $k_0 \neq 0$, then the *tangent at z_0* to any mapped curve passing through that point is rotated through *exactly* the angle α_0 into the tangent at w_0 to the image curve. Consequently, *the angle between two mapped curves intersecting at z_0 is the same as the angle between the two image curves at w_0 if $f(z)$ is analytic at z_0 and $f'(z_0) \neq 0$.* It is important to notice that this statement applies not only to the *magnitude*, but also to the *sense* of the angle.

At points where $f'(z) = 0$ the magnification factor is *zero* and the angle of rotation is indeterminate. Such points are often known as *critical points* of the mapping. It should be noticed that a critical point z_0 corresponds to a point w_0 at which the *inverse* function $z = F(w)$ is *not analytic*.

If we write

$$w = f(z) = u + iv = f_1(x, y) + if_2(x, y) \tag{28}$$

and notice that the straight lines $u = $ constant and $v = $ constant are orthogonal in the w plane, we see that the curves $u(x, y) = f_1(x, y) = $ constant and $v(x, y) = f_2(x, y) = $ constant in the z plane (of which these lines are the images) also must be orthogonal at points where $f'(z) \neq 0$. This is, *if $u(x, y)$ and $v(x, y)$ are the real and imaginary parts of an analytic function $f(z)$, then the curves $u = $ constant and $v = $ constant are orthogonal at points of analyticity where $f'(z) \neq 0$.* At singular points of $f(z)$ the u and v curves may or may not *exist* and, in any case, need not be orthogonal. Similarly, at points for which $f'(z) \neq 0$

the inverse function $z = F(w)$ is analytic (if single-valued), and the curve sets $x = $ constant and $y = $ constant in the z plane correspond to orthogonal sets of curves in the w plane. Thus, in general, small rectangles bounded by coordinate lines in either plane correspond to small "curvilinear rectangles" in the other plane except near points where $f'(z)$ is zero or nonexistent.

11.5. Application to Two-Dimensional Fluid Flow. Since the real part of an analytic function of z satisfies Laplace's equation (Section 10.4), it represents the *velocity potential* of an ideal fluid flow in the xy plane (Section 6.19). Thus, if we write

$$\Phi(z) = \varphi(x, y) + i\psi(x, y), \tag{29}$$

where Φ is an analytic function of z, and consider $\varphi(x, y)$ as a velocity potential corresponding to a flow velocity with components V_x and V_y in the x and y directions at a point, we then have

$$V_x = \frac{\partial \varphi}{\partial x}, \qquad V_y = \frac{\partial \varphi}{\partial y}. \tag{30}$$

Since the curves $\psi = $ constant are orthogonal to the equipotential lines $\varphi = $ constant, they must be identified with the streamlines of the flow. In fact, since the Cauchy–Riemann equations give

$$\frac{\partial \psi}{\partial x} = -\frac{\partial \varphi}{\partial y} = -V_y, \qquad \frac{\partial \psi}{\partial y} = \frac{\partial \varphi}{\partial x} = V_x, \tag{31}$$

it follows that $\psi(x, y)$ can be taken as the *stream function* of the flow. That is, since

$$\rho \, d\psi = \rho(-V_y \, dx + V_x \, dy) = \rho \mathbf{V} \cdot \mathbf{n} \, ds = df, \tag{32}$$

where ρ is the fluid density, $\mathbf{n}$ the unit normal vector to a curve C, and f the flux across C, the difference between the values of ψ at two points in the xy plane is numerically equal to the rate of mass flow of fluid with unit density across a curve joining these points.

The complex function $\Phi(z)$ is sometimes called the *complex potential*; its real part is the velocity potential and its imaginary part is the stream function. We notice that, from Equation (64) of Chapter 10, we have also

$$\Phi'(z) = \frac{d\Phi}{dz} = \frac{\partial \Phi}{\partial x} = \frac{\partial \varphi}{\partial x} + i \frac{\partial \psi}{\partial x} = V_x - iV_y. \tag{33}$$

The function $\Phi'(z)$ is frequently called the *complex velocity*; its real part is V_x and its imaginary part $-V_y$. It may be seen that the conjugate function

$$\overline{\Phi'(z)} = V_x + iV_y$$

can be considered as specifying the actual *velocity vector*.

In addition, we find that, since

$$\begin{aligned}
\Phi'(z) \, dz &= (V_x - iV_y)(dx + i \, dy) \\
&= (V_x \, dx + V_y \, dy) + i(V_x \, dy - V_y \, dx) \\
&= (\mathbf{V} \cdot \mathbf{u} + i\mathbf{V} \cdot \mathbf{n}) \, ds,
\end{aligned}$$

where **u** and **n** are unit tangent and normal vectors on a curve C, there follows

$$\oint_C \Phi'(z) \, dz = \oint_C \mathbf{V} \cdot \mathbf{u} \, ds + i \oint_C \mathbf{V} \cdot \mathbf{n} \, ds.$$

Thus we deduce that

$$\Delta_C \Phi = (circulation) + i \, (flux), \tag{34}$$

that is, that the change in Φ corresponding to a circuit about a simple closed curve C has as its real part the *circulation* of **V** around C and as its imaginary part the *flux* of **V** across C (when $\rho = 1$).

Now suppose that a second complex variable $w = u + iv$ is defined as an analytic function of the complex variable $z = x + iy$ by the relationship

$$w = f(z) \tag{35}$$

and suppose that this mapping gives a one-to-one correspondence between points in a region $\mathfrak{R}$ in the xy plane and a region $\mathfrak{R}'$ in the uv plane. Then the equipotential lines and streamlines corresponding to a flow in the region $\mathfrak{R}$ will be mapped onto a corresponding configuration in the region $\mathfrak{R}'$. If we write the inverse of (35) in the form

$$z = F(w), \tag{36}$$

the original potential function $\Phi(z)$ is then expressible as a function of w in the form $\Phi[F(w)]$. Since we have

$$\frac{d\Phi}{dw} = \frac{d\Phi}{dz} \frac{dz}{dw} = \frac{\Phi'(z)}{f'(z)} \tag{37}$$

and since $\Phi(z)$ and $f(z)$ are analytic, it follows that Φ is an analytic function of w except at points in the w plane which correspond to points in the z plane where $f'(z) = 0$. Hence, if we think of u and v as rectangular coordinates in the w plane, we see that the real and imaginary parts of $\Phi[F(w)]$ satisfy Laplace's equation in rectangular (u, v) coordinates, and hence the new configuration in the w plane represents a new flow pattern of an ideal fluid.

The complex velocity is of the form

$$V_u - iV_v = \frac{d\Phi}{dw} \tag{38}$$

or, equivalently,

$$V_u - iV_v = \frac{d\Phi}{dz} \frac{dz}{dw} = \frac{\Phi'(z)}{f'(z)} = \frac{V_x - iV_y}{f'(z)}. \tag{39}$$

From (39) it follows that

$$\sqrt{V_u^2 + V_v^2} = \frac{\sqrt{V_x^2 + V_y^2}}{|f'(z)|}, \tag{40}$$

so that *the absolute velocity at a point w_0 in the w plane is obtained by dividing the absolute velocity at the corresponding point z_0 in the z plane by $|f'(z_0)|$.* Moreover, the actual (vector) velocities at the corresponding points will be equal if and only if $f'(z_0) = 1$.

The streamlines $\psi(x, y) = $ constant and equipotential lines $\varphi(x, y) = $ constant map onto corresponding curves in the w plane, with equations obtained by replacing x and y by their equivalent expressions in terms of u and v, or determined from the transformation (35) or (36).

We may verify *directly* that the velocity potential and stream function satisfy Laplace's equation in rectangular (u, v) coordinates as follows. Since u and v satisfy the Cauchy–Riemann equations, we have

$$\frac{\partial}{\partial x} = u_x \frac{\partial}{\partial u} + v_x \frac{\partial}{\partial v} = u_x \frac{\partial}{\partial u} - u_y \frac{\partial}{\partial v},$$

and hence

$$\frac{\partial \varphi}{\partial x} = u_x \frac{\partial \varphi}{\partial u} - u_y \frac{\partial \varphi}{\partial v},$$

$$\frac{\partial^2 \varphi}{\partial x^2} = u_{xx} \frac{\partial \varphi}{\partial u} - u_{xy} \frac{\partial \varphi}{\partial v} + u_x \frac{\partial}{\partial x} \frac{\partial \varphi}{\partial u} - u_y \frac{\partial}{\partial x} \frac{\partial \varphi}{\partial v}$$

$$= u_{xx} \frac{\partial \varphi}{\partial u} - u_{xy} \frac{\partial \varphi}{\partial v} + u_x^2 \frac{\partial^2 \varphi}{\partial u^2} - 2u_x u_y \frac{\partial^2 \varphi}{\partial u \, \partial v} + u_y^2 \frac{\partial^2 \varphi}{\partial v^2}.$$

Similarly, we obtain

$$\frac{\partial^2 \varphi}{\partial y^2} = u_{yy} \frac{\partial \varphi}{\partial u} + u_{xy} \frac{\partial \varphi}{\partial v} + u_y^2 \frac{\partial^2 \varphi}{\partial u^2} + 2u_x u_y \frac{\partial^2 \varphi}{\partial u \, \partial v} + u_x^2 \frac{\partial^2 \varphi}{\partial v^2}.$$

When the last two results are added, there follows

$$\nabla^2 \varphi = \frac{\partial^2 \varphi}{\partial x^2} + \frac{\partial^2 \varphi}{\partial y^2} = (u_{xx} + u_{yy}) \frac{\partial \varphi}{\partial u} + (u_x^2 + u_y^2) \left(\frac{\partial^2 \varphi}{\partial u^2} + \frac{\partial^2 \varphi}{\partial v^2} \right).$$

The coefficient of $\partial \varphi / \partial u$ vanishes, since $u(x, y)$ satisfies Laplace's equation in xy coordinates. Also, since $u_x - i u_y = u_x + i v_x = f'(z)$ there follows $u_x^2 + u_y^2 = |f'(z)|^2$, and we have

$$\frac{\partial^2 \varphi}{\partial x^2} + \frac{\partial^2 \varphi}{\partial y^2} = |f'(z)|^2 \left(\frac{\partial^2 \varphi}{\partial u^2} + \frac{\partial^2 \varphi}{\partial v^2} \right). \tag{41}$$

(See also Problem 101, Chapter 6.) Thus, unless $f'(z) = 0$, the vanishing of the left-hand member implies the vanishing of the parentheses on the right. The same result naturally holds also when φ is replaced by ψ.

11.6. Basic Flows. In this section we investigate the flows corresponding to a few elementary complex potential functions.

Example 1. The potential

$$\Phi = V_0 z, \tag{42}$$

with V_0 real, corresponds to the complex velocity

$$V_x - i V_y = \frac{d\Phi}{dz} = V_0: \qquad V_x = V_0, \quad V_y = 0,$$

and hence represents uniform flow with velocity V_0 in the positive x direction (Figure 11.4). Similarly, if α is a real angle, the potential

$$\Phi = V_0 e^{-i\alpha} z \qquad (43)$$

corresponds to flow with velocity components

$$V_x = V_0 \cos \alpha, \qquad V_y = V_0 \sin \alpha,$$

and hence represents uniform flow with velocity V_0 in a direction making an angle α with the positive x axis (Figure 11.5). ∎

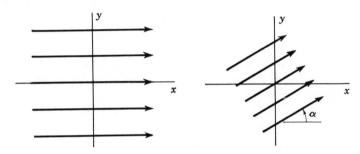

Figure 11.4   **Figure 11.5**

Example 2. For the potential

$$\Phi = k \log z = k(\log r + i\theta), \qquad (44)$$

with k real, the streamlines are

$$\psi = k\theta = \text{constant}.$$

Hence the flow is radial, with radial velocity $V_r = \partial\varphi/\partial r = k/r$ and with zero circumferential velocity $V_\theta = r^{-1}\partial\varphi/\partial\theta = 0$, so that there is a point *source* at the origin if $k > 0$ (Figure 11.6), or a point *sink* if $k < 0$. Its *strength m*, defined as the outward rate of mass flow of fluid with unit density across an arbitrary simple closed curve surrounding the origin, is given by the change in ψ corresponding to a positive circuit around such a curve, and hence has the value $m = 2\pi k$ [see also (34)]. Similarly, $\Phi = k \log (z - a)$ corresponds to the presence of an isolated source (or sink) at $z = a$ if k is real. ∎

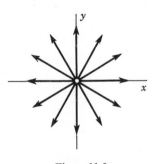

Figure 11.6

Example 3. For the potential

$$\Phi = -ik \log z = k(\theta - i \log r), \qquad (45)$$

with k real, the streamlines are

$$\psi = -k \log r = \text{constant}.$$

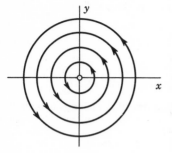

Figure 11.7

Hence the flow is circumferential, with zero radial velocity and with $V_\theta = r^{-1}\partial\varphi/\partial\theta = k/r$. The flow may be called a *circulatory flow*, about a point *vortex* at the origin, and is counterclockwise if $k > 0$ (Figure 11.7). It is conventional to define the *strength m* of the vortex to be $m = 2\pi k$. Reference to (34) (or direct calculation) shows also that $2\pi k$ is the value of the circulation $\oint_C \mathbf{V} \cdot d\mathbf{r}$ around an *arbitrary* simple closed curve C surrounding the origin.† Similarly, it follows that the complex potential $\mathbf{\Phi} = -ik \log(z - a)$ corresponds to a circulatory flow about a vortex at $z = a$ if k is real. ∎

Example 4. If a *sink* (negative source) of strength $-m$ is located at the origin and a *source* of strength m is located at a neighboring point $z = a$, the potential corresponding to the combination is given, according to Example 2, by

$$\Phi_a(z) = \frac{m}{2\pi}[-\log z + \log(z - a)] = \frac{m}{2\pi}\log\frac{z - a}{z}.$$

As the sink is brought into coincidence with the source, the two will cancel unless at the same time the strength m is continuously increased in inverse proportion to the separation $|a|$. If we write

$$m = \frac{K}{|a|}, \qquad a = |a|\,e^{i\alpha},$$

the potential can be put in the form

$$\Phi_a(z) = -\frac{K}{2\pi}\frac{\log z - \log(z - a)}{a}\frac{a}{|a|} = -\frac{K}{2\pi}e^{i\alpha}\frac{\log z - \log(z - a)}{a}.$$

Proceeding to the limit as $a \longrightarrow 0$, we obtain

$$\Phi(z) = -\frac{Ke^{i\alpha}}{2\pi}\frac{1}{z}. \tag{46}$$

The limiting combination of source and sink is called a *doublet* (or *dipole*) of strength K. The angle α, measured from the positive real axis, specifies the direction along which the coincidence was effected, and is called the *orientation* of the doublet. If we write $z = re^{i\theta}$, Equation (46) can be expressed in the form

$$\Phi(z) = -\frac{K}{2\pi}\frac{1}{r}e^{-i(\theta - \alpha)} = -\frac{K}{2\pi r}[\cos(\theta - \alpha) - i\sin(\theta - \alpha)].$$

Hence the streamlines are the circles

$$\psi = \frac{K}{2\pi r}\sin(\theta - \alpha) = \text{constant}.$$

The velocity at any point is of magnitude $K/2\pi r^2$, where r is distance from the doublet (Figure 11.8). ∎

---

†The term "circulation" is often used in the literature to refer to the *flow* (here called a "circulatory flow"), as well as to the quantity $\oint_C \mathbf{V} \cdot d\mathbf{r}$ associated with this flow (or with any other flow).

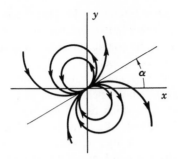

Figure 11.8

Example 5. By superimposing two or more of the basic flows described, various other flows of interest may be obtained. Thus, for example, if we superimpose upon a uniform flow with velocity V_0 in the positive x direction the flow corresponding to a doublet with orientation $\alpha = \pi$ and strength $2\pi c$, we obtain the potential

$$\Phi = V_0 z + \frac{c}{z}.$$

In terms of real and imaginary parts we then have

$$\varphi + i\psi = \frac{V_0 r^2 + c}{r}\cos\theta + i\frac{V_0 r^2 - c}{r}\sin\theta.$$

The streamlines

$$\psi = \frac{\sin\theta}{r}(V_0 r^2 - c) = \text{constant}$$

are seen to include the circle $r = a$ as well as the real axis $\sin\theta = 0$ if $c = V_0 a^2$, so that

$$\Phi = V_0\left(z + \frac{a^2}{z}\right). \tag{47}$$

In fact, it is easily seen that we have obtained in (47) a representation of the flow of an initially uniform stream around a circular barrier (Figure 11.9). The velocity is determined in the form

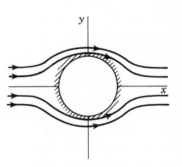

Figure 11.9

$$V_x = V_0\left(1 - \frac{a^2}{r^2}\cos 2\theta\right), \qquad V_y = -V_0\frac{a^2}{r^2}\sin 2\theta;$$

$$V = \sqrt{V_x^2 + V_y^2} = |\Phi'(z)| = \frac{V_0}{r^2}\sqrt{r^4 - 2a^2 r^2 \cos 2\theta + a^4}.$$

At the points $(r = a; \theta = 0, \pi)$ on the surface of the barrier, the velocity is zero. These points are known as *stagnation points*. By superimposing also multiples of the circulatory flow (45), we may obtain further flows in which the stream function ψ is again constant along the circular boundary and in which the flow tends toward uniformity with large distances from the boundary, but in which the stagnation points are displaced and the flow is no longer symmetrical with respect to the x axis. (See also Problems 48–50 of Chapter 9 and Problem 30 of this chapter.) ∎

It follows from the preceding examples that the presence of a *logarithmic singularity* in the complex potential function corresponds to the presence of a *source* (or sink) in the flow if the coefficient of the logarithmic term is *real*, and to the presence of a *vortex* if the coefficient is purely imaginary. Also, a *simple pole* at a point corresponds to a *doublet* at that point. If we recall (see Section 10.10) that a function $\Phi(z)$ behaves near z_∞ just as $\Phi(1/t)$ behaves near $t = 0$, we see that the potential $V_0 z$, which represents uniform flow ("streaming") in the finite plane, has a simple pole at z_∞, and hence this flow includes a doublet at z_∞. Similarly, the potential $k \log z$ represents a flow with a source at the origin and a sink of opposite strength at z_∞; the potential $ik \log z$ indicates equal and opposite vortices at the origin and at z_∞; and the potential c/z defines a flow with a doublet at the origin but with no singularity at z_∞.

It should be noted that, since the "two-dimensional" flows considered here are in fact only plane sections of flows imagined to not vary in the direction normal to that plane, it follows that the "point" sources, sinks, vortices, doublets, and *curves* (or *lines*) considered here in the xy plane truly are sections of infinite "line" sources, sinks, and so forth, and *cylinders* (or *planes*) in the three-dimensional flow.

11.7. Other Applications of Conformal Mapping. As has been seen in Chapter 9, it is frequently necessary in various fields to determine a function of x and y which satisfies Laplace's equation and which takes on prescribed values at points of a given curve C in the xy plane. Suppose that we have somehow obtained an analytic function $w = f(z) = u + iv$ which maps the curve C onto the real axis ($v = 0$) of the uv plane. The same mapping relation will transform the prescribed values of φ along C to corresponding values at points along the u axis in the w plane. If now we can find in the w plane a solution of Laplace's equation which takes on these values, this solution can be transformed back into xy coordinates to give the solution of the original problem. Thus, once the proper transformation is known, the problem is reduced from the determination of a function taking on prescribed values at points of a *curve* to the determination of one which takes on prescribed values along a *straight line*. This latter problem is considerably less difficult and, as a matter of fact, has been solved in Section 9.14 [Equation (230)].

In the case of ideal fluid flow, the problem just discussed can be considered as essentially reducing to the determination of a stream function $\psi(x, y)$ which satisfies Laplace's equation, takes on a *constant* value at points along a prescribed streamline, and behaves suitably at infinity. This is true since, once ψ is known, its conjugate $\varphi(x, y)$, the velocity potential, is determined by the Cauchy–Riemann equations (31), and the flow is completely determined.

Similar problems involving steady-state temperature distributions, electrostatic fields, and so on, are of frequent occurrence.

In general, the direct analytical determination of a suitable mapping function is not easily accomplished. However, if the curve C is *polygonal*, that is, made up of straight line segments, such a mapping can be obtained by methods to be presented in the following section.

In certain cases it may be more desirable to map the curve C instead onto the *unit circle* in the w plane and to solve the transformed problem by the method of Section 9.4 or 9.5. As a single illustration we consider the mapping represented by the relation

$$z = \frac{1}{2}\left(w + \frac{1}{w}\right). \tag{48}$$

Here the mapping function is conveniently expressed in a form solved for z. If we introduce coordinates (ρ, φ) in the w plane by writing

$$w = u + iv = \rho e^{i\varphi},$$
$$u = \rho \cos \varphi, \qquad v = \rho \sin \varphi, \tag{49}$$

Equation (48) becomes

$$x + iy = \frac{1}{2}\left(\rho e^{i\varphi} + \frac{1}{\rho}e^{-i\varphi}\right),$$

so that we have

$$x = \frac{1}{2}\left(\rho + \frac{1}{\rho}\right)\cos\varphi, \qquad y = \frac{1}{2}\left(\rho - \frac{1}{\rho}\right)\sin\varphi. \tag{50}$$

By eliminating ρ and φ successively from these equations, we obtain the relations

$$\frac{x^2}{\left(\rho + \frac{1}{\rho}\right)^2} + \frac{y^2}{\left(\rho - \frac{1}{\rho}\right)^2} = \frac{1}{4}, \tag{51a}$$

$$\frac{x^2}{\cos^2 \varphi} - \frac{y^2}{\sin^2 \varphi} = 1. \tag{51b}$$

Thus it follows that the circles $\rho = $ constant in the w plane correspond to the ellipses (51a) and parts of the radial lines $\varphi = $ constant correspond to the hyperbolas (51b), as represented in Figure 11.10. In particular, we verify that the unit circle $\rho = 1$ in the w plane is flattened into a two-sided "slit" in the z plane

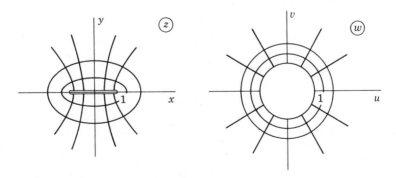

Figure 11.10

between $z = -1$ and $z = +1$. The other circles in the w plane correspond to ellipses which decrease in flatness with increasing size. We notice that the entire cut z plane maps onto the exterior of the unit circle.

Now suppose that it is required to determine a harmonic function $T(x, y)$, say steady-state temperature, which takes on prescribed values along the upper and lower edges of the slit, say

$$T(x, 0+) = f_1(x), \qquad T(x, 0-) = f_2(x) \qquad (-1 < x < 1).$$

For points on the slit (50) gives

$$\rho = 1, \qquad x = \cos \varphi,$$

where $0 < \varphi < \pi$ for the upper edge of the slit and $\pi < \varphi < 2\pi$ for the lower edge. Thus in the region $\rho \geq 1$ of the w plane we must determine a harmonic function $T(\rho, \varphi)$ which reduces when $\rho = 1$ to $f_1(\cos \varphi)$ when $0 < \varphi < \pi$ and to $f_2(\cos \varphi)$ when $\pi < \varphi < 2\pi$. This problem is solved in Section 9.4. The value of T determined at any point (ρ, φ) is also the required value of T at the corresponding point (x, y), as determined by (50).

Example. As another illustration of the use of the mapping (48), we notice that a *circulatory flow* with circulation m in the w plane, and specified by the complex potential $\Phi = -(im/2\pi) \log w$, is mapped onto a flow with complex potential

$$\Phi = -\frac{im}{2\pi} \log [z + (z^2 - 1)^{1/2}]$$

$$= -\frac{im}{2\pi} \cosh^{-1} z \tag{52}$$

about the slit in the z plane. The streamlines and equipotential lines of the two flows accordingly are the curves appearing in the z plane in Figure 11.10 and the velocity components are derivable from the relation

$$V_x - iV_y = \frac{d\Phi}{dz} = \frac{-(im/2\pi)}{(z^2 - 1)^{1/2}}. \tag{53}$$

Here, in order to specify a branch of $(z^2 - 1)^{1/2}$ having a cut on the real axis between $z = -1$ and $z = 1$, we may write

$$(z^2 - 1)^{1/2} = (p_1 p_2)^{1/2} e^{i(\omega_1 + \omega_2)/2}, \tag{54}$$

with the bipolar notation of Figure 11.11, where $p_1 \geq 0$ and $p_2 \geq 0$ and where $0 \leq \omega_1 < 2\pi$ and $0 \leq \omega_2 < 2\pi$. Then, in correspondence with a crossing of the real axis, the angle $(\omega_1 + \omega_2)/2$ is continuous when $x < -1$, has a jump of π when $-1 < x < 1$, and has a jump of 2π when $x > 1$, so that $\exp [i(\omega_1 + \omega_2)/2]$ is discontinuous along the cut but continuous elsewhere. With this definition of the interpretation of $(z^2 - 1)^{1/2}$, we verify easily that a positive circuit around any simple closed curve *enclosing* $z = -1$ and $z = 1$ increases Φ by m, so that, according to Equation (34), the circulation around any such curve is indeed m, the flow accordingly being counterclockwise when $m > 0$. Also, along the real axis $V_x = -(m/2\pi)(1 - x^2)^{-1/2}$ on the upper bank of the cut, with the sign reversed on the lower bank, while $V_y = (m/2\pi)(x^2 - 1)^{-1/2}$ when $x > 1$, with the sign reversed when $x < -1$.

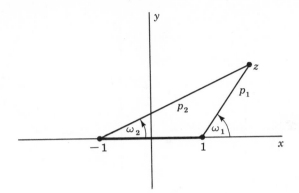

Figure 11.11

More generally, the complex potential

$$\Phi = -\frac{im}{2\pi} \cosh^{-1} \frac{z}{R} \tag{55}$$

would specify the result of expanding this flow pattern in a ratio R to 1, so that the points $z = \pm 1$, in particular, are replaced by the points $z = \pm R$. ■

11.8. The Schwarz–Christoffel Transformation. In this section we describe and illustrate a mapping which sets up a one-to-one correspondence between points on a polygonal boundary in the z plane and points on the real axis ($v = 0$) in the w plane. It will be convenient first to think of the mapping as *from the w plane to the z plane.*

For this purpose, we consider the mapping $z = F(w)$ for which

$$\frac{dz}{dw} = C(w - u_1)^{k_1}(w - u_2)^{k_2} \cdots (w - u_n)^{k_n}, \tag{56}$$

where $u_1, u_2, \ldots, u_n$ are any n points arranged in order along the real axis in the w plane (Figure 11.12), $u_1 < u_2 < \cdots < u_n$, and where the k's are real constants and C is a real or complex constant. By taking the logarithm of both sides, we may rewrite Equation (56) in the form

$$\log \frac{dz}{dw} = \log C + k_1 \log (w - u_1) + k_2 \log (w - u_2) + \cdots + k_n \log (w - u_n), \tag{57}$$

it being agreed that principal values are taken of all logarithms on the right (so that each associated angle is between 0 and 2π). We then consider the mapping of the u axis onto its image in the xy plane, remembering (see Section 10.9) that the *magnitude* of dz/dw at a point in the w plane is *the local magnification factor* in the mapping from the w plane to the z plane, and that the *angle* (or *argument*) of dz/dw, which can be expressed in the form

$$\triangle \frac{dz}{dw} = \text{Im} \left(\log \frac{dz}{dw} \right),$$

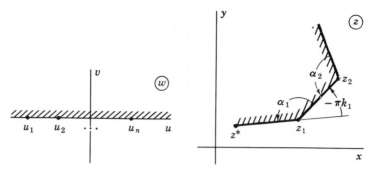

Figure 11.12   Figure 11.13

gives *the angle through which the tangent to a mapped curve passing through that point is rotated in the mapping*. Thus we may obtain from (57) the relation

$$\measuredangle \frac{dz}{dw} = \measuredangle C + k_1 \measuredangle(w - u_1) + k_2 \measuredangle(w - u_2) + \cdots + k_n \measuredangle(w - u_n) \quad (58)$$

by taking imaginary parts of both sides.

Let the point at infinity in the w plane be mapped onto the point z^* in the z plane, where z^* *may be* z_∞. Then, if we consider the image of a point $w = u$ moving from an infinite distance toward the right along the negative real axis in the w plane, we notice that so long as $w = u < u_1$ the numbers $w - u_1$, $w - u_2, \ldots, w - u_n$, are all *real and negative*. Hence their angles are all *constantly* equal to π in Equation (58), and so there follows

$$\measuredangle \frac{dz}{dw} = \measuredangle C + (k_1 + k_2 + \cdots + k_n)\pi \quad (w = u < u_1). \quad (59)$$

Thus the portion of the u axis to the left of the point u_1 is mapped onto a *straight line segment*, making the angle defined by (59) with the real axis in the z plane, and extending from z^* to z_1, the image of $w = u_1$ (Figure 11.13).

Now as the point w crosses the point u_1 on the real axis, the real number $w - u_1$ becomes positive, so that its angle abruptly changes from π to zero. Hence $\measuredangle(dz/dw)$ abruptly decreases by an amount $k_1 \pi$ and then remains constant as w travels from u_1 toward u_2. It follows that the image of the segment $(u_1 u_2)$ is a line segment $(z_1 z_2)$ in the z plane making an angle of $-k_1 \pi$ with the segment $(z^* z_1)$.

Proceeding in this way, we see that each segment $(u_p u_{p+1})$ is mapped onto a line segment $(z_p z_{p+1})$ in the z plane, making an angle of $-k_p \pi$ with the segment previously mapped. Thus, if the *interior* angle of the resultant polygonal contour at the point z_p is to have the magnitude α_p, with $0 \leq \alpha_p \leq 2\pi$,† we must set $\pi - \alpha_p = -k_p \pi$ or

$$k_p = \frac{\alpha_p}{\pi} - 1 \quad (60)$$

---

†The identification $\alpha_p = 0$ can apply only to a vertex at z_∞, as in Example 3 of this section.

in Equation (56). After an integration, we then conclude that *the mapping*

$$z = C \int (w - u_1)^{k_1}(w - u_2)^{k_2} \cdots (w - u_n)^{k_n} \, dw + K, \qquad (61)$$

where C and K are arbitrary (complex) constants, will map the real axis $(v = 0)$ *of the w plane onto a polygonal boundary in the z plane in such a way that the vertices* $z_1, z_2, \ldots, z_n$, *with interior angles* $\alpha_1, \alpha_2, \ldots, \alpha_n$, *are the images of the points* $u_1, u_2, \ldots, u_n$.

We notice that for the final segment $(w = u > u_n)$ the numbers $w - u_i$ are *all real and positive* and hence possess *zero* angles, so that this segment is rotated through the angle

$$\triangle \frac{dz}{dw} = \triangle C \qquad (w = u > u_n). \qquad (62)$$

For a *closed* polygon the sum of the interior angles is

$$\alpha_1 + \alpha_2 + \cdots + \alpha_n = (n - 2)\pi,$$

and hence also

$$k_1 + k_2 + \cdots + k_n = \frac{(n-2)\pi}{\pi} - n = -2.$$

Thus, according to Equations (59) and (62), the two infinite segments of the line $v = 0$ are rotated through the angles $\triangle C - 2\pi$ and $\triangle C$, as is clearly necessary for a closed figure.

If the constant C is written in the form $C = ce^{i\beta}$, we see that it comprises an arbitrary magnification factor c and a rotation β and, in fact, that according to (62), the angle β is the angle through which the infinite segment to the right of $w = u_n$ is rotated when it is mapped (onto a finite or infinite segment). If we write $K = u_0 + iv_0$, we see that the additive constant K represents an arbitrary translation of the polygon, without rotation or distortion, through the vector $u_0 + iv_0$.

It can be shown that the numbers $u_1, u_2, \ldots, u_n$ and the complex constants C and K can always be chosen so that *any* prescribed polygon in the z plane is made to correspond point by point to the real axis $(v = 0)$ in the w plane and, in fact, that the correspondence can be set up in infinitely many ways, in that *three of the numbers* $u_1, u_2, \ldots, u_n$ *can be determined arbitrarily* (provided that they are appropriately *ordered*).†

Further, the mapping can be shown to establish a one-to-one correspondence between points in the *interior* of the polygon in the z plane and points in the *upper half* of the w plane. In this connection, one must be careful to label the vertices on the polygonal boundary as $z_1, z_2, \ldots, z_n$ in the order in which they would be passed by an observer traversing the boundary *with the interior of the polygon to his left*. The corresponding points $u_1, u_2, \ldots, u_n$ on the u axis in

---

†When the polygon has a vertex at z_∞, the mapping of this vertex onto u_∞ amounts to specifying one of the three assignable u's. (See Problem 56.)

the w plane then are to be labeled in order from left to right, as has already been specified.

The propriety of this rule follows from the fact that the conformal property of the mapping guarantees that, if an observer traveling along the polygonal boundary turns to his left from the boundary (*into* the polygon), his image in the w plane will be traveling forward along the u axis and also will turn to *his* left (into the *upper* half-plane).

Thus, for example, if the open region *above* (and to the left of) the polygonal boundary in the z plane in Figure 11.14 is to be mapped onto the upper half of

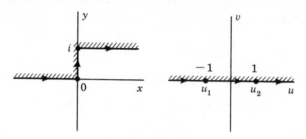

Figure 11.14

the w plane, with the two finite vertices mapped (say) onto $u_1 = -1$ and $u_2 = 1$, then the proper identification is $z_1 = 0$ and $z_2 = i$, in order that the vertex sequence $z = 0$, $z = i$, $z = z_\infty$ map onto the image sequence $w = -1$, $w = 1$, $w = u_\infty$. Consequently, there follows also $\alpha_1 = \pi/2$, $\alpha_2 = 3\pi/2$ and $k_1 = -1/2$, $k = 1/2$, and hence

$$\frac{dz}{dw} = C\left(\frac{w-1}{w+1}\right)^{1/2}.$$

After the integration, C and the constant of integration K would be determined so that $z = 0 \leftrightarrow w = -1$ and $z = i \leftrightarrow w = 1$. In order to map instead the region *below* (and to the right of) the polygonal boundary onto the upper half-plane, one would take $z_1 = i$, $z_2 = 0$ and $\alpha_1 = \pi/2$, $\alpha_2 = 3\pi/2$. In this special case it happens that the *same* differential equation for the mapping function would be obtained, but different values for C and K would result from then making $z = i \leftrightarrow w = -1$ and $z = 0 \leftrightarrow w = 1$.

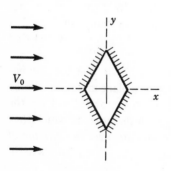

Figure 11.15

Certain artifices may also be used in obtaining mappings related to problems involving the *exterior* of a closed polygon. As an example, suppose that we wish to investigate the effect of the barrier of quadrilateral section, sketched in Figure 11.15, on an initially uniform flow of an ideal fluid. From the physical symmetry it is clear that, when no circulation is present, one streamline will consist of the part of the x axis exterior

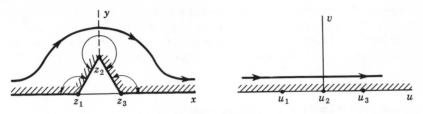

Figure 11.16

to the polygonal section and, say, the *upper* half of the boundary. Thus we are led to consider a mapping which transforms this streamline into the u axis of the w plane, as is indicated in Figure 11.16. Here we map the region *above* the polygonal contour onto the region $v > 0$, taking as the "interior angles" those indicated in the figure. The mapped region may be considered as an *infinite* polygon having the contour as its finite boundary. Once the flow is determined in the upper half of the z plane, by mapping an appropriate uniform flow in the upper half of the w plane, the flow is determined in the lower half-plane by symmetry. A superimposed flow, with circulation m, and with a velocity vector which also is tangential to the barrier but which tends to zero with increasing distance from it, would be obtained by effecting the same mapping on the complex potential

$$\Phi = -\frac{im}{2\pi}\cosh^{-1}\frac{w}{R}, \qquad (63)$$

provided that the upper boundary of the barrier is mapped onto the segment $-R \leq u \leq R$ of the real axis in the w plane. This fact follows immediately from the considerations at the end of Section 11.7.

A similar procedure clearly can be applied in all analogous cases in which there is an axis of symmetry parallel to the direction of flow.

A generalization of the Schwarz–Christoffel transformation permits the mapping of the complete exterior of a closed polygon in the z plane onto the upper half of the w plane.† In this connection, it should be noted that only a simply connected region can be mapped conformally onto another simply connected region, so that, for example, an *annular* region could not be mapped onto a half-plane. However, when the point z_∞ is included, the region outside a simple closed curve C is in fact simply connected since any simple closed curve enclosing C can be "shrunk to a point" at z_∞.

Many illustrations of the Schwarz–Christoffel transformation may be found in the literature. For *closed* polygons, even in the simplest cases of a triangle and a rectangle the use of *elliptic functions* is necessary, and in other cases resort usually must be had to numerical analysis. We here consider three applications to *open* polygons, in increasing order of complexity.

†See, for example, Reference 1 of Chapter 10.

Example 1. As a simple example, we consider first the case in which the polygon is merely the interior of the sector of the z plane bounded by the positive x axis ($\theta = 0$) and the line $\theta = \alpha$ (Figure 11.17). If we take the point u_1, corresponding to the single finite vertex at $z = 0$, at the origin ($u_1 = 0$), Equation (61) gives

$$z = C \int w^k \, dw,$$

where $k = (\alpha/\pi) - 1$ and C is an arbitrary constant, and hence $z = [C/(k + 1)]w^{k+1} + K$.

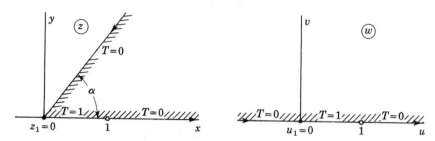

Figure 11.17

For $z = 0$ and $w = 0$ to correspond, the constant of integration must vanish, but C remains arbitrary. If, *for convenience*, we require that the point $z = 1$ map onto the point $w = 1$, we find that $C = k + 1$, and there then follows

$$z = w^{\alpha/\pi}, \qquad w = z^{\pi/\alpha}, \tag{64}$$

subject to the obvious branch specifications $0 \leq \angle z \leq \alpha$ and $0 \leq \angle w \leq \pi$.

Suppose now that we require the solution T of Laplace's equation in the infinite sector $0 \leq \theta \leq \alpha$ for which $T = 1$ when $0 < x < 1$ and $T = 0$ when $x > 1$, along the edge $\theta = 0$, and for which $T = 0$ everywhere along the edge $\theta = \alpha$ (Figure 11.17). The mapping (64) transforms this problem to that of determining the solution of Laplace's equation in the half-plane $v \geq 0$ which reduces, when $v = 0$, to 1 when $0 < u < 1$ and to 0 otherwise. But the solution of this problem is given by Equation (230) of Section 9.14 in the form†

$$T = \frac{1}{\pi} \int_0^1 \frac{v \, d\xi}{v^2 + (\xi - u)^2} = \frac{1}{\pi} \left(\tan^{-1} \frac{1 - u}{v} + \tan^{-1} \frac{u}{v} \right)$$

$$= \frac{1}{\pi} \cot^{-1} \frac{u^2 + v^2 - u}{v} \tag{65}$$

or, in polar coordinates (ρ, φ),

$$T = \frac{1}{\pi} \cot^{-1} \frac{\rho - \cos \varphi}{\sin \varphi}. \tag{66}$$

---

†It is useful to notice that the conventions $-\pi/2 \leq \tan^{-1} x \leq \pi/2$ and $0 \leq \cot^{-1} x \leq \pi$, when x is real, imply the relations

$$\tan^{-1} A + \tan^{-1} B = \begin{cases} -\cot^{-1}\left[(AB - 1)/(A + B)\right] & \text{if} \quad A + B \leq 0 \\ \tan^{-1}\left[(A + B)/(1 - AB)\right] & \text{if} \quad AB \leq 1 \\ \cot^{-1}\left[(1 - AB)/(A + B)\right] & \text{if} \quad A + B \geq 0 \end{cases}$$

when A and B are real.

From Equation (64) there follows

$$p = r^{\pi/\alpha}, \qquad \varphi = \frac{\pi\theta}{\alpha}$$

and hence the required solution is obtained in the convenient closed form

$$T = \frac{1}{\pi}\cot^{-1}\left[\frac{r^{\pi/\alpha} - \cos(\pi\theta/\alpha)}{\sin(\pi\theta/\alpha)}\right]. \tag{67}$$

The more general case in which T is prescribed arbitrarily along the two radial boundaries clearly can be treated in the same way. ∎

Example 2. As a second example, we investigate the effect of a plane barrier of finite length $2a$ on an initially uniform flow of an ideal fluid perpendicular to it (Figure 11.18), when no circulation is present. This is the limiting case of the problem discussed in connection with the barrier of Figure 11.15, when the interior angles of the quadrilateral tend to 0 and to 2π. Thus we are led to consider the mapping of Figure 11.19. Here we wish to map the negative x axis onto the segment $u < u_1$, the left-hand edge of the barrier onto the segment $u_1 < u < u_2$, the right-hand edge of the barrier onto the segment $u_2 < u < u_3$, and the positive x axis onto $u > u_3$. For purposes of sym-

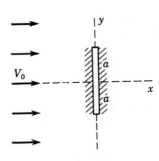

Figure 11.18

metry, we take $u_1 = -b$, $u_2 = 0$, $u_3 = b$, where b is as yet unspecified. With $\alpha_1 = \alpha_3 = \pi/2$, $\alpha_2 = 2\pi$, there follows, from (60), $k_1 = k_3 = -\frac{1}{2}$, $k_2 = 1$, and hence (61) gives

$$z = C\int\frac{w\,dw}{(w^2 - b^2)^{1/2}} = C(w^2 - b^2)^{1/2} + K, \tag{68}$$

where K, C, and b are yet to be determined. The desired correspondence

$$w = -b \leftrightarrow z = 0, \quad w = 0 \leftrightarrow z = ia, \quad w = +b \leftrightarrow z = 0 \tag{69}$$

then is obtained if we set

$$K = 0, \qquad Cb = a, \tag{70}$$

so that the mapping becomes

$$z = \frac{a}{b}(w^2 - b^2)^{1/2}, \qquad w = \frac{b}{a}(z^2 + a^2)^{1/2}. \tag{71}$$

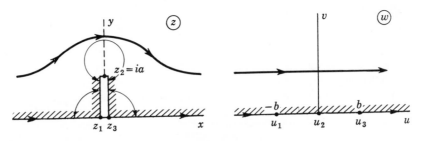

Figure 11.19

Finally, we notice that $z_\infty \longleftrightarrow w_\infty$, and hence if a uniform flow in the w plane is to be undistorted in the mapping at large distances from the barrier, we must have $dw/dz \longrightarrow 1$ as $z \longrightarrow z_\infty$, or

$$\lim_{|z| \to \infty} \frac{dw}{dz} = \lim_{|z| \to \infty} \left[\frac{b}{a} \frac{z}{(z^2 + a^2)^{1/2}} \right] = \frac{b}{a} = 1; \qquad b = a. \tag{72}$$

Thus the desired mapping takes the form

$$z = (w^2 - a^2)^{1/2}, \qquad w = (z^2 + a^2)^{1/2} \equiv f(z), \tag{73}$$

where the angle associated with each of the roots is to be between 0 and π.

It is readily verified that this mapping gives the specified correspondence. In particular, points in the interior of a quadrant in one half-plane correspond to points inside the same quadrant in the other half-plane.

A uniform flow $\Phi(w) = V_0 w$ in the upper half of the w plane then maps onto the desired flow in the z plane,† with complex potential $\Phi[f(z)]$, or

$$\Phi = V_0(z^2 + a^2)^{1/2} = \varphi(x, y) + i\psi(x, y). \tag{74}$$

The streamlines and equipotential lines are obtained by equating to constants the real and imaginary parts φ and ψ.

The complex velocity is given by

$$V_x - iV_y = \frac{d\Phi}{dz} = \frac{V_0 z}{(z^2 + a^2)^{1/2}} = \frac{V_0(x + iy)}{[(x^2 - y^2 + a^2) + i(2xy)]^{1/2}}, \tag{75}$$

where the sign of the denominator is most conveniently determined (without analytical specification of the relevant branch) by noticing that, for a given point z, the denominator represents the *image point* w. In particular, for points on the y axis there follows

$$V_x - iV_y = \frac{iV_0 y}{(a^2 - y^2)^{1/2}}, \tag{76}$$

with the proper interpretation. At any point on the *right-hand* edge of the barrier the denominator is real and *positive*, since the image point is on the positive u axis, and there follows

$$V_x = 0, \qquad V_y = -\frac{V_0 y}{\sqrt{a^2 - y^2}},$$

whereas along the *left-hand* edge the sign is reversed. Along the remainder of the positive y axis ($y > a$) we have

$$(a^2 - y^2)^{1/2} = +i\sqrt{y^2 - a^2},$$

and hence here

$$V_x = \frac{V_0 y}{\sqrt{y^2 - a^2}}, \qquad V_y = 0.$$

The flow velocity is seen to be infinite at the ends of the barrier.

A superimposable flow with circulation m, in which the flow is tangential to the barrier and the velocity tends to zero with increasing distance from the barrier, is

---

†Alternatively, b could have been assigned any convenient positive value (say, $b = 1$ or $b = a$) in advance, in which case the condition $(dw/dz)_{z_\infty} = 1$ could not have been imposed, unless $b = a$. Here one would write $\Phi = Aw$ and determine A in such a way that $(d\Phi/dz)_{z_\infty} = V_0$, and hence $A = V_0 a/b$, and would again obtain (74).

obtained from (63), with $R = a$ and $w = (z^2 + a^2)^{1/2}$, in the form

$$\Phi = -\frac{im}{2\pi} \cosh^{-1}\left(\frac{z^2}{a^2} + 1\right)^{1/2}$$

or, equivalently,

$$\Phi = -\frac{im}{2\pi} \sinh^{-1}\frac{z}{a}. \tag{77}$$

∎

Example 3. To illustrate additional features and techniques of the mapping, we consider the determination of the flow of an ideal fluid from a long channel with parallel walls. The width of the channel is denoted by $2\pi l$, and the coordinate axes in the z

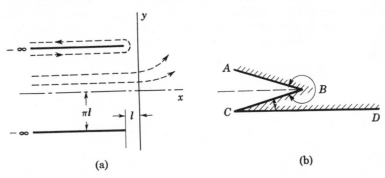

(a)   (b)

Figure 11.20

plane are located (for later convenience) as is indicated in Figure 11.20(a). We make use of the symmetry with respect to the x axis and consider only the flow in the upper half-plane. The polygonal boundary to be mapped then can be considered as the limit of the boundary indicated in Figure 11.20(b) as the points A and C tend to infinity and

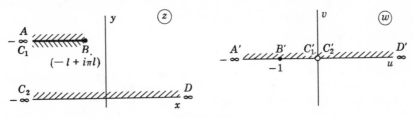

Figure 11.21

the indicated angles tend to 2π and 0. We next attempt to accomplish the mapping of Figure 11.21, in which the point B is mapped (for convenience) onto the point -1 in the w plane and the point $(C_1 C_2)$ at infinity is mapped onto the origin. We should notice, however, for future reference, that the desired flow around the barrier is in the direction $C_1 BA$, whereas that along the axis of symmetry is in the direction $C_2 D$. This condition suggests that we shall need to map a flow from a source at the origin in the w plane, rather than a uniform flow as in Example 2, onto the desired flow.

With $u_1 = -1$, $u_2 = 0$ and $k_1 = 1$, $k_2 = -1$, Equation (61) gives

$$\frac{dz}{dw} = C\frac{w+1}{w}, \tag{78}$$

and hence there follows

$$z = C(w + \log w) + K, \tag{79}$$

where K is a constant of integration. Since $\log w$ is multivalued, we must select a specific branch of that function. Since that branch must be analytic in the upper half of the w plane (and since only that half-plane is involved here), we adopt the definition

$$\log z = \log |z| + i\theta_P \qquad (0 \leqq \theta_P \leqq \pi).$$

We next determine the constants C and K.

It frequently is convenient to obtain conditions determining unknown constants by essentially integrating the equal members of the basic differential equation between corresponding limits. In illustration, we notice that transition from C_1 to C_2 in the z plane must correspond to transition from C_1' to C_2' in the w plane. The transition in the w plane must be considered as taking place along a semicircle of vanishing radius around the branch point at $w = 0$. If the radius is denoted by ρ, then near $w = 0$ we have, from (78),

$$dz = C\frac{w+1}{w}\,dw \sim C\frac{dw}{w} \qquad as \ |w| = \rho \to 0$$

and hence, on the small circle,

$$dz \sim C\frac{i\rho e^{i\varphi}\,d\varphi}{\rho e^{i\varphi}} = iC\,d\varphi. \tag{80}$$

Thus, in the limit as $\rho \to 0$, we have

$$z\Big|_{C_1}^{C_2} = iC\int_{\pi}^{0} d\varphi = -i\pi C, \qquad -i\pi l = -i\pi C,$$

and hence

$$C = l. \tag{81}$$

In order that the point B ($z = -l + i\pi l$) correspond to B' ($w = -1$), we must have, from (79),

$$-l + i\pi l = l[-1 + \log(-1)] + K = -l + i\pi l + K,$$

and hence

$$K = 0. \tag{82}$$

(The position of the origin in the z plane was chosen, after a preliminary analysis, in such a way that this convenient result would be obtained.)

Hence the desired mapping is given by

$$z = l(w + \log w). \tag{83}$$

To verify the correctness of this result, we may write $w = \rho e^{i\varphi}$. Then if $\varphi = \epsilon$, a simple calculation shows that $z \approx l[(\rho + \log \rho) + i(1 + \rho)\epsilon]$ except for higher powers of ϵ; and if $\varphi = \pi - \epsilon$, there follows similarly $z \approx l\{-[\rho - \log \rho] + i[\pi + (\rho - 1)\epsilon]\}$. From these results it is easy to verify the correspondence between the upper boundary of the u axis (as $\epsilon \to 0+$) and the path ABC_1C_2D.

Next it is necessary to determine a flow in the w plane, for which $C_1'B'A'$ and $C_2'D'$ are streamlines. Also it is necessary that the associated complex potential function Φ have the property that the velocity $d\Phi/dz$ shall reduce to a uniform velocity V_0 in the x direction at the point (C_1C_2) in the z plane. *These, however, are the only requirements to be satisfied.* Although Φ can be determined from these conditions by purely direct methods, it appears that in the present case a flow from a source at the origin in the w plane will have the desired properties. Hence we write

$$\Phi = k \log w$$

and attempt to determine k in such a way that the velocity $d\Phi/dz$ in the z plane will tend to V_0 as $z \longrightarrow (C_1 C_2)$ and, correspondingly, $w \longrightarrow 0$. Making use of (78) and (81), we obtain the general result

$$\frac{d\Phi}{dz} = \frac{dw}{dz}\frac{d\Phi}{dw} = \frac{1}{l}\frac{w}{w+1}\frac{d\Phi}{dw},$$

and hence, with $\Phi = k \log w$, there follows

$$\frac{d\Phi}{dz} = \frac{k}{l}\frac{1}{w+1}.$$

Setting $w = 0$, we obtain the condition

$$k = lV_0$$

so that the desired potential function takes the form

$$\Phi = lV_0 \log w. \tag{84}$$

The corresponding potential function in the z plane then is to be obtained by eliminating w between Equations (84) and (83), and hence is the solution of the equation

$$z = l\left(e^{\Phi/lV_0} + \frac{\Phi}{lV_0}\right). \tag{85}$$

Whereas Φ cannot be expressed as a function of z in explicit closed form, we may write

$$\Phi = \varphi + i\psi, \tag{86}$$

where, in consequence of (84) and the limitation of permissible values of the logarithm,

$$-\infty < \varphi < \infty, \qquad 0 \leq \psi \leq \pi lV_0, \tag{87}$$

and Equation (85) thus is equivalent to the two real equations

$$\frac{x}{l} = e^\alpha \cos \beta + \alpha,$$

$$\frac{y}{l} = e^\alpha \sin \beta + \beta, \tag{88}$$

with the abbreviations

$$\alpha = \frac{\varphi}{lV_0}, \qquad \beta = \frac{\psi}{lV_0} \qquad (-\infty < \alpha < \infty, 0 \leq \beta \leq \pi). \tag{89}$$

These parametric equations are convenient in plotting the streamlines $\beta = \text{constant}$ and the equipotential lines $\alpha = \text{constant}$. It can be verified that the streamlines for which $0 \leq \beta \leq \pi/2$ extend indefinitely in the x direction, whereas on the remaining streamlines a maximum value of x exists, so that these streamlines turn back around the channel walls. ∎

In each of the preceding examples, the region enclosed by the polygonal boundary could instead be considered as a homogeneous conducting sheet carrying an electric current, or as a plane section of an *electrostatic* field between long plane or cylindrical conductors. The *stream function* ψ and the *streamlines* ψ = constant in the problem of *fluid flow* then would correspond to the *potential function* and to the *equipotential lines* in the problem of *electrostatics*, whereas the *potential function* φ and the *equipotential lines* in the flow problem would correspond to the *stream function* and to the *lines of force* in the electrostatic problem.

In two-dimensional electrostatics, the *negative* gradient of the potential function is the electric field intensity **E**, whose *tangential* component must vanish on the surface of a perfect conductor. The *normal* component of **E** at such a surface is proportional to the surface charge density σ. The stream function then has the property that the difference between its values at two points on the conducting boundary is proportional to the total charge q, per unit height of the conductor, on the portion of the conductor joining those points. (See Problems 57–62 and Reference 11.)

11.9. Green's Functions and the Dirichlet Problem. In correspondence with a linear differential operator L, a related region $\mathfrak{R}$, and a specification of the nature of appropriate information to be prescribed along the boundary of $\mathfrak{R}$, there often exists a certain associated function, called a *Green's function*, which plays an important role in the solution (or formulation) of problems governed in $\mathfrak{R}$ by differential equations involving the operator L.

In this section and in Section 11.10 we identify L with the two-dimensional Laplacian operator ∇^2 and suppose that it is to be associated with a related problem of Dirichlet type, so that the solution is specified in advance along the boundary C of $\mathfrak{R}$, in which case the determination of the Green's function can be related to the problem of mapping the interior of $\mathfrak{R}$ conformally onto an upper half-plane, when $\mathfrak{R}$ is simple. Section 11.11 deals with some other two-dimensional examples, one of which is the Neumann problem associated also with $L = \nabla^2$, and with conformal mapping.

Green's functions also are defined over intervals for one-dimensional problems (see Problems 55–59 of Chapter 1 and Problems 65, 66, and 79–82 of this chapter) as well as over regions in three or more dimensions, in correspondence with appropriate linear differential operators and boundary conditions.

Each Green's function can be defined in a variety of equivalent ways, as will be illustrated. In addition, it should be noted that such definitions in the literature lead frequently to a function which is the *negative* of the one to be defined here, or occasionally to another constant multiple of that function.

For the two-dimensional operator $L = \nabla^2$, our starting point is the two-dimensional specialization of Green's theorem, Equation (123) of Chapter 6, as applied to a region $\mathfrak{R}$ of the xy plane, bounded by a curve C. Here it is convenient to think of the point $P(x, y)$ as a temporarily fixed point in $\mathfrak{R}$ and to use ξ and η to denote the "dummy variables" of integration, so that Green's theorem can be written in the form

$$\iint_{\mathfrak{R}} G \nabla^2 \varphi \, d\xi \, d\eta = \oint_C \left(G \frac{\partial \varphi}{\partial n} - \varphi \frac{\partial G}{\partial n} \right) ds + \iint_{\mathfrak{R}} \varphi \nabla^2 G \, d\xi \, d\eta, \quad (90)$$

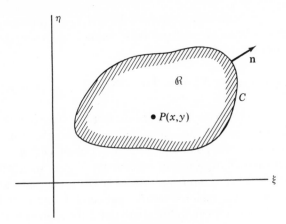

Figure 11.22

where here $\mathbf{V} = \mathbf{i}\,\partial/\partial\xi + \mathbf{j}\,\partial/\partial\eta$ in rectangular coordinates and $\partial/\partial n = \mathbf{n} \cdot \mathbf{V}$, with the unit normal vector $\mathbf{n}$ directed outward from $\mathfrak{R}$ at a point of C (Figure 11.22).

Suppose now that φ satisfies the Poisson equation

$$\mathbf{V}^2\varphi = h \qquad in \ \mathfrak{R} \tag{91}$$

and is to be such that

$$\varphi = f \qquad on \ C, \tag{92}$$

where h and f are prescribed functions. If a function G can be determined so that

$$\mathbf{V}^2 G = \delta(\xi - x)\delta(\eta - y) \qquad in \ \mathfrak{R}, \tag{93}$$

where $\delta(t - c)$ is the *delta function* of Section 2.6, located at $t = c$, and having the basic property that

$$\int_a^b F(t)\delta(t - c)\,dt = \begin{cases} 0 & (c < a), \\ F(c) & (a < c < b), \\ 0 & (c > b), \end{cases}$$

when $b > a$ [see Problem 31(a) of Chapter 2], then since $\delta(\xi - x)\delta(\eta - y)$, considered as a function of ξ and η, is a delta function located at the fixed point $P(x, y)$, the double integral on the right in (90) will reduce to $\varphi(x, y)$. If also G is such that

$$G = 0 \qquad on \ C, \tag{94}$$

Equation (90) will take the form

$$\iint_{\mathfrak{R}} Gh\,d\xi\,d\eta = -\oint_C f\,\frac{\partial G}{\partial n}\,ds + \varphi(x, y),$$

where both integrands involve only known functions.

The function G satisfying (93) and (94), which will depend upon the current variables (x, y) as well as the dummy variables (ξ, η), is in fact the *Green's function* associated with $\mathbf{V}^2$ in $\mathfrak{R}$ when the related boundary conditions prescribe the *solution* of a problem along the boundary of $\mathfrak{R}$, so that the problem is a *Dirichlet problem*. If the Green's function is denoted by $G(x, y; \xi, \eta)$, the above result transposes into the form

$$\varphi(x, y) = \oint_C f \frac{\partial G}{\partial n} ds + \int\int_{\mathfrak{R}} G(x, y; \xi, \eta) h(\xi, \eta) \, d\xi \, d\eta. \tag{95}$$

It should be noted that, in addition to the fact that formal operations involving the delta function need to be justified,† it is also true that since the function G is unbounded when the point $P(x, y)$ and the point $Q(\xi, \eta)$ approach each other (as is next shown), the rigorous mathematical proof that (95) does indeed satisfy (91) and (92) when h and f are sufficiently well behaved is rather formidable, particularly when the region $\mathfrak{R}$ also is unbounded, and is omitted here. In addition, in the examples which follow it will be tacitly assumed that the behavior of h and f is in fact satisfactory.

We notice that when $f = 0$, so that $\varphi = 0$ on C, the solution φ is given completely by the integral over $\mathfrak{R}$, and when $h = 0$, so that $\mathbf{V}^2\varphi = 0$ in $\mathfrak{R}$, the solution is provided by the integral along C. In addition, it is seen that if the differential equation (91) is generalized to the equation

$$\mathbf{V}^2\varphi + \lambda r\varphi = h, \tag{96}$$

where λ is a constant and r is a prescribed function, and if the boundary condition (92) is unchanged, then the additional term

$$-\lambda \int\int_{\mathfrak{R}} G(x, y; \xi, \eta) r(\xi, \eta) \varphi(\xi, \eta) \, d\xi \, d\eta \tag{97}$$

is introduced into the right-hand member of (95). The result is an *integral equation* to be solved for φ, with φ appearing in the integral (97) as well as outside it. Whereas such an equation usually cannot be solved analytically, in closed form, it often is preferable to an alternative formulation of a certain problem when an *approximate* solution is to be obtained, by numerical methods or otherwise.‡ (See, for example, Reference 7.)

In order to exhibit the behavior of the present Green's function as the point $Q(\xi, \eta)$ approaches the point $P(x, y)$, where P is an *interior* point of $\mathfrak{R}$, we consider the result of integrating $\mathbf{V}^2G$ over a circular disk D interior to $\mathfrak{R}$, with center at the point P (Figure 11.23). By making use of the two-dimensional

---

†The justification of the use of (90) when $\mathbf{V}^2G$ is identified with the delta function could consist of first defining $\mathbf{V}^2G$ to be a *continuous* function which (say) vanishes except in the circular disk $(x - \xi)^2 + (y - \eta)^2 \le \epsilon$, and which has the property that its integral over that disk is unity for all positive values of ϵ, then applying (90), and *afterward* taking the limit of the result as $\epsilon \longrightarrow 0$. Similar delayed limiting processes also are to be imagined in connection with subsequent operations involving the delta functions.

‡Treatments in which the principal interest in Green's functions centers on their use in integral equations generally define G with an algebraic sign opposite to that used here, so that the minus sign then is not present in (97).

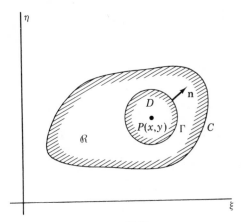

Figure 11.23

specialization of Equation (127), Chapter 6, we obtain the relation

$$\iint_D \nabla^2 G \, d\xi \, d\eta = \oint_\Gamma \frac{\partial G}{\partial n} \, ds, \tag{98}$$

where Γ is the circular boundary of D. Since G satisfies (93), it follows that the left-hand member of (98) has the value 1, regardless of the radius of Γ, so that (98) gives

$$\oint_\Gamma \frac{\partial G}{\partial n} \, ds = 1. \tag{99}$$

If we denote by R the distance PQ,

$$R = \sqrt{(x - \xi)^2 + (y - \eta)^2}, \tag{100}$$

we see that on Γ we may write

$$ds = R \, d\theta, \qquad \frac{\partial G}{\partial n} = \frac{\partial G}{\partial R}.$$

Hence (99) becomes

$$\int_0^{2\pi} R \frac{\partial G}{\partial R} \, d\theta = 1$$

and we deduce that *the mean value of $R \, \partial G/\partial R$ on Γ is $1/(2\pi)$*, so that for small values of R we must have $\partial G/\partial R \sim 1/(2\pi R)$, and hence

$$G \sim \frac{1}{2\pi} \log R \qquad (R \to 0). \tag{101}$$

Thus, in the present case, the Green's function must become logarithmically infinite as $P(x, y)$ and $Q(\xi, \eta)$ approach coincidence inside $\mathfrak{R}$. Consequently, an *alternative* definition of G consists of the requirement that

$$\nabla^2 G = 0 \qquad in \ \mathfrak{R} \quad when \ P(x, y) \neq Q(\xi, \eta), \tag{102}$$

that (101) hold as P and Q approach each other inside $\Re$, and that (94) be satisfied, so that G vanishes on C. The term $(2\pi)^{-1} \log R$ is sometimes called the *principal part* of G. Since $\mathbf{V}^2(\log R) = 0$ when $R \neq 0$, as is easily verified, it follows also that we may write

$$G(x, y; \xi, \eta) = \frac{1}{2\pi} \log R + g(x, y; \xi, \eta) \tag{103}$$

where

$$\mathbf{V}^2 g = 0 \quad in \ \Re \quad \text{and} \quad g = -\frac{1}{2\pi} \log R \quad on \ C. \tag{104}$$

It is found in Problem 63 that

$$G(x, y; \xi, \eta) = G(\xi, \eta; x, y) \tag{105}$$

in the present case, so that the Green's function is *symmetric* in (x, y) and (ξ, η). Thus the operator $\mathbf{V}^2$ in (93), (102), and (104) can be taken to be either $\partial^2/\partial\xi^2 + \partial^2/\partial\eta^2$ or $\partial^2/\partial x^2 + \partial^2/\partial y^2$ (or to be either of two equivalent alternative forms in terms of another coordinate system in the plane).

Example 1. When $\Re$ is the *upper half-plane*, the definition of (103) and (104) permits the determination of G by inspection. For since the function

$$\frac{1}{2\pi} \log R \equiv \frac{1}{2\pi} \log \sqrt{(x - \xi)^2 + (y - \eta)^2}$$

satisfies Laplace's equation in (x, y) for fixed (ξ, η), when $(x, y) \neq (\xi, \eta)$, the function

$$\frac{1}{2\pi} \log \sqrt{(x - \xi)^2 + (y + \eta)^2}$$

also satisfies Laplace's equation in (x, y) for any (x, y) and (ξ, η) in $\Re$, since then $y > 0$ and $\eta > 0$, and clearly the *difference* between these functions vanishes when $y = 0$ or $\eta = 0$. Since also that difference, which can be written conveniently in the form

$$G = \frac{1}{4\pi} \log \left[\frac{(x - \xi)^2 + (y - \eta)^2}{(x - \xi)^2 + (y + \eta)^2} \right], \tag{106}$$

tends to zero as $x^2 + y^2 \rightarrow \infty$ or $\xi^2 + \eta^2 \rightarrow \infty$ in the upper half-plane, it satisfies (103) and (104) in that region and is in fact the required Green's function.†

A simple calculation shows that along the boundary $\eta = 0$ there follows

$$\frac{\partial G}{\partial n} = \left[-\frac{\partial G}{\partial \eta} \right]_{\eta=0} = \frac{1}{\pi} \frac{y}{(x - \xi)^2 + y^2} \tag{107}$$

and hence the solution of the problem

$$\frac{\partial^2\varphi}{\partial x^2} + \frac{\partial^2\varphi}{\partial y^2} = h(x, y),$$

$$\varphi(x, 0) = f(x), \qquad \lim_{x^2+y^2\to 0} \varphi(x, y) = 0 \tag{108}$$

†A procedure of this type, in which superposition is used to achieve antisymmetry of a continuous function about a line when the function is to vanish on the line, or symmetry when the normal derivative of the function is to vanish there, is often called the *method of images*. (Compare Problems 83 and 84 of Chapter 9 and Problems 32 and 33 of Chapter 10, as well as Problem 33 of Chapter 8.)

in the half-plane $-\infty < x < \infty,\ 0 \leqq y < \infty$ can be written in the form

$$\varphi(x, y) = \frac{1}{\pi} \int_{-\infty}^{\infty} \frac{yf(\xi)\,d\xi}{(x - \xi)^2 + y^2} + \int_{-\infty}^{\infty} \int_{0}^{\infty} G(x, y; \xi, \eta)h(\xi, \eta)\,d\xi\,d\eta, \qquad (109)$$

where G is defined by (106), by virtue of (95). In the special case $h = 0$, this result properly reduces to Equation (230) of Section 9.14. Here, as in other examples to follow, it is assumed that $f(x)$ and $h(x, y)$ are such that the integrals in (109) exist and tend to zero as $x^2 + y^2 \longrightarrow \infty$ when $y > 0$. ∎

Example 2. In order to determine the Green's function of Example 1 by a more direct method, we can write $G = (2\pi)^{-1} \log R + g$, where $g(x, y; \xi, \eta)$ is the solution φ of the problem

$$\frac{\partial^2 \varphi}{\partial x^2} + \frac{\partial^2 \varphi}{\partial y^2} = 0,$$

$$\varphi(x, 0) = -\frac{1}{2\pi} \log R(x, 0; \xi, \eta), \qquad \lim_{x^2 + y^2 \to 0} \varphi = 0$$

in the upper half-plane. Here ξ and η are to be treated as *constants* while x and y vary. The solution can be obtained (for example) by use of the Fourier integral and is, in fact, given by Equation (230) of Section 9.14 in the form

$$g(x, y; \xi, \eta) = -\frac{1}{4\pi^2} \int_{-\infty}^{\infty} \frac{y \log \left[(u - \xi)^2 + \eta^2\right]}{y^2 + (u - x)^2}\,du.$$

This integral can be evaluated by first *differentiating* with respect to η, then using residue calculus (Section 10.13), and finally integrating with respect to η. A somewhat involved calculation thus yields the anticipated result

$$g(x, y; \xi, \eta) = -\frac{1}{4\pi} \log \left[(x - \xi)^2 + (y + \eta)^2\right]. \qquad ∎$$

Example 3. Another direct determination of the Green's function in Examples 1 and 2 can be based on the identification $G(x, y; \xi, \eta) = \varphi(x, y)$, with ξ and η again treated as constants, where φ is the solution of the problem

$$\frac{\partial^2 \varphi}{\partial x^2} + \frac{\partial^2 \varphi}{\partial y^2} = \delta(x - \xi)\delta(y - \eta),$$

$$\varphi(x, 0) = 0, \qquad \lim_{x^2 + y^2 \to 0} \varphi(x, y) = 0$$

in the upper half-plane. Perhaps the simplest method of solving this problem is by use of the Fourier transform (see Problem 64). ∎

In physically motivated problems, it is often possible to interpret h and φ in (91) (and in other such equations) as *cause and effect*, or as *input and output*, respectively. In such cases, the definition (93) states that $G(x, y; \xi, \eta)$ can be interpreted as *the effect at $P(x, y)$ due to a unit cause distribution concentrated at $Q(\xi, \eta)$*, so that G is sometimes referred to as an "influence function," and the symmetry property (105) then corresponds to what is often called a "reciprocity relation" in applications. The condition (94) requires here that no causes (or effects) be present along the boundary of the relevant region $\mathfrak{R}$ when G is observed.

Applications of Analytic Function Theory

For example, if φ is interpreted in (91) as *velocity potential* in a two-dimensional fluid flow, then $h(x, y)$ measures the intensity of a distribution of line sources and sinks and, as was shown in Section 11.6, the term

$$\frac{1}{2\pi} \log R = \frac{1}{2\pi} \sqrt{(x - \xi)^2 + (y - \eta)^2}$$

is the potential at a point $P(x, y)$ due to a *unit line source* (that is, a line source of unit strength per unit length normal to the plane of the flow) at a point $Q(\xi, \eta)$. The Green's function (106) accordingly corresponds to the introduction also of a *unit line sink* at the external point $Q'(\xi, -\eta)$ which is the "image" of Q relative to the boundary $\eta = 0$.

If φ is *steady-state temperature*, then the term $-(2\pi)^{-1} \log R$, with a prefixed *negative* sign, is the temperature at $P(x, y)$ due to a unit line heat source at $Q(\xi, \eta)$ when the medium has unit conductivity. Analogous interpretations are possible, for example, in terms of *gravitational potential* (*without* a negative sign) and electrostatic potential (*with* a negative sign).

It should be noted that the properties exhibited for the Green's function in this section have been with reference to the specific operator $L = \nabla^2$, subject to the specification of the solution of the associated problem along the boundary of $\mathfrak{R}$, and they do not necessarily persist in other cases. (See Section 11.11.)

11.10. The Use of Conformal Mapping. When the region $\mathfrak{R}$ is simply connected, we show next that the determination of the Green's function of Section 11.9 can be reduced to the problem of determining an analytic function which specifies a mapping of the region $\mathfrak{R}$ onto the upper half of the w plane.

For this purpose, we begin by noticing that the relevant Green's function for the upper half of the xy plane, as obtained in the preceding Example 1, can be written in the form.

$$G(x, y; \xi, \eta) = \frac{1}{2\pi} \log \left| \frac{z - \zeta}{z - \bar{\zeta}} \right| = \frac{1}{2\pi} \operatorname{Re} \left[\log \left(\frac{z - \zeta}{z - \bar{\zeta}} \right) \right]$$

where $$z = x + iy, \qquad \zeta = \xi + i\eta.$$

Accordingly, we may verify that *if $w = F(z)$ is an analytic function which maps the region $\mathfrak{R}$ in the z plane onto the upper half of the w plane, with $F'(z) \neq 0$ in $\mathfrak{R}$ so that the mapping is one-to-one, then the Green's function which satisfies (102), (101), and (94) is given by*

$$G(x, y; \xi, \eta) = \frac{1}{2\pi} \log \left| \frac{F(z) - F(\zeta)}{F(z) - \overline{F(\zeta)}} \right|$$

$$= \frac{1}{2\pi} \operatorname{Re} \left\{ \log \left[\frac{F(z) - F(\zeta)}{F(z) - \overline{F(\zeta)}} \right] \right\} \tag{110}$$

where again $$z = x + iy, \qquad \zeta = \xi + i\eta. \tag{111}$$

To confirm this assertion, we first observe that, since $\log |\alpha| = \frac{1}{2} \log (\alpha \bar{\alpha})$, Equation (110) also can be written in the form

$$G(x, y; \xi, \eta) = \frac{1}{4\pi} \log \left\{ \frac{[F(z) - F(\zeta)][\overline{F(z)} - \overline{F(\zeta)}]}{[F(z) - \overline{F(\zeta)}][\overline{F(z)} - F(\zeta)]} \right\}, \tag{112}$$

from which it is evident that the right-hand member is *symmetric* in z and ζ, and hence in (x, y) and (ξ, η). Second, since the real part of an analytic function of z satisfies Laplace's equation except at singular points of that function, and since $F(z) = F(\zeta)$ only when $z = \zeta$, and also $F(z) \neq \overline{F(\zeta)}$, when z is inside $\Re$,† it follows that the right-hand member of (110) satisfies Laplace's equation when z is in $\Re$ and $z \neq \zeta$. Third, since $F(\zeta)$ is real when ζ is on C, it follows that the right-hand member of (110) vanishes when ζ is on C and hence also, from the symmetry, when z is on C. Finally, since we have

$$F(z) - F(\zeta) = (z - \zeta)F'(\zeta) + \frac{(z - \zeta)^2}{2!}F''(\zeta) + \cdots,$$

when z and ζ are inside $\Re$, where $F'(\zeta) \neq 0$, we may deduce that the right-hand member of (110) behaves like $(2\pi)^{-1} \log |z - \zeta| = (2\pi)^{-1} \log R$ when $z - \zeta \to 0$, where R is given by (100). Thus the right-hand member of (110) satisfies (102), (101), and (94), and hence it is indeed identified with G.‡

We see that the Schwarz–Christoffel transformation (Section 11.8) is available for the determination of a mapping function F when C is polygonal. An additional useful fact is that the function

$$w = \frac{F(z) - F(\zeta)}{F(z) - \overline{F(\zeta)}}, \tag{113}$$

which appears in (110), is an analytic function which maps $\Re$ conformally onto the unit circular disk $|w| \leq 1$, in such a way that the point $z = \zeta$ maps onto the origin $w = 0$. (See Problem 69.)

Example 1. If $\Re$ is the *circular disk* of unit radius, $x^2 + y^2 \leq 1$, the result of Problem 23(c) gives $F(z) = i(1 - z)/(1 + z)$, in accordance with which (110) determines the Green's function in the form

$$G = \frac{1}{2\pi} \log \left| \frac{z - \zeta}{1 - \bar{\zeta}z} \right|, \tag{114}$$

with $z = x + iy$ and $\zeta = \xi + i\eta$. If polar coordinates are introduced, such that

$$z = re^{i\theta}, \qquad \zeta = \gamma e^{i\beta}, \tag{115}$$

a simple computation yields the expanded form

$$G(r, \theta; \gamma, \beta) = \frac{1}{4\pi} \log \left[\frac{r^2 - 2r\gamma \cos (\theta - \beta) + \gamma^2}{1 - 2r\gamma \cos (\theta - \beta) + r^2\gamma^2} \right]. \tag{116}$$

---

†Notice that the imaginary part of $F(z)$ is positive and the imaginary part of $\overline{F(\zeta)}$ is negative or zero.

‡The consequence that two different identifications of the mapping function $F(z)$ must produce the same G is independently verified in Problem 68.

In addition, we find that on the boundary $\gamma = 1$ there follows

$$\frac{\partial G}{\partial n} = \left[\frac{\partial G}{\partial \gamma}\right]_{\gamma=1} = \frac{1}{2\pi} \frac{1 - r^2}{1 - 2r\cos(\theta - \beta) + r^2}. \tag{117}$$

Thus the solution of the problem

$$\nabla^2 \varphi = h(r, \theta),$$
$$\varphi(1, \theta) = f(\theta) \tag{118}$$

in the unit disk $|r| \leq 1$ is given by

$$\varphi(r, \theta) = \frac{1}{2\pi} \int_0^{2\pi} \frac{(1 - r^2)f(\beta)}{1 - 2r\cos(\theta - \beta) + r^2}\, d\beta + \int_0^{2\pi}\int_0^1 G(r, \theta; \gamma, \beta)h(\gamma, \beta)\, d\gamma\, d\beta. \tag{119}$$

When $h = 0$, this result reduces to the Poisson integral formula of Section 9.5, with $a = 1$. ∎

Example 2. If $\Re$ is the *infinite strip* $-\infty < x < \infty$, $0 \leq y \leq b$, the result of Problem 23(a) combines with (110) to yield the expression

$$G = \frac{1}{2\pi} \log\left|\frac{e^{\pi z/b} - e^{\pi \zeta/b}}{e^{\pi z/b} - e^{\pi \bar{\zeta}/b}}\right| \tag{120}$$

or, after some manipulation,

$$G(x, y; \xi, \eta) = \frac{1}{4\pi} \log\left[\frac{\cosh\frac{\pi}{b}(\xi - x) - \cos\frac{\pi}{b}(\eta - y)}{\cosh\frac{\pi}{b}(\xi - x) - \cos\frac{\pi}{b}(\eta + y)}\right]. \tag{121}$$

Thus, for example, the solution of the problem

$$\nabla^2 \varphi = h(x, y),$$
$$\varphi(x, 0) = \varphi(x, b) = 0, \tag{122}$$

for $-\infty < x < \infty$, $0 \leq y \leq b$ is given by

$$\varphi(x, y) = \int_0^b \int_{-\infty}^{\infty} G(x, y; \xi, \eta)h(\xi, \eta)\, d\xi\, d\eta, \tag{123}$$

with G defined by (121). To solve the problem

$$\nabla^2 \varphi = 0,$$
$$\varphi(x, 0) = f(x), \qquad \varphi(x, b) = 0 \tag{124}$$

in that strip, we first make the computation

$$\left[-\frac{\partial G}{\partial n}\right]_{\eta=0} = \frac{1}{2b} \frac{\sin\frac{\pi y}{b}}{\cosh\frac{\pi}{b}(\xi - x) - \cos\frac{\pi y}{b}} \tag{125}$$

and deduce that

$$\varphi(x, y) = \frac{1}{2b} \sin\frac{\pi y}{b} \int_{-\infty}^{\infty} \frac{f(\xi)}{\cosh\frac{\pi}{b}(\xi - x) - \cos\frac{\pi y}{b}}\, d\xi, \tag{126}$$

in accordance with the result of Problem 82 of Chapter 9. ∎

11.11. Other Two-Dimensional Green's Functions. For a more general two-dimensional linear partial differential operator L and an associated region $\mathcal{R}$, one method of defining a Green's function G starts with the derivation of a related "Green's theorem" of the form

$$\iint_{\mathcal{R}} G\, L\varphi\, d\xi\, d\eta = \oint_C (\cdots)\, ds + \iint_{\mathcal{R}} \varphi\, L^*G\, d\xi\, d\eta, \qquad (127)$$

where L^* is a new differential operator, said to be the *adjoint operator* relative to L, and where $(\cdots)$ denotes an expression to be integrated along the boundary C. When $L^* = L$, as has been seen in (90) to be the case when $L = \mathbf{V}^2$, the operator L is said to be *self-adjoint*. In the special case $L = \mathbf{V}^2$, reference to (90) shows that the expression $(\cdots)$ is of the form $G(\partial\varphi/\partial n) - \varphi(\partial G/\partial n)$.

The relevant Green's function, *if it exists*, then is defined to be that function $G = G(x, y; \xi, \eta)$ which satisfies the differential equation

$$L^*G = \delta(\xi - x)\delta(\eta - y) \qquad in\ \mathcal{R}, \qquad (128)$$

together with certain specified homogeneous boundary conditions on C. In particular, if φ is to satisfy the equation

$$L\varphi = h(x, y) \qquad in\ \mathcal{R} \qquad (129)$$

and if, along each part of C, φ and/or $\partial\varphi/\partial n$ is prescribed, then along each part of C one requires that the corresponding G be such that the only nonvanishing terms in $(\cdots)$ are those which involve *prescribed* information relative to φ and/or $\partial\varphi/\partial n$. With G so defined, Equation (127) takes the form

$$\varphi(x, y) = -\oint_C (\cdots)\, ds + \iint_{\mathcal{R}} G(x, y; \xi, \eta)h(\xi, \eta)\, d\xi\, d\eta, \qquad (130)$$

where $(\cdots)$ now involves only known functions.

If L is *self-adjoint*, and if also it happens that the expression $(\cdots)$ in (130) vanishes everywhere on C whenever $\varphi(\xi, \eta)$ satisfies the same homogeneous conditions on C as does G, it is found that G is *symmetric* in (x, y) and (ξ, η), so that $G(x, y; \xi, \eta) = G(\xi, \eta; x, y)$. (See Problem 85.) In this case the operator L^* in (128) is identified with L and, because of the symmetry, the differentiations in LG may be either with respect to the coordinates of $Q(\xi, \eta)$, as was assumed in the derivation, or with respect to the coordinates of $P(x, y)$.[†] Except in the special cases when symmetry of G can be anticipated, the operator L^* in (128) *must* differentiate with respect to the coordinates of Q if (130) is to be valid.

In illustration, the symmetry of G in (95) could be deduced from the facts that, first, $L^* = L\ (= \mathbf{V}^2)$ and, second, G is to *vanish* on C and the boundary integrand in (95) will vanish if f (the boundary value of φ) *also* vanishes. Ex-

[†]In some references, the differential operator L is said to be "*formally* self-adjoint" when $L^* = L$. When also the homogeneous conditions which must be satisfied by an associated Green's function G on the boundary of the relevant region $\mathcal{R}$ are such that G is in fact symmetric, the term "self-adjoint" (without a modifier) then is used to describe the "domain operator" specified by L *and* by those homogeneous conditions.

ample 4 of this section will illustrate a situation in which $L^* = L$ but G is *not* symmetric.

The transformation of the left-hand member of (127) into the right-hand member can be effected by use of the two-dimensional analogies to integration by parts, yielded by the two-dimensional specialization of the divergence theorem, in the forms

$$\iint_{\mathfrak{R}} u \frac{\partial v}{\partial \xi} \, d\xi \, d\eta = \iint_{\mathfrak{R}} \frac{\partial}{\partial \xi}(uv) \, d\xi \, d\eta - \iint_{\mathfrak{R}} v \frac{\partial u}{\partial \xi} \, d\xi \, d\eta$$

or

$$\iint_{\mathfrak{R}} u \frac{\partial v}{\partial \xi} \, d\xi \, d\eta = \oint_{C} uv \cos \alpha \, ds - \iint_{\mathfrak{R}} v \frac{\partial u}{\partial n} \, d\xi \, d\eta \qquad (131a)$$

and, similarly,

$$\iint_{\mathfrak{R}} u \frac{\partial v}{\partial \eta} \, d\xi \, d\eta = \oint_{C} uv \cos \beta \, ds - \iint_{\mathfrak{R}} v \frac{\partial u}{\partial \eta} \, d\xi \, dy, \qquad (131b)$$

where
$$\mathbf{n} = \mathbf{i} \cos \alpha + \mathbf{j} \cos \beta \qquad (132)$$

is the unit outward normal vector at a point of C. One of the variables may represent time, in which case the region $\mathfrak{R}$ is a space–time domain.

Example 1. For the *modified Helmholtz operator*

$$L = \mathbf{\nabla}^2 - k^2,$$

with k real and positive, there follows

$$\iint_{\mathfrak{R}} G(\mathbf{\nabla}^2\varphi - k^2\varphi) \, d\xi \, d\eta = \oint_{C} \left(G \frac{\partial \varphi}{\partial n} - \varphi \frac{\partial G}{\partial n} \right) ds + \iint_{\mathfrak{R}} \varphi(\mathbf{\nabla}^2 G - k^2 G) \, d\xi \, d\eta.$$

Hence, for the self-adjoint operator $L = \mathbf{\nabla}^2 - k^2$, an associated Green's function must be such that

$$\mathbf{\nabla}^2 G - k^2 G = 0 \quad \text{in } \mathfrak{R} \quad \text{when } (x, y) \neq (\xi, \eta) \qquad (133)$$

and the requirement

$$1 = \lim_{R \to 0} \iint_{\Gamma_R} (\mathbf{\nabla}^2 G - k^2 G) \, d\xi \, d\eta = \lim_{R \to 0} \left(\int_{0}^{2\pi} R \frac{\partial G}{\partial R} \, d\theta - k^2 \iint_{\Gamma_R} G \, d\xi \, d\eta \right),$$

where Γ_R is a circular disk of radius R, with center at $P(x, y)$, leads to the condition

$$G \sim \frac{1}{2\pi} \log R \qquad (R \longrightarrow 0) \qquad (134)$$

as before, the term $\iint_{\Gamma_R} G \, d\xi \, d\eta$ then tending to zero with $\frac{1}{2}R^2 \log R$ as $R \longrightarrow 0$.

In the special case when $\mathfrak{R}$ is the entire plane, it is convenient to write

$$\xi - x = R \cos \psi, \qquad \eta - y = R \sin \psi$$

and to introduce R and ψ as new (polar) coordinates. Since G clearly will not depend upon ψ, Equation (133) becomes

$$\frac{d^2 G}{dR^2} + \frac{1}{R}\frac{dG}{dR} - k^2 G = 0 \qquad (R \neq 0)$$

and there follows

$$G = c_1 I_0(kR) + c_2 K_0(kR) \qquad (R \neq 0), \qquad (135)$$

where I_0 and K_0 are modified Bessel functions (Section 4.8). Since $I_0(x) \longrightarrow 1$ and $K_0(x) \sim -\log x$ as $x \longrightarrow 0$, the condition (134) requires that $c_2 = -1/(2\pi)$. In addition, since $I_0(x) \sim e^x/\sqrt{2\pi x}$ and $K_0(x) \sim e^{-x}/\sqrt{2x/\pi}$ as $x \longrightarrow \infty$, we must take $c_1 = 0$ in order that G tend to zero as $R \longrightarrow \infty$. Hence there follows

$$G = -\frac{1}{2\pi} K_0(kR). \qquad (136)$$

Thus, for example, the solution of the problem

$$\nabla^2 \varphi - k^2 \varphi = h(x, y),$$
$$\lim_{x^2 + y^2 \to \infty} \varphi = 0, \qquad (137)$$

for $-\infty < x < \infty$, $-\infty < y < \infty$, is given by

$$\varphi(x, y) = -\frac{1}{2\pi} \int_{-\infty}^{\infty} \int_{-\infty}^{\infty} K_0(k\sqrt{(x - \xi)^2 + (y - \eta)^2})\, h(\xi, \eta)\, d\xi\, d\eta. \qquad (138)$$

Clearly, the corresponding Green's function in the *upper half-plane* is

$$G = -\frac{1}{2\pi}[K_0(k\sqrt{(x - \xi)^2 + (y - \eta)^2}) - K_0(k\sqrt{(x - \xi)^2 + (y + \eta)^2})], \qquad (139a)$$

when $\varphi(x, 0)$ is to be prescribed, and is

$$G = -\frac{1}{2\pi}[K_0(k\sqrt{(x - \xi)^2 + (y - \eta)^2}) + K_0(k\sqrt{(x - \xi)^2 + (y + \eta)^2})], \qquad (139b)$$

when $\varphi_y(x, 0)$ is to be prescribed.

It is important to notice that conformal mapping, as used in Section 11.10 when $k = 0$, cannot be used when $k \neq 0$, to determine the Green's function in a region $\mathfrak{R}$ by mapping that region onto (say) a half-plane. This negative result follows from the fact that if $u + iv = F(x + iy)$, where F is an analytic function of $z = x + iy$, then

$$\frac{\partial^2 \varphi}{\partial x^2} + \frac{\partial^2 \varphi}{\partial y^2} + \lambda\varphi = |F'(z)|^2 \left(\frac{\partial^2 \varphi}{\partial u^2} + \frac{\partial^2 \varphi}{\partial v^2} \right) + \lambda\varphi.$$

[See Equation (41) of Chapter 11.] Thus, whereas $\varphi_{xx} + \varphi_{yy} = 0$ implies $\varphi_{uu} + \varphi_{vv} = 0$ when $|F'(z)| \neq 0$, the equation $\varphi_{xx} + \varphi_{yy} + \lambda\varphi = 0$ does *not* imply the equation $\varphi_{uu} + \varphi_{vv} + \lambda\varphi = 0$ unless $\lambda = 0$; that is, the Helmholtz (or modified Helmholtz) operator is not invariant under a conformal mapping unless $k = 0$.

In addition, we see that, whereas the Green's function (136) properly tends to zero as $x^2 + y^2 \longrightarrow \infty$, for any positive constant k, the function $(2\pi)^{-1} \log R + c$ ($c = $ constant) which results when $k = 0$ (see Problem 86) does *not* have this property. Correspondingly, it is found that the problem

$$\nabla^2 \varphi = h(x, y),$$
$$\lim_{x^2 + y^2 \to \infty} \varphi = 0, \qquad (137')$$

for $-\infty < x < \infty$, $-\infty < y < \infty$, generally does not possess a solution. When h is respectable, the function

$$\varphi(x, y) = \frac{1}{2\pi} \int_{-\infty}^{\infty} \int_{-\infty}^{\infty} \log \sqrt{(x - \xi)^2 + (y - \eta)^2} \, h(\xi, \eta) \, d\xi \, d\eta \tag{138'}$$

will satisfy $\nabla^2\varphi = h$, but it also will tend to zero as $x^2 + y^2 \longrightarrow \infty$ only if also h is such that

$$\int_{-\infty}^{\infty} \int_{-\infty}^{\infty} h(\xi, \eta) \, d\xi \, d\eta = 0, \tag{140}$$

in which case the addition of a constant c to $(2\pi)^{-1} \log R$ will not modify the definition (138'). ∎

Example 2. For the *Helmholtz operator*

$$L = \nabla^2 + k^2,$$

with k real and positive, we find that (135) is replaced by

$$G = c_1 J_0(kR) + c_2 Y_0(kR) \qquad (R \neq 0). \tag{141}$$

Since $J_0(x) \longrightarrow 1$ and $Y_0(x) \sim (2/\pi) \log x$ as $x \longrightarrow 0$, we must have $c_2 = \frac{1}{4}$. But since $J_0(x) \sim \sqrt{2/(\pi x)} \cos (x - \pi/4)$ and $Y_0(x) \sim \sqrt{2/(\pi x)} \sin (x - \pi/4)$ as $x \longrightarrow \infty$, the requirement that G (and/or $\partial G/\partial R$) tend to zero as $R \longrightarrow \infty$ permits c_1 to be arbitrary! Frequently the Helmholtz operator appears in a formulation as a result of requiring that the equation

$$\nabla^2 U - \frac{1}{c^2} \frac{\partial^2 U}{\partial t^2} = h(x, y)e^{i\omega t} \tag{142}$$

admit a solution of the form

$$U(x, y, t) = \varphi(x, y)e^{i\omega t},$$

in which case φ must satisfy the equation

$$\nabla^2\varphi + k^2\varphi = h(x, y), \tag{143}$$

with $k = \omega/c$. If the relevant region $\mathfrak{R}$ comprises the entire plane, the usual requirement imposed by the underlying problem is the specification that U satisfy the so-called *radiation condition*, that is, that it involve only "outward-traveling waves" as $R \longrightarrow \infty$ (see Section 9.13). The corresponding requirement that $Ge^{i\omega t}$ satisfy this condition demands that the right-hand member of (141) be a multiple of the Hankel function

$$H_0^{(2)}(kR) \equiv J_0(kR) - iY_0(kR)$$

(Section 9.12), and hence that $c_2 = -ic_1$ in (141). Thus we must have $c_1 = ic_2 = i/4$ and hence

$$G = \frac{i}{4} H_0^{(2)}(kR), \tag{144}$$

so that the solution of (143) for which $U = \varphi e^{i\omega t}$ satisfies the radiation condition as $x^2 + y^2 \longrightarrow \infty$ is

$$\varphi(x, y) = \frac{i}{4} \int_{-\infty}^{\infty} \int_{-\infty}^{\infty} H_0^{(2)}(k\sqrt{(x - \xi)^2 + (y - \eta)^2}) \, h(\xi, \eta) \, d\xi \, d\eta. \tag{145}$$

∎

Example 3. For the *Laplace operator*

$$L = \nabla^2,$$

when the associated boundary condition specifies the *normal derivative* of the solution on the boundary C of a region $\mathcal{R}$, so that the related problems are *Neumann problems*, it would appear from the related Green's theorem (90) that the relevant Green's function is the solution of (93) for which $\partial G/\partial n = 0$ on C. However, we notice that, if this were assumed to be the case, then the relation

$$\iint_{\mathcal{R}} \nabla^2 G \, d\xi \, d\eta = \oint_C \frac{\partial G}{\partial n} \, ds \qquad (146)$$

would provide a contradiction since the left-hand member would be 1 and the right-hand member would be 0. Thus G in fact *does not exist in this case.* This situation is a consequence of two related facts. First, if φ is to satisfy

$$\begin{aligned}\nabla^2 \varphi &= h &&\text{in } R, \\ \frac{\partial \varphi}{\partial n} &= g &&\text{on } C,\end{aligned} \qquad (147)$$

then the relation

$$\iint_{\mathcal{R}} \nabla^2 \varphi \, dx \, dy = \oint_C \frac{\partial \varphi}{\partial n} \, ds$$

requires that

$$\iint_{\mathcal{R}} h \, dx \, dy = \oint_C g \, ds. \qquad (148)$$

Unless g and h satisfy this compatibility condition, the problem (147) has *no solution.* In addition, as has been noted, if (147) does have a solution, then to that solution can be added any *constant,* so that the solution is not *unique.*

When (148) is satisfied (and g and h are respectable), the problem (147) is known to have a solution which is unique apart from this additive constant. (See Section 9.2 relative to the case $h = 0$.) In addition, a *modified Green's function* can be defined in either of two ways, to serve in place of the nonexistent G.

Neumann's function, denoted here by $N(x, y; \xi, \eta)$, is required to satisfy the equation

$$\nabla_{\varrho}^2 N = \delta(\xi - x)\delta(\eta - y) \qquad \text{in } \mathcal{R} \qquad (149)$$

and the boundary condition

$$\frac{\partial N}{\partial n_{\varrho}} = \alpha(s) \qquad \text{on } C, \qquad (150)$$

where $\alpha(s)$ is any function for which the condition replacing (146),

$$\iint_{\mathcal{R}} \nabla_{\varrho}^2 N \, d\xi \, d\eta = \oint_C \frac{\partial N}{\partial n_{\varrho}} \, ds,$$

is satisfied, and hence for which

$$\oint_C \alpha(s) \, ds = 1. \qquad (151)$$

Here, since (for the first time) symmetry in (x, y) and (ξ, η) is not necessarily present, ambiguity has been avoided by use of the notations ∇_{ϱ}^2 and $\partial/\partial n_{\varrho}$ to specify that the relevant differentiations are with respect to ξ and η, with x and y held fixed.

In terms of the function N, the associated Green's theorem

$$\iint_{\mathfrak{R}} \varphi\,\nabla_{\varrho}^2 N\,d\xi\,d\eta = \oint_C \left(\varphi\,\frac{\partial N}{\partial n_{\varrho}} - N\,\frac{\partial \varphi}{\partial n_{\varrho}}\right) ds + \iint_{\mathfrak{R}} N\,\nabla_{\varrho}^2\varphi\,d\xi\,d\eta \qquad (90')$$

then provides the formula

$$\varphi(x, y) = A - \oint_C N(x, y; \xi(s), \eta(s))\,g(s)\,ds + \iint_{\mathfrak{R}} N(x, y; \xi, \eta)h(\xi, \eta)\,d\xi\,d\eta, \qquad (152)$$

where $A = \oint_C \alpha\varphi\,ds$ is the anticipated arbitrary constant.

It is apparent that to any specific N can be added any function of x and y only, since the satisfaction of (149), (150), and (90') will not be affected. Moreover, by virtue of (148), this modification of N will not modify the right-hand member of (152). The additive function of x and y can be so chosen that N satisfies the supplementary condition

$$\oint_C \alpha(s)N(x, y; \xi(s), \eta(s))\,ds = \text{constant}, \qquad (153)$$

in which case Problem 88 shows that N becomes *symmetric* in (x, y) and (ξ, η). Problem 89 deals with the Neumann function when $\mathfrak{R}$ is a unit disk and hence also provides the basis for the determination of N in correspondence with other (simple) regions by use of conformal mapping.

Hilbert's function, denoted here by $H(x, y; \xi, \eta)$, is required to satisfy the equation

$$\nabla_{\varrho}^2 H = \delta(\xi - x)\delta(\eta - y) - \beta(\xi, \eta) \qquad in\ \mathfrak{R} \qquad (154)$$

and the boundary condition

$$\frac{\partial H}{\partial n_{\varrho}} = 0 \qquad on\ C, \qquad (155)$$

where $\beta(\xi, \eta)$ is any function for which the condition replacing (146) is satisfied, so that

$$\iint_{\mathfrak{R}} \nabla_{\varrho}^2 H\,d\xi\,d\eta = \oint_C \frac{\partial H}{\partial n_{\varrho}}\,ds,$$

and hence for which

$$\iint_{\mathfrak{R}} \beta(\xi, \eta)\,d\xi\,d\eta = 1. \qquad (156)$$

In terms of the function H, the relation (90), with G replaced by H, takes the form

$$\varphi(x, y) = B - \oint_C H(x, y; \xi(s), \eta(s))\,g(s)\,ds + \iint_{\mathfrak{R}} H(x, y; \xi, \eta)h(\xi, \eta)\,d\xi\,d\eta, \qquad (157)$$

where $B = \iint_{\mathfrak{R}} \beta\varphi\,d\xi\,d\eta$ now is the arbitrary constant. An arbitrary additive function of x and y, which clearly is present in H, but which does not affect (157) when (148) is satisfied, can be so determined that

$$\iint_{\mathfrak{R}} \beta(\xi, \eta)H(x, y; \xi, \eta)\,d\xi\,d\eta = \text{constant}, \qquad (158)$$

in which case H becomes symmetric in (x, y) and (ξ, η). (See Problem 91.) The Hilbert function is treated in Problem 92 when $\mathfrak{R}$ is the unit disk and in Problem 94 in other cases. ∎

Example 4. For the operator

$$L = c^2 \frac{\partial^2}{\partial x^2} - \frac{\partial^2}{\partial t^2},$$

associated with the *wave equation*, there follows

$$\int_0^T \int_a^b G(c^2\varphi_{\xi\xi} - \varphi_{\tau\tau}) \, d\xi \, d\tau = c^2 \int_0^T \left[G\varphi_\xi - G_\xi\varphi \right]_{\xi=a}^b d\tau$$
$$- \int_a^b \left[G\varphi_\tau - G_\tau\varphi \right]_{\tau=0}^T d\xi + \int_0^T \int_a^b \varphi(c^2 G_{\xi\xi} - G_{\tau\tau}) \, d\xi \, d\tau, \quad (159)$$

when $\mathfrak{R}$ is the space–time domain $a \leqq x \leqq b$, $0 \leqq t \leqq T$, where T is any positive constant. Thus $L^* = L$ and the Green's function must satisfy the equation

$$c^2 \frac{\partial^2 G}{\partial \xi^2} - \frac{\partial^2 G}{\partial \tau^2} = \delta(\xi - x)\delta(\tau - t) \quad in \ \mathfrak{R}. \quad (160)$$

We suppose here that $(a, b) = (-\infty, \infty)$ and that the associated problem is that of determining $\varphi(x, t)$ such that

$$c^2 \frac{\partial^2 \varphi}{\partial x^2} - \frac{\partial^2 \varphi}{\partial t^2} = h(x, t), \quad (161)$$

for $-\infty < x < \infty$, $0 \leqq t \leqq T$, subject to the initial conditions

$$\varphi(x, 0) = f(x), \qquad \varphi_t(x, 0) = g(x), \quad (162)$$

Accordingly, in order to suppress the boundary terms in (159) which involve unspecified information, we require that G satisfy the conditions

$$\lim_{\xi \to \pm\infty} G = \lim_{\xi \to \pm\infty} G_\xi = 0 \qquad (0 \leqq \tau \leqq T) \quad (163)$$

and

$$G = G_\tau = 0 \quad when \ \tau = T \quad (-\infty < x < \infty). \quad (164)$$

In terms of this function, the required solution φ is given by (159) in the form

$$\varphi(x, t) = \int_{-\infty}^\infty [G_\tau(x, t; \xi, 0)f(\xi) - G(x, t; \xi, 0)g(\xi)] \, d\xi$$
$$+ \int_0^T \int_{-\infty}^\infty G(x, t; \xi, \tau)h(\xi, \tau) \, d\xi \, d\tau. \quad (165)$$

For the purpose of determining G,† it is convenient to first take the Fourier transform, with respect to ξ, of the equal members of (160). If we write $\bar{G}$ for the transform of G, so that

$$\bar{G}(x, t; u, \tau) = \int_{-\infty}^\infty e^{-iu\xi} G(x, t; \xi, \tau) \, d\xi, \quad (166)$$

---

†Here, in spite of the fact that $L^* = L$, the Green's function will not be symmetric in (x, t) and (ξ, τ). It should be noticed that the condition required for symmetry in Problem 85 is not satisfied here, since G and its τ derivative are to vanish when $\tau = T$, whereas the boundary integrand in (165) will vanish identically only if f and g vanish, so that $\varphi(\xi, \tau)$ and its τ derivative vanish when $\tau = 0$.

then the transforms of $\partial^2 G/\partial \xi^2$, $\partial^2 G/\partial \tau^2$, and $\delta(\xi - x)$ are $-u^2 \bar{G}$, $\partial^2 \bar{G}/\partial \tau^2$, and e^{-iux}, respectively, and from (160) and (164) there follows

$$\frac{\partial^2 \bar{G}}{\partial \tau^2} + c^2 u^2 \bar{G} = -e^{-iux}\delta(\tau - t), \tag{167}$$

where

$$\bar{G} = \frac{\partial \bar{G}}{\partial \tau} = 0 \qquad \text{when } \tau = T. \tag{168}$$

The solution of (167) satisfying (168) is obtainable by the elementary method of variation of parameters [see Section 1.9 or Problem 110 of Chapter 9] in the form

$$\bar{G} = -e^{-iux} \int_T^\tau \frac{\sin cu(\tau - \tau_1)}{cu} \delta(\tau_1 - t) \, d\tau_1$$

$$= e^{-iux} \int_\tau^T \frac{\sin cu(\tau - \tau_1)}{cu} \delta(\tau_1 - t) \, d\tau_1$$

$$= \begin{cases} \dfrac{e^{-iux} \sin cu(\tau - t)}{cu} & (t > \tau), \\ 0 & (t < \tau). \end{cases} \tag{169}$$

Here the lower limit T in the first right-hand member is introduced to make $\bar{G} = 0$ when $\tau = T$ and the second right-hand member is written in such a way that the variable of integration is increasing, so that the contribution of the delta function becomes evident. When the transform (169) is inverted, there follows

$$G = -\frac{1}{2\pi} \int_{-\infty}^\infty \frac{e^{-iu(x-\xi)} \sin cu(t - \tau)}{cu} \, du$$

$$= -\frac{1}{2\pi c} \int_{-\infty}^\infty \frac{\cos u(x - \xi) \sin cu(t - \tau)}{u} \, du$$

when $\tau < t$ only. This integral can be evaluated by use of residue calculus [or by reference to Equation (247) of Chapter 5] to give the result

$$G(x, t; \xi, \tau) = \begin{cases} -\dfrac{1}{2c} & \text{when } |x - \xi| < c(t - \tau), \\ -\dfrac{1}{4c} & \text{when } |x - \xi| = c(t - \tau), \\ 0 & \text{otherwise.} \end{cases} \tag{170}$$

We see that G is independent of T, and hence can be considered to be associated with the domain $-\infty < x < \infty, 0 \leq t < \infty$.

Thus $G = -1/(2c)$ inside the sector

$$x - c(t - \tau) < \xi < x + c(t - \tau)$$

of the $\xi \tau$ plane and $G = 0$ outside that sector. With the Heaviside *unit step function* notation

$$H(u) = \begin{cases} 1 & (u > 0), \\ 0 & (u < 0), \end{cases} \tag{171}$$

Equation (170) also can be written in the form†

$$G = \begin{cases} -\dfrac{1}{2c}[H(x - \xi + c(t - \tau)) - H(x - \xi - c(t - \tau))] & (\tau < t), \\ 0 & (\tau > t), \end{cases} \tag{172}$$

and hence there follows further (see Problem 32 of Chapter 2)

$$\frac{\partial G}{\partial \tau} = \frac{1}{2}[\delta(x - \xi + c(t - \tau)) + \delta(x - \xi - c(t - \tau))] \tag{173}$$

when $\tau < t$. Accordingly, Equation (165) takes the form

$$\varphi(x, t) = \frac{1}{2}[f(x + ct) + f(x - ct)] + \frac{1}{2c} \int_{x-ct}^{x+ct} g(\xi)\,d\xi$$

$$- \frac{1}{2c} \int_0^t \int_{x-c(t-\tau)}^{x+c(t-\tau)} h(\xi, \tau)\,d\xi\,d\tau. \tag{174}$$

This relation is equivalent to the result of Problem 37 of Chapter 8, and it reduces when $h = 0$ to D'Alembert's formula [Equation (67) of Chapter 8]. ∎

Example 5. For the operator

$$L = \alpha^2 \frac{\partial^2}{\partial x^2} - \frac{\partial}{\partial t},$$

associated with the *heat-flow* (or *diffusion*) *equation*, there follows

$$\int_0^T \int_a^b G(\alpha^2 \varphi_{\xi\xi} - \varphi_\tau)\,d\xi\,d\tau = \alpha^2 \int_0^T \Big[G\varphi_\xi - G_\xi\varphi \Big]_{\xi=a}^b d\tau$$

$$- \int_a^b \Big[G\varphi \Big]_{\tau=0}^T d\xi + \int_0^T \int_a^b \varphi(\alpha^2 G_{\xi\xi} + G_\tau)\,d\xi\,d\tau, \tag{175}$$

when $\Re$ is the space–time domain $a \leq x \leq b$, $0 \leq t \leq T$. In particular, we see that

$$L^* = \alpha^2 \frac{\partial^2}{\partial x^2} + \frac{\partial}{\partial t},$$

in terms of x- and t-differentiation, so that L here is *not* self-adjoint. An associated Green's function correspondingly must satisfy the equation

$$\alpha^2 \frac{\partial^2 G}{\partial \xi^2} + \frac{\partial G}{\partial \tau} = \delta(\xi - x)\delta(\tau - t) \qquad in \ \Re, \tag{176}$$

together with appropriate homogeneous conditions along the space and time boundaries of $\Re$.

In illustration, in correspondence with the problem

$$\alpha^2 \frac{\partial^2 \varphi}{\partial x^2} = \frac{\partial \varphi}{\partial t} + h(x, t),$$

$$\varphi(x, 0) = f(x), \tag{177}$$

for $-\infty < x < \infty$, $0 \leq t \leq T$, we require that G satisfy the conditions

$$\lim_{\xi \to \pm\infty} G = \lim_{\xi \to \pm\infty} G_\xi = 0 \qquad (0 \leq \tau \leq T) \tag{178}$$

†It is easily verified that, if $f(u) = 1$ when $|u| < \alpha$ and $f(u) = 0$ when $|u| > \alpha$, then there follows $f(u) = H(u + \alpha) - H(u - \alpha)$.

and

$$G = 0 \quad when \ \tau = T \quad (-\infty < \xi < \infty), \tag{179}$$

so that (175) gives

$$\varphi(x, t) = -\int_{-\infty}^{\infty} G(x, t; \xi, 0) f(\xi) \, d\xi + \int_0^T \int_{-\infty}^{\infty} G(x, t; \xi, \tau) h(\xi, \tau) \, d\xi \, d\tau. \tag{180}$$

In order to determine G in this case, we again introduce the Fourier transform with respect to ξ, and find that $\bar{G}$ must be such that

$$\frac{\partial \bar{G}}{\partial \tau} - \alpha^2 u^2 \bar{G} = e^{-iux} \delta(\tau - t),$$

$$\bar{G} = 0 \quad when \ \tau = T, \tag{181}$$

so that

$$\begin{aligned}
\bar{G} &= e^{-iux} \int_T^\tau e^{-\alpha^2 u^2 (\tau_1 - \tau)} \delta(\tau_1 - t) \, d\tau_1 \\
&= -e^{-iux} \int_\tau^T e^{-\alpha^2 u^2 (\tau_1 - \tau)} \delta(\tau_1 - t) \, d\tau_1 \\
&= \begin{cases} -e^{-iux} e^{-\alpha^2 u^2 (t - \tau)} & (\tau < t), \\ 0 & (\tau > t). \end{cases}
\end{aligned} \tag{182}$$

Hence there follows

$$\begin{aligned}
G &= -\frac{1}{2\pi} \int_{-\infty}^{\infty} e^{-iu(x - \xi)} e^{-\alpha^2 u^2 (t - \tau)} \, du \\
&= -\frac{1}{2\pi} \int_{-\infty}^{\infty} e^{-\alpha^2 u^2 (t - \tau)} \cos u(\xi - x) \, du \\
&= -\frac{1}{2\alpha\sqrt{\pi(t - \tau)}} e^{-[(\xi - x)^2 / 4\alpha^2 (t - \tau)]}
\end{aligned} \tag{183}$$

when $\tau < t$ only, by virtue of Equation (241) of Chapter 9. The result of introducing (183) into (180) reduces to the result of Problem 85 of Chapter 9 when $h = 0$. As in the preceding example, G is independent of T, which accordingly can be taken to be $+\infty$.

It is easily seen that the corresponding Green's function for the region $0 \le x < \infty$, $0 \le t < \infty$ is given by

$$G = -\frac{1}{2\alpha\sqrt{\pi(t - \tau)}} [e^{-[(\xi - x)^2 / 4\alpha^2 (t - \tau)]} \mp e^{-[(\xi + x)^2 / 4\alpha^2 (t - \tau)]}], \tag{184}$$

when $\tau < t$ only, with the upper sign applying if $\varphi(0, t)$ is prescribed and the lower sign if $\varphi_x(0, t)$ is known. The appropriate specialization of (175) then solves the relevant modification of the problem (177). [See also Equation (242) of Chapter 9, as well as Problems 86 and 87 of that chapter, when $h = 0$.] ∎

REFERENCES

1. References at end of Chapters 8 and 10.

2. GREENBERG, M. D., *Application of Green's Functions in Science and Engineering*, Prentice-Hall, Inc., Englewood Cliffs, N.J., 1971.

3. KOBER, H., *Dictionary of Conformal Representations*, Dover Publications, Inc., New York, 1957.

4. LE PAGE, W. R., *Complex Variables and the Laplace Transform for Engineers*, McGraw-Hill Book Company, Inc., New York, 1961.

5. MACKIE, A. G., *Boundary Value Problems*, Hafner Publishing Company, Inc., New York, 1965.

6. McLACHLAN, N. W., *Modern Operational Calculus*, Dover Publications, Inc., New York, 1962.

7. MIKHLIN, S. G., *Integral Equations and Applications*, Pergamon Press, Inc., Elmsford, N.Y., 1957.

8. NEHARI, Z., *Conformal Mapping*, McGraw-Hill Book Company, Inc., New York, 1952.

9. ROTHE, R., F. OLLENDORF, and K. POHLHAUSEN, *Theory of Functions as Applied to Engineering Problems*, Dover Publications, Inc., New York, 1961.

10. SOMMERFELD, A., *Partial Differential Equations in Physics*, Academic Press, Inc., New York, 1949.

11. WALKER, M., *Schwarz–Christoffel Transformation and Its Applications: A Simple Exposition* (original title: *Conjugate Functions for Engineers*), Dover Publications, Inc., New York, 1964.

PROBLEMS

Section 11.2

1. Use Equation (8) to determine the inverse Laplace transform of each of the following functions, and compare with the appropriate formula given in Table 1, pages 67 and 68:

(a) $\dfrac{1}{s+a}$,   (b) $\dfrac{1}{s^2+a^2}$,   (c) $\dfrac{1}{s^2-a^2}$,

(d) $\dfrac{1}{s^4}$,   (e) $\dfrac{s}{(s^2+a^2)^2}$,   (f) $\dfrac{s}{s^4+4a^4}$.

2. Use Equation (8) to determine the inverse Laplace transform of each of the following functions (with n a positive integer):

(a) $\dfrac{1}{(s+a)^n}$,   (b) $\dfrac{s}{(s+a)^n}$,   (c) $\dfrac{1}{(s+b)^2+a^2}$,

(d) $\dfrac{s}{(s+b)^2+a^2}$,   (e) $\dfrac{1}{[(s+b)^2+a^2]^2}$,   (f) $\dfrac{1}{(s^2-a^2)^2}$.

3. Establish the following result: *If $F(z)$ is analytic except for a finite number of poles, and if $e^{t_0 z}F(z)$ tends uniformly to zero on the semicircle C_R: $|z| = R$ in the first and fourth quadrants as $R \longrightarrow \infty$, for some positive value of t_0, then*

$$\mathcal{L}^{-1}\{F(s)\} = 0 \qquad when\ t < t_0.$$

[Suppose that $t < t_0$, close the relevant Bromwich line from $c - Ri$ to $c + Ri$ to the *right* by an arc of $C_{R'}$, with $R' = \sqrt{R^2 + c^2}$, write

$$e^{tz}F(z) = e^{-(t_0-t)z}[e^{t_0 z}F(z)],$$

and use Theorem II.4 of Section 10.14.]

4. Use residue calculus to derive the formula

$$\mathcal{L}^{-1}\left\{\frac{e^{-as}}{s}\right\} = \begin{cases} 0 & (t < a), \\ 1 & (t > a), \end{cases}$$

when $a > 0$. (Use the result of Problem 3 when $t < a$.)

5. Use residue calculus to obtain the result

$$\mathcal{L}^{-1}\left\{\frac{1}{s}\frac{1 - e^{-\pi s}}{1 + e^{-\pi s}}\right\} = \frac{4}{\pi}\sum_{k \text{ odd}}\frac{1}{k}\sin kt,$$

where k takes on the values 1, 3, 5, ... in the infinite series, assuming that the formula (8) is applicable. [The series obtained is the Fourier expansion of a "square-wave function" $f(t)$ such that $f(t) = 1$ when $0 < t < \pi$, $f(t) = -1$ when $\pi < t < 2\pi$, and $f(t + 2\pi) = f(t)$ (see Problem 10 of Chapter 2). Here, in addition to justifying the presence of an infinite series, one must take into account the fact that $F(z)$ has its infinite set of poles $z = ki$ ($k = \pm 1, \pm 3, \ldots$) on the imaginary axis, so that $|F(z)|$ certainly is not even *bounded* on any semicircle C_R for which R is an odd integer. This difficulty can be avoided, in the present case, by letting $R \longrightarrow \infty$ through an appropriately chosen discrete *sequence* of values $R_1, R_2, \ldots$.]

6. Use residue calculus to obtain the result

$$\mathcal{L}^{-1}\left\{\frac{1}{s}\frac{\sinh (sx/c)}{\sinh (sl/c)}\right\} = \frac{x}{l} + \frac{2}{\pi}\sum_{k=1}^{\infty}\frac{(-1)^k}{k}\sin\frac{k\pi ct}{l}\sin\frac{k\pi x}{l},$$

assuming that (8) is applicable. (See Problems 102 and 106 of Chapter 9.)

7. Use residue calculus to obtain the result

$$\mathcal{L}^{-1}\left\{\frac{1}{s}\frac{\sinh (s^{1/2}x/\alpha)}{\sinh (s^{1/2}l/\alpha)}\right\} = \frac{x}{l} + \frac{2}{\pi}\sum_{k=1}^{\infty}\frac{(-1)^k}{k}e^{-k^2\pi^2\alpha^2t/l^2}\sin\frac{k\pi x}{l},$$

assuming that (8) is applicable. [See Problem 60 of Chapter 10. Also see Equation (168) and Problems 101 and 108 of Chapter 9.]

Section 11.3

8. Determine the inverse transform of

$$F(s) = \frac{e^{-\beta s^{1/2}}}{s} \qquad (\beta > 0)$$

in the form

$$f(t) = 1 - \frac{1}{\pi}\int_0^\infty e^{-rt}\frac{\sin \beta\sqrt{r}}{r}\,dr$$

and, by referring to Equation (247) of Section 9.14, deduce that

$$\mathcal{L}^{-1}\left\{\frac{e^{-\beta s^{1/2}}}{s}\right\} = 1 - \text{erf}\left(\frac{\beta}{2\sqrt{t}}\right).$$

9. Obtain the inverse transforms

(a) $\mathcal{L}^{-1}\left\{\dfrac{e^{-\beta s^{1/2}}}{s^{1/2}}\right\} = \dfrac{2}{\pi\sqrt{t}}\displaystyle\int_0^\infty e^{-u^2}\cos\dfrac{\beta u}{\sqrt{t}}\,du = \dfrac{1}{\sqrt{\pi t}}e^{-\beta^2/4t}$,

(b) $\mathcal{L}^{-1}\{e^{-\beta s^{1/2}}\} = \dfrac{2}{\pi t}\displaystyle\int_0^\infty e^{-u^2}\sin\dfrac{\beta u}{\sqrt{t}}u\,du = \dfrac{\beta}{2\sqrt{\pi t^3}}e^{-\beta^2/4t}$,

where $\beta > 0$. [Use Equation (241) of Section 9.14 to evaluate the integral in part (a) and differentiate that equation with respect to b to deal with part (b).]

10. Solve, *by use of Laplace transforms,* the heat-flow problem considered in Section 9.14, in which $T(x, t)$ satisfies the differential equation

$$\frac{\partial^2 T}{\partial x^2} = \frac{1}{\alpha^2} \frac{\partial T}{\partial t},$$

the initial condition $T(x, 0) = T_0 =$ constant, and the end condition $T(0, t) = 0$ for $0 \leq x < \infty, 0 \leq t < \infty.$ [Show that the transform $\bar{T}$ then satisfies the equation

$$\frac{\partial^2 \bar{T}}{\partial x^2} - \frac{s}{\alpha^2}\bar{T} = -\frac{T_0}{\alpha^2}$$

and hence must be of the form

$$\bar{T} = A(s)e^{-s^{1/2}x/\alpha} + B(s)e^{s^{1/2}x/\alpha} + \frac{T_0}{s}.$$

Determine A and B so that the end condition is satisfied, and so that $\bar{T}$ is bounded as $x \longrightarrow \infty$, and show that

$$\bar{T} = T_0\left(\frac{1}{s} - \frac{e^{-s^{1/2}x/\alpha}}{s}\right).$$

Thus, using the result of Problem 8, obtain the result

$$T = T_0 \operatorname{erf}\left(\frac{x}{2\alpha\sqrt{t}}\right),$$

in accordance with the result of Section 9.14.]

11. Use (11) to show that

$$\mathcal{L}^{-1}\left\{\frac{\log s}{s^2 + 1}\right\} = \frac{\pi}{2}\cos t - \int_0^\infty \frac{e^{-rt}}{r^2 + 1}\,dr$$

and deduce from the result of Problem 98, Chapter 10, that this result can be written

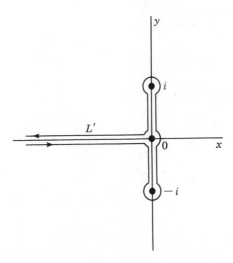

Figure 11.24

in the form

$$\mathcal{L}^{-1}\left\{\frac{\log s}{s^2 + 1}\right\} = \cos t \, Si(t) - \sin t \, Ci(t).$$

(Notice also that in this case the infinite Bromwich line would be equivalent to the modified loop contour L' indicated in Figure 11.24, in which the contributions of the segments along the imaginary axis would cancel in pairs, leaving the contributions of the integral on L and of the two poles.)

12. Use (11) to show that

$$\mathcal{L}^{-1}\left\{\frac{1}{s^{1/2}(s - a)}\right\} = \frac{e^{at}}{\sqrt{a}} - \frac{1}{\pi}\int_0^\infty \frac{e^{-tr}}{r^{1/2}(r + a)}\, dr \qquad (a > 0).$$

13. Express the result of Problem 12 in terms of tabulated functions as follows:

(a) Show that

$$\int_0^\infty \frac{e^{-tr}}{r^{1/2}(r + a)}\, dr = e^{at} g(t),$$

where

$$g(t) = 2\int_0^\infty \frac{e^{-t(v^2 + a)}}{v^2 + a}\, dv.$$

(Write $r = v^2$.)

(b) By differentiating under the integral sign (see Section 7.9), show that

$$g'(t) = -2e^{-at}\int_0^\infty e^{-tv^2}\, dv = -\sqrt{\frac{\pi}{t}}\, e^{-at}.$$

(c) Noticing that $g(\infty) = 0$, obtain the result

$$g(t) = -\sqrt{\pi}\int_\infty^t \frac{e^{-a\tau}}{\sqrt{\tau}}\, d\tau = 2\sqrt{\frac{\pi}{a}}\int_{\sqrt{at}}^\infty e^{-u^2}\, du$$

$$= \frac{\pi}{\sqrt{a}}(1 - \operatorname{erf}\sqrt{at}),$$

where $\operatorname{erf} x$ is the error function.

(d) Hence deduce from Problem 12 that

$$\mathcal{L}^{-1}\left\{\frac{1}{s^{1/2}(s - a)}\right\} = \frac{e^{at}}{\sqrt{a}}\operatorname{erf}\sqrt{at} \qquad (a > 0).$$

14. Generalize the results of Problems 12 and 13 as follows:

(a) Use Equation (11) to show that, when $0 < m < 1$ and $a > 0$, there follows

$$\mathcal{L}^{-1}\left\{\frac{1}{s^m(s - a)}\right\} = \frac{e^{at}}{a^m} - \frac{\sin m\pi}{\pi}e^{at} g(t),$$

where

$$g(t) = \int_0^\infty \frac{e^{-t(r+a)}}{r^m(r + a)}\, dr.$$

(b) Show that

$$g'(t) = -\Gamma(1 - m)e^{-at}t^{m-1}$$

and hence that

$$g(t) = \Gamma(1 - m)\int_t^\infty e^{-a\tau}\tau^{m-1}\, d\tau$$

$$= \frac{\Gamma(1 - m)}{a^m}[\Gamma(m) - \gamma(m, at)],$$

where $$\gamma(m, x) = \int_0^x e^{-u} u^{m-1}\, du \qquad (m > 0)$$

is the so-called *incomplete Gamma function* (and is rather well tabulated).

(c) Use Equation (59) of Chapter 2 to express the consequence of the preceding results in the form

$$\mathcal{L}^{-1}\left\{\frac{1}{s^m(s-a)}\right\} = \frac{e^{at}}{a^m}\frac{\gamma(m, at)}{\Gamma(m)} \qquad (0 < m < 1, a > 0).$$

(d) Verify that

$$\gamma(\tfrac{1}{2}, x) = \sqrt{\pi}\ \mathrm{erf}\ \sqrt{x}$$

and accordingly that the result of part (c) properly reduces to the result of Problem 13 when $m = \tfrac{1}{2}$. [Use Equation (57) of Chapter 2.]

15. Let

$$F(s) = \log\frac{s+1}{s}$$

and consider the integral $\oint_C e^{tz} F(z)\, dz$ around the closed contour of Figure 11.25(a), where $F(z)$ is defined to be that branch of $\log[(z+1)/z]$ which is analytic except along a cut joining the branch points at $z = 0$ and $z = -1$, and which is real on the positive real axis.

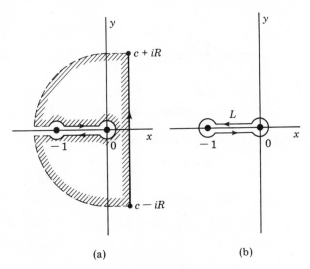

(a)   (b)

Figure 11.25

(a) Show that then

$$F(z) = \log\frac{p_2}{p_1} + i(\omega_2 - \omega_1),$$

where $z - a = p_1 e^{i\omega_1}$ and $z + a = p_2 e^{i\omega_2}$, with $-\pi < \omega_1 < \pi$ and $-\pi < \omega_2 < \pi$, and that $\omega_2 - \omega_1 = 0$ on the real axis except when $-1 < x < 0$, whereas $\omega_2 - \omega_1 = -\pi$ on the upper bank of the cut and $\omega_2 - \omega_1 = \pi$ on the lower branch.

(b) By letting $R \to \infty$ and letting the radii of the indentations tend to zero, show that the "loop integral" reduces to an integration along the closed loop L of Figure

11.25(b) where, in fact, the integrals around the small circles tend to zero and there follows

$$f(t) = \frac{1}{2\pi i} \left[\int_1^0 e^{-tr}(+\pi i)(-dr) + \int_0^1 e^{-tr}(-\pi i)(-dr) \right]$$

$$= \int_0^1 e^{-tr}\, dr = \frac{1 - e^{-t}}{t}.$$

[See also Problem 13(b) of Chapter 2.]

16. If $F(z)$ satisfies the conditions presumed in Section 11.2, except that it also has poles at a finite number of points $z = b_j$ on the negative real axis, show that the expression

$$\tfrac{1}{2} \sum_j [\operatorname{Res}\{e^{tz}F(z); |b_j|\, e^{-i\pi}\} + \operatorname{Res}\{e^{tz}F(z); |b_j|\, e^{i\pi}\}]$$

must be added to the formula for $f(t)$ and that the Cauchy principal value of the integral generally must be taken in that formula. [Indent the upper and lower segments of the loop L outward about each such pole (Figure 11.26) and use Theorem IV of Section 10.14. An alternative procedure (which may or may not be preferable) clearly consists of rotating the cut so that it avoids all poles.]

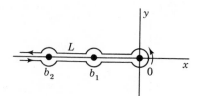

Figure 11.26

17. Use the method of Problem 16 to obtain the result

$$\mathcal{L}^{-1}\left\{\frac{1}{s^m(s + a)}\right\} = \frac{e^{-t}\cos m\pi}{a^m} + \frac{\sin m\pi}{\pi}\, \mathrm{P}\int_0^\infty e^{-rt}\frac{dr}{r^m(a - r)}$$

when $0 < m < 1$ and $a > 0$.

18. When $m = \tfrac{1}{2}$ in Problem 17, use the procedure suggested in Problem 13 to show that

$$\mathcal{L}^{-1}\left\{\frac{1}{s^{1/2}(s + a)}\right\} = \frac{e^{-at}}{\pi} g(t) \qquad (a > 0),$$

where

$$g'(t) = 2e^{at} \int_0^\infty e^{-tv^2}\, dv = \sqrt{\frac{\pi}{t}}\, e^{at}$$

and hence also

$$g(t) = \sqrt{\pi} \int_0^t \frac{e^{a\tau}}{\sqrt{\tau}}\, d\tau$$

$$= 2\sqrt{\frac{\pi}{a}} \int_0^{\sqrt{at}} e^{u^2}\, du,$$

when use is made of Equation (22) of Chapter 2. Thus deduce that

$$\mathcal{L}^{-1}\left\{\frac{1}{s^{1/2}(s+a)}\right\} = \frac{2}{\sqrt{\pi a}}e^{-at}\int_0^{\sqrt{at}} e^{u^2}\, du \qquad (a > 0).$$

[The function $(2/\sqrt{\pi})\int_0^x e^{u^2}\, du$ is sometimes called the *modified error function* and denoted by erfi x, and is tabulated.]

19. For the purpose of identifying the function

$$f(t) = \mathcal{L}^{-1}\left\{\frac{1}{(s^2+a^2)^{1/2}}\right\} \qquad (a > 0),$$

show that we may consider the limit of the integral $\oint_C e^{tz}F(z)\, dz$ around the contour of Figure 11.27(a), where $F(z)$ is that branch of the function $1/(z^2+a^2)^{1/2}$ for which

$$(z^2 + a^2)^{1/2} = \sqrt{p_1 p_2}\, e^{i(\omega_1 + \omega_2)/2},$$

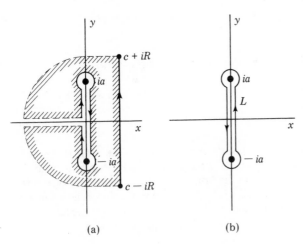

(a)   (b)

Figure 11.27

with $z - ia = p_1 e^{i\omega_1}$ and $z + ia = p_2 e^{i\omega_2}$, where $-\pi < \omega_1 < \pi$ and $-\pi < \omega_2 < \pi$, and where $F(z)$ is analytic except on the finite cut joining the branch cuts. Also show that $(\omega_1 + \omega_2)/2$ is 0 on the right bank of the cut and π on the left bank and that $\sqrt{p_1 p_2} = \sqrt{a^2 - y^2}$ on the cut. Hence, after proceeding to the limit, deduce that

$$f(t) = \frac{1}{2\pi i}\oint_L e^{tz}F(z)\, dz,$$

where L is the closed loop of Figure 11.27(b), and where the integrals around the circles tend to zero in the limit, so that

$$f(t) = \frac{1}{2\pi i}\left(\int_{-a}^{a}\frac{e^{ity}}{\sqrt{a^2-y^2}}i\, dy + \int_{a}^{-a}\frac{e^{ity}}{\sqrt{a^2-y^2}e^{i\pi}}i\, dy\right)$$

$$= \frac{1}{\pi}\int_{-a}^{a}\frac{e^{ity}}{\sqrt{a^2-y^2}}\, dy.$$

Finally, express this result in the form

$$f(t) = \frac{2}{\pi} \int_0^1 \frac{\cos atu}{\sqrt{1 - u^2}} \, du = J_0(at),$$

where J_0 is a Bessel function. [See Problem 65(c) of Chapter 5.]

Section 11.4

20. Consider the mapping $w = az + b$, where a and b are complex constants and $a \neq 0$.

(a) Show that the mapping is one-to-one everywhere.

(b) Verify that the mapping consists of a *rotation* of the z plane about the origin through the angle arg a and a homogeneous magnification or *stretching* by a factor $|a|$, followed by a *translation* through the vector whose components are the real and imaginary parts of b.

(c) Deduce, in particular, that circles map onto circles, and straight lines onto straight lines.

21. Consider the mapping $w = 1/z$.

(a) Show that the mapping is one-to-one everywhere when the points z_∞ and w_∞ are included.

(b) With the notations $w = \rho e^{i\varphi}$ and $z = re^{i\theta}$, show that $\rho = 1/r$ and $\varphi = -\theta$, and deduce that the mapping consists of an *inversion* with respect to the unit circle combined with a *reflection* about the real axis. [Notice that the point (r, θ) maps onto the point $(1/r, -\theta)$.]

(c) Show that the general straight line in the z plane has the polar equation $r \cos (\theta - \theta_0) = p$, where θ_0 is the angle made by its normal with the positive x axis and p is its distance from the origin, and deduce that its image in the w plane is the curve $p\rho = \cos (\varphi + \theta_0)$, where ρ and φ are polar coordinates in the w plane.

(d) Show that the general circle in the z plane, with center at the point with polar coordinates (r_0, θ_0) and radius R, has the polar equation

$$r^2 - 2r_0 r \cos (\theta - \theta_0) + (r_0^2 - R^2) = 0,$$

and deduce that its image in the w plane is the curve

$$(r_0^2 - R^2)\rho^2 - 2r_0\rho \cos (\varphi + \theta_0) + 1 = 0.$$

(e) Deduce from the results of parts (c) and (d) that, under the mapping $w = 1/z$, circles and straight lines in the z plane not passing through the origin map onto circles in the w plane, whereas circles and straight lines in the z plane passing through the origin map onto straight lines in the w plane.

22. *Bilinear transformations.* Consider the mapping corresponding to the general bilinear (or *linear fractional*) transformation

$$w = \frac{\alpha z + \beta}{\gamma z + \delta} \qquad (\Delta \equiv \alpha\delta - \beta\gamma \neq 0),$$

where α, β, γ, and δ are complex constants.

(a) Show that the mapping is one-to-one everywhere (when z_∞ and w_∞ are included) but that it would not be so if $\Delta = 0$. (Notice that the z plane maps onto a single point if $\Delta = 0$.)

(b) Verify that the mapping can be considered as the result of the three successive mappings

$$z' = \gamma z + \delta, \qquad z'' = \frac{1}{z'}, \qquad w = -\frac{\Delta}{\gamma} z'' + \frac{\alpha}{\gamma}$$

when $\gamma \neq 0$, and is of the form

$$w = \frac{\alpha}{\delta} z + \frac{\beta}{\delta}$$

when $\gamma = 0$.

(c) From the results of Problems 20 and 21, and the results of part (b), deduce that *any linear fractional transformation always maps circles onto circles*, with the convention that straight lines are to be considered as limiting forms of circles.

23. *Mappings onto a half-plane.* In each of the following mappings, a specified region in the z plane is mapped onto the upper half ($v \geq 0$) of the w plane, with an indicated correspondence of certain boundary points. Verify the correctness of each mapping and show also that dw/dz exists and is nonzero everywhere inside the mapped region. (As is shown in Problem 55, in each case the given mapping is only one of a triple infinity of available ones, chosen because of the convenience of the boundary-point correspondence.)

(a) *Infinite strip:*

$$(-\infty < x < \infty, 0 \leq y \leq b).$$

Mapping: $\qquad w = e^{\pi z/b}.$

Boundary points:

$$(+\infty, b) \rightarrow (-\infty, 0), \quad (0, b) \rightarrow (-1, 0), \quad (-\infty, y) \rightarrow (0, 0),$$
$$(0, 0) \rightarrow (1, 0), \quad (+\infty, 0) \rightarrow (+\infty, 0).$$

(b) *Semi-infinite strip:*

$$(0 \leq x \leq a, 0 \leq y < \infty).$$

Mapping: $\qquad w = -\cos\frac{\pi z}{a}.$

Boundary points:

$$(0, +\infty) \rightarrow (-\infty, 0), \quad (0, 0) \rightarrow (-1, 0), \quad (\tfrac{1}{2}a, 0) \rightarrow (0, 0),$$
$$(a, 0) \rightarrow (1, 0), \quad (a, +\infty) \rightarrow (+\infty, 0).$$

(c) *Interior of circle:*

$$(|z| \leq R).$$

Mapping: $\qquad w = i\frac{R - z}{R + z}.$

Boundary points:

$$(-R, 0-) \rightarrow (-\infty, 0), \quad (0, -R) \rightarrow (-1, 0), \quad (R, 0) \rightarrow (0, 0),$$
$$(0, R) \rightarrow (1, 0), \quad (-R, 0+) \rightarrow (+\infty, 0).$$

(d) *Interior of semicircle:*

$$(|z| \leq R, y \geq 0).$$

Mapping:
$$w = -\frac{1}{2}\left(\frac{z}{R} + \frac{R}{z}\right).$$

Boundary points:
$$(0+, 0) \longrightarrow (-\infty, 0), \quad (R, 0) \longrightarrow (-1, 0), \quad (0, R) \longrightarrow (0, 0),$$
$$(-R, 0) \longrightarrow (1, 0), \quad (0-, 0) \longrightarrow (+\infty, 0).$$

(e) *Half-plane indented by semicircle:*
$$(|z| \geq R, \, y \geq 0).$$

Mapping:
$$w = \frac{1}{2}\left(\frac{z}{R} + \frac{R}{z}\right).$$

Boundary points:
$$(-\infty, 0) \longrightarrow (-\infty, 0), \quad (-R, 0) \longrightarrow (-1, 0), \quad (0, R) \longrightarrow (0, 0),$$
$$(R, 0) \longrightarrow (1, 0), \quad (+\infty, 0) \longrightarrow (+\infty, 0).$$

24. *Mappings of sectors.* Verify each of the following mappings, showing also that dw/dz exists and is nonzero everywhere inside each specified mapped region.

(a) *Circular sector onto unit semicircle:*
$$(|z| \leq R, \, 0 \leq \theta \leq \alpha) \longrightarrow (|w| \leq 1, \, 0 \leq \varphi \leq \pi).$$

Mapping:
$$w = \left(\frac{z}{R}\right)^{\pi/\alpha}.$$

Boundary points:
$$0 \longrightarrow 0, \quad R \longrightarrow 1, \quad Re^{i\alpha/2} \longrightarrow i, \quad Re^{i\alpha} \longrightarrow -1.$$

(b) *Circular sector onto half-plane:*
$$(|z| \leq R, \, 0 \leq \theta \leq \alpha) \longrightarrow (v \geq 0).$$

Mapping:
$$w = -\frac{1}{2}\left[\left(\frac{z}{R}\right)^{\pi/\alpha} + \left(\frac{R}{z}\right)^{\pi/\alpha}\right].$$

Boundary points:
$$0e^{i0} \longrightarrow \infty e^{i\pi}, \quad R \longrightarrow -1, \quad Re^{i\alpha/2} \longrightarrow 0, \quad Re^{i\alpha} \longrightarrow 1, \quad 0e^{i\alpha} \longrightarrow \infty e^{i0}.$$

(c) *Infinite sector onto half-plane:*
$$(0 \leq \theta \leq \alpha) \longrightarrow (v \geq 0).$$

Mapping:
$$w = z^{\pi/\alpha}.$$

Boundary points:
$$\infty e^{i\alpha} \longrightarrow \infty e^{i\pi}, \quad 0 \longrightarrow 0, \quad \infty e^{i0} \longrightarrow \infty e^{i0}.$$

25. *Mappings onto a unit disk.*

(a) *Half-plane.* By inverting the mapping of Problem 23(c) with $R = 1$, and interchanging z and w, deduce that the mapping
$$w = \frac{i - z}{i + z}$$

transforms the half-plane $y \geq 0$ into the unit disk $|w| \leq 1$ with the following boundary-point correspondence:
$$(-\infty, 0) \longrightarrow (-1, 0-), \quad (-1, 0) \longrightarrow (0, -1), \quad (0, 0) \longrightarrow (1, 0),$$
$$(1, 0) \longrightarrow (0, 1), \quad (+\infty, 0) \longrightarrow (-1, 0+).$$

(b) *Other regions.* If the analytic function G is such that the mapping $w = G(z)$ maps a region $\mathfrak{R}$ onto the upper half of the w plane, show that the mapping

$$w = \frac{i - G(z)}{i + G(z)}$$

maps $\mathfrak{R}$ onto the unit disk $|w| \leq 1$. (First map $\mathfrak{R}$ onto the upper half of a complex t plane, then map that half-plane onto the unit disk in the w plane.)

26. *Examples of mappings onto a unit disk.* Use the mappings of Problems 23(a) and 24(c) to illustrate the result of Problem 25 in the two following cases:

(a) *Infinite strip:*

$$(-\infty < x < \infty, 0 \leq y \leq b).$$

Mapping: $\qquad w = \dfrac{i - e^{\pi z/b}}{i + e^{\pi z/b}}.$

Boundary points:

$$(+\infty, b) \longrightarrow (-1, 0-), \quad (0, b) \longrightarrow (0, -1), \quad (-\infty, y) \longrightarrow (1, 0),$$
$$(0, 0) \longrightarrow (1, 0), \quad (+\infty, 0) \longrightarrow (-1, 0+).$$

(b) *Infinite sector:*

$$(0 \leq \theta \leq \alpha, 0 \leq r < \infty).$$

Mapping: $\qquad w = \dfrac{i - z^{\pi/\alpha}}{i + z^{\pi/\alpha}}.$

Boundary points:

$$\infty e^{i\alpha} \longrightarrow (-1, 0-), \quad e^{i\alpha} \longrightarrow (0, -1), \quad (0, 0) \longrightarrow (1, 0),$$
$$(1, 0) \longrightarrow (0, 1), \quad \infty e^{i0} \longrightarrow (-1, 0+).$$

[Other mappings onto the unit disk follow similarly, for example, from the mappings of Problems 23(b, d, e) and 24(c).]

27. Determine a mapping of the *semi-infinite strip* $(0 \leq x \leq a, 0 \leq y < \infty)$ in the z plane onto the *infinite sector* $(0 \leq \varphi \leq \alpha, 0 \leq \rho < \infty)$ of the w plane, by first mapping the semi-infinite strip onto the upper half of a t plane [Problem 23(b)] and then mapping that half-plane onto the sector [Problem 24(c), inverted], in the form

$$w = \left(-\cos\frac{\pi z}{a}\right)^{\alpha/\pi}.$$

Also verify the correspondences $(0, +\infty) \longrightarrow \infty e^{i\alpha}$, $(0, 0) \longrightarrow e^{i\alpha}$, $(a/2, 0) \longrightarrow (0, 0)$, $(a, 0) \longrightarrow (1, 0)$, $(a, +\infty) \longrightarrow (+\infty, 0)$.

Section 11.5

28. With the notation of Section 6.17, show that the scale factors associated with the curvilinear coordinates defined by the relation $u + iv = f(x + iy)$ are given by $h_u = h_v = |df/dz|^{-1}$. Hence, by using the results of that section, deduce that

$$ds^2 = \left|\frac{df}{dz}\right|^{-2} (du^2 + dv^2)$$

and

$$\nabla\varphi = \left|\frac{df}{dz}\right| \left(\frac{\partial\varphi}{\partial u}\mathbf{u} + \frac{\partial\varphi}{\partial v}\mathbf{v}\right),$$

where **u** and **v** are unit vectors in the u and v directions, and also that

$$\nabla^2 \varphi = \left| \frac{df}{dz} \right|^2 \left(\frac{\partial^2 \varphi}{\partial u^2} + \frac{\partial^2 \varphi}{\partial v^2} \right),$$

where ∇^2 is the two-dimensional Laplacian operator, in accordance with the results of Section 11.5.

29. *Bernoulli's equation* [Equation (212) of Section 6.20] states that the pressure at a point in an incompressible fluid is given by

$$p = p_0 - \tfrac{1}{2}\rho V^2,$$

where p_0 is the equilibrium pressure, ρ is the density, and $V = |\mathbf{V}|$ is the absolute velocity at the point.

   (a) In a flow past a cylindrical surface whose cross section is bounded by a simple closed curve C in the z plane, show that the *normal force* per unit length of cylinder, exerted by the fluid, has components

$$-\oint_C p \, \frac{dy}{ds} \, ds \quad \text{and} \quad +\oint_C p \, \frac{dx}{ds} \, ds$$

in the x and y directions, respectively, and show that these components are the real and imaginary parts of the complex quantity $i \oint_C p \, dz$.

   (b) By making use of Bernoulli's equation, deduce the relation

$$-D + iL = -\frac{1}{2} i\rho \oint_C \left| \frac{d\Phi}{dz} \right|^2 dz,$$

where Φ is the complex potential function, D is the *drag* component of the force due to pressure in the negative x direction, and L is the *lift* component in the positive y direction, this terminology implying that the preponderant flow is in the positive x direction.

30. (a) Show that the complex potential $\Phi = Aw$ (A real and positive) corresponds to a steady flow, with velocity A in the positive u direction, in the w plane.

   (b) Use the result of Problem 23(e) to map the upper half of the w plane onto the upper half of the z plane, indented by a semicircle of radius a with center at the origin.

   (c) Show that the complex potential takes the form

$$\Phi = \frac{A}{2} \left(\frac{z}{a} + \frac{a}{z} \right).$$

By noticing that the real axis in the w plane corresponds to the portion of the x axis outside the circle $x^2 + y^2 = a^2$, and to the upper half of the boundary of that circle, deduce that the flow has this composite contour as a streamline.

   (d) Determine A in such a way that the velocity tends to V_0 at large distances from the origin, and hence obtain

$$\Phi = V_0 \left(z + \frac{a^2}{z} \right)$$

as the complex potential in the half-plane $y > 0$ of a flow, about a circular cylinder, which tends to a uniform flow in the positive x direction at large distances from the origin (see Figure 11.9).

31. For the flow obtained in Problem 30, show that the velocity components are of the form

$$V_x = V_0\left[1 - \frac{a^2(x^2 - y^2)}{(x^2 + y^2)^2}\right], \qquad V_y = -V_0\frac{2a^2xy}{(x^2 + y^2)^2}$$

and verify that, because of appropriate symmetry in these expressions, the flow is correctly specified by Φ over the entire z plane.

Section 11.6

32. Suppose that two sources of equal strength m are present at the points $z = \pm a$, where a is real and positive.

(a) Show that the complex potential is given by

$$\Phi = \frac{m}{2\pi} \log (z^2 - a^2),$$

and the complex velocity by

$$V_x - iV_y = \frac{m}{\pi} \frac{z}{z^2 - a^2}.$$

(b) Show (by symmetry or otherwise) that the stream function is constant along the y axis, and deduce that Φ is also the complex potential in the half-plane $x > 0$ corresponding to a source at $z = a$ and a fixed boundary along the y axis.

(c) Show that the velocity at a point on the boundary $x = 0$ is given by $V_y = (m/\pi)[y/(y^2 + a^2)]$, and at a point on the line $y = 0$ by $V_x = (m/\pi)[x/(x^2 - a^2)]$.

33. Suppose that a vortex of strength m is located at $z = a$, and that an equal and opposite vortex is at $z = -a$, where a is real and positive.

(a) Show that the complex potential is given by

$$\Phi = \frac{m}{2\pi i} \log \frac{z - a}{z + a},$$

and the complex velocity by

$$V_x - iV_y = \frac{m}{\pi i} \frac{a}{z^2 - a^2}.$$

(b) Show that the vortex at $z = -a$ may be replaced by a fixed boundary along the y axis in part (a), without affecting the flow.

(c) Show that the velocity at a point of the boundary $x = 0$ is given by $V_y = -(m/\pi)[a/(y^2 + a^2)]$, and that the velocity at a point on the line $y = 0$ is given by $V_y = (m/\pi)[a/(x^2 - a^2)]$.

34. Fluid is introduced at a steady rate m through a small opening at the origin into an infinite channel occupying the region $(-\infty < x < +\infty, 0 \le y \le b)$ of the z plane. The flow pattern in the channel is required.

(a) Use the mapping of Problem 23(a) to obtain a transformed problem in which fluid is introduced into the upper half of the w plane at the rate m at the point $w = 1$, and is withdrawn at the rate $m/2$ at the points $w = 0$ and w_∞. Noticing that a source of strength $2m$ then must be present at $w = 1$, and a sink of strength m at $w = 0$ (as well as at w_∞), since only half the fluid enters the upper half-plane from a source on its

Applications of Analytic Function Theory

boundary, obtain the complex potential in the w plane in the form

$$\Phi = \frac{m}{\pi}\left[\log{(w-1)} - \frac{1}{2}\log w\right] + \text{constant}$$

$$= \frac{m}{\pi}\log{(w^{1/2} - w^{-1/2})} + \text{constant}.$$

[The availability of an arbitrary additive constant is noted for the purpose of part (b).]

(b) Obtain the required complex potential in the form

$$\Phi = \frac{m}{\pi}\log\sinh\frac{\pi z}{2b},$$

neglecting an arbitrary additive constant.

35. (a) Show that the velocity vector in Problem 34 is of the form

$$V_x + iV_y = \frac{m}{2b}\coth\frac{\pi(x-iy)}{2b},$$

and that the velocity along the boundary $y = b$ is given by $V_x = (m/2b)\tanh{(\pi x/2b)}$, whereas the velocity along the boundary $y = 0$ is given by $V_x = (m/2b)\coth{(\pi x/2b)}$.

(b) Show that the stream function is of the form

$$\psi = \frac{m}{\pi}\cot^{-1}\frac{\tanh\pi x/2b}{\tan\pi y/2b},$$

and verify that ψ increases by m as a small semicircle is traversed about the origin in the counterclockwise direction. Deduce also that the streamlines are the curves

$$\tan\frac{\pi y}{2b} = c\tanh\frac{\pi x}{2b}.$$

36. (a) Show that the complex potential

$$\Phi = V_0\left(z + \frac{a^2}{z}\right) + i\frac{m'}{2\pi}\log z$$

defines a flow past the circle $|z| = a$, in which the velocity tends to V_0 as $|z| \to \infty$, for any constant real value of m'. (See also Problem 48 of Chapter 9.) (Notice that a positive value of m' corresponds to the presence of a *clockwise* circulation.)

(b) By making use of the result of Problem 29(b), show that the drag D and lift L, per unit length, exerted on a cylinder with $|z| = a$ as a cross section is then given by

$$-D + iL = -\frac{1}{2}i\rho\oint_C\left|V_0\left(1 - \frac{a^2}{z^2}\right) + i\frac{m'}{2\pi z}\right|^2 dz,$$

where C is the circle $|z| = a$.

(c) Show that the integral involved in part (b) can be expressed in the form

$$\int_0^{2\pi}\left[V_0(1 - e^{-2i\theta}) + i\frac{m'}{2\pi a}e^{-i\theta}\right]\left[V_0(1 - e^{2i\theta}) - i\frac{m'}{2\pi a}e^{i\theta}\right]iae^{i\theta}\,d\theta.$$

By noticing that all product terms of the form $e^{ir\theta}$ will integrate to zero when r is an integer other than zero, and to 2π when $r = 0$, obtain the value of the integral in the form $-2V_0m'$. Hence deduce that

$$D = 0, \qquad L = \rho V_0 m',$$

where m' is the clockwise circulation.

Section 11.7

37. Use the mapping (48) to obtain the solution $T(x, y)$ of Laplace's equation in the xy plane for which

$$T(x, 0+) = \sqrt{1 - x^2}, \qquad T(x, 0-) = -\sqrt{1 - x^2}$$

along a slit $|x| \leq 1$, in the form

$$T = \frac{\sin \varphi}{\rho},$$

where ρ and φ are to be determined in terms of x and y by use of the relations (50). In particular, show that $T(x, 0) = 0$ when $|x| \geq 1$ and that

$$T(0, y) = \begin{cases} \dfrac{1}{|y| + \sqrt{y^2 + 1}} & (y > 0), \\[4mm] \dfrac{-1}{|y| + \sqrt{y^2 + 1}} & (y < 0). \end{cases}$$

38. (a) If φ solves the problem

$$\nabla^2 \varphi = 0 \quad in \; \Re, \qquad \varphi = g \quad on \; C,$$

where C is the boundary of $\Re$, and if ψ is the *harmonic conjugate* of φ, so that $\varphi + i\psi$ is analytic in $\Re$, show that ψ solves the problem

$$\nabla^2 \psi = 0 \quad in \; \Re, \qquad \frac{\partial \psi}{\partial n} = -\frac{\partial g}{\partial s} \quad on \; C,$$

where $\partial/\partial s$ and $\partial/\partial n$ differentiate with respect to distance along C and in the direction of the outward normal, respectively.

(b) Verify the result of part (a) when

$$\varphi(x, y) = \frac{\sin x \sinh y}{\sin a \sinh b}$$

in the rectangle $(0 \leq x \leq a, \, 0 \leq y \leq b)$.

39. (a) *Mapping of exterior of ellipse onto exterior of circle.* Verify that the transformation

$$z = w + \frac{c^2}{4w}$$

maps the exterior of the ellipse

$$\frac{x^2}{a^2} + \frac{y^2}{b^2} = 1 \qquad (a > b)$$

onto the exterior of the circle $|w| = \frac{1}{2}(a + b)$, where c is the distance $\sqrt{a^2 - b^2}$ from the center of the ellipse to each focus, in such a way that the mapping is not distorted or rotated at large distances from the ellipse. [Notice that the point $w = \frac{1}{2}(a + b)e^{i\varphi}$ corresponds to the point $z = a \cos \varphi + ib \sin \varphi$.]

(b) *Mapping of plane with finite cut onto exterior of circle.* As a limiting case of part (a) as $b \longrightarrow 0$, show that the transformation

$$z = w + \frac{a^2}{4w}$$

maps the z plane with a cut along the real axis from $x = -a$ to $x = +a$ onto the exterior of the circle $|w| = a/2$, with no distortion or rotation at large distances from the cut. (Notice that the point $w = \frac{1}{2}ae^{i\varphi}$ corresponds to the point $z = a \cos \varphi$.)

40. *Rotation of mappings of Problem 39.*

(a) Suppose that the ellipse defined in Problem 39(a) is rotated about its center through the angle α, where α is positive when the rotation is counterclockwise. By considering the result of replacing z by $e^{-i\alpha}z$ and w by $e^{-i\alpha}w$ in the transformation of that problem, show that the exterior of the resultant ellipse is mapped onto the exterior of the circle $|w| = \frac{1}{2}(a + b)$ by the mapping

$$z = w + \frac{c^2 e^{2i\alpha}}{4w},$$

where again $c^2 = a^2 - b^2$, without distortion or rotation at z_∞.

(b) In the limiting case $b = 0$, show that the transformation

$$z = w + \frac{a^2 e^{2i\alpha}}{4w}$$

maps the z plane, with a cut of length $2a$ extending from the point $-ae^{i\alpha}$ to the point $ae^{i\alpha}$, onto the exterior of the circle $|w| = a/2$, without distortion or rotation at z_∞.

(c) Verify that the point $w = \frac{1}{2}(a + b)e^{i(\alpha+\beta)}$ corresponds to the boundary point $z = e^{i\alpha}(a \cos \beta + ib \sin \beta)$ in the transformation of part (a), and that the point $w = \frac{1}{2}ae^{i(\alpha+\beta)}$ corresponds to the point $z = ae^{i\alpha} \cos \beta$ on the cut in part (b).

41. *Flow about an elliptic cylinder.* Show that the complex potential for an ideal fluid flow in the xy plane about a section of an elliptic cylinder, with center at the origin, with semimajor axis of length a along the line $\theta = \alpha$, and with semiminor axis of length b, which tends to a uniform flow with velocity V_0 in the positive x direction at large distances from the ellipse, is obtained by eliminating w between the two equations

$$\Phi = V_0\left[w + \frac{(a + b)^2}{4w}\right] - i\frac{m}{2\pi} \log w, \qquad z = w + \frac{(a^2 - b^2)e^{2i\alpha}}{4w},$$

where m is the circulation. (Notice that the first equation defines the complex potential for the flow of Problem 36 in the w plane about a circle of radius $\frac{1}{2}(a + b)$, and use the result of Problem 40.)

42. By specializing the result of Problem 41 when $b = 0$ and $\alpha = \pi/2$, deduce that the complex potential for a steady flow about a plane barrier of width $2a$, at right angles to the flow and with center at the origin, is of the form

$$\Phi = V_0(z^2 + a^2)^{1/2} - \frac{im}{2\pi} \sinh^{-1} \frac{z}{a},$$

where V_0 is the limiting uniform velocity in the x direction. (An alternative derivation of this result is given in Section 11.8).

43. By using the result of Equation (230) of Section 9.14, show that the steady-state temperature distribution in the upper half of the w plane, which reduces to unity along the u axis when $|u| < 1$ and to zero elsewhere along the u axis, is given by

$$T = \frac{1}{\pi}\left(\tan^{-1} \frac{1 - u}{v} + \tan^{-1} \frac{1 + u}{v}\right) = \frac{1}{\pi} \cot^{-1} \frac{u^2 + v^2 - 1}{2v}.$$

(See footnote, page 646.)

44. The steady-state temperature distribution is required in the semi-infinite strip $(0 \leq x \leq a, 0 \leq y < \infty)$, subject to the requirements that it reduce to unity along the end $y = 0$ and to zero along the sides $x = 0$ and $x = a$. Use the results of Problems 23(b) and 43 to obtain this distribution in the form

$$T = \frac{1}{\pi} \cot^{-1}\left[\left(\sinh^2 \frac{\pi y}{a} - \sin^2 \frac{\pi x}{a}\right) \Big/ 2\sin\frac{\pi x}{a}\sinh\frac{\pi y}{a}\right] = \frac{2}{\pi}\tan^{-1}\left(\sin\frac{\pi x}{a}\Big/\sinh\frac{\pi y}{a}\right).$$

(Notice that the methods of Chapter 9 would give this solution in the form of an infinite series.)

45. The steady-state temperature is required in the semicircle bounded by the upper half of the circle $x^2 + y^2 = R^2$ and the x axis, subject to the requirements that it reduce to unity along the curved boundary and to zero along the straight boundary. Use the results of Problems 23(d) and 43 to obtain the distribution in the form

$$T = \frac{1}{\pi}\cot^{-1}\left\{\frac{\left[\frac{1}{2}\left(\frac{R}{r} - \frac{r}{R}\right)\right]^2 - \sin^2\theta}{\left(\frac{R}{r} - \frac{r}{R}\right)\sin\theta}\right\} = \frac{2}{\pi}\tan^{-1}\frac{2rR\sin\theta}{R^2 - r^2}.$$

46. The steady-state temperature $T(x, y)$ is required in the upper half-plane $y \geq 0$ subject to the conditions

$$T(x, 0) = 0 \quad (-\infty < x < -1), \qquad T(x, 0) = 1 \quad (1 < x < \infty)$$

and

$$\frac{\partial T(x, 0)}{\partial y} = 0 \quad (|x| < 1)$$

along the x axis.

   (a) Use the inverse of the transformation of Problem 23(b), in the form

$$z = -\cos \pi w$$

(taking $a = 1$), to map this problem into the problem of determining the solution T of Laplace's equation in the semi-infinite strip $(0 \leq u \leq 1, 0 \leq v < \infty)$ for which $T = 0$ along the edge $u = 0$, $T = 1$ along the edge $u = 1$, and $\partial T/\partial v = 0$ along the end $v = 0$.

   (b) Noticing that the solution of the transformed problem is $T = u$, by inspection, deduce that the required function is

$$T = \frac{1}{\pi} \text{Re} \left[\cos^{-1}(-z)\right].$$

   (c) Verify that

$$\sqrt{(x + 1)^2 + y^2} = \cosh \pi v - \cos \pi u, \qquad \sqrt{(x - 1)^2 + y^2} = \cosh \pi v + \cos \pi u$$

and deduce that T can be expressed in the explicit form

$$T = \frac{1}{\pi}\cos^{-1}\frac{1}{2}[\sqrt{(x - 1)^2 + y^2} - \sqrt{(x + 1)^2 + y^2}].$$

   (d) Verify the satisfaction of the prescribed conditions when $y = 0$.

Section 11.8

47. Derive the mapping of Problem 23(a).

48. Derive the mapping of Problem 23(b).

49. (a) Show that the mapping of Figure 11.14, in which the region *above* the indicated polygonal boundary is mapped onto the upper half of the w plane, is specified by the relation

$$z = \frac{i}{\pi}[(1 - w^2)^{1/2} - \cos^{-1} w + \pi].$$

(b) When the region *below* the polygonal boundary in Figure 11.14 is mapped onto the upper half of the w plane, obtain the mapping relation

$$z = \frac{i}{\pi}[\cos^{-1} w - (1 - w^2)^{1/2}].$$

(In both cases, the multivalued functions must be properly interpreted. The fact that neither relation can be inverted, to express w as an explicit function of z, unfortunately represents a situation of frequent occurrence. Text Example 3 illustrates the feasibility of using such relations in applications in spite of this inconvenience.)

50. Obtain a transformation which maps the region in the z plane, above the line $y = b$ when $x < 0$ and above the x axis when $x > 0$, onto the upper half of the w plane, in the form

$$z = \frac{b}{\pi}[(w^2 - 1)^{1/2} + \cosh^{-1} w],$$

by making the point $z = bi$ correspond to $w = -1$ and the point $z = 0$ correspond to $w = +1$.

51. The infinite strip $(-\infty < x < +\infty, 0 \leq y \leq b)$ has a semi-infinite cut extending in the positive x direction from the point $(0, b/2)$. It is required to map this cut region onto the upper half of the w plane, in such a way that the x axis maps onto the segment $u < -1$, the lower boundary of the cut onto $-1 < u < 0$, the upper boundary of the cut onto $0 < u < 1$, and the line $y = b$ onto the segment $u > 1$.

(a) Show that the mapping must be such that

$$\frac{dz}{dw} = \frac{Cw}{w^2 - 1}, \qquad z = \frac{C}{2} \log(w^2 - 1) + K.$$

(b) Show that the point $z = ib/2$ maps onto $w = 0$ if and only if

$$\frac{b}{2}i = \frac{\pi C}{2}i + K,$$

and that the line $z = x + ib$ maps onto $w = u$, where $u > 1$, if and only if $K = ib$. Hence obtain the mapping in the form

$$z = ib - \frac{b}{2\pi} \log(w^2 - 1), \qquad w^2 = 1 + e^{-2\pi z/b}.$$

52. Suppose that the temperature is maintained at zero along the boundaries $y = 0$ and $y = b$ of the strip of Problem 51, and at unity along the semi-infinite cut. Use the results of Problems 51 and 43 to show that the steady-state temperature distribution $T(x, y)$ in the strip is obtained by eliminating u and v from the relations

$$T = \frac{1}{\pi} \cot^{-1} \frac{u^2 + v^2 - 1}{2v},$$

$$u^2 - v^2 = 1 + e^{-2\pi x/b} \cos \frac{2\pi y}{b}, \qquad 2uv = -e^{-2\pi x/b} \sin \frac{2\pi y}{b}.$$

53. *Mapping of half-plane onto half-plane.* Show that the bilinear transformation

$$w = \frac{az + b}{cz + d}$$

defines a conformal one-to-one mapping of the upper half of the z plane onto the upper half of the w plane if and only if a, b, c, and d are real and $ad - bc > 0$. [Note that w must be real when z is real. Also notice that dw/dz must be real and positive when z is real, in order that as an observer travels in the positive direction along the x axis, with the mapped region to his left, his image will travel in the positive direction along the u axis, with the image region (to *his* left) accordingly the *upper* half of the w plane.]

54. Show that *only* the transformation defined in Problem 53 defines a conformal one-to-one mapping of the entire upper half of the z plane onto the entire upper half of the w plane. [Notice that the mapping function must have exactly one zero and one pole in the extended upper half-plane (including the point at infinity) and use results of Section 10.11.]

55. Deduce from Problems 53 and 54 that, if $w = F(z)$ maps the interior of a region $\mathcal{R}$ conformally onto the upper half of the w plane, then the same is true of the mapping

$$w = \frac{aF(z) + b}{cF(z) + d}$$

when a, b, c, and d are real and $ad > bc$, and *only* of that mapping. [Let $t = F(z)$ map the interior of R onto the upper half of a complex t plane and then introduce the mapping $w = (at + b)/(ct + d)$. Notice that, since the numerator and denominator can be divided by one of the four constants, there are *three* independent parameters. This accounts, in particular, for the assignability of three u's in the Schwarz–Christoffel transformation (56).]

56. (a) In the mapping of text Example 1, show that the three-parameter mapping function corresponding to the result of Problem 55 is of the form

$$w = \frac{az^{\pi/\alpha} + b}{cz^{\pi/\alpha} + d},$$

where a, b, c, and d are real and $ad > bc$, and that it yields the relation

$$\frac{dz}{dw} = \left[(ad - bc)\frac{\alpha}{\pi}\right](b - dw)^{(\alpha/\pi)-1}(cw - a)^{-(\alpha/\pi)-1}.$$

(b) If the requirements that $z = 0$ map onto $w = 0$ and $z = 1$ map onto $w = 1$ again are imposed, show that there results the one-parameter mapping function

$$w = \frac{z^{\pi/\alpha}}{1 - k(1 - z^{\pi/\alpha})},$$

from which

$$\frac{dz}{dw} = \frac{\alpha}{\pi}(1 - k)^{\alpha/\pi}w^{(\alpha/\pi)-1}(1 - kw)^{-(\alpha/\pi)-1},$$

where $k = c/a$ is to be a real constant such that $k < 1$.

(c) Verify the correctness of the mapping of part (b), showing in particular that the point z_∞ maps onto the point $w = 1/k$.

[Notice that accordingly the third degree of freedom promised in the text and in Problem 55 here corresponds to the choice of k. The derivation of the relation (64), which corresponds to $k = 0$, in effect ignores the presence of a vertex at z_∞ and causes it to be mapped onto w_∞. It can be seen also that when $k \neq 0$ the derivative formula in part (b) would correspond to mapping "a vertex at z_∞ with an interior angle $\alpha_\infty = -\alpha$" onto $w = 1/k$. Whereas this phrase is not readily subject to a geometric interpretation, it is in fact in accordance with the requirement that the sum of the interior angles of a closed polygon be $(n - 2)\pi$, where n is the number of vertices, if we here consider n to be two! If, more plausibly, we assert that there are *two* vertices at z_∞, so that $n = 3$, then the *sum* of the "interior angles at those vertices" must be $\pi - \alpha$; but the Schwarz–Christoffel transformation cannot accept this assertion. (A similar analysis of the mapping of text Example 2 defines in the same way "the interior angle $\alpha_\infty = -\pi$" at a single vertex at z_∞, in accordance with the requirement that the sum of the *four* angles $\alpha_1 = \alpha_3 = \pi/2$, $\alpha_2 = 2\pi$, and α_∞ be equal to $(n - 2)\pi$ when $n = 4$.)]

57. *Electrostatic fields about cylindrical conductors.* In a region free of charges, in which an electrostatic field is independent of position normal to the xy plane, the *electric field intensity* vector **E** is the negative gradient of an *electrostatic potential* $\varphi(x, y)$ which satisfies Laplace's equation,

$$\mathbf{E} = -\nabla\varphi \quad \text{where} \quad \nabla^2\varphi = 0.$$

(a) Show that the fact that the component of **E** tangential to a perfect conductor vanishes leads to the conclusion that φ is constant along the boundary of a perfect conductor.

(b) If $\psi(x, y)$ is the function conjugate to $\varphi(x, y)$, so that $\varphi + i\psi$ is an analytic function of $x + iy$, show that the *lines of force* then are the curves for which $\psi = \text{constant}$.

(c) The *surface charge density* σ on a conductor is given by $-K(\partial\varphi/\partial n)$, where K is the electric inductive capacity. Show that the *total charge* q on a cylindrical conductor, per unit length normal to the xy plane, is given by

$$q = \oint_B \sigma \, ds = -K \oint_B \frac{\partial\varphi}{\partial n} \, ds = -K \oint_B \frac{\partial\psi}{\partial s} \, ds = -K\Delta_B\psi,$$

where $\Delta_B\psi$ is the increment in ψ corresponding to a counterclockwise circuit of the boundary B of the section of the conductor in the xy plane. [See Problem 25(b) of Chapter 10.]

58. *Capacity of a condenser.* The *capacity* C, per unit length, of a condenser formed by two cylindrical conductors B_1 and B_2, is defined as the absolute value of the ratio of the total charge per unit length on either of the conductors to the potential difference between them. If $\Phi(z) = \varphi + i\psi$ is analytic in the region between the conductors, and if φ is constant on each conductor, show that

$$C = \left|\frac{q}{\varphi_{B_1} - \varphi_{B_2}}\right| = K\left|\frac{\Delta_B\psi}{\varphi_{B_2} - \varphi_{B_1}}\right|.$$

59. By making appropriate use of the function

$$\Phi = \varphi + i\psi = A \log z + B,$$

show that the electrostatic potential outside a *circular* cylindrical conductor is given by

$$\varphi = -\frac{q}{2\pi K} \log r + \text{constant},$$

where r is distance from the axis of the cylinder and q is the charge per unit length on the cylinder, whereas the function ψ is given by

$$\psi = -\frac{q}{2\pi K}\theta + \text{constant},$$

where θ is an angle measured around the axis of the cylinder.

60. Show that the capacity, per unit length, of a condenser formed by concentric *circular* cylinders of radii R_1 and R_2 is given by

$$C = \frac{2\pi K}{\log (R_2/R_1)}.$$

61. (a) By making use of the results of Problems 39 and 59, show that the result of eliminating w between the relations

$$\Phi = \varphi + i\psi = -\frac{q}{2\pi K} \log w + \text{constant}, \qquad z = w + \frac{c^2}{4w}$$

gives the potential φ and the function ψ for the electrostatic field outside an *elliptical* cylinder with cross section

$$\frac{x^2}{a^2} + \frac{y^2}{b^2} = 1 \quad \text{where} \quad a^2 - b^2 = c^2,$$

with charge q per unit length.

(b) Show that the elimination described in part (a) leads to the result

$$\varphi + i\psi = -\frac{q}{2\pi K} \cosh^{-1} \frac{z}{c} + \text{constant}.$$

62. By combining the results of Problems 61 and 60, show that the capacity, per unit length, of a cylindrical condenser whose cross section consists of confocal ellipses with focal length c and semi-major axes a_1 and a_2 is given by

$$C = \frac{2\pi k}{\log\left(\dfrac{a_2 + \sqrt{a_2^2 - c^2}}{a_1 + \sqrt{a_1^2 - c^2}}\right)} \equiv \frac{2\pi K}{\cosh^{-1} \dfrac{a_2}{c} - \cosh^{-1} \dfrac{a_1}{c}}.$$

[Notice that the ellipse $x^2/a^2 + y^2/(a^2 - c^2) = 1$ corresponds to the circle $|w| = \frac{1}{2}(a + \sqrt{a^2 - c^2})$ in the mapping $z = w + c^2/(4w)$.]

Section 11.9

63. Show that $G(x, y; \xi, \eta) = G(\xi, \eta; x, y)$ if $G(x, y; \xi, \eta)$ satisfies (93) and (94), by taking $G = G(a, b; \xi, \eta)$ and $\varphi = G(c, d; \xi, \eta)$ in (90) and deducing that $G(a, b; c, d) = G(c, d; a, b)$.

64. Determine φ such that

$$\frac{\partial^2 \varphi}{\partial x^2} + \frac{\partial^2 \varphi}{\partial y^2} = \delta(x - \xi)\delta(y - \eta),$$

$$\varphi(x, 0) = 0, \qquad \lim_{x^2+y^2\to\infty} \varphi(x, y) = 0,$$

in the upper half-plane, by the following steps:

(a) If $\bar\varphi$ is the Fourier transform of φ with respect to x, show that

$$\frac{\partial^2 \bar\varphi}{\partial y^2} - u^2\bar\varphi = e^{-iu\xi}\delta(y - \eta).$$

(b) If $\bar\varphi_s$ is the Fourier sine transform of $\bar\varphi$ with respect to y, show that

$$\bar\varphi_s(u, v) = -\frac{e^{-iu\xi} \sin v\eta}{u^2 + v^2}.$$

(c) Deduce that

$$\varphi_S(x, v) = -\frac{\sin v\eta}{2\pi} \int_{-\infty}^{\infty} \frac{e^{-iu(\xi-x)}\, du}{u^2 + v^2}$$

$$= -\frac{e^{-v|\xi-x|} \sin v\eta}{2v}$$

and hence that

$$\varphi(x, y) = -\frac{1}{\pi} \int_0^{\infty} \frac{e^{-v|\xi-x|} \sin v\eta\, dv}{v}.$$

(d) Evaluate this integral by first differentiating with respect to η, then evaluating the result and integrating with respect to η, noting that $\varphi = 0$ when $\eta = 0$, and so obtain (106).

65. Develop the one-dimensional analogy to the treatment of Section 11.9 for the operator $L = d^2/dx^2$ in the interval $(0, l)$, as follows:

(a) Derive the "Green's formula"

$$\int_0^l G\, Ly\, d\xi = \Big[Gy' - G'y\Big]_0^l + \int_0^l y\, LG\, d\xi.$$

(b) Deduce that the solution of the problem

$$\frac{d^2y}{dx^2} = h(x),$$

$$y(0) = A, \qquad y(l) = B,$$

for $0 \leq x \leq l$, is of the form

$$y(x) = B\frac{\partial G(x, l)}{\partial \xi} - A\frac{\partial G(x, 0)}{\partial \xi} + \int_0^l G(x, \xi)h(\xi)\, d\xi$$

if $G(x, \xi)$ is such that

$$\frac{\partial^2 G}{\partial \xi^2} = \delta(x - \xi),$$

$$G(x, 0) = 0, \qquad G(x, l) = 0$$

for $0 \leq \xi \leq l$.

(c) Show that $G(x, \xi) = G(\xi, x)$, so that x and ξ may be interchanged in the definition of G in part (b). (Use the method of Problem 63.)

66. Let $G(x, \xi)$ denote the Green's function of Problem 65.

(a) Use the results of parts (b) and (c) of that problem to deduce that

$$\lim_{\epsilon \to 0} \int_{\xi-\epsilon}^{\xi+\epsilon} \frac{\partial^2 G}{\partial x^2}\, dx = 1$$

and hence that

$$\left[\frac{\partial G}{\partial x}\right]_{\xi-}^{\xi+} = 1,$$

so that $\partial G/\partial x$, considered as a function of x for a fixed value of ξ, must have a *unit jump* as x increases through ξ.

(b) Use the formulation

$$\frac{\partial^2 G}{\partial x^2} = 0 \qquad (x \neq \xi),$$

$$G(0, \xi) = 0, \qquad G(l, \xi) = 0,$$

$$\left[\frac{\partial G}{\partial x}\right]_{\xi-}^{\xi+} = 1,$$

with G defined to be *continuous* in $(0, l)$, to determine G in the form

$$G(x, \xi) = \begin{cases} \dfrac{x}{l}(\xi - l) & (0 \le x \le \xi), \\[2mm] \dfrac{\xi}{l}(x - l) & (\xi \le x \le l), \end{cases}$$

and deduce that the solution of the problem

$$\frac{d^2y}{dx^2} = h(x),$$

$$y(0) = A, \qquad y(l) = B$$

is

$$y(x) = B\frac{x}{l} + A\left(1 - \frac{x}{l}\right) + \int_0^l G(x, \xi)h(\xi)\, d\xi.$$

[Compare Problem 59(a) of Chapter 1. Since, to a first approximation, a small deflection $y(x)$ of a flexible string under uniform tension T, and subject to a load distribution of intensity f, satisfies the differential equation $T\, d^2y/dx^2 = f$ when f is positive in the *negative* y direction (see Section 5.2), this Green's function can be interpreted as the deflection of a string fixed at $x = 0$ and at $x = l$, and subject to a unit concentrated load at $x = \xi$ when $T = 1$.]

67. Obtain relations analogous to those of Section 11.9 associated with $L = \nabla^2$ in a three-dimensional region $\mathcal{R}$. In particular, show that the relevant Green's function G is to be such that

$$\nabla^2 G = \delta(x - \xi)\delta(y - \eta)\delta(z - \zeta) \qquad in\ \mathcal{R}$$

and

$$G = 0 \qquad on\ S,$$

where S bounds $\mathcal{R}$, and that

$$G = G(x, y, z; \xi, \eta, \zeta) \sim -\frac{1}{4\pi R} \qquad (R \to 0)$$

where

$$R = \sqrt{(x - \xi)^2 + (y - \eta)^2 + (z - \zeta)^2}.$$

Section 11.10

68. Prove that if $F_1(z)$ and $F_2(z)$ are both analytic functions in $\mathcal{R}$, such that $w = F_1(z)$ and $w = F_2(z)$ each map the region $\mathcal{R}$ onto the upper half of the w plane, then both $F_1(z)$ and $F_2(z)$ yield the same Green's function (110). [Use the result of Problem 55 to deduce that $F_2 = (aF_1 + b)/(cF_1 + d)$, where a, b, c, and d are real and $ad - bc > 0$.]

69. If $w = F(z)$ maps the interior of a region $\mathcal{R}$ conformally onto the upper half of the w plane, with $F'(z) \ne 0$ in $\mathcal{R}$, show that

$$w = \frac{F(z) - F(\zeta)}{F(z) - \overline{F(\zeta)}}$$

maps the interior of $\mathcal{R}$ conformally onto the interior of the unit disk $|w| \le 1$, with

$dw/dz \neq 0$ in $\Re$ and with $z = \zeta$ mapping onto $w = 0$. [Write $F(z) = r + is$ and $F(\zeta) = \rho + i\sigma$ and show that $0 \leq |w| \leq 1$ when $s \geq 0$ and $\sigma \geq 0$.]

70. Suppose that the analytic function $w = K(z)$ maps the region $\Re$ onto the unit disk $|w| \leq 1$. Show that the associated Green's function is

$$G(x, y; \xi, \eta) = \frac{1}{2\pi} \log \left| \frac{K(z) - K(\zeta)}{1 - \overline{K(\zeta)}K(z)} \right|,$$

where $z = x + iy$ and $\zeta = \xi + i\eta$. [Verify that the conditions (96), (97), and (98) are satisfied.]

71. Derive the Green's function for the unit circular disk $|z| \leq 1$ from the result of Problem 70. [Take $K(z) = z$.]

72. Obtain the Green's function for the first quadrant $0 \leq x < \infty, 0 \leq y < \infty$ in the form

$$G = \frac{1}{2\pi} \log \frac{|z - \zeta||z + \zeta|}{|z - \bar{\zeta}||z + \bar{\zeta}|}$$

$$= \frac{1}{4\pi} \log \frac{[(x - \xi)^2 + (y - \eta)^2][(x + \xi)^2 + (y + \eta)^2]}{[(x - \xi)^2 + (y + \eta)^2][(x + \xi)^2 + (y - \eta)^2]}.$$

Also show that

$$\left[\frac{\partial G}{\partial n}\right]_{\eta=0} = \frac{4}{\pi} \frac{xy\xi}{[y^2 + (x - \xi)^2][y^2 + (x + \xi)^2]}$$

and

$$\left[\frac{\partial G}{\partial n}\right]_{\xi=0} = \frac{4}{\pi} \frac{xy\eta}{[x^2 + (y - \eta)^2][x^2 + (y + \eta)^2]}$$

and verify that the first of these two results is in accordance with the result of Problem 83, Chapter 9. [Notice that here $F(z)$ can be taken to be z^2.]

73. Interpret the result of Problem 72 in terms of "images" of a basic singularity at $z = \zeta$. Also write down the corresponding Green's functions associated with the *right half-plane* and with the *third quadrant*, by use of the "method of images."

74. Obtain the Green's function for the infinite sector $0 \leq \theta \leq \alpha, 0 \leq r < \infty$ in the polar form

$$G = \frac{1}{4\pi} \log \frac{r^{2\pi/\alpha} - 2r^{\pi/\alpha}\gamma^{\pi/\alpha} \cos \frac{\pi}{\alpha}(\theta - \beta) + \gamma^{2\pi/\alpha}}{r^{2\pi/\alpha} - 2r^{\pi/\alpha}\gamma^{\pi/\alpha} \cos \frac{\pi}{\alpha}(\theta + \beta) + \gamma^{2\pi/\alpha}},$$

where $\xi + i\eta = \gamma e^{i\beta}$. Also show that

$$\left[\frac{\partial G}{\partial n}\right]_{\beta=0} = \frac{1}{\alpha\gamma} \frac{r^{\pi/\alpha}\gamma^{\pi/\alpha} \sin \frac{\pi\theta}{\alpha}}{r^{2\pi/\alpha} - 2r^{\pi/\alpha}\gamma^{\pi/\alpha} \cos \frac{\pi\theta}{\alpha} + \gamma^{2\pi/\alpha}}$$

and

$$\left[\frac{\partial G}{\partial n}\right]_{\beta=\alpha} = \frac{1}{\alpha\gamma} \frac{r^{\pi/\alpha}\gamma^{\pi/\alpha} \sin \frac{\pi\theta}{\alpha}}{r^{2\pi/\alpha} + 2r^{\pi/\alpha}\gamma^{\pi/\alpha} \cos \frac{\pi\theta}{\alpha} + \gamma^{2\pi/\alpha}}.$$

75. (a) Specialize the results of Problem 74 to the case $\alpha = \pi/2$ and verify that they then are equivalent to the results of Problem 72.

(b) Specialize the results of Problem 74 to the case $\alpha = \pi$ and verify that they then are equivalent to the relations (106) and (107).

76. Use the results of Problem 74 to show that the solution of the problem

$$\nabla^2 T = 0,$$

$$T(r, 0) = f(r), \qquad T(r, \alpha) = g(r),$$

for $0 \le r < \infty$, $0 \le \theta \le \alpha$, can be written in the form

$$T(r, \theta) = \frac{1}{\pi} \left[\int_0^\infty \frac{f(u^{\alpha/\pi}) \, \rho \sin \varphi \, du}{\rho^2 \sin^2 \varphi + (u - \rho \cos \varphi)^2} + \int_0^\infty \frac{g(u^{\alpha/\pi}) \, \rho \sin \varphi \, du}{\rho^2 \sin^2 \varphi + (u + \rho \cos \varphi)^2} \right],$$

where $\rho = r^{\pi/\alpha}$ and $\varphi = \pi\theta/\alpha$, so that $z^{\pi/\alpha} = \rho e^{i\varphi}$.

77. (a) Verify that the mapping

$$w = -i \frac{1 - z}{1 + z}$$

transforms the *exterior* of the unit disk ($|z| \ge 1$) into the upper half of the w plane. [Note the reversed sign relative to the mapping of Problem 23(c).]

(b) Show that the Green's function for the exterior problem is the same as that given by Equation (116) for the interior problem.

(c) Account for the negative sign in the result of Problem 35, Chapter 9.

78. Use the result of Problem 45(b), Chapter 9, to obtain the Green's function for the *rectangle* $0 \le x \le a$, $0 \le y \le b$ in the form

$$G(x, y; \xi, \eta) = -\frac{4ab}{\pi^2} \sum_{m=1}^\infty \sum_{n=1}^\infty \frac{\sin \dfrac{m\pi x}{a} \sin \dfrac{n\pi y}{b} \sin \dfrac{m\pi\xi}{a} \sin \dfrac{n\pi\eta}{b}}{m^2 b^2 + n^2 a^2}.$$

(This result can be expressed in closed form only in terms of elliptic functions.)

Section 11.11

79. (a) Obtain the analogy to Equation (127) in the one-dimensional case when

$$Ly = py'' + sy' + qy$$

in an interval (a, b) and show that then

$$L^*y = py'' + (2p' - s)y' + (p'' - s' + q)y.$$

Thus deduce that L is *self-adjoint* [or "*formally* self-adjoint" (see footnote, page 661)] if and only if

$$s = p',$$

in which case there follows

$$Ly = (py')' + qy.$$

[Since any equation of the form $a_0 y'' + a_1 y' + a_2 y = f$ can be written in an equivalent form $(py')' + qy = h$ (see Section 5.6), there is no loss in generality if this form is presumed, so that the associated operator L is self-adjoint.]

(b) When

$$Ly = (py')' + qy,$$

show that the Green's formula of part (a) takes the form

$$\int_a^b G\, Ly\, d\xi = \left[p\left(Gy' - \frac{\partial G}{\partial \xi} y \right) \right]_a^b + \int_a^b y\, LG\, d\xi.$$

80. (a) Use the result of Problem 79 to show that the solution of the problem

$$\frac{d}{dx}\left[p(x)\frac{dy}{dx} \right] + q(x)y = h(x),$$

$$y(a) = A, \qquad y(b) = B,$$

for $a \leq x \leq b$, is of the form

$$y(x) = Bp(b)\frac{\partial G(x,\, b)}{\partial \xi} - Ap(a)\frac{\partial G(x,\, a)}{\partial \xi} + \int_a^b G(x,\, \xi)h(\xi)\, d\xi$$

if $G = G(x,\, \xi)$ is such that

$$\frac{\partial}{\partial \xi}\left[p(\xi)\frac{\partial G}{\partial \xi} \right] + q(\xi)G = \delta(\xi - x),$$

$$G(x,\, a) = 0, \qquad G(x,\, b) = 0,$$

for $a \leq \xi \leq b$.

(b) Use the method of Problem 63 to show that here $G(x,\, \xi) = G(\xi,\, x)$, so that x and ξ can be interchanged in the preceding definition of G.

(c) Use the method of Problem 66(a) to show that G must be such that

$$\left[\frac{\partial G}{\partial x} \right]_{\xi-}^{\xi+} = \frac{1}{p(\xi)},$$

under the assumption that $p(x) \neq 0$ for $a \leq x \leq b$, so that G can be specified by the alternative formulation

$$\frac{\partial}{\partial x}\left(p\frac{\partial G}{\partial x} \right) + qG = 0 \qquad (x \neq \xi),$$

$$G(a,\, \xi) = 0, \qquad G(b,\, \xi) = 0,$$

$$\left[\frac{\partial G}{\partial x} \right]_{\xi-}^{\xi+} = \frac{1}{p(\xi)},$$

with G defined to be *continuous* in (a, b). (Compare Problem 57 of Chapter 1.)

81. Use the result of Problem 80 to obtain the solution of the problem

$$\frac{d^2y}{dx^2} - k^2 y = h(x),$$

$$y(\pm\infty) = 0$$

in the form

$$y(x) = -\frac{1}{2k}\int_{-\infty}^{\infty} e^{-k|x-\xi|}h(\xi)\, d\xi$$

when $k > 0$. (Compare Problem 93 of Chapter 5.)

82. (a) Show that, if the condition $y(b) = B$ is replaced by $y'(b) = C$ in Problem 80, then the condition $G(b,\, \xi) = 0$ must be replaced by $\partial G(b,\, \xi)/\partial x = 0$.

(b) Use the result of part (a) to obtain the solution of the problem

$$\frac{d^2y}{dx^2} = h(x),$$

$$y(0) = A, \qquad y'(l) = C$$

in the form

$$y(x) = A + Cx + \int_0^l G(x, \xi)h(\xi)\, d\xi,$$

where

$$G(x, \xi) = \begin{cases} -x & (0 \le x \le \xi), \\ -\xi & (\xi \le x \le l). \end{cases}$$

[The results of Problem 80 generalize analogously to admit end conditions of the form

$$k_1 y(a) + k_2 y'(a) = A, \qquad k_3 y(b) + k_4 y'(b) = B.$$

In this more general case (but generally *not* when a condition is of *two-point* type) the Green's function again is symmetric.]

83. For the operator

$$L = A\frac{\partial^2}{\partial x^2} + C\frac{\partial^2}{\partial y^2} + D\frac{\partial}{\partial x} + E\frac{\partial}{\partial y} + F,$$

show that

$$\iint_\mathcal{R} G\,L\varphi\, dx\, dy = \oint_C (P\cos\alpha + Q\cos\beta)\, ds + \iint_\mathcal{R} \varphi\, L^*G\, dx\, dy,$$

where

$$P = AG\varphi_x - \varphi(AG)_x + DG\varphi,$$
$$Q = CG\varphi_y - \varphi(CG)_y + EG\varphi,$$

and

$$L^* = A\frac{\partial^2}{\partial x^2} + C\frac{\partial^2}{\partial y^2} + (2A_x - D)\frac{\partial}{\partial x} + (2C_y - E)\frac{\partial}{\partial y} + (A_{xx} + C_{yy} - D_x - E_y + F).$$

Deduce that L then is self-adjoint if and only if

$$D = A_x, \qquad E = C_y,$$

in which case

$$L\varphi = (A\varphi_x)_x + (C\varphi_y)_y + F\varphi$$

and

$$P = A(G\varphi_x - G_x\varphi), \qquad Q = C(G\varphi_y - G_y\varphi).$$

[When $B\,\partial^2/\partial x\,\partial y$ appears in L, the self-adjointness conditions become $D = A_x + \frac{1}{2}B_y$, $E = C_y + \frac{1}{2}B_x$, in which case $L\varphi = (A\varphi_x)_x + \frac{1}{2}(B\varphi_x)_y + \frac{1}{2}(B\varphi_y)_x + (C\varphi_y)_y + F\varphi$.]

84. Show that

$$\frac{\partial}{\partial x} = \cos\alpha\frac{\partial}{\partial n} - \cos\beta\frac{\partial}{\partial s}, \qquad \frac{\partial}{\partial y} = \cos\alpha\frac{\partial}{\partial s} + \cos\beta\frac{\partial}{\partial n},$$

on a curve C with unit normal vector $\mathbf{n} = \mathbf{i}\cos\alpha + \mathbf{j}\cos\beta$, and use this result to express the boundary integral in Problem 83 in the form

$$\oint_C \left[(A\cos^2\alpha + C\cos^2\beta)\left(G\frac{\partial\varphi}{\partial n} - \frac{\partial G}{\partial n}\varphi\right) - (A - C)\cos\alpha\cos\beta\left(G\frac{\partial\varphi}{\partial s} - \frac{\partial G}{\partial s}\varphi\right)\right] ds$$

when L is self-adjoint.

85. Use the method of Problem 63 to prove that

$$G(x, y; \xi, \eta) = G(\xi, \eta; x, y)$$

if it is true that both $L^* = L$ and also the boundary integral in (130) vanishes when $\varphi(\xi, \eta)$ satisfies the same homogeneous conditions on the curve C in the $\xi\eta$ plane as does G. [Notice that then it is true that

$$\iint_\mathcal{R} G\,L\varphi\,d\xi\,d\eta = \iint_\mathcal{R} \varphi\,LG\,d\xi\,d\eta$$

when G and φ satisfy the same homogeneous conditions on C.]

86. Show that, if G is to be a Green's function for $L = \nabla^2$ associated with the entire plane, and accordingly must be independent of *angle* about the point (ξ, η), then it must satisfy the equation

$$\frac{d}{dR}\left(R\frac{dG}{dR}\right) = 0 \qquad (R \neq 0)$$

and hence cannot be made to approach zero as $x^2 + y^2 \longrightarrow \infty$.

87. Use the Fourier transform to determine the Green's function for the modified Helmholtz operator $L = \nabla^2 - k^2$ in the entire plane, as follows:

(a) Starting with the formulation

$$\frac{\partial^2 G}{\partial x^2} + \frac{\partial^2 G}{\partial y^2} - k^2 G = \delta(x - \xi)\delta(y - \eta),$$

$$\lim_{x^2+y^2\to\infty} G = 0,$$

take the repeated Fourier transform, relative to both x and y, and deduce that

$$\bar{\bar{G}} = -\frac{e^{-iu\xi}e^{-iv\eta}}{u^2 + v^2 + k^2}$$

and hence that

$$G = -\frac{1}{2\pi^2}\int_{-\infty}^{\infty}\int_{-\infty}^{\infty}\frac{e^{i[u(x-\xi)+v(y-\eta)]}}{u^2 + v^2 + k^2}\,du\,dv.$$

(b) By introducing polar coordinates (r, θ) such that $u = r\cos\theta$, $v = r\sin\theta$, and writing

$$x - \xi = R\cos\psi, \qquad y - \eta = R\sin\psi,$$

show that the preceding result can be written in the form

$$G = -\frac{1}{4\pi^2}\int_0^\infty\left[\int_0^{2\pi} e^{irR\cos(\theta-\psi)}\,d\theta\right]\frac{r\,dr}{r^2 + k^2}$$

and, by referring to Problem 65 of Chapter 5, deduce that

$$G = -\frac{1}{2\pi}\int_0^\infty \frac{rJ_0(rR)\,dr}{r^2 + k^2}.$$

[The integral obtained here is known to have the value $K_0(kr)$ when $k > 0$, in accordance with the result of text Example 1.]

88. By replacing G by $N(a, b; \xi, \eta)$ and φ by $N(c, d; \xi, \eta)$ in (90), and using (149) and (150), deduce that

$$N(a, b; c, d) = \oint_C N(a, b; \xi(s), \eta(s))\alpha(s)\, ds$$

$$- \oint_C N(c, d; \xi(s), \eta(s))\alpha(s)\, ds + N(c, d; a, b).$$

Hence show that the satisfaction of (153) ensures that $N(x, y; \xi, \eta) = N(\xi, \eta; x, y)$.

89. Verify that the function

$$N = \frac{1}{2\pi} \log |(z - \zeta)(1 - z\bar{\zeta})|$$

serves as a Neumann function for the unit disk $x^2 + y^2 \leqq 1$, with $\alpha(s) = 1/(2\pi)$ in Equations (150) and (151). [Writing $z = re^{i\theta}$ and $\zeta = \gamma e^{i\beta}$, show that

$$N = \frac{1}{4\pi}\{\log [r^2 - 2r\gamma \cos (\theta - \beta) + \gamma^2] + \log [1 - 2r\gamma \cos (\theta - \beta) + r^2\gamma^2]\}$$

and verify, in particular, that the condition $[\partial N/\partial \gamma]_{\gamma=1} = 1/(2\pi)$ is satisfied.]

90. (a) Verify that, if $w = K(z)$ maps a region $\mathcal{R}$ conformally onto the unit disk $|w| \leqq 1$, then

$$N = \frac{1}{2\pi} \log |[K(z) - K(\zeta)][1 - K(z)\overline{K(\zeta)}]|$$

serves as a Neumann function for $\mathcal{R}$. [The fact that

$$\oint_C \frac{\partial N}{\partial n}\, ds = \frac{1}{2\pi} \oint_C |K'(\zeta)|\, ds = \frac{1}{2\pi} \int_0^{2\pi} d\beta = 1,$$

where C is the boundary of $\mathcal{R}$, follows from results of Problems 89 and 28.]

(b) By interchanging z and w in Problem 23(c) and taking $R = 1$, deduce from part (a) with $K(z) = (i - z)/(i + z)$ that the function

$$N = \frac{1}{2\pi} \log \left| \frac{(z - \zeta)(z - \bar{\zeta})}{(z + i)^2(\bar{\zeta} + i)^2} \right|$$

is a Neumann function for the upper half-plane. (Here an irrelevant additive constant was discarded.) Also verify directly that

$$\left[-\frac{\partial N}{\partial \eta} \right]_{\eta=0} = \frac{1}{\pi(\xi^2 + 1)} \equiv \alpha(\xi)$$

and hence also that (150) and (151) are indeed satisfied.

(c) Deduce that, if $w = F(z)$ maps a region $\mathcal{R}$ conformally onto the upper half-plane, then the function

$$N = \frac{1}{2\pi} \log \left| \frac{[F(z) - F(\zeta)][F(z) - \overline{F(\zeta)}]}{[F(z) + i]^2[F(\zeta) + i]^2} \right|$$

is a Neumann function for $\mathcal{R}$.

91. Modify the treatment of Problem 88 to show that the satisfaction of (158) ensures that $H(x, y; \xi, \eta) = H(\xi, \eta; x, y)$.

92. Verify that the function

$$H = \frac{1}{2\pi} \log |(z - \zeta)(1 - z\bar{\zeta})| - \frac{1}{4\pi}(|z|^2 + |\zeta|^2)$$

is a Hilbert function for the unit disk $x^2 + y^2 \leq 1$, with $\beta(\xi, \eta) = 1/\pi$ in (154) and (156). [Notice that $H = N - (r^2 + \gamma^2)/(4\pi)$, where $x + iy = re^{i\theta}$ and $\xi + i\eta = \gamma e^{i\beta}$, and where N is the Neumann function of Problem 89, and recall that $\nabla^2 N = 0$ when $R \neq 0$ and $\partial N/\partial n = 1/(2\pi)$ on the circular boundary.]

93. Let $w = f(z)$ map a region $\Re$ conformally onto a region $\Re^*$. With the abbreviations $\nabla^2 = \partial^2/\partial x^2 + \partial^2/\partial y^2$ and $\nabla^{*2} = \partial^2/\partial u^2 + \partial^2/\partial v^2$, where $w = u + iv$, show that

$$\nabla^{*2}\varphi = \frac{\nabla^2\varphi}{|f'(z)|^2}$$

and that

$$\iint_{\Re^*} \nabla^{*2}\varphi \, du \, dv = \iint_{\Re} \nabla^2\varphi \, dx \, dy.$$

[Use Equations (19) and (41), together with (46) of Chapter 7.]

94. Use the result of Problem 93 to justify the use of conformal mapping with reference to the Hilbert function. In particular, proceed as in Problem 90(b) to deduce that

$$H = \frac{1}{2\pi} \log \left| \frac{(z - \zeta)(z - \bar{\zeta})}{(z + i)^2(\zeta + i)^2} \right| - \frac{1}{4\pi} \left(\left| \frac{z - i}{z + i} \right|^2 + \left| \frac{\zeta - i}{\zeta + i} \right|^2 \right)$$

is a Hilbert function for the upper half-plane and that, if $w = F(z)$ maps $\Re$ conformally onto the upper half-plane, then the result of replacing z by $F(z)$ and ζ by $F(\zeta)$ in this function is a Hilbert function for $\Re$.

95. Develop the analogy $N(x, \xi)$ to Neumann's function of text Example 3 in correspondence with the problem

$$\frac{d^2y}{dx^2} = h(x),$$

$$y'(0) = C, \qquad y'(l) = D,$$

showing that the condition (148) corresponds to the compatibility requirement

$$D - C = \int_0^l h(x) \, dx,$$

that the conditions (149), (150), and (151) correspond to the formulation

$$\frac{\partial^2 N}{\partial \xi^2} = \delta(\xi - x),$$

$$N_\xi(x, 0) = -\alpha_1, \qquad N_\xi(x, l) = \alpha_2,$$

with

$$\alpha_1 + \alpha_2 = 1,$$

and that, when the compatibility condition is satisfied, the problem solution is

$$y(x) = A + CN(x, 0) - DN(x, l) + \int_0^l N(x, \xi)h(\xi) \, d\xi,$$

where $A = \alpha_1 y(0) + \alpha_2 y(l)$ is an arbitrary constant. Show also that the symmetry-ensuring condition (153) corresponds to the requirement

$$\alpha_1 N(x, 0) + \alpha_2 N(x, l) = \text{constant}.$$

Finally, with the special choice $\alpha_1 = \alpha_2 = \frac{1}{2}$, show that

$$N(x, \xi) = \tfrac{1}{2}|x - \xi| + \text{constant}.$$

[Notice that an arbitrary function of x can be added to $N(x, \xi)$ if symmetry is not required.]

96. Proceed as in Problem 95 with the analogy $H(x, \xi)$ to Hilbert's function in correspondence with the specified problem, showing that H is to be determined by

$$\frac{\partial^2 H}{\partial \xi^2} = \delta(\xi - x) - \beta(\xi),$$

$$H_\xi(x, 0) = 0, \qquad H_\xi(x, l) = 0,$$

where

$$\int_0^l \beta(\xi)\, d\xi = 1,$$

and that then, if the compatibility condition is satisfied, there follows

$$y(x) = B + CH(x, 0) - DH(x, l) + \int_0^l H(x, \xi)h(\xi)\, d\xi,$$

where $B = \int_0^l \beta(\zeta)y(\zeta)\, d\zeta$ is an arbitrary constant. Show also that the symmetry-ensuring condition (158) corresponds to the requirement

$$\int_0^l H(x, \xi)\beta(\xi)\, d\xi = \text{constant}.$$

Finally, with the special choice $\beta(x) = 1/l$, show that

$$H = \begin{cases} x & (\xi \leq x) \\ \xi & (\xi \leq x) \end{cases} - \frac{x^2 + \xi^2}{2l} + \text{constant}.$$

[Notice that an arbitrary function of x can be added to $H(x, \xi)$ if symmetry is not required.]

97. Deduce Equation (174) from (165), (172), and (173).

98. Derive the Green's function (184) (with the *upper* ambiguous sign) for the heat-flow operator in the region $0 \leq x < \infty$, $0 \leq t < \infty$, with $\varphi(0, t)$ assumed to be prescribed, by use of the method of variation of parameters (Section 9.17) as follows:

(a) For fixed x and t, write

$$G = \int_0^\infty C(u, \tau) \sin u\xi\, du$$

and show that

$$\frac{\partial C}{\partial \tau} - \alpha^2 u^2 C = \frac{2}{\pi} \delta(\tau - t) \sin ux.$$

(b) Deduce that

$$C(u, \tau) = -\frac{2}{\pi} e^{-\alpha^2 u^2 (t - \tau)} \sin ux \qquad (\tau < t)$$

and hence that

$$G = -\frac{1}{2\alpha\sqrt{\pi(t - \tau)}}[e^{-[(\xi-x)^2/4\alpha^2(t-\tau)]} - e^{-[(\xi+x)^2/4\alpha^2(t-\tau)]}]$$

when $\tau < t$, by virtue of Equation (241) of Chapter 9.

99. Use the method of variation of parameters (Section 9.17) to determine the Green's function for the heat-flow operator $L = \alpha^2(\partial^2/\partial x^2) - \partial/\partial t$ when $\mathcal{R}$ is the domain $0 \leq x \leq l, 0 \leq t < \infty$, if the solution φ of a related problem is assumed to be specified when $x = 0$, $x = l$, and $t = 0$, in the form

$$G = -\frac{2}{l} \sum_{n=1}^{\infty} e^{-n^2\pi^2\alpha^2(t-\tau)/l^2} \sin\frac{n\pi x}{l} \sin\frac{n\pi\xi}{l} \qquad (t > \tau).$$

[For fixed values of x and t, assume

$$G = \sum_{n=1}^{\infty} C_n(\tau) \sin\frac{n\pi\xi}{l}$$

in Equation (176) and deduce that

$$C_n' - \frac{n^2\pi^2\alpha^2}{l^2} C_n = \frac{2}{l}\,\delta(\tau - t) \sin\frac{n\pi x}{l}$$

and accordingly

$$C_n = -\frac{2}{l} e^{-n^2\pi^2\alpha^2(t-\tau)/l^2} \sin\frac{n\pi x}{l}$$

when $\tau < t$.]

100. With the abbreviation

$$S(x - \xi, t - \tau) = \frac{1}{2\alpha\sqrt{\pi(t-\tau)}} e^{-(\xi-x)^2/4\alpha^2(t-\tau)},$$

for the *negative* of the Green's function associated with the heat-flow operator in the region $-\infty < x < \infty$, $0 \leq t < \infty$, use the method of images to show that the corresponding Green's function in the region $0 \leq x \leq l, 0 \leq t < \infty$ can be expressed as the sum

$$G = -\sum_{n=-\infty}^{\infty} [S(x - \xi + 2nl, t - \tau) - S(x + \xi + 2nl, t - \tau)]$$

if G is to vanish when $\xi = 0$ and when $\xi = l$, as the sum

$$G = -\sum_{n=-\infty}^{\infty} [S(x - \xi + 2nl, t - \tau) + S(x + \xi + 2nl, t - \tau)]$$

if $\partial G/\partial\xi$ is to vanish when $\xi = 0$ and when $\xi = l$, and as the sum

$$G = -\sum_{n=-\infty}^{\infty} [-S(x - \xi + 2l + 4nl, t - \tau) + S(x + \xi + 4nl, t - \tau)$$
$$+ S(x - \xi + 4nl, t - \tau) - S(x + \xi - 2l + 4nl, t -\!-\tau)]$$

if $\partial G/\partial\xi = 0$ when $\xi = 0$ and $G = 0$ when $\xi = l$. [Notice that S is symmetric in x and ξ (but not in t and τ). These expansions may be preferred to series of the type obtained in Problem 99 when t is small.]

Answers to Problems

Chapter 1

1. (a) $y' - y = 0$.
   (b) $y' - 2xy = 2 - 4x^2$.
   (c) $y' - y^2 = 0$.
   (d) $xy' - y \log y = 0$.
   (e) $yy' + x = 0$.
   (f) $x^2 y'^2 = (4y + 1)(xy' - y)$.

2. (a) $y'' - y = 0$.
   (b) $y'' - 2y' + y = 0$.
   (c) $y'' + y = 0$.
   (d) $yy'' - y'^2 = 0$.
   (e) $(y - 1)y'' - 2y'^2 = 0$.
   (f) $y''^2 = (1 + y'^2)^3$.

3. (a) $y = Ce^{x^2}$.
   (b) $x - C = \sqrt{1 - y^2}, y = \pm 1$.
   (c) $(y - 1)(x - 1) = C(y + 1)(x + 1)$.
   (d) $\sqrt{1 - y^2} - \sqrt{1 - x^2} = C, x = \pm 1, y = \pm 1$.

6. (b) Dependent.

7. (a) Dependent.
   (b) Dependent.
   (c) Independent.
   (d) Dependent.

13. (a) $y = x^2/(2 - k) + Cx^k$ if $k \neq 2$; $y = x^2 \log x + Cx^2$ if $k = 2$.
    (b) $y = x \tan x + 1 + C \sec x$.
    (c) $y = \sin x + C \cos x$.
    (d) $y = 2 \sin x - 2 + Ce^{-\sin x}$.
    (e) $y = (x + 1)(x + C)/(x - 1)$.
    (f) $y = (Cx - 1)/(x \log x)$.
    (g) $y^2 = x(x + C)$.
    (h) $x = y(y + C)$.

18. (a) $y = c_1 e^{-x} + c_2 e^{2x}$.

(b) $y = c_1 e^{-x} + (c_2 + c_3 x)e^x$.

(c) $y = e^x(c_1 \cos x + c_2 \sin x)$.

(d) $y = e^x(c_1 + c_2 x + c_3 \cos x + c_4 \sin x)$.

(e) $y = c_1 e^x + e^{-x/2}[c_2 \cos (\sqrt{3}x/2) + c_3 \sin (\sqrt{3}x/2)]$.

(f) $y = c_1 e^{(1+i)x} + c_2 e^{-(1+i)x}$.

19. (a) $y = c_1 e^{kx} + c_2 e^{-kx} + c_3 \cos kx + c_4 \sin kx$.

(b) $y = e^{kx}(c_1 \cos kx + c_2 \sin kx) + e^{-kx}(c_3 \cos kx + c_4 \sin kx)$.

(c) $y = e^{kx}(c_1 + c_2 x) + e^{-kx}(c_3 + c_4 x)$.

20. (a) $y = c_1 \cos kx + c_2 \sin kx + (\sin x)/(k^2 - 1)$ if $k^2 \neq 0, 1$.

(b) $y = c_1 \cos x + c_2 \sin x - \frac{1}{2}x \cos x$.

(c) $y = c_1 e^x + c_2 e^{-x} - \frac{1}{2} \sin x$.

(d) $y = c_1 e^x + c_2 e^{-x} + \frac{1}{2}xe^x$.

(e) $y = c_1 e^x + c_2 e^{-x} + \frac{1}{4}(x^2 - x)e^x$.

(f) $y = e^x(c_1 \cos x + c_2 \sin x) + \frac{1}{5}(2 \cos x + \sin x)$.

(g) $y = e^x(c_1 \cos x + c_2 \sin x) - \frac{1}{2}xe^x \cos x$.

(h) $y = c_1 e^{4x} + c_2 e^{5x} + (2x^2 + 6x + 7)e^{3x}$.

(i) $y = c_1 e^{2x} + c_2 e^{-x} - xe^{-x} - 3 \sin x + \cos x + 2x - 1$.

21. (a) $y = c_1 x^k + c_2 x^{-k}$ if $k \neq 0$; $y = c_1 + c_2 \log x$ if $k = 0$.

(b) $y = x(c_1 \cos \log x + c_2 \sin \log x)$.

(c) $y = c_1 x^2 + c_2 x^{-1}$.

(d) $y = x(c_1 + c_2 \log x)$.

(e) $y = c_1 x^{-1} + x(c_2 + c_3 \log x)$.

(f) $y = c_1 x^n + c_2 x^{-n-1}$.

(g) $y = c_1 x + c_2 x^{-1} + \frac{1}{3}x^2$.

(h) $y = c_1 x + c_2 x^{-1} + \frac{1}{2}x \log x$.

(i) $y = c_1 x^3 + c_2 x^2 + 3x + 2$.

22. (a) $y = c_1(a + x)^2 + c_2(a + x)^{-1}$.

(b) $y = c_1(a + x)^2 + c_2(a + x)^{-1} + (a + x)^2 \log (a + x) - \frac{1}{2}$.

(c) $y = c_1(a + x)^2 + c_2(a + x)^{-1} + \frac{1}{3}(a + x)^2 \log (a + x) - 2ax - a^2$.

28. $y = c_1 e^x + c_2 e^{-x} + \frac{1}{2}xe^x$.

30. $y = c_1 x^2 + c_2 x^{-1} + \frac{1}{4}x^3$.

32. $x = c_1 t + c_2, y = -\frac{1}{2}c_1 t^2 - (c_1 + c_2)t + c_3$.

33. (a) $x = c_1 e^{2t} + c_2 e^t, y = -c_1 e^{2t} + c_2 e^t$.

(b) $x = c_1 e^t + c_2 \cos t + c_3 \sin t$,

$y = 2(c_1 - 1)e^t + (c_2 - c_3) \cos t + (c_2 + c_3) \sin t$.

(c) $x = e^t(c_1 + c_2 t) + c_3 e^{-3t/2} - \frac{1}{2}t$,

$y = e^t(6c_2 - 2c_1 - 2c_2 t) - \frac{1}{3}c_3 e^{-3t/2} - \frac{1}{3}$.

(d) $x = c_1 \cos \omega_1 kt + c_2 \sin \omega_1 kt + c_3 \cos \omega_2 kt + c_4 \sin \omega_2 kt$,

$y = \sqrt{2}(c_1 \cos \omega_1 kt + c_3 \sin \omega_1 kt - c_3 \cos \omega_2 kt - c_4 \sin \omega_2 kt)$,

where $\omega_1 = \sqrt{2 + \sqrt{2}}$ and $\omega_2 = \sqrt{2 - \sqrt{2}}$.

(e) $x = e^{2t} + 1, y = e^{2t} - 1$.

(f) $x = c_1 e^{-2t}, y = 5c_1 e^{-2t} + c_2 e^{-t}$.

(g) $x = c_1 e^{2t}, y = \frac{3}{4}c_1 e^{2t} + c_2 e^{-2t}, z = -\frac{3}{2}c_1 e^{2t} - \frac{2}{3}c_2 e^{-2t} + c_3 e^{3t}$.

34. (a) $x = c_1 t + c_2 t^2, y = c_1 t - c_2 t^2$.

(b) $x = c_1 t^2, y = \frac{3}{4}c_1 t^2 + c_2 t^{-2}, z = -\frac{3}{2}c_1 t^2 - \frac{2}{3}c_2 t^{-2} + c_3 t^3$.

35. (a) $py = \int ph \, dx + c$, where $p = e^{\int a_1 dx}$.

(b) $y = c_1 \cos x + c_2 \sin x + (\sin x) \log (\tan \frac{1}{2}x)$.

(c) $y = c_1 \cos x + c_2 \sin x + (\cos x) \log (\cos x) + x \sin x$.

(d) $y = c_1 \cos x + c_2 \sin x + \int^x \sin (x - \xi) \log \xi \, d\xi$.

(e) $y = e^x[c_1 \cos x + c_2 \sin x - (\cos x) \log (\sec x + \tan x)]$.

(f) $y = c_1 x^3 + c_2 x^2 + x^3 \int \dfrac{\sin x}{x^2}\, dx - x^2 \int \dfrac{\sin x}{x}\, dx.$

(g) $y = c_1 x + c_2 x^2 - x \log x - \tfrac{1}{2} x (\log x)^2.$

39. $y = c_1 x + c_2 e^x + 1.$

40. $y = c_1(1 + x \tan x) + c_2 \tan x.$

41. $y = c_1 x + c_2\left(\dfrac{x}{2}\log\left|\dfrac{1+x}{1-x}\right| - 1\right) + \dfrac{3}{2}(x^2 + 1).$

49. $u_0 = \dfrac{1}{2}\left(\dfrac{x}{a} + \dfrac{a}{x}\right),\ u_1 = \dfrac{1}{2}\left(x - \dfrac{a^2}{x}\right)\ (a \neq 0).$

51. $u_0 = 1,\ u_1 = \sin x,\ u_2 = 1 - \cos x.$

52. $u_0 = \tfrac{1}{2}(\cosh x + \cos x),\ u_1 = \tfrac{1}{2}(\sinh x + \sin x),$
$u_2 = \tfrac{1}{2}(\cosh x - \cos x),\ u_3 = \tfrac{1}{2}(\sinh x - \sin x).$

54. (a) $k = n\pi/a$, n a nonzero integer.
(b) $k \neq m\pi/a$, m an odd integer.

60. (a) $x^2 y^3 + xy - y^2 = c.$
(b) $x^3 y - xy^3 = c.$
(c) $ye^x + xe^y = c.$

61. (a) $y^2 + 2xy - x^2 = c.$
(b) $y^2 + xy = cx.$
(c) $y + \sqrt{x^2 + y^2} = c.$
(d) $y = x \sin^{-1}(cx).$

62. (a) $xy = e^{cx}.$
(b) $x^3 - 3xy = c.$
(c) $y^3 - 3xy = c.$
(d) $(y - 2)^2 + 2(x - 1)(y - 2) - (x - 1)^2 = c.$
(e) $xy(x + c) = 1.$
(f) $(x^2 + 1)^2(2y^2 - 1) = c.$

64. (a) $x^2 - y^3 = cx^4.$
(b) $x^m + y^n = cy^p.$

65. (a) $y = \log(x + c_1) + c_2.$
(b) $y = c_1 \tan^{-1}(c_1 x) + c_2.$
(c) $y^2 = c_1 x + c_2.$
(d) $y = c_2 - [1 - (x - c_1)^2]^{1/2}.$
(e) $y = c_1 \cos x + c_2 \sin x.$
(f) $x = 2p^3 + p + c_1,\ y = \tfrac{3}{2}p^4 + \tfrac{1}{2}p^2 + c_2.$

Chapter 2

3. (a) $(s - a)/[(s - a)^2 + k^2].$
(b) $n!/(s + a)^{n+1}.$
(c) $(1 + e^{-s\pi})/(s^2 + 1).$
(d) $(e^{-sa} - e^{-sb})/s.$

4. (a) $6/s^4.$ (b) $2/(s + 3)^3.$ (c) $a(s^2 - 2a^2)/(s^4 + 4a^4).$
(d) $4(s - 1)/(s^2 - 2s + 5)^2.$ (e) $(6as^2 - 2a^3)/(s^2 + a^2)^3.$
(f) $(s - a)/[(s - a)^2 - b^2].$

5. (a) $s^3 \bar{f}(s) - s^2 f(0) - s f'(0) - f''(0).$ (b) $-\bar{f}'(s - 1).$
(c) $\displaystyle\sum_{n=0}^{N}(n!\,a_n)/s^{n+1}.$ (d) $\displaystyle\sum_{n=0}^{N} a_n s/(s^2 + n^2).$

11. $(\tanh \tfrac{1}{4}as)/s^2.$

20. (a) $1 - \dfrac{t^2}{2!3!} + \dfrac{t^4}{4!5!} - \dfrac{t^6}{6!7!} + \cdots$.

(b) $1 - \dfrac{(t/2)^2}{(1!)^2} + \dfrac{(t/2)^4}{(2!)^2} - \dfrac{(t/2)^6}{(3!)^2} + \cdots$.

26. (a) $(1 - \cos at)/a$.   (b) $(e^{at} - 1 - at)/a^2$.   (c) $(e^{at} - e^{bt})/(a - b)$.
(d) $(b \sin at - a \sin bt)/(b^2 - a^2)$.

28. (a) $\displaystyle\int_0^t e^{-u} f(t - u)\, du = \int_0^t e^{-(t-u)} f(u)\, du$.

(b) $\dfrac{1}{a} \displaystyle\int_0^t f(t - u) \sin au\, du = \dfrac{1}{a} \int_0^t f(u) \sin a(t - u)\, du$.

(c) $\displaystyle\int_0^t ue^{-au} f(t - u)\, du = \int_0^t (t - u)e^{-a(t-u)} f(u)\, du$.

(d) $\dfrac{1}{a - b} \displaystyle\int_0^t (e^{-bu} - e^{-au}) f(t - u)\, du = \dfrac{1}{a - b} \int_0^t [e^{-b(t-u)} - e^{-a(t-u)}] f(u)\, du$.

33. (a) $e^{2t} - e^t$.   (b) $e^t \cos 2t + \tfrac{1}{2} e^t \sin 2t$.
(c) $\sin T \sinh T + (\sin T \cosh T - \cos T \sinh T)/\sqrt{2}$, where $T = t/\sqrt{2}$.
(d) $(1 + 2e^{-t} - 3e^{-2t})/2$.   (e) $t - \sin t$.
(f) $e^{-(t-1)}$ when $t > 1$; 0 when $0 \leq t < 1$.

34. (a) $(\sin at + at \cos at)/(2a)$.   (b) $(\sinh at + at \cosh at)/(2a)$.
(c) $-e^{-t} + 5e^{-2t} - 4e^{-3t}$.
(d) $[e^{-at} - e^{at/2}(\cos \tfrac{1}{2}\sqrt{3}\, at - \sqrt{3} \sin \tfrac{1}{2}\sqrt{3}\, at)]/(3a^2)$.
(e) 1 when $0 \leq t < 1$; 0 when $t > 1$.
(f) $(b \sin at - a \sin bt)/[ab(b^2 - a^2)]$ if $b^2 \neq a^2$; $(\sin at - at \cos at)/(2a^3)$ if $b = a$.

35. (a) $e^{-t} \cos t$.   (b) $(\sin at \cosh at + \cos at \sinh at)/(2a)$.
(c) 0 when $0 \leq t \leq \pi$; $-\sin t$ when $t \geq \pi$.   (d) $\tfrac{1}{2}(2 - 4at + a^2 t^2)e^{-at}$.
(e) $\tfrac{1}{8}[(3 - t^2) \sin t - 3t \cos t]$.   (f) $\tfrac{1}{8}[(3 + t^2) \sinh t - 3t \cosh t]$.

36. (a) Staircase function: b $(0 < t < a)$, $2b$ $(a < t < 2a)$, $3b$ $(2a < t < 3a)$, $\ldots$.
(b) Square-wave function: 1 $(0 < t < a/2)$, -1 $(a/2 < t < a)$, 1 $(a < t < 3a/2)$, $\ldots$.
(c) Rectified sine: $\sin \omega t$ $(0 < t < \pi/\omega)$, 0 $(\pi/\omega < t < 2\pi/\omega)$,
$\sin \omega t$ $(2\pi/\omega < t < 3\pi/\omega)$, $\ldots$.
(d) $|\sin \omega t|$.

39. 0 $(0 < t < P/8)$, 1 $(P/8 < t < 3P/8)$, 2 $(3P/8 < t < 5P/8)$, 1 $(5P/8 < t < 7P/8)$,
0 $(7P/8 < t < P)$, $\ldots$.

40. (a) $y = e^{-kt}$.   (b) $y = (1 - e^{-kt})/k$.
(c) $y = e^{-kt}$ when $0 \leq t < 1$; $y = (1 + e^k)e^{-kt}$ when $t > 1$.
(d) $y = e^{-kt} \left[y_0 + \displaystyle\int_0^t f(u)e^{ku}\, du \right]$.

41. (a) $y = e^{-t} \cos t$.   (b) $y = 1 - e^{-t} \cos t$.
(c) $y = e^{-t} \cos t$ when $0 \leq t \leq 1$;
$y = e^{-t} \cos t + e^{-(t-1)} \sin (t - 1)$ when $t \geq 1$.
(d) $y = e^{-t} \left[y_0 \cos t + (y_0 + y_0') \sin t + \displaystyle\int_0^t f(u)e^u \sin (t - u)\, du \right]$.

42. (a) $y = \tfrac{1}{4}(\sin t \cosh t - \cos t \sinh t)$.
(b) $y = 1 - \cos t \cosh t$.

43. $y = -[\sin t \sin (b - a)]/(\sin b)$ when $0 \leq t \leq a$;
$y = -[\sin a \sin (b - t)]/(\sin b)$ when $a \leq t \leq b$.

44. (b) $x = \dfrac{f_0}{k} \left[1 - e^{-\alpha t} \left(\cos \beta t + \dfrac{\alpha}{\beta} \sin \beta t \right) \right]$ $(\beta^2 = \omega_0^2 - \alpha^2 > 0)$,

$x = \dfrac{f_0}{k} [1 - (1 + \alpha t)e^{-\alpha t}]$ $(\alpha = \omega_0)$,

$x = \dfrac{f_0}{k} \left[1 - \dfrac{\alpha + \gamma}{2\gamma} e^{-(\alpha-\gamma)t} + \dfrac{\alpha - \gamma}{2\gamma} e^{-(\alpha+\gamma)t} \right]$ $(\gamma^2 = \alpha^2 - \omega_0^2 > 0)$.

(c) $x = ae^{-\alpha t}\left(\cos \beta t + \dfrac{\alpha}{\beta} \sin \beta t\right)$ $(\beta^2 = \omega_0^2 - \alpha^2 > 0)$,

$\quad x = ae^{-\alpha t}(1 + \alpha t)$ $(\alpha = \omega_0)$,

$\quad x = a\left[\dfrac{\alpha + \gamma}{2\gamma}e^{-(\alpha - \gamma)t} - \dfrac{\alpha - \gamma}{2\gamma}e^{-(\alpha + \gamma)t}\right]$ $(\gamma^2 = \alpha^2 - \omega_0^2 > 0)$.

45. $x = -e^{-t}(\cos t + \sin t)$, $y = e^{-t}(1 + \sin t)$.

50. (a) $\sqrt{\pi}/2$.   (b) 1.54.   (c) 3.33.   (d) 1.43.   (e) -4.33.   (f) $3\sqrt{\pi}/4$.

54. (a) 0 when $\alpha > 0$; 1 when $\alpha = 0$; no finite limit when $\alpha < 0$.
  (b) $1/\sqrt{\pi}$.

Chapter 3

15. $y(0.1) \approx 0.9949$, $y(0.2) \approx 0.9785$.

19. Rounded true values: (a) 1.5527; (b) 1.5841; (c) 0.07842; (d) 1.6082.

21. Rounded true values: (a) 0.3152; (b) 0.3121.

26. Rounded true values: (a) 1.4049; (b) 1.4333; (c) 0.07096; (d) 0.07024; (e) 2.8394;
  (f) 5.6856.

Chapter 4

1. (a) All values of x.   (b) $-3 < x < 1$.   (c) $1 - \sqrt{3} \leq x \leq 1 + \sqrt{3}$.
  (d) $-1 < x < 1$; also $x = -1$ if $k < 1$ and $x = +1$ if $k < 0$.
  (e) All values of x.   (f) $x = 0$.
  (g) $a - 1 < x < a + 1$; also $x = a + 1$ if $\alpha > 1$ and $x = a - 1$ if $\alpha > 0$.
  (h) $-4 < x < 4$.   (i) $x > 1$ and $x \leq -1$.   (j) $x > -1$.

5. (a) $y = c_1\left(1 + \dfrac{x^3}{2 \cdot 3} + \dfrac{x^6}{2 \cdot 3 \cdot 5 \cdot 6} + \cdots\right) + c_2\left(x + \dfrac{x^4}{3 \cdot 4} + \dfrac{x^7}{3 \cdot 4 \cdot 6 \cdot 7} + \cdots\right)$.

  (b) $y = c_1\left(1 + \dfrac{x^2}{2!} - \dfrac{x^4}{4!} + \dfrac{1 \cdot 3 x^6}{6!} - \dfrac{1 \cdot 3 \cdot 5 x^8}{8!} + \cdots\right) + c_2 x$.

  (c) $y = c_1\left(1 + \dfrac{x^4}{2!} + \dfrac{x^8}{4!} + \cdots\right) + c_2\left(x^2 + \dfrac{x^6}{3!} + \dfrac{x^{10}}{5!} + \cdots\right)$
    $= c_1 \cosh x^2 + c_2 \sinh x^2$.

6. (a) $y = A_0\left(1 - \dfrac{x^2}{2!} + \dfrac{x^4}{4!} - \cdots\right) + A_1\left(x - \dfrac{x^3}{3!} + \dfrac{x^5}{5!} - \cdots\right)$.

  (b) $y = A_0\left(1 - \dfrac{3}{2}x^2 + \dfrac{1}{6}x^3 + \cdots\right) + A_1\left(x - \dfrac{1}{2}x^3 + \dfrac{1}{12}x^4 + \cdots\right)$.

  (c) $y = A_0\left(1 + \dfrac{1}{2}x^2\right) + A_1 x$.

  (d) $y = A_0\left(1 + x + \dfrac{1}{2}x^2 + \cdots\right)$.

  (e) $y = A_0\left(1 + \dfrac{x}{1!} + \dfrac{x^2}{2!} + \cdots\right)$.

  (f) $y = A_0$.

7. (a) No singular points.
  (b) No singular points.
  (c) Regular singular points at $x = \pm\sqrt{2}$.
  (d) Irregular singular point at $x = 0$.
  (e) Regular singular points at $x = 0$ and $x = -1$.
  (f) Irregular singular point at $x = 0$.

8. (a) Irregular singular point at $x = 0$, regular singular point at $x = 1$.
  (b) Irregular singular point at $x = 0$.

(c) No singular points.

(d) Irregular singular points at $x = \pm 1$.

(e) No singular points.

11. (a) $y = c_1\left(1 + x + \dfrac{x^2}{2!} + \dfrac{x^3}{3!} + \cdots\right) + c_2 x^{1/2}\left(1 + \dfrac{2x}{1 \cdot 3} + \dfrac{(2x)^2}{1 \cdot 3 \cdot 5} + \dfrac{(2x)^3}{1 \cdot 3 \cdot 5 \cdot 7} + \cdots\right)$

(coefficient of c_1 is e^x).

(b) $y = c_1 x^{-1/2}\left(1 - \dfrac{x^2}{2!} + \dfrac{x^4}{4!} - \cdots\right) + c_2 x^{1/2}\left(1 - \dfrac{x^2}{3!} + \dfrac{x^4}{5!} - \cdots\right)$

$= x^{-1/2}(c_1 \cos x + c_2 \sin x)$.

(c) $y = c_1 x^{-1}\left(1 - \dfrac{x^2}{2!} + \dfrac{x^4}{4!} + \cdots\right) + c_2\left(1 - \dfrac{x^2}{3!} + \dfrac{x^4}{5!} - \cdots\right)$

$= x^{-1}(c_1 \cos x + c_2 \sin x)$.

(d) $y = c_1(1 + x + x^2) + c_2 x^3(1 + x + x^2 + x^3 + \cdots)$

$= c_1(1 + x + x^2) + c_2 x^3/(1 - x) \quad (|x| < 1)$

$= (C_1 + C_2 x^3)/(1 - x) \quad (|x| < 1)$.

12. (a) $y = c_1 x\left(1 + \dfrac{x^2}{2!} + \dfrac{x^4}{4!} + \cdots\right) + c_2 x^2\left(1 + \dfrac{x^2}{3!} + \dfrac{x^4}{5!} + \cdots\right)$.

(b) $y = c_1\left(1 + \dfrac{1}{2}x^2 + \dfrac{1}{6}x^3 + \cdots\right) + c_2 x$.

(c) $y = c_1\left(1 - \dfrac{x^4}{2!} + \dfrac{x^8}{4!} - \cdots\right) + c_2\left(x^2 - \dfrac{x^6}{3!} + \dfrac{x^{10}}{5!} - \cdots\right)$.

(d) $y = c_1\left(x - \dfrac{1}{6}x^3 + \cdots\right) + c_2\left(x^2 - \dfrac{1}{12}x^4 + \cdots\right)$.

13. (a) $(-\infty, \infty)$.   (b) $(-1, 1)$.   (c) $(-\infty, \infty)$.   (d) $(-2\pi, 2\pi)$.

14. $M(a, c; x) = 1 + \displaystyle\sum_{k=1}^{\infty} \dfrac{a(a + 1)(a + 2) \cdots (a + k - 1)}{c(c + 1)(c + 2) \cdots (c + k - 1)} \dfrac{x^k}{k!}$

$= 1 + \dfrac{a}{c}x + \dfrac{a(a + 1)}{c(c + 1)}\dfrac{x^2}{2!} + \cdots$.

15. $\alpha = 0$: $y = A_0 + A_2\left(x^2 - \dfrac{2x^3}{3!} + \dfrac{3x^4}{4!} - \cdots\right)$

$= A_0 + 2A_2[1 - (x + 1)e^{-x}] = c_1 + c_2(x + 1)e^{-x}$.

$\alpha = 1$: $y = A_0(1 - x) + A_2\left(x^2 - \dfrac{x^3}{3} + \dfrac{x^4}{3 \cdot 4} - \dfrac{x^5}{3 \cdot 4 \cdot 5} + \cdots\right)$

$= A_0(1 - x) + 2A_2(e^{-x} - 1 + x) = c_1(1 - x) + c_2 e^{-x}$.

16. (b) $u_1(x) = \displaystyle\sum_{k=0}^{\infty} \dfrac{x^k}{(k!)^2}$.

(d) $v(x) = -\left(2x + \dfrac{3}{4}x^2 + \dfrac{11}{108}x^3 + \cdots\right)$

$= -2\left[\dfrac{1}{(1!)^2}x + \dfrac{1 + \frac{1}{2}}{(2!)^2}x^2 + \dfrac{1 + \frac{1}{2} + \frac{1}{3}}{(3!)^2}x^3 + \cdots\right]$.

17. (b) $u_1(x) = \displaystyle\sum_{k=0}^{\infty} \dfrac{x^{k+1}}{k!} = xe^x$.

(d) $v(x) = 1 - x\left(\dfrac{1}{1!}x + \dfrac{1 + \frac{1}{2}}{2!}x^2 + \dfrac{1 + \frac{1}{2} + \frac{1}{3}}{3!}x^3 + \cdots\right)$.

19. (a) $y = \dfrac{x^2}{2!} + \dfrac{x^3}{3!} + \dfrac{x^6}{6!} + \dfrac{x^7}{7!} + \cdots$.

(b) $y = \dfrac{x^2}{2!} - \dfrac{3x^5}{5!} + \dfrac{3 \cdot 6 x^8}{8!} - \dfrac{3 \cdot 6 \cdot 9 x^{11}}{11!} + \cdots$.

(c) $y = \dfrac{x^2}{2(1!)^2} + \dfrac{x^3}{3(2!)^2} + \dfrac{x^4}{4(3!)^2} + \dfrac{x^5}{5(4!)^2} + \cdots$.

(d) $y = x^{-1/2}\left(\dfrac{4}{7} + \dfrac{4x}{3} + \dfrac{2x^2}{7} + \dfrac{2x^3}{57} + \cdots\right)$.

20. See Problem 16.

21. See Problem 17.

22. $y = c_1 u_1(x) + c_2[u_1(x) \log x - v(x)]$,

where $u_1(x) = 1 + x + \dfrac{x^2}{2!} + \dfrac{x^3}{3!} + \cdots = e^x$,

$$v(x) = \sum_{k=1}^{\infty} \frac{\left(1 + \dfrac{1}{2} + \cdots + \dfrac{1}{k}\right)}{k!} x^k = x + \frac{3}{4}x^2 + \frac{11}{36}x^3 + \cdots.$$

23. $y = c_1 u_1(x) + c_2[u_1(x) \log x + 1 - v(x)]$,

where $u_1(x) = x\left(1 + x + \dfrac{x^2}{2!} + \dfrac{x^3}{3!} + \cdots\right) = xe^x$,

$$v(x) = \sum_{k=2}^{\infty} \frac{\left(1 + \dfrac{1}{2} + \cdots + \dfrac{1}{k-1}\right)}{(k-1)!} x^k = x^2 + \frac{3}{4}x^3 + \frac{11}{36}x^4 + \cdots.$$

27. (a) 0.1483. (b) -1.081. (c) 0.1924. (d) 0.1357.
   (e) $0.9900 - 1.081i$. (f) 0.4539.

28. $x = c_1 J_0(t) + c_2 Y_0(t)$, $y = c_1 t J_1(t) + c_2 t Y_1(t)$.

29. $y = c_1 J_0(Ax + B) + c_2 Y_0(Ax + B)$.

30. (a) $\sqrt[3]{2}/\Gamma(\frac{2}{3})$. (b) $-2/\pi$. (c) 2. (d) $1/(2^n n!)$. (e) $2/\pi$.

46. (a) $y = x^2[c_1 J_2(x) + c_2 Y_2(x)]$.
   (b) $y = x[c_1 J_{1/2}(x^2) + c_2 J_{-1/2}(x^2)] = C_1 \cos x^2 + C_2 \sin x^2$.
   (c) $y = c_1 I_{1/2}(x) + c_2 I_{-1/2}(x) = x^{-1/2}(C_1 \cosh x + C_2 \sinh x)$.
   (d) $y = e^{-x}[c_1 J_0(x) + c_2 Y_0(x)]$.
   (e) $y = x[c_1 I_1(x) + c_2 K_1(x)]$.
   (f) $y = c_1 x \cos (a/x) + c_2 x \sin (a/x)$.
   (g) $y = c_1 x^{1/2} I_{1/4}(x^2/2) + c_2 x^{1/2} I_{-1/4}(x^2/2)$.
   (h) $y = c_1 e^{-x} I_0(x) + c_2 e^{-x} K_0(x)$.
   (i) $y = c_1 e^{-x^2} J_0(x) + c_2 e^{-x^2} Y_0(x)$.

47. (a) $y = x^{1/2} Z_{(2m+1)/2}(i\alpha x)$.
   (b) $y = x^{(2m+1)/2} Z_{(2m+1)/2}(i\alpha x)$.

48. $y = x^{-1} J_1(2x)$.

49. $y = c_1 J_1(x) - \frac{1}{4} P Y_1(x)$.

50. (a) $y = x^p Z_p(x/l)$, where $p = (1 - n)/2$.
   (b) $y = x^{ps} Z_p(x^s/ls)$, where $p = (1 - n)/(2 - n)$ and $s = (2 - n)/2$, $n \neq 2$.
   (c) $y = x^{-1/2}(c_1 + c_2 \log x)$.

51. (a) 0.9844. (b) 0.2496. (c) 1.016. (d) 0.2483.

55. $U = r^n[c_1 P_n(\cos \varphi) + c_2 Q_n(\cos \varphi)]$.

71. (a) $y = c_1 + c_2\left(1 + \dfrac{1}{1!\,x} + \dfrac{1}{2!\,x^2} + \cdots\right) = c_1 + c_2 e^{1/x}$.

   (b) $y = c_1\left(1 - \dfrac{1}{2!\,x^2} + \dfrac{1}{4!\,x^4} - \cdots\right) + c_2\left(\dfrac{1}{x} - \dfrac{1}{3!\,x^3} + \dfrac{1}{5!\,x^5} - \cdots\right)$

   $= c_1 \cos \dfrac{1}{x} + c_2 \sin \dfrac{1}{x}$.

72. $y = c_1\left(1 - \dfrac{1}{3!\,x^2} + \dfrac{1}{5!\,x^4} - \cdots\right) + c_2\left(x - \dfrac{1}{2!\,x} + \dfrac{1}{4!\,x^3} - \cdots\right)$

   $= c_1 x \sin \dfrac{1}{x} + c_2 x \cos \dfrac{1}{x}$.

Chapter 5

2. $k = n\pi/l$, where n is any integer other than zero.

4. $y = CJ_0(\lambda^{1/2}x)$, where $J_0(\lambda^{1/2}a) = 0$.

10. $\omega_n = \sqrt{\dfrac{EI}{\rho}}\dfrac{\mu_n^2}{l^2}$, where $\cos \mu_n = \mathrm{sech}\ \mu_n$;

$$\varphi_n = \frac{\sinh (\mu_n x/l) - \sin (\mu_n x/l)}{\sinh \mu_n - \sin \mu_n} - \frac{\cosh (\mu_n x/l) - \cos (\mu_n x/l)}{\cosh \mu_n - \cos \mu_n}.$$

11. $\omega_n = \sqrt{\dfrac{EI}{\rho}}\dfrac{\mu_n^2}{l^2}$;

Symmetrical modes: $\tan (\mu_n/2) = -\tanh (\mu_n/2)$,

$$\varphi_n = \frac{\cosh (\mu_n x/l)}{\cosh (\mu_n/2)} - \frac{\cos (\mu_n x/l)}{\cos (\mu_n/2)}.$$

Antisymmetrical modes: $\tan (\mu_n/2) = +\tanh (\mu_n/2)$,

$$\varphi_n = \frac{\sinh (\mu_n x/l)}{\sinh (\mu_n/2)} - \frac{\sin (\mu_n x/l)}{\sin (\mu_n/2)}.$$

19. Exact value rounds to 6.55.

22. (a) $(x^2 y')' + (x^2 + \lambda x)y = 0$.
    (b) $(y' \sin x)' + \lambda y \sin x = 0$.
    (c) $(e^{ax}y')' + e^{ax}(b + \lambda)y = 0$.
    (d) $(x^c e^{-x}y')' + x^{c-1}e^{-x}(-a + \lambda)y = 0$.

31. (a) $A_n = 2(1 - \cos n\pi)/(n\pi)$.
    (b) $A_n = (-1)^{n+1}2/n$.
    (c) $A_n = \left(1 - \cos \dfrac{n\pi}{2}\right)\Big/\dfrac{n\pi}{2}$.

34. $1 = \displaystyle\sum_{n=1}^{\infty} A_n \sin \mu_n \frac{x}{l}\ (0 < x < l)$, where $\mu_n + k \tan \mu_n = 0$;

$$A = \frac{2(1 - \cos \mu_n)}{\mu_n - \frac{1}{2}\sin 2\mu_n} = \frac{2(1 - \cos \mu_n)}{\mu_n(1 + k^{-1}\cos^2 \mu_n)}.$$

35. (c) $f(x) = 1$: $A_n = \dfrac{2}{n\pi}(1 - \cos n\pi)$.

$$f(x) = x:\quad A_n = \frac{2n\pi}{n^2\pi^2 + (\log b)^2}(1 - b \cos n\pi).$$

37. $\displaystyle\int_0^l h(x) \sin \frac{p\pi x}{l}\, dx = 0$.

$$y = \sum_{n=1}^{\infty} a_n \sin \frac{n\pi x}{l} \quad \text{with} \quad a_n = \frac{2l}{(p^2 - n^2)\pi}\int_0^l h(x) \sin \frac{n\pi x}{l}\, dx\ (n \neq p),\ a_p \text{ arbitrary}.$$

49. (a) $\dfrac{8l^2}{\pi^3}\left(\dfrac{1}{1^3}\sin \dfrac{\pi x}{l} + \dfrac{1}{3^3}\sin \dfrac{3\pi x}{l} + \dfrac{1}{5^3}\sin \dfrac{5\pi x}{l} + \cdots\right)$.

(b) $\dfrac{4l}{\pi^2}\left(\dfrac{1}{1^2}\sin \dfrac{\pi x}{l} - \dfrac{1}{3^2}\sin \dfrac{3\pi x}{l} + \dfrac{1}{5^2}\sin \dfrac{5\pi x}{l} - \cdots\right)$.

(c) $\dfrac{2}{\pi}\left(\sin \dfrac{\pi x}{l} + \sin \dfrac{2\pi x}{l} + \dfrac{1}{3}\sin \dfrac{3\pi x}{l} + \cdots + \dfrac{1 - \cos (n\pi/2)}{n}\sin \dfrac{n\pi x}{l} + \cdots\right)$.

(d) $\dfrac{8}{\pi}\displaystyle\sum_{n=1}^{\infty}\dfrac{(-1)^{n+1}n}{4n^2 - 1}\sin \dfrac{n\pi x}{l}$.

(e) $\dfrac{1}{2}\sin \dfrac{\pi x}{l} - \dfrac{2}{\pi}\displaystyle\sum_{n=2}^{\infty}\dfrac{n \cos (n\pi/2)}{n^2 - 1}\sin \dfrac{n\pi x}{l} = \dfrac{1}{2}\sin \dfrac{\pi x}{l} + \dfrac{4}{\pi}\displaystyle\sum_{m=1}^{\infty}\dfrac{(-1)^{m+1}m}{4m^2 - 1}\sin \dfrac{2m\pi x}{l}$.

(f) $\dfrac{2}{\pi}\displaystyle\sum_{n=1}^{\infty}\dfrac{1 - \cos (n\pi\epsilon/l)}{n\epsilon}\sin \dfrac{n\pi x}{l}$.

50. (a) $\dfrac{l^2}{6} - \dfrac{4l^2}{\pi^2}\left(\dfrac{1}{2^2}\cos\dfrac{2\pi x}{l} + \dfrac{1}{4^2}\cos\dfrac{4\pi x}{l} + \dfrac{1}{6^2}\cos\dfrac{6\pi x}{l} + \cdots\right).$

  (b) $\dfrac{l}{4} - \dfrac{8l}{\pi^2}\left(\dfrac{1}{2^2}\cos\dfrac{2\pi x}{l} + \dfrac{1}{6^2}\cos\dfrac{6\pi x}{l} + \dfrac{1}{10^2}\cos\dfrac{10\pi x}{l} + \cdots\right).$

  (c) $\dfrac{1}{2} + \dfrac{2}{\pi}\left(\dfrac{1}{1}\cos\dfrac{\pi x}{l} - \dfrac{1}{3}\cos\dfrac{3\pi x}{l} + \dfrac{1}{5}\cos\dfrac{5\pi x}{l} - \cdots\right).$

  (d) $\dfrac{2}{\pi}\left(1 - 2\displaystyle\sum_{n=1}^{\infty}\dfrac{1}{4n^2-1}\cos\dfrac{n\pi x}{l}\right).$

  (e) $\dfrac{1}{\pi}\left(1 + \cos\dfrac{\pi x}{l} + 2\displaystyle\sum_{n=2}^{\infty}\dfrac{n\sin(n\pi/2)-1}{n^2-1}\cos\dfrac{n\pi x}{l}\right).$

  (f) $\dfrac{1}{l}\left(1 + \dfrac{2l}{\pi}\displaystyle\sum_{n=1}^{\infty}\dfrac{\sin(n\pi\epsilon/l)}{n\epsilon}\cos\dfrac{n\pi x}{l}\right).$

54. $y = 2\dfrac{\sin\omega l}{l^2}\displaystyle\sum_{n=1}^{\infty}\dfrac{(-1)^{n+1}n\pi\sin\dfrac{n\pi x}{l}}{\left(\Lambda - \dfrac{n^2\pi^2}{l^2}\right)\left(\dfrac{n^2\pi^2}{l^2} - \omega^2\right)}.$

55. $\Lambda = p^2\pi^2/l^2$: no solution unless also $\omega = r\pi/l$ (r a positive integer, but $r \neq p$), in which case infinitely many solutions.
  $\omega = p\pi/l$: unique solution unless also $\Lambda = p^2\pi^2/l^2$, in which case no solution.

57. (a) $\dfrac{l^2}{12} - \dfrac{2l^2}{\pi^2}\left(\dfrac{1}{2^2}\cos\dfrac{2\pi x}{l} + \dfrac{1}{4^2}\cos\dfrac{4\pi x}{l} + \dfrac{1}{6^2}\cos\dfrac{6\pi x}{l} + \cdots\right)$

$+ \dfrac{4l^2}{\pi^3}\left(\dfrac{1}{1^3}\sin\dfrac{\pi x}{l} + \dfrac{1}{3^3}\sin\dfrac{3\pi x}{l} + \dfrac{1}{5^3}\sin\dfrac{5\pi x}{l} + \cdots\right).$

  (b) $\dfrac{l}{8} - \dfrac{4l}{\pi^2}\left(\dfrac{1}{2^2}\cos\dfrac{2\pi x}{l} + \dfrac{1}{6^2}\cos\dfrac{6\pi x}{l} + \dfrac{1}{10^2}\cos\dfrac{10\pi x}{l} + \cdots\right)$

$+ \dfrac{2l}{\pi^2}\left(\dfrac{1}{1^2}\sin\dfrac{\pi x}{l} - \dfrac{1}{3^2}\sin\dfrac{3\pi x}{l} + \dfrac{1}{5^2}\sin\dfrac{5\pi x}{l} - \cdots\right).$

  (c) $\dfrac{3}{4} + \dfrac{1}{\pi}\left(\dfrac{1}{1}\cos\dfrac{\pi x}{l} - \dfrac{1}{3}\cos\dfrac{3\pi x}{l} + \dfrac{1}{5}\cos\dfrac{5\pi x}{l} - \cdots\right)$

$- \dfrac{1}{\pi}\left(\sin\dfrac{\pi x}{l} - \sin\dfrac{2\pi x}{l} + \dfrac{1}{3}\sin\dfrac{3\pi x}{l} + \cdots\right.$

$\left. + \dfrac{\cos(n\pi/2)-\cos(n\pi)}{n}\sin\dfrac{n\pi x}{l} + \cdots\right).$

  (d) $\dfrac{8}{\pi}\displaystyle\sum_{n=1}^{\infty}\dfrac{(-1)^{n+1}n}{4n^2-1}\sin\dfrac{n\pi x}{l}.$

  (e) $\dfrac{1}{2\pi}\left(1 + \cos\dfrac{\pi x}{l} + \dfrac{\pi}{2}\sin\dfrac{\pi x}{l}\right)$

$+ \dfrac{1}{\pi}\displaystyle\sum_{n=2}^{\infty}\dfrac{1}{n^2-1}\left[\left(n\sin\dfrac{n\pi}{2}-1\right)\cos\dfrac{n\pi x}{l} - n\cos\dfrac{n\pi}{2}\sin\dfrac{n\pi x}{l}\right].$

  (f) $\dfrac{1}{2l} + \dfrac{1}{\pi\epsilon}\displaystyle\sum_{n=1}^{\infty}\dfrac{1}{n}\left[\sin\dfrac{n\pi\epsilon}{l}\cos\dfrac{n\pi x}{l} + \left(1 - \cos\dfrac{n\pi\epsilon}{l}\right)\sin\dfrac{n\pi x}{l}\right].$

58. (a) $\left(a + \dfrac{bP}{2}\right) - \dfrac{bP}{\pi}\left(\dfrac{1}{1}\sin\dfrac{2\pi x}{P} + \dfrac{1}{2}\sin\dfrac{4\pi x}{P} + \dfrac{1}{3}\sin\dfrac{6\pi x}{P} + \cdots\right).$

  (b) $\dfrac{1}{2} + \dfrac{2}{\pi}\left(\dfrac{1}{1}\sin\pi x + \dfrac{1}{3}\sin3x + \dfrac{1}{5}\sin5\pi x + \cdots\right).$

  (c) $\dfrac{2}{\pi} - \dfrac{4}{\pi}\left(\dfrac{1}{3}\cos2x + \dfrac{1}{15}\cos4x + \cdots + \dfrac{1}{4n^2-1}\cos2nx + \cdots\right).$

  (d) $\dfrac{3}{2} - \dfrac{1}{\pi}\left(\dfrac{1}{1}\sin2\pi x + \dfrac{1}{2}\sin4\pi x + \dfrac{1}{3}\sin6\pi x + \cdots\right).$

60. $y = \dfrac{1}{4\Lambda} + \displaystyle\sum_{n=1}^{\infty}\dfrac{\sin\dfrac{n\pi}{2}\cos nx + \left(1 - \cos\dfrac{n\pi}{2}\right)\sin nx}{n\pi(\Lambda - n^2)}$

63. (a) $a_0 = 1, a_n = 0 \quad (n \neq 0)$.

  (b) $a_n = \dfrac{l \sin l \cos n\pi}{l^2 - n^2\pi^2}$.

  (c) $a_n = \dfrac{\sinh al \cos n\pi}{al - n\pi i}$.

  (d) $a_0 = 0, a_n = \dfrac{1 - \cos n\pi}{n\pi i} \quad (n \neq 0)$.

71. $x = 2l^p \displaystyle\sum_{n=1}^{\infty} \dfrac{J_p\left(\alpha_n \dfrac{x}{l}\right)}{\alpha_n J_{p+1}(\alpha_n)}$, where $J_p(\alpha_n) = 0$.

72. $x^p = 2l^p \displaystyle\sum_{n=1}^{\infty} \dfrac{\mu_n l J_{p+1}(\mu_n l)}{(\mu_n^2 l^2 - p^2)[J_p(\mu_n l)]^2} J_p(\mu_n x)$

  $= 2l^p \displaystyle\sum_{n=1}^{\infty} \dfrac{p}{\mu_n^2 l^2 - p^2} \dfrac{J_p(\mu_n x)}{J_p(\mu_n l)}$, where $J'_p(\mu_n l) = 0$, when $p \neq 0$.

  When $p = 0$, expansion is $1 = 1$.

76. $f(x) = \tfrac{1}{4}P_0(x) + \tfrac{1}{2}P_1(x) + \tfrac{5}{16}P_2(x) + \cdots$.

77. $f(\varphi) = \tfrac{1}{4}P_0(\cos \varphi) + \tfrac{1}{2}P_1(\cos \varphi) + \tfrac{5}{16}P_2(\cos \varphi) + \cdots$.

85. $0 \ (x < -a), \ -\pi/4 \ (x = -a), \ -\pi/2 \ (-a < x < 0), \ 0 \ (x = 0), \ \pi/2 \ (0 < x < a),$
  $\pi/4 \ (x = a), \ 0 \ (x > a)$.

86. (b) $A(u) = B(u) = e^{-\alpha u}$.

87. $e^{-ax} \cos bx = \displaystyle\int_0^{\infty} \left[\dfrac{2}{\pi} \dfrac{u(a^2 - b^2 + u^2)}{(a^2 - b^2 + u^2)^2 + 4a^2b^2}\right] \sin ux \, du \quad (0 < x < \infty)$,

  $e^{-ax} \sin bx = \displaystyle\int_0^{\infty} \left[\dfrac{4}{\pi} \dfrac{abu}{(a^2 - b^2 + u^2)^2 + 4a^2b^2}\right] \sin ux \, du \quad (0 < x < \infty)$.

99. $A(u) = \dfrac{2}{\pi} \displaystyle\int_0^{\infty} f(x) \sin ux \, dx$.

Chapter 6

1. $3; (\tfrac{2}{3}, -\tfrac{2}{3}, -\tfrac{1}{3})$.

2. $\mathbf{c} = (n\mathbf{a} + m\mathbf{b})/(m + n)$.

6. (a) $\pm(\tfrac{1}{3}, \tfrac{2}{3}, -\tfrac{2}{3})$.   (b) $\pm(\tfrac{1}{3}\mathbf{i} + \tfrac{2}{3}\mathbf{j} - \tfrac{2}{3}\mathbf{k})$.

8. (a) $\tfrac{2}{3}$.   (b) $\cos^{-1} \tfrac{2}{15}$.

9. $\tfrac{8}{3}$.

12. $45°$.

13. (a) $\pm(\sqrt{2}/10)(4\mathbf{i} - 3\mathbf{j} - 5\mathbf{k})$.   (b) $5\sqrt{2}/2$.

14. (a) $\pm(2\mathbf{i} - \mathbf{j} - \mathbf{k})/\sqrt{6}$.   (b) $\sqrt{6}/2$.

15. $\sqrt{\tfrac{2}{3}}\,\omega$.

16. -5.

20. $\tfrac{1}{6}$.

24. $\pm(\mathbf{i} - \mathbf{k})/\sqrt{2}$.

28. $\mathbf{F} \cdot \dfrac{d\mathbf{F}}{dt} \times \dfrac{d^3\mathbf{F}}{dt^3}$.

29. (a) $\mathbf{v} = -\omega\mathbf{a} \sin \omega t + \omega\mathbf{b} \cos \omega t; \ \mathbf{r} \times \mathbf{v} = \omega(\mathbf{a} \times \mathbf{b})$.
  (b) $\mathbf{a} = -\omega^2\mathbf{r}$.

35. (c) $\mathbf{u} = (-a\mathbf{i}\sin t + a\mathbf{j}\cos t + c\mathbf{k})/\sqrt{a^2 + c^2}$,
$\quad \mathbf{n} = -(\mathbf{i}\cos t + \mathbf{j}\sin t)$,
$\quad \mathbf{b} = (c\mathbf{i}\sin t - c\mathbf{j}\cos t + a\mathbf{k})/\sqrt{a^2 + c^2}$.

42. $1/\rho = \sqrt{266}/98$, $1/\tau = \frac{3}{19}$.

44. (a) $\pm(1, 1, 1)/\sqrt{3}$; $2\sqrt{3}$. (b) $-\frac{2}{3}$.

45. $\text{grad } r = \mathbf{i}\cos\theta + \mathbf{j}\sin\theta$; $\text{grad } \theta = -\mathbf{i}(\sin\theta)/r + \mathbf{j}(\cos\theta)/r$.

46. $(l, m, n) = (\varphi_x, \varphi_y, \varphi_z)/\sqrt{\varphi_x^2 + \varphi_y^2 + \varphi_z^2}$.

47. $\pm(2\mathbf{i} - \mathbf{j} + 2\mathbf{k})/3$.

48. $\text{div } \mathbf{V} = 0$, $\text{curl } \mathbf{V} = 0$.

52. (a) $2lx + 2my + 2nz$. (b) $2lx\mathbf{i} + 2my\mathbf{j} + 2nz\mathbf{k}$.

60 (a) No solution. (b) $\varphi = x^2 y + y^2 z + z + C$.

61. (a) $-\frac{4}{13}$. (b) 3.

62. (a) 0. (b) 0.

65. (a) $\frac{4}{3}\sqrt{4\pi^2 + 1}$.
$\quad$ (b) 2π.
$\quad$ (c) $\dfrac{270}{7}$.

67. $\frac{3}{4}$.

68. $\pi/2$.

69. 24.

70. 0.

71. (a) $\sqrt{3}\pi/4$.
$\quad$ (b) $(2 + \sqrt{2})/3$.

73. (a) 4π.
$\quad$ (b) $-\pi/2$.
$\quad$ (c) $\pi/2$.
$\quad$ (d) 0.

88. Both sides of Equation (133) reduce to π.

89. $-\pi$.

90. (a) π.
$\quad$ (b) $\frac{3}{2}$.
$\quad$ (c) 0.

104. (a) $\mathbf{u}_\theta/r$. (b) $nr^{n-1}\mathbf{u}_r$. (c) $\mathbf{u}_z/r$. (d) 0. (e) $3\cos\theta$. (f) 0.

106. (a) $\mathbf{u}_\varphi/r$. (b) $\mathbf{u}_\theta/(r\sin\varphi)$. (c) 0. (d) 0.

110. 0.

111. $\frac{4}{3}\pi$.

113. (b) $\mathbf{V} = 3(x^2 - y^2)\mathbf{i} - 6xy\mathbf{j}$; $V = 3(x^2 + y^2)$.
$\quad$ (c) $\psi = 3x^2 y - y^3$; $3x^2 y - y^3 = \text{constant}$.
$\quad$ (d) 14ρ.

117. (b) $\mathbf{V} = -2r\mathbf{u}_r + 4z\mathbf{k}$; $V = 2\sqrt{r^2 + 4z^2}$.
$\quad$ (c) $\psi = 2r^2 z + c$.

Chapter 7

1. (a) $\cos \theta$.   (b) $-r \sin \theta$.   (c) $\sin \theta$.   (d) $r \cos \theta$.
   (e) $\cos \theta$.   (f) $\sin \theta$.   (g) $-r^{-1} \sin \theta$.   (h) $r^{-1} \cos \theta$.

2. (a) $-\cot \theta$.   (b) $-\tan \theta$.   (c) $\tan \theta$.   (d) $\cot \theta$.
   (e) $r \tan \theta$.   (f) $r^{-1} \cot \theta$.   (g) $-r \cot \theta$.   (h) $-r^{-1} \tan \theta$.

10. (a) $\dfrac{\partial(f, g)}{\partial(u, v)} \neq 0, h_z \neq 0$.
    (e) $z_x = 4x^3 - 4xy^3 - 2x,\ z_y = 6y^5 - 6x^2y^2 - 3y^2$.

16. Yes.

17. Yes.

29. $1 + x + y + \frac{1}{2}x^2 + xy + \frac{1}{2}y^2 + \cdots$.

30. $1 - x + y + x^2 - 2xy + y^2 + \cdots$.

32. 0.906.

33. (a) Min at $(0, 0)$ if $b^2 < 1$, saddle point at $(0, 0)$ if $b^2 > 1$; min along line $y = -x$ if $b = 1$; min along line $y = x$ if $b = -1$.
    (b) Min at $(1, 0)$.
    (c) Rel. max at $(0, 0)$; saddle points at $(1, 1), (1, -1), (-1, 1), (-1, -1)$.
    (d) Saddle point at $(1, 1)$.

34. (a) Min at $(0, 0)$.
    (b) Min at $(1, -1)$.
    (c) Min along lines $y = \pm x$.
    (d) Max at $(1, -1)$; min along circle $(x - 1)^2 + (y + 1)^2 = 2$.

35. (a) Min at $(0, 0)$.
    (b) Min along lines $y = \pm x$.
    (c) Saddle point at $(0, 0)$.
    (d) Min at $(0, 0)$.

36. Max $= 3$ at $(-1, 0)$; min $= -\frac{3}{2}$ at $(\frac{1}{2}, \pm\frac{1}{2}\sqrt{3})$.

37. Min at $(1, 1), (-1, -1)$.

38. Max at $(1, 1), (-1, -1)$; min at $(1, -1), (-1, 1)$.

39. (a) Min $= 12$ when $x^2 = 4, y^2 = 1, z^2 = \frac{1}{4}$.
    (b) Min $= 4$ when $x^2 = 2, y^2 = \frac{1}{2}, z = 0$.
    (c) Min $= 1$ at $(1, 0, 0)$.

40. Furthest at $(1, 0, 0)$; nearest at $(\frac{1}{3}, \frac{2}{3}, 0)$.

41. (a) $(py')' + qy = f$.   (b) $(Ay')' + (B' - C)y = E - D'$.

42. (a) $(Au_1')' + (Bu_2')' = 0,\ (Bu_1')' + (Cu_2')' = 0$.
    (b) $(a_{11}u_1')' + (a_{12}u_2')' = b_{11}u_1 + b_{12}u_2$,
    $(a_{12}u_1')' + (a_{22}u_2')' = b_{12}u_1 + b_{22}u_2$.

43. (a) $z = c_1\theta + c_2$.   (b) $\theta = c_3z + c_4$.   (c) $z = c_5t + c_6, \theta = c_7t + c_8$.
    [Here a geodesic is either a helix, a circular arc, or a straight line. Part (a) ignores the straight lines $\theta =$ constant and part (b) ignores the circles $z =$ constant, while part (c) includes *all* possibilities.]

44. $\theta = \sin^{-1}(c_1 \cot \varphi) + c_2$.

48. (a) Value of integral always $\frac{1}{2}$.
    (b) Only candidate $(y = 1/x)$ inadmissible.

50. π^2, corresponding to $y = \sqrt{2} \sin n\pi x$ with $n = 1$.

53. $y = 1 + \sin x - \cos x,\ z = 1 - \sin x - \cos x;\ I = 4$.

58. $y''' = h,\ y(a) = y'(a) = y''(a) = 0$.

60. 2.0945515.

61. 4.493.

62. 3.927.

63. 1.31.

64. $x \doteq 0.883$, $y \doteq 0.469$.

65. $x \doteq 1.01941$, $y \doteq 1.03062$.

Chapter 8

1 (a) $(x - y)(z_x - z_y) - 2z = 0$.
 (b) $bdz_{xx} - (ad + bc)z_{xy} + acz_{yy} = 0$.
 (c) $b^2 z_{xx} - 2abz_{xy} + a^2 z_{yy} = 0$.

5. (a) $z = cy/b + f(bx - ay)$ $(b \neq 0)$.
 (b) $z = e^{cx/a} f(bx - ay)$ $(a \neq 0)$.
 (c) $z = f(x^2 + y^2)$.
 (d) $z = e^{-x^2} f(y - x)$.
 (e) $z = xf(xy)$.
 (f) $z^{-1} = x^{-1} + f(x^{-1} - y^{-1})$.

6. (b) $z = f(2x + y) + x^2 + x + \frac{1}{2}e^y$.

11. $z = (x + y)^2$.

12. (a) $z = [(b - c)x + (c - a)y]/(b - a)$ $(b \neq a)$;
    $z = x + f(x - y)$, where $f(0) = 0$ if $a = b = c$.
 (b) $z = e^{c(y-x)/(b-a)}(bx - ay)/(b - a)$ $(b \neq a)$.
 (c) $z = \sqrt{(x^2 + y^2)/2}$.
 (d) No solution.
 (e) $z = x$.
 (f) $z = x\left/\left[1 + xf\left(\dfrac{y - x}{xy}\right)\right]\right.$, where $f(0) = 0$.

13. (a) $z = [(bx - ay)/(b - 2a)]^2 + c(y - 2x)/(b - 2a)$ $(b \neq 2a)$.
 (b) $z = e^{c(y-2x)/(b-2a)}[(bx - ay)/(b - 2a)]^2$.
 (c) $z = \frac{1}{5}(x^2 + y^2)$.
 (d) $z = (y - x)^2 e^{y(y-2x)}$.
 (e) $z = (\frac{1}{3}x^3 y)^{1/2}$.
 (f) $z = (x^2 y^2)/(4x^2 - 8xy + 4y^2 - xy^2 + 2x^2 y)$.

14. (a) $z = \sqrt{x + y}\, f(x - y) + 1$.
 (b) $z = \frac{1}{2}x^2 y + f(y)$.
 (c) $z^2 = xy + f(y/x)$.
 (d) $z = e^{3xy^2 - y^3} f(x - y)$.

15. (b) No solution unless $\varphi(x) = x + a$ (a constant), in which case $z = x + f(y - x)$, where f is any differentiable function for which $f(b) = a$.

23. (a) $z = f(3x + y) + g(x - y)$.
 (b) $\varphi = f(x + y + ix) + g(x + y - ix)$.
 (c) $\varphi = f(x + y) + g(y)$.
 (d) $rw = f(r + ct) + g(r - ct)$ $(c \neq 0)$.
 (e) $\varphi = f[x - (V + U)t] + g[x + (V - U)t]$ $(V \neq 0)$.
 (f) $\varphi = f_1(x + y) + f_2(x - y) + f_3(x + iy) + f_4(x - iy)$.
 (g) $z = f_1(x + y) + xf_2(x + y) + f_3(x - y) + xf_4(x - y)$.
 (h) $\varphi = f_1(x + iy) + f_2(x - iy) + xf_3(x + iy) + xf_4(x - iy)$.

27. (a) $z = f(y + x) + g(y - x) + \frac{1}{6}x^3$.
 (b) $z = f(y + x) + g(y + 2x) - \frac{1}{2}\cos y$.

29. (a) $z = f(y - x) + \frac{1}{2}x^2 y^2$.

(b) $z = f(x + iy) + g(x - iy) + \frac{1}{12}x^4 + \frac{1}{6}x^3 y$.

(c) $z = f(y + x) + g(y - x) + \frac{1}{6}x^3(y + 1)$.

(d) $z = f(y + x) + xg(y + x) + \frac{1}{12}x^4 + \frac{1}{6}y^3$.

31. $z = f(y + x) + g(y - x) + \sin xy$.

32. $\varphi = x^2 + c^2 t^2 + c^{-1} \cos x \sin ct$.

37. (b) $\varphi = x^2 + (c^2 - \frac{1}{2})t^2 + t$.

44. $\varphi = x^m y^n e^{Ax + By}[f(x + iy) + g(x - iy)]$.

54. $\alpha = 2, \beta = 1, \gamma = 2$.

55. (b) $h = 1$: $\varphi = f(x + ct) + g(x - ct) + \frac{1}{4}(x^2 - c^2 t^2)$.

$h = \cos(x + ct)$: $\varphi = f(x + ct) + g(x - ct) + \frac{1}{4}(x - ct)\sin(x + ct)$.

Chapter 9

14. (a) $T = \dfrac{400}{\pi} \displaystyle\sum_{n \text{ odd}} \dfrac{\sinh(n\pi y/100)}{n \sinh n\pi} \sin \dfrac{n\pi x}{100}$.

(b) Exact value is $25°$.

15. $T = \displaystyle\sum_{n=1}^{\infty} \left[a_n \sinh \dfrac{n\pi y}{l} + b_n \sinh \dfrac{n\pi(d - y)}{l} \right] \sin \dfrac{n\pi x}{l}$,

where $a_n \sinh \dfrac{n\pi d}{l} = \dfrac{2}{l} \displaystyle\int_0^l f(x) \sin \dfrac{n\pi x}{l} \, dx$

and $b_n \sinh \dfrac{n\pi d}{l} = \dfrac{2}{l} \displaystyle\int_0^l g(x) \sin \dfrac{n\pi x}{l} \, dx$.

22. $T = \alpha_1 + (\alpha_2 - \alpha_1)\dfrac{x}{l} + \beta_1 y + (\beta_2 - \beta_1)\dfrac{xy}{l} + \displaystyle\sum_{n=1}^{\infty} a_n e^{-n\pi y/l} \sin \dfrac{n\pi x}{l}$,

where $a_n = \dfrac{2}{l} \displaystyle\int_0^l f(x) \sin \dfrac{n\pi x}{l} \, dx - \dfrac{2}{n\pi}(\alpha_1 - \alpha_2 \cos n\pi)$.

23. $T = \dfrac{x}{l}\left(T_0 + c\dfrac{y}{d}\right) + \displaystyle\sum_{n=1}^{\infty} \left[a_n \sinh \dfrac{n\pi y}{l} + b_n \sinh \dfrac{n\pi(d - y)}{l} \right] \sin \dfrac{n\pi x}{l}$,

where $a_n \sinh \dfrac{n\pi d}{l} = \dfrac{2}{l} \displaystyle\int_0^l f(x) \sin \dfrac{n\pi x}{l} \, dx + \dfrac{2(T_0 + c) \cos n\pi}{n\pi}$

and $b_n \sinh \dfrac{n\pi d}{l} = \dfrac{2T_0 \cos n\pi}{n\pi}$.

24. $T = \dfrac{\sin \omega l}{\omega^2} \dfrac{x}{l} - \dfrac{\sin \omega x}{\omega^2} + \displaystyle\sum_{n=1}^{\infty} a_n e^{-n\pi y/l} \sin \dfrac{n\pi x}{l}$,

where $a_n = \dfrac{2n\pi \cos n\pi \sin \omega l}{\omega^2}\left(1 + \dfrac{1}{\omega^2 l^2 - n^2 \pi^2}\right)$ if $\omega \neq \dfrac{k\pi}{l}$ $(k = 1, 2, \ldots)$.

$T = -(1 - e^{-\omega y})\dfrac{\sin \omega x}{\omega^2}$ if $\omega = \dfrac{k\pi}{l}$.

25. $AX'' + DX' + (F_1 + \lambda G)X = 0$, $CY'' + EY' + (F_2 - \lambda H)Y = 0$;

$AX'' + (D_1 + \lambda G)X' + F_1 X = 0$, $Y' + (D_2 - \lambda H)Y = 0$;

$X' + (E_1 - \lambda G)X = 0$, $CY'' + (E_2 + \lambda H)Y' + F_2 Y = 0$.

27. $T = \dfrac{2}{r}(r^2 - 100) \sin \theta + \dfrac{1}{2r}(400 - r^2) \cos \theta$.

28. (a) $T = \dfrac{1}{2} + \dfrac{2}{\pi} \displaystyle\sum_{n \text{ odd}} \dfrac{r^n}{n} \sin n\theta$ $(r \leq 1)$.

(b) $T = \dfrac{1}{2} + \dfrac{2}{\pi} \displaystyle\sum_{n \text{ odd}} \dfrac{r^{-n}}{n} \sin n\theta$ $(r \geq 1)$.

29. (a) $T = \sum_{n=1}^{\infty} A_n \left(\dfrac{r}{a}\right)^{n\pi/\alpha} \sin \dfrac{n\pi\theta}{\alpha}$, where $A_n = \dfrac{2}{\alpha} \displaystyle\int_0^{\alpha} f(\theta) \sin \dfrac{n\pi\theta}{\alpha} d\theta$.

36. (a) $T = T_0 \left(1 - \dfrac{r}{a} \cos \varphi\right)$.

(b) $T = T_0 \left[\dfrac{1}{2} + \dfrac{3}{4} \dfrac{r}{a} P_1 (\cos \varphi) - \dfrac{7}{16} \left(\dfrac{r}{a}\right)^3 P_3 (\cos \varphi) + \cdots \right]$.

46. (c) $\varphi = \displaystyle\sum_{m=1}^{\infty} \sum_{n=1}^{\infty} \dfrac{A_{mn}}{\Lambda - \pi^2(m^2/l_1^2 + n^2/l_2^2)} \sin \dfrac{m\pi x}{l_1} \sin \dfrac{n\pi y}{l_2}$,

if $\Lambda \neq \pi^2(p^2/l_1^2 + q^2/l_2^2)$, where p and q are integral, with A_{mn} as defined in Problem 45(b).

53. (b) $w = \displaystyle\sum_{m=1}^{\infty} \sum_{n=1}^{\infty} b_{mn} \sin \dfrac{m\pi x}{l} \sin \dfrac{n\pi y}{l} \sin \omega_{mn} t$,

where $b_{mn} = \dfrac{4}{\omega_{mn} l^2} \displaystyle\int_0^l \int_0^l v(x, y) \sin \dfrac{m\pi x}{l} \sin \dfrac{n\pi y}{l} dx \, dy$.

55. $T = 100 - x - \dfrac{200}{\pi} \displaystyle\sum_{m=1}^{\infty} \dfrac{1}{m} \sin \dfrac{m\pi x}{50} e^{-m^2\pi^2\alpha^2 t/2500}$.

56. (a) $T_{T,p} = b_n \cos \dfrac{n\pi x}{l} e^{-n^2\pi^2\alpha^2 t/l^2}$ $(n = 0, 1, 2, \ldots)$.

(b) $T = 50 - \dfrac{400}{\pi^2} \displaystyle\sum_{n \, \text{odd}} \dfrac{1}{n^2} \cos \dfrac{n\pi x}{100} e^{-n^2\pi^2\alpha^2 t/10^4}$.

64. (a) $T = T_0 \left[\dfrac{x}{l} \sin \omega t + \dfrac{2\lambda\omega}{\pi} \displaystyle\sum_{n=1}^{\infty} \dfrac{(-1)^n n}{n^4 + \lambda^2\omega^2} \left(\cos \omega t + \dfrac{\lambda\omega}{n^2} \sin \omega t - e^{-n^2 t/\lambda}\right) \sin \dfrac{n\pi x}{l} \right]$.

77. $\varphi = \cos \dfrac{\omega}{c} (x - ct)$.

93. $z = \begin{cases} e^t & (t \leq x), \\ e^x & (t \geq x). \end{cases}$

94. $z = e^t$.

95. $\varphi = \begin{cases} 0 & \left(t < \dfrac{x}{c}\right), \\ \cos \omega \left(t - \dfrac{x}{c}\right) & \left(t > \dfrac{x}{c}\right). \end{cases}$

96. $\varphi = \begin{cases} 1 + t - \frac{1}{2}t^2 & (t \leq x), \\ 1 + x - tx + \frac{1}{2}x^2 & (t \geq x). \end{cases}$

98. $z = e^t \varphi(x - t)$.

111. $C_n(y) = \dfrac{2(1 - \cos n\pi)}{n\pi} \left[\left(\dfrac{l^2}{n^2\pi^2} + 1\right) e^{-n\pi y/l} - \dfrac{l^2}{n^2\pi^2} \right]$.

112. $\varphi = 1 + \displaystyle\sum_{n=1}^{\infty} C_n(y) \sin \dfrac{n\pi x}{l}$, where

$C_n(y) = \dfrac{2(1 - \cos n\pi)}{n\pi} \left[\left(\dfrac{l^2}{n^2\pi^2} - 1\right) e^{-n\pi y/l} - \dfrac{l^2}{n^2\pi^2} \right]$.

113. (c) $C_n(y) = h_n \displaystyle\int_0^{\infty} G(y, \eta) \, d\eta + a_n e^{-n\pi y/l}$, where

$h_n = a_n = \dfrac{2}{n\pi} (1 - \cos n\pi)$, $\displaystyle\int_0^{\infty} G(y, \eta) \, d\eta = -\dfrac{l^2}{n^2\pi^2} (1 - e^{-n\pi y/l})$.

120. $\dfrac{V_x}{U} = 1 + \dfrac{1}{2} \left\{ \epsilon[x - (V_S + U)t] + \epsilon[x + (V_S - U)t] \right\}$,

$\dfrac{p}{p_0} = 1 + \dfrac{U}{V_S} \left\{ \epsilon[x - (V_S + U)t] - \epsilon[x + (V_S - U)t] \right\}$.

Chapter 10

5. (a) $-2 + 2i$. (b) $0 + i$. (c) $0 + i$. (d) $\dfrac{e^2}{\sqrt{2}} + \dfrac{e^2}{\sqrt{2}}i$.

(e) $\dfrac{\cosh 2}{\sqrt{2}} + \dfrac{\sinh 2}{\sqrt{2}}i$. (f) $\dfrac{\cosh 2}{\sqrt{2}} + \dfrac{\sinh 2}{\sqrt{2}}i$.

7. $\sin z$, $\cos z$, $\sec z$, and $\csc z$ have period 2π; $\tan z$ and $\cot z$ have period π; $\sinh z$, $\cosh z$, $\operatorname{sech} z$, and $\operatorname{csch} z$ have period $2\pi i$; $\tanh z$ and $\coth z$ have period πi.

13. (a) $\log \sqrt{2} + i\left(\dfrac{\pi}{4} + 2k\pi\right)$, where k is any integer;

principal value is $\log \sqrt{2} + (\pi i/4)$.

(b) $\cos \dfrac{3\pi}{8}(1 + 4k) + i \sin \dfrac{3\pi}{8}(1 + 4k)$, where $k = 0, 1, 2, 3$;

principal value is $\cos \dfrac{3\pi}{8} + i \sin \dfrac{3\pi}{8}$.

(c) $\sqrt[4]{2}\left[\cos \dfrac{\pi}{8}(1 + 8k) + i \sin \dfrac{\pi}{8}(1 + 8k)\right]$, where $k = 0, 1$;

principal value is $\sqrt[4]{2}\left(\cos \dfrac{\pi}{8} + i \sin \dfrac{\pi}{8}\right)$.

14. $z = \sqrt[4]{2}(\cos \alpha + i \sin \alpha)$, where $\alpha = 3\pi/8, 5\pi/8, 11\pi/8, 13\pi/8$.

15. $z^\pi = r^\pi[\cos \pi(\theta_P + 2k\pi) + i \sin \pi(\theta_P + 2k\pi)]$, where $z = re^{i\theta}$; $\cos (\pi^2/4) + i \sin (\pi^2/4)$.

16. $f(i) = e^{-(\pi/2 + 2k\pi)}$ $(k = 0, \pm 1, \pm 2, \ldots)$; $g(i) = e^{-\pi/2}$.

18. (a) $\left(\dfrac{\pi}{2} + 2k\pi\right) \pm i \log (2 + \sqrt{3})$, where k is any integer.

(b) $(2k + 1) \dfrac{\pi}{2} + i \log \sqrt{3}$, where k is any integer.

23. (a) $3xy^2 - x^3 + C$. (c) It is not an analytic function.

24. (b) $f(z) = iz/(z^2 + 1)$.

(b) Replace iy by z in $f(iy) = u(0, y) + iv(0, y)$ when $f(z)$ is analytic on part of the imaginary axis.

27. (a) $v = r^{-2} \cos 2\theta$.

(b) $u = -r^3 \cos 3\theta$.

28. $f(z) = U(z, 0) + iV(z, 0)$.

29. $f(z)$ must be a branch of $\log z + iz$.

33. (a) 0.

(b) $2\pi i$.

35. (a) $2\pi i$.

(b) $2\pi i$.

36. $2\pi i$; $2\pi a^2 i$.

37. 0; 0.

50. (a) 1. (b) $\frac{1}{2}$. (c) 1. (d) -1.

51. (a) $1 + z + z^2 + \cdots + z^n + \cdots$.

(b) $-\dfrac{1}{z} - \dfrac{1}{z^2} - \dfrac{1}{z^3} - \cdots - \dfrac{1}{z^n} - \cdots$.

(c) $\dfrac{1}{2} + \dfrac{z + 1}{2^2} + \dfrac{(z + 1)^2}{2^3} + \cdots + \dfrac{(z + 1)^n}{2^{n+1}} + \cdots$.

(d) $-\dfrac{1}{z + 1} - \dfrac{2}{(z + 1)^2} - \dfrac{2^2}{(z + 1)^3} - \cdots - \dfrac{2^{n-1}}{(z + 1)^n} - \cdots$.

52. (a) $\dfrac{1}{z} + 1 + z + z^2 + \cdots + z^n + \cdots$.

(b) $-\dfrac{1}{z^2} - \dfrac{1}{z^3} - \dfrac{1}{z^4} - \cdots - \dfrac{1}{z^n} - \cdots$.

(c) $-\dfrac{1}{z-1} + 1 - (z-1) + (z-1)^2 - \cdots + (-1)^n(z-1)^n + \cdots$.

(d) $-\dfrac{1}{(z-1)^2} + \dfrac{1}{(z-1)^3} - \dfrac{1}{(z-1)^4} + \cdots + (-1)^{n+1}\dfrac{1}{(z-1)^n} + \cdots$.

(e) $-\frac{1}{2} - \frac{3}{4}(z+1) - \frac{7}{8}(z+1)^2 - \cdots - (1 - 2^{-n-1})(z+1)^n - \cdots$.

(f) $\cdots + \dfrac{1}{(z+1)^n} + \cdots + \dfrac{1}{(z+1)^2} + \dfrac{1}{z+1} + \dfrac{1}{2} + \dfrac{z+1}{2^2}$
$$+ \dfrac{(z+1)^2}{2^3} + \cdots + \dfrac{(z+1)^n}{2^{n+1}} + \cdots.$$

(g) $-\dfrac{1}{(z+1)^2} - \dfrac{3}{(z+1)^3} - \dfrac{7}{(z+1)^4} - \cdots - \dfrac{2^{n-1}-1}{(z+1)^n} - \cdots$.

53. $2 - z + 2z^3 - z^4 + 2z^6 - z^7 + \cdots$ $(|z| < 1)$,
$\dfrac{1}{z^2} - \dfrac{2}{z^3} + \dfrac{1}{z^5} - \dfrac{2}{z^6} + \dfrac{1}{z^8} - \dfrac{2}{z^9} + \cdots$ $(|z| > 1)$.

54. $1 < |z - 1| < \sqrt{5}$.

55. $\dfrac{1}{z} + \dfrac{1}{6}z + \dfrac{7}{360}z^3 + \cdots$.

61. (a) Simple poles at $z = \pm i$.
(b) Simple poles at $z = -1$, $z = \frac{1}{2}(1 \pm i\sqrt{3})$.
(c) Branch points at $z = \pm i$.
(d) Branch points at $z = 1$, $z = 2$.
(e) Simple poles at $z = (2k+1)\pi/2$, where k is any integer.
(f) Branch points at $z = 1 \pm i$.

65. (a) Inside the circle $|z - 1| = \sqrt{3}$.
(c) Launch another series expansion from (say) $z = 1 + 1.7i$.

67. (a) $|x| < 1$.   (b) $|x| < 1$.   (c) $|x| > 0$.
(d) $|x| < \pi$.   (e) $|x| < \sqrt{2}$.   (f) $|x| < \pi/2$.

71. (a) Pole of order two.   (b) Regular point.   (c) Regular point.
(d) Branch point.   (e) Simple pole.   (f) Branch point.

78. (a) $\text{Res}\,(\pm ai) = \pm e^{\pm ai}/2ai$.
(b) $\text{Res}\,(\pm a) = \pm 1/4a^3$, $\text{Res}\,(\pm ai) = \pm i/4a^3$.
(c) $\text{Res}\,(0) = 1$.
(d) $\text{Res}\,(0) = 0$.
(e) $\text{Res}\,(0) = 1$, $\text{Res}\,(1) = 0$.
(f) $\text{Res}\,(\pm ai) = \pm 1/4a^3 i$.
(g) $\text{Res}\,(2) = \frac{1}{3}e^{2a}$, $\text{Res}\,(\frac{1}{2}) = -\frac{1}{3}e^{a/2}$.
(h) $\text{Res}\,(-1) = \frac{1}{2}(e^{-2} - 1)$.
(i) $\text{Res}\,(0) = -a^8/(8!)$.
(j) $\text{Res}\,(2k\pi i) = 2$ $(k = 0, \pm 1, \pm 2, \ldots)$.
(k) $\text{Res}\,(k\pi) = 1$ $(k = 0, \pm 1, \pm 2, \ldots)$.
(l) $\text{Res}\,(0) = -\frac{1}{24}$.

80. (a) $-\pi/2$.
(b) $+\pi/2$.

81. (a) $-ie^{ai/\sqrt{2}} \sinh(a/\sqrt{2})$.
(b) $e^{a/\sqrt{2}} \sin(a/\sqrt{2})$.

82. $2\pi i[\text{Res}\,(1) - \text{Res}\,(-1)] = 2\pi i$.

86. $-2\pi i$.

88. (a) Res $(0) = 1$.
    (b) Res $(0) = 0$.
    (c) Res $(\pi) = 0$.
    (d) Res $(0) = -$Res $(z_\infty) = \frac{7}{6}$.

93. (b) $(\pi/128a^7)e^{-am}[(3 + 2am)\sin am + 3\cos am]$ $(a > 0, m \geqq 0)$.

96. $f(t) = (1/2a)e^{-a|t|}$ $(a > 0)$.

110. (a) $\pi(1 - e^{-a})/a^2$ $(a > 0)$.
     (b) $2/\pi$.

115. (b) $\mathrm{P}\displaystyle\oint_C f(z)\,dz = \alpha i\,\mathrm{Res}\,(a_0) + 2\pi i \sum_{k=1}^{n} \mathrm{Res}\,(a_k)$.

Chapter 11

1. (a) e^{-at}.   (b) $\dfrac{1}{a}\sin at$.   (c) $\dfrac{1}{a}\sinh at$.

    (d) $\dfrac{1}{6}t^3$.   (e) $\dfrac{1}{2a}t\sin at$.   (f) $\dfrac{1}{2a^2}\sin at \sinh at$.

2. (a) $\dfrac{t^{n-1}e^{-at}}{(n-1)!}$.

    (b) $\dfrac{(n-1) - at}{(n-1)!}t^{n-2}e^{-at}$ $(n > 1)$.

    (c) $(1/a)e^{-bt}\sin at$.

    (d) $(1/a)\,e^{-bt}(a\cos at - b\sin at)$.

    (e) $(1/2a^3)e^{-bt}(\sin at - at\cos at)$.

    (f) $(1/2a^3)(at\cosh at - \sinh at)$.

73. Right half plane: $G = \dfrac{1}{2\pi}\log\dfrac{|z - \zeta|}{|z + \bar{\zeta}|}$.

    Third quadrant: G same as for first quadrant.

86. $G = (1/2\pi)\log R + $ constant.

Index

Index